MARINE DESIGN XIII

PROCEEDINGS OF THE 13TH INTERNATIONAL MARINE DESIGN CONFERENCE (IMDC 2018), 10–14 JUNE 2018, ESPOO, FINLAND

Marine Design XIII

Editors

Pentti Kujala & Liangliang Lu

Marine Technology, Department of Mechanical Engineering, School of Engineering, Aalto University, Finland

VOLUME 2

CRC Press is an imprint of the
Taylor & Francis Group, an **informa** business
A BALKEMA BOOK

Cover photo: Meyer Turku shipyard

CRC Press/Balkema is an imprint of the Taylor & Francis Group, an informa business

Typeset by V Publishing Solutions Pvt Ltd., Chennai, India

Published by: CRC Press/Balkema
Schipholweg 107C, 2316 XC Leiden, The Netherlands
e-mail: Pub.NL@taylorandfrancis.com
www.crcpress.com – www.taylorandfrancis.com

ISBN: 978-1-138-54187-0 (set of 2 volumes + CD in volume 1)
ISBN: 978-1-138-34069-5 (Vol 1)
ISBN: 978-1-138-34076-3 (Vol 2)
ISBN: 978-1-351-01004-7 (eBook set of 2 volumes)
ISBN: 978-0-429-44053-3 (eBook, Vol 1)
ISBN: 978-0-429-44051-9 (eBook, Vol 2)

Marine Design XIII – Kujala & Lu (Eds)
© 2018 Taylor & Francis Group, London, ISBN 978-1-138-34076-3

Table of contents

Ship concept design

VOLUME 2

Risk and safety

Marine Design XIII – Kujala & Lu (Eds)
© 2018 Taylor & Francis Group, London, ISBN 978-1-138-34076-3

Preface

This book collects the contributions to the 13th International Marine Design Conference, IMDC 2018, held in Espoo, Finland between 10 and 14 June 2018. This is the thirteenth in the IMDC conference series. In spring 1982, the first of the IMDC series of conferences was held in London (United Kingdom). Successive conferences were held every three years, namely 1985 in Lyngby (Denmark), 1988 in Pittsburgh (USA), 1991 in Kobe (Japan), 1994 in Delft (The Netherlands), 1997 in Newcastle (United Kingdom), 2000 in Kyongju (Korea), 2003 in Athens (Greece), 2006 in Ann Arbor-Michigan (USA), 2009 in Trondheim (Norway), 2012 in Glasgow (United Kingdom) and 2015 in Tokyo (Japan).

The aim of IMDC is to promote all aspects of marine design as an engineering discipline. The focus of this year is on the key design challenges and opportunities in the area of current maritime technologies and markets, with special emphasis on:

- Challenges in merging ship design and marine applications of experience-based industrial design
- Digitalisation as technological enabler for stronger link between efficient design, operations and maintenance in future
- Emerging technologies and their impact on future designs
- Cruise ship and icebreaker designs including fleet compositions to meet new market demands

To reflect on the conference focus, the book covers the following research topic series from worldwide academia and industry:

- State of the art ship design principles – education, design methodology, structural design, hydrodynamic design
- Cutting edge ship designs and operations – ship concept design, risk and safety, Arctic design, autonomous ships
- Energy efficiency and propulsions – energy efficiency, hull form design, propulsion equipment design
- Wider marine designs and practices – navy ships, offshore and wind farms and production

In total, the book contains 111 papers, including 2 state of the art reports related to the design methodologies and cruise ships design and 4 keynote papers related to the new direction for vessel design practices and tools, digital maritime traffic, naval ship designs and new tanker design for the Arctic.

The articles in this book were accepted after peer-review process, based on the full text of the papers. Many thanks are sincerely given to the reviewers of IMDC 2018 who helped the authors deliver better papers by providing constructive comments. Meanwhile, we also would like to thank the sponsors of IMDC 2018: ABB Marine, Aker Arctic, Arctech Helsinki shipyard, Elomatic, Meyer Turku shipyard, Royal Caribbean Cruise Ltd.

Hope the proceedings of IMDC 2018 contribute to marine design research and industry.

Pentti Kujala
Local Chairman, IMDC2018
Vice Dean, Professor, Aalto University

Marine Design XIII – Kujala & Lu (Eds)
© 2018 Taylor & Francis Group, London, ISBN 978-1-138-34076-3

Committees

INTERNATIONAL COMMITTEE

David Andrews (Chairman), *Professor, University College London, United Kingdom*
Apostolos Papanikolaou, *Professor, Hamburgische Schiffbau-Versuchsanstalt GmbH, Germany*
Makoto Arai, *Professor, Yokohama National University, Japan*
Richard Birmingham, *Professor, University of Newcastle, United Kingdom*
Stein Ove Erikstad, *Professor, Norwegian University of Science and Technology, Norway*
Sheming Fan, *Professor, Marine Design and Research Institute of China, China*
Stefan Krüger, *Professor, Technical University of Hamburg, Germany*
Patrik Rautaheimo, Dr., *Managing Director, Elomatic Oy, Finland*
Hiroyuki Yamato, *Professor, The University of Tokyo, Japan*
David Singer, *Associate Professor, University of Michigan, United States of America*
Dracos Vassalos, *Professor, University of Strathclyde, United Kingdom*
Hans Hopman, *Professor, Delft University of Technology, The Netherlands*
Per Olaf Brett, Dr., *Deputy Managing Director, Ulstein International AS, Norway*
Kelly Cooper, *Program Manager, Ship Systems and Engineering Research, US Navy, United States of America*
Chris Mckesson, Dr., *University of British Columbia, Canada*

LOCAL ORGANIZING COMMITTEE (FINLAND)

Pentti Kujala (Chairman), *Professor, Aalto University*
Patrik Rautaheimo, *Managing Director, Elomatic*
Mervi Pitkänen, *Head of External Funding, Rolls-Royce*
Reko-Antti Suojanen, *Managing director, Aker Arctic*
Niko Rautiainen, *Senior Vice President, Design, Arctech Helsinki shipyard*
Riku-Pekka Hägg, *Vice-President Ship Design, Wärtsilä*
Mikko Ilus, *Head of Ship Theory, Meyer Turku shipyard*
Tommi Arola, *Head of Unit, Finnish Transport Safety Agency*
Elina Vähäheikkilä, *Secretary General, Finnish Maritime Industries*
Marjo Keiramo, *Senior Program Manager, Royal Caribbean Cruises Ltd*
Andrei Korsstrom, *Product Manager, ABB Marine*
Teemu Manderbacka, *Senior R&D Engineer, NAPA Shipping Solutions*
Jani Romanoff, *Professor, Aalto University*
Otto Sormunen, *Postdoctoral Researcher, Aalto University*
Liangliang Lu, *Doctoral Researcher, Aalto University*
Sophie Cook, *Project Manager, HRG Nordic*

ADDITIONAL AALTO TEAM

Heikki Remes, *Professor, Aalto University*
Kari Tammi, *Professor, Aalto University*
Markus Ahola, *Project Manager, Experience Platform, Aalto University*
Tommi Mikkola, *Lecturer, Aalto University*

Floris Goerlandt, *Lecturer, Aalto University*
Osiris A. Valdez Banda, *Postdoctoral Researcher, Aalto University*
Martin Bergström, *Postdoctoral Researcher, Aalto University*
Mihkel Korgesaar, *Postdoctoral Researcher, Aalto University*
Jakub Montewka, *Postdoctoral Researcher, Aalto University*
Mikko Suominen, *Doctoral Researcher, Aalto University*
Fang Li, *Doctoral Researcher, Aalto University*
Lei Du, *Doctoral Researcher, Aalto University*

Risk and safety

Marine Design XIII – Kujala & Lu (Eds)
© 2018 Taylor & Francis Group, London, ISBN 978-1-138-34076-3

Collision accidents analysis from the viewpoint of stopping ability of ships

M. Ueno
National Maritime Research Institute, Tokyo, Japan

ABSTRACT: Collisions account for a major part of ship accidents even after the adoption of the standards for ship manoeuvrability of the International Maritime Organization. This paper presents an analysis of 24 serious collision accidents in Japan of passenger ships, cargo ships, and tankers excluding collisions with small boats. The author estimates distances between the collision points and where ships recognize the danger of collisions first. Since the initial ship speed at the danger recognition varies, he normalized the distances by converting them into those at 12 kn initial speed based on a simplified mathematical model. He, then, deduces the relation between the stopping criterion and reduction rates of collision accidents. Discussion on the required engine power for better stopping ability is provided. The study presented here excluding human factors should give an idea on how to determine a criterion of stopping ability to decrease ship collisions.

1 INTRODUCTION

Ship collisions dominate in ship accidents by over half the numbers. They often attribute the causes of the accidents to human errors and many researchers have dedicated their effort to prevent ship collisions from the viewpoint of man-machine systems. On the other hand, the International Maritime Organization (IMO) adopted in 2002 the standards for ship manoeuvrability (IMO 2002a, b) that aims to prevent ship accidents from the viewpoint of ship performance regardless of human factors.

The IMO standards for ship manoeuvrability include the stopping criterion as follows. "The track reach in the full astern stopping test should not exceed 15 ship lengths. However, this value may be modified by the Administration where ships of large displacement make this criterion impracticable, but should in no case exceed 20 ship lengths". "The test speed used in the Standards is a speed of at least 90% of the ship's speed corresponding to 85% of the maximum engine output", which is regarded as a navigation full speed.

However, the criterion of 15 ship lengths has no clear reasoning (Yoshimura 1994), and ship collisions have seen no obvious sign of decrease even after the adoption of the IMO standards. Moreover, the Energy Efficiency Design Index (EEDI) of IMO (IMO 2011, 2012) that regulates the air pollution from ships is stimulating the increase of ships with small engines, which leads the interim guidelines for determining minimum propulsion power to maintain the manoeuvrability of ships in adverse weather conditions (IMO 2013). Although it seems not to be recognized well comparing with the problems of manoeuvrability in adverse weather conditions, the ships with smaller engines are the ships with inferior stopping ability. These facts imply the needs of careful and continuous study for the stopping ability of ships and its criterion to prevent and decrease collisions of ships.

The author presents here an analysis of collision accidents for clarifying the relation between ship collisions and stopping criterion. The source data of accidents are the reports of the Japan Transport Safety Board (JTSB 2017) from 2008 to 2016. Serious collision accidents of passenger ships, cargo ships, and tankers that do not include those with small boats are for the analysis. The ship speeds when recognizing the danger of collisions first and when colliding in the end, and the duration between them estimate the distances needed for the ships to stop before reaching collision points. The author normalizes the estimated distances of which initial speeds are various by converting them into those at 12 kn initial speed. He, then, assumes a stopping criterion as the distance for ships to stop at 12 kn initial speed by full astern. Considering different situations in overtaking, crossing, and facing, the analysis tells the relation between the stopping criterion and the reduction rate of ship collisions. The required engine power for improving the stopping ability to comply with the arbitrary criterion is also discussed. The study presented here from the viewpoint of stopping ability of ships excluding human factors provides an insight how the ship stopping ability with its criterion accorded contributes to prevent and decrease ship collisions.

2 SOURCE DATA

The source data are JTSB reports of ship accidents (JTSB 2017) from April 2008 to August 2016. Accident data of which kind is "Collision", category is "Serious", and ship types are "Passenger ship", "Cargo carrier", and "Tanker" are for the analysis. The resultant number of accidents is 52 consisting of 4 of passenger ships, 38 of cargo carriers, and 10 of tankers, including duplication in searching by the ship types. Excluding the duplicates; those involving small ships such as fishing vessels, pleasure boats, commuter boats, sightseeing boats, water taxies, patrol boats and sailboats; and a case with little information makes the total accidents number 26. Moreover, exclusion of a case with a large containership with tugs and a case with a barge with a pusher results in total 24 cases. An accident with two collisions involving three ships neglects the second collision. Therefore, the 24 collisions involve 48 ships.

The 24 collisions consist of 6 overtaking cases, 7 crossings, and 11 facings. Eighteen cases are in routes or harbors. Eight ships among the 48 ships ordered propellers astern.

Table 1 shows data numbers; maximum, minimum, and mean values of principal particulars of the 48 ships. Note that JTSB reported navigation full speed for only 22 ships among the 48 ships. Figure 1 is a histogram of ship lengths, L, of the 48 ships of which mean is 100.8 m. Twenty-one ships are over 100 m that IMO standards for ship manoeuvrability should apply to. Histograms of breadth and depth of the ships are omitted here. Figures 2 and 3 are the histograms of gross tonnages, GT, and maximum continuous ratings of engines, MCR, respectively. Note that the abscissas are not linear scale in Figs. 2 and 3. The reason why the number of ships under 500 t is large is related to the Japanese rule on the ship tonnage. Figure 4 is a histogram of navigation full speeds.

Table 1. Data numbers and ranges of principal particulars of sample ships.

	Data numb.	Max.	Min.	Mean
Ship length, L (m)	48	276.0	50.0	100.8
Breadth (m)	48	43.4	8.3	16.1
Depth (m)	48	25.5	4.5	8.9
Gross tonnage (t)	48	94446	170	5600
MCR (kW)	48	51485	404	4609
Nav. full speed (kn)	22	24.7	10.5	15.0

*MCR; Maximum continuous rating.

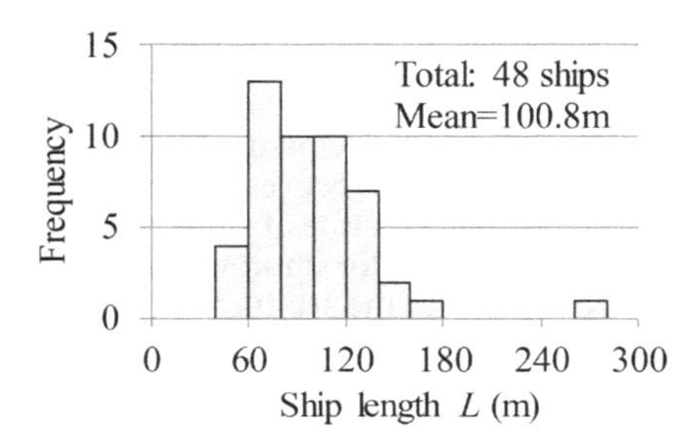

Figure 1. Histogram of ship lengths.

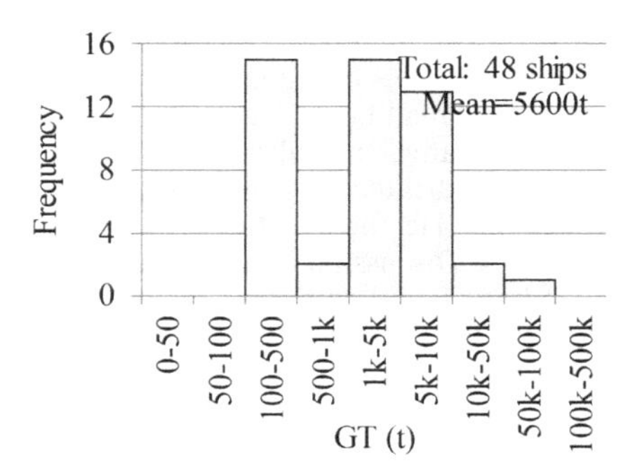

Figure 2. Histogram of gross tonnages.

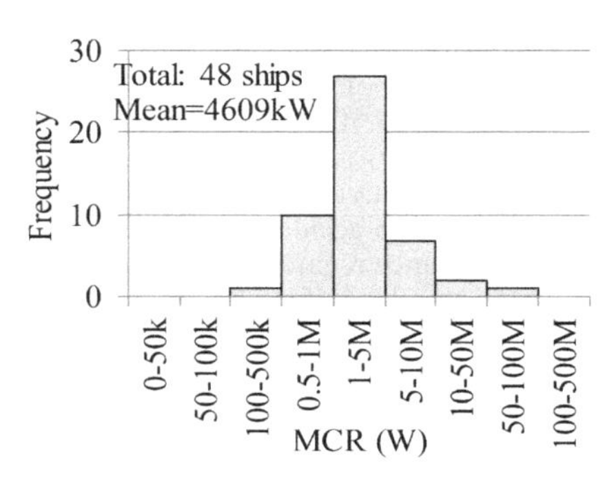

Figure 3. Histogram of maximum continuous ratings.

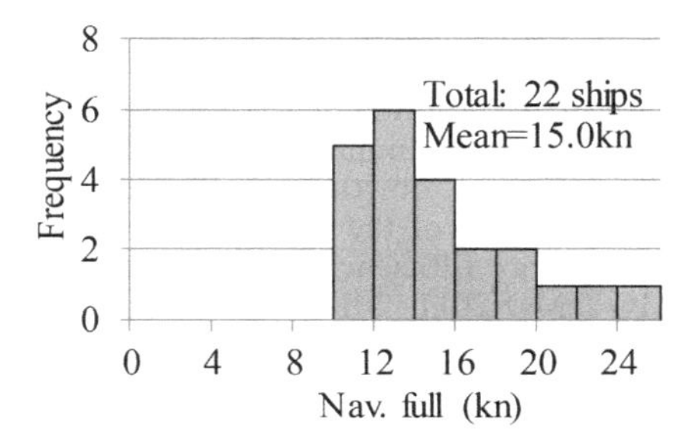

Figure 4. Histogram of navigation full speeds.

The mean value of navigation full speeds for the 22 ships is 15.0 kn. Since the minimum navigation speed is 10.5 kn and the median is 13.8 kn, over the half ships' navigation full speeds are around 12 kn as Fig. 4 shows.

3 COLLISION SITUATIONS

Table 2 shows data numbers; maximum, minimum, and mean values of characteristics of collision accidents of which histograms are in Figs. 5 through 10. Figure 5 is a histogram of initial ship speeds V_i. V_i is a speed when a ship recognizes danger firstly and exert an initial action to avoid a collision. The number of V_i is 36 since V_i of 12 ships are unknown. V_i distributes in a wide range from 4.1 kn to 18.0 kn of which mean is 11.2 kn. Figure 6 is a histogram of collision ship speeds V_c at which ships collide. V_c also distributes in a wide range and its mean is 10.2 kn, less than the mean of V_i. Figure 7 is a histogram of speed decreases from V_i to V_c, $V_i - V_c$, of which mean is 0.9 kn. The speed decreases are small, and even negative values or acceleration cases are seen. The author considers that this is one of reasons why they resulted in the "Serious" accidents.

Figure 8 is a histogram of time from the recognitions of danger to the collisions, t_c. In cases that V_i is unknown, V_i is assumed to be equal to V_c, which leads to t_c equal to zero. The mean t_c is 38s but the distribution of t_c leans to zero as Fig. 8 shows since t_c in 17 cases are equal to zero. Figure 9 is a histogram of distances from a recognition point of danger to a collision point D_S, that is estimated by the following equation.

Table 2. Data numbers and ranges of characteristics of collision accidents.

	Data numb.	Max.	Min.	Mean
Initial speed, V_i (kn)	36	18.0	4.1	11.2
Collision speed, V_c (kn)	48	18.0	2.3	10.2
Speed decrease (kn)	36	6.2	−1.0	0.9
Time to collision, t_c	48	3′27″	0′0″	0′38″
Estimated dist. to collision point, D_S (L)	48	7.1	0.0	1.9
Normalized distance at 12 kn, $D_{S[12]}$ (L)	48	12.2	0.0	2.6

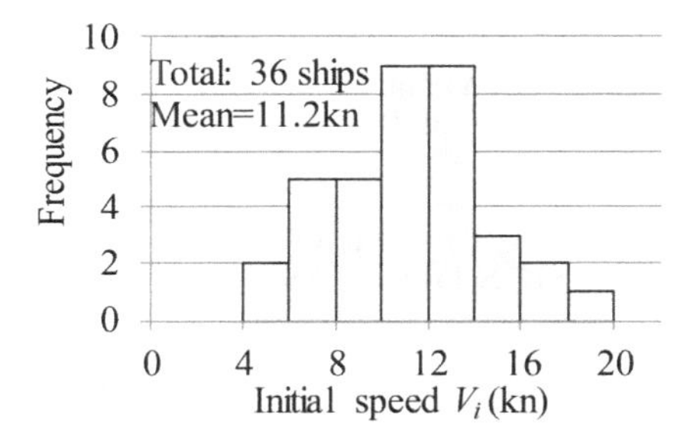

Figure 5. Histogram of initial ship speeds.

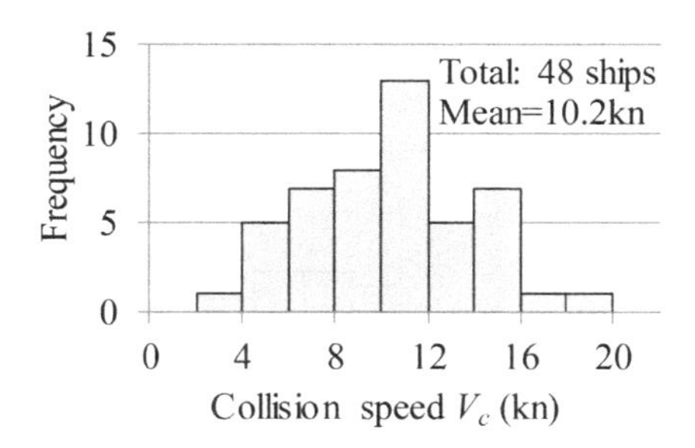

Figure 6. Histogram of collision ship speeds.

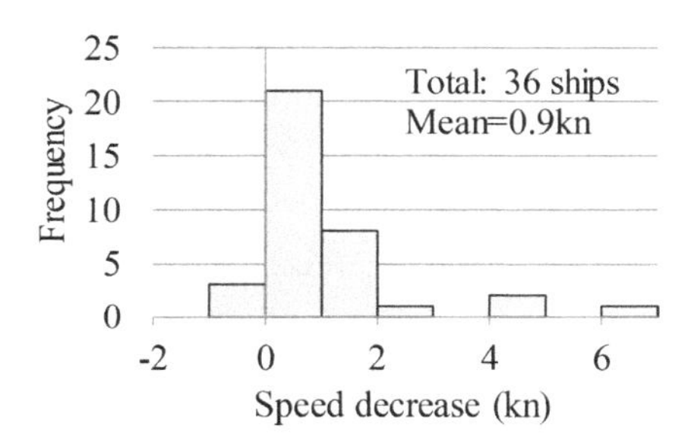

Figure 7. Histogram of speed decreases before collisions.

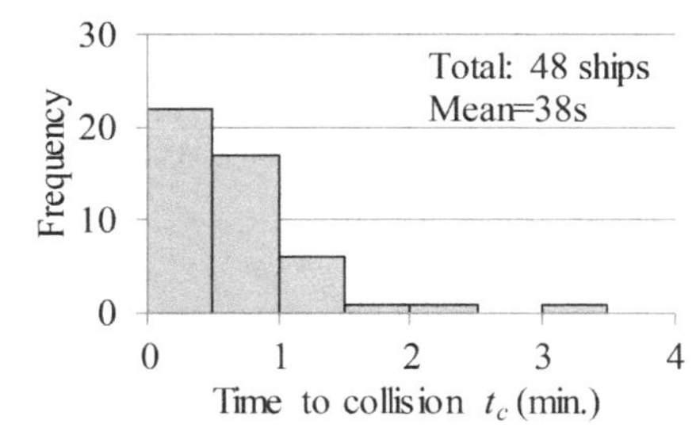

Figure 8. Histogram of time to collision.

$$D_S = \frac{(V_i + V_c)}{2} t_c \tag{1}$$

Equation (1) assumes ship motion from a recognition of danger to a collision point is monoto-

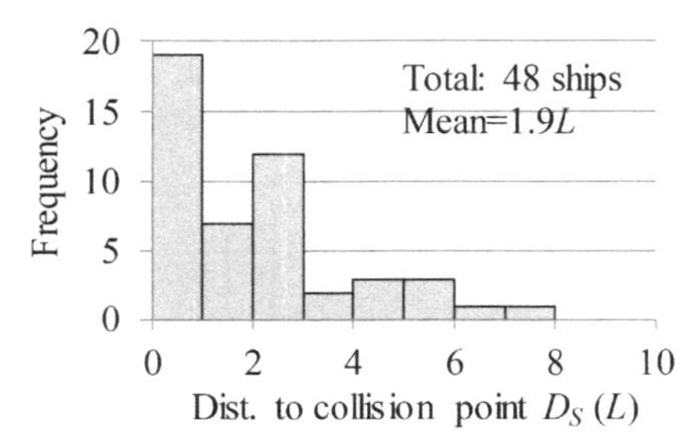

Figure 9. Histogram of distances to collision point.

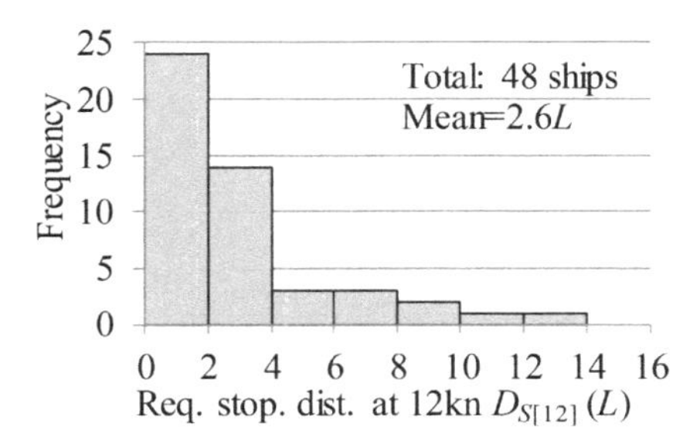

Figure 10. Histogram of stopping distances at 12 kn before reaching collision points $D_{S[12]}$.

nous in a straight course in which the ship speed varies from V_i to V_c proportionally to time. Effects of external forces due to wind, waves, and current are neglected. Including the 17 cases of D_S equal to 0.0L that corresponds to those of t_c is equal to zero, the mean D_S is 1.9L.

4 NORMALIZATION OF STOPPING DISTANCES

Ships do not necessarily stop in straight courses by reversing propellers even in calm water because the reversing propeller generates not only the astern force but also sway force and yaw moment. However, large astern propeller rates relative to small initial ship speeds do not induce significant sway and yaw motions. In other words, if hull advance ratios $|J_{Hai}|$ defined by the following equation (Yamazaki 1978) were small, stopping motion of ships should be approximately in straight courses (Yoshimura & Nomoto 1978, Ueno & Tsukada 2017).

$$J_{Hai} = \frac{V_i}{n_a D_P} \tag{2}$$

In Eq. (2), n_a is an astern propeller rate and D_P is a propeller diameter. J_{Hai} is a parameter characterizing the stopping motion by propeller astern (Yamazaki 1978). Small $|J_{Hai}|$ corresponds to an emergency stop where n_a is large relative to V_i.

If, therefore, a ship ordered full astern immediately on recognition of danger and the ship could stop within a distance D_S defined by Eq. (1), the ship might avoid the collision. The stopping distance in a straight course is equal to both the track reach and the head reach, which is the situation the discussion hereafter assumes.

D_S defined by Eq. (1) is the distance within which a ship must stop to avoid a collision in each case. However, D_S depends on the initial speed V_i. Therefore, D_S cannot be compared with each other as a parameter representing a stopping ability of each ship.

To discuss the stopping ability of each ship to avoid the collision, the author converts D_S to those at a designated initial speed. The converted D_S can represent a stopping ability on a common basis. A possible basis is the navigation full speed as in the IMO standards for ship manoeuvrability, though they vary depending on specifications of each ship. However, the number of collisions in which the both ships' navigation full speeds in a collision are known are only 9 out of the 24 cases, which is too small to deduce any feature. The author, therefore, chooses 12 kn as a common initial ship speed because the speeds in most of ship routes are around 12 kn. The stopping distance at 12 kn initial speed could be a practical scale to measure the stopping ability of ships since they most represent the situations of collision accidents as Fig. 5 suggests.

The stopping distance would be approximately proportional to the square of initial ship speed V_i if J_{Hai} was small (Appendix A). Therefore, the following equation in which V_i is in knot converts the distances D_S in Fig. 9 at various V_i to the stopping distances at 12 kn, $D_{S[12]}$.

$$D_{S[12]} = D_S \left(\frac{12}{V_i} \right)^2 \tag{3}$$

Figure 10 is a histogram of $D_{S[12]}$ of which maximum, minimum, and mean values are in Table 2. The distribution of $D_{S[12]}$ in Fig. 10 is more continuous than that of D_S in Fig. 9, probably because of the normalization. The mean value of $D_{S[12]}$ grows to 2.6L from 1.9L of D_S.

5 STOPPING CRITERION AND COLLISION DECREASE

Let us suppose $D_{S[12]c}$ as a stopping criterion representing the allowable maximum stopping distance at 12 kn and consider how $D_{S[12]c}$ could work to prevent the collisions if all the ships fulfilled the criterion.

Figure 11 shows the three patterns of two ships to collide; overtaking, crossing, and facing. Either stopping of the two ships before reaching the collision point prevent the collision in cases of overtaking and crossing. On the other hand, both stoppings of the two ships are needed in cases of facing to prevent the collision. Therefore, $D_{S[12]c}$ for the three patterns are defined as follows.

$$D_{S[12]c} = \begin{cases} max\left(D_{S[12]\mathbf{A}}, D_{S[12]\mathbf{B}} \right), \\ \quad \textit{for overtaking and crossi} \\ min\left(D_{S[12]\mathbf{A}}, D_{S[12]\mathbf{B}} \right), \\ \quad \textit{for facing} \end{cases} \tag{4}$$

$D_{S[12]A}$ and $D_{S[12]B}$ are $D_{S[12]}$ of the two ships, ship A and ship B, in a collision, respectively.

The total 24 cases consist of 6 overtaking cases, 7 crossings, and 11 facings. The 48 $D_{S[12]}$ shown in Fig. 10 and the judge of the right-hand side of Eq. (4) tell $D_{S[12]c}$ for each case to prevent the collision. Figure 12 is a histogram of collision avoidance depending on $D_{S[12]c}$ of which increment is $2L$. Figure 12 tells, for example, if $D_{S[12]c}$ was $2L$, an extreme assumption, and all the ships fulfill the criterion, 12 cases at least out of the 24 could avoid the collisions by crush astern of the both ships. Figure 12 also tells, for example, that 1 case at least could avoid the collision if $D_{S[12]c}$ was $12L$.

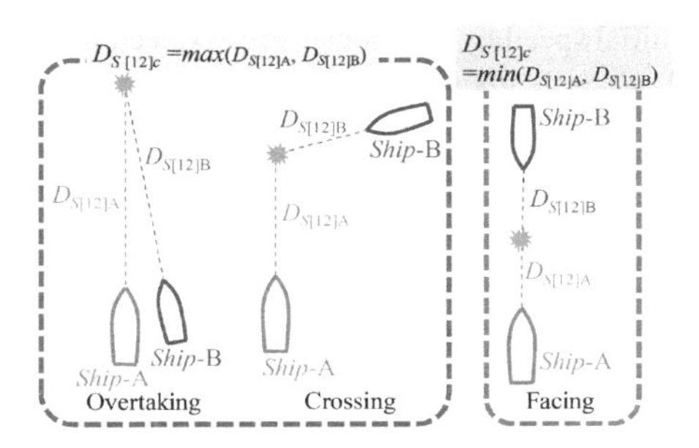

Figure 11. Stopping criterion at 12 kn $D_{S[12]c}$ to avoid collisions and stopping distance at 12 kn before reaching collision points $D_{S[12]}$ in overtaking, crossing, and facing.

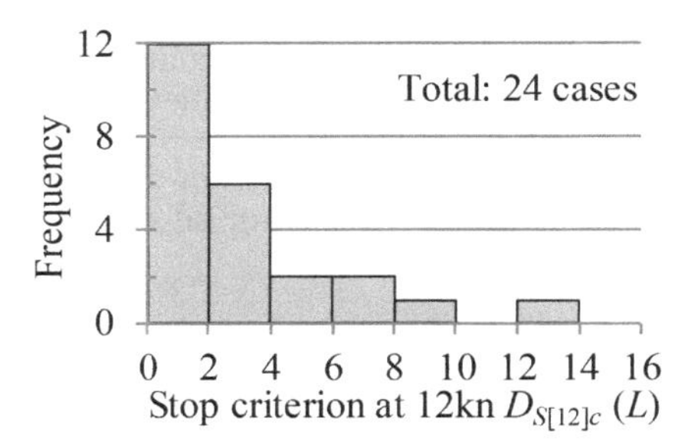

Figure 12. Histogram of collision avoidances depending on stopping criterion at 12 kn $D_{S[12]c}$.

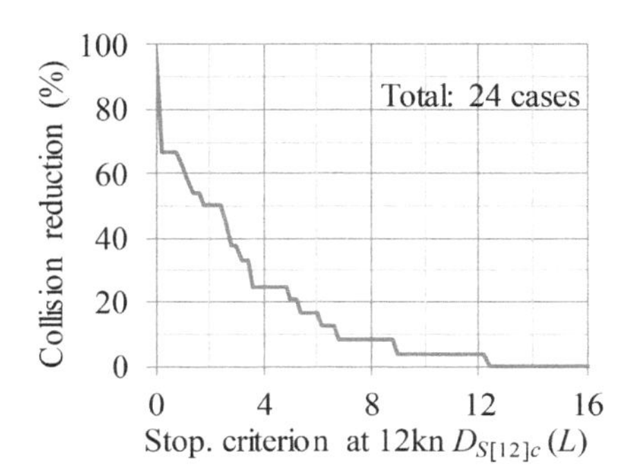

Figure 13. Relation between collision reduction rate and stopping criterion at 12 kn $D_{S[12]c}$

The data of Fig. 12 can tell the relation between collision reduction rates and $D_{S[12]c}$ as in Fig. 13 where the abscissa is continuous. Figure 13 reveals that $D_{S[12]c}$ over $12.8L$ has no effect on collision decrease. Around 33% of the accidents cannot be avoided by any strict $D_{S[12]c}$. In these cases, no one could notice the danger until the collision occurred. Figure 13 also tells, for examples, that $D_{S[12]c}$ equal to $10L$ and $6L$ reduce collisions by about 4% and 17%, respectively. Note that the reduction rate is expected if all the ships exerted full astern on the danger recognition.

The stopping distance $10L$ at 12 kn corresponds to approximately $15L$ at 14.7 kn according to Eq. (3). The mean navigation full speed is 15.0 kn, close to 14.7 kn, as shown in Table1 and Fig. 4. Complying the IMO standards for ship manoeuvrability ensures that ships can stop within $15L$. Therefore, the 4% reduction rate for $D_{S[12]c}$ equal to $10L$, mentioned above, implies that one of the 24 cases could avoid the collision by the full astern on danger recognition.

6 INFLUENCE ON SHIP DESIGN

The tighter the stopping criterion is for decreasing collision accidents, the higher the required stopping ability of ships must be. An option to cope with the improvement of stopping ability is to increase the engine power or *MCR*.

Yoshimura (1994) approximated the relation between *MCR* and the stopping distance D_S as follows.

$$\frac{D_S}{L} \propto \frac{\Delta}{MCR} V_{MCR} F_{ni}^{\ 2} \tag{5}$$

Δ is displacement of a ship, V_{MCR} is speed corresponding to *MCR*, and F_{ni} is Froude number for

V_i. Equation (5) implying D_S proportional to V_i^2 is consistent with Eq. (3).

Assumption that MCR is proportional to $V_{MCR}{}^3$ leads Eq. (5) to the following relation.

$$\frac{D_S}{L} \propto MCR^{-\frac{2}{3}} V_i^2 \tag{6}$$

Let us consider how a ship having the stopping ability equal to $15L$ at her navigation full speed must increase the engine power from MCR_0 to MCR for satisfying a supposed stopping criterion $D_{S[12]c}$. According to Eq. (6) the following equation tells the ratio of MCR to MCR_0.

$$\frac{MCR}{MCR_0} = \left\{ \frac{D_{s[12]c}}{15} \left(\frac{V_{NF}}{12} \right)^2 \right\}^{-\frac{3}{2}} \tag{7}$$

V_{NF} in Eq. (7) is the navigation full speed.

Figure 14 demonstrates Eq. (7) for ships with V_{NF} equal to 12 kn, 15 kn, 18 kn, and 21 kn. The dotted horizontal line represents MCR/MCR_0 equal to 1. Since ships with high V_{NF} have high MCR_0, they can endure stricter or lower $D_{S[12]c}$ without increasing MCR. Ships with low V_{NF} must increase MCR significantly for satisfying low $D_{S[12]c}$ if they narrowly stopped with $15L$ distance at their V_{NF}. Ships of which V_{NF} is 15 kn must increase their MCR to fulfill $D_{S[12]c}$ lower than $9.6L$, around $10L$. This confirms the description in the former section that $D_{S[12]}$ equal to $10L$ corresponds to approximately $15L$ at V_i equal to 14.7 kn, around 15 kn. Note that Eqs. (5) through (7) and Fig. 14 are estimates that do not take into account the duration between the setting off of full-astern order and the starting of propeller reversing.

A goal-based standard aiming at decreasing collision accidents would promote ships having higher stopping ability. One measure to cope with this requirement is an increased MCR as described above. However, since a larger MCR degrades EEDI, the ship design should go through an iterative process for the stopping ability as for the requirement for the minimum propulsion power to maintain the manoeuvrability in adverse weather. Among other measures, spoilers mounted on a ship hull and parachutes thrown into water, for example, are to reduce ship speed in emergency. Although they are not prevalent in practical use, a potential revision of the standard could stimulate innovative technologies for improving the stopping ability, in which assessment the discussions on the stopping criterion presented here should contribute.

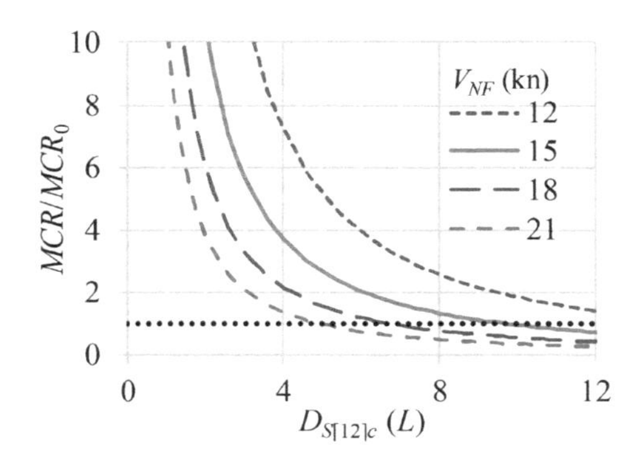

Figure 14. MCR increase ratio to original MCR_0 of ships with stopping ability equal to $15L$ at navigation full speed V_{NF} for satisfying $D_{S[12]c}$, dotted horizontal line for MCR/MCR_0 equal to 1.

7 CONCLUSIONS

The author analyzed data of 24 serious collision accidents of passenger ships, cargo ships, and tankers excluding collisions with small boats. The data source is the reports from 2008 to 2016 of the Japan Transport Safety Board. The estimated distances from the recognition points of danger to the collision points at various initial ship speeds are normalized by converting them to the distances at 12 kn initial speed. The author assumed a stopping criterion that represents a stopping distance at 12 kn initial speed by full astern. He deduced the relation between the stopping criterion and collision reduction rates by considering different situations in overtaking, crossing, and facing for all the collision cases. The reduction rates are, for example, about 4% by a criterion of 8 ship lengths and about 17% by a criterion of 10 ship lengths, while about 33% of collisions are unavoidable by any criterion. The discussion on the required engine power to satisfy a supposed stopping criterion clarified the ratio of increased engine power to the original depending on the navigation full speed.

Since these reduction rates assume the immediate full astern on the recognition of danger, no human factor is concerned and only stopping ability of ships is taken into consideration. The data number is not sufficient to have sound reduction rates of collisions. The analysis, however, should give an aspect how to consider the stopping criterion to prevent ship collisions from the viewpoint of ship performance required for safety.

REFERENCES

Hewins, E.F. et al. 1950. The backing power of geared-turbine-driven vessels. *SNAME transactions* 58.

IMO 2002a. Annex 6, Resolution MSC. 137(76), Standards for ship manoeuvrability. *MSC 76/23/Add. 1.*

IMO, 2002b. Explanatory notes to the standards for ship manoeuvrability. *Ref. T4/3.01, MSC/Circ. 1053*.

IMO, 2011. Annex 19, Resolution MEPC. 203(62), Amendments to the annex of the protocol of 1997 to amend the international convention for the prevention of pollution from ships, 1973, as modified by the protocol of 1978 relating thereto. *MEPC 62/24/Add. 1*.

IMO, 2012. Annex 8, Resolution MEPC. 212(63), 2012 guidelines on the method of calculation of the attained energy efficiency design index (EEDI) for new ships. *MEPC 63/23*.

IMO, 2013. Annex 16, Resolution MEPC. 232(65), 2012 interim guidelines for determining minimum propulsion power to maintain the manoeuvrability of ships in adverse conditions. *MEPC 65/22*.

JTSB, 2017. http://jtsb.mlit.go.jp/jtsb/ship/.

Ueno M. & Tsukada Y. 2017. Estimation of stopping ability of full-scale ship using free-running model. *Ocean engineering* 130: 260–273.

Yamazaki Y. 1978. A fundamental study on the stopping ability of ships. *J. of the kansai society of naval architectures, Japan* 168: 17–27.

Yoshimura Y. & Nomoto, K. 1978. Modeling of manoeuvring behaviour of ships with a propeller idling, boosting and reversing. *J. of the society of naval architects, Japan* 144: 57–69.

Yoshimura Y. 1994. Studies on the stopping ability of a manoeuvring standard. *J. of the society of naval architects, Japan*, 176: 259–265.

APPENDIX A: SIMPLIFIED 1-DOF ANALYSIS OF STOPPING MANOEUVRE

A large astern propeller rate n_a relative to a small initial ship speed V_i leads to a small initial hull advance ratio $|J_{Hai}|$ defined by Eq. (2). In such cases, sway and yaw motions are assumed small or negligible (Yoshimura & Nomoto 1978, Ueno & Tsukada 2017). Therefore, the ship motion by full astern propeller in emergency conditions could be approximated by the following one-degree-of-freedom equation of motion.

$$(m+m_x)\dot{V} = X_{VV}V^2 + X_{P(J_{Ha})} \tag{A1}$$

M and m_x stand for the ship mass and the added mass in longitudinal direction, respectively. X_{VV} that is usually negative is the resistance coefficient in calm water and assumed a constant. $X_{P(JHa)}$ is the force originated in a reversing propeller including the interaction with the ship hull. According to Yamazaki (1978), and Yoshimura & Nomoto (1978), $X_{P(JHa)}$ is a function of the apparent propeller advance ratio or the hull advance ratio, J_{Ha}, defined as follows.

$$J_{Ha} = \frac{V}{n_a D_P} \tag{A2}$$

Equation (2) is the initial value of J_{Ha}.

In a range of small $|J_{Ha}|$, $X_{P(JHa)}$ can be assumed constant (Hewwins et al. 1950, Yoshimura & Nomoto 1978). This assumption makes Eq. (5) the following equation where X_{P1} stands for the constant value of $X_{P(JHa)}$.

$$(m+m_x)\dot{V} = X_{VV}V^2 + X_{P1} \tag{A3}$$

Although the first term in the right-hand side of Eq. (A3) could have a larger effect than the second term at the initial phase of stopping motion, it diminishes rapidly as the ship speed decreases and the second term should dominate. Neglecting the hull resistant term or the first term reduces Eq. (A3) to the following equation.

$$(m+m_x)\dot{V} = X_{P1} \tag{A4}$$

Equation (A4) is a simplified differential equation representing uniformly decelerated motion of a ship in emergency stop.

The integration of Eq. (A4) from zero to stopping time t_s gives the following equation.

$$-(m+m_x)V_i = X_{P1}t_S \tag{A5}$$

Equation (A5) tells the momentum change of a ship and fluid is equal to the impulse of X_{P1} during stopping motion.

The twice integration of Eq. (A4) from zero to stopping time t_s gives the following equation.

$$-(m+m_x)H_R = \frac{1}{2}X_{P1}t_S{}^2 \tag{A6}$$

H_R stands for a head reach or a stopping distance.
Substituting Eq. (A5) into Eq. (A6) leads to the following equation.

$$-X_{P1}H_R = \frac{1}{2}(m+m_x)V_i^2 \tag{A7}$$

Equation (A7) tells the kinetic energy of a ship and fluid at the initial condition is equal to the work done by X_{P1} during the stopping motion.
Equations (A7) and (A5) give representations for H_R and t_s, respectively, as follows.

$$\begin{cases} H_R = \dfrac{1}{2}\dfrac{(m+m_x)V_i^2}{-X_{P1}} \\ t_S = \dfrac{(m+m_x)V_i}{-X_{P1}} \end{cases} \tag{A8}$$

The first equation of Eq. (A8) tells that H_R is proportional to V_i^2.

Equations (A5) and (A7), or Eq. (A8), also tell a simple relation of H_R and t_s as follows.

$$H_R = \frac{1}{2}V_i t_S \tag{A9}$$

Equations (A8) and (A9) are equivalent to those derived by Yoshimura (1994) in a following way. He assumed a constant X_{P1}, resolved Eq. (A3) using the analytical solution of a Riccati's differential equation with constant coefficients, expanded the solution in series, and approximated it by the leading term. His procedure, as the result, neglects the first term of the right-hand side of Eq. (A3), which leads to the same results presented here.

Marine Design XIII – Kujala & Lu (Eds)
© 2018 Taylor & Francis Group, London, ISBN 978-1-138-34076-3

Collision risk factors analysis model for icebreaker assistance in ice-covered waters

M.Y. Zhang, D. Zhang & X.P. Yan
Intelligent Transportation Systems Research Center, Wuhan University of Technology, China
National Engineering Research Center for Water Transport Safety (WTS), WoTS, China

F. Goerlandt & P. Kujala
School of Engineering, Department of Mechanical Engineering, Marine Technology, Aalto University, Espoo, Finland

ABSTRACT: With the global warming and a large amount of sea ice melting, the available Arctic Sea Route has greatly enhanced the value of Arctic shipping. Ship operations under icebreaker assistance have become an essential way to facilitate the safe navigation of merchant vessels sailing through the Arctic Sea Route in ice-covered waters, but they can also put the crew and the ship in danger caused by a possible collision between the assisted ship and the icebreaker. In this paper, a dedicated Human and Organizational Factors (HoFs) model of ship collision accidents between an assisted ship and an icebreaker is developed and analyzed with the aim to identify and classify collision risk factors. A modified model of the Human Factors Analysis and Classification System (HFACS) for collision accidents between a ship and an icebreaker in ice-covered waters is proposed, which helps to analyze ship collision reports. An important guidance for the risk control of ship collisions during icebreaker assistance in ice-covered waters is provided for policy makers and shipping companies.

1 INTRODUCTION

With the global warming and a large amount of sea ice melting, the extremely valuable Northern Sea Route (NSR) has led to an increased interest in Arctic activities of ships (Fu et al., 2017). In this area, navigational operations under icebreaker assistance are key to the success of the safe navigation of merchant vessels (Zhang et al., 2017; Montewka et al., 2015; Valdez Banda et al., 2015). It is very difficult to ensure the safety of navigation in Arctic waters when vessels sail independently facing harsh conditions, such as the presence of sea ice, low temperatures, electromagnetic interference, and other complex environmental conditions (Stoddard et al. 2015, Goerlandt et al., 2016; Fu et al., 2017; Khan et al., 2017; Ostreng et al., 2013). At the same time, many merchant vessels lack the capability of ice-breaking, so they are unable to sail independently in a harsh ice environment, which can easily lead to ice accidents (Kum et al., 2015; Zhang et al., 2017; Fu et al., 2016). Hence, navigational operation under icebreaker assistance represents a typical model of ship operation in ice-covered waters. In 2016, 62.5% of General Cargo Carriers were under icebreaker assistance in the ice-covered waters of the NSR which provided by Northern Sea Route Information Office (Transit Statistics 2011–2016). In addition, icebreaker assistance operations also play an essential role in the ice-covered waters of the Baltic Sea in winter. The numbers of vessels under icebreaker assistance during the icebreaking season in the Baltic Sea in different years are shown in the following picture taken from *Baltic Sea Icebreaking Report 2007–2016* provided by the *Baltic Organization.*

Icebreaker assistance is a widespread method used in navigation in ice-covered waters. Navigational operations under icebreaker assistance are organized into four identified icebreaker operations: *Escort operations, Convoy operations, Breaking a ship loose operations* and *Towing operations* (Goerlandt et al., 2017; Valdez Banda et al., 2015), where escort operations and convoy operations are key to the success of the safe navigation of merchant ships. Convoy operations are similar to escort operations, where several ships follow an icebreaker at a short distance in case the ice channel is filled with ice cakes (Zhang et al., 2017). Escort and convoy operations under icebreaker assistance reduce the risk of frequent accidents, such as ice collisions and propeller or rudder damage. Nevertheless, collision accidents do occur between icebreakers and assisted ships. The statistics of accidents occurred in ice-covered waters of Russian sea area (Loanov et al., 2013) and Finnish

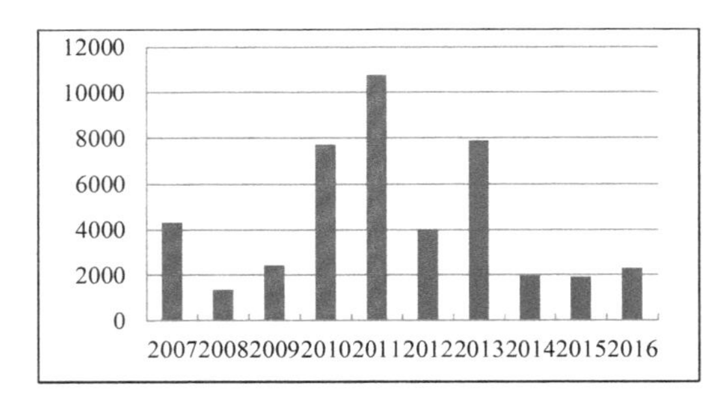

Figure 1. The number of assisted vessels under icebreaker assistance in the Baltic Sea in wintertime (2007–2016). Source: Baltic Sea Icebreaking Report (2007–2016).

sea area (Valdez Banda O.A. 2017) are presented in Figure 2. It can be seen that in Finnish sea area the percentage of collisions is 48% out of all accidents, and 95% out of all accidents under icebreaker assistance, the analysis of statistics and ship collision risk factors in ice-covered waters also indicates that ship collision accidents under icebreaker assistance should be avoided.

Collision risks between icebreakers and assisted ships sailing within a close distance cannot be ignored in ice-covered waters. In the scientific literature, the risks of collisions under icebreaker assistance are different from other ship collision accidents. Accidents in icebreaker assistance are more likely to occur than in open water conditions, but typically have lower severity in terms of consequences (Zhang et al., 2014; Franck and Holm Roos, 2013; Sulistiyono et al., 2015; Goerlandt et al. 2017). Accordingly, it is meaningful to investigate collision risk factors under specific conditions. There exists literature on accidents analysis in open waters and ice-covered waters. The risks of ship collisions are assessed, which is of signification importance for narrow, shallow and busy waterways (Qu et al., 2011; Klanac et al., 2010; Zhang et al., 2016). Furthermore, the analysis of the risks of navigational operations in ice-covered waters suggests that escort and convoy operations under icebreaker assistance are quite dangerous operations performed in the ice-covered waters. Overall, collisions between assisted ships and icebreakers present the most significant risk (Valdez Banda et al., 2016; Goerlandt et al. 2017). A root cause analysis method is presented to analyze the risks of collisions and grounding in Arctic waters, which aims at proposing a recommendation to reduce the occurrence probability based on fuzzy fault tree analysis (Kum et al., 2015). An Arctic shipping accident scenario is analyzed to identify essential accident risk factors in a potential accident scenario (Afenyo et al., 2017). Risk analysis

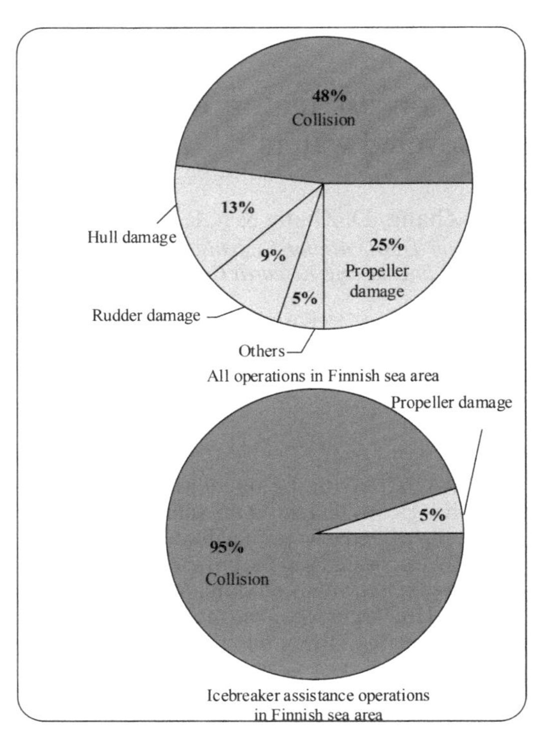

Figure 2. The statistics of accidents occurred in ice-covered waters in Finnish Sea area (Valdez Banda O.A. 2017).

models of ships stuck in ice are proposed (Fu et al., 2014, 2016; Montewka et al., 2015). Another line of work focuses on the application of risk-based design principles to Arctic shipping (Bergström et al. 2015, Ehlers et al. 2017).

However, these studies are limited in terms of the risk analysis of typical operational conditions or accidents, such as collisions between ships or a ship and ice, grounding accidents, and ship stuck incidents in ice-covered waters, not focusing specifically on collision risk factors during icebreaker assistance operations in ice-covered waters.

Icebreaker assistance operations in ice-covered waters refer to a team navigation system consisting of an icebreaker and an assisted ship. The detuning of the navigational conditions between the icebreaker and the assisted ship is the cause of a collision accident after a change in the team navigational system. Accordingly, human and organizational factors are main factors contributing to the occurrence of collision accidents. In the scientific literature, a number of human error analysis models and frameworks have been proposed to aid in the understanding of faults and errors related to human and organizational factors in complex systems where such accidents occur, such as the four stage information processing model presented by Wickens et al. (1988), the SHEL model (Edwards, 1972), the multiple SHEL model (IMO, 1999), and the GEMS (Generic Error Modeling System) proposed by Reason et al. (1990). These models and frameworks focus on the human errors of operators.

The framework of the Human Factors Analysis and Classification System (HFACS) was presented by Wiegmann et al., 2003, which was used to classify and identify contributing factors in accident factors analysis. The HFACS-framework was used to analyze maritime accidents (Chauvin et al., 2013; Chen et al., 2013), grounding accidents (Mazaheri et al., 2015) and road traffic accidents (Baysari et al., 2008; Reinach et al., 2006; Patterson et al., 2010), and classify and identify fundamental risk factors based on accident reports.

In this paper, an analysis model of ship collision risk factors under icebreaker assistance is established based on the HFACS to solve the problem of accurate classification and identification of collision risk factors under icebreaker assistance in ice-covered waters. In particular, the systematic and multi-factorial analysis of collision factors under icebreaker assistance is presented, which aim at identifying and classifying collision risk factors. The research relies on the HFACS, which are utilized to identify and classify collision risk factors that are mentioned in reports on accidents between icebreakers and assisted ships in ice-covered waters. In this paper, a dedicated Human and Organizational Factors (HoFs) model for collision accidents between assisted ships and icebreakers is developed and analyzed, by reviewing collision accident reports, and identifying and classifying collision risk factors. A collision risk factors analysis method for ship collision accidents between assisted ships and icebreakers in ice-covered waters, named the HFACS-SIBCI model, is proposed.

The rest of this paper is organized as follows. Section 2 describes the theoretical framework along with the problem statement. The HFACS-SIBCI model of collision accidents under icebreaker assistance in ice-covered waters are presented to classify and identify ship collision risk factors based on accident reports in Section 3. We examine ship collision risk factors using the statistics of factors mentioned in accident reports in order to evaluate the proposed method's performance in Section 4. Section 5 presents our conclusions.

2 THEORETICAL FRAMEWORK

According to accident statistics and scientific literature, accidents caused by human and organizational factors (HoFs) account for 90% of the total number of maritime accidents (Chauvin et al., 2011; Hetherington et al., 2006). At the same time, the lack of system coordination after detuning the navigational conditions between the icebreaker and the assisted ship resulting from human and organizational factors without effective Risk Control Options (RCOs) leads to ship collision accidents. Therefore, human errors and hidden organizational factors play vital roles in the ship collision risk factors classification and identification under icebreaker assistance. In view of this, this paper proposes a HFACS-based framework of ship collision risk factors under icebreaker assistance to solve the problem of the identification of human and organizational factors in ship collision risk factors classification under icebreaker assistance. We discuss the initial framework of the HFACS in what follows. The research flowchart is presented in Figure 3.

The initial Human Factors Analysis and Classification System (HFACS) framework (Wiegmann et al., 2003) consists of four layers: *organizational factors*, *unsafe supervision*, *preconditions for unsafe acts*, and *unsafe acts*. Reinach and Viale (2006) proposed a fifth layer, *External factors*. They believed that the identification of accident risk factors should consider the economy, law and policy as a supplement in the HFACS. Chauvin et al. (2013) developed a model to analyze human and organizational factors in maritime accidents using the five layers of the HFACS. The framework of the HFACS-Ground with five layers was presented to classify and identify contributing factors using accident reports and incident reports by Mazaheri et al. (2015). In these applications of the HFACS framework to specific contexts, the contributing factors of each layer are interpreted in specific situations considering accidents. Overall, the HFACS framework is approved by scientific literature for risk assessment and risk analysis, where the factors of each layer change continuously based on the research object (Chauvin et al., 2013; Chen et al., 2013; Mazaheri et al., 2015). This paper constructs

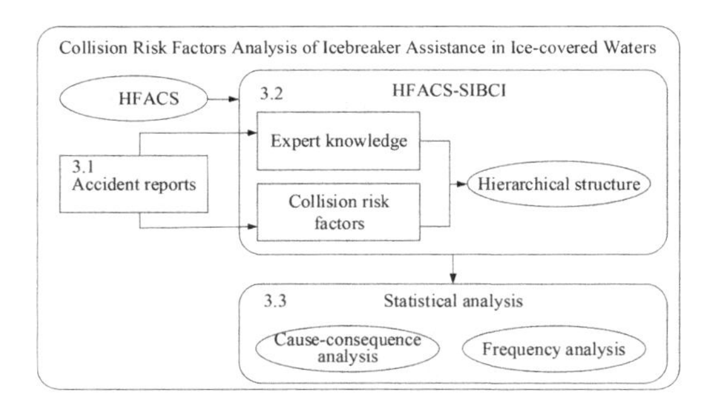

Figure 3. The flowchart of the collision risk factors analysis of icebreaker assistance in ice-covered waters.

a ship collision risk factors analysis model of an icebreaker and an assisted ship in ice-covered water based on the HFACS framework.

3 METHODOLOGY

3.1 *Collision accident reports*

Official accident reports play an essential role in risk factors analysis and being analyzed by an accident investigation board usually present valuable and detailed information about accidents (Mazaheri et al., 2015). We processed 17 accident reports on ship collision accidents between icebreakers and assisted ships in ice-covered waters. We utilized 14 ship collision reports on icebreaker assistance selected from *Swedish Accident Investigation Board (SHK)*, two collision reports from the UK *Marine Accident Investigation Branch (MAIB)* and one collision report from Russian *FleetMon*. A total of 17 accidents during 1989–2017 freely accessible to the public were analyzed using the proposed approach. 16 collision reports contained detailed information including the summary, general description of the ship, external conditions and conclusion of the investigation. One collision report only described the process of the collision accident.

3.2 *HFACS-Ship-Icebreaker collision in Ice-covered waters (HFACS-SIBCI)*

In this paper, the 17 accidents during 1989–2017 are used for detailed analysis. The HFACS-based ship collision risk analysis model of the HFACS-SIBCI (HFACS-**S**hip-**Ic**e**b**reaker **C**ollision in **I**ce-covered waters) is established to classify and identify ship collision factors. Only the collision risk factors mentioned in the accident reports were considered and further classified based on the proposed model. The HFACS-SIBCI model consists of five ship collision risk analysis levels and 28 classification categories, as shown in Figure 3. The HFACS-SIBCI is established as a five levels framework, which is similar to the HFACS-Coll, HFACS-Grounding and HFACS-MA (Chauvin et al., 2013; Chen et al., 2013; Mazaheri et al., 2015). In particularly, the 28 classification categories contain collision fundamental risk factors affecting collision accidents under icebreaker assistance.

The systematic methodology of the HFACS and the corresponding risk factor classification can assist in reducing the shortcomings of the subjective bias, experience restrictions and accident information omission in investigations and analyses of ship collision risk factors under icebreaker assistance in ice-covered waters. Accordingly, this paper constructs the ship collision risk factors identification and classification model of ship collisions under icebreaker assistance using the HFACS-SIBCI model, as shown in Figure 4, to solve the problem of ship collision risks under icebreaker assistance. This paper retains the four original levels of the HFACS framework: *(1) Unsafe acts of the operator; (2) Preconditions for unsafe acts; (3) Unsafe supervision; (4) Organizational factors*, and supplements them with *(5) External factors* causing icebreaker-ship collision accidents during icebreaker assistance; and further proposes the HFACS-SIBCI model with five levels.

3.2.1 *Analysis on risk factors of ship collisions under icebreaker assistance*

The contributing factors mentioned in the accident reports were identified as risk factors. In this paper, in order to establish collision risk factors classification, the collision accident risk factors analysis model is presented based on the HFACS-SIBCI. First, ship collision risk factors under icebreaker assistance are identified using the five-layer HFACS-SIBCI model introduced in Section 3.2. The classification model is utilized to classify ship collision factors based on the classification categories that contain the five collision risk analysis levels, and 28 classification categories, such as *decision errors*, *technical errors*, *legislation gaps* and so on. The HFACS-SIBCI accident risk factors classification model is shown in Figure 5.

In order to classify accurately ship collision risk factors under icebreaker assistance, the understanding of the HFACS-SIBCI accident risk factor analysis model is required.

3.2.2 *Ship collision risk factors classification and hierarchical structure model*

Icebreaker assistance operations are typical team operations in ice-covered waters involving icebreakers and assisted ships sailing in complex environments with harsh weather conditions. Ship collision accident reports under icebreaker assistance in ice-covered waters and accident research literature are analyzed according to experts' knowledge. The collision contributing factors mentioned in the accident reports are considered and classified based on the proposed model. Then, the HFACS-SIBCI model constructed in Section 3.2 is used to classify and identify the collision risk factors, namely, *External factors, Organizational factors, Unsafe supervision, Preconditions for unsafe acts* and *Unsafe actors*, of the navigational system in ice-covered waters. At the same time, the hierarchical structure of ship collision risk factors is constructed based on the HFACS-SIBCI.

We classify ship collision accidents between icebreakers and assisted ships and establish the hierarchical structure of ship collision risk factors.

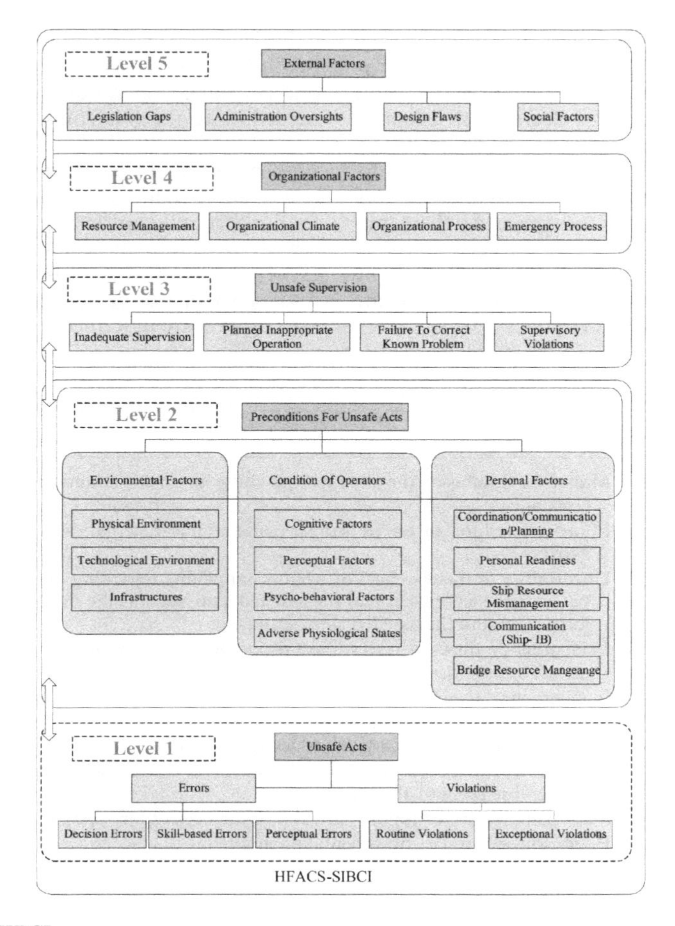

Figure 4. HFACS-SIBCI.

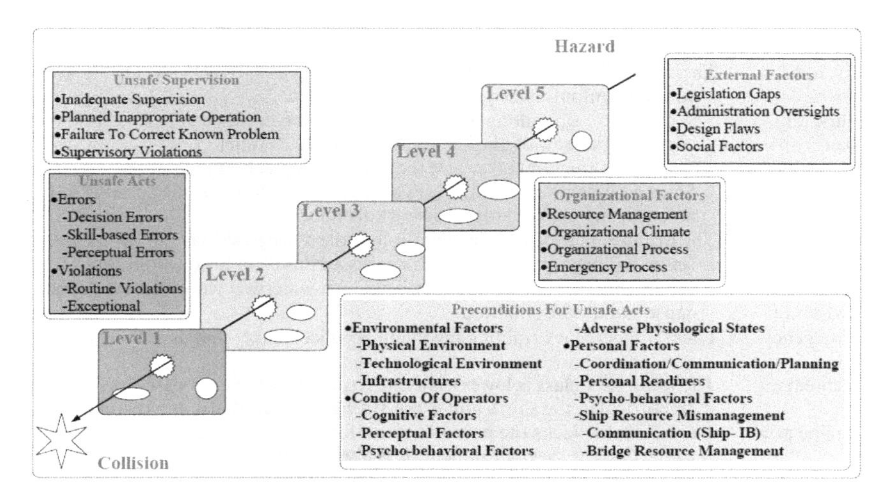

Figure 5. Accident risk factors classification model based on HFACS-SIBCI.

The collision risk classification procedure is presented as follows. First, we preliminary select ship collision contributing factors mentioned in the accident reports described in Section 3.1. Second, ship collision factors are identified and some other risk factors are presented by experts. Even if we do not have many accident reports, this way we will not miss ship collision risk factors. At the same time, literature is referenced to check the results regarding ship collisions in open water (Chauvin et al., 2013) and in Arctic ice-covered waters (Kum et al., 2015), as shown in Table 1.

The results on ship collision factors are classified by four experts who have experience to carry out assistance operations in ice-covered waters, according to the five collision risk levels, namely, *unsafe acts, preconditions for unsafe acts, unsafe supervision, organizational factors and external factors*. If more than three classification results of the four experts are consistent, they are adopted. Otherwise, we adopt the classification results of the expert who is always consistent with other experts' results, which is discussed by the fours experts. The four experts are described as follows.

Table 1. Description of ship collision risk factors under icebreaker assistance.

	Factors	Description
1	Maneuver failures of the assisted ship	Maneuver failures of the assisted ship cause an unsafe situation during icebreaker assistance.
2	Maneuver failures of the icebreaker	Maneuver failures of the icebreaker cause an unsafe situation during icebreaker assistance.
3	Lack of situational awareness	Uncertainty or unawareness of what is happening regarding the dangerous situation between the icebreaker and the assisted ship, such as the distance between the icebreaker and the assisted ship, speed etc.
4	Negligence	In emergency, the crews fail to take proper actions preventing the process of the unsafe situation.
5	Judgment failures	In emergency, the chosen action is inadequate or wrong, resulting in an undesired state.
6	Ice conditions	Ice conditions ranging from slush ice to solid pack. Ice conditions can be defined by ice concentration, ice thickness and ice type. Such as POLARIS, EGG CODE, etc.
7	Ice ridge	The edge of the ice is superimposed, which is called ice ridge, and it is easy to cause sudden breaking of the icebreaker.
8	Bad visibility	Poor visibility due to fog or snow that influences radar visibility.
9	Snow or rain weather	Hazardous natural environmental phenomena.
10	Engine failure	Mechanical failure related to the power of the icebreaker.
11	Steering gear failure	Mechanical failure related to the control of the course.
12	Anti-collision system failure	Mechanical failure related to the equipment of anti-collision, such as ECDIS, ARPA etc.
13	Communication equipment failure	Mechanical failure related to the equipment of communication between the icebreaker and the assisted ship.
14	Poor communication between ships	Given emergency, there is a communication gap between the icebreaker and the assisted ship regarding cooperative actions. Or there is misunderstanding.
15	Improper route selection	The design of the route makes it hard or dangerous to navigate. Or the severe ice environment influences the navigation system.
16	Wrong course of icebreaker	The icebreaker sails along a wrong course, which causes a collision between the ice edge and the bow of the icebreaker in the ice channel.
17	Over safety speed	The speed of the icebreaker and the assisted ship is higher than the standard set by Sailing Directions, which causes an unsafe situation. Or the speed is so high that the assisted ship cannot stop in a short distance.
18	Unmaintained safety distance	The distance between the icebreaker and the assisted ship is shorter than the standard set by Sailing Directions, which causes an unsafe collision situation.
19	Deviation from suggested route	Not being in the planned route or being in a waterway with severe ice conditions causes an unsafe situation.
20	Lack of emergency operation	Lack of emergency training involving both icebreaker and assisted ship.
21	Lack of icebreaking ability	The icebreaker class is lower than one required by the current ice environment. In particular, the ship's hull is not strong enough. It also is a relative parameter.
22	Lack of engine power	The icebreaker lacks the power to move forward, which causes a sudden break related to the current ice environment. It is also a relative parameter.
23	Anti-collision rule gap	Given emergency, there is no unified ship collision avoidance rule during icebreaker assistance, which results in an unsafe collision situation.

- One captain: the captain of a research ship with more than fifteen years' navigation experience in ice-covered waters, including the experience of carrying out icebreaker assistance operations.
- One professor: a professor engaged in navigation safety in Arctic ice-covered waters for more than twenty years, conducting research on communication equipment for navigation in polar conditions.
- Two pilots: each of them with more than five years of experience in icebreaker assistance operations in Bohai Bay of China during wintertime.

When the ship collision factors are classified, the hierarchical structure model of ship collision risk factors is established, as shown in Figure 6.

3.3 *Statistical analysis*

Each ship collision accident is caused by many factors, where main contributing factors are different for different accidents. The proposed statistical analysis procedure can be divided into two steps: (a) the contributing collision factors are selected one by one in the 17 accidents, and (b) hazard situations caused by the contributing collision factors are defined and analyzed, where this step similar to Cause-Consequence Analysis (Chen et al., 2013). At the same time, the consequences, such as hazard situations, caused by the contributing collision factors are elaborated and defined. For example, on 20th of January 2011 at 0057 LT, a ship collision happened between an icebreaker and an assisted ship during an icebreaker assistance operation at 65°05.1'5, 026°41.0'1. The icebreaker was damaged to the rubber fender of the towing notch. The escorted ship got a 1.5 m hole in the port bow, and was also damaged to the plating and frames in ice-covered waters. This ship collision accident is analyzed based on the HFACS-SIBCI model according to the two steps shown in Figure 7.

According to the proposed procedure, the statistical analysis of the collision risk factors present in the 17 accidents is conducted according to the accident reports, as shown in Figure 8.

The statistics on the occurrence frequency of the corresponding accidents shows that *Preconditions for unsafe acts* and *unsafe acts* exert greater impact on ship collision accidents under icebreaker assistance, where the factor *ice ridge [L2–2]* is related to all 17 accidents. In addition, some collision risk factors are not mentioned in the 17 accident

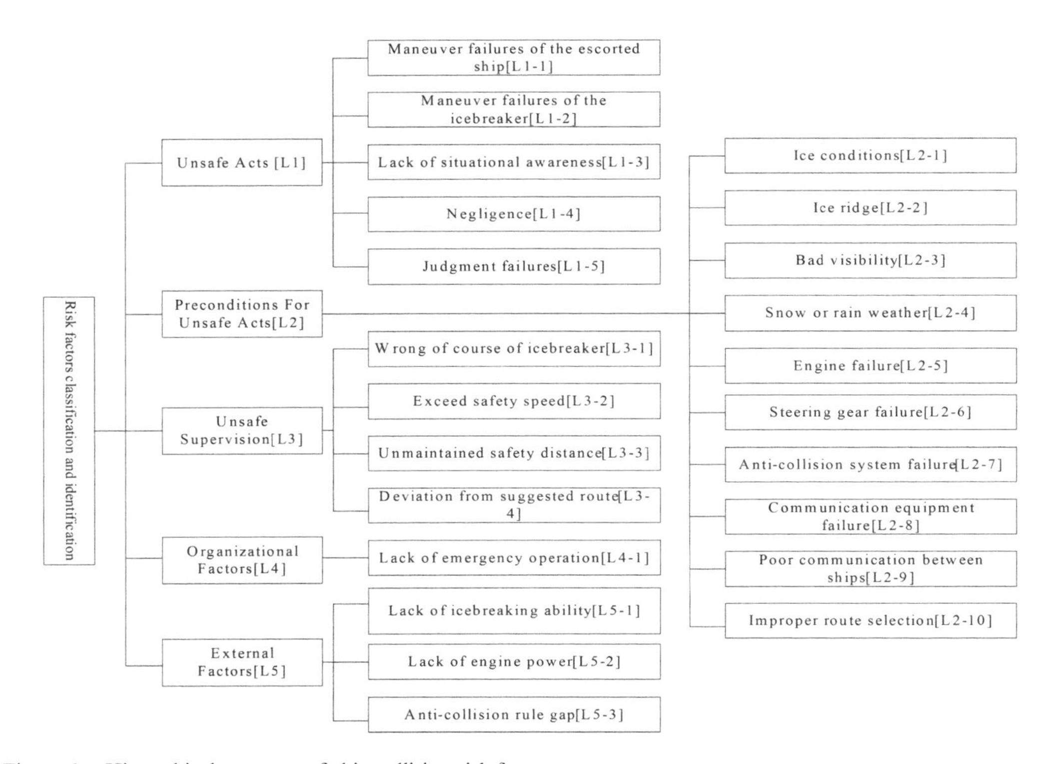

Figure 6. Hierarchical structure of ship collision risk factors.

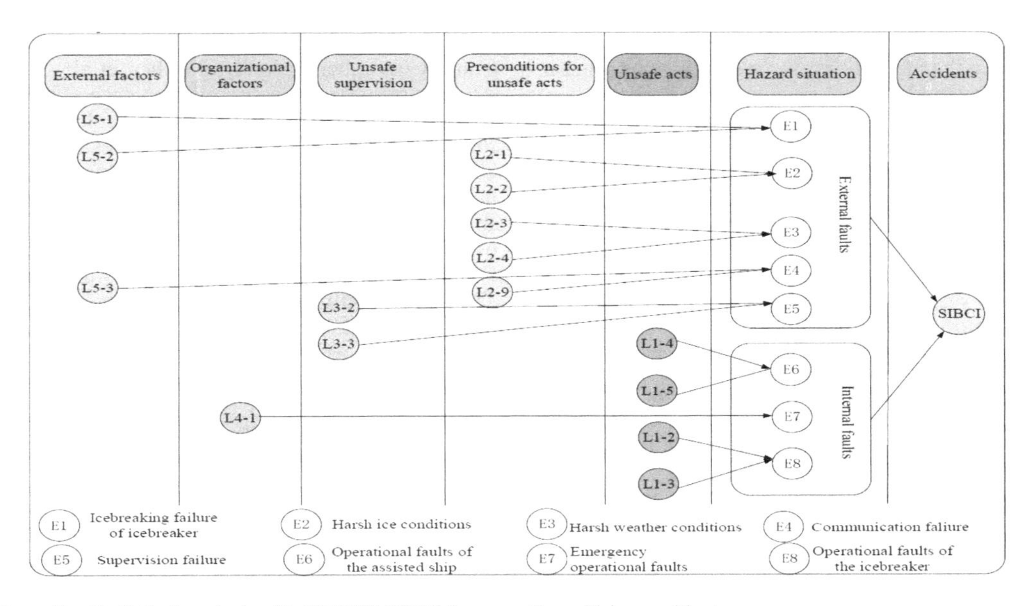

Figure 7. Statistical analysis with HFACS-SIBCI focus on the collision accident.

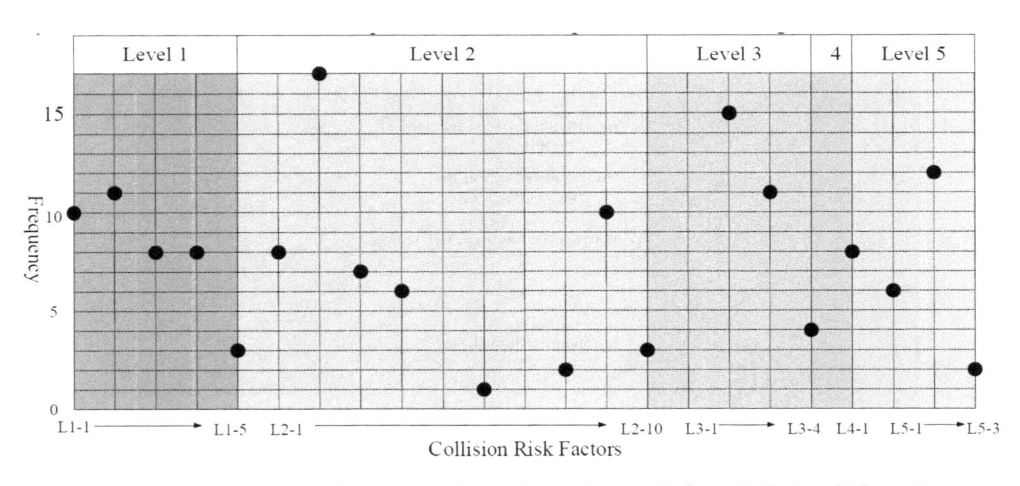

Figure 8. Frequency of collision risk factors under icebreaker assistance (Refer to Table 1 and Figure 6).

reports, such as *engine failure [L2–5], anti-collision system failure [L2–7] etc.*, but they are also main collision risk factors according to the experts and literature described in Section 3.2.2. So, the frequencies of collision factors are different. But they all can cause ship collision accidents. In brief, the classification and identification of the risk factors of collisions between icebreakers and assisted ships can be realized using the proposed model, which is a basis for the risk analysis modeling of collision accidents.

4 DISCUSSIONS

In the paper, the HFACS-SIBCI model is established, and the contributing factors of collision accidents are identified and classified. Then, the hierarchical structure model of the ship collision risk factors is presented according to the results of the classification. At the same time, statistical analysis is carried out. The results of the research indicate that the most frequent active failure mentioned in the accident reports is *"Preconditions for unsafe acts"*. *"Unsafe acts"* contribute second, followed by *"External factors"*, *"Unsafe supervision"* and *"Organizational factors"* (Figure 7), which can provide theoretical guidance to policy makers and shipping companies regarding the prevention of ship collision accidents between icebreakers and assisted ships in ice-covered waters.

The ship collision accident reports were taken from three different accident investigation boards. At the same time, the number of the ship collision reports is small, which is a limitation of the research, but it is enough for an exploratory qualitative analysis. However, this may question the reliability of the ship collision accident reports when used to determine ship collision risk factors in accident risk modeling. Besides, the research involves uncertainties. The research relies on accident reports formulated in a specific format, where a larger number of accident reports would reduce the uncertainty of the collision risk factors. We will improve the research in the terms of uncertainty analysis in the future. The proposed method of collision risk factors analysis has also some potential benefits to the analysis of collision risks between icebreakers and assisted ships in ice-covered waters.

5 CONCLUSIONS

The paper provides comprehensive analysis of ship collision risk factors under icebreaker assistance using the HFACS-SIBCI model based on accident reports. Ship collision risk factors are identified and classified into five levels, including *"Preconditions for unsafe acts", "Unsafe supervision", "External factors", "Organizational factors" and "Unsafe acts"*, where the HFACS-SIBCI model-based hierarchical structure of ship collision risk factors is shown in Table 1 and Figure 6. The results obtained are promising as they can help to understand ship collision factors during icebreaker assistance. In addition, several advantages of the proposed method for ship collision risk factors analysis are elaborated with the focus on collision risk analysis under icebreaker assistance in ice-covered waters.

ACKNOWLEDGMENT

This study was supported by the National Science Foundation of China (NSFC) under Grant No. 51579203 and 51711530033 and the Excellent Dissertation Cultivation Funds of Wuhan University of Technology under Grant No. 2016-YS-041. This work has also been supported by the H2020-MCSA-RISE project "RESET—Reliability and Safety Engineering and Technology for large maritime engineering systems" (No.730888). We thank professor Wu from Wuhan University of Technology, Captain Wang from Polar Research Institute of China and pilot Wang and Tao from Pilot Station of Tianjin. We are also grateful to Chengpeng Wan, Chi Zhang and Kai Zhang for their assistance in this research.

REFERENCES

Afenyo, M., Khan, F., Veitch, B., and Yang, M. (2017). Arctic shipping accident scenario analysis using Bayesian Network approach. *Ocean Engineering*, 133, 224–230.

Baltic Organization. Baltic Sea Icebreaking Report 2007–2016. Retrieved from <http://baltice.org/> on 09th October, 2017.

Baysari, M. T., Mcintosh, A. S., & Wilson, J. R. (2008). Understanding the human factors contribution to railway accidents and incidents in australia. *Accident Analysis & Prevention*, 40(5), 1750.

Bergström M., Erikstad S.O.. Ehlers S. 2015. Applying risk-based design to arctic ships. Proceedings of the International Conference on Offshore Mechanics and Arctic Engineering—OMAE, Vol. 8, code 116005.

Chen, S. T., & Chou, Y. H. (2013). Examining Human Factors for marine casualties using HFACS—maritime accidents (HFACS-MA). *International Conference on ITS Telecommunications*, 391–396.

Chen, S. T., Wall, A., Davies, P., Yang, Z., Wang, J., & Chou, Y. H. (2013). A human and organisational factors (hofs) analysis method for marine casualties using hfacs-maritime accidents (hfacs-ma). *Safety Science*, 60(12), 105–114.

Edwards, E., (1972). Man and machine: systems for safety. In: Proceedings of British Airline Pilots Association Technical Symposium. British Airline Pilots Association, London, UK, pp. 21–36.

Ehlers S., Cheng F., Jordaan I., Kuehnlein W., Kujala P., Luo Y., Freeman R., Riska K., Sirkar J., Oh Y.-T., Terai K., Valkonen J. 2017. Towards mission-based structural design for arctic regions. Ship Technology Research 64(3):115–128.

FleetMon. Retrieved from <https://www.fleetmon.com/> on 09th October, 2017.

Franck, M., and Holm Roos, M. (2013). Collisions in ice: a study of collisions involving swedish icebreakers in the baltic sea. *Swedish: Linnaeus University.*

Fu, S., Yan, X., Zhang, D., Shi, J., & Xu, L. (2015). Risk factors analysis of Arctic maritime transportation system using structural equation modeling. *Port and Ocean Engineering under Arctic Conditions*.

Fu, S., Zhang, D., Montewka, J., Yan, X., and Zio, E. (2016). Towards a probabilistic model for predicting ship besetting in ice in arctic waters. *Reliability Engineering & System Safety*, 155, 124–136.

Fu, S., Zhang, D., Montewka, J., Zio, E., and Yan, X. (2017). A quantitative approach for risk assessment of a ship stuck in ice in Arctic waters. *Safety Science*, https://doi.org/10.1016/j.ssci.2017.07.001.

Goerlandt, F., Goite, H., Banda, O. A. V., Höglund, A., Ahonen-Rainio, P., & Lensu, M. (2017). An analysis

of wintertime navigational accidents in the northern baltic sea. *Safety Science*, 92, 66–84.
Goerlandt, F., Montewka, J., Zhang, W., and Kujala, P. (2016). An analysis of ship escort and convoy operations in ice conditions. *Safety Science*, 95, 198–209.
International Maritime Organization. (1999). Resolution A.884(21): Amendments to the Code for the Investigation of Marine Casualties and Incidents (resolution A.849(20)). International Maritime Organization (IMO), London, UK.
International Maritime Organization. (2015). International Code for Ship Operating in Polar Waters (POLAR CODE). Retrieved from <https://edocs.imo.org/FinalDocuments/English/MEPC68-21-ADD.1(E)>. on 17th July 2016.
Khan, B., Khan, F., Veitch, B., Yang, M., Khan, B., & Khan, F., et al. (2018). An operational risk analysis tool to analyze marine transportation in arctic waters. Reliability Engineering System Safety.169, 485–502.
Kum, S., and Sahin, B. (2015). A root cause analysis for Arctic Marine accidents from 1993 to 2011. *Safety Science*, 74, 206–220.
Lobanov, V.(2013). "Ice performance and ice accidents fleet inland and river-sea navigation. " *Internet Journal "Naukovedenie"*, 67(4), 12–16.
Marine Accident Investigation Branch (MAIB). Retrieved from <https://www.gov.uk/government/organisations/marine-accident-investigation-branch> on 09th October, 2016.
Maritime Safety Administration of the People's Republic of China. (2014). Guidance on Arctic navigation in the northeast route. China communications press.
Mazaheri, A., Montewka, J., Nisula, J., & Kujala, P. (2015). Usability of accident and incident reports for evidence-based risk modeling—a case study on ship grounding reports. *Safety Science*, 76, 202–214.
Montewka, J., Goerlandt, F., Kujala, P., and Lensu, M. (2015). Towards probabilistic models for the prediction of a ship performance in dynamic ice. *Cold Regions Science & Technology*, 112, 14–28.
Montewka, J., Hinz, T., Kujala, P., and Matusiak, J. (2010). Probability modelling of vessel collisions. *Reliability Engineering & System Safety*, 95(5), 573–589.
Northern Sea Route Information Office (NSR), Retrieved from <http://www.arctic-lio.com/nsr_transits> on 19th October, 2017.
Ostreng W, Eger KM, Floistad B, Jorgensen-Dahl A, Lothe L., et al. (2013) Shipping in Arctic waters: a comparison of the Northeast, Northwest and trans polar passages. Heidelberg: Springer.
Patterson, J. M., & Shappell, S. A. (2010). Operator error and system deficiencies: analysis of 508 mining incidents and accidents from queensland, australia using hfacs. *Accident Analysis & Prevention*, 42(4), 1379.
Qu, X., Meng, Q., and Suyi, L. (2011). Ship collision risk assessment for the Singapore Strait. *Accident Analysis & Prevention*, 43(6), 2030–2036.
Reason, J. (1990). Human Error. Cambridge University Press, New York.
Reinach, S., & Viale, A. (2006). Application of a human error framework to conduct train accident/incident investigations. *Accident Analysis & Prevention*, 38(2), 396–406.
Stoddard M.A., Etienne L., Fournier M., Pelot R., Beveridge L. 2015. Making sense of Artcic maritime traffic using the Polar Operational Limits Assessment Risk Indexing System (POLARIS). IOP Conference Series: Earth and Environmental Science 34(1).
Sulistiyono, H., Khan, F., Lye, L., and Yang, M. (2015). "A risk-based approach to developing design temperatures for vessels operating in low temperature environments." *Ocean Engineering*, 108, 813–819.
Swedish Accident Investigation Board (SHK). Retrieved from <http://www.havkom.se/en/> on 09th May, 2016.
Transport Canada. Arctic ice regime shipping system (1988). User assistance package for the implementation of Canada's Arctic ice regime shipping system (AIRSS).
Valdez Banda O.A. (2017). Maritime risk and safety management with focus on winter navigation. *Finland: Aalto University*.
Valdez Banda, O. A., Goerlandt, F., Kuzmin, V., Kujala, P., and Montewka, J. (2016). Risk management model of winter navigation operations. *Marine Pollution Bulletin*, 108(1–2), 242.
Valdez Banda, O. A., Goerlandt, F., Montewka, J., and Kujala, P. (2015). A risk analysis of winter navigation in Finnish sea areas. *Accident Analysis & Prevention*, 79, 100.
Wickens, C., Flach, J., (1988). Information processing. In: Wiener, E.L., Nagel, D.C. (Eds.), Human Factors in Aviation. Academic, San Diego, CA, 111–155.
Wiegmann, D. A., and Shappell, S. A. (2003). A human error approach to aviation accident analysis: The human factors analysis and classification system. Reference & Research Book News.
Zhang M., Zhang D., Fu S., Yan X. (2017) A method for Arctic sea route planning under multi-constraint conditions. Proceedings of International Conference on Port and Ocean Engineering under Arctic Conditions, June 10–16, 2017, Busan, Korea.
Zhang M., Zhang D., Fu S., Yan X., Luo J., Goncharov V. (2017). Safety distance modelling for vessels under the icebreaker assistance: taking "yong sheng" and "50 let pobedy" as an example. Transportation Research Board 96th Annual Meeting, January 8–12, 2017. Washington, D.C., USA.
Zhang, D., Yan, X., Yang, Z., et al., (2014). An accident data based approach for congestion risk assessment of inland waterways: A Yangtze River case. Journal of Risk & Reliability, 228(2): 176–188.
Zhang, D., Yan, X., Zhang, J., Yang, Z., Wang, J. (2016). Use of fuzzy rule-based evidential reasoning approach in the navigational risk assessment of inland waterway transportation systems. Safety Science, 82, 352–360.
Zhang, M., Zhang D., Fu S., Yan X., Goncharov, V. (2017). Safety distance modeling for ship escort operations in Arctic ice-covered waters. Ocean Engineering, 146, 202–216.

Marine Design XIII – Kujala & Lu (Eds)
© 2018 Taylor & Francis Group, London, ISBN 978-1-138-34076-3

Collision risk-based preliminary ship design—procedure and case studies

X. Tan & J. Tao
Maritime Institute at Nanyang Technological University, Singapore

D. Konovessis
Singapore Institute of Technology, Singapore

H.E. Ang
School of Mechanical and Aerospace Engineering, Nanyang Technological University, Singapore

ABSTRACT: This paper proposes a procedure for collision-risk-based preliminary ship design. The procedure is a synthesis of a probabilistic ship-ship collision damage model, common ship design software, and several analytical parametric models. The procedure is implemented in a parametric way such that by systematically varying the parameters related to the conceptual design of the ship, e.g., the main particulars and the inner tank arrangement, the performances of large amount of design variants can be evaluated. Risk is treated as a design objective, alongside other conventional ship design merits, as opposed to a design constraint. Thus trade-offs among different objectives become explicit and can be evaluated quantitatively across the design space to achieve an optimum at an early design phase. The application of the established procedure is discussed through two case studies, one with the preliminary design of a Suezmax oil tanker, and the other one with that of a container ship. The risks of oil spill and in-hold container damage are quantified for the two ship types respectively. By using the utility function technique with several sets of predefined weighing factors, favored designs under different settings can be identified. The procedure can facilitate decision making when specific design requirements and operational profile are given.

1 INTRODUCTION

Ship-ship collision is a complicated physical process where uncertainties are present at every single stages of the interaction between the two ships (Goerlandt and Montewka, 2015), (Sormunen et al., 2014). In order to quantitatively assess the collision risk and to obtain a statistically sound cargo loss distribution, extensive amount of collision cases have to be evaluated for each design variant. Since even the preliminary design optimization can easily involve hundreds of hull variants, methods such as performing integrated dynamic simulations with the finite element method and rigid body dynamics to collect the damage data to establish the damage distributions directly are considered unsuitable for the purpose of risk-based ship design and optimization. Although historical accident statistics are harvested from the real world, it reflects only fragments of reality. Relying solely on accident statistics may lead to biases either because they cannot reflect the future or because they do not take into consideration of specific conditions, e.g. ship structural design, for different cases (Brown et al., 2000), (DNV, 2003).

This paper introduces a scheme for the preliminary design and optimization of tankers and containerships considering the risk of cargo loss due to ship-ship collision accidents, expressed as oil outflow (for the case of tankers) or in-hold cargo loss due to water ingress (for the case of containerships). Recent developments in the area of risk-based design procedures for tankers and containerships, for example, (Papanikolaou et al., 2010) and (Priftis et al., 2016), have demonstrated the viability of the approach, which is being extended to the examination of new design criteria, as above. The procedure is a synthesis of various first-principle-based analysis models and common naval architecture software. Risk is treated as a design objective as opposed to a constraint to achieve an optimal design that is a result of balancing it with other conventional ship design merits such as steel weight, calm water resistance and cargo capacity. The variables that define each design variant are the main particulars of the vessel and the inner tank arrangement parameters. In order to evaluate and compare safety level across the design variants, quantification of the risk of accidental in-hold cargo loss due to water ingress has become an important subject.

Risk is in the present work expressed as an expected cargo loss (ISO, 2002):

$$R = \sum_i p_i \cdot C_i \tag{1}$$

where, i is the index of hazardous events leading to consequences, which represents the different magnitudes of oil outflow volumes (for the case of tankers) or the different numbers of cargo loss, represented by number of damaged TEUs due to water ingress (for the case of containerships);

C is the oil outflow volume (for the case of tankers) or the number of damaged TEUs due to water ingress in cargo tanks (for the case of containerships);

p is the probability that a certain amount of outflow occurs (for the case of tankers) or a certain number of TEU damage occurs (for the case of containerships).

In the following sections of this paper, first the methodology is briefly introduced. Then its intended application is examined on two cases studies, on the preliminary design of a Suezmax crude oil tanker, and a 7,500 TEU containership, in order to gain an insight into the trade-offs among design metrics including steel weight, cargo capacity and the risk metrics as considered (oil outflow volume for the case of tankers and number of damaged TEUs for the case of containerships).

2 METHODOLOGY

2.1 *Overview*

The procedure consists of a number of analysis models, common naval architecture software and a probabilistic collision damage model. The idea is to establish a procedure that has high computational efficiency and at the same time can still take into consideration of details such as ship's main particulars and randomly distributed collision scenario variables.

Parametric ship hull model was setup using NAPA. Conventional ship performance aspects such as the intact and damage stability, calm water resistance, cargo capacity and steel weight were calculated using NAPA too. In parallel, Monte Carlo simulations were conducted to obtain the probabilistic collision damage for each design variant. The main particulars and inner tank arrangement parameters were fed into the probabilistic damage model together with the randomly generated collision scenario variables.

2.2 *Resistance, steel weight and cargo capacity*

At the initial stage of ship design, empirical parametric models for resistance and steel weight are employed to calculate the global performance of the vessel. The weight component accounted for are the hull steel, outfit and machinery weights (Watson and Gilfillan, 1977). The resistance at the design speed was calculated using Holtrop and Mennen's formulae (Holtrop and Mennen, 1982). The cargo volume of each tank is given by the geometric model of NAPA.

2.3 *Probabilistic collision damage*

The probabilistic ship-ship collision damage model given by (Van de Wiel and Van Dorp, 2011) was adopted and implemented numerically in the present study to evaluate the probabilistic collision damage and subsequent oil outflow (for the case of tankers) or in-hold cargo loss due to water ingress (for the case of containerships). The simulations were conducted by using a deterministic numerical procedure of ship-ship collisions developed by (Brown, 2001).

The inputs of the model are: the velocities of the two colliding vessels right before collision, the masses of the two vessels, the collision angle and location and the half bow entrance angle of the striking vessel. The outputs of the model are the damage properties including the length, depth and start and end location along the ship side. By knowing the damage properties and inner tank arrangement, the breached tanks can then be identified and statistics of cargo loss due to water ingress can be calculated accordingly.

3 CASE STUDIES

In this section of the paper, we present two cases studies to examine the applicability of the risk-based design procedure developed.

The first case study is on the application of the procedure in the preliminary design of a Suezmax tanker ship considering oil outflow risk. Main particulars and parameters that control the inner tank arrangement were varied systematically so that the design space was explored efficiently and effectively. For each feasible design variant, the expected oil outflow volume was calculated using the probabilistic collision damage model introduced in Section 2.3. Intact and damage stability, steel weight and calm water resistance were calculated in NAPA. Finally, the trade-offs among the design merits including risk were discussed.

The second case study is on the preliminary design and optimization of a container ship considering the risk of cargo loss as well as other conventional ship design merits such as steel weight, TEU capacity and calm water resistance. Main particulars and parameters that control the cargo tank arrangement were varied systematically so that the design space was explored efficiently and effectively. For each feasible design variant, the expected in-hold cargo loss was calculated using the probabilistic collision damage model introduced in Section 2.3. Intact and damage stability, steel weight and calm water resistance were calculated in NAPA. Finally, the trade-offs among the design merits including risk were discussed.

3.1 *Tanker case study*

A representative double hull Suezmax tanker given in (IMO, 2008a) was chosen as the reference ship. The main particulars of the ship are: $L_{OA} = 273.64$ m; $L_{BP} = 264$ m; B = 48 m; D = 23.1 m; $C_B = 0.86$; LCB (to AP) = 133 m; HDS = 2.5 m (width of double skin sides); HDB = 2.8 m (double bottom height); T = 17.11 m; Displ. = 190,243 ton; DWT = 158,982 ton; cargo tank block length = 213.14 m.

Design constraints & design space

The following design constraints were applied to either setting the bounding limits of the design variables or filtering out the designs that do not fulfill the general requirement of ship design, e.g. stability, to ensure a design space of feasible designs.

- Suezmax are the maximum marine vessels that meet the restrictions of the Suez Canal. The maximum permissible draught of Suezmax is 20.1 m to pass the canal. From 2010, the wetted surface cross sectional area of the ship is limited by 1,006 m^2, which means 20.1 meters of draught for ships with the beam no wider than 50.0 m or 12.2 meters of draught for ships with maximum allowed beam of 77.5 meters.
- The ratios *L*/*B*, *B*/*D* and B/T are limited to between 5.5 and 6.0, 1.9 and 2.1 and 2.25 and 3.75, respectively (Lamp, 2003).
- Damage stability of each design is checked against IMO's damage stability code (IMO, 1974).

In addition, the examination is restricted to the feasible designs that have a displacement falls in the range of ±10% the displacement of the reference design, so that the variants are relatively comparable in terms of ship size. In this respect, the ranges of ship design variables considered are as follows:

- L_{BP} – 248.7 m ÷ 289.5 m
- B – 45 m ÷ 50 m
- LCB – 123.1 m ÷ 147 m
- C_B – 0.818 ÷ 0.896
- D – 21.5 m ÷ 25.2 m
- HDB – 1.8 m ÷ 3.2 m
- HDS – 1.8 m ÷ 3.2 m

In order to explore the design space effectively and efficiently, low variance sequence was used to fill the design space by generating the design variables within the bounds given above. In the present study, originally 10,000 designs were generated, of which 4,023 are feasible ones that fulfill the constraints listed above.

Parametric ship model

A parametric ship hull model, Figure 1, was established with NAPA. The design variables of the ship model were varied according to the considered ranges given above.

The tank arrangement is illustrated in Figure 2.

Probabilistic distributions of collision damages

The simulated probabilistic distributions of oil outflow volume and damage extents are exemplified by the results of the reference ship as given in Figures 3a, 3b and 4. Figures 3a and 3b illus-

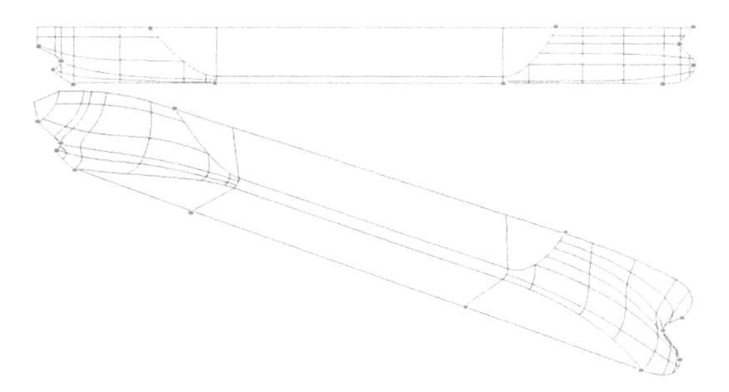

Figure 1. Parametric hull model in NAPA (the points that control the geometry of the hull are highlighted).

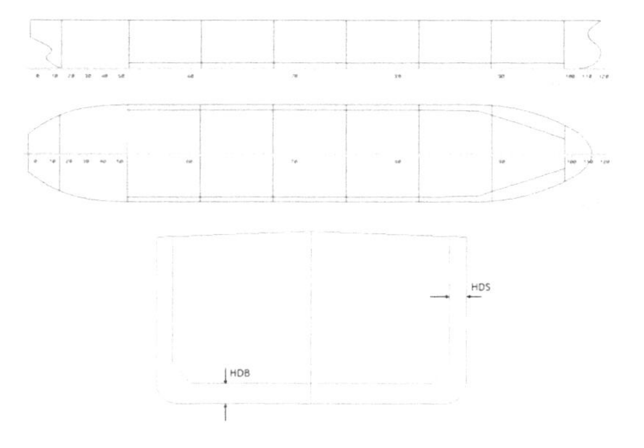

Figure 2. Tank arrangement.

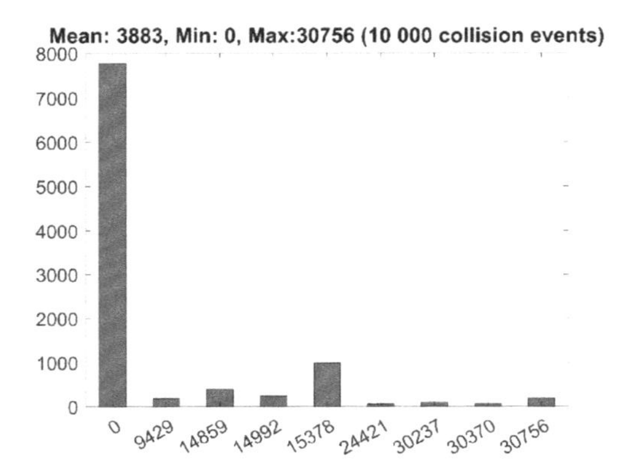

Figure 3a. Oil outflow volume for the reference design (number of collision cases simulated versus oil outflow in m³). Cases with zero oil outflow included in the distribution.

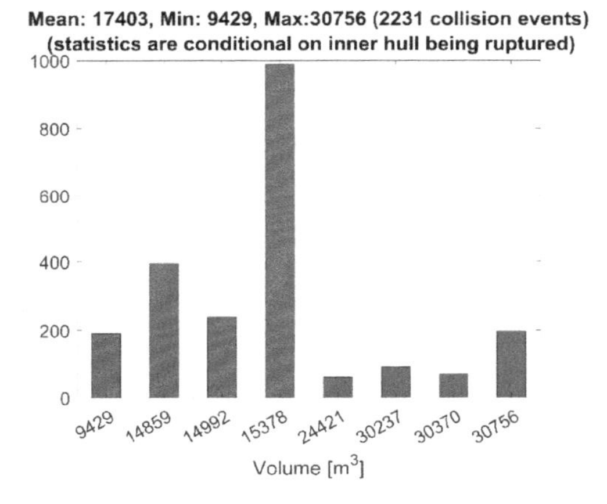

Figure 3b. Oil outflow volume for the reference design (number of collision cases simulated versus oil outflow in m³). Cases with zero oil outflow excluded from the distribution.

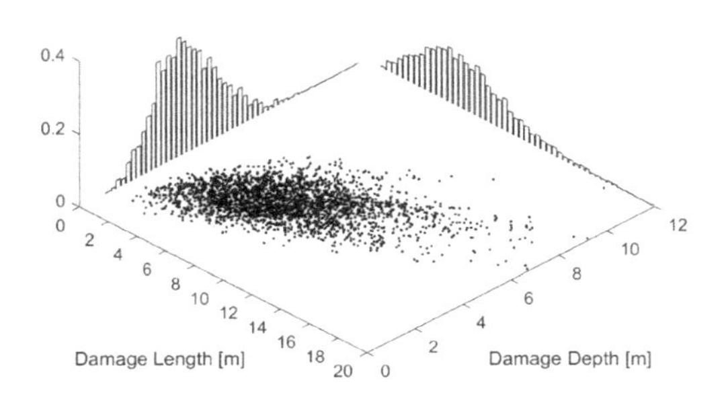

Figure 4. Joint and marginal (normalized) distributions of damage length and depth for the reference ship.

trate the distribution of number of collision cases examined versus the corresponding calculated oil outflow, when collisions with zero outflow are included (Figure 3a) or excluded (Figure 3b) from the distribution. The probability of zero outflow for the reference ship is 77.69%. The statistics are conditional on tanker being struck. Figure 4 illustrates the joint and marginal (normalized) distributions of damage length and depth for the reference tanker ship. It is clearly shown in Figure 4 that the damage length and depth are correlated.

Results

The relationships among the steel weight, calm water resistance, cargo capacity and mean oil outflow as obtained from the procedure are summarized in Figures 5 to 7. The figures indicate the trade-offs between each two conflicting objectives. Also marked in the figures are the reference design, the "Pareto front" of cargo capacity versus oil outflow, and the five most favorable designs in terms of utility levels which combines the effects of all the four objectives with serval sets of predefined weighing factors.

The designs fall on the Pareto front are the "non-dominated" set, which means that specific preference and weighting among the criteria examined should be introduced. How a favored design is subsequently to be chosen depends on the specific settings of the intended service profile of the ship, which may include, e.g. the sailing area, design life span etc. For example, it was found that within the examined design space, there did not exist a design that is better than the reference design in terms of the absolute values for all four merits. This is as expected since the reference design is believed to be an optimized one at least for the three conventional merits that do not include oil outflow risk. Nevertheless, depending on the operational profile of the vessel, relatively better designs can still be pursued by putting more weights on the dominant factor(s) and evaluating the overall performance.

In Figure 5, equal weights, i.e., no preference, were given to the four design merits, and the five most favorable designs given by ranking the utility levels of all the designs are in general concentrated in the smaller size segments. This is reasonable because all of the four merits but the cargo capacity favor a smaller ship in general.

In environmentally sensitive areas, more preferences should be given to the designs with relatively low oil outflow level with some sacrifice of economic aspects in order to avoid or mitigate the devastating consequences from oil spill. In contrast to the scenario in Figure 5, when more weights are given to oil outflow and cargo capacity, as shown in Figure 6 and Figure 7, some larger vessels can become competitive too.

3.2 *Containership case study*

A representative container ship with a nominal TEU capacity of 7,500 is adopted in the present study as reference design. The main particulars of the ship are: LOA = 323.83 m; LBP = 293.90 m; B = 42.85 m; D = 24.71 m; CB = 0.65; LCB (to AP) = 137.48 m; HDS = 2.3 m (width of double skin sides); HDB = 2.1 m (double bottom height); T = 12.05 m; T = 12.05 m; Displ. = 101,851 ton; Nominal TEU Capacity = 7,500.

Design space & feasible designs

The bounding limits for the design variables which define the design space are summarized as follows:

- L_{BP} – 290.5 m ÷ 293.9 m
- B – 15 rows ÷ 16 rows
- LCB – 131.1 m ÷ 139.6 m

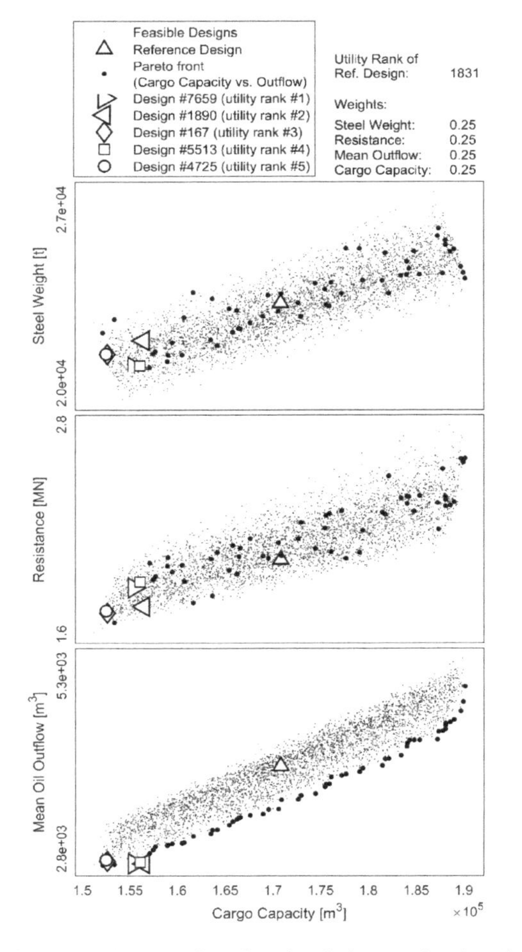

Figure 5. Scatter plots for the design merits (equal weights).

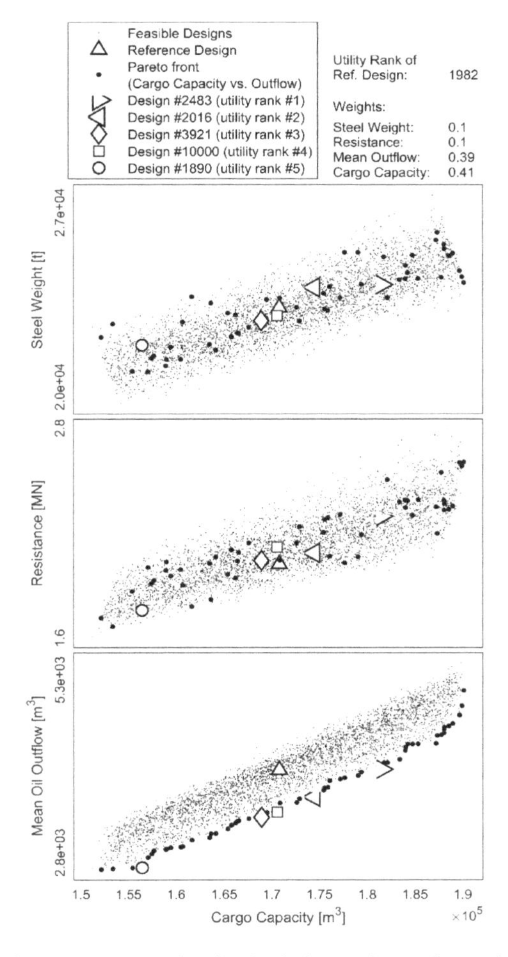

Figure 6. Scatter plots for the design merits (preference is firstly given to cargo capacity and then to oil outflow risk).

- C_B – 0.6 ÷ 0.8
- D – 9 tiers ÷ 10 tiers
- HDB – 1.8 m ÷ 2.5 m
- HDS – 2 m ÷ 2.5 m
- BHW – 0.8 m ÷ 1 m (gap size between two cargo tanks)

The beam and the depth of the vessels are represented by the number of containers in each corresponding direction. In the present study, the length of the ship in terms of container number in the longitudinal direction was not varied. The variation in L_{BP} as shown is due to the change in the gap sizes, BHW, between each two adjacent cargo tanks.

In order to explore the design space effectively and efficiently, low variance sequence was used to fill the design space by generating the design variables within the bounds given above.

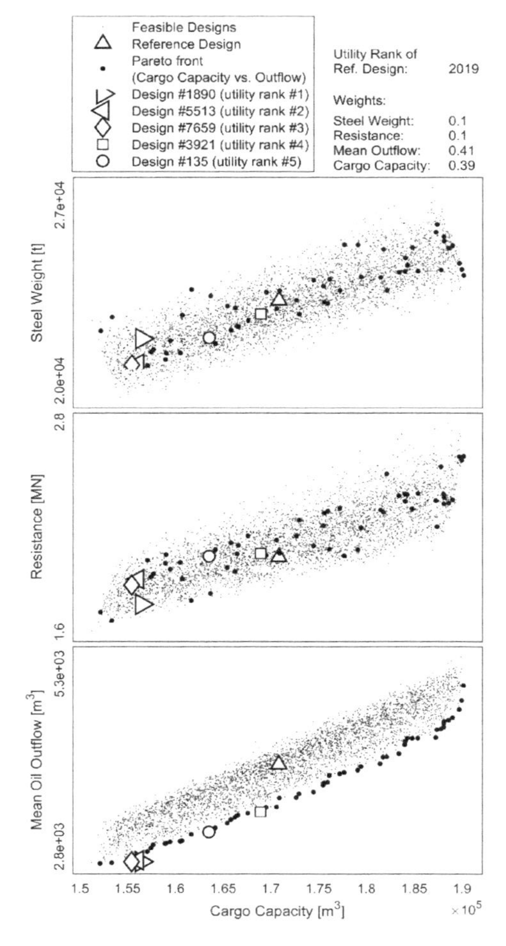

Figure 7. Scatter plots among the design objectives (preference is firstly given to oil outflow and then to cargo capacity).

The design constraints, with which feasible designs in the investigated design space can be identified, are listed below:

- The ratio B/D is limited to between 2.5 and 3.75;
- IMO Intact stability criteria (IMO, 2008b);
- IMO damage stability code (IMO, 1974);

In the present study, originally 6,500 design variants were generated, of which 4,826 are feasible ones that fulfill the constraints listed above.

The designs that have been filtered out have a large draft relative to the beam, which can lead to a less stable geometry under intact conditions. In addition, the draft of those designs are also relatively large compared with the depth, implying a smaller freeboard and hence higher probability of losing stability under damaged conditions due to water ingress through the hatches on deck.

Parametric containership model

A parametric ship hull model, Figure 8, was established with NAPA.

Loading condition

In the present study, a homogenous weight of 10 tons for the containers was used. The vessel was first loaded up to the visibility line, then intact and damage stability criteria were checked. If the criteria were not fulfilled, then a tier of on-deck containers were removed, this process is repeated until the criteria were fulfilled. Ballast tanks are empty except for the need to adjust the vessel to an upright position.

Results

In order to achieve an overall optimal design, multiple ship design merits were evaluated simultaneously, namely, the steel weight, TEU capac-

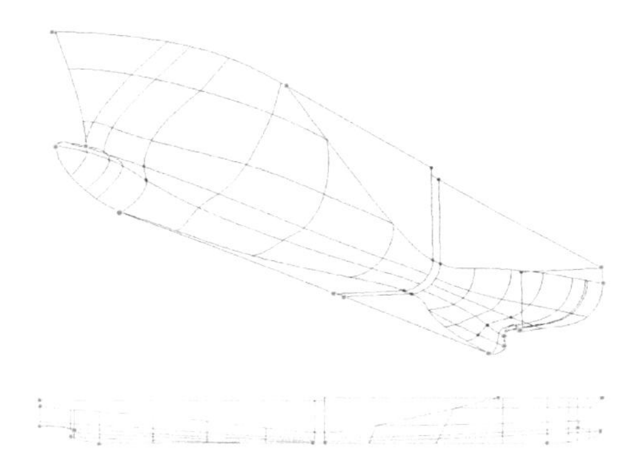

Figure 8. Parametric hull model in NAPA (the points that control the geometry of the hull are highlighted).

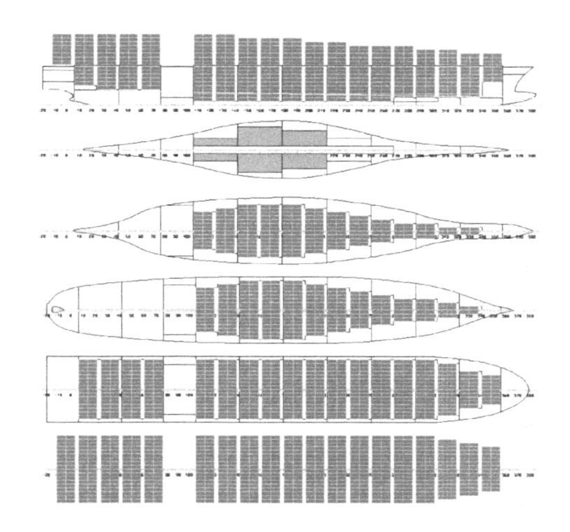

Figure 9. Side and plan views of the vessel with stacked containers (red blocks).

ity, calm water resistance and the risk of cargo loss due to water ingress following a collision accident. This is discussed in this section over the simulated results.

Figures 11 through 13 shows the scatter plots among the design merits. The designs were ranked according to their utility values combined with different weights. It can be seen that the resistance

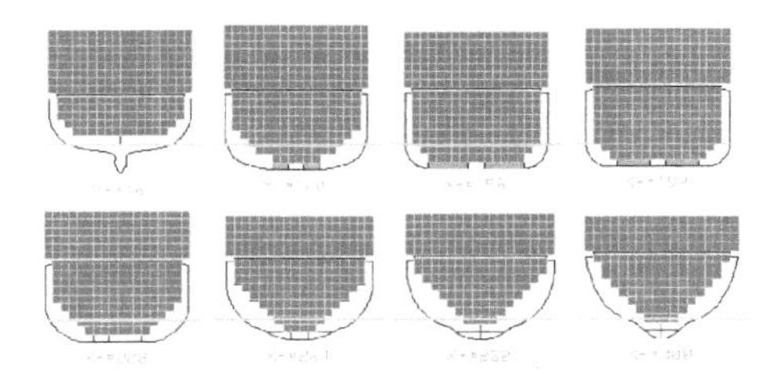

Figure 10. Front view of the vessel at different cross sections with stacked containers (red blocks).

and TEU capacity values are clustered around several discrete values. This is due to the fact that the sizes of the ship are clustered around four categories, two different beams and two different depths, and the ship size for container ships is primarily a staircase function of the beam and depth.

In Figure 11, equal weights are put on the four design merits. The five most highly ranked designs for this setting are concentrated in the smaller size segment, which is reasonable as all the merits but TEU capacity favor a smaller ship in general.

In Figures 12 and 13, more weights are put on TEU capacity and TEU damage risk. It is then seen that the top ranked designs now lie on the "Pareto front" between the TEU capacity and the TEU damage risk, which is the non-dominated set of designs. Also can be seen is that when preferences are given, larger designs can become competitive.

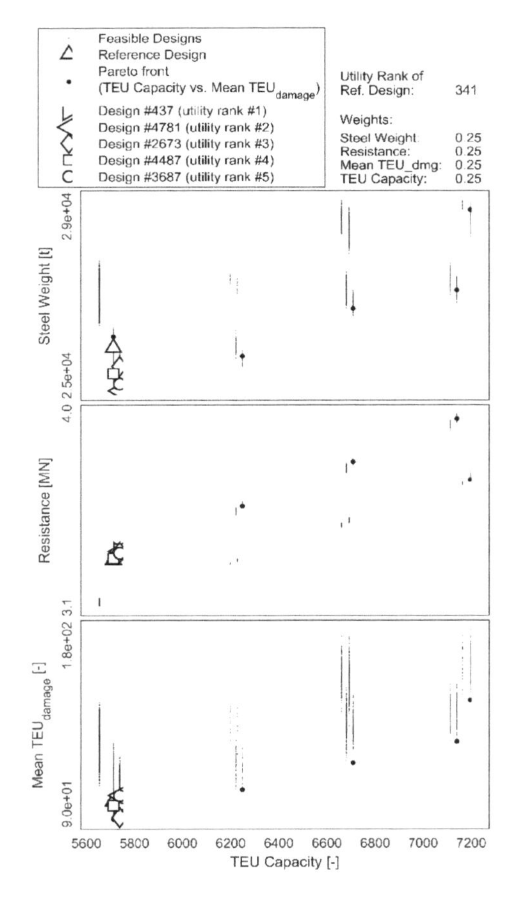

Figure 11. Scatter plots for the design merits (equal weights).

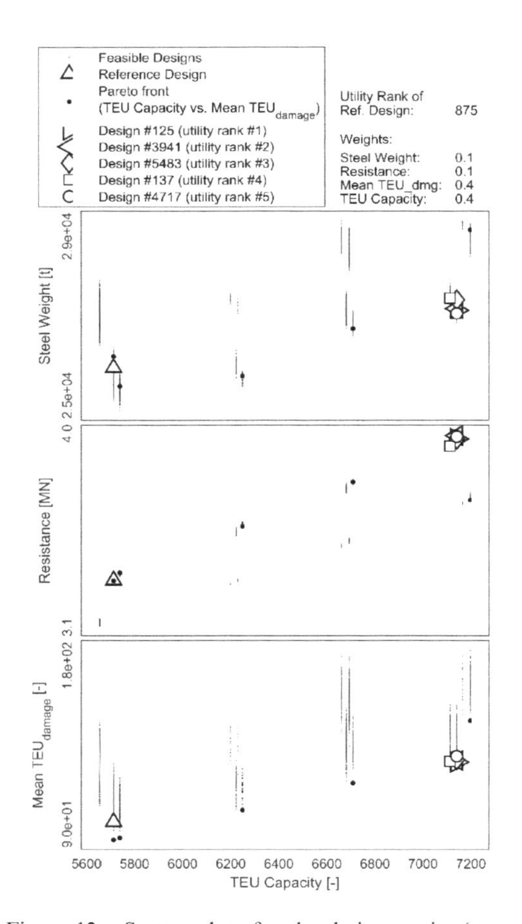

Figure 12. Scatter plots for the design merits (more weights on TEU damage risk and TEU capacity).

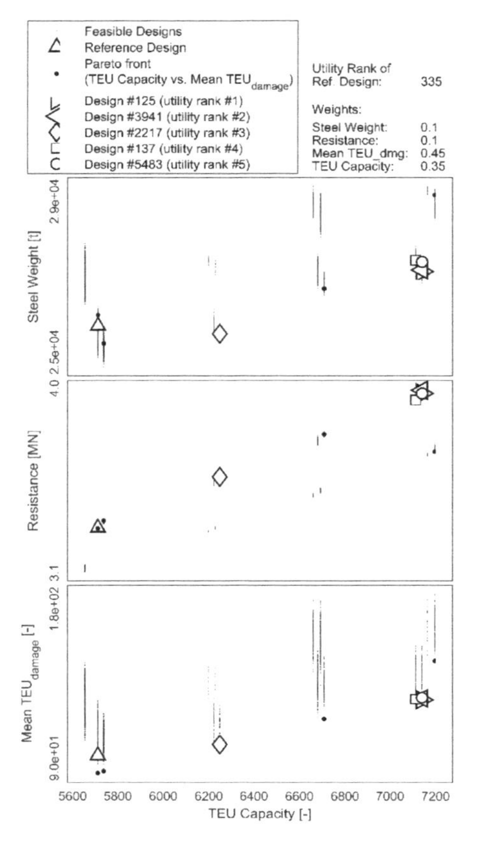

Figure 13. Scatter plots for the design merits (more weights on TEU damage risk).

4 CONCLUSIONS

The paper proposes a computational procedure for risk-based preliminary design and optimisation of tankers and containerships. The procedure is a synthesis of statistical and analytical models and common ship design software. The application of the procedure was examined by conducting cases studies with a representative Suezmax tanker and a containership with a 7,000 to 8,000 TEU nominal capacity. Different design metrics, including appropriate risk metrics (oil outflow for the case of tankers and number of damaged TEUs for the case of containerships) and ones as normally treated in typical ship design process, were evaluated simultaneously to achieve an overall optimal design.

It is shown that as the ship size grows, steel weight, resistance, cargo loss risk as well as cargo capacity grows, as expected. Since only the increase of the cargo capacity is the desirable feature among the four design metrics, trade-off needs to be made to arrive at the optimal design. The study showed that by using the utility function technique with different weight specifications, different design variants can surface up as optimal. Also several designs have exhibited better overall performance as compared with the reference design. The procedure can help facilitate decision making when a particular operational setting and design requirement is given.

ACKNOWLEDGEMENTS

This research is funded by Singapore Maritime Institute under the Simulation and Modelling R&D Programme—Project SMI-2014-MA-04.

REFERENCES

Brown, A. 2001. Alternative Tanker Designs, Collision Analysis, NRC Marine Board Committee on Evaluating Double-Hull Tanker Design Alternatives.

Brown, A., Tikka, K., Daidola, J. C., Lutzen, M., and Choe, I-H. 2000. Structural Design and Response in Collision and Grounding, In Annual meeting. Vancouver: SNAME.

DNV, 2003. Risk Assessment—Large Passenger Ships—Navigation, DNV Report 2003-0277, Annex II.

Goerlandt, F., Montewka, J. 2015. A Framework for Risk Analysis of Maritime Transportation Systems: A Case Study for Oil Spill from Tankers in A Ship-Ship Collision, Safety Science, Volume 76, Pages 42–66.

Holtrop, J., and Mennen, G.G.J. 1982. An Approximate Power Prediction Method, International Ship Building Progress.

IMO, 1974. International Convention for the Safety of Life at sea, Chapter II-1, Part B-1, Regulation 7.

IMO, 2008a. Marine Environmental Protection Committee. FSA—Crude Oil Tankers, MEPC 58/INF.2.

IMO, 2008b. International Code on Intact Stability, Resolution MSC.267(85).

ISO, 2002. Risk Management Vocabulary, Guidelines for Use in Standards, ISO/IEC Guide. International Standards Organisation, Geneva. ISO Guideline vol. 73.

Lamb, T. 2003. Ship Design and Construction. Jersey City, NJ, Society of Naval Architects and Marine Engineers.

Papanikolaou, A., Zaraphonitis, G., Boulougouris, E., Langbecker, U., Matho, S. and Sames, P. 2010. Multi-Objective Optimization of Oil Tanker Design, Journal Marine Science and Technology, Springer Verlag, Tokyo.

Priftis, A., Papanikolaou, A. and Plessas, T. 2016. Parametric Design and Multi-Objective Optimization of Containerships, Journal Ship Production and Design, SNAME.

Sormunen, O. E., Goerlandt, F., Hakkinen, J., et al. 2014. Uncertainty in Maritime Risk Analysis: Extended Case Study on Chemical Tanker Collisions, Proceedings of the Institution of Mechanical Engineers, Part M: Journal of Engineering for the Maritime Environment, Volume: 229 issue: 3, page(s): 303–320.

Van de Wiel, G. and Van Dorp, J. R. 2011. An Oil Outflow Model for Tanker Collisions and Groundings, Annals of Operations Research, 187, 279–304.

Watson, D.G.M., and GILFILLAN, A.W. 1977. Some Ship Design Methods, Transactions RINA, Vol. 119.

Marine Design XIII – Kujala & Lu (Eds)
ISBN 978-1-138-34076-3

Using FRAM to evaluate ship designs and regulations

D. Smith, B. Veitch, F. Khan & R. Taylor
Memorial University of Newfoundland, St. John's, Canada

ABSTRACT: Ship design decisions and regulations are made with the intentions of improving ship performance and safety. While design decisions and regulations may have strong theoretical merits, there can be uncertainty regarding if the intended improvements are experienced in practice. The approach put forth in this paper demonstrates how it can be checked and understood if these design decisions and regulations are having the intended impact on ship operations. The performance of a shipping operation can be tracked using system performance measurement techniques, which help understand if improved performance is being achieved. Then, by using the Functional Resonance Analysis Method (FRAM) to monitor the functional dynamics of each performance measurement, an understanding the processes that produce that measurement can be obtained. The combined understanding obtained from using system performance measurement and the FRAM over time, provide a framework to evaluate, justify, adjust, and make better ship designs and regulations, informed by practical ship performance.

1 INTRODUCTION

Shipping operations are continuously evolving to accommodate changing markets, different environmental conditions, new technologies, and dynamic social and political climates. As ship operations evolve, so too should ship designs and regulations. In order for ship designs and regulations to keep pace with the evolution of ship operations, they should be evaluated to understand their effect on the operation, which will inform the next generation of design and regulation practices. This has been traditionally done by using the shared knowledge of the shipping industry regarding the performance of ship designs to inform future designs. It starts with the design specification, which is informed by the operational profile and the owner's requirements. Then the ship design is performed by synthesizing first principles, rules, regulations, and construction to meet the design specification. The success of the as-built ship can subsequently be evaluated through sea trials, operational feedback, and economics. The performance evaluations are then used to inform future designs and specifications through lessons learned. This process of using shared knowledge to create improved designs and regulations has been effective and can be seen by comparing modern ships to their predecessors.

Areas that can be improved upon include (1) how the design evaluations are understood and (2) the rate at which evaluations take place. As ship designs and operations evolve, they become more complex and there is more uncertainty regarding how changes might affect ship performance. Additionally, the information that traditionally comes from the design evaluations can be unorganized, dissociated from operational practices, and uncertain in terms of economic impact. Given that the rate of change within organizations and in technological systems is continually increasing, it is essential that new adaptive approaches be developed to ensure rapid, comprehensive evaluations, to help clarify links between design and operations, while also supporting evolution of the design process.

This paper will illustrate how to use the functional resonance analysis method (FRAM) and system performance measurement to evaluate ship design parameters and regulations.

2 BACKGROUND

2.1 *Ship design and safety*

The major design methodology that is practiced by naval architects is the design spiral (Taggart, 1980). The design spiral is an iterative process where designers select, evaluate, and optimize predicted ship performance components. Over time, performance is evaluated and fed back to ship designers and owners and they use this information to make decisions for future ship designs. This process can be seen Figure 1.

Ship design is a crucial part of ship performance and safety. The performance of a ship is dependent on a combination of factors, including the design, operation, and environmental conditions. Sometimes these contributing factors can have complex relationships and high uncertainty. In particular, socio-technical relations—relationships between humans and

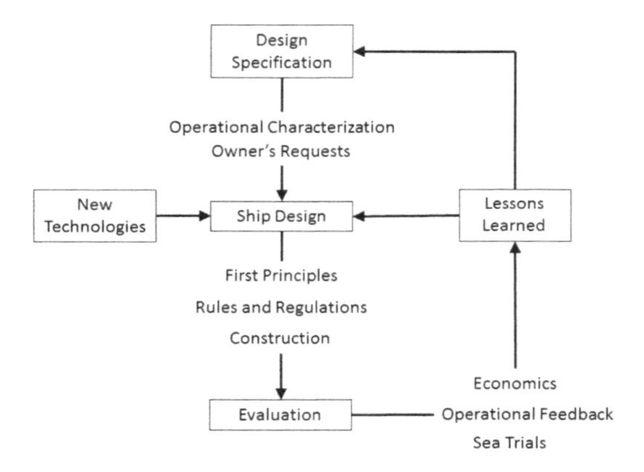

Figure 1. Traditional ship design and regulation evaluation method.

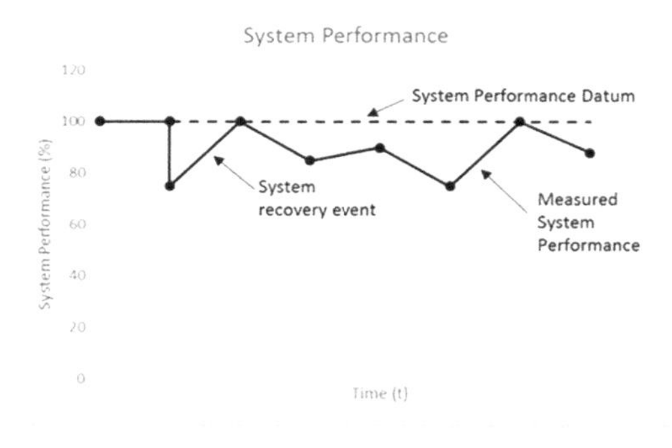

Figure 2. Measuring system performance over time.

technologies—have not been well understood in traditional safety methodologies (Perrow, 1984). Sociotechnical relationships can have a major impact on ship performance. To achieve higher performance, design decisions should not only be made on the basis that a technology exhibits better performance when assessed by itself, but also how it performs in assessments, including inter-related components of the ship operation. More concisely, technologies should have an affinity with the operators who will be using them (Vicente, 2004).

Understanding the relationships that exist between technical design parameters and operational practices can improve design and management decisions for shipping operations. A way to understand the impact that interactions between design parameters and operational practices influence ship performance is to recursively monitor the ship performance over time (Park et al., 2013). This includes monitoring the performance and the intermediate processes that contribute to the overall ship performance.

Traditionally, the ship design evaluation process has been informed by lessons learned, usually from intermittent monitoring of processes that have poor outcomes. For example, accident investigation is an intermittent monitoring technique used to learn lessons. Accidents make up a small portion of the entire operational profile of a ship. By monitoring a larger portion of the operational profile, greater understanding of complicated accident mechanisms could be achieved. For example, 70–90% of marine accidents can be attributed to human error (Rothblum, 2000). Human error is the leading contributor to marine accidents, and is difficult to manage for ship operators. In order to manage the operation for human error, understanding is required of what errors are occurring, when they are occurring, and why they are occurring. It is difficult to fully understand this from a single isolated event.

The concept of Safety II promotes understanding of successes in addition to failures, as a means to manage safety (Hollnagel, 2014b). Safety II can be thought of as a modern version of safety that builds on the traditional approach to safety (Safety I), which focuses on accidents as a way to inform safety management. This new approach can help better inform safety by helping operators to understand the things that should be done in addition to the things that should be avoided to promote safety (Hollnagel, 2014a).

The functional resonance analysis method (FRAM) allows an operation to be modelled and assessed for its full operational profile. The method is focused on understanding functional execution for a variety of outcomes of an operation. This method also allows monitoring of the intermediate processes of a system (Hollnagel, 2012).

Ayyub (2014) proposes that resilience principles should be used to monitor the overall performance of a system. A metric should be chosen that represents the main objective of the system and then tracked recursively. This provides an understanding of the system's performance over time and the overall system response to certain events. The system performance should be monitored over time as illustrated in Figure 2 (Ayyub, 2015).

2.2 *FRAM*

The FRAM is a method for assessing system functionality. The method produces a model of interconnected functions that represents the potential pathways that an operation could take. Cases (or variations) of the operation can be examined to understand the functional pathways that are actually taken to produce an operational outcome. Once a number of variations of the model and the outcomes have been observed and recorded, it is possible to look for operational conditions that produce functional resonance. Functional resonance is the idea that an operation requires small adjustments in response to variability in order to function, and that from time to time certain combinations of these small adjustments can result in an unexpected outcome. These unexpected outcomes can be unfavorable (accidents), but can also produce good outcomes.

In order to build a FRAM model, the potential functions should be presented as nodes, as seen in Figure 3 (Hollnagel, 2012). The function should represent the work being done.Each function has 6 aspects that represent the way the work is connected, which should be states within the system. The six aspects for each function are Input(s), Output(s), Preconditions, Resources, Time, and Control(s).

Once the model is constructed, the variability of the operation can be assessed. Variability is assessed at the functional level by the variability seen in the output(s). Functional outputs are typically not produced exactly the same each time the function is executed, so should be monitored. Because the functional output is connected to other downstream functions, variability has an opportunity to propagate through the model. The combined effects of the variability in the system can lead to variable system outputs, and in some cases, to functional resonance.

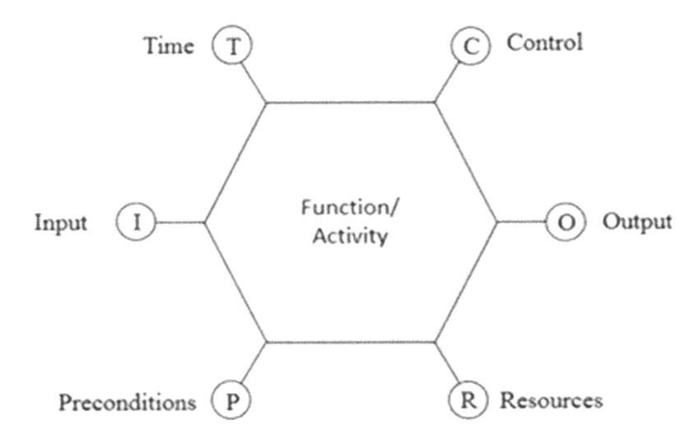

Figure 3. FRAM function diagram.

Monitoring an operation using this method can allow assessors to gain insight into the functional relationships and conditions that promote certain outcomes for an operation. Understanding what promotes good outcomes and what promotes poor outcomes enables operators to adjust management decisions and have better outcomes more often, and avoid poor outcomes. This understanding can help inform operational quality and safety.

3 METHODOLOGY

The FRAM can be a useful tool to help organize and collect information on ship designs and evaluate them. By using the FRAM to recursively monitor the performance of ships, it will serve as a framework for ship design assessment. This framework allows for learning from all operational outcomes and maps out the functional processes that lead to each outcome.

The methodology as seen in Figure 4 shows the integration of the FRAM into the traditional ship design evaluation process. The FRAM is used to organize information, understand operational practices, and evaluate practical performance. The FRAM provides a functional map that acts as an organizational aid for information. The variability of in the FRAM model provides an opportunity to understand operational practices that exist in the operation. The system performance is a signal by which the variable outcomes of the operation can be monitored. By connecting the functional understanding of the operation to the variety of operational practices in the operation to the variable performance of the operation, there is framework to inform operational management and ship design decisions.

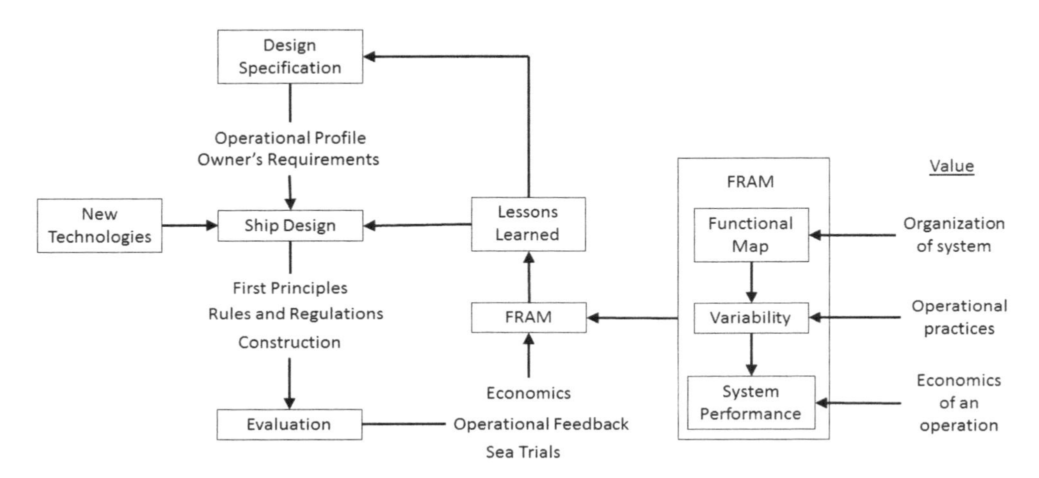

Figure 4. Ship design and regulation evaluation method including the FRAM.

Ship operations can occur many different ways. While the functional map outlines the many pathways that might be available for the operation to function, only a portion of the full potential may be active at a given time. This represents the variability of the operation. Each variation that produces an outcome is a functional signature for that outcome.

Since outcomes are also variable, the performance of the operation should be monitored for every case to understand this variability. A metric should be defined that represents the main objective of the operation and its functionality. This will allow performance to be monitored regularly.

3.1 *System performance*

To monitor system performance, the performance of each voyage should be tracked using a metric that represents the main objective of the operation and its functionality. This performance should be monitored over time, as illustrated in Figure 2. For a shipping operation, for example, the main objective is to transport a payload from one port to another. The functionality can be measured by the speed at which that payload is being transported. The performance metric could be payload × distance/time. This metric can help characterize the variability in the performance of the operation and identify high and low performance voyages, which may be of interest to managers. This particular metric does not consider other factors, such as costs to the shipping operation, which is an important consideration when making design decisions. There are ways to incorporate additional factors, like costs, in the performance metric of a FRAM analysis, although in the example here, a simple metric is used. Nevertheless, it serves to illustrate the method.

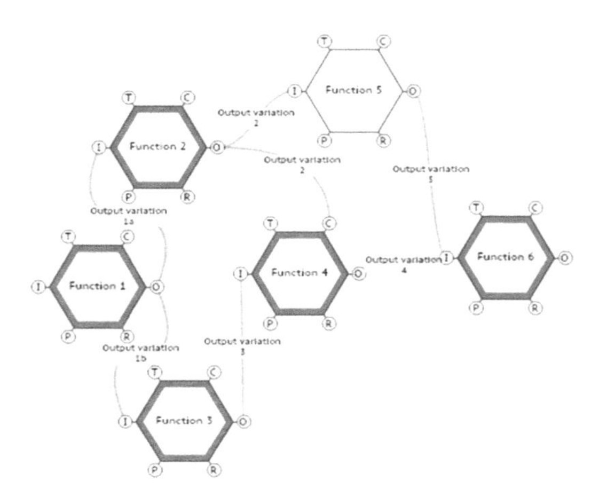

Figure 5. Example of functional signature at time (t).

3.2 *Functional signatures*

Each voyage performed by a vessel has a functional signature, which can be described using the FRAM model. The functional signature represents a variation of the functional model that has produced an outcome. The outcome also has an associated performance measurement. The signature then can be investigated to determine contributors to the operational outcome.

The FRAM model describes the many ways that the operation can be performed. Each functional signature for the operation many only require that portions of the FRAM model are active at a given time (t). Figure 5 shows a functional signature for time (t). The nodes outlined in bold represent the functions that are active at that time (t) and the specific outputs of this signature are shown on the connecting lines.

4 DISCUSSION

A methodology has been outlined for using the FRAM to evaluate shipping operations. Two hypothetical cases are used to demonstrate the application of the method: one examines a design parameter and another a regulation. While ship design decisions and regulations are well-intended and may have merits, using the FRAM allows assessors to check if those merits translate to practical improvements for an operation.

4.1 *Ship design*

Consider a design decision such as selecting a propulsor for a new ship that will regularly enter ice

Table 1. Performance metrics for Vessel 1.

Voyage (–)	Payload ($)	Distance (km)	Time (hours)	P*D/T ($*km/hours)
1	217999	2300	102	4915663
2	207261	2300	108	4413891
3	215390	2300	81	6116012
4	223311	2300	85	6042532
5	217880	2300	86	58270236
6	205399	2300	91	5191403
7	207944	2300	73	6551660
8	221162	2300	91	5589808
9	200410	2300	71	6492154
10	208020	2300	85	5628776
11	219440	2300	102	4948156
12	211081	2300	107	4537255
13	221388	2300	88	5786277
14	213871	2300	109	4512874
15	209677	2300	71	6792353
16	220679	2300	68	7464142
17	212990	2300	105	4665495

infested waters. Imagine that you have a fleet of 3 vessels that are currently operating in ice: vessel 1 has podded propulsors and vessels 2 and 3 have shafted propellers with rudders. You would like to use their past performance to guide the propulsor selection for the new ship design. You have been monitoring the fleet's performance for a while and will use the FRAM to help understand this information.

First the performance of the vessels must be monitored. Table 1 to 3 display the recorded performance of vessels 1 to 3 for their last 17 voyages.

Table 2. Performance metrics for Vessel 2.

Voyage (–)	Payload ($)	Distance (km)	Time (hours)	P*D/T ($*km/hours)
1	221143	1900	94	4469911
2	224906	1900	85	50273109
3	215222	1900	82	4986851
4	204479	1900	81	47964209
5	220770	1900	81	51785556
6	215359	1900	109	3753964
7	208438	1900	91	4352002
8	204650	1900	200	1944175
9	201212	1900	108	3539840
10	200934	1900	104	3670909
11	210234	1900	85	4699348
12	216266	1900	95	4325320
13	224108	1900	107	3979487
14	221922	1900	81	5205577
15	203002	1900	101	3818849
16	217117	1900	106	3891719
17	219972	1900	107	3906044

Table 3. Performance metrics for Vessel 3.

Voyage (–)	Payload ($)	Distance (km)	Time (hours)	P*D/T ($*km/hours)
1	207452	2200	89	5128026
2	224300	2200	104	4744807
3	222117	2200	108	4524605
4	224235	2200	96	5138718
5	200124	2200	93	4734116
6	217667	2200	113	4237764
7	205282	2200	105	4301146
8	215439	2200	100	4739658
9	212107	2200	102	4574856
10	205143	2200	350	1289470
11	217140	2200	99	4825333
12	206593	2200	117	3884654
13	208305	2200	111	4128567
14	222117	2200	95	5143762
15	224493	2200	116	4257625
16	202634	2200	113	3945086
17	206389	2200	104	4365921

The vessels have variable payloads, depending on the value of the cargo for each voyage. It is assumed the distance is constant for each vessel as they sail the same route each time. The time it takes to complete the voyage is variable and as a result the performance metric is variable.

Figure 6 shows the time history of the vessel's performance. There were two cases where there were mechanical failures that prolonged the duration of the voyage, thus producing the two lowest performance measurements. The remaining performance measurements would typically be seen as successes, but it can be seen that not all successes are equal.

The shipping operation is so dynamic that it is difficult to attribute lower measurements of performance to a particular area of the operation. Lower performance could be influenced by having less payload, mechanical failures, operating in harsher ice conditions, having a certain design feature, crew performance, or any combination of factors. This is where the FRAM can help assess contributors to low performance.

By having a detailed map of the functional pathways for a shipping operation, it allows the functional signatures to be tracked for each voyage. The functional signatures contain information about the operating conditions, operational practices, and ship design performance. An example of a functional signature for one particular voyage can be seen in Figure 7.

By grouping lower performance measurements together and looking for common trends in the functional signatures, the factors contributing to the low performance may be determined. In this hypothetical example, it is seen that vessel 1 generally has better performance than the other 2 vessels. Since vessel 1 is the only vessel with podded propulsors it might suggest that podded propulsors would be the best design choice for a new ship. However, other parameters may be influencing performance, such as hull form, engine power, payload capacity, and operating conditions.

It was seen in 15 of the lower performance measurements that ice ridges were encountered, which impeded the vessels progress. An example of one instance can be seen in Figure 7. When the vessels encounter ice ridges, one technique used to transit them is to ram the ridge repeatedly until the vessel eventually breaks through. When ramming the ridge, the vessel must reverse from the ridge back a distance to be able to ram forward with enough momentum to weaken the ridge. It was seen that reversing into an ice field with rudders was challenging and took longer to make each subsequent ram. This ultimately increased the duration of the voyage and in turn lowered the performance of the voyage. This is a mechanism that explains, at least in part, why podded propulsors might be a better design choice for a new ship.

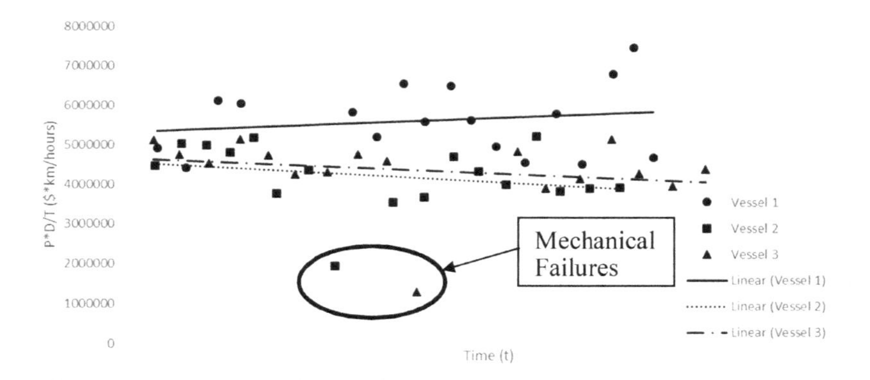

Figure 6. Vessel performance measurements over time.

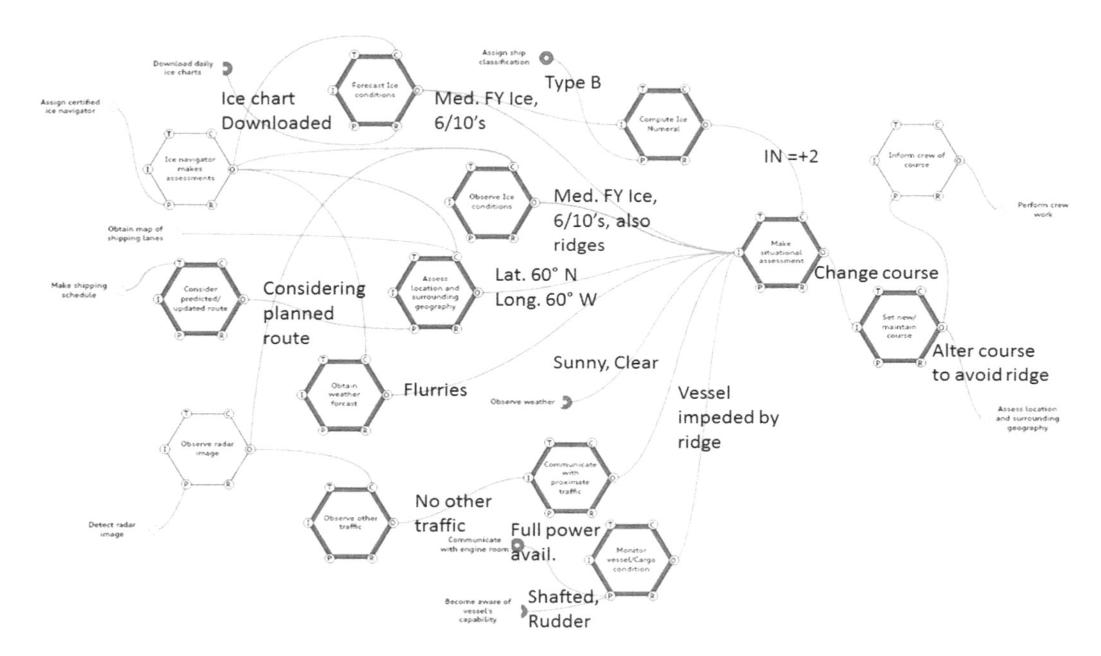

Figure 7. Functional signature for time (t) of one voyage for a vessel.

This case is hypothetical and does not prove that podded propulsors would be a better design choice than shafted propellers with rudders. However, it does illustrate how performance measurement and the FRAM can be used to inform ship design decisions.

4.2 *Regulations*

Consider the same vessel performance measurements as presented in section 4.1. This time, instead investigating how a ship design parameter can influence operational performance, the influence of a regulation on operational performance will be examined. In this case, the Arctic Ice Regime Shipping System (AIRSS) regulation will be assessed. The AIRSS is a regulatory decision making aid intended to be used by ships entering Canadian ice covered waters. AIRSS uses the ice conditions and ship classification to determine whether or not entry to a certain route should be prohibited for a ship. Ice Multipliers are assigned for different ice types and ship classifications. An ice numeral is computed from the summation of the products of ice concentration and ice multiplier for each ice type present along the route being assessed.

The sign of the ice numeral will determine the ice regime and whether or not entry should be prohibited. If the ice numeral is zero or positive, the risk is considered acceptable for the ship to navigate the route. If the ice numeral is negative, then the risk is considered to be too high for the ship to operate on the route. (Transport Canada, 2012).

In this hypothetical case, there are 7 voyages where the Ice Numeral was calculated to be in the positive regime, but harsher ice conditions than expected were encountered after entry to the route. In 2 of the 7 voyages, vessels had mechanical failures as seen in Figure 6. The nature of the mechanical failures were structural damage. Using the FRAM in Figure 7, it can be seen that the ice numeral was computed to be + 2 based on the forecasted ice conditions. The presence of ice ridges was missed in the forecast, and when ridges were noticed by the ice observer, they were assumed to be small enough to be neglected in the ice numeral calculation. In this case, the ice ridge caused damage to the ship's structure.

Here the ice numeral regulation did not achieve its purpose of preventing structural damage to ships transiting ice infested waters. In many cases, the ice numeral calculation does achieve its intended results, but by closely monitoring its influence on operational performance it can be seen that there is some room to improve the regulation.

Again, this hypothetical case does not prove that the ice numeral calculation does or does not need improvements. The example illustrates how performance measurements and the FRAM can be used to evaluate the practical effectiveness of regulations.

5 CONCLUSIONS

Ship design decisions and regulations are made to promote performance and safety of shipping operations. It is important to assess if operators are actually reaping the benefits of these such decisions. In this paper, a methodology is presented that enables a practical evaluation of ship design decisions and regulations. This method uses system performance measurement to monitor the ship's overall performance, and FRAM to understand the mechanisms that produce each performance measurement. This method can help formulate an understanding of how ship designs and regulations actually impact a shipping operation, and help inform future designs and regulations. A hypothetical example was used for illustration purposes, and thus conclusions reached are also hypothetical. However, the examples do illustrate the application of this method for practitioners who would like to apply this to their own operation.

ACKNOWLEDGEMENTS

The financial support of the Lloyd's Register Foundation is acknowledged with gratitude. Lloyd's Register Foundation helps to protect life and property by supporting engineering-related education, public engagement and the application of research.

REFERENCES

Ayyub, B.M. (2014). Systems resilience for multihazard environments: definition, metrics, and valuation for decision making. Risk Analysis: An Official Publication of the Society for Risk Analysis, 34(2), 340–355. https://doi.org/10.1111/risa.12093.

Ayyub, B.M. (2015). Practical Resilience Metrics for Planning, Design, and Decision Making. Asce-ASME Journal of Risk and Uncertainty in Engineering Systems, Part A: Civil Engineering, 1(3). http://dx.doi.org/10.1061/AJRUA6.0000826.

Hollnagel, E. (2012). FRAM: The Functional Resonance Analysis Method. Ashgate Publishing Ltd.

Hollnagel, E. (2014a). Is safety a subject for science? Safety Science, 67, 21–24. https://doi.org/10.1016/j.ssci.2013.07.025.

Hollnagel, E. (2014b). Safety-I and Safety-II: The Past and Future of Safety Management (1st ed.). Farnham, Surrey, UK England ; Burlington, VT, USA: Ashgate Publishing Ltd.

Park, J., Seager, T.P., Rao, P.S.C., Convertino, M., & Linkov, I. (2013). Integrating Risk and Resilience Approaches to Catastrophe Management in Engineering Systems. Risk Analysis, 33(3),356–367.

Perrow, C. (1984). Normal Accidents: Living with High-Risk Technologies. Princeton, N.J: Princeton University Press.

Rothblum, A.M. (2000). Human Error and Marine Safety. Presented at the National Safety Council Congress and Expo, Orlando, USA.

Taggart, R. (1980). Ship Design and Construction. The Society of Naval Architects and Marine Engineers.

Transport Canada. (2012). Arctic Ice Regime Shipping System.

Vicente, K. (2004). The Human Factor: Revolutionizing the Way People Live with Technology. New York: Routledge.

Marine Design XIII – Kujala & Lu (Eds)

ISBN 978-1-138-34076-3

Using enterprise risk management to improve ship safety

S. Williams
Maritime Safety Research Centre, Department of Naval Architecture, Ocean and Marine Engineering, University of Strathclyde, Glasgow, Scotland, UK

ABSTRACT: Every safety critical industry grapples with implementing an effective corporate risk management approach suited to their particular operational environment. A literature review revealed that Airlines and trains, as well as other safety critical industries, have adopted Enterprise Risk Management (ERM) to help improve the safety of their operations, but identified a gap in the application for cruise and ferry companies. Incidents like the sinking of the Costa Concordia highlight the need for improved risk management in passenger carrying ships used in cruise and ferry operations. This paper explains the creation of a holistic risk framework, using an idealized template as a guide for implementing and improving current risk approaches employed by operators of passenger carrying ships. Based on the positive results observed in other industries from implementing Enterprise Risk Management, passenger ship operators can expect an improved safety record and corresponding better financial performance of their companies.

1 INTRODUCTION

1.1 *Risk management background*

> "Risk is like fire: If controlled it will help you; If uncontrolled it will rise up and destroy you."
>
> *Theodore Roosevelt*

The 26th President of the United States captured the essence of why corporations adopted risk management. This paper focuses on an apparent lack of holistic risk management being used by cruise and ferry companies. This gap is apparent from the lack of articles that address the use of Enterprise Risk Management (ERM) or other holistic methods to address the full range of risks evident in the safe operation of the current generation of highly complex passenger carrying ships.

In other safety critical industries like airlines and trains, the literature shows a clear evolution of risk management to the current concept of Enterprise Risk Management (ERM). ERM is a holistic approach that grew out of the financial sector's desire to integrate risk management across the full breadth of their businesses. ERM's roots are in the general application of risk management within corporations that began shortly after the end of World War II and steadily evolved for the past seventy years. The goal for applying a more holistic approach to risk management for cruise and ferry companies is to improve their safe operation while increasing the profitability of the ship operating company, by reducing the incidents of serious accidents and loss of life.

1.2 *Hypothesis*

Historically, maritime businesses tend to be slow in adopting new technologies and new management approaches. My hypothesis is that if a cruise or ferry company implements ERM as its risk management system, then their safety performance and financial performance should move them to be the top performers in their segment of the market.

1.3 *Purpose*

The goal of the new Maritime Safety Research Centre at the University of Strathclyde is to conduct a broad range of research into improving ship safety. The Centre has two corporate sponsers: Royal Caribbean Cruise Lines (RCCL) and Det Norske Veritas-Germanischer Lloyd (DNV-GL). This paper captures one area being pursued to improve safe ship operations.

In other safety critical industries like airlines, trains and offshore oil and gas the adoption of a more holistic approach to risk management is evident from a large number of papers and articles documenting the use of ERM. Since passenger carrying ships operate in an unforgiving, high risk environment, the use of a more rigorous risk management approach merits study.

The first step in this process was to create a generic ERM model based on similar implementations in other safety critical industries. This model can then be tuned to the unique nature of passenger carrying ships and reflect the latest ERM research. This model will then be validated by surveying a number of cruise and ferry operating companies via a series of interviews and responses to questionnaires. This data will then form the basis for the development of a risk model.

1.4 *Preliminary findings/contributions*

A number of cruise and ferry operating companies have agreed to participate in this project. Data on each company's risk management approach, safety management and incident data has provided the following initial set of findings:

- None of the companies have fully embraced a holistic approach to risk management
- The linkage between the safety management system and the risk management system is not tightly integrated
- Incident data helps drive the creation of Key Performance Indicators (KPIs), which in turn can provide linkage back to the design process to focus on operational issues and link to create Key Risk Indicators (KRIs).

1.5 *Research paper structure*

The research article is structured as follows: firstly, the background literature is summarized in greater detail; secondly, the methodology used is outlined; thirdly, preliminary results and analysis are displayed and lastly the paper discusses the implications of the findings.

2 DETAILS OF LITERATURE REVIEW

Most safety critical industries have had a seminal event that triggers a more comprehensive use of risk management to try and avoid a repeat of that type of issue from occurring in the future. The Piper Alpha fire in 1988 (Drysdale, 1998) served as the driver behind the North Sea oil and gas industry moving to an approach that uses risk-based design and operations as the preferred method of reducing the likelihood of any future incidents. The Piper Alpha rig had a large loss of life (167) and huge financial impact on Occidental Petroleum (Caledonia) Limited. Risk based design for ships (Lloyds, 2016), is now an accepted design approach that analyses the risk implications of alternative designs.

The change in design (Turner, 2013) after Piper Alpha included moving away from prescriptive rules to analysing the design for risks and then proceeding to eliminate those risks via the design process for each offshore installation. The legislation passed after Piper Alpha included developing a safety case for each installation to provide details of health and safety management and major accident hazard control systems on the installation. The safety cases are living documents that are tested and revised over the lifecycle of the rig with the intention of ensuring the rig remains safe during operations.

In a similar fashion, the 2012 Costa Concordia accident has been a wake-up call to the passenger ship industry, and has served as the trigger for starting the implementation of more advanced risk management in cruise and ferry ship companies.

One of the continuing difficulties embedded with practitioners of risk in the marine area is the lack of a standard set of definitions, for even the most basic elements of a risk program. A specific definition for risk is available from maritime law (Mandaraka-Shepard, 2014):

> Risk is the possibility of harm or loss associated with an activity, or the likelihood of an incident happening that may result in danger to life, property or the environment, or may lead to commercial disputes and litigation.

For ERM, the Committee of Sponsoring Organizations of the Treadway Commission (COSO, 2004) created the ERM Integrated Framework. Over time it has become an accepted definition of ERM:

> Enterprise Risk Management is a process, effected by an entity's board of directors, management, and other personnel, applied in *strategy setting and across the enterprise*, designed to identify potential events that may affect the entity and manage risks to be within its risk appetite to provide reasonable assurance regarding the achievement of entity objectives.

The International Safety Management Code (ISM, 2010), as amended, sets the cornerstone for risk management process for ships;

Paragraph 1.2.2 states:

Safety management objectives of the company should, inter alia;

1. Provide for safe practices in ship operations and safe working environment;
2. *Assess all identified risks to its ship, personnel, and the environment, and establish appropriate safeguards*; and
3. Continuously improve safety management skills of personnel ashore and aboard ships.

Part (2) is the only specific mention of risk management in the regulation, which therefore gives a very broad avenue for interpretation by each ship operator.

One area of concentration for my research is to look at how different ship operators have crafted their safety management systems from the bottom up to meet the intent of the ISM requirement to identify the risks to the ships it operates. The interweaving of risk and safety creates a beneficial tension within the organization to accommodate both the desire to reduce risks, while improving ship safety. An important concept is that ERM integrates across the various corporate silos reflected in the various organizational elements. Without this integrating function, risks will not be shared between parts of the organization and lessons learned in one area will not benefit other parts of the organization. The understanding gained from analyzing a wide range of ship operating companies will help drive the development of an idealized risk management approach. Since other safety critical industries have already adopted ERM as their solution to managing risk, the goal is to use the research done for those industries to understand the fundamental approaches and then apply them to create a generic template for passenger ship operators. Rail and airline experience is there are limitations to implementing holistic risks because of the stovepipe nature of their organizations.

3 METHODOLOGY

3.1 *Overview*

The initial step, highlighted above, was to conduct a literature review to determine the state-of-the-art in risk and ERM in the marine sector. For ships, risk based design and risk based operations exhibit a reasonable number of papers over this period, but there is a general lack of papers on financial and operational risk management approaches for ship and ferry operating corporations. In fact, only one reference to ERM dealing with cruise ships was uncovered during an extensive literature review.

The methodology to prove the hypothesis is to conduct field work by visiting the top ten cruise ship companies (this represents 92% of all cruise ships worldwide) and a representative number of ferry operators.

With agreement from each company to supply their approach to risk management so that their maturity level can be determined, then incidents and accidents data for the past decade will be researched, and publically available financial performance information will be captured.

Safety and financial performance over the past decade captures the period with a very large expansion in the cruise sector, and a steady increase in ferry operations. During the first year of research, six cruise ship companies and two ferry operators have been visited and agreed to participate in this research.

3.2 *The case for ERM*

Figure 1, shows a very favorable trend in safety improvement over the past fifteen years for passenger trains in the United Kingdom (Abbott, 2017).

In discussion with Network rail, they feel the improvement shown from 2005 to 2012 is attributed to the adoption of risk management and a better safety culture. The steep slope shows significant improvements, with no fatal accidents on UK rail networks for the past nine years. The implementation of ERM occurred over the past five years and is credited with handling the really difficult phase of safety improvement. In essence the one-two punch of safety culture and ERM have proven to be very effective.

From a PhD thesis (Missura, 2015) focused on the implementation of ERM at airlines, Figure 2, shows the transformation in risk management over

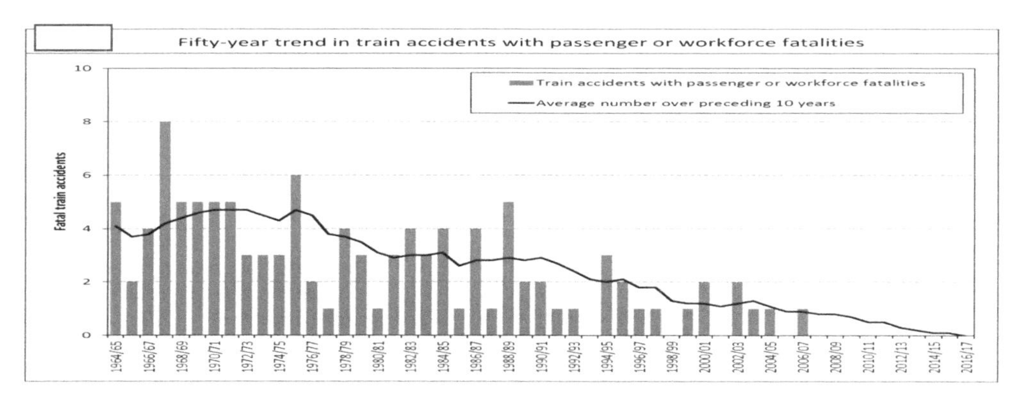

Figure 1. Fatal accidents on United Kingdom rail system.

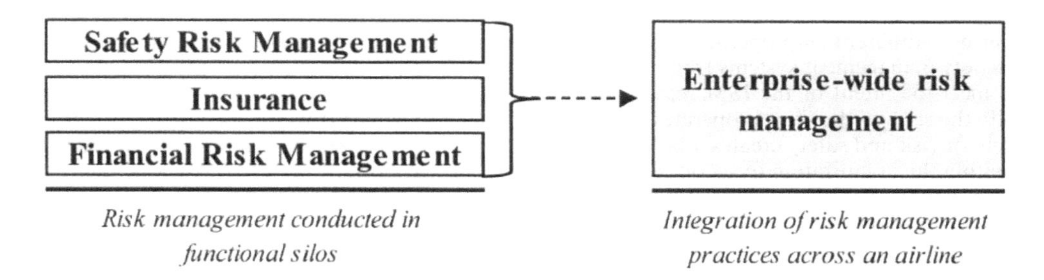

Figure 2. Change in risk approach at airlines.

the course of the last ten years in airline companies, another safety critical industry.

A majority of the airlines, seven of the ten surveyed, adopted ERM and have used it to break down the old silos of risk management.

3.3 *Generic ERM maturity model*

When comparing the maturity level of risk management at corporations, a number of different maturity indices have been created. One standard was created by the Risk and Insurance Management Society (Minksy, 2008). The five-level maturity model to assess corporate risk management maturity is defined in Table 1.

A recent paper, (Florio, 2017), incorporated more detailed measurements of ERM maturity by determining whether a Chief Risk Officer (CRO) has been appointed and whether a separate internal control and risk committee function exists. The paper analysed mid-sized Italian companies, establishing the level of ERM maturity using factors similar to Table 1.

The results show that the financial performance and market evaluation of companies with high ERM maturity are consistently higher than those with lower ratings.

3.4 *Validation of ERM model*

Ideally as the set of companies expands, one or more of them may have already implemented a form of holistic risk management, even if they have not use the term ERM. If these companies have a higher maturity level, then the generic model proposed in this paper can be validated by analyzing the performance of these companies. Initial findings show that many of the companies have implemented risk management for their various projects. This project based risk management is then integrated with the overarching risk approach for the company.

Table 1. Corporate ERM maturity levels.

Level 1 Ad Hoc	No coordinated focus on risk management
Level 2 Initial	Some risks identified, silo focused, audit focused
Level 3 Repeatable	Risks are tracked, enterprise risks identified, risk management plan
Level 4 Managed	Business planning and investments are linked to risk, board of directors briefed periodically
Level 5 Leadership	Corporation understands its risk tolerance/appetite, risk is part of the day to day management of the organization with a strategic focus

3.5 *ERM implementation aspects*

A key finding from many sources (Lam, 2017, Florio, 2017) is that successful ERM implementations show that the process is supported by both the Chief Executive Officer (CEO) and the Board of Directors (BOD). Key indicators of this are the appointment of a Chief Risk Officer (CRO) by the CEO and the establishment of a Risk Committee as one of the sub-committees of the BOD.

4 PRELIMINARY RESULTS/ANALYSIS

4.1 *Initial incident data implications for design*

In the first year of research six cruise ship and two ferry operating companies have been visited and agreed to participate in this project. As a first step, Key Performance Indicators (KPIs) derived from incident data were reviewed. From these, the subset that cause safety issues and could be linked to risks were created. Another use of this data is to determine which of these incidents are influenced by design decisions and deserve more attention in early and detailed design phases. Table 2 lists a number of operational risk drivers that flow from design decisions and merit a focus for ship designers.

4.2 *Generic ERM model*

Using other safety critical industries as a guide, the Ship Lifecycle Risk Management Framework (Figure 3) was created to capture the holistic implementation of risk across a ship operating organization, including the unique nature of the lifecycle of ships.

The fact that a major ship operator will have ships in all phases of the lifecycle means that the risks in each area must be identified and tracked. Further, it is clear in this template that with the advent of risk based design, the implications of that across the other phases of a ship's life grow in importance. For example, if a new cruise ship coming on line now used the alternative approval method to allow for novel materials or larger than normal fire zone boundaries, then there is a direct linkage between the risk-models created in design and the risk of operations. In recognition of this change, DNV GL has extended the concept of a "digital twin" first proposed at the University of Michigan in 2003 (Grieves, 2013) by applying this concept to offshore structures and ships (DNV GL, 2017). In addition to risk models, the approach is to capture software intensive system models, like machinery controls, and integrate them into a virtual representation of the physical offshore platform or ship. A digital model should be created and maintained for the lifecycle of each platform or ship. For a typical new cruise or ferry designed using a risk based process, the risk model of the ship used must be maintained and updated with the as-built changes that occur during construction. This updated model then serves as the basis for development of the operational risk models.

Table 2. Key Performance Indicators (KPIs) linked to design.

1. Hard to maintain life-saving appliances
2. Control issues that cause a loss of main propulsion power
3. Control issues that cause a complete loss of electric plant operation
4. Fire suppression system failures
5. Incinerator fires
6. Stabilizer failures
7. Release of refrigerants to atmosphere
8. Fuel leaks from flexible hoses

5 RELATED WORK

Recent research (Misiura, 2015) analyzed the use of ERM by the airlines as the basis for a PhD. In this work, the author created a survey and analyzed the maturity of ERM at approximately a dozen airlines. Although useful for capturing the extent to which ERM is embraced as a risk management approach by the airlines, no attempt was made to establish the maturity of each company's ERM, and no linkage was attempted to their safety records or financial performance.

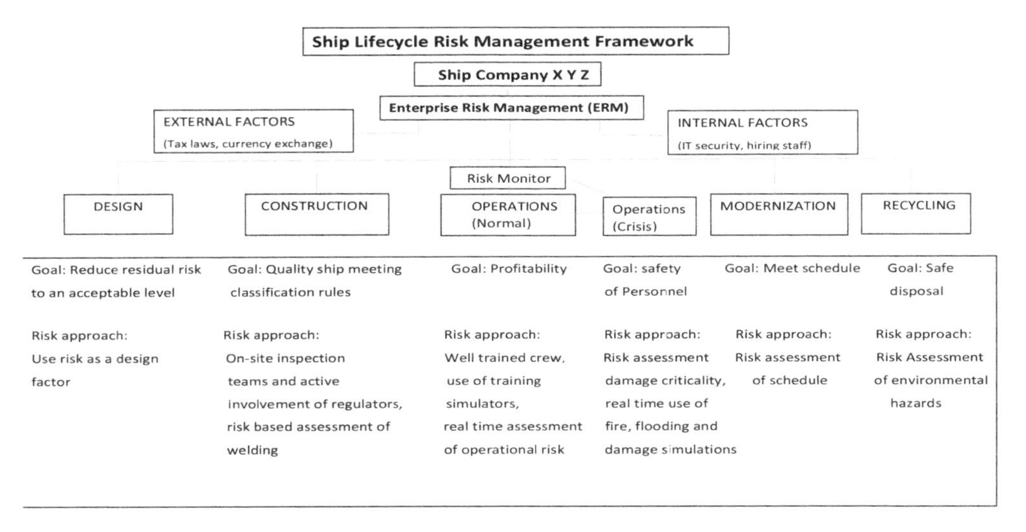

Figure 3. Ship lifecycle risk management framework.

6 CONCLUSIONS

Cruise and ferry ship operators are entering an era where risk will play a more major role throughout the lifecycle of the ownership of their fleet. Enterprise Risk Management is the glue that will help analyze the risks consistently and holistically, yielding improved safety, better linkage to design and better profitability. By integrating all risks effecting the business ERM helps the Board and company management understand the impediments to success and also, the opportunities created by these risks available for the company to exploit. The concept of a risk model as part of the "digital twin" that every new construction ship will need is here now and needs to be embraced by industry. New ships that use a Risk Based Design approach must retain the assumptions that allowed the variations from the prescriptive rules so that when changes are made during construction and modifications are proposed during modernizations, safe decisions can be made that do not cause an increase in the risk of operations.

7 ADDITIONAL WORK

The next phase of this research is to refine the development of the risk template and create a model of the interaction of risks across entire organizations. The plan is to use the System Theoreitc Accident Modelling and Process (STAMP) developed at MIT (Leveson, 2004) to validate the implementation of risk at each of the companies investigated.

REFERENCES

Abbott, John, Safety on Britian's railway 1804–2017, the story so far!, Safety Culture at Sea Conference, U.K. Chamber of Shipping, 22–24 September, 2017.

Boyd, Stephen, Moolman Johannes, Nwosu, Nkemjika, Risk Reporting and Key Risk Indicatores, a Case Study Analysis, NC State 2016.

DNV GL, Det Norske Veritas Germascher Lloyd, Making your asset smarter with the digitaik twin, retrieved 14 August, 2017. www.dnvgl.com/article/making-your-asset-smarter-with-digitial-twin/63328.

Drysdale, D.D., Sylvester-Evans, R., The Explosion and Fire on the Piper Alphas Platform, 6 July 1988, *Proceedings of the Royal Society,* Vol. 356, issue 1748, pp 2929–2951, 1998.

Committee of Sponsoring Organization of the Treadway Commission (COSO), Enterprise Risk Management—Integrated Framework, September 2004.

Florio, Cristina, Leoni, Giulia, Enterprise risk management and firm performance: The Italian Case, *British Accounting Review*, 49 (2017) pp. 56–74.

Grieves, Michael, Digitial Twin: Manufacturing Excellence Through Virtual Factory Replication, 2013.

International Safety Management Code (ISM) and guidelines on the implementation of the ISM Code, 20110 Edition, International Maritime Organizations (IMO).

Lam, James, Implementing Enterprise Risk Management: From Methods to Applications, 2017.

Leveson, Nancy, A New Accident Model for Engineering Safer Systems, Safety Science, Vol. 42, No. 4, April 2004.

Lloyds Register, ShipRight Design and Construction, additional design procedures, Risk Based Designs (RBD), May 2016.

Mandaraka-Sheppard, Aleka, Modern Maritime Law, Vol 2: Managing Risk and Liability, 2013.

Minsky, Stephen, Risk and Insurance Management Society (RIMS), State of ERM Report 2008.

Misiura, Anna, 2015. *Enterprise Risk Management in the Airline Industry—Risk Management Structures and Practicds.* PhD thesis, Bruenel Business School, Bruenel Univerity, London, United Kingdom.

Marine Design XIII – Kujala & Lu (Eds)
© 2018 Taylor & Francis Group, London, ISBN 978-1-138-34076-3

Using system-theoretic process analysis and event tree analysis for creation of a fault tree of blackout in the Diesel-Electric Propulsion system of a cruise ship

V. Bolbot, G. Theotokatos & D. Vassalos
Department of Naval Architecture, Ocean and Marine Engineering, Maritime Safety Research Centre, University of Strathclyde, Glasgow, Scotland, UK

ABSTRACT: Diesel-Electric Propulsion (DEP) has been widely used for propulsion of LNG carriers, icebreakers, drilling units, warships and cruise ships. It is important that every blackout is prevented, especially on cruise ships, considering the possible consequences of such an event. In this work, hazard analysis of a simplified DEP system of a cruise ship is implemented to identify the hazardous scenarios leading to a blackout. This is achieved by combining System-Theoretic Process Analysis (STPA) and Event Tree Analysis (ETA). The STPA is used to identify the hazards and the possible control actions leading to hazards along with their causal factors, whilst the ETA is used to determine the propagation of hazards into the other hazards or accident. Next, the results of STPA and ETA are mapped into a Fault Tree for better representation of results. In this way, the relationship between accident, hazards and unsafe control actions is explicitly described and a more comprehensive picture of the potential accidental scenarios in the system is provided, rendering possible allocation of quantitative performance requirements as per IEC 61508.

1 INTRODUCTION

The cruise ship industry has been rapidly developing, with both the vessels size and number constantly growing up (CLIA 2016), and ensuring the passengers, crew and ship safety is always a paramount necessity. Collision, contact and grounding are reported among the most important accidents on cruise ships with blackout and propulsion loss being included in the top 10 hazards leading to these accidents (Nilsen 2005). Considering the potential consequences of blackouts, it is crucial to prevent such undesirable incidences (Nilsen 2005, MAIB 2011).

Diesel-Electric Propulsion (DEP) systems have been widely used for propulsion and power generation of a number of ships including LNG carriers, icebreakers, drilling units, naval vessels and cruise ships (Hansen & Wendt 2015). The DEP can be viewed as a complex system consisting of a number of subsystems including physical components, hardware and software, which interact with each other to ensure the continuous power generation and distribution, covering the ship power demand (Ådnanes 2003, Rokseth et al. 2017). Although these components provide adequate flexibility, improved operational efficiency and increased reliability, they suffer from a number of hazards which can lead to undesired incidences. Despite the redundancy incorporated in each DEP system design, a number of blackouts have been reported for cruise ships with DEP or ships with similar systems (MAIB 2011, Hossain et al. 2013, Rokseth et al. 2017). These hazards should be also investigated in the light of increasing connectivity capabilities and software dependence of the cruise ships and their systems.

In practice, the avoidance of these incidences can be achieved by accurate identification and control of the possible scenarios leading to blackout during the design and operation. The current regulations require implementation of Failure Mode Effects Analysis (FMEA) on ships with Dynamic Position System (DPS) (Rokseth et al. 2017). A modified version of FMEA is also required for assessment of the availability of propulsion and other systems on cruise ships after flooding and fire (IMO 2010, DNV GL 2016). Several other approaches focused on the implementation of Fault Tree Analysis for studying of the system reliability (Buja et al. 2010, Vedachalam & Ramadass, 2017). However, FMEA and FTA cannot capture properly the interactions between the system components, especially the interactions between the control components and the physical components, and therefore, additional methods need to be employed to cover this gap (Leveson 2011, Thomas 2013).

System-Theoretic Process Analysis (STPA) is a top-down approach for hazard identification and has been referred as an effective method for capturing the interactions between the system components (Ishimatsu et al. 2010, Thomas 2013, Dawson et al. 2015). Results of STPA and FMEA can guide the implementation of testing, like Hardware-in-the-loop or Software-in-the-loop testing (Rokseth et al. 2018). The deficiency of STPA, however, is that it provides a cause-effect description of how the accident, hazards, Unsafe Control Actions (UCAs) and causal factors are related to each other, which is quite ambiguous. This impedes the allocation of testing resources wisely, focusing not just on safety-related but on safety-critical test cases and allocation of requirements based on the desired performance of the system according to IEC 61508 during the design of the system (BSI, 2010). This can be overcome by combining STPA with Event Tree Analysis (ETA) and by using the Fault Tree structure, which can illustrate explicitly how the causal factors, UCAs and hazards are related to each other and to the accident.

The paper is organized as following. In the next section, the adopted approach is described in its general form. After a brief introduction to the case study, the results of implementing the adopted methodology are provided. The paper finishes with some discussion on the application of the adopted methodology which paves the way for future research.

2 DESCRIPTION OF ADOPTED APPROACH

2.1 *The overall picture*

The implementation of the hazard identification, analysis and structuring of results in the present work was carried out according to the following steps:

0. Preparatory step including the gathering of the information on the system and the associated hazards.
1. The identification of accidents and system level hazards and development of hierarchical control structure.
2. Identification of UCAs.
3. Identification of causal factors.
4. Implementation of ETA.
5. Combining the Event Trees to create a Fault Tree.
6. Populating the Fault Trees with UCAs.
7. Refinement of common causal factors of STPA in the Fault Tree.

The first three steps are almost the same as the classical STPA approach. In the fourth step, the ETA uses the information from the first step of STPA on the hazards and available constraints in the system and follows the typical steps of the ETA. The fifth step is based on the constructed Event Trees and synthesizes them into one Fault Tree. During the sixth step, the hazards in the Fault Tree are broken down into the UCAs. During the seventh, the causal factors common to the UCAs are refined to avoid their duplicate contribution to the hazard. The steps followed are explained in detail in the following sections and the flowchart of the adopted approach is shown in Figure 1.

2.2 *The preparatory step of the STPA*

2.2.1 *Accidents, system hazards and safety constraints definition*

STPA defines the accident as: “an undesired and unplanned event that results in a loss, including a

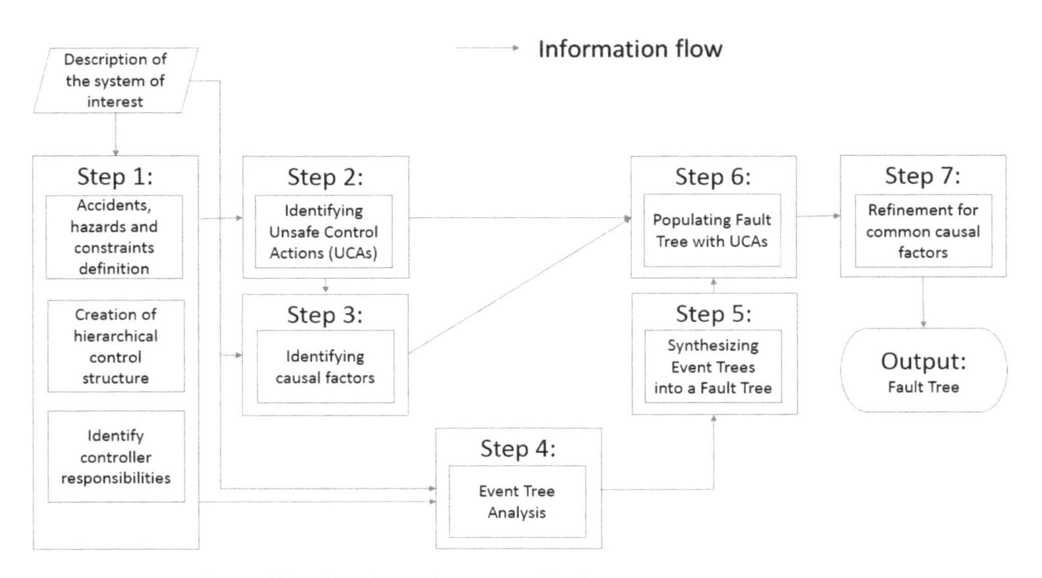

Figure 1. The steps of hazard identification and creation of fault tree.

loss of human life or human injury, property damage, environmental pollution, mission loss, financial loss, etc." (Leveson & Thomas, 2015). Yet, the decision on what constitutes an undesired event is dependent solely on the situation and goals of the safety analysis (Leveson 2011, Leveson & Thomas 2015).

The hazards in the STPA framework are understood as: "system states or set of conditions that together with a worst-case set of environmental conditions, will lead to an accident" (Leveson & Thomas, 2015). The hazards in STPA are viewed on a system level, so they go beyond the single failures that may occur in the system and should be referred to a specific state of the system. Herein, in contrast with STPA, as hazards are considered also some triggering events and some of them on a subsystem level. Based on the hazards then, the necessary safety constraints and requirements that should be implemented are established.

2.2.2 *Construction of the system control structure*
The construction of functional control structure is one of the differentiating points of the STPA analysis, compared with the other methods (Leveson & Thomas, 2015). Usually, it starts with a high level of abstraction and proceeds to a more detailed level in the process. The initial control structure consists of the high-level controller, the human operator and the controlled process with the basic control, feedback and communication links. A more detailed description would incorporate a hierarchy of controllers. Both types of the control structure that can be used for the analysis are shown in Figure 2. In the present work, a more detailed description has been directly used to incorporate more details in the analysis.

2.2.3 *Refinement of the control loop*
The next step after the development of the basic control structure is its refinement. The necessary

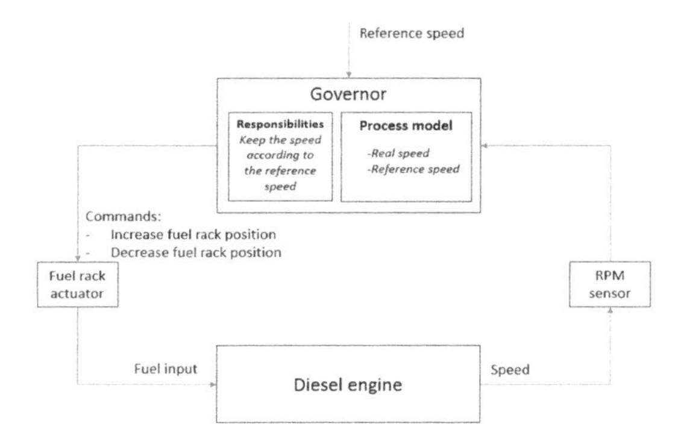

Figure 2. Refinement of control structure.

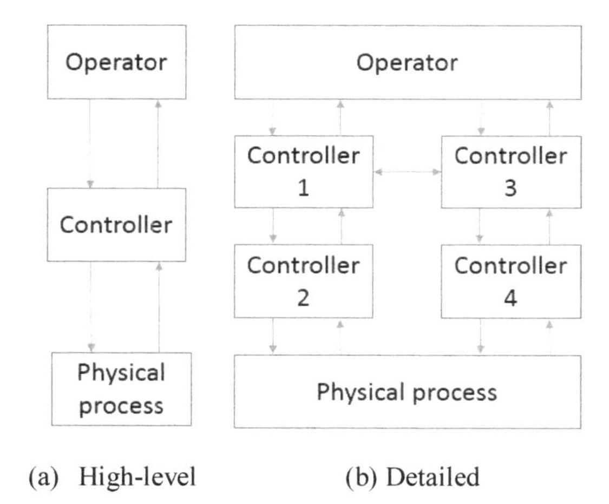

Figure 3. The types of control structure.

activities include the definition of responsibilities of each controller, of the process model with process variables and potential process variable values, of the control actions, the behaviour of the actuators, the information coming from the sensors and the information coming from the other controllers. The output of the analysis should be in the form as shown in Figure 2 for the governor of a Diesel Generator (D/G).

2.3 *Unsafe control actions identification*

The previous three steps are identical with the preparatory steps of the STPA. The actual hazards identification starts by finding the Unsafe Control Actions (UCAs). The possible ways to proceed are either by using the control actions types as initially proposed for the STPA (Leveson 2011) or by using the context tables as proposed by (Thomas 2013). The second approach is more exhaustive and complete but more tedious. Unfortunately, the capabilities of the supporting software didn't facilitate its application. As a result, the first is used here. According to that, the possible UCAs can be of the following four types (Leveson & Thomas 2015):

- Not providing the action leads to a hazard.
- Providing of a UCA that leads to a hazard.
- Providing the safe control action too late, too early or out of sequence.
- A safe control action is stopped too soon or applied for too long.

The control actions are viewed as safe in general and become unsafe only under specific conditions. According to the STPA, there is also another type of UCA, when the safe control action is provided but is not followed. This type of failure mode is addressed during the identification of causal factors in the next step. Similarly, with the system hazards, safety constraints can be derived for the UCAs, aiding the identification of possible risk control options.

2.4 *Identification of causal factors*

The second step in the hazard identification of the STPA has the purpose to determine all the causal factors leading to the UCAs. This is done by examining the hazardous scenarios including software and physical failures and design errors but the approach can degenerate to the identification only of failures and deviations leading to the UCAs. There is a number of ways to organize the results of the hazardous scenarios by using tables or lists. In this work, the process was augmented by usage of a modified tree structure proposed in the work of (Blandine 2013) as shown in Figure 4. This

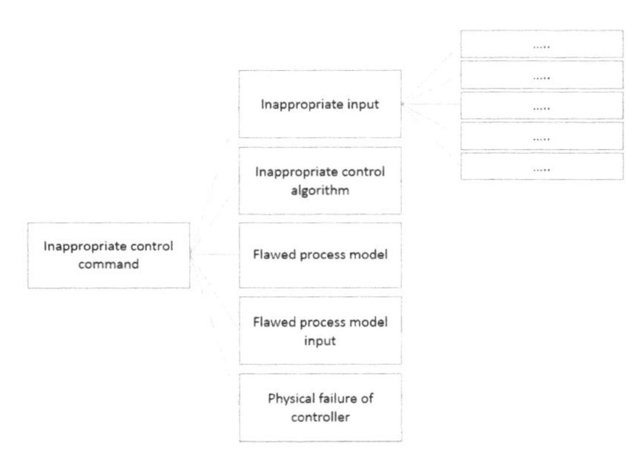

Figure 4. The causal factors according to (Blandine, 2013).

allowed easy transition from the STPA to the Fault Tree structure, as in this way the causal factors can be connected by using the OR gate of a Fault Tree. Similarly, for the hazards and the UCAs, the necessary safety constraints can be defined based on the hazardous scenarios.

2.5 *Event tree analysis*

The ETA can commence earlier, as soon as, the hazards, the accidents of interest and the relevant constraints have been defined. Each one of the hazards is used as an initiating event and the propagation of hazards is studied by considering the sequence of events taking into account the responses of the system, the presence of protective barriers and the combinatory faults. The safety constraints that have been found during the preparatory step of STPA are used as barriers herein. The ETA completes when all the outcomes stop either in the safe condition or at another hazard or at the investigated accident. The ETA is implemented qualitatively in the present work. A symbolic Event Tree is shown in Figure 5.

2.6 *Fault tree creation from event trees*

Since not all hazards lead directly to the accident, but there are some interactions between hazards, the Event Trees are restructured so as to show how a hazard may propagate into another hazard. This will allow the unification of Event Trees and the final outflow of this structure is either an accident or a safe condition. Furthermore, the ETA is quite independent of the STPA, so it can give an explicit information on whether there is a link between the

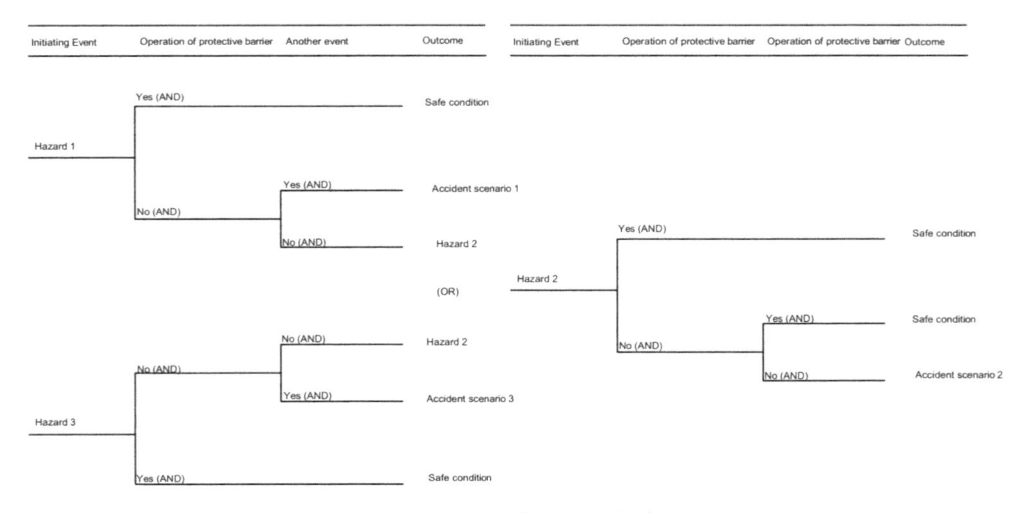

Figure 5. The combination of event trees and transformation into a fault tree.

hazards. Afterwards, the Event Trees are transformed into a Fault Tree by translating each of the branches of the Event Tree into AND gate. The different accident scenarios are connected by using the OR gates. If there is a path from a hazard to another hazard, then this path is connected using the OR gate to the hazards along with contributory factors as described below. This is also shown in Figure 5.

2.7 *Populating fault tree with UCAs*

The development of the Fault Tree is not completed by the conversion of the Event Trees into a Fault Tree as it must be populated with the UCAs. The causal relationship between UCAs and hazards has been already found during the STPA analysis. For each of the hazards, all the relevant UCAs are considered initially independent and thus connected with the OR gate. In case a UCA represents a failure of the protective barrier, as coming from ETA, then it is connected with AND gate with the hazard. If a UCA is leading to more than one hazard, then it is repeated for each and every hazard. The causal factors for each of the UCAs are considered independent and thus connected with the OR gate. If the right control action is provided but is not followed, then this type of control action is connected with OR gate with other UCAs.

2.8 *Refinement of common causal factors and UCAs*

Quite often UCAs are sharing common causal factors; the one causal factor will cause two or more UCAs to occur. This type of refinement is necessary primary for those functions that are implemented by the same physical entities, where the same type of failure or error will compromise several functions. Since UCAs are control actions that become unsafe under specific conditions, the control actions are connected with AND gate with these conditions.

3 CASE STUDY

The methodology described above has been applied for the case study of a cruise ship DEP system. For the purposes of the study, a simplified version is used which is shown in Figure 6. The analysis has been restricted to the components in the box. The actual existing redundancy has not been taken into account, as it was not necessary for the qualitative analysis. Also, the focus was only on one accident in the system, which in the specific case is the blackout. The considered vessel operational mode is the sailing mode.

A number of assumptions have been used in the analysis, which primarily affected the ETA. They are listed below:

- The engines start based on load demand from the consumers and switchover is implemented based on the running hours of D/G sets in the system.
- Other consumers are considered to be excluded from this analysis, even their contribution in terms of short circuits to the network, except for tripping of heavy consumers.
- The human factor neither reduces nor introduces new hazards.
- The D/G sets operate in the droop mode.

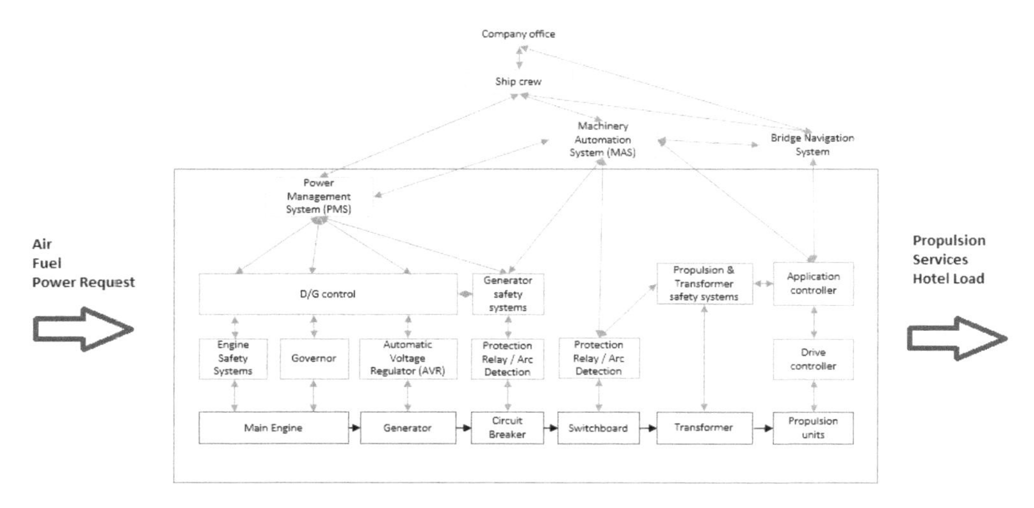

Figure 6. A simplified version of the system.

- No hazards are created in the system because of breaches in cybersecurity system.
- Failure to start a D/G set in time during abnormal situation will result in sudden loss of a D/G set.
- An uncontrolled arc failure in the Switchboard will cause a total blackout.
- An uncontrolled short circuit in propulsion units will lead to the overload of the D/G sets and a consequent blackout.
- Any load sharing imbalance in the system can be safely handled only by the intelligent diagnosis systems.

The STPA analysis is implemented using the open source software XSTAMPP (Abdulkhaleq and Wagner 2016).

4 RESULTS

4.1 *The results of preparatory steps of STPA*

4.1.1 *Accidents, system hazards and safety constraints definition*

A number of accidents could be considered for the DEP system such as partial blackout, excessive emissions, electric shock (Rokseth et al. 2017), but the primary goal of this study is to tackle the blackout. The list of hazards and the related safety constraints that have been found for the system is shown in Table 1.

4.1.2 *Construction of the system control structure*

Following the steps of STPA, the control structure has been developed. On the contrary to the (Blandine, 2013), the safety functions have been included in the control structure as the interest is to identify those physical failures that provoke their realization. Also, the adopted structure was more consistent with the physical structure. The D/G control could be also broken down into the functions related to the start of a D/G set and those related to the control during operation since this is implemented by one unit, it is represented as one controller. Similarly, for application controller, the safety functions for propulsion unit could be distinguished as a separate controller. In any case, the impact on the result from not using a functional structure is negligible.

4.1.3 *Refinement of the control structure*

For each of the control loop, the process goes on and the responsibilities of each controller, the possible control actions, the relevant process variables and their possible values are defined. In total 34 control actions and 92 process variables have been specified. Most of the control actions were discrete like start a D/G set and all the process variables had a value either acceptable or non-acceptable (true or false). The various multiplicities coming from the fact that the network is 3 phase and the fact that more than one D/G set are employed in the network was not required to be considered. For some of the higher controllers, the process variables have been considered to be containing a number of other process variables for simplicity.

4.2 *Unsafe control actions identification*

Based on the previous steps, the UCAs have been found using the possible failure modes of control actions. Totally 74 UCAs have been identified. In comparison with (Rokseth et al. 2017), additional control actions are given. A typical example is provided in Table 2.

Table 1. The hazards for the system.

a/a	Hazards	Safety constraints
[H-1]	D/G sets unavailable.	There must be always D/G sets available to be connected when requested by safety functions.
[H-2]	Inability to connect a D/G set in time to the network, when requested.	The standby/available D/G set must always be connected in time to the network.
[H-3]	Tripping of a D/G set.	The D/G set must be never disconnected without a standby/available D/G set is connected to the network.
[H-4]	Imbalanced power generation.	The system must always avoid imbalance in power generation.
[H-5]	Overloading of D/G sets.	The system must always avoid working at conditions where the power demand is higher than power can be given by the connected D/G sets.
[H-6]	Too fast increase in the power demand.	The system must always avoid speed of increase in power demand that cannot be accepted by the D/G sets.
[H-7]	Tripping of heavy consumers in the network.	The system must avoid inadvertent tripping of the propulsion units.
[H-8]	Disturbances in the network like a short circuit and arcing faults.	The system must prevent the occurrence of short circuits and not allow the short circuit and arc fault to be uncontrolled in the system.

Table 2. Example of UCAs in the system.

Control action	Type of UCA	Description
Start a D/G set (given by PMS)	Not providing causes hazard	If the order to start a D/G set is not given when there is a request for higher power production it will result in overload. [H-5]
	Providing incorrect causes hazard	Giving an order to start a D/G set, when it is faulty, will cause disturbances to the network. [H-4]
		Trying to start-up an already running D/G set will result in the unavailability of production power when the power demand is high and failure to implement a switchover to a faulty D/G set. [H-2] [H-5]
	Wrong timing or order causes a hazard	A delayed order to start a D/G set will cause a delay in change over when there is a faulty D/G set. This will result in tripping of D/G set. [H-3]
		A delayed order to start a D/G set, when the power demand is higher than provided safely by other D/G sets will result in overload. [H-2]

4.3 *Identification of causal factors*

The second step of the STPA is the determination of the causal factors contributing to the system performance. For each of the UCAs from 1 to 10 causal factors have been identified. This task was repeated for the 74 UCAs.

4.4 *Event tree analysis*

The 8 hazards that have been defined for the system are used as an initiating event in the Event Tree Analysis. A typical example is given in Figure 7. For the 8th hazard, subcases are used for each type of fault (short circuit & arcing fault) in the system and taking into account the locations of the faults.

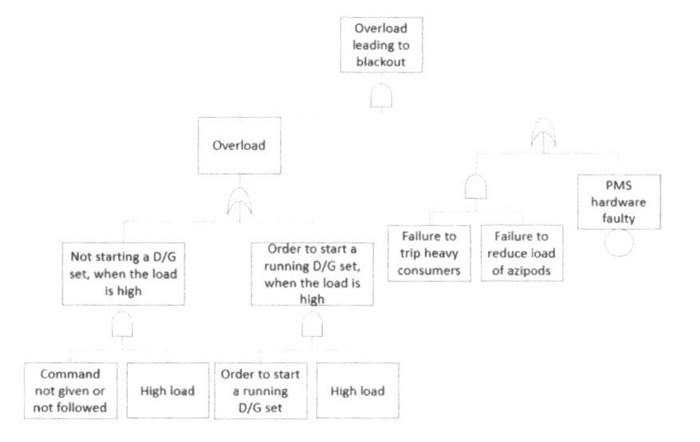

Figure 7. An extract from generated fault tree.

4.5 *Creation of fault tree*

In this step, all the developed Event Trees are merged into one Fault Tree. Most of the hazards are found to be correlated with each other. For instance, [H-1], [H-2], [H-7] & [H-8] are leading to the [H-3]. [H-3] in turn is leading to the [H-5]. In some cases [H-4], [H-5], [H-6] & [H-8] are leading directly to the blackout. So in this way, the primary accidental scenarios are eliminated into four large categories:

- Instability in power generation.
- Overload of D/G sets.
- The too high slope of limiters of requested propulsion demand.
- Uncontrolled disturbances.

These four categories of hazards ([H-4], [H-5], [H-6], [H-8]) are states for the system, in contrast with [H-1], [H-2], [H-3], [H-7] which can be viewed as triggering events at subsystem level. This result justifies the use of the definition of hazards as states in the system according to STPA (Leveson & Thomas, 2015), as they directly lead to an accident, without any intermediate hazard. Still, a looser definition of hazards can add more flexibility and freedom to the analysis and ETA can be used to clarify the setting.

4.6 *Populating the fault tree with UCAs & refinement for common cause factors*

The constructed Fault Tree incorporates only the data from ETA. Based on the information from the STPA, these hazards have been broken down into corresponding UCAs. This is applied to each of the hazards determined during the first steps of STPA. For instance, an attempt to start an already running D/G set or failure to start a D/G set when the power demand is higher than the available power generation will result in both cases into overload. The refinement of the common causal factors is applied to the Power Management

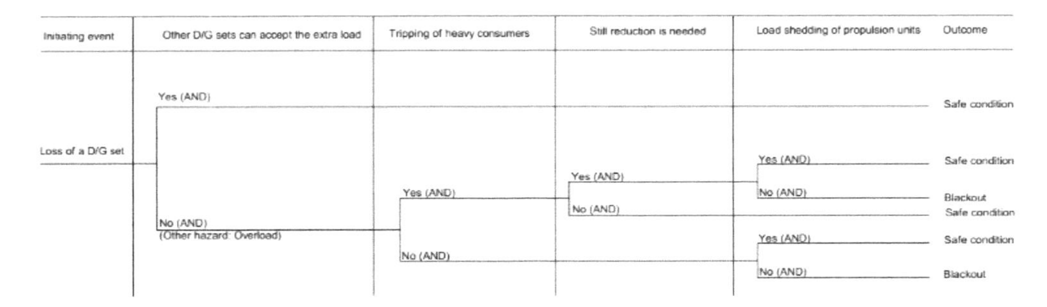

Figure 8. The event tree for one of the hazards.

System (PMS) safety system and engine safety systems, as both PMS and engine safety implement a number of functions. UCAs were broke down into the corresponding conditions and control actions. An extract from the generated Fault Tree is shown in Figure 8.

5 DISCUSSION AND CONCLUSIONS

In this work, an approach has been proposed that combines two hazard identification and analysis techniques, the classical ETA methods and the systemic STPA method and transforms the results into a Fault Tree. In this way, it is possible to define the relationship between the causal factors, UCAs, hazards and the accident in the system using Fault Tree syntax. The constructed system model can incorporate the hazards more systematically and can also incorporate the design errors as a result of the use of STPA. The constructed Fault Tree can be used for the implementation of Quantitative Risk Assessment (QRA) of the system, provided the availability of necessary data. The generated Fault Tree can be also used for allocation of requirements based on target performance as in IEC 61508. The implementation of this approach, although requires significant domain knowledge, can be applied without the exact details of the systems. The constructed Fault Tree is generic and can be easily adjusted to any existing power generation systems on ships. Due to the incorporation of the safety functions in the analysis, it is also possible to focus on critical physical failures.

On the other hand, such an approach has proved tedious, resulting in a quite large number of UCAs, process variables, causal factors. The switching between the methods also has required specific care to avoid any mistake and error during handling of the process. Despite the effort to account for common cause failures, a verification for exhaustiveness of refinement is required. The approach is disconnected from system model and system architecture, which also impedes analysis. The STPA has been also proved as a weak method when it comes to the incorporation of physical failures and single failure accidents (Dawson et al., 2015), (Rokseth et al., 2017) & (Sulaman et al., 2017) and its implementation should be augmented by the use of an FMEA. Since the STPA focused primary on UCAs identification and the relevant contributing factors, the constructed Fault Tree involved mostly the controllers, software and actuators failures. The implemented ETA also used several assumptions. All these issues show directions for future research.

ACKNOWLEDGMENT

The authors are thankful to Dr Luminita Manuela Bujorianu, Dr Romanas Puisa from Maritime Safety Research Centre and to George Psarros, Ole Christian Astrup, Knut Erik Knutsen and Pierre C Sames from DNV GL for their valuable comments and support.

REFERENCES

Abdulkhaleq, A. & Wagner, S. 2016. XSTAMPP 2.0: new improvements to XSTAMPP Including CAST accident analysis and an extended approach to STPA. *5th STAMP Workshop.* United Kingdom, Cambridge.

Ådnanes, A.K. 2003. *Maritime electrical installations and diesel-electric propulsion.* Lecture notes, Norway, Trondheim.

Blandine, A. 2013. *System theoretic hazard analysis applied to the risk review of complex systems: an example from the medical device industry.* PhD thesis, Massachusetts Institute of Technology, Cambridge.

British Standard Institute (BSI). 2010. Functional safety of electrical/electronic/programmable electronic safety-related systems. BS EN 61508

Buja, G., Rin, A. d., Menis, R. & Sulligoi, G. 2010. Dependable design assessment of Integrated Power Systems for All-Electric Ships. *Electrical Systems for*

Aircraft, Railway and Ship Propulsion, Bologna, 19–21 Oct. 2010. 1–8.
Cruise Lines International Association (CLIA). 2016. 2017 cruise industry outlook.
Dawson, L.A., Muna, A.B., Wheeler, T.A., Turner, P.L., Wyss, G.D. & Gibson, M.E. 2015. Assessment of the Utility and Efficacy of Hazard Analysis Methods for the Prioritization of Critical Digital Assets for Nuclear Power Cyber Security.
DNV GL. 2016. Guidance for safe return to port projects. *Class Guideline, DNVGL-CG-0004*
Hansen, J.F. & Wendt, F. 2015. History and State of the Art in Commercial Electric Ship Propulsion, Integrated Power Systems, and Future Trends. *Proceedings of the IEEE,* 103(12), pp 2229–2242.
Hossain, M.A., Kelly, S.J., Ahmed, M.F. & Roa, M.J. 2013. Cause and effect of the catastrophic failure of shipboard and offshore vessel/platform power sources. *Petroleum and Chemical Industry Technical Conference (PCIC), 2013 Record of Conference Papers Industry Applications Society 60th Annual IEEE, IEEE,* 1–8.
International Maritime Organisation (IMO). 2010. Interim explanatory notes for the assessment of passenger ship system's capabilities after a fire or flooding casualty. *MSC.1/Circ. 1369*
Ishimatsu, T., Leveson, N.G., Thomas, J., Katahira, M., Miyamoto, Y. & Nakao, H. 2010. Modelling and hazard analysis using STPA. *Fourth IAASS Conference, 2010 USA, Alabama.*
Leveson, N. & Thomas, J. 2015. An STPA Primer.
Leveson, N. 2011. *Engineering a safer world: Systems thinking applied to safety.* Cambridge, MA: MIT Press.
MAIB. 2011. Report on the investigation of the catastrophic failure of a capacitor in the aft harmonic filter room on board RMS Queen Mary 2 while approaching Barcelona 23 September 2010, United Kingdom, Southampton.
Nilsen, O.V. 2005. FSA for Cruise Ships—Subproject 4.1 Task 4.1.1 - Hazid identification.
Rokseth, B., Utne, I.B. & Vinnem, J.E. 2017. A systems approach to risk analysis of maritime operations. *Proceedings of the Institution of Mechanical Engineers, Part O: Journal of Risk and Reliability,* 231(1), pp 53–68.
Rokseth, B., Utne, I.B. & Vinnem, J.E. 2018. Deriving verification objectives and scenarios for maritime systems using the systems-theoretic process analysis. *Reliability Engineering & System Safety,* 169(18–31).
Sulaman, S.M., Beer, A., Felderer, M. & Höst, M. 2017. Comparison of the FMEA and STPA safety analysis methods–a case study. *Software Quality Journal,* 1–39.
Thomas, J. 2013. *Extending and automating a systems-theoretic hazard analysis for requirements generation and analysis.* PhD thesis, Massachusetts Institute of Technology, Cambridge.
Vedachalam, N. & Ramadass, G.A. 2017. Reliability assessment of multi-megawatt capacity offshore dynamic positioning systems. *Applied Ocean Research,* 63(251–261).

Marine Design XIII – Kujala & Lu (Eds)
© 2018 Taylor & Francis Group, London, ISBN 978-1-138-34076-3

Safe maneuvering in adverse weather conditions

S. Krüger
Hamburg University of Technology, Hamburg, Germany

H. Billerbeck
Flensburger Schiffbau—Gesellschaft, Flensburg, Germany

A. Lübcke
Pella Sietas, Hamburg, Germany

ABSTRACT: The paper describes the interaction between the IMO instrument Energy Efficieny Design Index (EEDI) and the ability of a ship to be able to maneuver safely in adverse weather conditions. A method based on first principles is developed which can compute the equilibrium in yaw and sway if the ship is exposed to wind and waves. From the external and the steering forces, the power requirement to keep the speed can be computed. The results obtained with this method can be used to quantify the steering capability of a ship in heavy weather. The method was applied to some twin screw designs taking into account the future alteration of propulsive power due to the EEDI. It was found that the EEDI had little effect on future steering abilities of the investigated ships, and as a result it can be concluded that properly designed twin screw ships will still have sufficient maneuverability in the future. This suggests that possible future regulatory attempts for safe maneuvering in adverse weather conditions should not focus on twin screw ships. It was also found that when then course keeping problem is to be investigated, the focus should be on following seas.

1 INTRODUCTION

Since the Energy Efficiency Design Index (EEDI) has been adopted, the installed power of ships is subject to a continuous reduction. Because the EEDI regime supports those ships which have a large capacity and a slow ship speed, and the fulfilment of the EEDI is hardly possible without a substantial reduction of the installed engine power. Therefore, the question has been put forward whether ships with reduced installed power can still be safely maneuvered in adverse weather conditions. The European research project SHOPERA has intensively studied this matter. Currently it is under discussion at IMO whether further regulations for minimum installed power shall be developed or not. In this context it must be mentioned that already two such rule sets exist: The minimum required power demands for safe navigation in ice and the minimum power requirements for safe return to port.

From the ship design point of view it clearly leads to severe design problems if one regulation prescribes a maximum installable power and a second regulation prescribes a minimum installable power. The EEDI formulates a maximum installable power, and the two above mentioned rule sets clearly describe a minimum power requirement. As it is presently not intended that a minimum power requirement (from safety considerations) will overrule the EEDI (environmental view point), ship design is confronted with two conflicting requirements concerning the selection of the main engine power. This problem will become really severe if in 2020, the next EEDI reduction phase will enter into force.

In the SHOPERA project, regulations have been suggested which focus on an achievable minimum speed against waves, which will result in a minimum power demand. The problem has been applied to single screw vessels and fixed pitch propellers.

As the German Shipbuilding Industry mainly produces twin screw vessels which have a high maneuverability as key design feature, the question was put forward whether these ships are affected by the problem or not. For this purpose, it was found useful to apply a first principle based direct computational procedure to analyze maneuvering in adverse weather conditions, which is in our analysis defined as course keeping against wind and waves. The computational procedure must at the same time be fast enough for application during the basic design phase. It must also ensure that all maneuvering devices can be designed for a prescribed course keeping procedure.

2 COURSE KEEPING PROBLEM

The principal definitions of the course keeping problem are shown in Fig. 1. The ship travels with the ship speed v_S and it is exposed to environmental forces, which may be a combination of wind, waves and current. These external forces result at first in a longitudinal force, which is called added resistance. This added resistance directly leads to an increase of the propulsion power. The external forces do also result in a hull cross force and a hull yawing moment. These are generated by a combination of drift motion of the ship (drift angle β) and a rudder yawing moment (rudder angle δ). Both drift motion and the rudder action result in an additional longitudinal force which also increases the resistance. The propulsion efficiency of the ship is affected by the fact that the propulsor(s) now work(s) in oblique flow due to the drifting motion. From the view point of maneuverability in heavy weather, the most important question is the ability of the rudder(s) to produce sufficient yaw moment to enable the ship to keep its course. As the inflow to the rudder(s) is strongly influenced by the propeller(s) (especially under the heavy loading), it is of interest whether the installed power is sufficient to guarantee the required inflow to the rudder(s).

The calculation is therefore divided into two steps: At first, two non-linear equations need to be solved to obtain the rudder angle and the drift angle. If the ship cannot keep its course, there exists no solution for the problem. During the second step, the added resistance, the propeller and rudder inflow need to be calculated. This results in propeller rpm and pitch setting and the delivered power of the drive train(s), if the torque/rpm combination is still located in the engine load diagram. The aim of the procedure is to identify limiting situations for the course keeping.

As the course keeping problem is a maneuvering problem, we use our maneuvering model for the computational procedures. This model is briefly described in the following section.

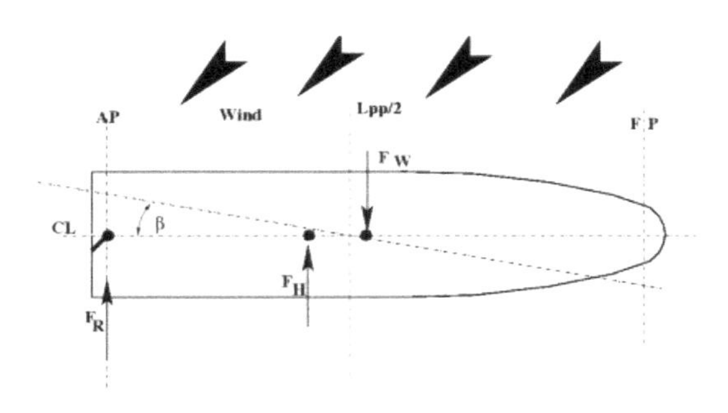

Figure 1. Principal definition of course keeping.

3 MANEUVERING MODEL

3.1 *General*

For the present investigations, we use the body fore maneuvering model which is implemented in the ship design system E4. The model was developed by Söding and it is continuously developed by Krüger et al. on the basis of full scale measurements of ships during sea trials and during the operation. The model is used during the initial design phase for the layout of the hull and the maneuvering devices as well as during the delivery of ship for the generation of the maneuvering booklet and the wheel house poster (see e.g. Fig. 2). The typical accuracy for full scale ships is about ship's breath for distances such as tactical diameters or advances and about 2 degree for overshoot angles. This maneuvering model is now adapted to the course keeping problem.

3.2 *Body forces*

Body Forces are computed according to the slender body theory developed by Söding (1993), see Eqn. 1, where X_2 denotes the longitudinal force, Y_2 the cross force and N_2 the yawing moment:

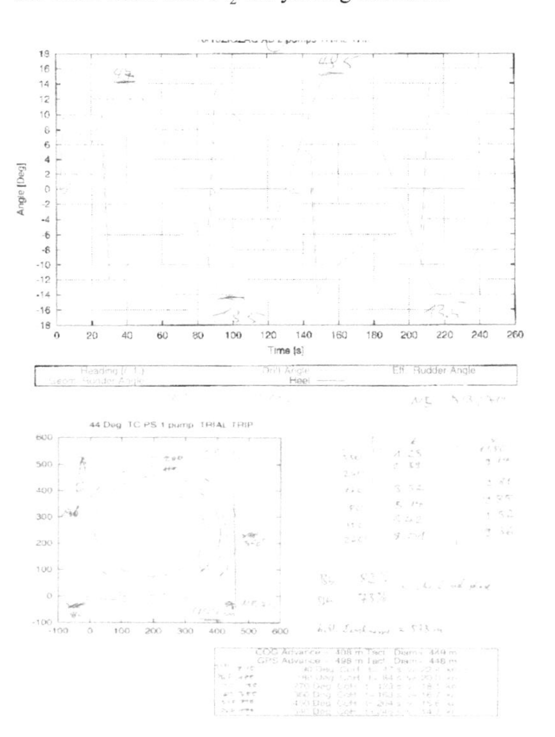

Figure 2. Two examples for the prediction of nautical maneuvers and their first visual validation during sea trial. Top: 10/10 Zig Zag, Bottom: Hard rudder turning circle. Example Ship 1.

$$\begin{pmatrix} X_2 \\ Y_2 \\ N_2 \end{pmatrix} =$$

$$\begin{pmatrix} R_i + X_{vr} v_S r + X_{rr} r^2 \\ R_F \left(v_S + x_T \dot{\psi} \right) \backslash u_s - 0.5\rho \int_L \left(v_S + x\dot{\psi} \right) \backslash v_s + x\dot{\psi} \backslash c_D \cdot d \cdot dx \\ R_F L^2 \dot{\psi} / \left(6u_s \right) - 0.5\rho \int_L x\left(v_s + x\dot{\psi} \right) \backslash v_s + x\dot{\psi} \backslash c_D \cdot d \cdot dx \end{pmatrix}$$

The cross force Y_2 and the yawing moment N_2 consist of two major contributions: The first part is due to the acceleration of the section added mass and the second part is due to the viscous cross resistance of a frame. Both contributions have to be integrated over ship length. For the course keeping problem, the drift angle can be assumed to be small, and therefore, the contribution of the section added mass part is dominating the hull forces. The drag coefficient C_D can be identified form full scale measurements or from viscous calculations (Lübcke 2014) and it lies in the range of 0.3 (for frames with large bilge radii) to 1 (for V – shaped frames). But for the course keeping problem, the viscous contribution is small against the added mass part. It should be noted that the course keeping of a ship strongly depends on the longitudinal distribution of section added mass, where the section added mass depends on the hull form and on the floating condition. Fig. 3 shows the longitudinal distribution of the section added mass for 0 and 15 Degree heel floating condition for the sea trial design floating condition of example ship 1.

Our body force model includes the roll degree of freedom indirectly: From hydrostatic considerations, the heel angle is found, and section added mass as well as viscous forces are corrected for the heel (Söding 1993).

The coefficient X_{vv} which gives the longitudinal force per drift angle can be identified from full scale measurements.

3.3 *Propeller and rudder forces*

Propeller forces are obtained from lifting line calculations. This includes computations in oblique inflow to include the propeller cross forces and the alterations of thrust and torque. These calculations include alterations of the propeller blade angle if the propeller is a controllable pitch propeller. Rudder forces are computed by a direct panel method by Söding (Söding 1998) where the rudder inflow comes from a lifting line theory. Fig. 4 shows the propeller/rudder model of example ship 1 for zero degree rudder angle and design load.

3.4 *Interaction effects*

Most important interaction effects are the alteration of the effective wake and the cross flow to the propeller. The cross flow v_c to the propeller due to the drifting and turning motion can be modelled according to the Kose approach (Söding 1993) as:

$$V_c = k_1 \, v_y + k_2 \, \text{rot} \, x_P$$

where v_y and rot denote the rigid body lateral speed and turning rate of the ship. x_P denotes the distance between propeller and pivoting point. K_1 and k_2 are factors which are below 1, because due to flow separations in the aft body the cross flow is smaller than the rigid body motion. For single screw vessels, k_1 and k_2 have been proposed as 0.36 and 0.66, but for twin screws this is not applicable as the cross flow depends on the combination of signs of turning rate and location of propeller. Furthermore, for twin screw vessels these factors depend

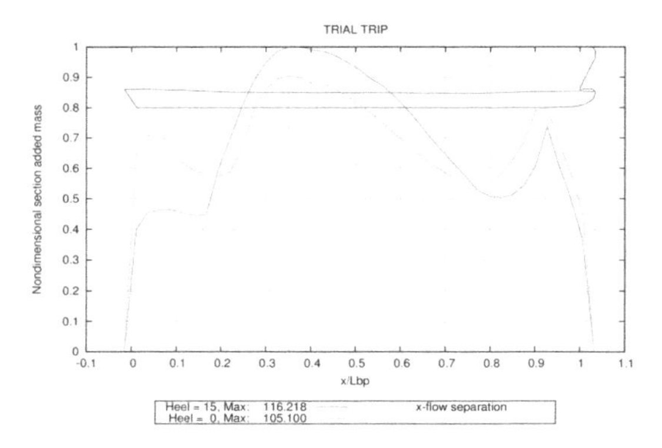

Figure 3. Longitudinal distribution of section added mass for 0 and 15 degree heel. Example Ship 1.

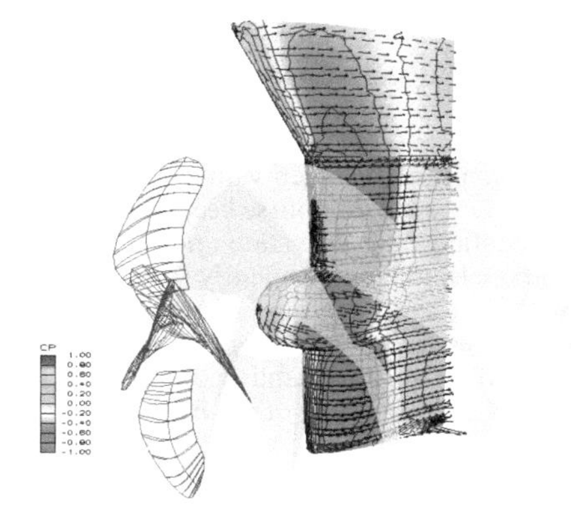

Figure 4. Panel model for rudder forces in the propeller slipstream for Example Ship 1.

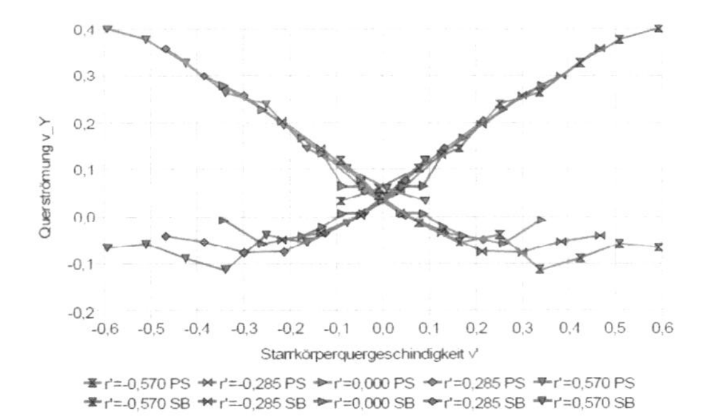

Figure 5. Kose factor k_1 computed for Ship 1. Red: port side propeller, green: starboard side propeller.

also on the ratio of v_y and rot, where for course keeping problems, the rate of turn can be assumed as zero. Fig. 5 shows computed values for the Kose factor k_1 for example ship 1 by a computational RANS—procedure developed by Vorhölter.

3.5 *Propulsion control system*

The model includes a propulsion control system for the calculation of propeller pitch and rpm settings during each time step. The lever command sets initial values for rpm and pitch according to the predefined combinatory curve, and both overload and wind milling switches modify the rpm or pitch value. In case of transient maneuvers such as crash stop astern, there are predefined engine slowdown and startup ramps. For the course keeping analysis in the present context, only the predefined combinatory curve is relevant for pitch and rpm control, as we compute time averaged values only.

4 EXTERNAL LOADS

4.1 *General*

Wave loads, wind and current are acting on the ship, but in the present context, current is disregarded. Although all force modules work in time domain, we have taken time averaged values for the external forces and we search the course keeping equilibrium for these static values. See state and wind parameters can be selected independently from each other, but for the present analysis we have selected wave period and wave height depending on the wind force according to the established ERN—concept by DNV GL, if not stated otherwise.

4.2 *Wind forces*

Wind forces are computed on the basis of Blendermann's coefficients (Blendermann 2013).

4.3 *Wave drift forces*

Second order wave drift forces are computed in the present context by means of a strip theory according to Augener. The investigated ships have a very low block coefficient and a relatively high forward speed, which justifies the application of a strip method. Alternatively, we can use the panel method NEWDRIFT by Papanikolau for these purposes in case it is doubtful whether the strip method is applicable.

5 VALIDAITION

Measurements of the course keeping ability could be performed during the sea trials of ship 1. Ship 1 is a five Deck RoRo ferry with abt. 200 m in length, beam 31 m, draft 7.40. During the sea trial, the draft was 6.47 m aft and 5.75 m fore. The windage area of the ship during sea trial amounted to 5900 m^2. The vessel is a twin screw ship with controllable pitch propellers of 6.30 m in diameter and 10800 kW per shaft line. The two full spade rudders are located in the propeller slipstream.

Designation	Measured	Calculated
Ship speed	17.6–17.8	17.7
Wind speed	16.3–17.8	16.8
Wind direction	228–234	230
Propeller rpm	102.9	102.9
Propeller pitch	77	73
Drift angle	1.8–2.0	2.2
Rudder angle	2.7–2.9	3.2
Shaft power PS	4680	4810
Shaft power STB	4785	4920

During the sea trials, wind forces were between BFT 7 and BFT 8, and the sea was rough with wave heights abt. 2.5 m. The period was estimated to be abt. 5 s. The measurements were performed in the Baltic Sea. The ships heading was abt. 118 Degree, wind and waves were coming from about 230 degree, which lead to a beam wind scenario. Measurements were taken from the bridge automation system (see Fig. 7)[1].

The measured values for the drift angle, rudder angle and wind speed varied slightly due to the unsteady nature of the environmental forces, due to the auto pilot and the propulsion control system.

The computed drift and rudder angle are roughly in line with the measured values, it seems to be that our method slightly overestimates drift and ruder angle. The computed power values are

[1]During the measurement, the rpm bridge display calibration was not finished, the rpm value in Fig, 6 should be 103 instead of 126.

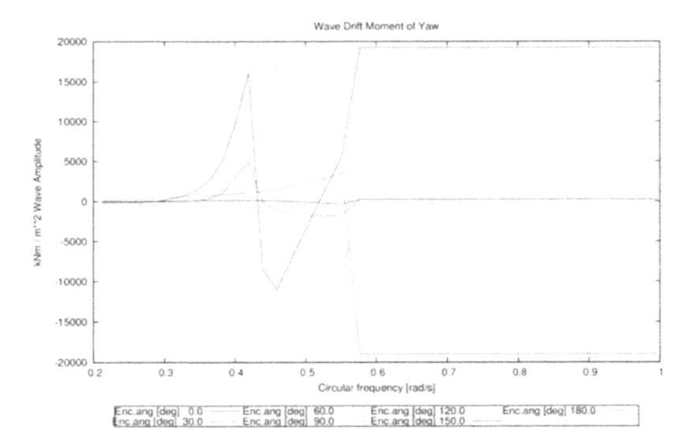

Figure 6. Second order wave yawing moment computed for ship 1 as function of circular frequency. Ship speed 17 knots.

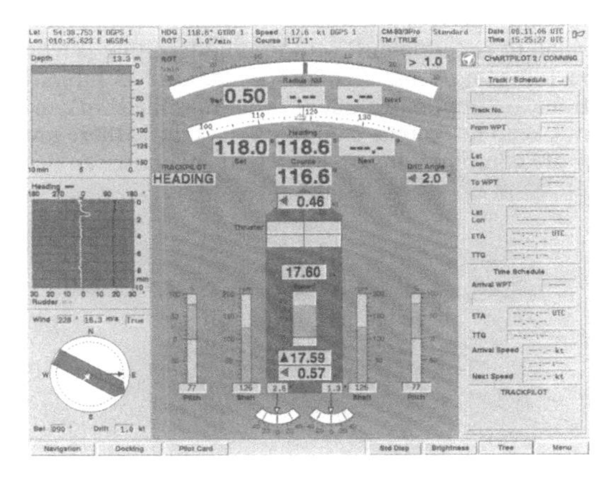

Figure 7. Screen shot of the bridge display during yaw checking measurements of ship 1.

slightly smaller compared to the measurements, this can be explained by the fact that the vessel performed better during sea trial compared to the model basin prognosis, and for the present analysis we have not made a correction to the model basin values. Most likely is that the full scale wake fraction differs from the model value, as there are also slight differences in the computed pitch settings. Although further validations would be useful, we assume for the moment that our model predicts the correct trends and we will apply this model for course keeping investigations.

6 COURSE KEEPING INVESTIGATIONS

6.1 *General remarks*

Currently, there is no agreed setup for the course keeping analysis. For the present analysis, we follow the safe-return-to port regulations which define environmental conditions as BFT 8 and sea state 7. According to the DNV GL ERN concept, this results in a significant wave height $H_{1/3}$ of 5.2 m and a wave period T_1 of 9.6 s. We will use these conditions during the following analyses. We further assume that wind and waves have the same direction. It is obvious that the critical condition for course keeping is the lightest seagoing condition, therefore we will carry out our analyses for the ballast arrival condition (which is close to the sea trial condition).

6.2 *Ship 1*

Although ship 1 does not need to comply with EEDI due to keel laying date, we use ship 1 as a first example because of its very large windage area. Fig. 8 shows the results of our analysis in form of polar plots of the required rudder angle for course keeping (top) and the resulting drift angle. The calculation speed was set between 6 and 20 knots. Due to the limited engine power, the achievable speed in head seas is smaller, that is why the polar plots have been cut accordingly. In following seas, the problem arises that at low ship speeds, there is no inflow to the rudders and consequently no

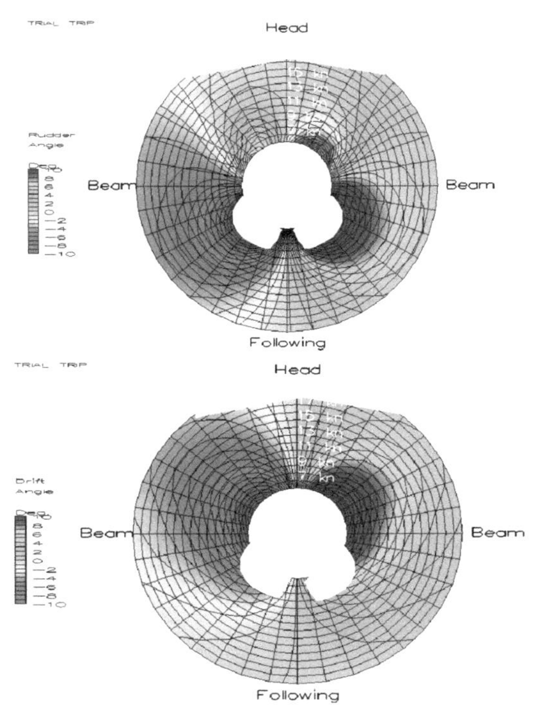

Figure 8. Course keeping polar diagrams for ship 1, trial draft, BFT 8 SS7. Top: Rudder angle, bottom: Drift angle.

rudder yaw moments, that is why certain areas in following seas between 20 and 40 degree can never be reached at slow speeds because the ship is unstable there. In head seas, the ship has not any problems to maintain its course. The polar plots do also show that the ship can reach more than 6 knots speed in all head and beam sea courses. If the speed is above 10 knots, the ship can reach all courses in following seas.

It has been mentioned that ship 1 does not need to comply with the EEDI. To study at least roughly the effect of future power limitations to the ship, we have computed the power demand of the ship during course keeping and plotted the results in Fig. 9.

Fig. 9 shows the required power during course keeping for ship 1 in a BFT 8 SS 7 scenario. The color scale was selected such that all power values below 10 MW get the same color index. It can easily be seen that even if the EEDI would reduce installed power by more than 50%, the ship would still be maneuverable in all head sea condition, but now only speeds between 9 and 11 knots can be reached in head seas. For the following sea scenarios, our assumed power reduction has no influence. The ship is even with reduced power fully maneuverable due to the fact that she is a twin screw vessel with two spade rudders which were designed for high maneuverability and due to the fact that the controllable pitch propellers can absorb 100% MCR at positive inflow to the propellers.

6.3 *Ship 2*

Ship 2 is a four—deck RoRo Ferry which complies with the current EEDI—requirements. She was developed from a proven FSG design by adding a parallel mid body of 16.80 m to increase the capacity. The layout of the maneuvering devices remained unchanged. The length of the ship is 209 m, beam 26 m, draft 6.45 m. The windage area amounts to 4600 m^2. The ship has two main engines with a total power of 19200 kW. The twin full spade rudders ale located in the slipstream of the two controllable pitch propellers of 5 m diameter. This ship was chosen for one reason that she fulfills the existing EEDI regulations and for the other that the capacity of the ship was increased by keeping the maneuvering devices of the original design.

Fig. 10 shows the results of our course keeping analysis. The computations were performed for ship speeds between 6 and 20 knots where the vessel was running on the light ballast draft. In head seas, a maximum speed of abt. 15.5 knots is possible with the installed power. In following seas, we found the same situation as for ship 1: When the ship speed is low, the power requirement is too small to generate inflow to the rudders, that is why in following seas, certain speed/course combinations are impossible due the fact that the ship cannot keep its course. Compared to ship 1, there are more combinations of speed and encounter angles

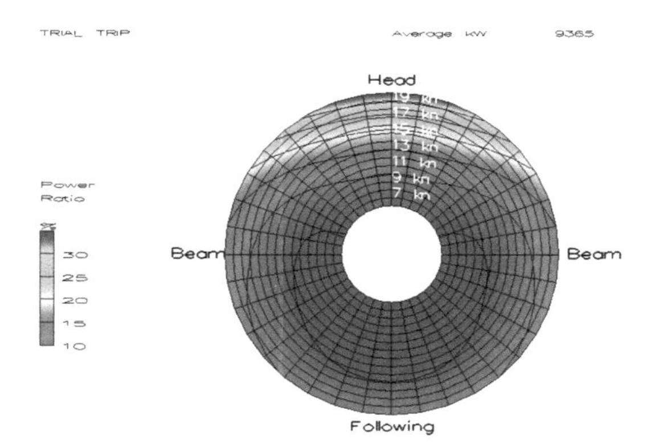

Figure 9. Required power for course keeping for ship 1, trial draft, BFT 8 SS7. Cutoff power was set to 10 MW.

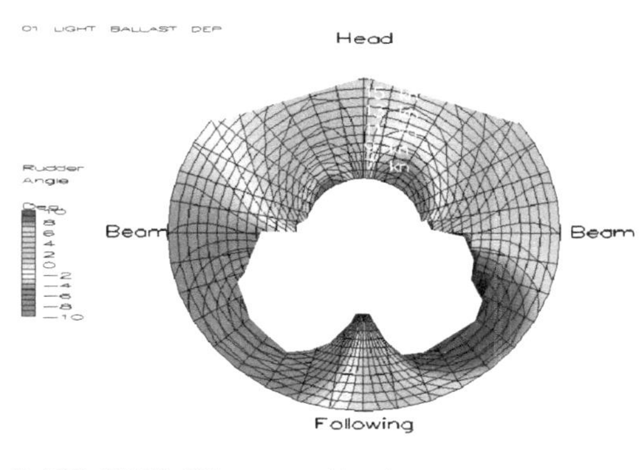

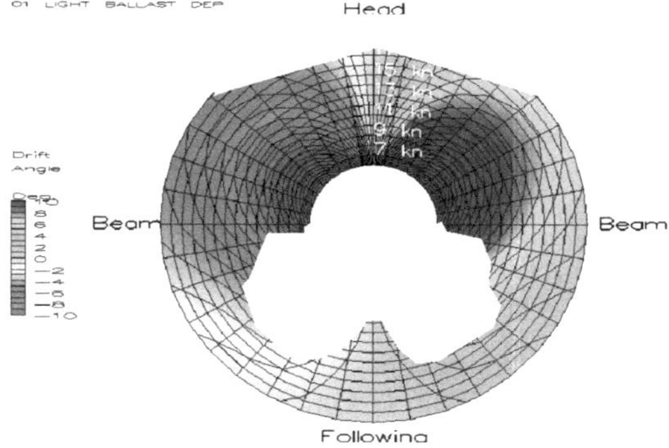

Figure 10. Course keeping polar diagrams for ship 2, light ballast draft, BFT 8 SS7. Top: Rudder angle, bottom: Drift angle.

where the ship is not able to steer. This is reasonable, because the maneuvering devices were taken from the basic design and they remained unchanged. If the ship speed is more than 13 knots, the ship is maneuverable on all courses.

From operational view, one can say that in the assumed scenario, the ship is well maneuverable, because from practical considerations, the ship will not operate at very low speeds in following seas. One can further conclude that even after the lowering of the EEDI baseline in 2020 the ship will be fully maneuverable, because the EEDI has an effect on the installed power, and the installed power has no impact on the situation on following seas. A reduced power would also reduce the achievable speed in head seas, but even with reduced power the speed will be abt. 13–14 knots.

The results do also show that if an EEDI—compliant ship shall have the same maneuverability compared to a pre—EEDI ship, the effectiveness of the rudders must be adapted. Critical in this respect is maneuvering in following seas with slow speeds. If future ships are designed for very low speeds following the EEDI reduction scheme, than it could be possible that operation in following seas is more limited or even hardly possible.

6.4 *Ship 3*

Ship 3 is built at Pella Sietas Yard in Hamburg. The type of the ship is a hopper dredger which is called Sietas Type 180. The ship has a length over all of 118.47 m, a breadth of 21.10 m and the draft is limited by 7,10 m. The hopper dredger has a diesel-electric drive concept with two electric drive motors with a total power of 3400 kW.

The ship does not need to comply with the EEDI-requirements due to the propulsion concept and the ship type. But it is a good example to show the course keeping capability of a small twin screw vessel with a relative low engine power and a fixed pitch propeller.

The results of the course keeping calculation in heavy weather are presented in Fig. 11 for a ship speed between 6 and 13 knots. The figure above gives the required rudder angle to hold the course of the ship and below the resulting drift angle is shown. Similar to the ship 1 and 2 the capability of course keeping is limited by the available power and the maximum rudder angle. In case of head sea, the ship speed is reduced to 7.5 knots by the available power. The more critical situation is for small ship speeds in following seas. In this condition the rudder angle limits the course keeping capability which is caused by the reduced inflow velocity of the rudder. The results of the calculation confirm the previous investigations of ship 1 and 2.

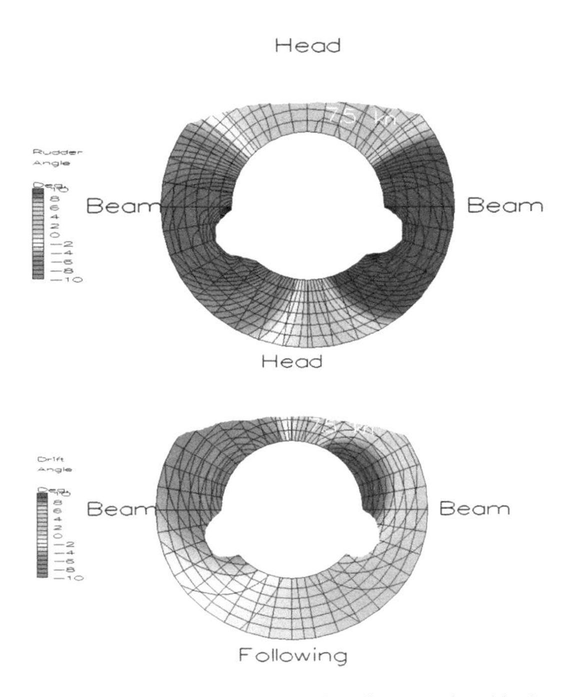

Figure 11. Course keeping polar diagrams for ship 3, design draft, BFT 8 SS7. Top: Rudder angle, bottom: Drift angle.

7 CONCLUSIONS

A method was presented which allows to compute the course keeping problem in heavy weather. The method is based on our existing maneuvering code and it determines drift and rudder angle in the equilibrium of cross forces and yawing moments. Afterwards, the required power can be determined. The comparison of measurements for a sea trial in bad weather and the computed results was reasonable. The application to three twin screw designs showed the following results: In head or beam seas, all ships are fully maneuverable in all possible courses and speeds. Reasonable ship speeds could be reached. It could be shown that all investigated ships would still be fully maneuverable even if they would have to comply with EEDI after 2020. This is due to fact that all ships have been designed for good maneuverability, therefore they all have twin controllable pitch propellers and twin rudders. It could also be shown that critical situations with respect to course keeping will at first occur in following seas if the ship speed falls below a critical speed. This is due to the fact that in these situations, the effectiveness of the rudders is too low because there is not enough inflow from the propellers to the rudders. It might be possible for future ship designs that this problem becomes more severe if ships are designed for significantly lower speeds. It can also be concluded from our investigations that EEDI—compliant twin screw ships which are equipped with CPPs and twin rudders will always have sufficient maneuvering ability in adverse weather conditions if the design ship speed is still reasonable.

REFERENCES

Augener, P. 2016. *Computation of Wave Drift Forces for Dynamic Positioning within the Early Design Stage.* PhD-Thesis, Hamb. Univ of Techn.

Blendermann. W. 2013. *Practical Ship and Offshore Areodynamics.* Schriftenreihe Schiffbau, Rep. 669. Hamburg.

Isay, W. H 1963. *Propellertheorie.* Springer, Berlin.

Krüger, S.: *Maneuvring—Which results can be obtained from strip methods?* PROC. JSTG 2018, Vol. 108, Hamburg.

Lübcke, A.: *Dynamic Positoning as a New Aspect in the Early Design Stage of Ships—Calculations, Model Tests and Results.* PROC. JSTG, 2014 Vol. 107, Hamburg.

Papanikolau, A., Zaraphonitis, G., Bitner-Gregersen, E., Shigunov, V., EL Moctar, O., Guedes Soares, C., Reddy, D. N., Sprenger, F. 2015. *Minimum Propulsion and Steering Requirements for Efficient and Safe Operation (SHOPERA).* Motorship P&EC Conf., Hamburg, 4-5/3/2015.

Söding, H.: 1993. *Body Forces.* Manoeuvring Technical Manual. Seehafen Verlag, Hamburg.

Söding, H.: 1998. *Limits of potential theory in rudder flow predictions.* Ship. Techn. Res. 45,3.

Vorhölter, H.: 2011. *Numerische Analyse des Nachstroms und der Propellereffektivität am manövrierenden Schiff.* PhD-Thesis, Hamb. Univ of Techn.

Marine Design XIII – Kujala & Lu (Eds)

Design method for efficient cross-flooding arrangements on passenger ships

P. Ruponen
NAPA, Helsinki, Finland

A.-L. Routi
Meyer Turku, Turku, Finland

ABSTRACT: Asymmetric flooding condition increases the risk of capsizing for a damaged ship. For this reason, efficient cross-flooding arrangements are needed. This is important, especially for passenger ship designs with double skin or large U-shaped voids in several watertight compartments. This paper presents an advanced method, utilizing time-domain flooding simulation, for design and analysis of cross-flooding devices. Special attention is paid on the selection of the damage scenarios that are used in the assessment of the efficiency of the cross-flooding arrangements. The presented method can be used to confirm the assumption of instantly flooded rooms for probabilistic SOLAS damage stability calculations.

1 INTRODUCTION

The increasing demand for improved damage stability is a challenge for designers of passenger ships. Especially since after the Costa Concordia accident, ship operators have started to require a double skin arrangement in the engine room compartments for new ships. Although this design enhances survivability in side groundings with small penetration, it also increases the risk of transient asymmetric flooding. In addition, longitudinal non-watertight bulkheads are often needed for fire integrity, but also these structures provide a challenge for ensuring high survivability in the event of flooding.

Asymmetric flooding should be equalized rapidly through self-acting cross-flooding devices in order to reduce extensive heeling. Furthermore, in the void spaces, air compression can be notable, and the counter pressure can delay the equalizing cross-flooding. Therefore, sufficient ventilation must be ensured, especially in the void spaces.

Efficient cross-flooding arrangements provide a larger attained subdivision index in the SOLAS (Safety of Life at Sea) damage stability calculations. If complete equalization occurs in 60 s or less, also the connected rooms can be assumed as instantly flooded. A simplified method for estimation of the cross-flooding time was initially presented by Solda (1961). This has later become an industry standard and a recommendation MSC.362(92) from the International Maritime Organization (IMO). The use of time-domain flooding simulation for design of cross-flooding arrangements of navy ships was introduced by Peters et al. (2003). Later, more advanced simulation methods have been developed to account for the air compression, Ruponen et al. (2012).

The present study introduces a design method for efficient and feasible cross-flooding arrangements, within the regulatory framework of the SOLAS Convention. Instant flooding is analyzed for realistic collision damages by using time-domain flooding simulation, accounting for air compression effects. The presented approach is demonstrated with a practical example, using a large passenger ship design with double skin in the engine room compartments. The focus is on two watertight compartments, where the risk of asymmetric flooding stages is obvious.

2 REGULATORY REQUIREMENTS

2.1 *SOLAS framework*

Requirements for sufficient reserve stability after a damage are specified in SOLAS Chapter II–1. The calculations are mainly based on a probabilistic method (Reg. 7). In practice, this means that a large group of damage cases needs to be calculated. Each case has a p-factor that denotes the probability that only the defined set of rooms and compartments is damaged. The actual breach extent is not defined, only the extent of damaged compartments. With increased number of more extensive damages, the sum of the p-factors approaches to unity. For this reason, several multi-zone damages with damage length up to the maximum of 60 m are often calculated for large passenger ships in order to maximize the attained subdivision index.

2.2 *Assumption of instant flooding*

An essential part of a damage stability analysis is to decide the first flooding stage to be calculated. Namely, what rooms are considered open to sea (i.e. lost buoyancy). The Explanatory Notes to the SOLAS Chapter II-1 (IMO, 2017) state that:

> *"The case of instantaneous flooding in unrestricted spaces in way of the damage zone does not require intermediate stage flooding calculations. Where intermediate stages of flooding calculations are necessary in connection with progressive flooding, flooding through non-watertight boundaries or cross-flooding, they should reflect the sequence of filling as well as filling level phases. Calculations for intermediate stages of flooding should be performed whenever equalization is not instantaneous, i.e. equalization is of a duration greater than 60 s. Such calculations consider the progress through one or more floodable (non-watertight) spaces, or cross-flooded spaces. Bulkheads surrounding refrigerated spaces, incinerator rooms and longitudinal bulkheads fitted with non-watertight doors are typical examples of structures that may significantly slow down the equalization of main compartments."*

This means that for efficient design, a calculation method is needed to check if the equalization takes place in 60 s or less.

2.3 *Cross-flooding*

The heeling due to asymmetric flooding is equalized by adding passive cross-flooding devices that allow floodwater to flow to the undamaged side.

A schematic cross-flooding case is illustrated in Figure 1. The damaged side of a U-shaped void is open to sea, and the ship is heeled with an initial waterline WL_0. The center of the cross-flooding duct is used as the reference point, and half of the duct is considered as lost buoyancy. The initial effective pressure head is H_0. After the cross-flooding process, volume W_f has flown through the duct, and the effective pressure head has decreased to h_f.

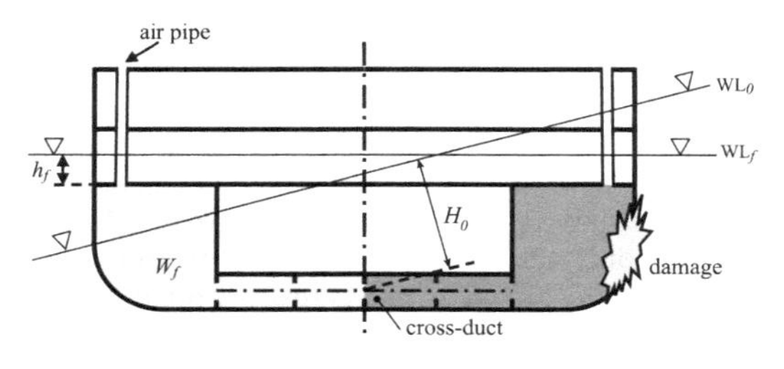

Figure 1. Principles of cross-flooding in a large U-shaped void (dark shaded area is treated as lost buoyancy).

The IMO Resolution MSC.362(92) provides a simple formula for calculating the cross-flooding time. However, this formula is valid only for one void space. If there is simultaneous cross-flooding in several compartments, the simplified method can only be used to get very conservative estimates, as pointed out by Ruponen and Lindroth (2016). On the other hand, time-domain flooding simulation provides realistic time-to-flood analysis, even in the most complex scenarios.

3 DESIGN OF CROSS-FLOODING ARRANGEMENTS

3.1 *Means of cross-flooding*

Asymmetric flooding can be equalized with various different arrangements, as illustrated in Figure 2. In double bottom, cross-flooding ducts are fitted to the structures, but also pipes can be used. On tank top and above, cross-flooding is arranged by transverse corridors. If fire integrity is needed, blowout panels or flaps need to be installed in the longitudinal bulkheads. In addition, collaps-

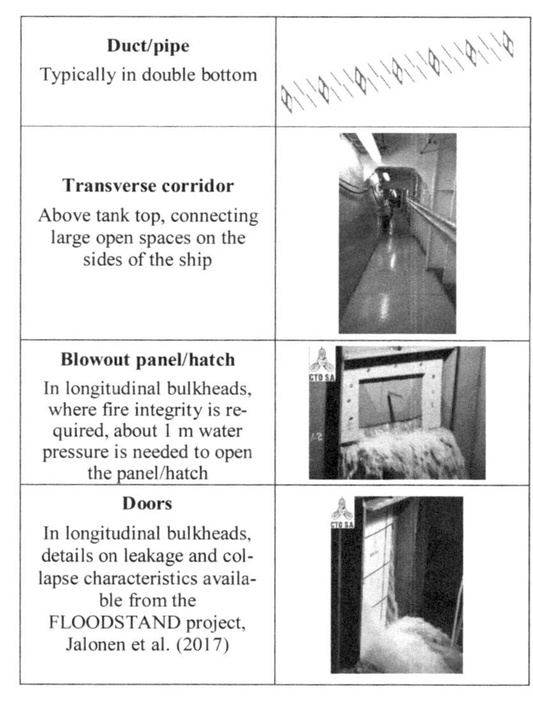

Duct/pipe Typically in double bottom	
Transverse corridor Above tank top, connecting large open spaces on the sides of the ship	
Blowout panel/hatch In longitudinal bulkheads, where fire integrity is required, about 1 m water pressure is needed to open the panel/hatch	
Doors In longitudinal bulkheads, details on leakage and collapse characteristics available from the FLOODSTAND project, Jalonen et al. (2017)	

Figure 2. Typical cross-flooding arrangements.

ing non-watertight doors can provide equalizing flooding.

For doors and blowout panels, the leakage and collapse characteristics need to be modelled realistically. Results of dedicated full-scale tests and FEM analysis within the EU FP7 project FLOODSTAND, Jalonen et al. (2017), may be useful. However, these are only guideline values, and a study on the effects of random variation in both critical pressure head for collapse and leakage area ratio for complex flooding cases have been presented in Ruponen (2017).

In practice, the maximum possible area for the cross-ducts is limited by the structural design, i.e. number frames, double bottom height and required web frames and stiffeners. In some cases, the only available option to decrease the cross-flooding time is to add a cross-pipe through the double bottom tanks, as illustrated in Figure 3. Therefore, it is important that the efficiency of the cross-flooding devices is evaluated already in the early design phase.

3.2 *Ventilation*

Air compression can significantly delay the cross-flooding. In arrangements, where the total air pipe sectional area is 10% or more of the cross-flooding sectional area, the restrictive effect of any air back pressure may be neglected, according to MSC.362(92). Interestingly, the regulation recognizes only the cross-sectional area of the air pipes and completely ignores the pressure losses in the pipes due to friction, bends and valves.

In practice, very large air pipes will restrict the general arrangement above the void. Preferably, the pipe should fit between the web frames, as illustrated in Figure 4. Therefore, if improved ventilation is needed, the solution can be to add more pipes with the same diameter.

Recently CFD methods have been introduced to calculate air compression effects during flooding, Gao et al. (2018) and Cao et al. (2018), but the simplified method, Ruponen et al. (2013), is more suitable for design purposes due to enormously faster computational performance.

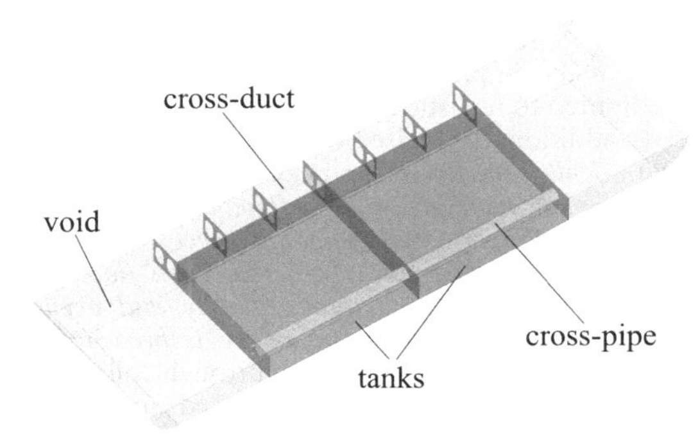

Figure 3. Example of cross-duct and cross-pipe connecting the same void pair Principles.

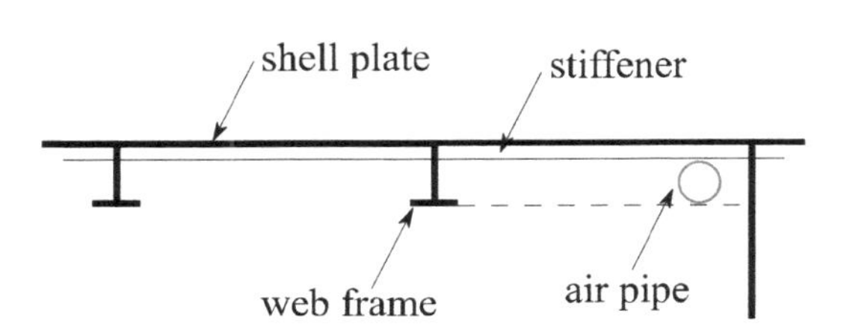

Figure 4. Deck plan view of structural limitations for the ventilation pipes from the void spaces below.

4 TIME-DOMAIN FLOODING SIMULATION

4.1 *Quasi-stationary hydraulic model*

The time-domain flooding simulation in the NAPA software is used. The applied method is based on implicit time integration with a pressure-correction algorithm. This has proven to be an efficient and accurate approach for calculation of damage scenarios with either extensive progressive flooding to several compartments or notable air compression.

At each time step, the conservation of mass must be satisfied in each flooded room:

$$\int_{\Omega} \frac{\partial \rho}{\partial t} d\Omega = -\int_{S} \rho \mathbf{v} \cdot d\mathbf{S} \tag{1}$$

where ρ is density, $\mathbf{v}$ is the velocity vector and $\mathbf{S}$ is the surface that bounds the control volume Ω. The normal vector of the surface points outwards from the control volume.

The velocities in the openings are calculated by applying Bernoulli's equation for a streamline from point A that is in the middle of a flooded room to point B in the opening, see Figure 5:

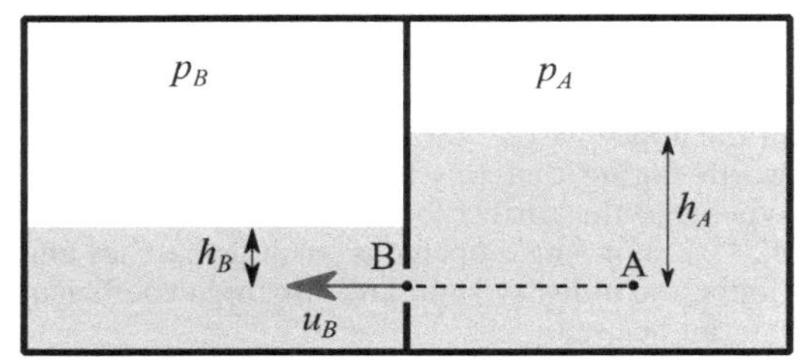

Figure 5. Application of Bernoulli's equation for a streamline from point A to point B.

$$\int_A^B \frac{dp}{\rho} + \frac{1}{2}(u_B^2 - u_A^2) + g(h_B - h_A) + \frac{1}{2}k_L u_B^2 = 0 \quad (2)$$

where p is air pressure, u is flow velocity, g is acceleration due to gravity and h is the water height from the common reference level. The flow velocities in the centers of the rooms are assumed to be negligible, and consequently $u_A = 0$. All losses in the opening are represented by the non-dimensional pressure-loss coefficient k_L that equals to the k-sum in MSC.362(92).

The calculation within a time step is iterative, based on linearized Bernoulli's equation. The algorithm changes the pressures in the flooded rooms until both Bernoulli's equation for each opening and continuity for each room are satisfied with sufficient accuracy. After that, the floating position of the ship is calculated based on the distribution of floodwater in the compartments. In the present study, all ship motions are considered quasi-stationary. The floodwater volumes at each time step are constant in the calculation of the righting lever curve.

A more detailed description of the applied simulation method is given in Ruponen (2014), and with focus on the application of Bernoulli's equation for airflows in Ruponen et al. (2013).

4.2 *Discharge coefficients*

Discharge coefficient for the flow through an opening can be evaluated based on the total sum of pressure loss coefficients:

$$C_d = \frac{1}{\sqrt{1 + k_L}} \quad (3)$$

which is equivalent to the flow reduction factor F in the simplified formula for cross-flooding time in MSC.362(92). Therefore, in this respect there is no difference between the calculation methods.

The effective discharge coefficient for the cross-flooding ducts can be evaluated with model tests, Stening et al. (2011), or by CFD analyses, Ruponen et al. (2012) and Ohashi et al. (2012).

In practice the simplified regression equations for various typical duct and pipe arrangements in the appendix of MSC.362(92) are applied. It is worth noting that this regulation contains some typos and the correct formulae are given in MSC 92/11/2. For single openings, such as hatches and doors, the industry standard discharge coefficient 0.6 can be used.

The IMO Resolution MSC.362(92) provides a simple equation for estimation of the total pressure losses, when the device consists of several successive openings with area S_i and pressure loss coefficient k_i:

$$\sum k = k_1 + k_2 \frac{S_1^2}{S_2^2} + k_3 \frac{S_1^2}{S_3^2} + \ldots \quad (4)$$

This can be applied to any typical cross-flooding duct. However, it should be noted that the first opening to consider should be the one with the smallest area, so that $S_1 = S_{min}$. In addition, the smallest area should be used as the effective area of the cross-flooding device.

5 ASSESSMENT OF INSTANT FLOODING

5.1 *Background*

The conventional method for cross-flooding analyses is limited to flooding between two rooms, MSC.362(92). Thus, for consistency the same assumption can be used also when using time-domain simulation. In practice, this means that each device would be studied independently, while the other compartments with cross-flooding devices are assumed to be instantly flooded.

However, the actual purpose of using simulation should be more realistic assessment of the flooding progression than with the conventional methods. In order to fully take advantage of simulation, the performance of different cross-flooding devices should be studied simultaneously. However, this is not possible in practice, unless the damage extents are limited to realistic collision damages.

In addition, the revised Explanatory Notes state that: *"Unless the flooding process is simulated using time-domain methods, when a flooding stage leads to both a self-acting cross-flooding device and a non-watertight boundary, the self-acting cross-flooding device is assumed to act immediately and occur before the non-watertight boundary is breached".* Thus in simulation, the flooding through collapsed non-watertight doors can be calculated simultaneously with the cross-flooding analyses.

5.2 *Selection of the studied damage scenarios*

The analysis of cross-flooding arrangement by using a time-domain flooding simulation is a deterministic procedure that can be done for certain damage cases. The current SOLAS damage stability requirements are based on a probabilistic method. In principle, fast cross-flooding should be verified for all damage cases contributing to the attained subdivision index. However, SOLAS II-1 Reg. 7 is based on zonal approach, where each damage case has a combined probability of all possible damages that would breach exactly the same set of compartments. Therefore, the actual size of the breach is not considered. For long collision damages this will inevitably result in unrealistically large breach sizes, meaning that the initial condition before cross-flooding commences is not survivable. Therefore, assessment of instant flooding due to efficient cross-flooding arrangement needs to be done only for a limited set of damage cases.

In general, the relative damage length in a collision is smaller for larger ships, Pedersen and Zhang (2000). This is logical since the damage length is mainly affected by the characteristics of the striking ship, as well as the structures of the struck ship. Furthermore, the relative damage lengths are smaller when using only the more recent accidents, Papanikolaou et al. (2013). The reason for this is evidently the increased size of the ships. The applied probability distributions in SOLAS can be very conservative, especially for long passenger ships with several decks. Consequently, it is not feasible to base the rapid cross-flooding evaluation on all damage scenarios to be included in the probabilistic damage stability assessment.

The worst realistic damage scenario from the cross-flooding point of view is a large collision damage. Raking side damages, like the one in the Costa Concordia accident, are not that critical for cross-flooding point of view, since the breach area is smaller. Consequently, also the water inflow is smaller. It is worth noticing, that in reality cross-flooding starts immediately when the device is submerged. This means that for small breach areas the initial condition, where the damaged side is fully flooded before cross-flooding is very unrealistic.

For a typical collision damage, the damage length of SOLAS Chapter II-1 Reg. 8 is considered as a suitable limitation for cross-flooding analysis. Thus the damage length is:

$$L_{dam} = 0.03 \cdot L_s \tag{5}$$

where L_s is the subdivision length of the ship. In practice, this means that damage cases involving more than two WT compartments can usually be ignored.

Cruise ships usually have 2–3 decks below the intact waterline. This will affect the penetration of the damage in collisions, Pedersen & Zhang (2000). The probabilistic SOLAS calculations in Reg. 7 consider damage extents up to *B*/2, but this is not very realistic, especially for modern passenger ships. Therefore, analogically to the damage length, also the penetration can be based on the Reg. 8 requirements, so that the rooms inside the *B*/10 limit are always intact.

Also the horizontal WT subdivision should be accounted for. Damages with a lesser vertical extent, e.g. only above the double bottom, may be more serious since the center of gravity of the floodwater is located higher. On the other hand, there are usually several cross-flooding devices in the double bottom, thus resulting in more asymmetry of flooding in damage cases that extent vertically to the baseline.

5.3 *Initial conditions*

SOLAS calculations are performed for three initial conditions, namely, the deepest subdivision draught (DS), light service draft (DL) and one partial condition (DP). Dry ship (no liquid loads) is assumed for all calculations. This is a conservative approach since the mass (and GM reduction) of the liquid loads is accounted for, but the tanks are empty and can be flooded without outflow. In principle, the same initial conditions are suitable also for cross-flooding studies.

Each simulation starts from a condition, where the rooms in way of the damage, including the damaged side of cross-flooding connections, are open to sea, i.e. treated as lost buoyancy.

5.4 *Criteria for rapid cross-flooding*

SOLAS regulations set criteria for the cross-flooding arrangements. In Chapter II-1 Reg. 7–2 Paragraph 26 it is stated that: "*For passenger and cargo ships, where cross-flooding devices are fitted, the time for equalization shall not exceed 10 min*".

The Explanatory Notes provide a more detailed description on how this should be evaluated: "*If complete fluid equalization occurs in 60 s or less, it should be treated as instantaneous and no further calculations need to be carried out. Additionally, in cases where s_{final} = 1 is achieved in 60 s or less, but equalization is not complete, instantaneous flooding may also be assumed if s_{final} will not become reduced*".

For passenger ships, the *s*-factor, representing the survivability level, is defined as:

$$s_{final} = K \cdot \left(\frac{GZ_{max}}{0.12} \cdot \frac{range}{16} \right)^{\frac{1}{4}} \tag{6}$$

where GZ_{max} is limited to 0.12 m and *range* to 16°. The effect of the heel angle ϕ is accounted for with the coefficient:

$$K = \sqrt{\frac{15° - \phi}{15° - 7°}} \tag{7}$$

when the heeling angle ϕ is between 7° and 15°. If the heeling exceeds 15° then K = 0.0, and if heeling is less than 7° then K = 1.0. The upper threshold limit is supported by the SOLAS requirement to be able to lower the lifeboats with heeling up to 15°.

For cross-flooding the heeling angle is usually the determinant factor, and in practice the use of s_{final}, equation (6), means that heeling must be reduced to 7° or less within 60 s.

In addition, reserve to external heeling moments is accounted for:

$$s = s_{mom} s_{final} \tag{8}$$

where

$$s_{mom} = \min\left(\frac{(GZ_{max} - 0.04) \cdot disp}{M_{heel}}, 1.0\right) \tag{9}$$

where *disp* is the intact displacement at subdivision draft and M_{heel} is the maximum external heeling moment, caused either by wind, passenger crowding or launching of the survival craft.

It is worth noticing that different *s*-factor formulation, with stricter requirements for GZ_{max} and *range*, are applied for RoPax ships in the damage cases, where a ro-ro room (vehicle deck) is flooded.

5.5 *Procedure*

The previously presented approach for assessment of rapid cross-flooding can be summarized as the following procedure:

1. Model all rooms with a cross-flooding connection as two separate rooms
2. Generate the damage cases to be calculated (same cases as SOLAS II-1, Reg. 8)
3. Evaluate the effective discharge coefficient for each cross-flooding devices and air pipes
4. Model the cross-flooding devices as openings and define the ventilation levels and air pipes
5. Calculate each damage case with time-domain flooding simulation
6. Analyze the *s*-factor for the calculated cases
7. Check which cross-flooding devices do not provide "instant flooding"
8. If needed, modify cross-flooding devices (e.g. larger area if possible) or add air pipes; then continue from step 3
9. Combine the room parts, where cross-flooding devices provide "instant flooding"
10. Continue with SOLAS damage stability calculations

A cross-flooding device is considered to provide instantaneous flooding between the two connected rooms if in all cases, where the device is involved:

- Within 60 s the survivability factor s_{final} = 1.0 (and is not reduced thereafter), or
- The cross-flooded room is either filled up, or the water level equals to the sea level

If instant flooding is not achieved, the cross-flooding device, or the ventilation arrangement, needs to be improved. Alternatively, the flooding stage before cross-flooding can be included in the SOLAS damage stability calculations.

6 PRACTICAL EXAMPLE

6.1 *Arrangement*

The presented method is demonstrated with a large passenger ship design, which is a slightly modified version of the FLOODSTAND Sample Ship A, Kujanpää & Routi (2009). For example, a double skin has been added to the engine room compartments.

In principle, all one and two WT compartment damages need to be studied. These are illustrated in Figure 6. Of course, not all compartments have cross-flooding devices, and these can be ignored. As an example, the focus is on two compartments in the aft part of the ship, with significant risk of notable transient asymmetric flooding due to the void spaces surrounding the double bottom tanks and machinery areas. These are illustrated in Figure 7. There are cross-flooding ducts in the double bottom and an open transverse corridor on

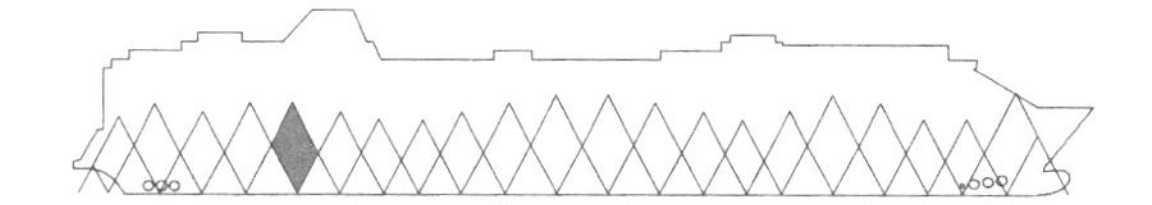

Figure 6. Studied damage cases for cross-flooding analyses; the triangles at the bottom line indicate single zone damages and the parallelograms indicate adjacent zones damages; the case where detailed results are presented is highlighted.

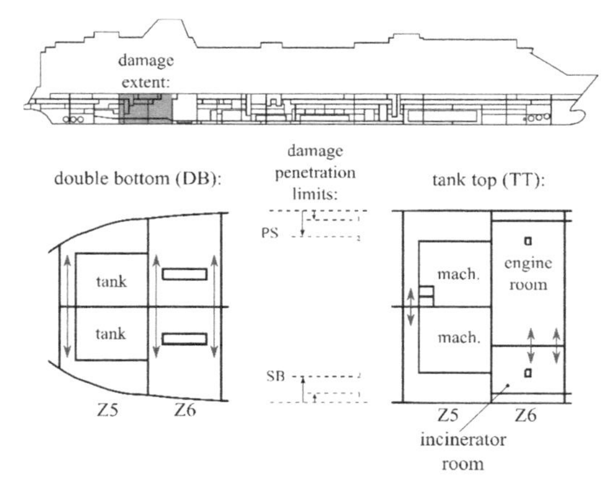

Figure 7. Arrangement and damage extension limits for the studied cases in Zones 5 and 6; the double-headed arrows mark the cross-flooding devices.

tank top in Zone 5. In Zone 6, there are both blowout panels and A-class hinged fire door connecting the incinerator and engine rooms.

Flooding simulations were carried out for the three initial conditions in SOLAS, namely DS, DP and DL (see Table 1), considering damage penetrations up to $B/10$ on both sides (PS and SB).

In addition to normal unprotected openings in SOLAS calculations, critical points have been modelled on the top of the buoyant hull. The effective range of positive stability in the calculation of the s-factor is limited to the heel angle when the first critical opening is immersed.

The tank top level is considered watertight, and the lesser extent damages, where the double bottom is not breached are also calculated.

The cross-duct and air pipe configurations are illustrated in Figure 8. The outlets of the air pipes are extended up to the boat deck level in order to minimize the pressure losses by avoiding the need for a non-return valve, Ruponen et al. (2012). The parameters of the pipes and calculation of the k-sum are shown in Table 2.

Each cross-duct in the double bottom, as illustrated in Figures 7–8, consists of seven individual girders, each with two equal sized manholes. Using the simplified method, equation (4), for pressure losses results in k-sum = 12.446, and correspondingly the effective discharge coefficient is 0.272.

The machinery spaces are considered fully vented and air compression is calculated only in the void spaces. The air pipe configuration and details of the calculation of the pressure losses in the pipes are presented in Figure 9 and Table 2, respectively.

The ducts are modelled at center line and the part of the duct on the damaged side is treated lost buoyancy. This approach notably simplifies calculations, but it is less conservative in cases, where the volume of the ducts is large when compared to the size of the void spaces.

Table 1. Studied initial conditions (as defined in SOLAS).

Initial condition	DS	DP	DL
Draft (m)	8.80	8.52	8.10
Trim (m)	0.00	0.00	0.00
Metacentric height, GM (m)	2.40	2.25	1.90

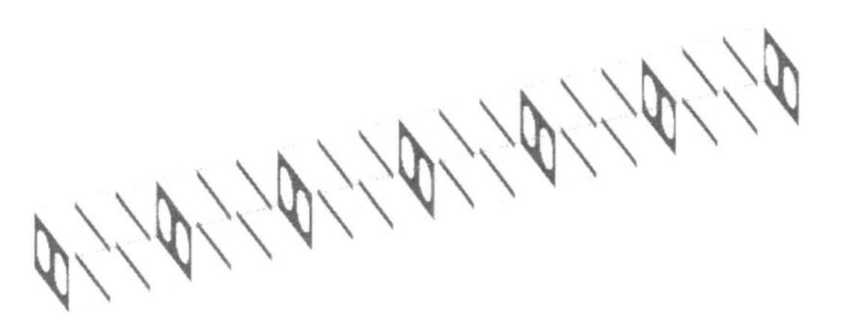

Figure 8. Cross-duct design.

Table 2. Calculation of pressure losses for the air pipes.

	Z5 DB	Z6 DB
No. of air pipes on each side	2	4
Diameter	0.25 m	0.25 m
Length	22.0 m	16.0 m
Pipe inlet (k_{inlet})	0.44	0.44
Frictional losses ($k_{friction}$)	1.76	1.28
3 double mitre bends (k_{bends})	1.20	1.20
k-sum	3.40	2.92
Discharge coefficient	0.477	0.505

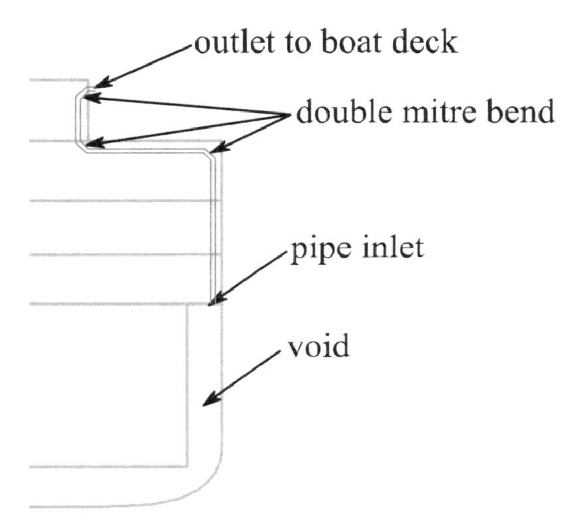

Figure 9. Air pipe configuration.

A rather short time step of 0.5 s is used since also air compression is calculated. The righting lever curve is evaluated at each time step, using constant volumes of floodwater at each heel angle, Ruponen et al. (2018). The maximum simulation time is set to 10 min.

6.2 *Results*

A summary of the simulation results for all cases involving the watertight zones 5 and 6 are presented in Table 3. An example of detailed results is shown in Figure 10. In addition, a comparison of the effect of the initial condition in the most critical damage case is presented in Figure 11. It is especially noteworthy that the initial condition DL is the most critical one. The reason for this is that with a smaller draft the pressure heads, acting on the cross-flooding devices, are also smaller. For the

Table 3. Results for damages to zones 5–6.

Case (init/damage)	Cross-flooding connection (zone-deck)			
	Z5-DB	Z6-DB	Z5-TT	Z6-TT
DS/D_5–6_SB_1	s = 1	s = 1	s = 1	–
DS/D_5–6_SB_2	s = 1	s = 1	s = 1	s = 1
DS/D_5–6_PS_1	s = 1	s = 1	s = 1	–
DS/D_5–6_PS_2	s = 1	s = 1	s = 1	s = 1
DP/D_5–6_SB_1	s = 1	s = 1	s = 1	–
DP/D_5–6_SB_2	s = 1	s = 1	s = 1	s = 1
DP/D_5–6_PS_1	s = 1	s = 1	s = 1	–
DP/D_5–6_PS_2	s = 1	s = 1	s = 1	s = 1
DL/D_5–6_SB_1	s = 1	s = 1	s = 1	–
DL/D_5–6_SB_2	ok	ok	ok	failed
DL/D_5–6_PS_1	s = 1	s = 1	s = 1	–
DL/D_5–6_PS_2	s = 1	s = 1	s = 1	s = 1
DS/D_5–6_SB_1_LE	–	s = 1	s = 1	–
DS/D_5–6_SB_2_LE	–	s = 1	s = 1	s = 1
DS/D_5–6_PS_1_LE	–	s = 1	s = 1	–
DS/D_5–6_PS_2_LE	–	s = 1	s = 1	s = 1
DP/D_5–6_SB_1_LE	–	s = 1	s = 1	–
DP/D_5–6_SB_2_LE	–	s = 1	s = 1	s = 1
DP/D_5–6_PS_1_LE	–	s = 1	s = 1	–
DP/D_5–6_PS_2_LE	–	s = 1	s = 1	s = 1
DL/D_5–6_SB_1_LE	–	s = 1	s = 1	–
DL/D_5–6_SB_2_LE	–	ok	ok	failed
DL/D_5–6_PS_1_LE	–	s = 1	s = 1	–
DL/D_5–6_PS_2_LE	–	s = 1	s = 1	s = 1

s = 1 means that s_{final} = 1.0 within 60 s ok means that cross-flooding is complete failed means that device does not provide instant flooding case is in the form: init/D_zones_side_penetration suffix LE means lesser extent damage above the tank top level.

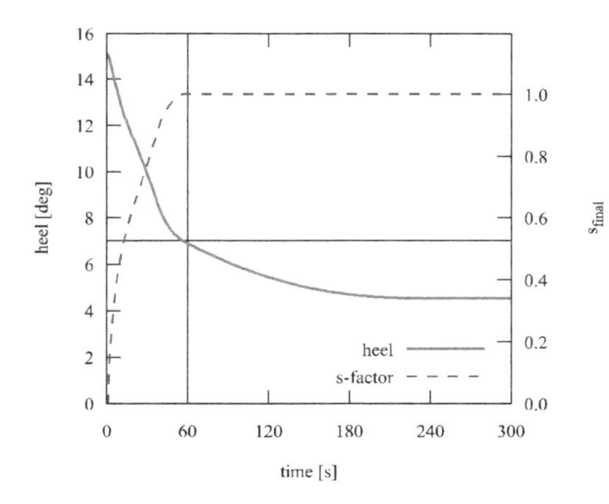

Figure 10. Example of simulation results, where s_{final} = 1 is reached within 60 s.

rooms extending above the waterline after flooding, the required volume to cross-flood is smaller as well, but this seems to have a much smaller effect on the results. Moreover, the metacentric height (GM) in the initial condition DL is smaller, which results in larger initial heeling before the cross-flooding starts.

In the most critical damage case, the lesser vertical extent causes smaller initial heel angle but the time to reduce the heeling to 7° is slightly longer, Figure 12.

The cross-flooding ducts in the double bottom and the transverse corridor on tank top in zone 5 provide s_{final} = 1.0 or full equalization within 60 s in all studied cases. However, in two cases the flooding between the incinerator room and the engine room in zone 6 is not equalized fast enough, as illustrated in Figure 13.

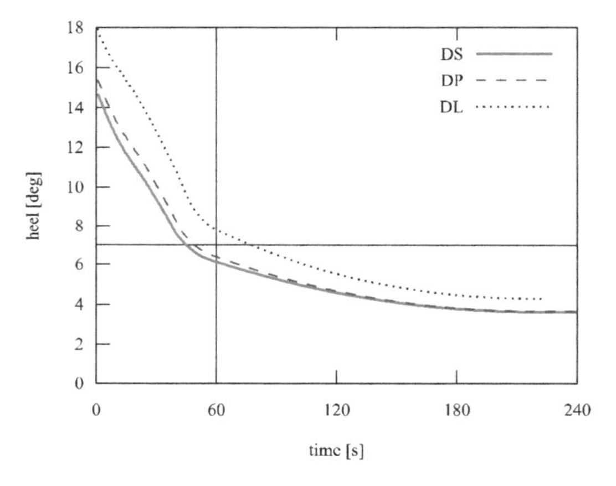

Figure 11. Comparison of the heel angle in the damage case D_5–6_SB_2 with different initial conditions.

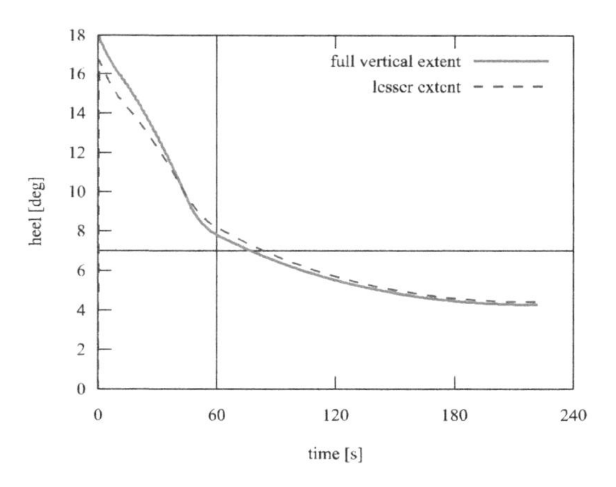

Figure 12. Effect of lesser vertical extent in the damage case DL/D_5–6_SB_2.

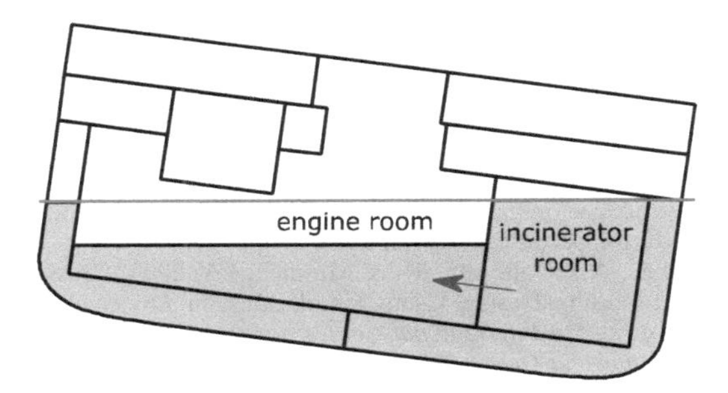

Figure 13. Example of incomplete cross-flooding at 60 s for the case DL/D_5–6_SB_2; $s_{final} < 1$, and thus instant flooding cannot be assumed.

It could be possible to add more blowout panels to the longitudinal bulkhead between the engine room and the incinerator room, in order to ensure complete equalization of flooding within 60 s. On the other hand, this A-class boundary can also be included in the calculation of the attained subdivision index of SOLAS.

7 DISCUSSION

Intuitively the most conservative approach would be to always include the stage before cross-flooding in the calculation of the attained subdivision index. However, the SOLAS probabilistic damage stability calculations usually contain also very long damages. Such damages result from either inclusion of raking side grounding damages in the applied statistics, or due to the use of non-dimensional data. In practice, this means that several damage cases that are included in the calculations are very theoretical. Furthermore, the inclusion of the stage before cross-flooding does not urge for efficient cross-flooding arrangement. Consequently, the resulting design may be even more prone to capsize during the transient flooding since the heeling is not equalized rapidly.

It is also worth noticing that the SOLAS calculations are based on the assumption that the rooms in way of the damage are always instantly flooded. In reality, the dynamics of the collision are very complex, Tabri et al. (2010), and the transient flooding stage is very much affected by the size of the breach in respect to the internal openings, Manderbacka & Ruponen (2016).

The proposed procedure for assessment of instant flooding through cross-flooding devices could also be included in an optimization framework, e.g. Manderbacka et al. (2016), with suitable constraints, for the size of the devices and air pipes.

8 CONCLUSIONS

Asymmetric flooding is a dangerous condition that should be equalized using self-acting cross-flooding devices. In order to minimize the risk of capsizing, this should be done rapidly. This is vital especially for large cruise ships with dense and complex non-watertight subdivision inside the watertight compartments. An advanced method, using time-domain flooding simulation within the SOLAS framework, has been presented to help in confirming that the cross-flooding devices allow rapid enough equalization of asymmetric flooding.

The SOLAS interpretation of the damage extent is extremely conservative for long damages that in reality are more likely caused by side grounding than a collision. Therefore, it is unfeasible to evaluate the cross-flooding arrangements with such damages. The presented methodology for evaluation of the cross-flooding arrangements is based on deterministic cases, where the damage length is limited in a similar way as in the SOLAS II-1 Reg. 8 (special requirements concerning passenger ship stability). This approach, where damage length is limited to 3% of ship length, covers the realistic large collision damages. It is noted that for small ships, it may be necessary to set a minimum applied damage length.

If the flooding is equalized to a condition, where $s_{final} = 1.0$, in practice meaning that the heel angle is less than 7°, in all cases within 60 s, then the cross-flooding arrangement is confirmed to be efficient enough. Consequently, the cross-flooded rooms can be considered as instantly flooded. If the condition of instant flooding is not met, the design of the cross-flooding arrangement can be improved, e.g. by increasing the area of the device, or the number of air pipes. Alternatively, the particular devices can be included in the SOLAS damage stability calculations.

The computation time for full assessment of instant flooding depends on the complexity of the arrangement, and most notably on the applied simulation parameters, such as time step and maximum time. In practice, for a large passenger ship the total computation time can be several hours.

The presented method for assessment of instant flooding through cross-flooding devices is considered a practical tool for design of efficient, yet feasible, arrangements. Especially for vessels with large U-voids, such as modern passenger ships, the assessment of the cross-flooding devices is very important. In future, such calculations become more relevant also for cargo ships due to updates in SOLAS, requiring analysis of the intermediate stages of flooding if cross-flooding devices are used. The possibility to achieve larger attained subdivision index urges for design of efficient devices, thus improving the safety.

REFERENCES

Cao, X.Y., Ming, F.R., Zhang, A.M. & Tao, L. 2018. Multi-phase SPH modelling of air effect on the dynamic flooding of a damaged cabin, *Computers and Fluids*, Vol. 163, pp. 7–19.

Gao, Z., Wang, Y., Su, Y. & Chen, L. 2018. Numerical study of damaged ship's compartment sinking with air compression effect, *Ocean Engineering*, Vol. 147, pp. 68–76.

IMO MSC 92/11/2 Modification of the revised regression formulae for cross-flooding through a series of structural ducts with 1 and 2 manholes in resolution MSC.245(83), submitted by Japan and Finland, 23 April 2013.

IMO MSC.362(93) Revised Recommendations on a Standard Method for Evaluating Cross-Flooding Arrangements, Adopted on 14 June 2013.

IMO MSC.429(98) Revised Explanatory Notes to the SOLAS Chapter II-1 Subdivision and Damage Stability Regulations, Adopted on 9 June 2017.

Jalonen, R., Ruponen, P., Weryk, M., Naar, H. & Vaher, S. 2017. A study on leakage and collapse of non-watertight ship doors under floodwater pressure, *Marine Structures*, Vol. 51, pp. 188–201.

Kujanpää, J & Routi, A-L. 2009. Concept Ship Design A, FLOODSTAND deliverable D1.1a.

Manderbacka, T. & Ruponen, P. 2016. The impact of the inflow momentum on the transient roll response of a damaged ship, *Ocean Engineering*, Vol. 120, pp. 346–352.

Manderbacka, T, Ruponen, P, Lindroth, D & Tompuri, M. 2016. Subdivision Optimization of LNG Fueled Ropax Ship, in *Proceedings of the 6th International Maritime Conference on Design for Safety*, Hamburg, Germany, 28–30 November 2016, pp. 90–97.

Ohashi, K., Ogawa, Y. & Shiraishi, K. 2012. Study on the Evaluation for Performance of the Cross-Flooding Arrangements by means of the Computational Fluid Dynamics, in *Proceedings of the 11th International Conference on the Stability of Ships and Ocean Vehicles, STAB2012*, 23–28 September 2012, Athens, Greece, pp. 381–389.

Papanikolaou, A, Hamann, R., Lee, BS, Mains, C, Olufsen, O, Vassalos, D & Zaraphonitis, G. 2013. GOAL-DS—Goal Based Damage Ship Stability and safety standards, *Accident Analysis & Prevention*, Vol. 60, pp. 353–365.

Pedersen, P.T. & Zhang, S. 2000. Effect of ship structure and size on grounding and collision damage distributions, *Ocean Engineering*, Vol. 27, pp. 1161–1179.

Peters, A.J, Galloway, M. & Minnick, P.V. 2003. Cross-Flooding Design Using Simulations, in *Proceedings of the 8th International Conference on the Stability of Ships and Ocean Vehicles, STAB2003*, 15–19 September 2003, Madrid, Spain, pp. 743–755.

Ruponen, P., Queutey, P., Kraskowski, M., Jalonen, R. & Guilmineau, E. 2012. On the calculation of cross-flooding time, *Ocean Engineering*, Vol. 40, pp. 27–39.

Ruponen, P., Kurvinen, P., Saisto, I. & Harras, J. 2013. Air compression in a flooded tank of a damaged ship, *Ocean Engineering*, Vol. 57, pp. 64–71.

Ruponen, P. 2014. Adaptive time step in simulation of progressive flooding, *Ocean Engineering*, Vol. 78, pp. 35–44.

Ruponen, P. & Lindroth, D. 2016. Time-Domain Simulation for Regulatory Flooding Analysis, in *PRADS 2016 – Proceedings of the 13th International Symposium on PRActical Design of Ships and Other Floating Structures*, 4–8 September 2016, Copenhagen, Denmark.

Ruponen, P. 2017. On the effects of non-watertight doors on progressive flooding in a damaged passenger ship, *Ocean Engineering*, Vol. 130, pp. 115–125.

Ruponen, P., Manderbacka, T. & Lindroth, D. 2018. On the calculation of the righting lever curve for a damaged ship, *Ocean Engineering*, Vol. 149, pp. 313–324.

Solda, G.S. 1961. Equalisation of Unsymmetrical Flooding, *RINA Transactions*, Vol. 103, pp. 219–225.

Stening, M., Järvelä, J., Ruponen, P. & Jalonen, R. 2011 Determination of discharge coefficients for a cross-flooding duct, *Ocean Engineering*, Vol. 38, pp. 570–578.

Tabri, K., Varsta, P. & Matusiak, J. 2010. Numerical and experimental motion simulations of nonsymmetric ship collisions, *J. Mar. Sci. Technol.*, Vol. 15, pp. 87–101.

Marine Design XIII – Kujala & Lu (Eds)
 ISBN 978-1-138-34076-3

Pro-active damage stability verification framework for passenger ships

Yu Bi
Maritime Safety Research Centre, University of Strathclyde, Glasgow, UK

Dracos Vassalos
University of Strathclyde, Glasgow, UK

ABSTRACT: Under the drive of nurturing a safety culture for the maritime industry, this paper presents a Life-Cycle Risk (damage stability) Management (LCRM) framework, which aims at controlling and managing risk through analysing the relevant risk control measures in design, operation and emergency phases. Affiliated to LCRM framework and inherent with important potential meanings, a pro-active damage stability verification framework for damage control measure is one of the essential elements for the realisation of LCRM. This paper elaborates this concept step by step along with a case study example.

1 INTRODUCTION

Accidents of passenger ships, involving thousands of lives on broad, are a matter of grave concern, consequences of which from time to time irritate and astonish the public. As a result, industry and academia's endeavour to improve the safety of passenger ships never stops and much of it targets the inadequate damage stability, the Achilles heel of passenger ships. For centuries, traditional passive ways of establishment and modification of safety regulations and rules in the aftermath of tragic accidents stay as the dominant method to help control the risk but nowadays it becomes difficult to catch up with the unrelenting pace of ship technology. In contrast, pro-active risk reduction ideas were put forward and various related methods are under development and tentatively expanding into the shipping industry. The typical representative, risk-based ship design method, integrating safety assessment procedure into the ship design process, widens the design envelope and inspires innovations on the new specifications while proactively controlling the risk. From all those new-born pro-active methodologies in the field of maritime safety, the idea was borne of establishing a life-cycle risk management (LCRM) (Vassalos, 2015) system to ensure risks are managed and controlled in systematic and comprehensive ways. Various kinds of LCRM ideas have been proposed in different perspectives due to the plethora of different types of risk and risk control methods proposed. This paper will also introduce an LCRM framework, which specifically targets management of damage stability. As an important building, a damage stability verification framework for damage control will then be elaborated in this paper.

2 LIFE-CYCLE RISK (DAMAGE STABILITY) MANAGEMENT FRAMEWORK

2.1 *Drives and aims*

Although it is always of great interest to see progress concerning damage stability, achieved through design measures, excessive focus on design measures to improve damage stability and negligence of risk control measures in other life phases of the ship could never be a good/cost-efficient plan, especially when a lot of improvements through design measures tend to be ineffective and expensive. Moreover, the improvement of damage stability through design measures serves only the new-buildings, leaving thousands of existing ships still confronted with uncontrollable risk and with state-of-art knowledge on damage stability wasted.

In the aforementioned background, a superior LCRM system should not only include design damage stability improvement measures but also pay more attention to the active measures in operation and emergencies, as an alternative approach to improving damage survivability. This kind of LCRM system can not only help the new-buildings but also thousands of existing ships, which are still in danger of uncertain damage stability. The life-cycle risk (damage stability) management framework presented in this paper is established following this idea. In general, it is a holistic approach aiming at controlling and managing damage stability through relevant risk control measures in all relevant life phases of passenger ships.

2.2 *Foundation (Safety Management System)*

As the common and well-known guide for safety management, Safety Management System (Health and Safety Executive guide—Successful health and safety management (HSG65) has been employed in various industry fields providing a holistic method of management of risks and safety problems. It systematically and reasonably classifies complex safety management processes into five steps, namely, policy, organizing, planning and implementing, measuring performance, reviewing and auditing (shown in Figure 1). A brief explanation of these five steps is presented as below.

- "Policy" describes the corporate approach to safety;
- 'Organizing' describes the management hierarchy relating to safety with responsibilities defined at each level;
- 'Planning' shows the safety tasks to be targeted at each stage and 'Implementing' is to conduct measures to reduce or mitigate risks;
- 'Measuring performance' refers to measurement and verification of the effectiveness the implemented measures;
- 'Reviewing and auditing' belong to the system of continuous improvement, ensuring that new hazards are identified, near-miss incidents are considered and the Safety Management System (SMS) is kept up to date.

Based on the HSE SMS guideline, the main part of LCRM for damage stability would be

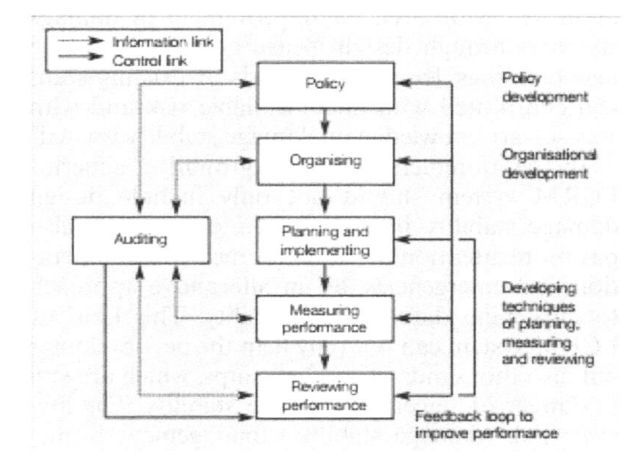

Figure 1. Key elements of successful safety management (after HSG65).

established following the last three steps of SMS, namely planning and implementing, measuring performance, reviewing and auditing. Specifically, they are planning and implementation of risk control measures, measuring the performance and effectiveness of these risk control measures and reviewing and offering recommendations for the former two processes based on the performance results of the risk control measures. In other words, risk control options in each ship's life phase are the "things" to be organised, planned and implemented. Following this, quantitative measurement of the effectiveness of the risk control options in each of the ship life phase is conducted to acquire the information of on risk reduction and control. Reviews of the measurement results will systematically present the safety status of the ship under different circumstances and in different life stages, and moreover, summarise the data of the more useful risk reduction or mitigation measures. Finally, suggestion and feedback to the risk reduction and mitigation measures or to relevant ship arrangement and design could be concluded at the last phase to help risk control through the life-cycle or future designs.

2.3 *Framework of life-cycle risk (damage stability) management*

Following the aforementioned contents of the LCRM framework, life-cycle risk (damage stability) management framework comprises risk (damage stability) control measures throughout ship life cycle, which would be the design phase, operational phase and emergency phase. Related to damage stability, operational phase is defined as the phase in which the ship is involved in a flooding (collision) accident but still able to be saved and the emergency phase is defined as the phase when the ship needs to be abandoned and evacuation is taking place.

Correspondingly, a variety of risk control measures are encompassed in these three phases. In the design phase, traditionally rules serve as passive risk control measures for damage stability improvement. Operational measures relate to active risk control measures, abundant in SOLAS Ch. II-2, e.g. damage control, emergency floatation and buoyancy devices, boundary control and so on. In emergencies, effective risk control measures relate to systems and measures focusing on emergency response, such as Decision Support Systems for Crisis Management, Evacuation, LSA, Escape and Rescue.

At each phase, risk control measures will be implemented, reducing the risk of the damaged ship like barriers (shown in Figure 2). Assess-

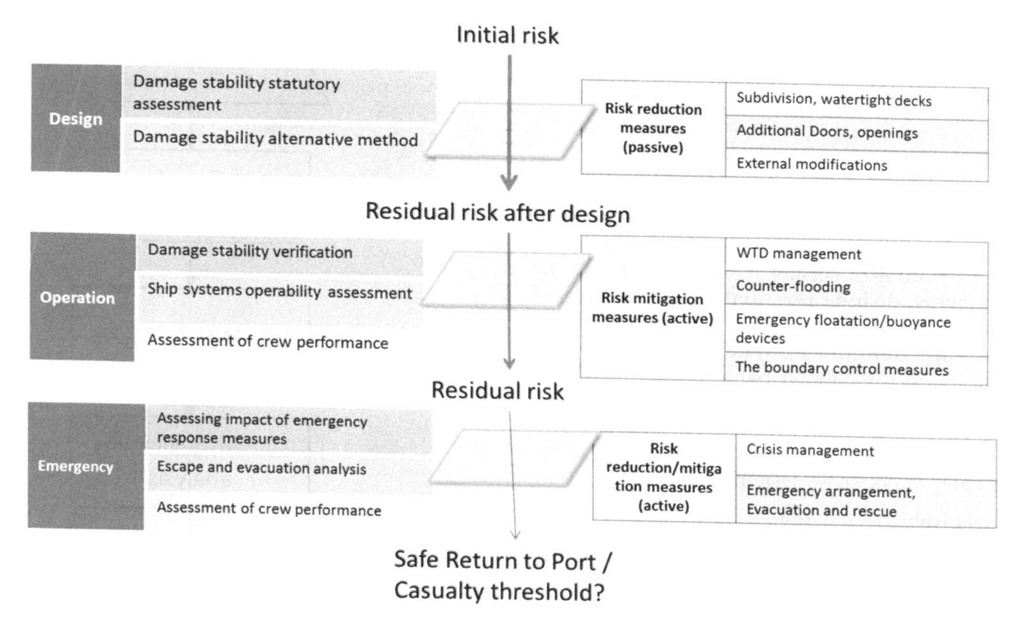

Figure 2. Image of life-cycle risk (damage stability) management.

ment and verification of each type of risk control measures are conducted to attain the safety status of the ship after the implementation of those risk reduction or mitigation measures, i.e. the value of residual risk. The ultimate target is to ensure the ship and people onboard survived the accident and safely return to port.

Although there are plenty options for planning and implementation of risk control measures, measurement of the performance and effectiveness of these risk control measures still remains as a big gap in this approach. In spite of passive design solutions, which stayed as a primary research target for centuries, operational measures have not been rigorously validated as alternative or supplementing design measures. The cost-effectiveness of emergency risk reduction potential has never been measured nor verified as 'residual' risk, which in this stage is perceived to be small by definition. These problems all need to be overcome before the overall process can be formalized and adopted.

Therefore, it is of primamy importance for a LCRM system to establish a damage stability verification frameworks (DSVF), which aims at quantitative measuring and verifying the effectiveness of risk (damage stability) reduction or mitigation measures. The following sections will elaborate one DSVF, which specifically works as damage control measure in the operation phase.

3 DAMAGE STABILITY VERIFICATION FRAMEWORK FOR THE DAMAGE CONTROL MEASURE

3.1 *Description of damage control measures*

The generic sequence of events that may occur after a flooding accident (typical muster list) relates to the following.

Incident → Detection & Alarm → Damage control → Muster of Passengers → Preparation of LSA → Abandon ship → Rescue

From the sequence given above, we can see that the damage control is the dominating measure conducted in the operation phase to improve damage stability. Various damage control measures would be taken by the master, either passive or active. Typical measures may comprise closure of the water-tight doors that are used at sea, cross-flooding and counter-flooding. Passive damage control measures infer lack of flexibility and the effectiveness of most of them has already been studied along with other design measures. Among active damage control measures, apart from the measures that only exists in a few ships, counter-flooding is the only damage control measure that can be commonly conducted but has not been verified in the perspective of risk reduction. Therefore, in the context of damage stability verification framework for damage control, verification of damage stability following implementation of counter-flooding measures, will be presented in the next section.

3.2 *The counter-flooding measure*

The counter-flooding (also called as counter-ballasting) measure refers to using pumping system to actively transmit water inside the ship or the sea in order to neutralize heeling moments and give the ship a better floating position (which may retard progressive flooding) and improve damage stability.

The water transmission systems onboard, which participate in the counter-ballasting measures are usually ballasting water (BW) system, fresh water (FW) system, heeling water (HW) system, and heavy fuel oil (HFO) system. Compared to other fluid water tanks, these systems have:

- A relatively low location in the ship;
- A relatively large volume, which could affect change to stability;
- Their independent and flexible and efficient piping and pumping facilitates/equipments.

3.3 *Framework components and structure*

The flowchart in Figure 3 presents the basic components and overall structure of the DSVF for counter-flooding measures established in this research. As shown in the flowchart, the whole damage stability verification framework mainly contains five core parts, from probabilistic damage stability analysis to hydrodynamic analysis and testing. Specifically, the framework starts from an existing ship, which means that the ship is already compliant with regulations and standards of subdivision. Probabilistic damage analysis is then conducted to obtain detailed information of the ship flooding vulnerability and survivability to all statistical damage scenarios (sampled from distributions as defined in SOLAS 2009). These two steps offer the basic database for following quantitative analyses of the effectiveness of counter-ballasting measures.

Based on flooding vulnerability of each damage scenario acquired from the probabilistic damage analyses, a filtration (or selection) process is conducted to obtain the critical scenarios that need application of the counter-flooding measures, as most of 2-compartment/zone and 3-compartment/zone damage scenarios already possess a high level of survivability. In addition, some damage cases are primarily caused by sinkage, which could not be affected by counter-flooding. After the critical scenarios have been identified, a series of analyses will then be implemented for each critical scenario, first of which would be the system availability assessment of the counter-flooding system. It will

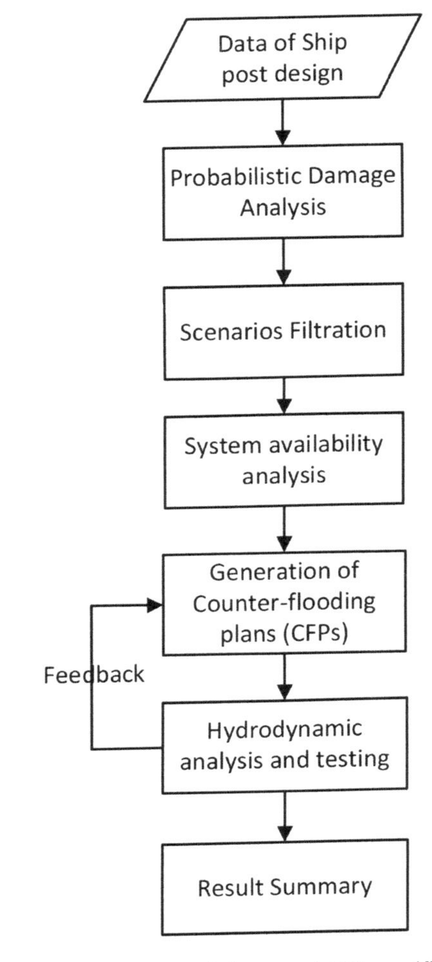

Figure 3. Flowchart of damage stability verification framework for the counter-flooding measure.

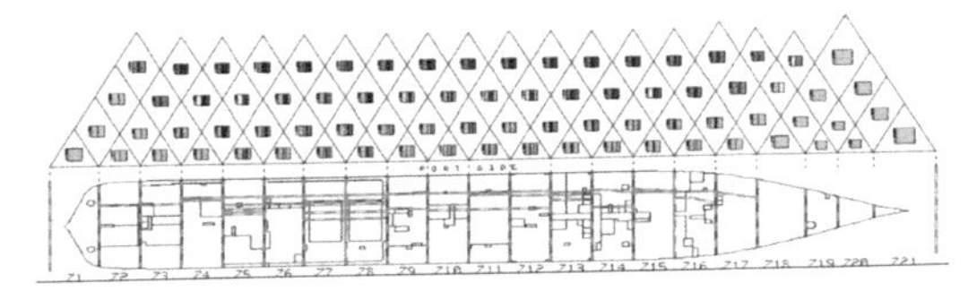

Figure 4. Result example of probabilistic damage stability analysis.

reveal the real working status of every component in the counter-flooding system and show the limitations of applying counter-flooding plans (CFPs). Based on the availability information, the process of generation of the CFPs comprises an optimization process to identify the best counter-ballasting measures for each damage case. The evaluation objectives employed in the optimization process are based on the hydrostatic performance of the damaged ship with CFPs as it takes both processing time and accuracy into account.

Because of the potential uncertainty and limitation of hydrostatics-based analysis, the hydrodynamic response of the damaged ship with counter-flooding measures in waves is also assessed. It comprises the numerical simulation of the whole execution process of the counter-flooding measures. In this respect, the predicted motion response will be computed in time-domain for each damage case.

After quantitative hydrostatic and hydrodynamic analyses of the damage control measures, quantitative results indicating the safety status of the damaged ship could be presented in various ways, one of which is via the updated damage consequence diagram (which is used in the damage control manual) showing the survivability status of the ship under different extents of damage and certain loading condition, following application of the most effective counter-ballasting measure. Details of each process in this framework will be elaborated in following sections along with a case study.

3.4 *Case study and process details*

In order to interpret the process of the aforementioned damage stability verification framework more directly, a case study is presented in the following to help with further explanation of each process.

3.4.1 *Probabilistic damage stability analysis*

Geometry data of ship post design, including hull, compartments, and subdivision information are the preparation data for the implementation of the framework. With these data, probabilistic damage stability can be conducted according to SOLAS 2009 Chapter II-1, Part B and Part B-1. S factor and p factor for each damage extent need to be acquired.

3.4.2 *Critical case filtration*

Damage cases are selected relative to damage extent. On account of inherent features of counter-flooding, CFPs may have a dominant benefit on the damage case with asymmetrical flooding and probably also avail damage cases leading to with high centre of gravity. Therefore, the filtration standards for the critical case of counter-flooding are:

- Survivability index—s factor of the damage case: $s < 1$ means there is a need and possibility to improve the stability in this certain damage case.
- Heel angle at equilibrium θ: while $|\theta|>0.5°$ which means there is relatively explicit asymmetry in flooding, the case will be selected.
- GM: With $|\theta|\leq 0.5°$ and $s < 1$, the causes of poor damage stability could be high centre of gravity (flooding in high deck) or sinkage. While $GMT < 0$, there is a high probability that flooding happened in high decks, one of the dominant reasons for poor damage stability.

3.4.3 *System availability assessment*

The results of counter ballasting systems availability assessment will indicate the available tanks, pumps, etc. which could work in the counter-flooding process in each damage scenario and if any limitations exist when applying counter-flooding measures because of the system failure. As the overall counter-flooding system comprises four subsystems, i.e. BW, HW, HFO, and FW, the system availability assessment also involves the assessment of these four subsystems, which are similar but also a slightly distinct from each other.

Logic dependency trees are used to assist the assessment. (Figure 5 shows the connection dependency of the ballast water and heeling water combined system.) System component location information (penetration and zone) and damage information will then be employed to obtain the availability of every system component. This part of the assessment is based on the conservative assumption (Cichowicz, 2010) that systems located within spaces directly affected by the flooding casualty are unavailable.

Through system availability assessment, detailed information of the transmission availability of each undamaged tank either with other tanks or with sea will be acquired, serving as critical input data for both the generation of CFP and hydrodynamic analysis. Moreover, the information of the availability of every pump in the system is also a requisite for the following processes in the framework.

3.4.4 *Generation of CFPs*

Constrained non-dominated sorting genetic algorithm-II (constrained NSGA-II) is adopted here to generate the optimal CFPs for each critical damage case. The constrained NSGA-II mainly employ the method described in (Deb, 2002), the flowchart of which is depicted in Figure 6. The constraints are set based on the inherent features of counter-flooding systems and also limitations caused by system availability as well. Some basic constraints (Martins, 2011) comprise:

- Fresh water tanks must have the same amount of water or less (sea discharge) since no ballast was considered to fill them up, i.e. the total volume of fresh water in the fresh water system should either stay the same or decrease.
- Fuel volume can be changed between tanks but it cannot be discharged overboard, i.e. the total

volume of heavy fuel oil in the HFO system should always remain unchanged.

Besides, another two points have been assumed while modelling:

- No watertight compartments that were not flooded due to damage are going to be flooded.
- Drainage is not considered because of the uncertainty and complexity of the flow of water ingress. This may cause a conservative result.

The evaluation objectives used in the genetic algorithm are adaptive corresponding to the convergence situation. The potential evaluation objectives comprise s factor, equilibrium heel angle and trim angle, deepest draught, time to apply CFPs, GM.

An example results of the optimization process is presented in the appendix, which proves feasibility of the proposed method which can be used to obtain the effectiveness levels of counter-flooding measures on different damage scenarios.

3.4.5 *Hydrodynamic analysis and testing*

Hydrodynamic simulation of application of CFPs is developed under the environment of PROTEUS 3, which is a state-of-the-art time-domain damage stability simulation tool. Ship hydrodynamics are based on asymmetrical strip theory formulation with Rankine source distribution (Jasionowski, 2002). The motion of counter-flooding water and floodwater are both modelled as Free-Mass-on-Potential-Surface (FMPS) system in an acceleration field. Assumption correlating to counter-flooding process is that the time derivative of the mass is proportional to the velocity of the centre

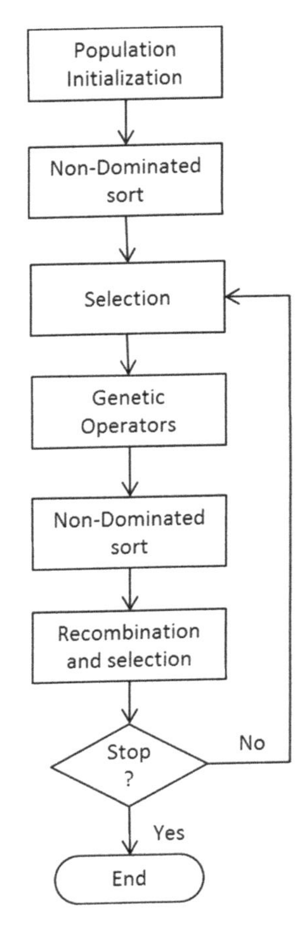

Figure 6. Flowchart of the optimization process.

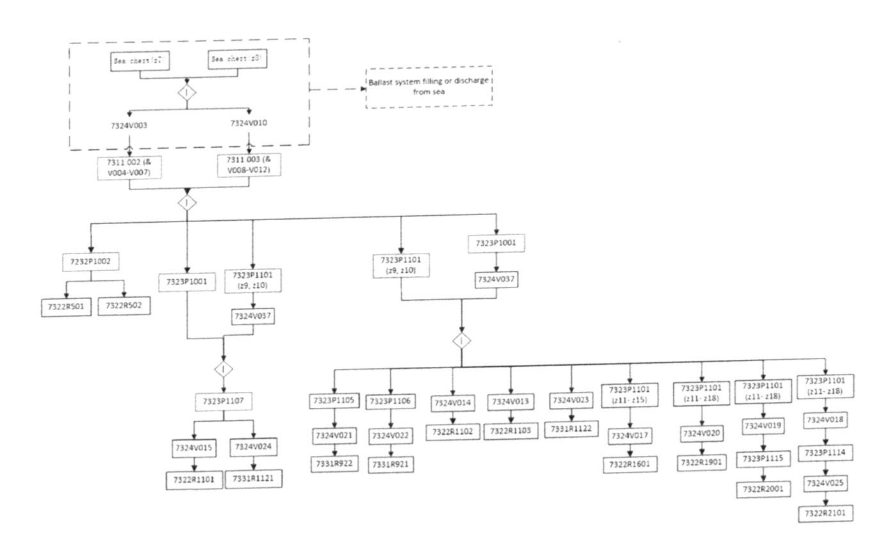

Figure 5. Connection dependency tree of BW and HW combined system.

of gravity of the ship, which implies that the mass that enters or leaves the system has zero velocity.

The hydrodynamic analysis serves as a supplement of hydrostatic analysis to double check the performance of the optimal CFPs.

3.5 *Potential impact*

This damage stability verification framework is meaningful and worthy of application because of its potential meanings, parts of which are listed below.

- Give a quantitative idea of the effectiveness of damage control measures;
- Give counter-ballasting suggestions to the master for different damage scenarios; The suggestions include the final volume distribution in the counter-ballasting system and execution sequence priority suggestions;
- Offering database to the real-time onboard damage control decision system if the ship is equipped with one;
- All this knowledge of damage control could help alleviate panic in the situation of a ship flooding accident and restore ship stability.

4 CONCLUSION

In summary, this paper proposes a life-cycle risk (damage stability) management framework, which has applied the SMS methodology and involves the relevant risk control measures in design, operation and emergency phases. As the main building block of the LCRM framework, a damage stability verification framework for the damage control measure is elaborated along with a case study example which proves the feasibility of the proposed methodology. Based on the potential meaning of these frameworks, it is worthwhile to develop and implement them in order to proactively control and manage the risk pertaining to damage stability.

Furthermore, potential future work based on the research presented in this paper could include:

- Integration of more operational measures into current damage stability verification framework, e.g. emergency floatation/buoyancy devices, boundary control measures.
- Development of risk verification framework for risk mitigation measures during emergencies.
- Integration of framework into regulation.

REFERENCES

Cichowicz, J., Vassalos, D. 2010, Onboard Systems Availability for Ship Life-Cycle Design; Proceedings of 4th International Conference on Design for Safety, Trieste, Italy.

Deb, K., Pratap, A., Agarwal, S., Meyarivan, T. 2002, A Fast and Elitist Multi-Objective Genetic Algorithm: NSGA-II. *IEEE Transactions on Evolutionary Computation*, 6(2):182–197.

HSG65, Managing for health and safety, Health and Safety Executive.

Jasionowski, A. 2002 An Integrated Approach to Damage Ship Survivability Assessment. PhD thesis, University of Strathclyde.

Martins, P.T. 2011, Real-time Decision Support System for Managing Ship Stability under damage. *OCEANS 2011 IEEE—Spain*, Santander, 2011, pp. 1–7.

Vassalos, D., Bi, Y. 2015, Life-Cycle Risk (Damage Stability) Management of Passenger Ships. *Proceedings of the 12th International Conference on the Stability of Ships and Ocean Vehicles*, pp. 635–642, 14–19 June 2015, Glasgow, UK.

APPENDIX

A result example from optimization process of generation of CFPs is presented in the appendix. The damage extent in the example is port side 4-compartment damage (zone 2 to zone 5) with 10.451 metres penetration. Four loading conditions are considered for this damage extent, which are:

- Load 1: 10% bunkers and stores, without passengers;
- Load 2: 10% bunkers and stores, with passengers
- Load 3: Draught at summer load line, with passengers
- Load 4: 50% bunkers and stores, with passengers

Table 1 presents various hydrostatic and execuation time information of the original damage case and one possible counter-flooding solution yielded from the optimization process for each aforementioned damage scenario.

The evoluation process of optimiation is depicted in Figure 7. The horizonal coordinate denotes the generation number, and the vertical coordinates represent six potential evaluation objectives respectively. The curves in the figure present the developing trend of the value of evaluation objectives of final selected solution in each generation. The final selected solutions are picked in the population at end of each generation by comparison of the value of evalution objectives in a priority sequence which is s factor, CFP excuation time, deepest draught, equalibrium trim angle, equalibrium heel angle, GM.

Table 1.

Evaluation objectives	Load 1		Load 2		Load 3		Load 4	
	Original	CFPs	Original	CFPs	Original	CFPs	Original	CFPs
S factor	0.8376	1	0	0.9332	0.8674	1	0.4933	1
Equilibrium heel angle (degree)	4.8957	0.0915	90	0.0069	7.1413	2.7950	9.1110	0
Equilibrium trim angle (degree)	0.9603	0.7627	90	0.5469	1.1089	0.6800	1.0918	0.8611
Deepest draught (metre)	12.8383	11.0423	NA	10.6572	14.7986	12.7122	14.9545	11.6641
GM (metre)	0.6811	0.9254	0.3201	0.6772	0.6270	1.0854	0.4047	0.5804
Time to apply CFPs (hour)	0	2.2384	0	3.2196	0	4.3018	0	2.1483
Draught at midship (metre)	11.3168	11.0050	NA	10.6492	12.5455	11.8446	12.0717	11.6547

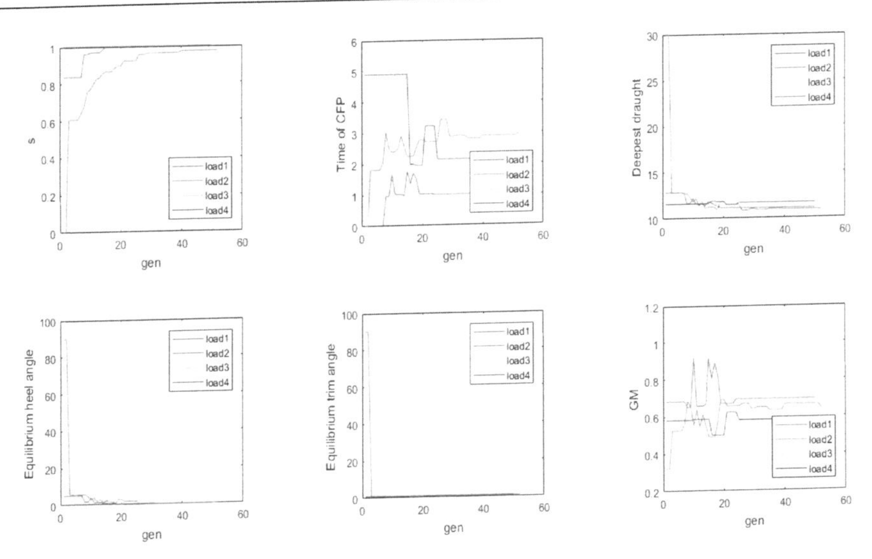

Figure 7. Evolution trend of evaluation objectives along optimization.

Marine Design XIII – Kujala & Lu (Eds)
 ISBN 978-1-138-34076-3

Weight and buoyancy is the foundation in design: Get it right

K.B. Karolius & D. Vassalos
Department of Naval Architecture, Ocean and Marine Engineering, Maritime Safety Research Centre, University of Strathclyde, Glasgow, Scotland, UK

ABSTRACT: Stability and cargo carrying capacity are fundamental in vessel design and are governed by the Vertical Centre of Gravity (VCG), identified through performing the inclining experiment. It is well known that the current method for calculating the VCG following inclining experiments is limited by assuming an unchanged metacentre position when the vessel is heeled and may produce error prone results. This paper will present alternative calculation methods which have been tested together with the *Classical* method on a range of vessels to highlight their accuracy and flexibility. The worst-case result using the *Classical* method will be utilised to highlight the implications such errors may have on stability and cargo carrying capacity performance. The results demonstrate the importance of correct VCG calculation for a safe and optimal vessel design, as even minor errors in the order of millimeters may translate into extensive weights and moments compromising stability and cargo carrying capacity.

1 INTRODUCTION

Modern ships are complex systems and their design comprise of a range of design variables and constraints. However, there are two main aspects that could be considered most critical in ship design, namely the cargo carrying capacity and the level of safety. The cargo carrying capacity is a measure of payload that can safely be carried in excess of the vessels lightweight, while safety is highly linked to the ability of the vessel to remain afloat and upright in any condition of loading including damaged condition. As such, these are closely linked parameters and will clearly affect one another.

Buoyancy and weight are the two key parameters governing both stability and carrying capacity of the vessel due to their well-known interplay in terms of flotation, heeling and righting moments. Buoyancy is solely dependent on the underwater hull shape, and is today easily assessed using modern software tools that replicate the hull shape with great accuracy. The vessel's design lightweight, however, is more difficult to assess in any detail and is only roughly calculated through the design process. It is not until the vessel is close to completion that it can be identified more accurately by performing the classic inclining experiment.

It is a well-known fact that the so called *Classical* method, in which the vertical centre of gravity (VCG) is calculated following inclining experiments, has its limitations on performance in terms of applied heel angle magnitude, applied loading condition and accuracy for certain hull forms. This is due to the assumption made of unchanged metacentre position when the vessel is heeled. In an attempt to ensure the correct application of the *Classical* method, various requirements have been set out in the 2008 IS Code Part B Ch. 8 and Annex I (IMO, 2008).

Recently, as a result of the limiting assumptions in the *Classical* calculation method, more accurate and flexible calculation methods have been proposed. A detailed study on such methods has been presented by Karolius & Vassalos (2018), highlighting possible dangers inherent in the *Classical* method whilst demonstrating due flexibility and higher accuracy through the use of the new methods. In this paper, higher focus will be placed on design implications in terms of stability and cargo carrying capacity.

The test undertaken in the aforementioned study, enabled establishment of an error potential for each method using a purely technical software-simulated inclining experiment. Using the established error potential, a corrected operational VCG could be calculated from actual inclining VCG values, which were evaluated against the loading conditions for each vessel to see if the stability margins had been compromised. Only the two calculation methods showing highest accuracy and flexibility are addressed in the following; namely the *Generalised* method, and the *Polar* method. This paper will in addition utilize the vessel identified in the study as having highest error potential in the operational VCG values to show how this may affect the stability and loading capacity, and highlight the importance in achieving correct VCG value following the inclining experiment for a safe and optimal vessel design.

2 DESIGN IMPLICATIONS

2.1 *Wrongful assumptions in the Classical method*

To assess possible implications on stability and carrying capacity the, the limitations of the *Classical* method need first to be reviewed. The *Classical* method validity is based on the assumption that the position of the metacentre is unchanged when the vessel is heeled, as illustrated in Figure 1.

The position of the metacentre can be represented by the metacentre-radius (BM) given by the well-known relationship (1) between the transverse second-moment of the waterplane area I_{XX} and the vessel displaced volume (∇):

$$BM = \frac{I_{XX}}{\nabla} \tag{1}$$

As the vessel displaced volume is constant during the incline, the change in the metacentre position is proportional to the change in the second moment of the waterplane area and consequently the waterplane area itself. For any heel angle, there will be a change in the waterplane area, which is crucial in obtaining a righting lever arm and subsequent righting moment, as it is directly related to the movement of the buoyancy position. This is highlighted by the fact that only a completely circular hull-shape will have unchanged waterplane area but also unchanged buoyancy position when heeled. A more realistic movement of the metacenter with increased waterplane area is illustrated in Figure 2.

The assumption in the *Classical* method, however, relates to smaller heel angles, which in more traditional vessel designs may hold to a satisfactory level. This is also the main reason for the IMO requirements set out in the 2008 IS Code, setting requirements in terms of heel angle magnitude, initial heel angle and loading condition used to ensure a minimal change in waterplane area.

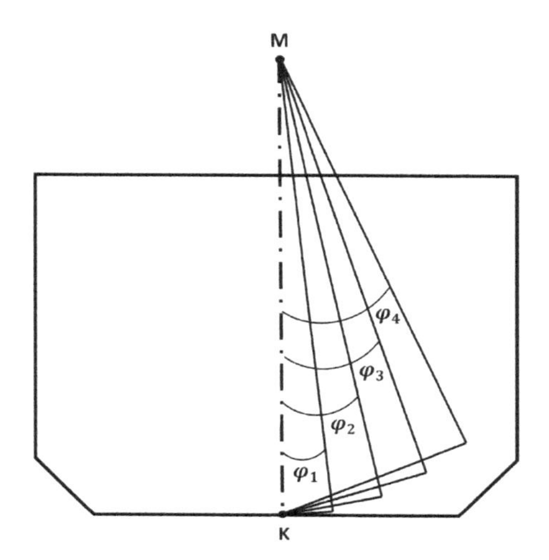

Figure 1. Assumption of unchanged metacentre position.

There is further a misconception in the industry that the assumptions hold for completely wall-sided vessels, and this appears to have given rise to the so-called "wall-sided" assumption in relation to the inclining experiment. This is not the case, as even a completely box-shaped vessel will experience change in the waterplane area when heeled and a subsequent movement of the metacenter as is illustrated in Figure 3.

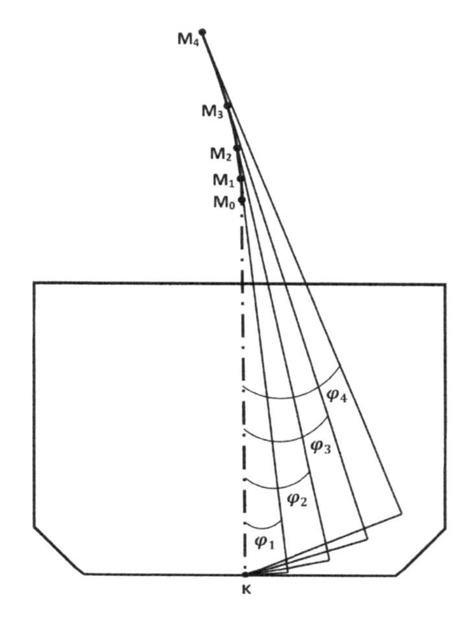

Figure 2. More realistic movement of metacentre position.

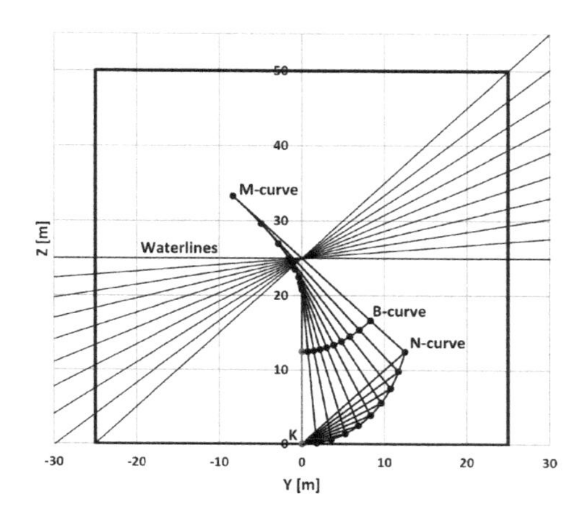

Figure 3. Actual movement of metacentre for box-shaped vessel, 0–45° heel.

The movement of the metacentre shown in Figure 3 is for extensive heel angles ranging from 0° to 45° heel for a box-shaped vessel with length and breadth of 100 and 50 meters respectively. If smaller heel angles, in the range of 0°–4° are considered in line with the maximum allowed heel in accordance with the IMO requirements, a significant smaller movement can be seen, as illustrated in Figure 4 represented by an increase in KM of 0.3% in this specific case.

The main reason for the assumptions in the *Classical* method is to utilise a simplified trigonometric relationship as illustrated in Figure 5.

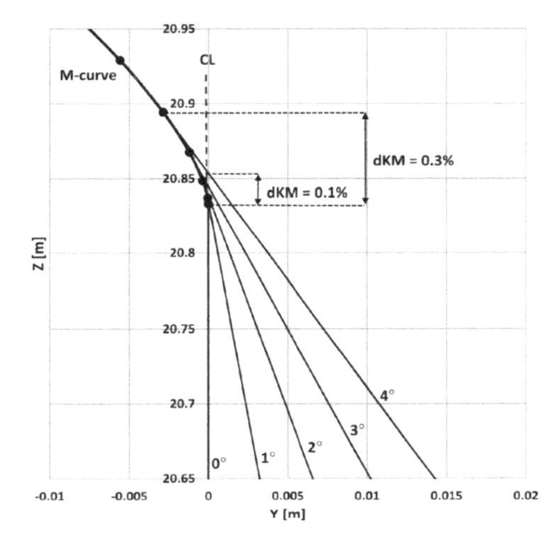

Figure 4. Actual movement for box-shaped vessel, 0°–4° heel.

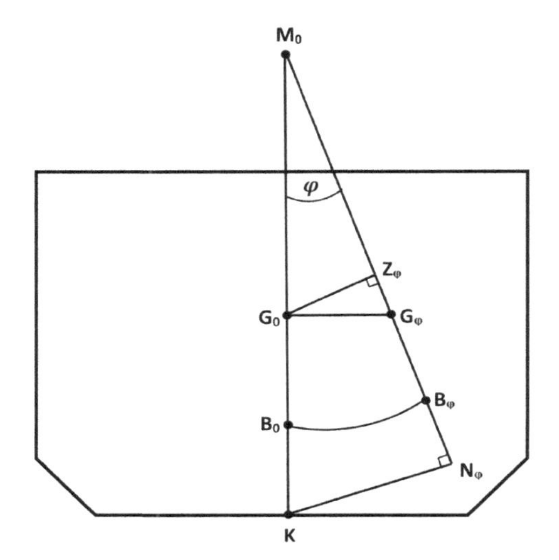

Figure 5. Simplified trigonometric relationship for deriving the *Classical* formula for GM.

This facilitate a formula for VCG to be derived as shown next.

$$\tan(\varphi) = \frac{G_0G_\varphi}{G_0M_0} \tag{2}$$

$$G_0M_0 = \frac{G_0G_\varphi}{\tan(\varphi)} \tag{3}$$

$$G_0G_\varphi = \frac{w \cdot d}{\Delta} \tag{4}$$

$$GM = G_0M_0 = \frac{w \cdot d}{\Delta \cdot \tan(\varphi)} \tag{5}$$

$$\tan(\varphi) = \frac{r}{L} \tag{6}$$

$$VCG = KM - GM \tag{7}$$

where w = inclining weight, d = movement distance, Δ = displacement, r = pendulum reading, and L = pendulum length. Remaining parameters are explained by Figure 5.

The assumption of unchanged waterplane area further enables the use of upright hydrostatics in the calculation for GM for every weight shift. As a simplified trigonometric relationship is used, it is clear that the rise in KM of 0.3% mentioned earlier is not the actual increase inherent in the formula. The increase in question is rather found at the intersection with the centerline, which results in an increase in KM of 0.1%. Based on this, it may be argued that the assumptions may be considered valid for a completely box-shaped vessel and may well be the reason for the emergence of the wall-sided assumption in relation with the inclining experiment.

It is important, however, to note than no vessel is completely box-shaped, nor circular and it is rather a combination of the two, with various design features that may result in much higher change in the water-plane area than should be accepted even for smaller heel angles. This is the main reason for the *Classical* method being subjected to scrutiny and debate. Such design features may include:

- Chine lines and knuckles
- Large fore- and aft flare
- Misc. appendages
- Large change in trim during heel
- Other unconventional hull forms

As the *Classical* calculation method was developed in the late 17th century (Hoste, 1693) when detailed software models were not available, the limiting assumptions makes sense as it enables upright hydrostatics to be utilised. Today, however, the strife is towards higher accuracy and there exists a range of tools for this purpose, making such simplifications and requisite assumptions obsolete.

2.2 *Implications on stability and cargo carrying capacity*

In the aforementioned examples, a clear increase in the KMT is seen as a result of an increase in the waterplane area. In reality the waterplane area may both increase or decrease depending on heel magnitude and which specific design features are emerged or submerged. As a general rule the following is true:

Case 1: Increase in waterplane area:

$$BM_0 < BM_\varphi \tag{8}$$

$$GM_0 < GM_\varphi \tag{9}$$

$$VCG_0 > VCG_\varphi \tag{10}$$

Case 2: Decrease in waterplane area:

$$BM_0 > BM_\varphi \tag{11}$$

$$GM_0 > GM_\varphi \tag{12}$$

$$VCG_0 < VCG_\varphi \tag{13}$$

It is clear that Case 1 above will by using the *Classical* method, overestimate vessel stability, thus producing a lower VCG value than is the actual case, while Case 2 will underestimate the vessel stability leading to a higher VCG than is the actual case. Cases 1 and 2 therefore translates directly into possible implications for stability and cargo carrying capacity respectively. These are explained in the following.

2.2.1 *Stability*

The VCG of a vessel is the parameter of highest importance in assessing intact and damage stability, this being the baseline for any condition of loading. It is also governing in other important aspects such as vessel motion behaviour through its effect on rolling period and hence the new second generation intact stability criteria. The VCG is utilised in most intact and damage stability requirements through enforcing requirements on the GZ righting curve. The GZ curve is represented by (14) and it is clear that any error in VCG will lead to subsequent errors in the GZ-curve, and therefore also incorrect assessment against relevant stability criteria.

$$GZ(\varphi) = KN(\varphi) - VCG \cdot \sin(\varphi) \tag{14}$$

From the formula above, it is clear that over- and underestimation of the VCG will result in under- and overestimation in the GZ-curve respectively.

For most vessels, it is the damage stability requirements that is governing and limiting the operational envelope. Using the probabilistic damage stability and the attained index A as a basis, it is possible to gauge the impact using (15). This approach is utilised and presented in section 7.3 to highlight implications in stability.

$$Risk = 1 - A \tag{15}$$

2.2.2 *Cargo carrying capacity*

A vessel's cargo carrying capacity is limited by volumetric constraints, but also by its safety requirements, especially those related to intact and damage stability. To best illustrate this, a traditional VCG stability limit curve can be used as shown in Figure 6.

The limit curve serves as the safe operational envelope for a vessel and represents the operational conditions for which the relevant intact and damage stability requirements are fulfilled. The black curve represents the limit curve prepared using the light-weight VCG, as obtained from the *Classical* method. For the sake of argument, an error of 1% overestimation in VCG is assumed. The actual curve is then represented by the stapled line as the overestimation of the lightweight VCG has resulted in a more stringent operational limit.

It is clear that more restrictions will be imposed to the operational envelope if the VCG is overestimated. Another way to illustrate the implications in cargo carrying capacity is to translate the overestimated VCG into potential reduction in cargo carrying capacity. This is the cargo that could have been carried if the VCG were calculated correctly. As most vessels have a maximum summer draught decided by load-line and strength requirements, the additional draught would have to be maintained whilst ballast-water with a lower VCG can

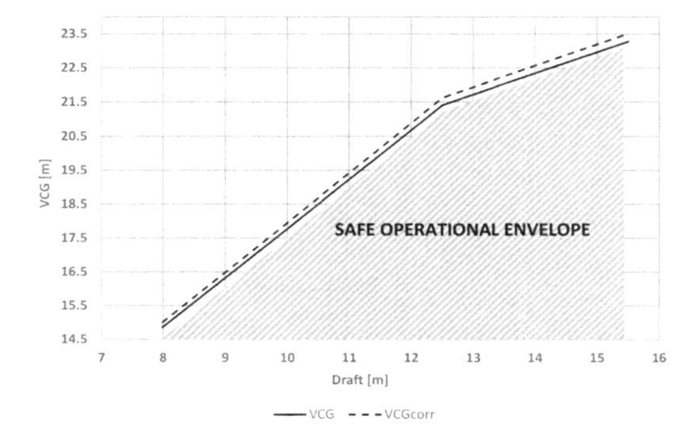

Figure 6. Underestimation of stability limit curve and subsequent reduction in operational envelope.

be exchanged by additional cargo with a higher VCG. This approach is utilised and presented in section 7.4 to highlight implications in cargo carrying capacity.

3 ASSESSMENT OF CALCULATION METHODS

The study assessing the various calculation methods comprises two parts. Firstly, identifying the potential errors inherent in each method. This is achieved using a purely technical, software-simulated inclining experiment using a stability model with known lightweight parameters. Each calculation method is then used in an attempt to replicate the actual VCG values, given as known input parameters to the software. An error potential is then developed using the percentage difference in actual and calculated VCG values using (16). The various methods are applied for 2, 4 and 10 degrees of maximum inclining heel angles.

$$Error[\%] = \frac{VCG_{actual} - VCG_{calculated}}{VCG_{actual}} \cdot 100\% \qquad (16)$$

Secondly, by calculating the error potential using an identical floating position and heel magnitude as was used in actual inclining experiments for the particular test vessels as approved by the administration, the ensuing error is assumed to be present in the operational VCG values of the test vessels. The operational VCG values for each test vessel can therefore be corrected for either over- or underestimation using the error potential and to assess whether the vessels stability or loading capacity are affected.

4 UNCERTAINTY AND ERRORS

The inclining experiment is subject to a range of sources of uncertainties and errors originating from external influences such as wind, waves, current and human measurement errors. This paper focuses only on the error originating from the choice of calculation method. Other sources of uncertainty and errors have been reviewed and discussed in many publications, such as Shakshober & Montgomery (1967) and Woodward et al. (2016).

For more detailed information on the range of uncertainties related to the inclining experiment, the above mentioned publications are recommended but in order to highlight the most common sources of uncertainty, Figure 7 has been borrowed from Woodward et al. (2016). The figure shows the various sources of component uncer-

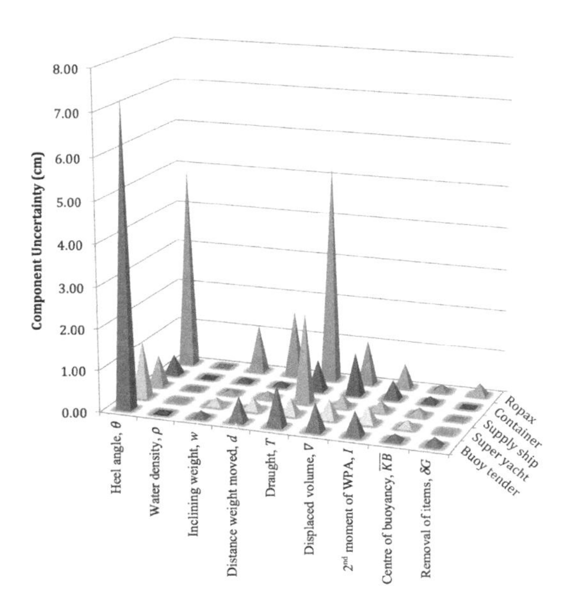

Figure 7. Component uncertainty contribution for various inclining experiment parameters. Reprinted by permission from Woodward (2016, Fig. 2).

tainty contribution in the vertical centre of gravity for various inclining experiment parameters for five case-study vessels.

The figure clearly indicates that the highest contribution is originating from the heel angle and draught, in terms of uncertainties related to pendula and draught marks.

5 CALCULATION METHODS

The *Classical* method has already been reviewed in section 2.1. In this section, the alternative methods will be covered. Among these, there were two methods that showed highest accuracy and flexibility from the aforementioned study; namely the *Generalised* and the *Polar* method, both described in the following.

5.1 *The Generalised method*

The *Generalised* method was initially proposed by R.J. Dunworth (2013) and further expanded by Dunworth (2014, 2015) and Smith, Dunworth & Helmore (2016). The method utilizes the fact that in equilibrium position for each weight shift, the vessel's righting arm GZ and heeling arm HZ must be equal. Using the trigonometric relationships, as illustrated in Figure 8, the following can be derived:

$$HZ = KN - VCG \cdot \sin(\varphi) - TCG \cdot \cos(\varphi) \qquad (17)$$

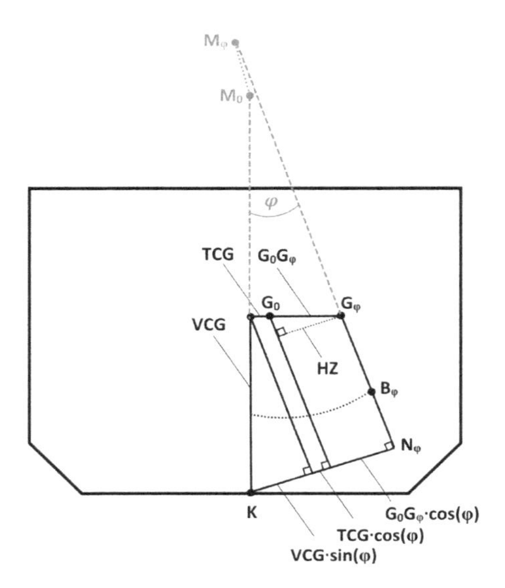

Figure 8. Main parameters for the *Generalised* method. Adopted from Dunworth (2014).

$$VCG \cdot \sin(\varphi) = KN - HZ - TCG \cdot \cos(\varphi) \tag{18}$$

$$HZ = \frac{w \cdot d \cdot \cos(\varphi)}{\Delta} \tag{19}$$

From the above, it is apparent that this method does not make any reference to the metacentre in the calculations and should, therefore, not be influenced by any change in the waterplane area during the weight shifts, as is the case for the *Classical* method. To this end, actual KN values are needed from a stability software model, corresponding to equilibrium floating position for each weight shift. The heeling angle is determined from pendulum deflection readings similar to the *Classical* method.

For each shift VCG · sin(φ) is calculated using equation (18) and plotted against sin(φ). The final value of VCG can be directly calculated as the regression slope using a least squares fit similar to the *Classical* method. Dunworth (2013) further suggests an alternative method for calculating the TCG off-set, by plotting the calculated HZ values corresponding to each weight shift against the heeling angle. The TCG offset for the whole system comprising ship and inclining weights are then found at the y-axis intercept, i.e. HZ for $\varphi = 0$. The curve fitting is suggested to be obtained using a 3rd order polynomial fit.

5.2 *The Polar method*

The *Polar* method was presented in the study by Karolius & Vassalos (2018), and is derived utilising the line PL illustrated in Figure 9. This line can be represented in polar coordinates using (20), and

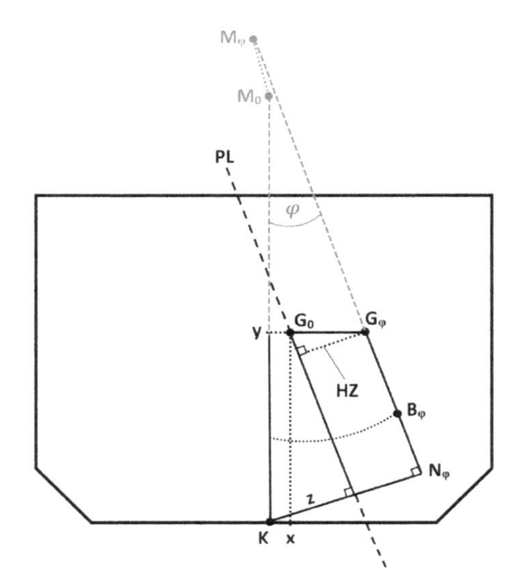

Figure 9. Main parameters from the *Polar* method.

if corrected for actual KN and HZ values for each weight shift using (21), the line will pass through the point (x, y) for any arbitrary weight shift from the neutral position, which results in (22). Knowing that the x-coordinate is equal to TCG, and the y-coordinate equal to VCG, equation (23) is obtained.

$$z = x \cdot \cos(\varphi) + y \cdot \sin(\varphi) \tag{20}$$

$$z = KN - HZ \tag{21}$$

$$KN - HZ = x \cdot \cos(\varphi) + y \cdot \sin(\varphi) \tag{22}$$

$$KN - HZ = VCG \cdot \cos(\varphi) + TCG \cdot \sin(\varphi) \tag{23}$$

This method takes advantage of the fact that both VCG and TCG need to be located on the PL line in the initial condition and to be kept constant in this position for each individual weight shift, i.e. the initial VCG_0 and TCG_0 are kept constant on this line, while the overall system TCG is shifted a distance G_0G_i for each shift *i* as represented by (24) and (25).

$$TCG_i = TCG_0 \tag{24}$$

$$VCG_i = VCG_0 \tag{25}$$

There are, as a result, two equations to derive the two unknown parameters and by using (23) and following some deduction, (24) results in a solution for VCG given by (26), and (25) in a solution for TCG given by (27) in their most general form:

$$VCG = \frac{(KN_i - HZ_i) \cdot \cos(\varphi_0) - (KN_0 - HZ_0) \cdot \cos(\varphi_i)}{\cos(\varphi_0) \cdot \sin(\varphi_i) - \sin(\varphi_0) \cdot \cos(\varphi_i)} \tag{26}$$

$$TCG = \frac{(KN_i - HZ_i)\cdot \sin(\varphi_0) - (KN_0 - HZ_0)\cdot \sin(\varphi_i)}{\cos(\varphi_i)\cdot \sin(\varphi_0) - \sin(\varphi_i)\cdot \cos(\varphi_0)} \quad (27)$$

The equations can further be simplified using the trigonometric relations in (28) and (29) and knowing that the heeling arm resulting from weight movement in the neutral position HZ_0 needs to be zero, this results in (30) and (31).

$$\cos(\varphi_0)\cdot \sin(\varphi_i) - \sin(\varphi_0)\cdot \cos(\varphi_i) = \sin(\varphi_i - \varphi_0) \quad (28)$$

$$\cos(\varphi_i)\cdot \sin(\varphi_0) - \sin(\varphi_i)\cdot \cos(\varphi_0) = \sin(\varphi_0 - \varphi_i) \quad (29)$$

$$VCG = \frac{(KN_i - HZ_i)\cdot \cos(\varphi_0) - KN_0 \cdot \cos(\varphi_i)}{\sin(\varphi_i - \varphi_0)} \quad (30)$$

$$TCG = \frac{(KN_i - HZ_i)\cdot \sin(\varphi_0) - KN_0 \cdot \sin(\varphi_i)}{\sin(\varphi_0 - \varphi_i)} \quad (31)$$

The *Polar* method can in theory be used to calculate VCG directly for any arbitrary shift from the neutral position, but to account for other sources of error and uncertainty as was discussed in Section 4, it is recommended to utilize a least squares linear regression also for the *Polar* method by plotting the denominator against the numerator.

6 TEST VESSELS

The test vessels used in the study comprise 9 vessels of various type, size and hull form in an attempt to account for the ship specific problematic design features, such as knuckles, large flare angles, sharp chine lines. More conventional wall-sided hull forms have been included as well for comparison. The vessels main particulars are presented in Table 1.

Table 1. Test vessel particulars.

Vessel type	L_{BP} [m]	B [m]	D [m]	C_B [m]
Fishing vessel	40.20	12.00	7.50	0.73
Yacht	36.60	7.70	4.20	0.54
RoPax	195.30	25.80	14.80	0.79
Bulk carrier	223.50	32.30	20.20	0.92
Passenger vessel	320.20	41.40	11.60	0.74
Naval I	54.10	10.60	5.00	0.65
Naval II	71.00	12.00	6.20	0.58
Container vessel	320.00	48.20	27.20	0.76
Supply vessel	76.80	19.50	7.75	0.69

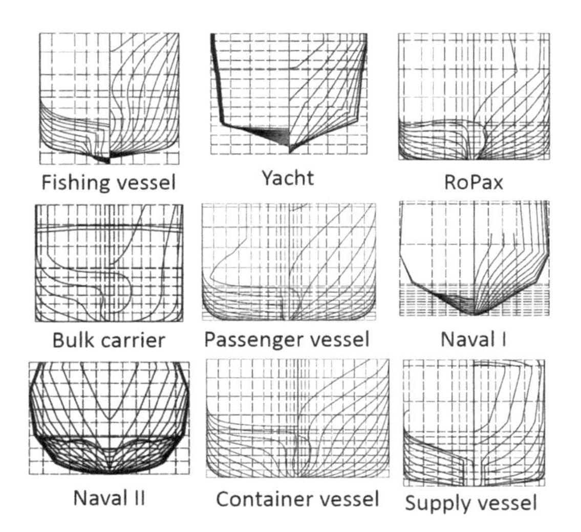

Figure 10. Lines plan of test vessels.

Despite covering the problematic design features, the chosen designs are still fairly conventional and it is clear that much larger errors would be evident for even more unusual hull shapes, especially for vessels under 24 m in length or of higher novelty such as high-speed or leisure craft. Line plans of the test vessels can be seen in Figure 10.

7 RESULTS

In the following section, only a summary of the results obtained in the study is presented to highlight the errors obtained. For more comprehensive result, including results for various initial heel angles, reference is made to Karolius & Vassalos (2018). In order to compare the methods against each other, the results from the software simulated inclining experiment are represented by the absolute value of the error potentials, irrespectively of over-, or underestimation. The errors have then been used to correct the operational VCG values for the vessels and are presented in section 7.2. Finally, sections 7.3 and 7.4 illustrates how the worst case may translate to stability and capacity implications.

7.1 *Software simulated inclining experiment*

In the following, results obtained for each method are presented together for comparison. In the presented result, initial heel of 0 degrees was used, i.e. vessel upright in neutral position. From Figure 11, it is clear that all methods produce accurate results, with the highest error below 0.5%, obtained by the *Classical* method for the Naval II vessel. In Figure 12, maximum heel angle of 4 degrees is presented.

The results still show good accuracy for all methods but the error using the *Classical* method is now increased to 1.5% for the Naval II vessel. In Figure 13, maximum heel angle of 10 degrees is presented. As expected, the results show much lower accuracy for the *Classical* method, with a maximum error above 6% for the Naval II vessel. The other methods still show high accuracy, with only an error of below 0.02%.

To summarize, Figure 14 presents the error potential averaged over all vessel types for all

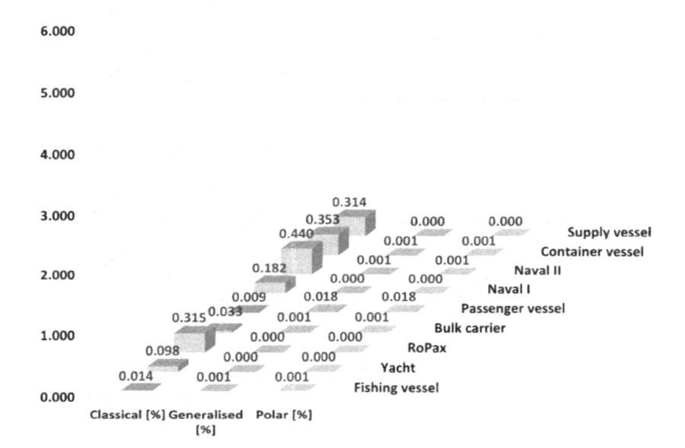

Figure 11. Percentage error for *VCG*, 2 degrees maximum heel angle and 0 degrees initial heel angle.

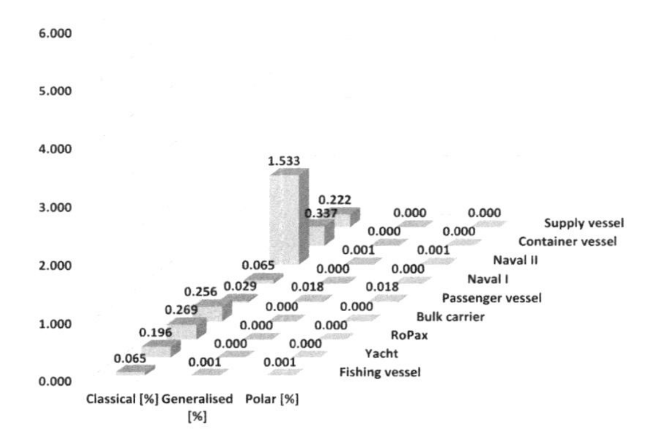

Figure 12. Percentage error for VCG, 4 degrees maximum heel angle and 0 degrees initial heel angle.

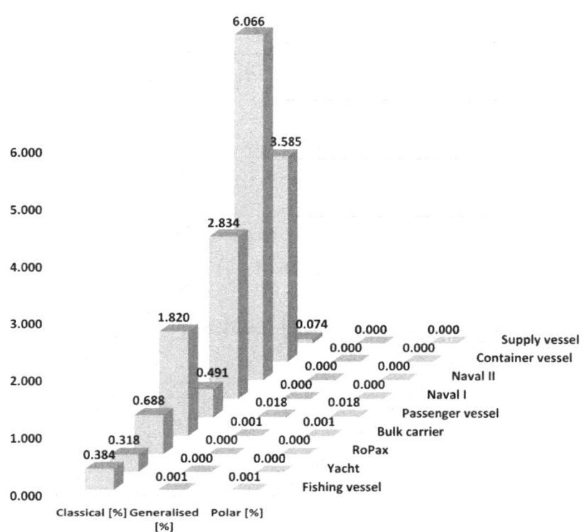

Figure 13. Percentage error for *VCG*, 10 degrees maximum heel angle and 0 degrees initial heel angle.

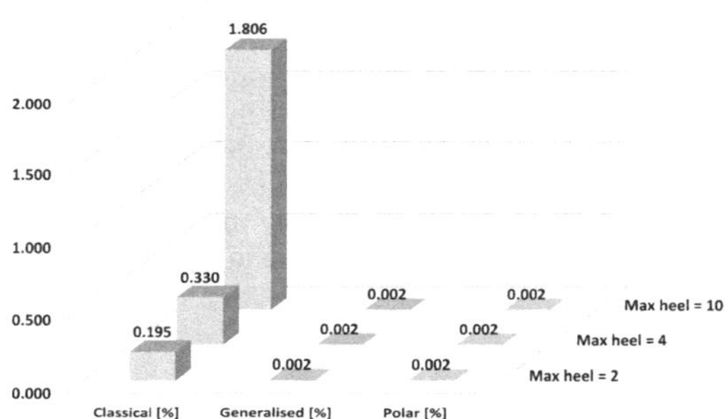

Figure 14. Error potential averaged over vessel types for various heel angles and methods.

methods for the various inclining heel angles. It is clearly shown that the *Classical* method is highly dependent on the inclining heel angles compared to the other methods, and produces results with increasing errors.

7.2 *Actual inclining experiment corrections*

Corrected operational VCG values are presented in Table 2. It is clear that there are potential errors in the operational VCG values as a result of using the *Classical* method. The highest error is obtained for the RoPax, Naval II and Container vessel, with 41, 22 and 61 mm errors respectively. The VCG of the Naval II vessel is overestimated, while the VCG for the RoPax and Container vessels are underestimated. As the container vessel shows the highest error, this has been used for illustrating possible design implications.

Table 2. Corrected *VCG* values obtained using error potentials for the *Classical* method.

	VCG	Correction	VCGcorr
Vessel type	[m]	[mm]	[m]
Fishing vessel	5.753	0.373	5.754
Yacht	3.747	4.207	3.752
RoPax	13.171	41.478	13.213
Bulk carrier	11.632	2.789	11.634
Passenger vessel	22.221	−3.084	22.218
Naval I	4.500	8.605	4.509
Naval II	4.934	−21.705	4.912
Container vessel	17.228	60.813	17.288
Supply vessel	7.592	7.897	7.600

7.3 *Error implication on stability*

The container vessel in question should comply with the probabilistic damage stability requirements in accordance with SOLAS Reg. II-1/7-8 (IMO, 2009), and for the sake of illustrating possible implications on stability, the attained index A for the operational VCG and the corrected VCG have been calculated. Knowing that 1-A can be regarded as a representation of the ensuing risk, this allows obtaining a measure of false safety inherent in the vessel as a result of the inaccurate VCG calculated using the *Classical* method as is seen in Table 3.

It is seen that an error in safety estimation of 3% is seen due to the error in VCG. The table further presents the difference in number of capsize cases and it is seen that the vessel actually has 37 additional capsize cases not accounted for due to the error in VCG.

7.4 *Error implication on carrying capacity*

Despite having an underestimated VCG as presented in the former sections, the Container vessel will be utilised in order to examine what the error of 61 mm in lightship VCG yields in terms of lost cargo carrying capacity if, for the sake of argument, this were an overestimated error. As highlighted in section 2.2.2, a vessel draught is often restricted by its maximum summer draught governed by load-line-, and strength requirements. As such, the draught and subsequent displacement needs to be maintained, while ballast-water can be replaced with cargo until the maximum permissible VCG is achieved.

For simplicity, the ballast water is subtracted from the global ballast VCG, while the additional cargo is added to the global container load VCG. The added cargo and reduced ballast is presented in Table 4. This difference, if considering an average weight of 12 tonnes per TEU, corresponds to over 11 TEU's.

If the vessel had available margins in terms of draught it may be possible to keep the ballast whilst adding cargo until the maximum permissible VCG is achieved. This results in a quite extensive loss in cargo carrying capacity as is seen in Table 5. Again, this difference, if considering an average weight of 12 tonnes per TEU, corresponds to over 56 TEU's.

Table 3. Underestimated VCG translated to overestimated probabilistic damage stability performance, i.e. false safety.

Lightweight case	A	Risk = 1-A	Capsize cases
VCG	0.711	0.289	896
VCG_{corr}	0.702	0.298	933
Difference [%]	1.28	3.02	3.96

Table 4. Overestimated VCG translated to underestimated cargo carrying capacity, maintained draught.

	Cargo	BW	Draught
Lightweight case	[t]	[t]	[m]
VCG	103716	9530.6	15.25
VCG_{corr}	103851	9395.6	15.25
Difference	135	135	0.00

Table 5. Overestimated VCG translated to underestimated cargo carrying capacity, increased draught.

	Cargo	BW	Draught
Lightweight case	[t]	[t]	[m]
VCG	103716	9530.6	15.25
VCG_{corr}	104399	9530.6	15.30
Difference	683	0	0.05

8 CONCLUDING REMARKS

As can be seen in the results from the software simulated inclining experiment, the *Classical* method is highly dependent on heel angle magnitude and may produce unacceptable errors that could affect important design parameters such as cargo carrying capacity and stability. This is highlighted by the corrected VCG values presented in the actual inclining experiment corrections seen in Table 2. It is important to highlight that all test vessels have been approved by class, and that IMO requirements have been adhered to. If for some reason the requirements in terms of heel magnitude, initial heel angle and loading condition were not followed, significant higher error would be expected. This also applies to the vessel designs. There exist numerous and more unconventional designs that would produce increasingly higher errors.

The additional measures imposed by IMO are unnecessary when applying the alternative methods, as they do not make reference to the metacentre in the equations. They produce very accurate results for any floating position, in terms of draught, heel magnitude and initial heel as they utilise actual KN values corresponding to each floating position. This reduces the possibility of making mistakes and they can therefore be considered to be more reliable and flexible than the *Classical* method.

The results further highlight the importance of achieving correct VCG value following the inclining experiment for a safe and optimal vessel design, as even minor errors in the order of millimeters may translate into extensive weights and moments compromising both stability/safety and cargo carrying capacity. The most common argument for maintaining the *Classical* method is that the errors are small and insignificant in comparison with the random errors mentioned in section 4, but the validity of this argument can be questioned.

As designers, it is our responsible to reduce errors that may compromise or undermine safety to the lowest possible degree, especially when the means to do so are available. Considering the results from this study, the industry should be more critical when applying the *Classical* method and it may even be time to replace it with better and more flexible calculation methods. It is at least important for the industry to know that there are other more reliable alternatives to the *Classical* method and should be accounted for in the regulations and guidelines in use today.

REFERENCES

Dunworth, 2013, "Up Against the Wall", International Maritime Conference, Pacific 2013 IMC, Sidney, Australia.

Dunworth, 2014, "Back Against the Wall", RINA Transactions (International Journal of Small Craft Technology), 2014, 156(B2), p. 99–106.

Dunworth, 2015, "Beyond the Wall", Proceedings of the 12th International Conference on the Stability of Ships and Ocean Vehicles, 14–19 June 2015, Glasgow, UK.

Hoste, Paul, (1697), "Théorie de la Construction des Vaisseaux" ("Theory of the Construction of Vessels"), Arisson & Posule, Lyon.

International Maritime Organization, 2008, "Part B Annex I of the International Code on Intact Stability 2008", as adopted in IMO Res. MSC.267(85), 2008.

International Maritime Organization, 2009, "Reg. II-1/7–8 of SOLAS Consolidated Edition 2009", as adopted in IMO Res. MSC 216(82)), 2006.

Karolius, Kristian & Vassalos, Dracos, 2018, "Tearing down the wall – The inclining experiment", Ocean Engineering Vol. 148, pp. 442–475.

Shakshober & Montgomery, 1967, "Analysis of the inclining experiment", Hampton Road Section of the Society of Naval Architects and Marine Engineers.

Smith, Dunworth & Helmore, 2016, "Towards the Implementation of a Generalised Inclining Method for the Determination of the Centre of Gravity", International Maritime Conference, Pacific 2015 IMC, Sidney, Australia.

Woodward, Rijsbergen, Hutchinson & Scott, 2016, "Uncertainty analysis procedure for the ship inclining experiment", Ocean Engineering 114 (2016), p. 79–86.

Marine Design XIII – Kujala & Lu (Eds)
© 2018 Taylor & Francis Group, London, ISBN 978-1-138-34076-3

SmartPFD: Towards an actively controlled inflatable life jacket to reduce death at sea

M. Fürth, K. Raleigh, T. Duong & D. Zanotto
Stevens Institute of Technology, Hoboken, New Jersey, USA

ABSTRACT: Personal Flotation Devices (PFD)/Life vests can considerably prolong life expectancy of a person that has gone overboard, yet they are not often worn by recreational users. Inflatable PFD are popular because of their slimmer profile that enhances user movement and comfortability, however, current models on the market are not without limitations. The SmartPFD has been developed to tackle problems associated with involuntary activation and to study the effect of multiple air pockets. The proposed PFD inflates when a pressure level has been maintained for a certain time; this will stop vests from inflating accidently. The vest has two separate air pockets that can deploy sequentially or simultaneously. The users' orientation and acceleration are measured using an embedded Inertia Measurement Unit to track the ability of the PFD to turn a person floating face down upright. The designed PFD has been experimentally tested and initial results show good promise.

1 PROBLEM STATEMENT

In marine contexts, drowning is an ever-present threat. Whether for recreation or commercially, people who engage in activities near the water, even those who can swim, face the risk of succumbing to the ocean. Drowning is cited as the third leading cause of unintentional injury death worldwide at 7% of all injury-related deaths. (World Health Organization, 2017) Between 2005 and 2014 there were an average of 332 boating-related drowning deaths in the United States each year (CDC Injury Center, 2016). Reasons for drowning include lack of swimming ability, improper use or no use of personal flotation devices, alcohol use, loss of consciousness, and unsafe environmental conditions.

Even trained swimmers can succumb to environmental conditions in water such as temperature, waves and currents. These conditions increase the chance of injuries. For example, hypothermia can occur even after a short time in moderately cold water, further impeding a swimmer's ability to stay afloat or seek help.

Use of a personal flotation device (PFD) could potentially prevent approximately half of all drowning deaths (Quistberg et al., 2014). PFDs serve to keep users afloat thereby reducing the energy expenditure of a swimmer so that they can better assess what must be done to improve chances of survival and to give rescuer more time to locate them. A study by McCormack et al.(2008) on the survival times of individuals accidentally immersed in water showed that the median time of survival for an immersed individual in a PFD is ten times greater compared to an individual without a PFD. In this study, it was reported that in spite of these statistics and regulations requiring the use of life vests, many people still do not wear them regularly.

According to a United States Coast Guard study, the life jacket wear rate for all boaters in 2016 was only 21.6% (Braynard, 2018) This number fluctuates year to year, but 2016 marks a lower rate than both 2015 and 2014. Various reasons are cited for low use rates of life jackets. Among these reasons are discomfort, cost, and improper judgement of conditions in which a personal flotation device is necessary.

Improper fit or use of uncomfortable vests leads users not to wear their vest, or remove it once it has been worn for some time. In either case, this leaves a user without a device they could need at a moment's notice. Many older life vests were made of buoyant material such as foam. This led to bulky vests which impeded a wearer's movement and reduced their comfort. To combat this, inflatable life vests were developed which allowed for a smaller vest profile in an uninflated state. This design helps eliminate the discomfort associated with inherent buoyancy life vests, offering a lighter suspender-like harness system to attach the vest. Inflatable vests are operated manually, with pull tabs, or automatically upon contact with water. The major drawback of inflatable designs is their cost. In addition, the inflatable design requires a re-arming kit after they have been inflated.

Inherently buoyant life vests are typically priced in the range of $35 to $60. Conversely, inflatable vests typically cost between $70 and $150, with higher prices for higher performance models. The strict relationship between cost and functionality often leads boat owners to opt for cheaper life vests which they are less likely to wear (International Maritime Organization, 2017).

The Coast Guard reports that when determining whether or not to wear a personal flotation device, it is common for users to only find them necessary in the case of foul weather. Users often assume that due to their self-perceived swimming ability, wearing a life vest is redundant or useless (Braynard, 2018).

In the United States, nearly 75 million people are involved in recreational boating. In 2016 alone, the US Coast Guard reported 4,463 recreational boating related accidents. These accidents lead to 701 deaths, 73% of which were due to drowning. The number of accidents represents an increase of 11% from 2015 (Quistberg et al., 2014).

Every PFD must meet certain codes, standards, and regulations that ensure the users' safety. The US Coast Guard classifies PFD into five different types. Type I includes offshore life jackets that are geared for situations where rescue is not immediate. They are bulky but offer increased buoyancy over other types of PFDs and can turn unconscious wearer face-up in the water. Near-shore vests are classified as Type II, and are designed for inland waters and fast rescue. They are less bulky than Type I vests, but will only turn some unconscious users face-up. Type III classification is for flotation aids where quick rescue is likely. Wearers must position themselves in the upright position in order to avoid being face-down in the water. Type IV is reserved for throw-able devices such as cushions or ring buoys and act as backups for PFDs. Finally, a Type V PFDs are special-use devices for specific activities such as kayaking, waterskiing, and windsurfing. Also included in type V are any flotation aids with additional functionality, such as immersion suits with hypothermia prevention (Boater Exam, 2017).

The International Standards Organization provides guidance for the design on PFDs through the ISO Standard 12402 ("*Personal Flotation Devices*"). The standard classifies PFDs into lifejackets, which help keep the user face up in the water, and buoyancy aids, which require the user to move themselves in order to float face up. The standard discusses safety and testing requirements for PFDs (ISO, 2006).

Many of the vests on the market comply with US Coast Guard regulations, however there are situations in which they are lacking (Braynard, 2018). Automatic inflatable life vests have the ability to detect water mechanically or chemically, using a substance which dissolves after contact with water causing a CO_2 canister to open and inflate the vest. Despite their higher costs compared to inherently buoyant vests, these systems might have malfunctions. Firstly, they can inflate even when the wearer is out of water due to water sprays, thus causing users to take them off. Secondly, unwanted deployments can be dangerous for users' safety. For example, if a user is trapped under a floating object such as an overturned boat, conventional automatic vests will deploy when they detect water, potentially trapping the wearer underneath the object.

Thirdly, in 2013 during the Islands Race, an Offshore Sailboat race leaving from the San Diego Yacht Club, a 32 foot sailboat had a structural failure and was blown into the rough surf of San Clemente Island. One member of the crew lost his life in the incident and the other members reported a particular malfunction with the life jacket. US sailings incident report came to the following conclusion: Four of five Spinlock deck vests failed to work properly, allowing the flotation chamber to pull over the wearer's head to one side of the body. The deceased was found floating face down with the flotation chamber pulled over his head. Given that the crew had to swim through large surf to reach the shore this was a life threatening failure (US Sailing, 2013). This incident brought light to the fact that even the most advanced life jacket technology needs improvement to be trusted in all conditions.

2 DESIGN REQUIREMENTS

This project seeks to provide a PFD solution that efficiently fills some of the gaps in the current recreational-use PFDs on the market. The proposed design can be utilized by users such as racing sailors or others who partake in activities on or around water, that offer some danger of accidental submergence and potential unconsciousness in the case of an accident. The designed life jacket is of the inflatable type II PFD and should therefore provide a minimum of 34 lbs (151.24 N) of buoyancy (Braynard, 2018). This type of PFD is only authorized for use by persons 16 years of age and older. A Type II PFD is approved for near-shore use; in calm or inland waters where there is a good chance of quick rescue.

Further, the PFD should inflate automatically when submerged in a depth of at least 6 ft (1.83 m) of water to prevent accidental inflations (e.g. the user is standing in the water, or has fallen into the water but can still easily swim to the surface). Additionally, it should turn an unconscious wearer to a face-up position in the water and keep the wearers head above the water. The vest itself should be easily adjustable to fit a variety of users and should allow freedom of movement. The automatic inflation system should be electronically controlled, to allow tracking of the user orientation in order to facilitate and evaluate the PFDs ability to turn a user face up.

3 DESIGN AND FABRICATION

3.1 *System architecture*

The PFD consists of a horseshoe vest with two separate compartments (anterior and posterior inflatable bladders) attached to a harness. The battery pack and electronics are housed in a waterproof box designed to be worn at the user's waist. Both of the two inlets of the box are connected to a 16 g CO_2 canister through a manually adjustable pressure regulator, which also integrates a pre-puncturing mechanism for the canister. Each of the two box outlets is connected to one bladder through polyurethane tubing.

Figure 1 illustrates the architecture of the system. A microcontroller continuously monitors environmental pressure through an analogue sensor. When a predefined threshold level is reached and maintained for a sufficient amount of time, the microcontroller activates two NC fast-response pneumatic valves, allowing the CO_2 to fill the bladders. The activation of the two valves can be modulated independently, based on the user's current depth and orientation, as estimated by the onboard inertial measurement unit (IMU).

3.2 *Electromechanical design*

The system is powered by a Li-Po battery pack (3000 mAh, 11.4 V) through step-up and step-down voltage regulators that serve the valves (24 V) and the logic circuit (5 V), respectively.

A 32-Bit ARM Cortex-M4 microcontroller samples data from the analog pressure sensor (U86B, by TE connectivity) and from the 9-degree-of-freedom IMU (3-Space Embedded, Yost Labs). Data is logged to an on-board mini-SD card and also sent to a host computer using an on-board Wi-Fi module.

The microcontroller triggers two fast-response 3/2 NC pneumatic valves (Festo MHE series) by means of power Darlington transistors (TIP120). The outlet of each valve is connected to one inflatable bladder through polyurethane tubing (OD = 4 mm, ID = 2.5 mm), while the NC port is connected to one of the pressure regulators. The

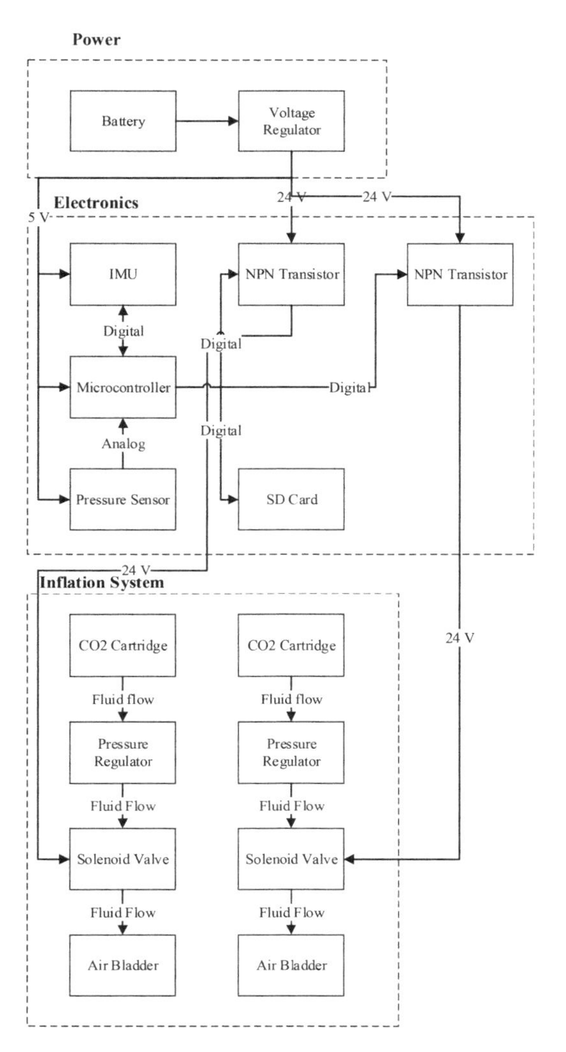

Figure 1. System architecture diagram.

NO ports are connected together to allow for air pressure distribution between the two inflatable bladders after inflation is complete.

For testing purposes, the electronics was housed inside an off-the-shelf watertight plastic box. Holes were drilled to accommodate the outlet ports of the pressure regulators, the sensing port of the pressure sensor, and the polyurethane tubing connecting the vest to the box. Silicone sealant and o-rings were used to ensure waterproofness.

Components were rigidly connected to the box using plastic standoffs. Custom electronic boards were attached to the internal walls of the box using industrial-strength Velcro straps as seen in Figure 2.

Figure 2. Waterproof box with electronics.

3.3 *System operation*

The system is automatically initialized after 4 seconds of being powered on. During this interval, the user is expected to stand in an upright position, to determine the relative orientation between the IMU and the user. Any bias in the pressure sensor is also zeroed at this point.

Alternatively, the system can be initialized and tared remotely from a host computer, though a custom-made graphical user interface (GUI) implemented in MATLAB (The Mathworks), which was designed to better control the PFD during the experimental validation.

IMU data include the user's orientation (in terms of yaw, pitch, and roll Euler angles) and 3D acceleration. These data are sampled at 310 Hz, along with the raw voltage readings from the pressure sensor. These raw data are then converted into pressure values using experimentally determined calibration coefficients.

Calibrated pressure is monitored using a smoothening FIR filter (moving average) with adjustable window size. When the filtered pressure is above a predefined threshold value for a specified time interval, the valves are activated. A time delay can be set between the activation of the anterior and posterior valve, to actively modulate the righting moment.

The duration of the time interval can be adjusted to let users swim to the surface (if they can), thus preventing unwanted inflations.

Orientation, acceleration, and pressure data, along with two variables indicating the states of the two valves, are stored to an onboard mini-SD card at 310 Hz. These data are also sent to the GUI running in a host computer at a lower frequency (25 Hz), using UDP over WLAN.

Besides real-time data visualization, the GUI also allows to fine-tune the adjustable parameters mentioned above.

Figure 3. Inflated vest.

3.4 *Inflatable bladders*

The inflatable air pockets designed for the PFD are horseshoe shaped, common to many PFDs on the market (Figure 3). The airbladder is divided into two distinct pockets: one behind the head and one on the chest. The chest pocket is larger than the one behind the neck, as is standard for similar designs in order to keep the wearer face up. After reaching surface mode, the air pocket behind the head will keep the head supported and out of the water.

Overall, the fully-inflated air bladder has an approximate volume of 3.57 gallons (13.5 liters), which was determined by filling up the bladders with water. This produces about 31 lbs. (14.1 kg) of buoyancy, which is close to the 34 lbs. requirement for a Type II vest. To keep the bladders secured to the wearer, a strap is sewn on to the sides of the vest with an adjustable buckle for increased comfort.

4 TESTING

4.1 *Pressure sensor calibration*

In order to turn the raw output from the pressure sensor into useful depth estimates, a calibration test was performed to determine the relationship between raw sensor data and measured depth.

The watertight box was attached to a measure line weighted at one end. At measured intervals, the rope was looped to allow it to be fixed to a crane positioned above the water tank used for testing.

Each interval between loops reflected a change in depth between measurements.

Once this test set up was established, the PFD controller was activated to start logging raw pressure data to the onboard mini-SD card. Then, the box was lowered at fixed depth increments until it reached the lowest depth, and brought up to the surface following a similar procedure. The full cycle was repeated twice to test reliability of the data (Figure 4).

Linear regression was then run to determine the best-fit (in the least squares sense) linear model approximating the relationship between sensor raw data and measured depth (Figure 5). The calibration model was then implemented into the PDF controller.

4.2 *Preliminary test*

A proof-of-concept test was carried out in the towing tank at the Stevens Institute of Technology, Davidson Laboratory. The goal was to verify the feasibility of the smart PFD's onboard electronics to trigger the inflation of the air bladders at a pre-programmed depth.

A 40-liter roll-top dry bag was used to simulate a human torso. First, two 2-lb weights (total 1.8 kg) were secured to the inner side surface of the dry bag, plastic sheets were wrapped around a cable reel placed inside the bag and the bag was filled with water and sealed. A rigid plastic plate was fixed to the external side of the bag to facilitate positioning of the electronics box. Secondly, the vest was strapped around the bag, and the electronic box was attached to the posteroinferior side of the bag, as shown in Figure 6.

Then, the IMU was initialized with the bag fixed in the upright position, and the pressure sensor was zeroed just above the free surface. Afterwards, the water-filled bag fitted with the PFD was lowered into the tank and let sink until the bladders were automatically inflated, bringing the device back to the surface.

Depth and orientation data, as well as the status of the valves, were logged to the on-board SD card for analysis.

The experiment consisted of two sessions, each including two tests. The first session simulated a user who fell in water "facing up". The second session simulated a user who fell in water "facing down". Because the heaviest side of the dry bag (i.e., the one with the weight attached) assumed a downward oriented configuration while sinking, before reaching the bottom of the tank, the difference between the two sessions consisted in simply

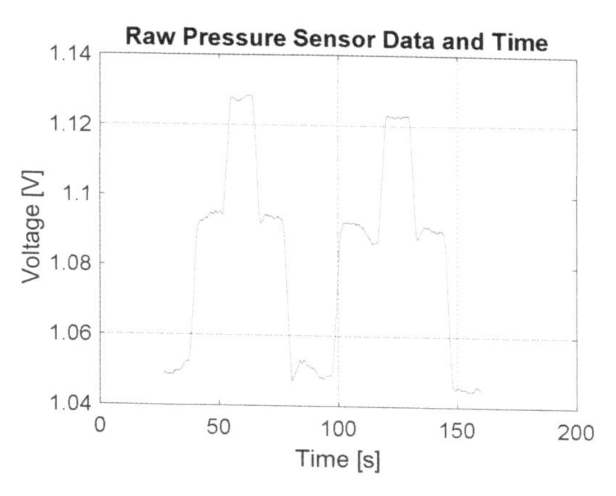

Figure 4. Pressure sensor calibration data.

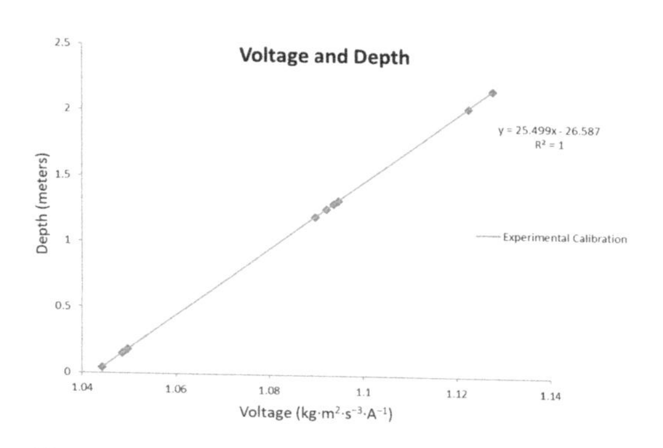

Figure 5. Voltage as a function of depth.

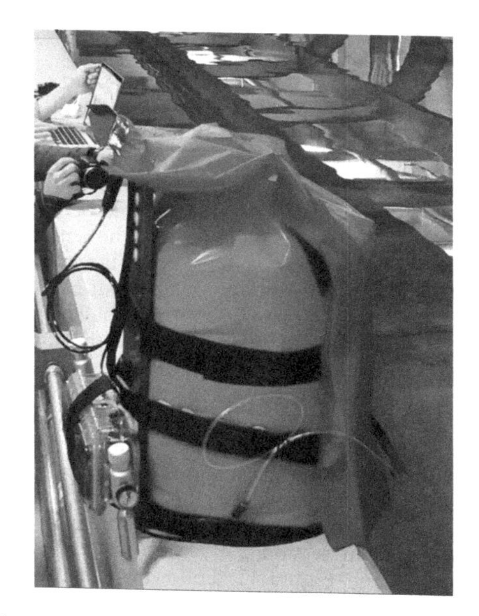

Figure 6. Bag used at testing with the PFD on.

Figure 7. Bag face down at the bottom of the tank.

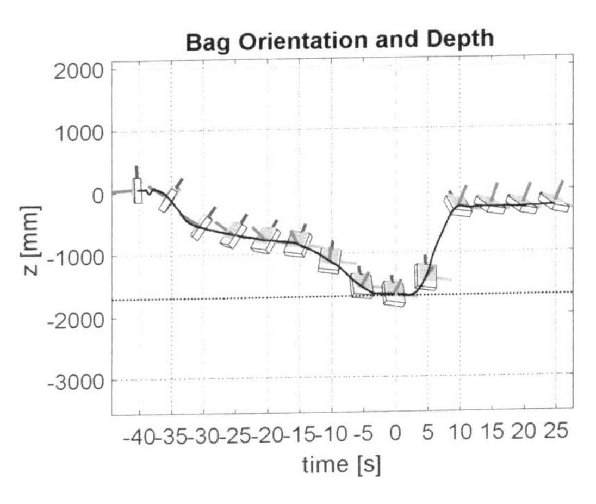

Figure 8. Session 1, test 1; Face up drop position with simultaneous activation of the air pockets. Red, green and blue segments indicate the anteroposterior, medi-olateral, and superior-inferior axes of the SmartPDF.

flipping the anterior/posterior parts of the vest relative to the bag.

Within each session, test 1 and test 2 differed in the timing of activation of the anterior and posterior valves. In test 1, the two valves were activated simultaneously. In test 2, the posterior valve was activated 2 seconds after the anterior valve.

Before starting each test, the GUI was used to remotely adjust the PFD activation parameters. In session 1, the activation depth was set to 1.7 m. In session 2, this parameter was reduced to 1.5 m to account for the bag laying "face down" on the bottom with the electronics box at the back of the bag (Figure 7).

The time interval between reaching the target depth and activating the first valve was set to 5 seconds. This value was chosen to give the bag enough time to rest at the bottom of the tank before inflating the PFD.

4.2.1 *Results*

Figure 8 and Figure 9 illustrates the depth and orientation of the bag as a function of time, and t = 0 s indicates the time instant when the anterior valve was activated. It can be noticed that the bag reached the bottom of the tank before the valves were activated. As expected, the activation occurred 5 s after the bag reached the target depth. The buoyancy of the inflated system was sufficient to bring the bag to the surface, and the bag's orientation stayed approximately vertical throughout the test. It can also be seen that the time required to bring the bag to the surface is similar in both "face up" tests. The dashed vertical lines marks the time of activation of the valves.

The difference in time required to bring the bag to the surface between test 1 (simultaneous activation) and test 2 (sequential activation) is further reduced in session 2 (face down) as seen in Figures 10 and Figure 11. Following the red axis pointing in the anterior direction (i.e., forward from the "chest"), it is clear that in both test cases during session 2 the bag is turned face up as it ascends towards the free surface.

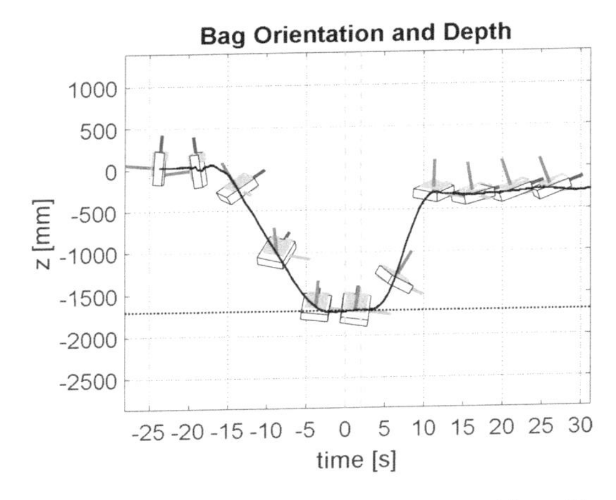

Figure 9. Session 1, test 2: Face up drop position with sequential activation of the air pockets.

Figure 12 and Figure 13 shows the inclination of the bag during the face down tests. In both tests the bag sinks "face down", laying on its "chest" at the bottom. The dashed vertical lines mark the time of activation of the air pockets (simultaneous in Figure 12 and sequential in Figure 13). Values above 90° (horizonral dotted line) indicate that the bag is facing up. It should be noted that it takes somewhat longer for the bag to turn around when the air pockets are activated sequentially. However, as seen in Figure 10 and Figure 11, the bag is turned face up in both tests once it reaches the free surface.

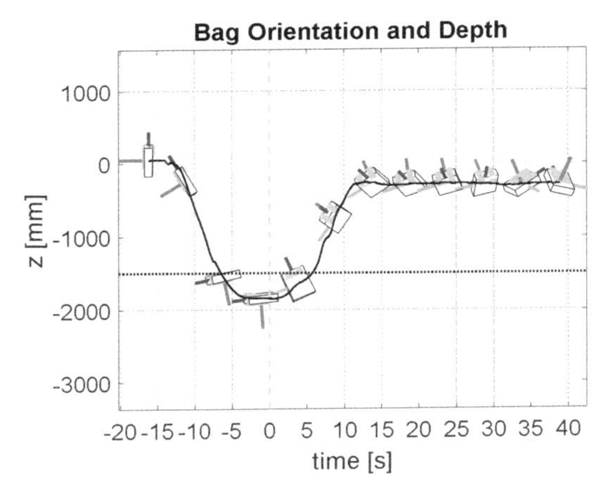

Figure 10. Session 2, test 1; Face down drop position with simultaneous activation of the air pockets.

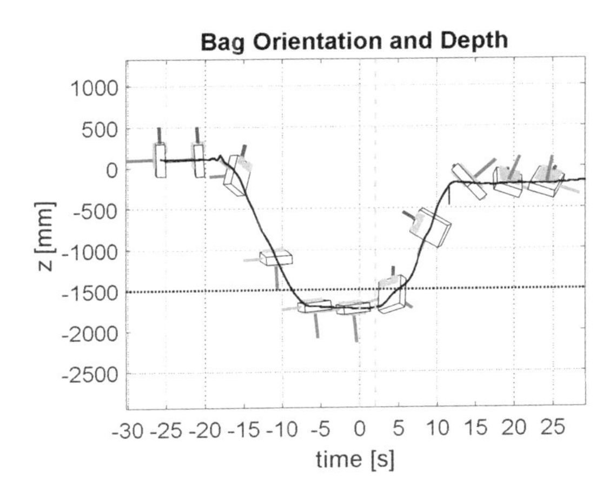

Figure 11. Session 2, test 2: Face up drop position with sequential activation of the air pockets.

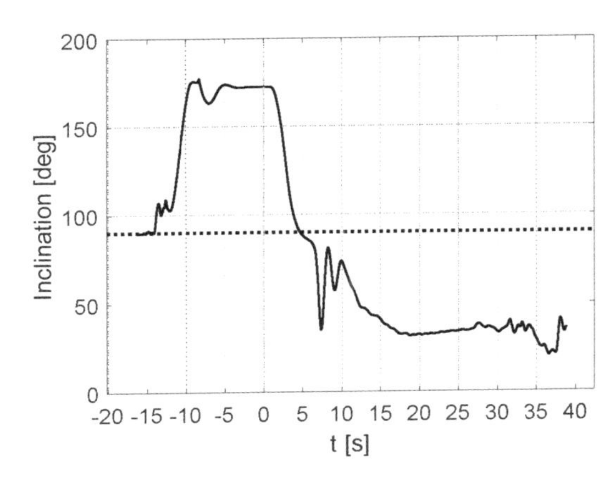

Figure 12. Session 2, test 1; Inclination of the bag.

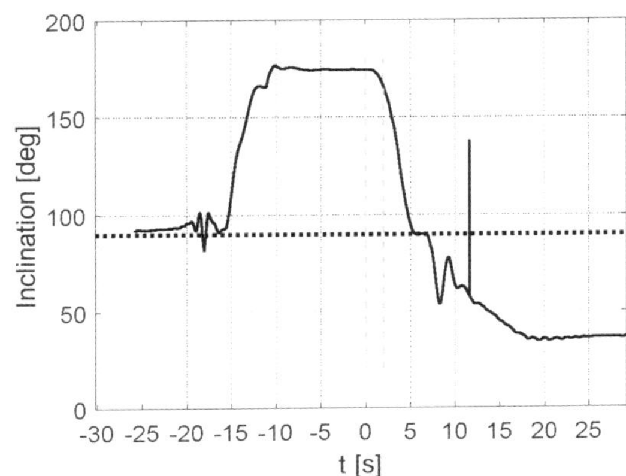

Figure 13. Session 2, test 2; Inclination of the bag.

5 CONLUSIONS AND FUTURE WORK

An electronically controlled PFD has been developed at the Wearable Robotic Systems Lab and the Davidson Lab at Stevens Institute of Technology. The PFD consists of a vest with two separate compartments and a waterproof box housing the electronics. The PFD inflates when the user has been below a specified depth for 5 seconds. Further, the user's orientation is tracked by an IMU.

Experimental proof of concept testing shows that the PFD performs as required and will bring the user (in this case, a bag simulating a human torso) to the surface face up. Further, the PFD successfully turns a face down user face up prior to reaching the free surface.

It should be noted sequential activation caused a small delay in bringing the bag to the surface in the "face up" session. However, this delay was reduced in the face down session. The sequential activation also took longer to turn the bag "face up", however, in both tests the bag was turned "face up" when it reached the free surface.

5.1 *Future work*

To better assess the effect of sequential activation, repeatability must be ensured during the testing. Thus, further testing should be conducted.

Dividing the vest into separate left- and right-side air pockets instead of anterior and posterior pockets could decrease the time required to turn the user "face up" in sequential activation, since this would facilitate rotation around the longitudinal axis, which corresponds to a smaller moment of inertia.

The bag used for testing represented only a torso, attaching a head might be needed to more accurately evaluate the effect of the air pocket behind the head.

The smart PFD is in an early design phase, and as such the authors prioritized proof-of-concept validation of the effects of staggered activation over a systematic comparison with PFDs currently available in the market, in order to control for other aspects of the design (e.g., vest geometries and volumes, activation depth, etc.) which are less critical at this stage. In the future, a comparison with market leaders is recommended.

ACKNOWLEDGEMENTS

The authors would like to acknowledge the assistance of all the students involved in this project throughout 2017, especially, Karine Jansen, Kiril Manchevski, Reed Oberlander, Julian Fraize and Juan Sanchez. We also wish to thank Megan Clifford at the Stevens' Systems Engineering Research Center for her support and valuable discussions. This project was partially supported by U.S. Special Operations Command.

REFERENCES

Boater Exam, 2017. Personal Flotation Device Types, Designs & Uses. URL https://www.boaterexam.com/boating-resources/personal-flotation-device-types.aspx (accessed 1.10.18).

Braynard, L.K., 2018. "Wear It!": Life jackets save lives. Coast Guard Compass. URL http://coastguard.dodlive.mil/2015/05/wear-it-life-jackets-save-lives/ (accessed 9.19.17).

CDC Injury Center, 2016. Unintentional Drowning: Get the Facts - Home and Recreational Safety. URL https://www.cdc.gov/homeandrecreationalsafety/water-safety/waterinjuries-factsheet.html (accessed 1.10.18).

International Maritime Organization, 2017. Life-saving appliances, 2017 edition. ed. London.

ISO, 2006. ISO 12402–5:2006(en), Personal flotation devices Part 5: Buoyancy aids (level 50) Safety requirements. URL https://www.iso.org/obp/ui/#iso:std:iso:12402:-5:ed-1:v1:en (accessed 1.12.18).

McCormack, E., Elliott, G., Tikuisis, P., Tipton, M., 2008. Search and Rescue (SAR) Victim Empirical Survival Model. University of Portsmouth.

Quistberg, D.A., Quan, L., Ebel, B.E., Bennett, E.E., Mueller, B.A., 2014. Barriers to Life Jacket Use Among Adult Recreational Boaters 20, 244–250. https://doi.org/10.1136/injuryprev-2013-040973.

US Sailing, 2013. US Sailing Releases Report on 2013 Islands Race Tragedy in Southern California. Cruis. World. URL http://www.cruisingworld.com/how/us-sailing-releases-report-2013-islands-race-tragedy-southern-california (accessed 9.12.17).

World Health Organization, 2017. Drowning. URL http://www.who.int/mediacentre/factsheets/fs347/en/ (accessed 9.19.17).

Arctic design

Marine Design XIII – Kujala & Lu (Eds)
© 2018 Taylor & Francis Group, London, ISBN 978-1-138-34076-3

Numerical simulation of interaction between two-dimensional wave and sea ice

Wen-jin Hu, Bao-yu Ni, Duan-feng Han & Yan-zhuo Xue
College of Shipbuilding Engineering, Harbin Engineering University, Harbin, China

ABSTRACT: The polar vessels sailing in Marginal Ice Zone (MIZ) are subjected to external loads including wave load and ice load. Therefore, the study of wave and sea ice interaction is of significance for ship design and safety from the perspective of external loads. Based on Computational Fluid Dynamics (CFD) software FLUENT, a Two-Dimensional (2D) numerical water tank is established by using Volume of Fluid (VOF) model and *k*-epsilon turbulence model. Based on wave theory, the waves are generated by a push pedal at one end of the water tank and are absorbed by using the method of momentum source wave elimination at the other end. Sea ice is simplified into a one-end rigid fixed plate for shore-connecting ice cover or a floating rigid plate for an ice floe. The responses of the plate under wave load are simulated and analyzed, including the pressure and force as well as the stress of the fixed plate and the motion of the floating plate. For a given incident wave, the threshold value of ice length for the fracture of the sea ice is discussed, referring to the physical and mechanical properties of sea ice in the Bohai Bay.

1 INTRODUCTION

Polar vessels sailing in the Arctic Passage generate waves, which may influence the motion of sea ice around the navigation area. In addition, in the marginal ice zone (MIZ), waves are one of the main reasons why sea ice is deformed and broken under the effect of wave-sea ice interaction (Shen, 2001; Vaughan and Squire, 2011). The sea ice obtains kinetic energy under the wave impact, so a large area of ice cover near the harbor may deform and break into pieces in the waves. These small floes move along with the waves, and collide with each other and possibly a passing ship or port facilities also. As a result, the polar vessels sailing in marginal ice zone (MIZ) may be subjected to external loads including wave load and ice load. Therefore, the study of wave and sea ice interaction is of significance for ship design and safety from the perspective of external loads.

In terms of the theoretical and numerical study of interaction between wave and sea ice, most of the existing researches focused on propagation and attenuation behavior of the wave in the water under the ice. Li (2000) used the flux energy theory to analyze the deformation effects of the ice sheet. Elastic energy transmitted from wave into the ice sheets, and the forces that the regular waves exert on the floating sea-ice blocks were calculated. Yang et al. (2017) developed a three-dimensional (3D) numerical wave flume. Regarding the sea ice as a rigid plate, they studied the effects of wave height, ice length and ice thickness on the vertical impact force, horizontal impact force, bending moment and maximum bending stress of the sea ice. Bai et al. (2017) used the potential flow model HydroSTAR and the viscous flow computational fluid dynamics (CFD) model OpenFOAM to investigate the kinematic response of rigid sea ice floes in waves. The velocities in surge, heave and drift motion were analyzed for various ice floe shapes. On the other hand, in terms of the model experiment and observation, researches concerned about the motion and fracture of sea ice more. Li (2000) analyzed the mechanism of ice breakup under tidal fluctuation and waves according to the characteristics of sea ice in Bohai. Physical model tests were carried out to confirm the numerical model of the rupture of ice sheet under the action of regular waves. The results indicated that bending failure happened in sea ice under waves. McGovern and Bai (2014) conducted an experiment in water tank to investigate the motion response of an ice floe to the floe and wave characteristics, where the ice floe is made of wax with density 890 kg/m^3. The results showed that Froude number, floe size, floe orientation and surface roughness did not affect the heave and surge motion much. Wavelength affected motion of the floe predominantly. Bennettsa et al. (2015) presented an experimental study on transmission of ocean waves through an ice floe. Thin plastic plates with different material properties and thicknesses were used to simulate the floe. Results showed that transmitted waves were regular for gently-sloping incident waves but irregular for storm-like incident waves. Guo et al. (2017) adopted paraffin wax to simulate the ice and carried out an experiment to study the motion of a single paraffin wax sheet floating in a wave tank. Longitudinal motion including drift and surge motions of the wax was recorded under some wave condition.

Upon reviewing the previous research, we find few researches pay attention to the force of the semi-infinite sea ice under wave actions, and it is difficult to make large areas of sea ice in the laboratory. Besides, studies are still in lack on numerical simulation of 2D rigid floating ice floe under waves. This forms the primary motivation for the present work and provides its key focus. Herein, we adopted the CFD software FLUENT to simulate the responses of ice cover and ice floe under action of incident waves.

2 ESTABLISHMENT AND VALIDATION OF NUMERICAL WATER TANK

The main types of wave maker include the paddle type, push-pedal type, punching box type and air type. The paddle type and the push-pedal type wave makers are widely used because of their simple structure, convenient control and easy maintenance. The paddle wave maker is generally used in deep water, while the push-pedal type is generally used in relatively shallow water. Considering that the numerical water tank is relatively shallow, we adopt push pedal to generate waves herein.

2.1 *Establishment of 2D numerical model*

By adding the FLUENT module in the ANSYS Workbench, we establish a 2D water wave tank with VOF method to capture the air-water interface. The length of the water tank is 12 m, the height is 2 m and the water depth is 1 m, as shown in Fig. 1. The upper boundary of the water tank is pressure outlet boundary, the left boundary is moving boundary for pushing plate to make waves, and the remaining boundaries are non-slip wall boundaries. As shown in Fig. 1, the water tank can be divided into 3 zones: wave making zone, working zone and wave elimination zone. The lengths of these three zones are 0.5 m, 6.5 m and 5 m, respectively, in this paper.

The left boundary is set as a push pedal and its motion velocity equation satisfies a cosine function:

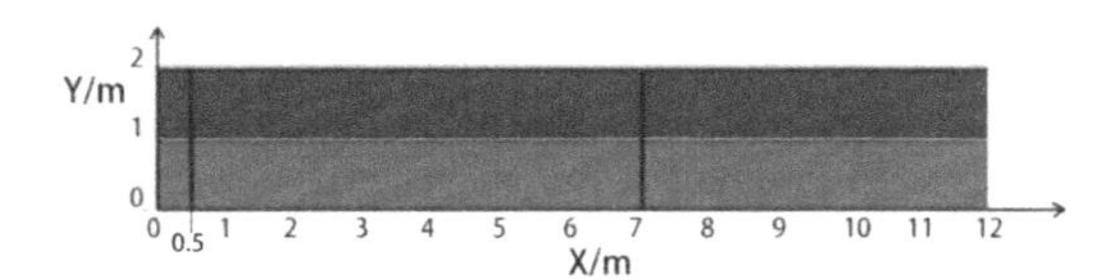

Figure 1. Sketch of the 2D water tank.

$$V = \omega \cdot S \cdot \cos(\omega \cdot t), \tag{1}$$

where V is velocity of the pedal, ω is angular frequency, S is a stroke, t is time. The equilibrium position is taken as $X = 0$ m.

On the wave elimination zone, the source of momentum attenuation is introduced as the artificial damping to reduce the reflection of the incident wave. User defined function (UDF) macro-programming (Gu, 2009 and Li, 2009.) is adopted to achieve the source of momentum attenuation. In the damping zone, momentum equation is:

$$\frac{\partial u}{\partial t} + u\frac{\partial u}{\partial x} + v\frac{\partial u}{\partial y} = -\frac{1}{\rho}\frac{\partial P}{\partial x} + \upsilon\left(\frac{\partial^2 u}{\partial x^2} + \frac{\partial^2 u}{\partial y^2}\right) - S_x, \tag{2}$$

$$\frac{\partial v}{\partial t} + u\frac{\partial v}{\partial x} + v\frac{\partial v}{\partial y} = g - \frac{1}{\rho}\frac{\partial P}{\partial x} + \upsilon\left(\frac{\partial^2 v}{\partial x^2} + \frac{\partial^2 v}{\partial y^2}\right) - S_y, \tag{3}$$

where ρ is fluid density, u and v are the velocity components in the direction of X and Y, respectively, P is pressure, υ is kinetic viscosity coefficient, S_x and S_y are the source of momentum in the direction of X and Y, respectively. The expressions of the source momentum attenuation are:

$$S_x = \frac{\mu}{\alpha}u + C \cdot \frac{1}{2}\rho|u|u, \tag{4}$$

$$S_y = \frac{\mu}{\alpha}v + C \cdot \frac{1}{2}\rho|v|v, \tag{5}$$

where μ is dynamic viscosity coefficient, $|u|$ and $|v|$ are the absolute values of the velocities, $1/\alpha$ and C are constants. The first term on the right-hand side of Eq. (5) is the viscous loss and the second term is the inertial loss. The proper viscous drag coefficient $1/\alpha$ is determined to eliminate the wave completely and no wave will be reflected from this zone as a result. In the wave elimination zone, viscous drag coefficient $1/\alpha$ is set to be linearly increase with x from zero,

$$\left(\frac{1}{\alpha}\right)_j = e^6 \cdot \frac{x_j - x_0}{x_e - x_0}(x_0 < x_j < x_e), \tag{6}$$

to ensure the continuity of the potential function in the whole flow field. In Eq. (6), $(1/\alpha)_j$ is the viscous drag coefficient at $x = x_j$, x_0 and x_e are the x-coordinates of the front and back edges of the wave elimination zone, respectively.

2.2 *Validation of numerical model*

According to the theory of linear push-pedal wave maker (Dong, 2009), under the motion of Eq.(1), the wave equation η is:

$$\eta = \frac{S}{2}\left[\frac{4\sin^2 k_p d}{2k_p d + \sinh 2k_p d}\cos(k_p x - \omega t) + \Sigma \frac{4\sin^2 k_s(n)d}{2k_s(n)d + \sinh 2k_s(n)d} e^{-k_s(n)x}\sin\omega t\right], \tag{7}$$

where d is the depth of tank, k_p is the progressive wave number, k_s is the decaying vertical wave number, and k_p satisfies the equation:

$$\omega^2 = gk_p \tanh k_p d, \tag{8}$$

$k_s(n)$ satisfies the equation

$$\omega^2 = -gk_s \tanh k_s d, \tag{9}$$

with root n.

In this example, the stroke of push-pedal S is 0.1 m and the angular frequency ω is π. In the Eq. (7), the first term is the progressive wave generated by the push-pedal, and the second one is the decaying vertical wave, which decays exponentially with the increase of distance. When the distance from the push-pedal is 2~3 times than the depth of water, the second term can be neglected, so the wave height curve of the following theory takes only the first term.

For the water depth 1 m, the distance between the monitoring points and the push-pedal is set in 2 m and 2.5 m respectively, which is convenient for observation. After calculating the wave number and wavelength of the plate, the theoretical waveform can be obtained.

The comparison curves of theoretical and simulation results are shown in the Fig. 2. The results of above figures show that the numerical result agrees with the theoretical solution. The wave cycle is basically consistent. Because of the influence of viscosity in the propagation process, the wave heights will decline along the X direction. Therefore, under the condition of 2~3 times water depth from the push-pedal, the waves making simulation and the method of momentum source wave elimination in our model are validated and can be used herein.

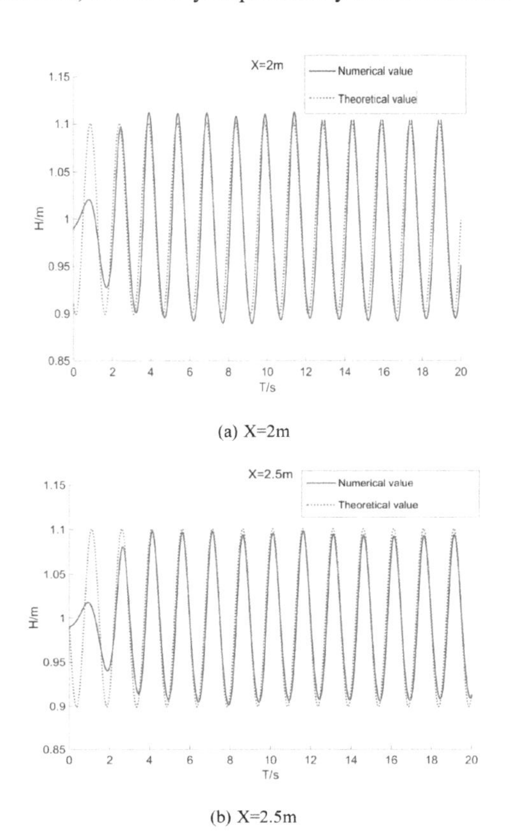

(a) X=2m

(b) X=2.5m

Figure 2. Comparison of numerical and theoretical results when X = 2.0 m and X = 2.5 m, respectively.

3 NUMERICAL SIMULATION OF INTERACTION BETWEEN WAVE AND A SHORE-CONNECTING ICE SHEET

3.1 *Establishment of numerical model*

Based on the 2D water wave tank established in section 2, we firstly simulate the interaction between wave and a shore-connecting ice sheet. Some assumptions need to be given to simplify this problem. Firstly, the ice sheet is assumed to have a uniform thickness. Secondly, considering that the deformation of the shore-connecting ice is very small under wave, we assume its deformation neglectable and simplify the shore-connecting ice as a rigid plate when calculating its external force and moment. Thirdly, although the ice sheet is assumed to be rigid when calculating external force and moment, the ice sheet has its internal stress which is calculated based on pure bending theory.

The numerical model is established in ANSYS Workbench, as shown in Fig. 3. To enhance calculation efficiency, the length of water tank in this section is taken as 6 m, the height and water depth are still 2 m and 1 m, respectively. The ice sheet is rigid fixed at the right end of water tank, whose length is L thickness is h and draught is d. Considering that the ice density is 0.9 times of water density and the buoyancy of ice sheet is equal to its gravity before encountering the wave load, the draught d = 0.9 h. In this paper, L may change from 0.5 m to 2.5 m with an interval 0.5 m and h may change from 0.06 m to 0.14 m with an interval 0.02 m. The incident wave is constant with wave height 0.2 m, cycle 1.5 s, wavelength 3.14 m and

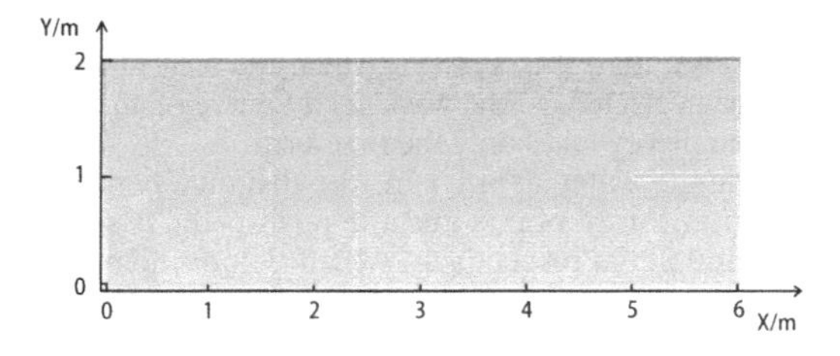

Figure 3. The schematic diagram of mesh.

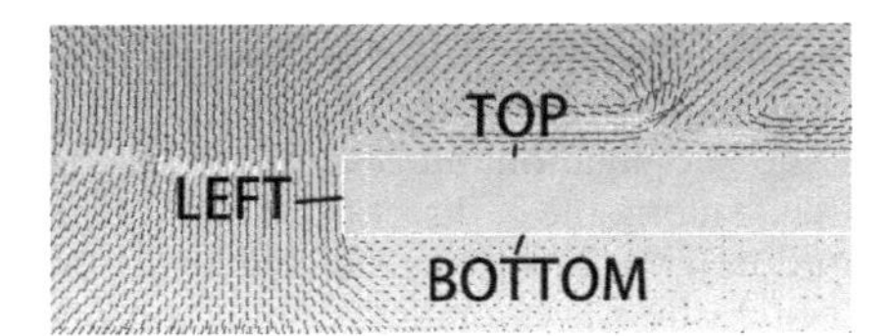

Figure 4. The local map of velocity vector of the fluid around the ice sheet.

wave number 2. The computational domain is gas-liquid two phases flow, and the grid is encrypted near the free surface to improve the calculation accuracy. Based on the convergence study of mesh and time step, about 10 grids are arranged in a wave height range and about 150 grids in a wavelength range, while the time step size is taken as 0.01 s. The schematic diagram of mesh is shown in Fig. 3.

During the interaction between wave and ice sheet, wave may wash up the top of the ice sheet, as shown in Fig. 4. As a result, all the pressure of the top, left and bottom faces of the ice sheet should be monitored. On this basis, the fluid force and moment can be obtained easily through calculation.

3.2 *Analysis of calculation results*

3.2.1 *Results at different ice thickness h*

Firstly, we keep the length of the ice sheet L as 1.0 m constant, and change the ice thickness h from 0.06 m to 0.14 m with an interval 0.02 m. The variation of the fluid pressure, fluid force and moment of the ice sheet will be investigated.

Fig. 5 provides the time histories of the fluid force on the top surface of the ice sheet with different thickness. It can be seen that the fluid force on the top surface of ice sheet increases with time. As shown in Fig. 4, the incident wave washes up on the ice and accumulate gradually with time, so the fluid force on the top face of ice rises gradually. After 14 s, the water amount accumulated on the top surface is almost stable so the curves of fluid force tends to be slightly oscillate around some constant value then. On the other hand, as the ice sheet get thinner, more water would be easier to wash up to the top surface of ice sheet, so the fluid force increases slightly.

Fig. 6 presents the time histories of the fluid force on the left surface of the ice sheet with different thickness. It can be seen the fluid force oscillates as the wave impacts the left surface of the ice periodically. It can be understood that a thick ice, which has a larger cross-sectional area, has a larger fluid force.

Fig. 7 gives the time histories of the fluid force on the bottom surface of the ice sheet with differ-

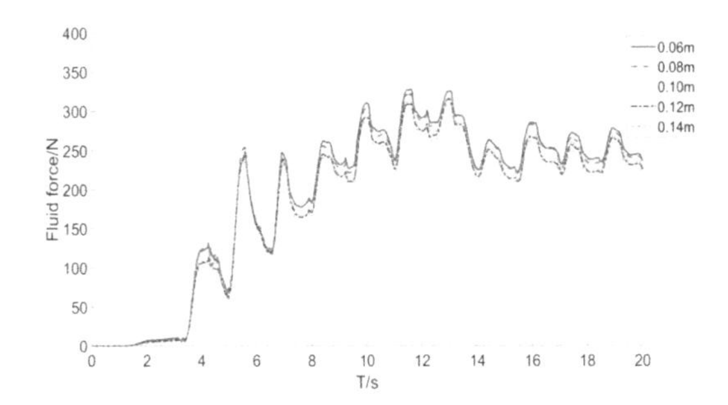

Figure 5. Comparison of fluid force curves in 5 kinds of thickness on the top surface of the ice model.

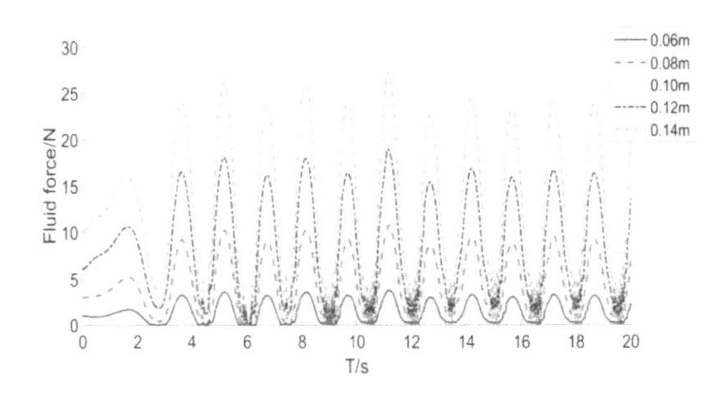

Figure 6. Comparison of fluid force in 5 kinds of thickness on the left surface of the ice model.

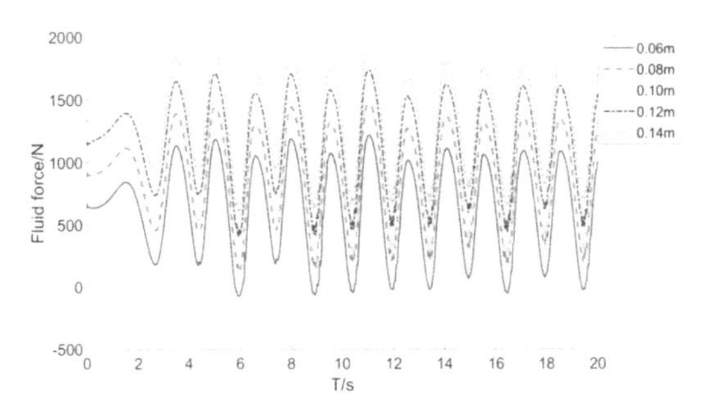

Figure 7. Comparison of fluid force in 5 kinds of thickness on the bottom surface of the ice model.

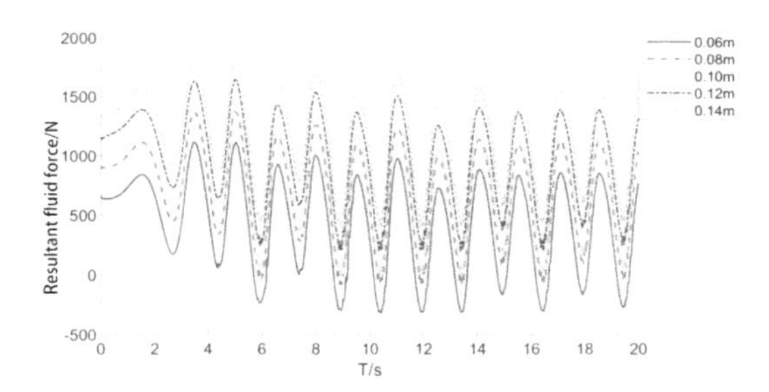

Figure 8. Comparison of vertical force in 5 kinds of thickness of ice model.

ent thickness. It can be seen the fluid force oscillates as the wave impacts the bottom surface of the ice periodically. All the curves have similar trends because the propagations of the wave are similar for different cases. It can be understood that a thicker ice, which has a larger hydrostatic pressure, has a larger fluid force. The resultant fluid force F can be obtained by subtracting the top fluid force F_t in Fig. 5 from the bottom fluid force F_b in Fig. 7, as is shown in Fig. 8. Because F_t is quite small relative to F_b, the trends of the resultant fluid force F are similar to those of F_b.

Because the problem is 2D, the bending of the ice sheet can be seen as the cylindrical bending and the moment can be calculated on a cantilever lath beam, or a strip beam with a unit breadth. When calculating the bending moment, we simplify the fluid force on the beam as a uniform distributed force. Under this assumption, the bending moment at the fixed end can be calculated by:

$$M_{\max} = -\frac{1}{2}\left(\frac{F}{L} - \rho_i hg\right) \cdot L^2, \tag{10}$$

where F is resultant fluid force, ρ_i is the density of sea ice, which is 900 kg/m³, L and h are the length and thickness of ice sheet, respectively.

Eq. (10) can be further converted into:

$$\begin{aligned} M_{\max} &= -\frac{1}{2}\left(-\int_{S_w} P n_y ds - \rho_i hgL\right) \cdot L^3 \\ &= -\frac{1}{2}\left(-\int_{S_w} (P_S + P_d) n_y ds - 0.9\rho hgL\right) \cdot L^3 \\ &= \frac{1}{2}\int_{S_w} P_d n_y ds \cdot L^3 \end{aligned} \tag{11}$$

where P is fluid pressure and P_S and P_d are hydrostatic and hydrodynamic pressures, respectively; S_w is the wetted surface of the ice sheet; n_y is the y-component of the unit normal vector of the ice boundary, pointing outwards from the ice surface. The last equality is because the buoyancy of ice sheet is equal to its gravity before encountering the wave. Thus, one can see the maximum bending moment of the ice sheet is related to the hydrodynamic pressure of the fluid on the ice surface.

Fig. 9 shows the hydrodynamic pressure of the fluid at the point 0.4 m away from the free end on the bottom surface of the ice. It can be seen that the hydrodynamic pressures of the fluid on the bottom surface of the ice are very close with each other at different ice thickness. On the other hand, the hydrodynamic forces on the top surface of the ice are also very close with each other at different ice thickness, as shown in Fig. 5. Thus, it can be predicted that the maximum bending moment at the fixed end of the ice sheet will be close with each other at different ice thickness by using Eq. (11), just as shown in Fig. 10. However, there are still small differences from enlarged view of Fig. 9 and Fig. 10, which denotes the hydrodynamic pressure and maximum bending moment are not sensitive to ice thickness in the range in this paper.

3.2.2 *Results at different ice length L*

Secondly, we keep the thickness of the ice sheet h as 0.10 m constant, and change the ice length L from 0.5 m to 2.5 m with an interval 0.5 m. The variation of the bending moment and stress of the ice sheet will be investigated.

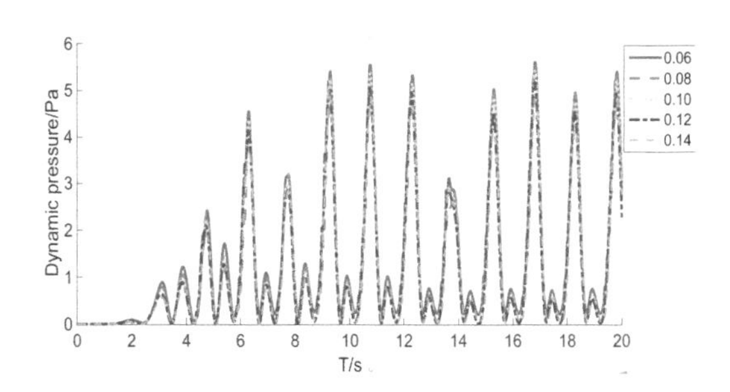

Figure 9. Hydrodynamic pressure of the fluid at the point 0.4 m away from the free end on the bottom surface of the ice.

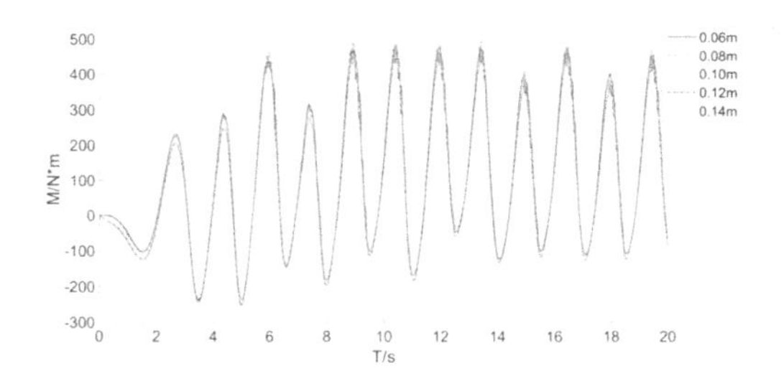

Figure 10. Comparison of maximum bending moments in 5 kinds of thickness of ice model.

According to the bending theory of elastic beam, the maximum bending stress can be obtained by:

$$\sigma_{\max} = M_{\max} / W, \tag{12}$$

where W is the elastic section modulus, which can be obtained by

$$W = bh^2 / 6, \tag{13}$$

for a rectangular section with a height of h and a width of b, which is taken as 1 for a 2D problem.

Fig. 11 provides the time histories of the maximum bending moments at the fixed end of the ice sheet and the maximum bending stress of the ice sheet with different length. From Fig. 11, the peak of bending moments and stresses increase as the ice length rises.

To predict whether the ice sheet would fracture under the wave or not, we compare the maximum bending stress $\sigma_{\max}$ with the bending strength σ_{str} of the sea ice. When

$$\sigma_{\max} \geq \sigma_{str}, \tag{14}$$

the ice sheet would break under wave loads. As the bending strength of sea ice are affected by many

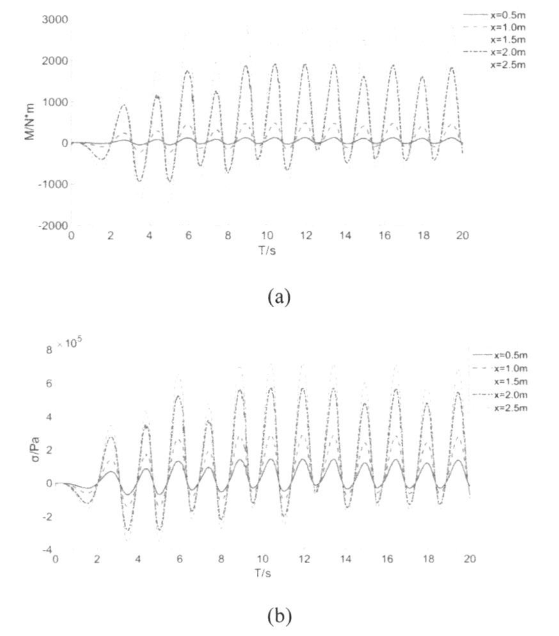

(a)

(b)

Figure 11. (a) Comparison of maximum bending moments at different lengths of ice and (b) Comparison of maximum bending stresses at different lengths of ice.

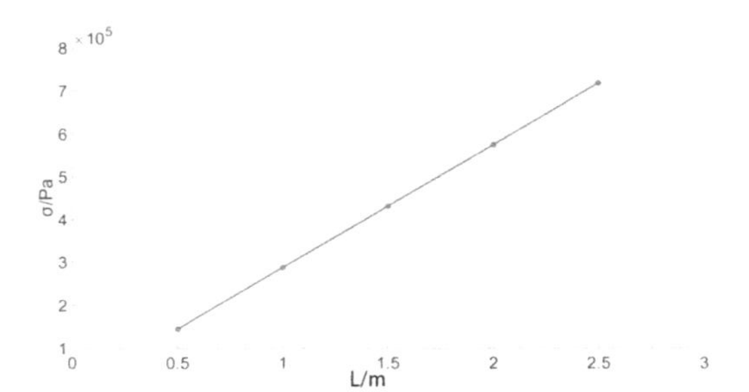

Figure 12. The curve of bending stresses of ice changing with length.

aspects, there is a range of σ_{str}. Here bending strength of sea ice σ_{str} in the Laizhou Bay of Bohai is selected as a reference (Ji et al., 2011), whose maximum σ_{str} is 1.94 MPa, the minimum σ_{str} is 0.19 MPa, and the average σ_{str} is 0.84 MPa.

On the other hand, the maximum bending stresses $\sigma_{\max}$ of the ice sheet with different lengths are shown in Fig. 12 under the action of a wave with wave height 0.2 m and cycle 1.5 s. It can be seen that the maximum bending stress has an almost linear rising relationship with ice length. Therefore, by fitting the curve of Fig. 12, one can predict the range of ice length when the fracture happens corresponding to the range of σ_{str}. It is predicted that the maximum ice length of the sea-ice fracture is 6.75 m, the minimum value is 0.66 m, and the average value is 2.92 m.

3.2.3 *Results of the pressure distribution on the sheet in the longitudinal direction*

To study the pressure distribution on the sheet in the longitudinal direction, we take a case with ice length L = 1 m and thickness h = 0.1 m. The incident wave parameter is same as above. Figure 4 presents the pressure on the top and bottom surfaces of ice in the longitudinal direction at 4, 7, 10, and 13 seconds, respectively.

As seen in Fig. 13 the pressure at the bottom is larger than that at the top in most of the time, except when the impact of the wash-up wave is severe as shown in Fig. 13 (c). The pressure at the bottom changes periodically with time, but changes slightly in the longitudinal direction. The wavelength is 3.14 m in our model, which is much larger than the ice length 1 m, so the spatial variation of fluid pressure on the ice bottom is small. On the other hand, the pressure on the top surface changes greatly in the longitudinal direction. At T = 4 s, the incident wave just washes up on the sheet partially. Then at T = 7 s, wave has washed up on the ice sheet totally and some water accumulates at the right hand of ice while more wave washes up from the left hand of ice. Around T = 10 s, the

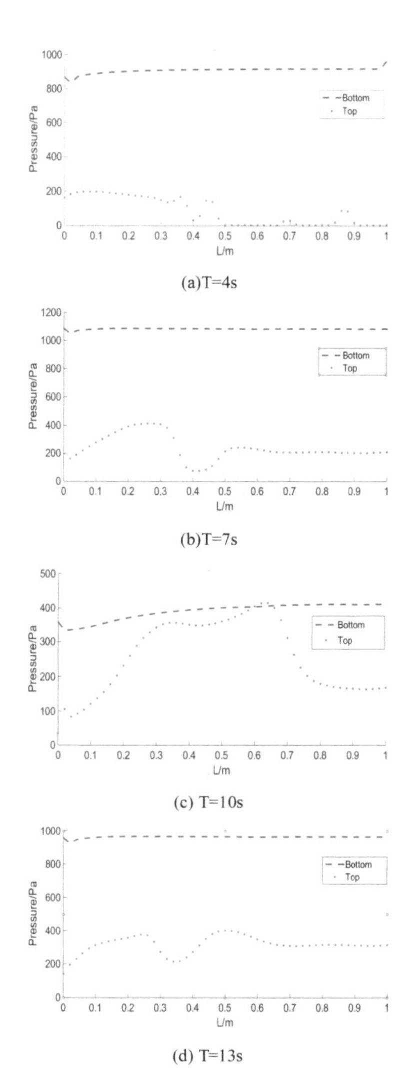

Figure 13. Pressure on the top and bottom surfaces of ice sheet in the longitudinal direction at different time.

impact of the wash-up wave on the middle part of the ice top peaks before it gets small again around T = 13 s. The pressure on the top face also has a periodic tendency along with wave motion, as shown in Fig.13 (b) and (d).

4 NUMERICAL SIMULATION OF RIGID ICE FLOE MOTION

4.1 *Establishment of numerical model*

When the scale of ice floe is very small relative to the wavelength, the small deformation of the ice caused by the wave can be neglected, so the small-scale ice floe can be simplified as a rigid body for kinematic analysis.

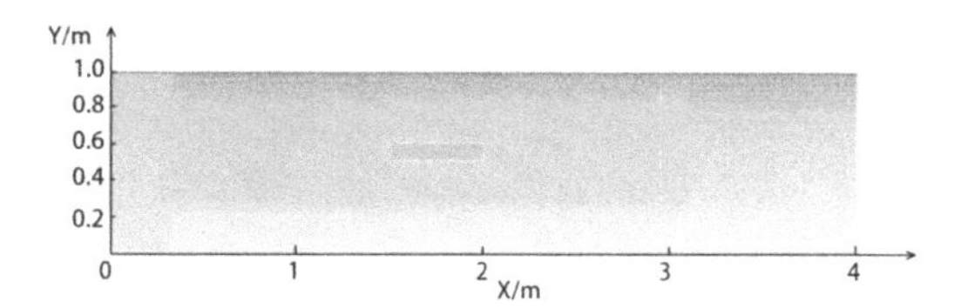

Figure 14. The schematic diagram of mesh.

A numerical tank model is established as shown in the Fig. 14, where the length of the tank is 4 m, the height is 1 m, and the water depth is 0.6 m. The length of the ice floe is 0.5 m, the thickness is 0.1 m and the draught is 0.09 m. Because the ice floe has a 3 DOF(degree of freedom) motion, more refined meshes and time steps are added to improve the calculation accuracy. Based on the convergence study of mesh and time step, about 20 grids are arranged in a wave height range and about 300 grids in a wavelength range in working zone, while the time step size is taken as 0.0001s. The coordinates of the mass center of the ice flow is (1.75 m,0.56 m). The density of water is taken as 1000 kg/m^3. The density of sea ice is taken as 900 kg/m^3, so the sea-ice mass is 45 kg per unit breadth. The moment of inertia of the sea ice around the Z-axis is taken as 0.975 kg*m^2. The wave period in tank is 1 s, and the height is 0.1 m.

4.2 *Analysis of calculation results*

Fig. 15 gives evolution of the wave and the motion of the floe at 0.00, 2.00, 4.00 and 6.00 s, respectively. It can be seen that initially the ice is floating on the water statically, as shown in Fig. 15 (a). Around $t = 2.00$ s in Fig. 15 (b), the incident wave just propagates to the left hand of the ice. As the wave propagates further, it reaches the ice floe and over washes it as shown in Fig. 15 (c). The ice moves, rolls in particular, under the wave loads. At the same time, the diffracted wave and the radiated wave will be superimposed after the ice floe as shown in Fig. 15 (d). To check the deformation of the waves, the wave heights at two points, before and after ice floe respectively, are monitored. These two points are 1.2 m and 2.5 m away from the push pedal on the initial flat free surface, respectively. The results are shown in Fig. 16.

From Fig. 16, the wave height is about 0.09 m before the floe, while about 0.05 m after the floe, reducing about 44%. Thus, the floe plays a certain effect on wave elimination, because the wave energy is transformed into the kinetic energy of the ice floe and also the viscous dissipation energy under interaction between wave and ice.

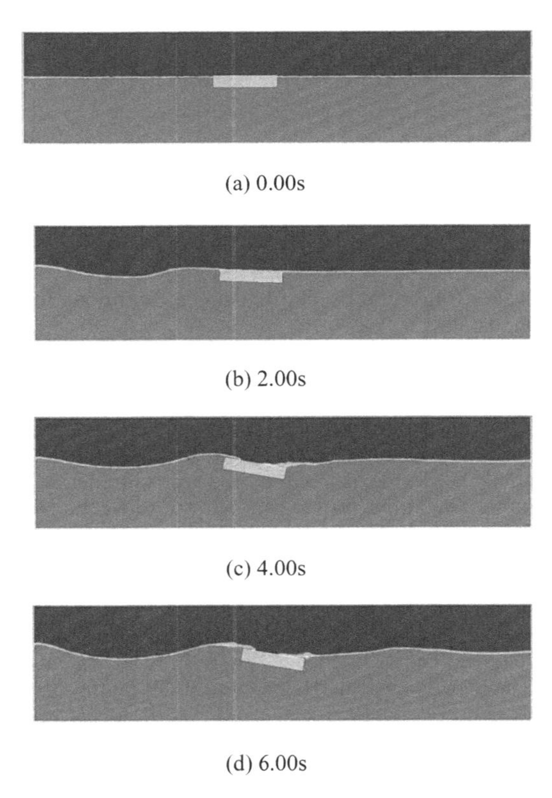

(a) 0.00s

(b) 2.00s

(c) 4.00s

(d) 6.00s

Figure 15. Evolution of the wave and motion of the floe in the water tank.

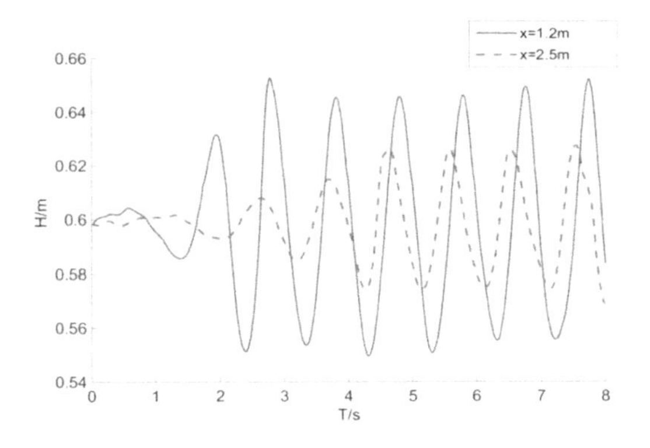

Figure 16. The contrast curves of wave height before and after the floe.

The displacement of the floe centroid in the direction of X and Y is monitored, as shown in Fig. 17. It can be seen that the ice floe just oscillates in the Y direction but both drifts and oscillates in the X direction. The drift may come from

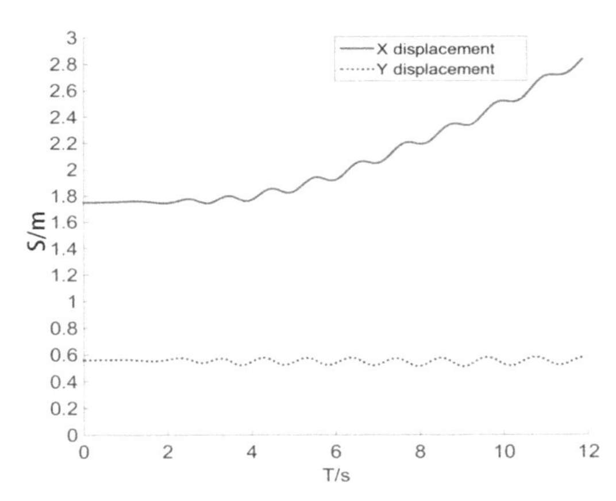

Figure 17. The displacement of the center of the floe.

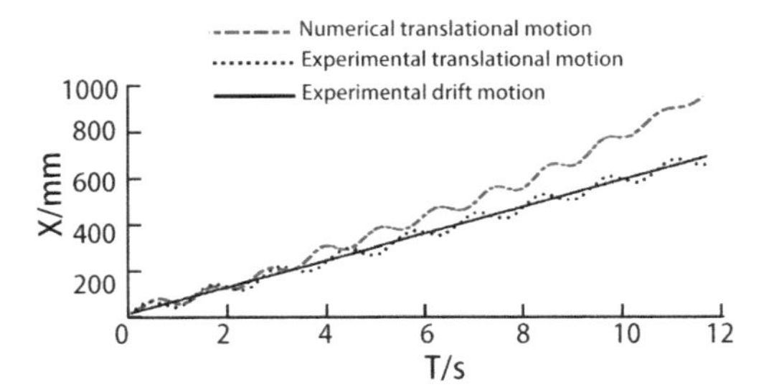

Figure 18. Comparison of 2D numerical and 3D experimental translational motion of a floating ice sheet, where the experimental data come from Guo et al (2017).

the two-order drifting force of the wave to the body (Dai and Duan, 2008). This phenomena was also observed from the 3D experiments of McGovern and Bai (2014) and Guo et al. (2017). To compare with the 3D experimental data, we simulate the same conditions of the experiments of Guo et al. (2017) by using our 2D numerical model.

To compare with the 3D experimental data, we simulate the same conditions of the experiments of Guo et al. (2017) by using our 2D numerical model. The wavelength and wave height of the incoming wave are 2.6 m and 0.08 m, respectively. The ice sheet is 0.6 m long and 0.09 m thick. The breadth of the ice sheet in experiment is 0.6 m, which is taken as unit breadth in the 2D simulation.

Fig. 18 provides the ice translation distance in our numerical simulation with comparison with experimental data from Guo et al. (2017). It can be seen that the numerical result has similar tendency with experimental data, except that the displacement growth rate of ice in 2D numerical simulation is higher than that in 3D experimental condition. This is because the wave can flow around the ice sheet in 3D experiment, but only flow through the top and bottom surfaces of the ice sheet in 2D simulation. As a result, incident wave energy is confined and more energy is transformed into the kinetic energy of the ice sheet in 2D model, which induces a larger growth rate of the displacement.

5 CONCLUSIONS

By using the CFD software FLUENT, we simulated the interaction between 2D wave and ice. Firstly, the accuracy of numerical wave making simulation and the method of momentum source wave elimination were verified by comparing numerical results and theoretical solutions. On this basis, two types of ice, a shore-connecting ice sheet and an ice floe, were simplified and studied. Then the interaction between the wave caused by push pedal and these two types of ice were investigated. Through a series of calculation, conclusions below can be drawn.

1. The shore-connecting ice sheet could be simplified into a one-end rigid fixed plate. Wave would wash up and accumulate at the top surface of the ice sheet. Although the total fluid pressure increases with ice thickness, the hydrodynamic pressure is not sensitive to ice thickness. As a result, the resultant external force including the gravity force, the bending moment and the stress of the ice sheet are all slightly affected by the ice thickness.
2. The bending moment and the stress of the ice sheet are sensitive to the ice length. For a given incident wave, the maximum bending stress has an almost linear rising relationship with increasing ice length. Take properties of the sea ice in Bohai Bay as a reference, it is predicted that the maximum ice length for the sea-ice fracture is 6.75 m, the minimum value is 0.66 m, and the average value is 2.92 m.
3. The ice floe could be simplified as a rigid body for its kinematic analysis. Wave would over wash the ice floe. On the one hand, the wave is eliminated by the ice through transforming its energy into kinetic energy of the ice floe and the viscous dissipation energy and so on. On the other hand, the 2D ice floe would occur 3 degree of freedom of motion under the action of waves, including roll, sway and heave. In addition, the ice floe would drift along the incident-wave direction, under the two-order drifting force of the wave to the body. Compared with 3D experiment, the 2D model has a larger translational displacement growth rate.

Although the result cannot be used directly in ship design, it would provide some references for relative researches in terms of the external loads of polar ships. In the next step, work will be extended to 3D model.

ACKNOWLEDGEMENTS

This work is supported by the National Natural Science Foundation of China (Nos. 51639004, 51579054 and 11472088), the Fundamental Research Funds for the Central Universities (No. HEUCFP201701, HEUCFP201777), the 111 Project in HEU, to which the authors are most grateful.

REFERENCES

Bai W., Zhang T., McGovern D., 2017. Response of small sea ice floes in regular waves: A comparison of numerical and experimental results. Ocean Engineering. 129, 495–506.

Bennetts L.G., Alberello A., Meylan M.H., Cavaliere C., Babanin A.V., Toffoli A., 2015. An idealise experimental model of ocean surface wave transmission by an ice floe. Ocean Modelling. 96, 85–92.

Dai Y.S., Duan W.Y., 2008. Potential flow theory of ship motion in waves. National Defend Industry Press.

Dong Z., Zhan J.M., 2009. Comparison of existing methods for wave generating and absorbing in VOF-based numerical tank. Journal of hydrodinamics. 24(01), 15–21.

Gu T.F., 2009. Simulation of Wave Generation in Ocean Engineering Basin. Harbin Engineering University.

Guo C.Y., Song M.Y., Lon W.Z., Tian T.P., 2017. Experimental study on longitudinal motion of sea ice in waves. Journal of Huazhong University of Science & Technology (Natural Science Edition). 45(06), 85–90.

Ji S.Y., Wang A.L., Su J., Yue Q.J., 2011. Test and characteristic analysis of the bending strength of sea ice in Bohai ring. Advances in water science. 22(02), 266–272.

Li C.H., 2000. The mechanism study of ice breakup on tidal fluctuation and waves. Dalian University of Technology.

Li H.W., 2009. Research on wave-generating method of numerical tank. Harbin Engineering University.

McGovern D., Bai W., 2014. Experimental study on kinematics of sea ice floes in regular waves. Regions Science and Technology. 103, 15–30.

Shen H.H., 2001. Wave-Ice interaction. Marine forecasts. S1, 22–34.

Vaughan G.L., Squire V.A., 2011. Wave induced fracture probabilities for arctic sea-ice. Cold Regions Science and Technology. 67, 31–36.

Yang C.Z., Liu C.G., Fang H.Y., Ma H.M., 2017. Numerical simulation of wave impact on sea ice in the Bohai sea. Chinese journal of hydrodynamics. 32(02), 141–147.

Marine Design XIII – Kujala & Lu (Eds)

Azimuthing propulsion ice clearing in full scale

P. Kujala, G.H. Taimuri & Jakke Kulovesi
Aalto University, Espoo, Finland

P. Määttänen
ABB Marine, Helsinki, Finland

ABSTRACT: Propeller jets can be used to break level ice, when the ship is stationary or moving. The amount and capacity of breaking or clearing the ice is based on the thrust of the propeller, angle between jet axis and water plane, and thickness of the ice. Current analysis is based on the full scale experiment image data and propeller flow parameters. The experiments performed in the month of March 2017, in which ice is being cleared at different thrust, pod angles and area is approximated from images, which represents that how much a particular thrust and pod angle of propeller is capable of breaking and clearing ice at stern of the ship. The data is of icebreaker "Polaris" which is a Finnish icebreaker of ice class PC4, Diesel-electric propulsion having three Azimuth propeller two at stern and one at bow. Study of 2 × 6500 kW (stern)—ABB Azipod propeller jet and its effect on clearing ice sheet is examined in this analysis. Starting from the image and power data area, thrust, open water and mix water distances of each frames were calculated. Ice clearing area depends on the power, ice thickness, heel of the ship and angle between jet axis and water plane. The ice thickness that are used in experiments were confirmed from surveillance videos. Only those results are considered to be valid in which experiments starts from the initial state (unbroken complete ice sheet). Polar plots were made which describes the variation of ice clearing area with respect to radius and angles of the region.

1 INTRODUCTION

Propeller jets can be used to break level ice, when the ship is stationary or moving. Propeller wash effect is utilized to widen the channels, washing the ice rubbles as well as floes away from the ship in the direction of the flow and clearing level ice from the frozen harbor (Tsarau et al. 2016). Application of propeller jet has also been used in oil and gas exploration in assisting drill ships to clear icy water, since other vessels are prone to thickness of the ice beyond their ice class specifications which causes damage to their hulls, therefore assistance of ice breaker is required to clear the ice. The amount and capacity of breaking or clearing the ice is based on the thrust of the propeller, angle between jet axis and water plane, and the thickness of the ice.

Full scale experiment (based on image data) to clear level ice was conducted in bollard pull condition. Using images capture at the time of experiment, ice clearance area was investigated. At the time of experiment propeller parameters were also measured. Full scale measurement data consist of podded propeller power, rate of rotation, orientation of propeller and images of the ice clearance area around the ship vicinity.

Experiment was performed in the month of March 2017, in the region of Gulf of Bothnia on the Baltic Sea. In full scale experiment ice is being cleared at different thrust, and pod angles as well as area is approximated from images, to determine the propellers capability of breaking and clearing the ice, at the stern of the ship. Data is measured onboard icebreaker "Polaris" which is a Finnish icebreaker of ice class PC4, see Table 1.

IB "Polaris" consist of diesel-electric propulsion having three Azimuth propeller two at stern and one at bow. Study of 2 × 6500 kW (stern)-ABB Azipod propeller jet and its effect on clearing ice sheet is examined in this analysis.

Table 1. Particulars of IB Polaris.

Ice class	PC4
Length	110 m
Beam	24 m
Operational draft	8 m
Displacement	10961 at design draft
LNG tanks	2 X 400 m^3
Speed	17 knots
Speed at 1.2 m ice	6 knots

2 EXPERIMENTAL STATISTICS

2.1 *Collection of data*

Full scale data is based on images that were taken at the time of experiment. In addition to it, ship heading direction, wind speed and direction, propeller power, and rotation speed along with the angle of the Azipod were noted. Moreover, time and location was also collected during experiment. Ice thickness data is compiled based on the surveillance videos recordings and visual observations. In some of the experiment there is a slight movement in ship's position, therefore only those data were studied in which ship was stationary or shifted slightly.

Table 2, summarizes various ice conditions at different pod angles and power. These powers are the average of the every experiment that runs for approximately less than 3 minutes, with fluctuating power having variation of less than 1%. Column of portside and starboard side power in Table 2, shows respective powers on each side for a particular experiment.

2.2 *Propeller thrust calculation*

From the power and thrust data provided by ABB for stern Azipod unit represented in Table 3, empirical factor of bollard pull K_E is calculated, using equation 1 (Juva & Riska, 2002). In order to calculate K_E from the data provided by ABB, diameter of the propeller is required and a value of is 4.2 m is used. The value of K_E obtained from calculation is 0.81. K_E is then used to calculate the thrust of propeller in respect to the power data of full-scale experiment.

Table 2. Summary of used power during various experiments.

Thickness	Angle	Power-PS	Power-SB
cm	°	kW	kW
60	0	1166, 1794	1174, 1799
	30	947, 3530	950, 3510
50	30	1176, 2613	1184, 2560
	60	1207	1253
	90	1221, 2652	1187, 2593
40	90	1233, 2834	1206, 2831
		4586, 6491	4635, 6491
30	60	1243, 2789	1177. 2779
10	30	1220, 2540, 2835	1165, 2560, 2732
	60	1238, 2629	1172, 2580
		4666, 6251	4620, 6285
	90	1231, 2650	1246, 2643
		4556, 6511	4632, 6489

Table 3. Stern Azipod power, thrust, torque and rotation speed data.

Power	Torque	Thrust	RPM
kW	kN	kN	rev/min
1239	125	245	95
1924	167	329	110
3176	233	460	130
4879	311	612	150
5921	353	696	160
6494	376	740	165

$$T_B = K_E\,(P_D.D)^{2/3} \tag{1}$$

where T_B is thrust at bollard pull condition; K_E is quality constant of the bollard pull; P_D is power delivered at bollard pull condition; and D is diameter of the propeller.

3 ICE CLEARING AREA ASSESSMENT

From the sets of image data area of level ice clearance due to propeller jet was estimated. Full scale image data was first modified in accordance to the level of the ice, since the location of camera is not directly above the ice sheet (camera was located on the bridge of the ship) therefore all images were adjusted accordingly with respect to the level of the ice. Ice clearance area was distributed into two regions. In first region there is complete visualization of open water and in second region there is a mixture of both ice and water or floes and water.

Full scale images are converted into binary data and then to RGB image to calculate open water area, mixed area (that is mixture of water and ice) and their maximum distances. Figure 1, shows one of the example of the full scale image and converted image.

The top picture in Figure 1, is of actual image and the bottom picture is converted image. Image in Figure 1, is of an experiment in which average power of Azipod at portside is 1177 kW and on starboard side 1243 kW, moreover pod angle is of 60° and ice thickness is 30 cm. The black region in converted image signifies the geometry of the ship, blue color (dark region) is showing open water and green color (light region) shows mixture of water and ice. It has been observed that, the mixture region consist of ice chunks and fragments that are formed in open water and moves towards the far region and attached to level ice. In addition to it some fraction of the water also flows on the top of the level ice.

In the analysis of ice cleared area formed with the method of propeller jet, particularly those

Figure 1. Conversion of full scale Image to RGB image, showing aft region of the ship, where right side is portside and left is starboard side.

images were consider where there is a complete visibility of the region around the ship. Furthermore, only those experiment were considered that begins with complete level ice. Images begins with zero cleared ice area and as the power of the pod increases, it develops open water region. The rotation speed of the pod remains constant for few minutes and after noticing no increase in area of open water, than rotation speed was again increased and same procedure followed for each experiment.

There is an increase of area as the power of the propeller increases. Figure 2, shows the trend of increasing area with the increment of propeller thrust. Figure 2, indicates the collective results of all the test preformed at 60° pod angles. In 60° pod angle experimental data, level ice has a thickness of 30 cm and 10 cm. It can be seen from the plots that when ice thickness reduces, propeller jet clear more area. In case of starboard side when the ice thickness reduces to 10 cm from 30 cm, there is an increase in area of about 1.25–3.8 times the open water area in comparison with 30 cm thick ice. On the contrary portside increment varies from 2–2.5 times the open water area at 30 cm.

Similarly, calculated thrust and area from image data and pod data for the case of 90° pod angle is illustrated in Figure 3. At 90° pod angle, data have a level ice has a thickness of 50 cm, 40 cm

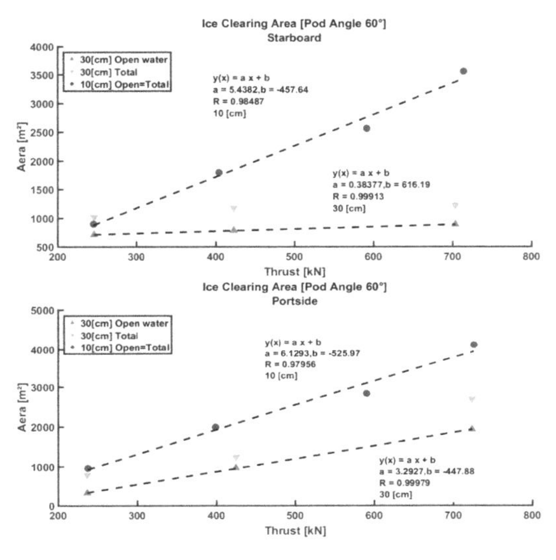

Figure 2. Estimated thrust and ice clearing area from image data at pod angle of 60°. Top plot starboard side and bottom right plot is shows portside.

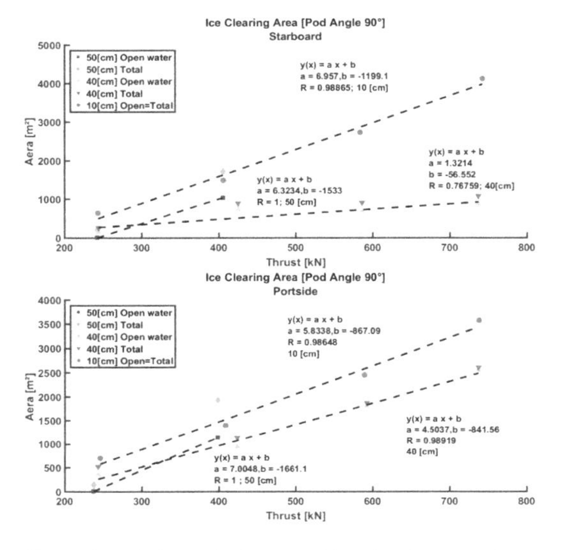

Figure 3. Estimated thrust and ice clearing area from image data at pod angle of 90°. Top plot starboard side and bottom plot is shows portside.

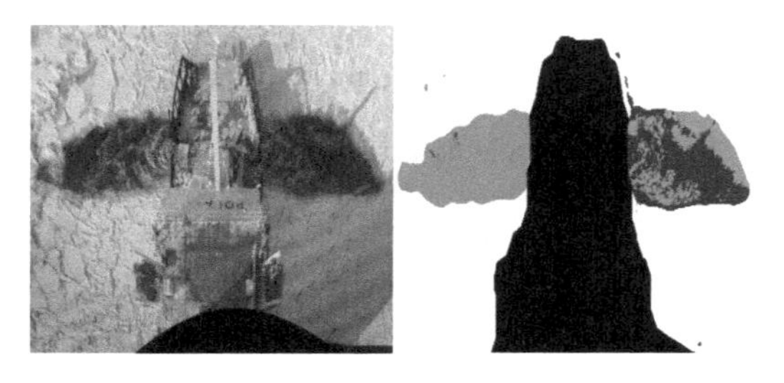

Figure 4. Full scale (left) and converted image (right). Ice thickness of 10 cm and pod angle 90°.

and 10 cm. Although there is overall an increasing trend but the key thing to notice here is that the open water area cover in 50 cm level ice is larger than 40 cm level ice on the starboard side of the ship. On the starboard side, 10 cm level ice clearance area is 2–4 times higher in comparison to 40 cm level ice and 1.5–2.25 times greater than 50 cm level ice. In case of portside level ice clearance area of 10 cm thick ice is 1.25–3 times higher in comparison to 50 cm level ice and 1.5–2 times higher than 40 cm thick level ice.

Looking at Figure 2, and Figure 3, the difference in the starboard and portside could possibly appears due to the tilting of the ship. Heeling of the ship is not recorded while performing experiment. Furthermore, the longitudinal and transversal distance of covered area are not the same for different ice thickness. In 10 cm thick level ice there is no formation of floes displayed in Figure 4, the green region in case of 10 cm thick ice depicts open water. Even though the ice break on both sides but due to formation of foam, the converted image shows green region.

4 AXIAL DISTANCE OF ICE CLEARANCE

Using image data, coordinates of the open water distance and mix water distance from the mid (separating starboard side and portside) location of the ship is measured and polar plots were generated. Polar plots describe the point of the maximum distance which a propeller jet can reach at different angles and directions. Figures 5–14, shows the plots of level ice clearance at approximately constant pod powers and for different ice thickness and pod angles. In Figures 5–14, angles that are mentioned shows the approximate section of pod rotation. Since the pods are few meters away from the center point therefore, the line of pod axis will lie in between the two angles. The configuration of pod is of puller type.

The transversal and longitudinal distance varies with the power and thickness of the ice.

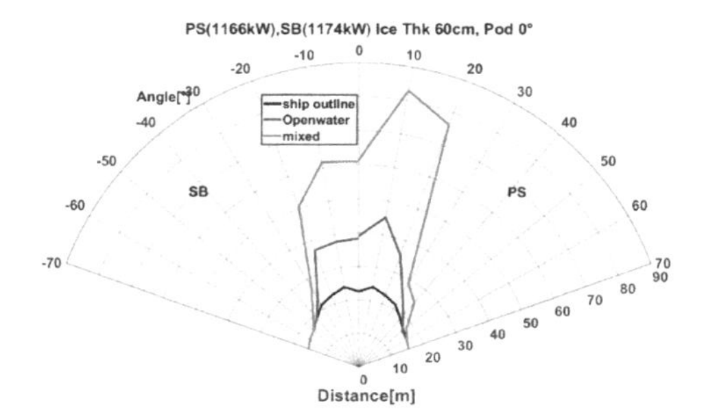

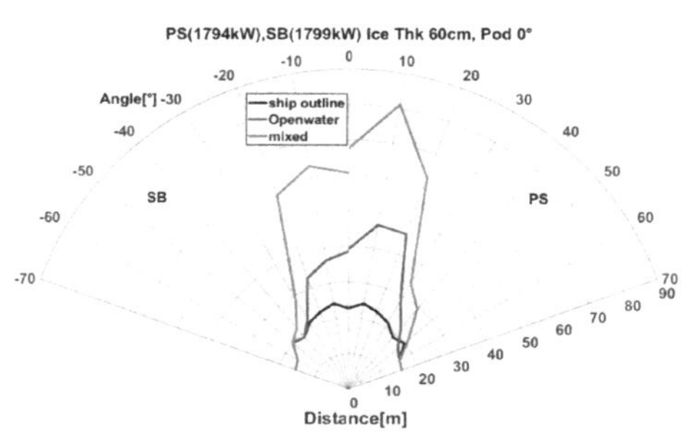

Figure 5a. Level ice thickness 60 cm, 0° pod angle, transversal and longitudinal distance of the open water area. Power of starboard side pod and portside pod is given in the plot.

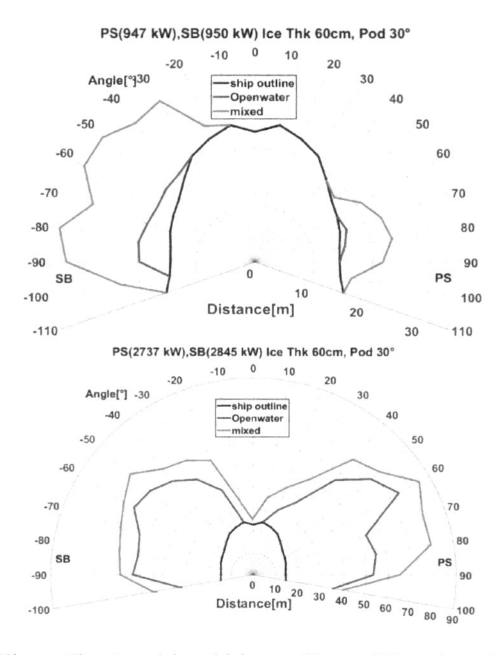

Figure 5b. Level ice thickness 60 cm, 30° pod angle, transversal and longitudinal distance of the open water area. Power of starboard side pod and portside pod is given in the plot.

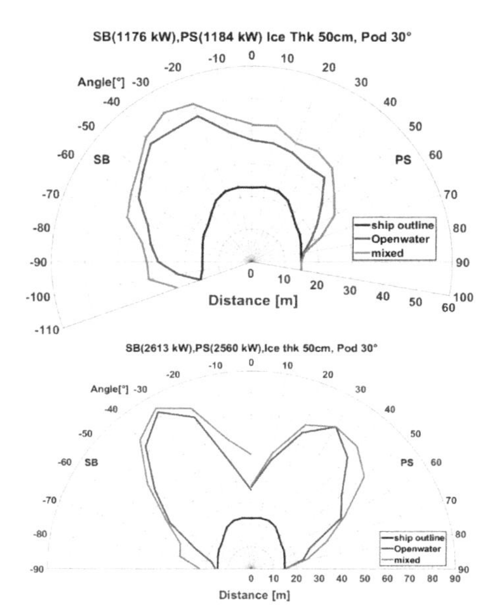

Figure 6a. Level ice thickness 50 cm, 30° pod angle, transversal and longitudinal distance of the open water area.

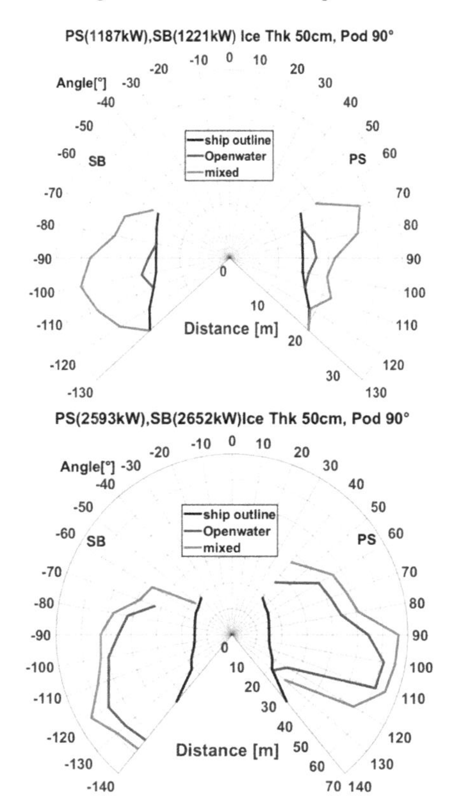

Figure 6b. Level ice thickness 50 cm, 90° pod angle, transversal and longitudinal distance of the open water area.

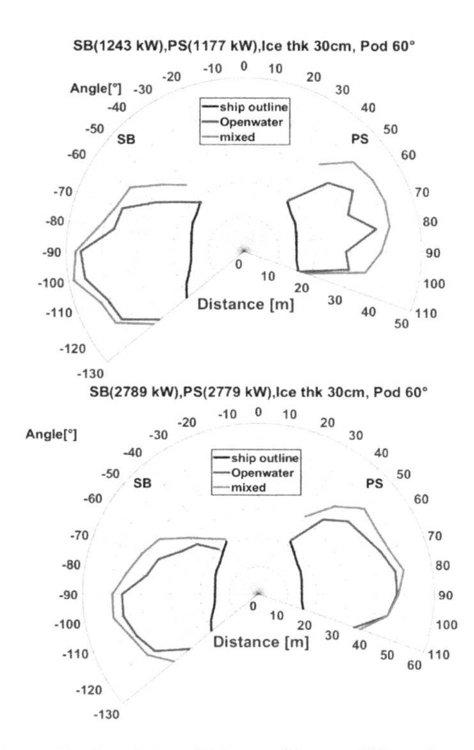

Figure 7. Level ice thickness 30 cm, 60° pod angle, transversal and longitudinal distance of the open water area. Power of starboard side pod and portside pod is given in the plot.

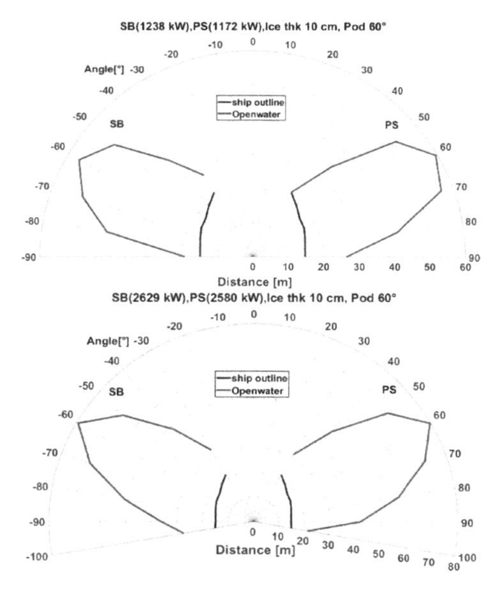

Figure 8. Level ice thickness 10 cm, 60° pod angle, transversal and longitudinal distance of the open water area.

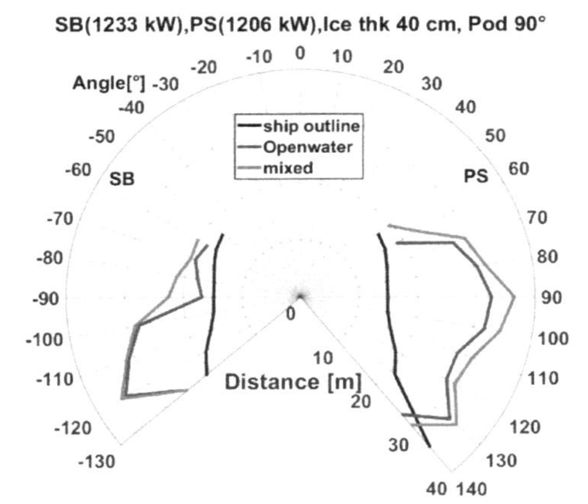

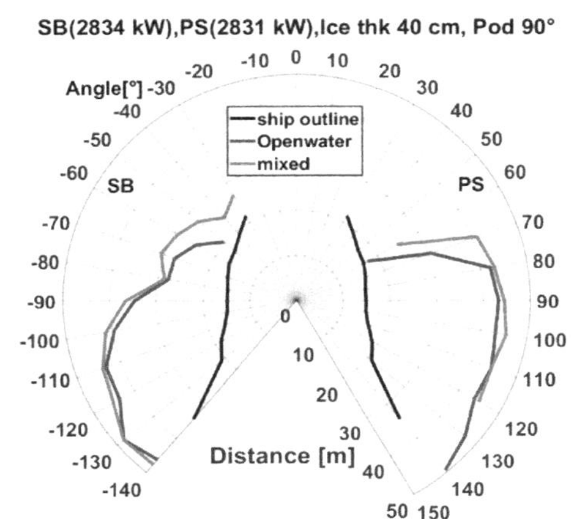

Figure 9. Level ice thickness 40 cm, 90° pod angle, polar plots.

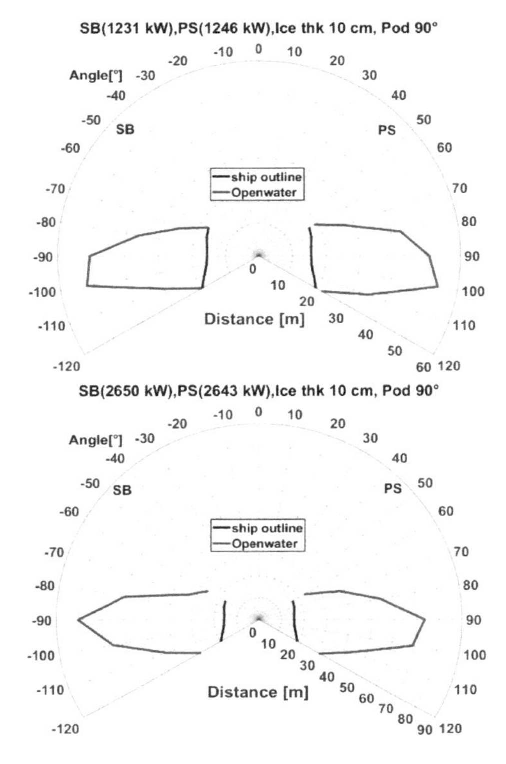

Figure 10. Level ice thickness 10 cm, 90° pod angle, polar plots.

Figure 5a, b gives the plot of ice thickness of 60 cm at pod angle of 0° and 30° respectively. With reference to 0° pod angle, maximum 5% of increase in the distance of portside is observed, while maximum transversal distance remains the same. As far as Figure 5b, is concern, increasing power on starboard side to 2850 kW which is 3 times as much as 950 kW; the longitudinal distance increase to 2 times as much from initial distance, whereas change in maximum transversal distance on starboard side is 1.85 times the initial distance. On the contrary port side parameters vary significantly which may be due to the tilting of the ship.

Figure 6a, b illustrate 50 cm thick level ice results at 30° and 90° pod angle. Increase of maximum axial and transversal distance is observed with increase in power to approximately 2.25 times from the initial value. The variation in longitudinal distance is 1.6–2 times and transversal is 1.1–1.8 times the starting open water distance for 30° pod angle. When pod angle is 90° Figure 6b, maximum axial open water distance ranges from 3.25–3.5 and transversal from 4–5 times the initial value. Figure 7, demonstrates that at ice thickness of 30 cm and pod angle of 60°, there is negligible change in the maximum longitudinal and transversal distance at the border of open water on starboard side. On the contrary portside shows 26% increase in longitudinal maximum distance and 36% increase in maximum transversal distance.

Considering top plots of Figures 7–8, power of the pod is nearly same for portside and starboard side and moving from 30 cm thick ice to 10 cm thick level ice, increase in the axial limit of open water can be observed to 24% and 58% on starboard and portside respectively.

In Figures 9–10, level ice thickness of 40 cm and 10 cm at pod angle of 90°, at different thrust is plotted. Increasing propeller thrust to approximately 1.7 times yields 33% increase in maximum axial distance and 40% in transversal direction.

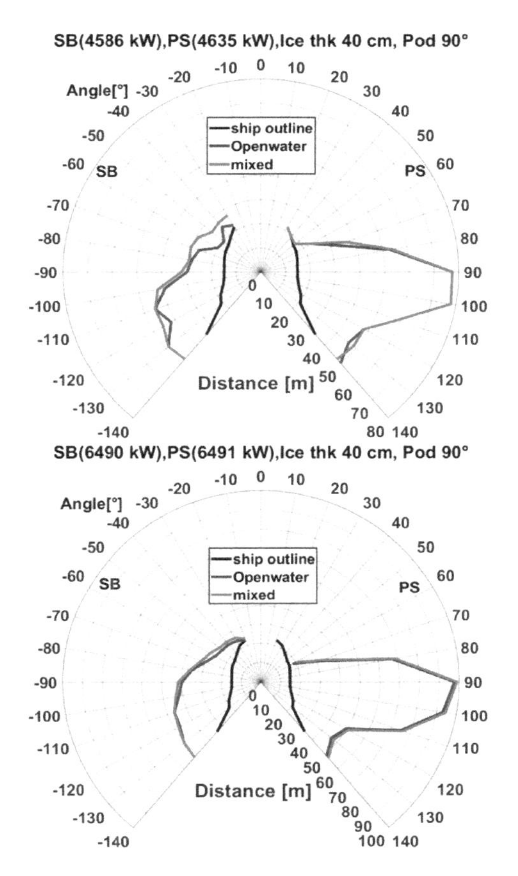

Figure 11. Level ice thickness 40 cm, 90° pod angle, polar plots at higher power.

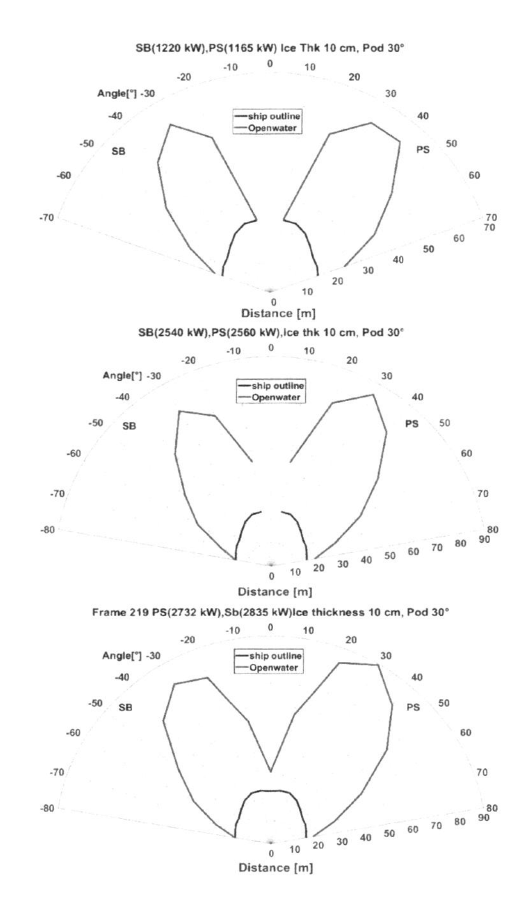

Figure 12. Level ice thickness 10 cm, 30° pod angle.

In case of 10 cm thick ice Figure 9, maximum longitudinal distance and maximum transversal distance increases to 45% and 42% respectively, having thrust to be 1.65 times larger than the first experiment. Furthermore, at nearly same power of propellers comparison of 40 cm and 10 cm thick ice reveals, roughly 50% increase in maximum axial distance of ice cleared open water area.

Rest of the polar plots representing remaining experiment of ice thickness 40 cm and 10 cm at higher power based on Table 2, are presented from Figure 11–13. As can be seen from Figure 11, that there is no symmetry on port and starboard side but increasing trend can be notice.

At the level ice thickness of 10 cm and pod angle of 30°, experiment with three different power was performed. The plot shows butterfly shape profile of the open water area Figure 12. Top Figure 12, is test 1, middle is test 2 and bottom is final test. Increasing the power of the pod from test 1 to test 2, maximum ice clear distance reach is 24% and 26% on starboard side and port side correspondingly. Further increase in the thrust just slightly increase the distance within 6% of the limit reach at test 2. Same kind of change is notice when pod angle is 60° and 90° as demonstrated in Figure 13a, b.

5 ANALYSIS OF PROPELLER JET

Using Propeller thrust it is possible to break ice and to wash away ice rubbles and floes. When ice breaker makes channel in ice it covers more area in comparison with the breadth of the ship. It is due to the fact that propeller jet acts with a force to ice sheet, rubble or floes, which causes ice to break or floes to wash away from the ship. Additionally propeller wash technique is used to break and clear ice away from quay (Tsarau et al. 2016).

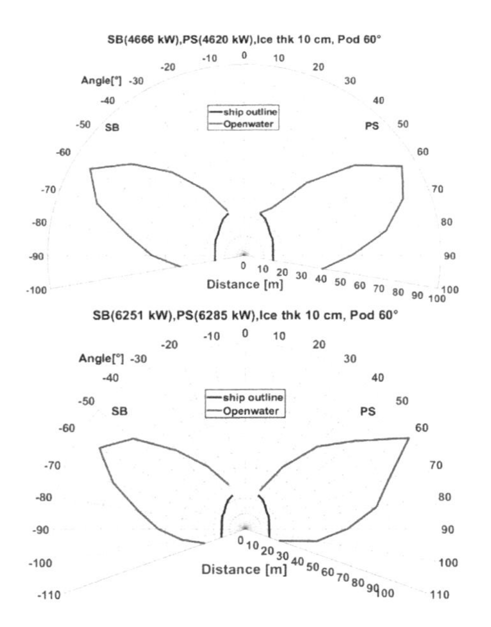

Figure 13a. Level ice thickness 10 cm, pod angle 60° at higher power.

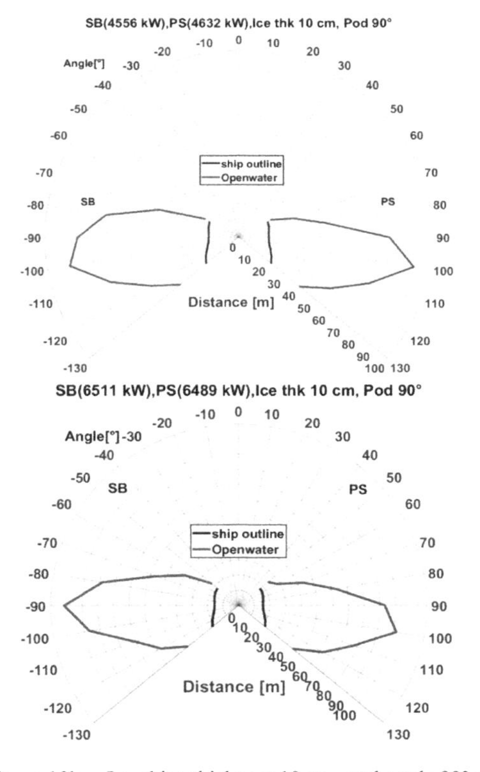

Figure 13b. Level ice thickness 10 cm, pod angle 90° at higher power.

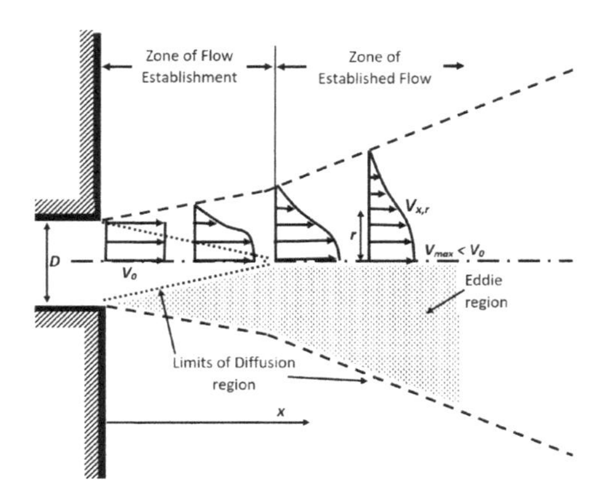

Figure 14. Representation of submerged jet from an orifice (Albertson et al. 1950).

5.1 *Flow features of a propeller jet*

In ice management the first step in defining the propeller flow, is the flow velocity of the jet. The starting point to define velocity field, is to use axial momentum theory which is most commonly used to inspect propeller flows. Recently many authors are using laser Doppler Anemometry (LDA) in their experimental setups to measure velocity of the flow and comparing the results with empirical equations, as well as modifying the equations.

(Albertson et al. 1950) Proposes that slipstream in propeller flow varies very little and same methodology can be implemented in submerged jet flow. In addition to it he also mentioned that velocity profile follows Gaussian normal probability distribution in submerge jets. In submerged flow as the velocity decrease flow area will increase and surrounding fluid will join the flow increasing the amount of total flow.

Fluid will have lateral diffusion and deceleration, moreover instability in the flow arises due to the difference in discharge velocity and surrounding velocity. Representation of jet flow proposed by (Albertson et al. 1950) is shown in Figure 14. Another region after zone of establish flow is fully establish flow, where jet becomes turbulent in nature and surrounding fluid entrainment remains uniform and velocity get reduced.

Propeller generates complex flow which differs from the assumptions of axial momentum theory. There are many factors in propeller that influence the discharge flow. These include the complex geometry of the propeller blade, propeller body and hub diameter which is not accounted in jet flows by (Albertson et al. 1950). (Lam et al. 2011) reviews the equation of propeller jet velocity in a study of sea bed scouring and mentioned that momentum theory assumptions are not effective to describe propeller flow. Moreover, since the major component that contributes most in propeller jet is axial component of velocity, therefore in image data analysis, only axial component is taken into account. Axial component is 10 times the magnitude of other components (PIANC, 2015).

5.2 *Zone of flow establishment*

5.2.1 *Axial efflux velocity equation, V_0*

The maximum velocity taken from a time-averaged velocity distribution along the initial propeller plane is termed as efflux velocity. Two most credible equations used to predict the axial efflux velocity are proposed by Fuehrer & Römisch (1977) Equation 2, and Hamill (1987) Equation 3, as mentioned below:

$$V_0 = 1.59\, n\, D_P \sqrt{C_t} \quad (2)$$

$$V_0 = 1.33\, n\, D_P \sqrt{C_t} \quad (3)$$

where V_0 = axial efflux velocity; n = rate of rotation of propeller in revolution per second; D_P = diameter of propeller in meters; and C_t = thrust coefficient of propeller. According to (Lam et al. 2012), least variation is obtained in comparison to experimental results and equation proposed by Fuehrer & Römisch (1977) Equation 2. Therefore, in calculation of axial velocity action on level ice Equation 2 is used.

5.2.2 *ZFE Axial velocity distribution*

(Albertson et al. 1950) work was modified according to the experimental investigation of propeller flows by many researchers in the zone of flow establishment. McGarvey (1996) derive an equation for axial velocity distribution based on the propeller geometry. In our case we do not have detail characteristics of propeller geometry therefore equation proposed by Hamill (1987) is used to calculate the axial velocity distribution in ZFE.

Hamill (1987) Equation 4, is the modified equation of normal probability distribution function as suggested by (Albertson et al. 1950).

$$\frac{V_{x,r}}{V_{max}} = \exp\left(-\frac{(r - R_{m0})^2}{2\sigma^2}\right) \quad (4)$$

where $V_{x,r}$ = axial velocity at location x,r; V_{max} = maximum axial velocity; R_{m0} = radial distance from propeller axis to location of maximum axial velocity, assumed to be $D_p/4$; σ = standard deviation measured by Hamill (1987) as a constant value and evaluated from Equations 5–6, as follows:

$$\sigma = \frac{R_{m0}}{2} \; for \; \frac{X}{D_p} < 0.5 \quad (5)$$

$$\sigma = \frac{R_{m0}}{2} + 0.075\left(X - \frac{D_p}{2}\right) \; for \; \frac{X}{D_p} > 0.5 \quad (6)$$

where X/D_P = distance from propeller plane. In accordance to experimental data as reported by Hamill & Kee (2016), maximum transition limit from ZFE to ZEF take place at $X/D_P = 3.5$. Maximum axial velocity within ZFE is computed from Equation 7, formulated by Hamill & Kee (2016).

$$\frac{V_{max}}{V_0} = 1.51 - 0.175\left(\frac{X}{D_p}\right) - 0.46P' \quad (7)$$

wehre P' = pitch to diameter ratio.

Calculated and assumed parameters of Azipod propeller are presented in Table 4, due to lack of availability of propeller characteristics pitch to diameter ratio is assumed as 1.

5.3 *Zone of established flow*

5.3.1 *Maximum axial velocity decay*

In the region of zone of establish flow decay of maximum velocity is quick because there is only external entrainment of the fluid in this region and internal diffusion is not taking place. For this region velocity distribution proposed by (Berger et al. 1981); Equation 8, Hashmi (1993); Equation 9, and Hamill & Kee (2016); Equation 10, is taken into account for calculating velocity decay in the region of ZEF.

$$\frac{V_{max}}{V_0} = 1.025\left(\frac{X}{D_P}\right)^{-0.6} \quad (8)$$

$$\frac{V_{max}}{V_0} = 0.638\, e^{-0.097\frac{X}{D_P}} \quad (9)$$

$$\frac{V_{max}}{V_0} = 0.96 - 0.039\left(\frac{X}{D_p}\right) - 0.344P' \quad (10)$$

Table 4. IB Polaris, approximated propeller parameters.

Propeller diameter, D_P	4.2 m
Hub diameter, D_h	1.4 m
Blade number	4
Pitch ratio, P'	1
Thrust coefficient, C_t	0.3

5.3.2 *Axial velocity distribution*

On the basis of different propeller investigation by Hamill & Kee (2016), Equation 11 proposed by Fuehrer & Römisch (1977) agrees within good approximation of experimental results.

$$\frac{V_{x,r}}{V_{max}} = \exp\left(-22.2(\frac{r}{x})^2\right) \tag{11}$$

where r = the varying radial distance from propeller line of axis; and x = the distance from propeller plane.

Full scale image study (action of propeller jet on level ice)

In image analysis vertical force acting on the level ice sheet is calculated using equation 12, suggested by (Tsarau et al. 2016). When the jet strikes the level ice it produces vertical force on the ice sheet. When the resultant vertical force is sufficiently high it causes bending failure and it breaks the ice. Propeller jet strikes the ice sheet with an angle as illustrated in Figure 15.

$$F_v = \rho sin^2\alpha \int V^2\, dS \tag{12}$$

where F_v = vertical force on ice due to propeller jet; V = jet axial velocity; α = angle between jet axis and water plane; ρ = density of water; and S = area of the ice sheet.

In estimation of vertical force, area 'dS' is considered as the open water area.

To calculate the vertical force that is contributing to break the ice; velocity of the flow, jet angle and area of the ice sheet is required. Angle between jet axis and water plane is 5° and the density of water is taken to be 1025 kg/m³.

Equation 4 and 11, defined axial velocity at a distance 'r' from the jet axis. Pictorial representation of the jet axis, ice sheet and varying radial distance from propeller is shown in Figure 16, where radial distance from jet axis to level ice sheet is calculated and then used in equations mentioned above.

From image data area is approximated and limits of integration of area in equation 12 is taken

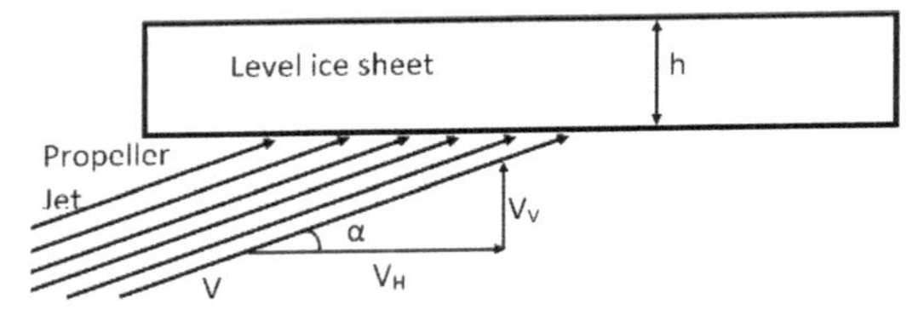

Figure 15. Representation of propeller jet flow on level ice.

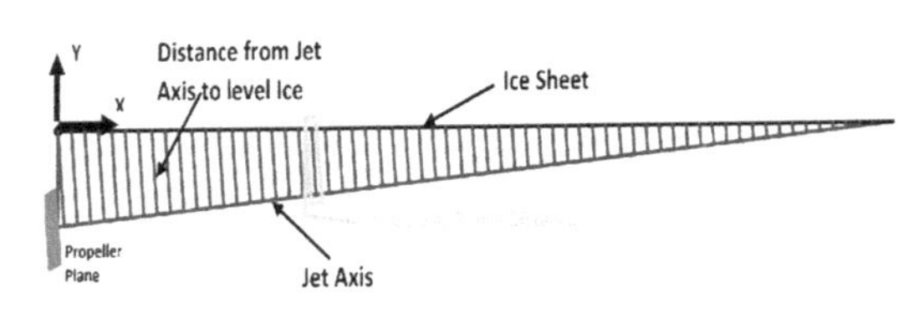

Figure 16. Representation of varying radial distance of velocity from jet axis to ice sheet.

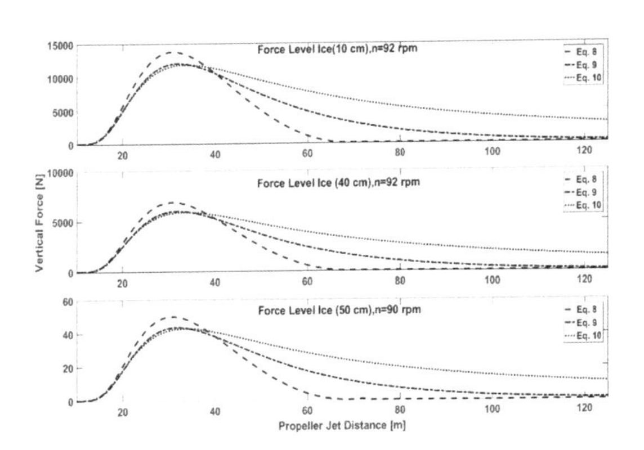

Figure 17. Propeller jet distance and vertical force plot at 90° pod angle having ice thickness of 10 cm, 40 cm and 50 cm.

from the beginning of the experiment to the end of experiment, performed at constant power. Plot of vertical force at 90° pod angle at almost constant propeller rotation speed for three different ice thickness is shown in Figure 17. It can be notice from the plot that maximum fore is acting at a distance of roughly 32 m from the propeller plane. Moreover equation proposed by Hamill & Kee (2016), gives the maximum force, but unable to predict the force after the distance of 70 m, it also contradict with the general trend of decreasing velocity. (Berger et al. 1981) Equation 9, on the other hand decays slowly as compare to the equations mentioned by Hashmi (1993) and Hamill & Kee (2016). After reaching the maximum point all three equations shows exponential decay of vertical force.

From image data the maximum distance of open water for 10 cm thick level ice is 54 m, for 40 cm thick ice open water distance is 32 m and at 50 cm thick level ice zero(since no ice is break), when propeller operates at a rotation speed of 92 rpm. Looking into Figure 17, Table 5, highlights the minimum allowable force required to break level ice as per Equations 8–10.

It has to be noted here that Equation 12, proposed by (Tsarau et al. 2016) is meant for ice floes,

Table 5. Minimum force needed to clear level ice for different ice thickness.

Ice thickness	Force. Eq. 8	Force. Eq. 9	Force. Eq. 10
cm	N	N	N
10	3056	6198	8703
40	6858	5973	5811
50	Not Break	Not Break	Not Break

but in our case we consider open water area to obtain the force.

Only few frames or experiments are consistent in showing icebreaking process from the initial (unbroken ice) to end of experiment (when propeller jet clears the ice and shows open water). In rest of the tests either experiments was started when there is already an open water or angle of the pod changes while the location of the ship remains the same.

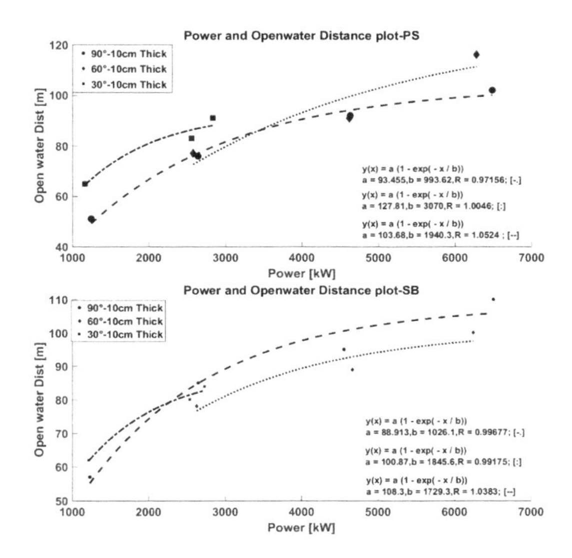

Figure 18. Open water distance and power trend of 10 cm thick level ice, at different pod angles.

6 DISCUSSION

Analysis of the ice clearing area using image data and propeller operation details is reasonable in examining the breaking of level ice and maximum ice clear distance that can be obtained. Based on the Equation 12, increasing the angle between the propeller jet axis and water plane by 1°, will increase the vertical force acting on ice sheet to 56%. Equation 12 that is used to estimate the vertical force on level ice is applicable but required further development in numerical model to predict accurate force on level ice sheet.

Full scale measurement shows that the general trend of power and maximum open water distance is an exponential decay toward a limiting value as illustrated in Figure 18, in which 10 cm thick level ice results are plotted at different pod angles and each experiment shows decay towards limiting value as the power of the propeller increases.

Uncertainties in the result could be mostly related to the heeling of the ship while performing experiment. Wind speed will only influence in drifting of ice floes and broken ice in the direction of the wind. Full scale images shows that wind-direction and speed are not contributing considerably in breaking of level ice. Furthermore, variation will also occur due to the interaction of the two jets with each other and flow intake by propeller because propeller are close to each other. Also, there would be some flow return when jet hits the level ice.

Furthermore approximated propeller geometry parameters are consider in analysis of force acting on ice sheet. Also in some of the images it was noticed that even at close thrust on portside and starboard side, one side clears more ice in comparison to other and this could be due to the tilting or heeling effect of the ship, which is not taken into consideration.

7 CONCLUSION

Starting from the image and power data area of the broken ice, thrust of the propeller, maximum open water and mix water distances of each frames were calculated. It is observed that ice clearing area depends on the power, ice thickness, heel of the ship and angle between jet axis and water plane. The ice thickness that are used in experiments were confirmed from surveillance videos. Only those results are considered to be valid in which experiments starts from the initial state (unbroken complete ice sheet). Polar plots were made which describes the variation of ice clearing area with respect to radius and angles of the region. Vertical force acting on the ice sheet shows exponential decay after reaching the maximum value at a distance of 32 m from propeller plane. Exponential decay to a limiting value is identified from the collection of image data examination. Moreover, Vertical force acting on the ice sheet is calculated from equation proposed by researchers, these results can be compared by doing model testing or simulating propeller jet action to a flat plane representing level ice sheet and analyzing pressure distribution over it. More detail explanation and comments can be made by validating the results using model scale testing.

ACKNOWLEDGEMENTS

This work has been enabled by the funding available in the TEKES project ARAJÄÄ and the support from the participants here gratefully acknowledged i.e. Aker Arctic, ABB Marine, Arctia, Hyötytuuli and Eranti Engineering. In addition Markus Karjalainen from the Finnish Traffic Agency is specially thank to get financial support for the full scale trial.

REFERENCES

Albertson, M.L., Dai, Y.B., Jensen, R.A., & Rouse, H. (1950). Diffusion of Submerged Jets. Transactions of the American Society of Civil Engineers, Vol. 115(Issue 1), 639–664.

Berger, W., Felkel, K., Hager, M., Oebius, H., & Schale, E. (1981). Courant provoquè par les bateaux. Protection des berges et solution pour èviter l' erosion du lit Haut Rhin. Proceedings of the 25th Congress Permanent International Association of Navigation Congresses (PIANC).Edinburgh, Section I-1, 51–69.

Fuehrer, M., & Romisch, K. (1977). Effects of modern ship traffic on islands and ocean waterways and their structures. In: proceedings of 24th Congress P.I.A.N.C., Leningrad, 1977 Sections 1–3

Hamill, G. (1987). Characteristics of the screw wash of a maneuvering ship and the resulting bed scour.. Thesis submitted to the Queen's University of Belfast for the degree of Doctor of Philosophy.

Hamill, G., & Kee, C. (2016). Predicting axial velocity profiles with in a diffusing marine propeller jet. Ocean Eng.

Hamill, G., Kee, C., & Ryan, D. (2015). 3D Efflux velocity characteristics of marine propeller jets. Proceedings of the ICE-maritime Engineering 168(2), 62–75.

Hamill, G., McGarvey, J., & Hughes, D. (2004). Determination of the efflux velocity from a ship's propeller. Proceedings of the Institution of Civil Engineers: 83–91.

Hashmi, H. (1993). Erosion of a granular bed at a quay wall by a ship's screw wash. Thesis submitted to the Queen's University of Belfast for the degree of Doctor of.

Juva, M., & Riska, K. (2002). On the power requirement in the Finnish-Swedish ice class rules. Winter navigation Research Board, Res. Rpt. No. 53, Helsinki, 80.

Lam, W., Hamill, G., Song, Y., Robinson, D., & Raghunathan, S. (2011). Review of the equations used to predicts the velocity distribution within a ship's propeller jet. Ocean Eng. 38 (1), 1–10.

Lam, W., Hamill, G., Robinson, D., Raghunathan, S. & Song, Y., (2012). Analysis of 3D zone of flow establishment from a ship's propeller. 465–477.

McGarvey, J. (1996). The influence of the rudder on the hydrodynamics and the resulting bed scour, of a ship's screw wash. Thesis submitted to the Queen's University of Belfast for the degree of Doctor of Philosophy.

PIANC. (2015). Guidelines for Protecting Berthing Structures from Scour Caused by Ships (ReportNo.180). Maritime Navigation Commission, ISBN: 978-2-87223-223, 9.

Stewart, D. (1992). Characteristics of a ships screw wash and the influence of Quay wall proximity. Thesis submitted to the Queen's University of Belfast for the degree of Doctor of Philosophy.

Tsarau, A., Lubbad, R., & Løset, S. (2016). A numerical model for simulating the effect of propeller flow in ice management. Cold Regions Science and Technology.

Marine Design XIII – Kujala & Lu (Eds)
© 2018 Taylor & Francis Group, London, ISBN 978-1-138-34076-3

Removable icebreaker bow with propulsion

H.K. Eronen
ILS Ltd., Turku, Finland

ABSTRACT: The development of removable icebreaker bow with propulsion also in the bow, not only in the pusher vessel, is described in the paper. Also a summary of removable bow icebreaking vessels used in the past is given. The concept development and model tests done in connection with WINMOS I and WINMOS II projects as well as the pilot vessel project for Lake Saimaa icebreaking are described. The different alternative propulsion systems are summarized.

An analysis and detailed description of the selected propulsion system (pat.pend.) with conventional shafts at the reamer area of wider bow is given. Performance comparison between the removable bow icebreaker and conventional icebreaker is also included and economic feasibility of the concept is summarized.

1 INTRODUCTION

1.1 *Development of icebreaking devices in front of a ship*

Using removable icebreaking bow in front of a vessel is an old invention. In 1893 a patent was granted to Weedermans invention of an icebreaking device in front of a ship, see Figure 1. This type of flat bow was used e.g. in Dutch waterways. The bow device was attached to the conventional cargo vessel bows to brake ice.

Also bow extension type which bends ice upwards was developed in 1960'. So called Alexbow and versions of it were tested in Canada and Soviet Union without great success.

To test spoon shaped icebreaking bow shape Wärtsilä built in 1985 an experimental removable bow which is still used successfully in Lake Saimaa icebreaking. The 11.1 m wide bow is attached to 1A ice class single screw tug Protector in winter time, see Figure 2.

Descriptions of different removable bow concepts are given e.g. in Johansson et al. (1994) and Jones (2008). A concept of icebreaker attachment with aft azimuth thrusters is presented in Gulling et al. (1997).

1.2 *WINMOS studies*

The further development of the old idea has taken place in connection with two WINMOS projects: Winter navigation motorways of the sea, a co-operation program with the aim to create infrastructure for future winter navigation is co-financed by the European Union (Trans European Transport Network, TEN-T).

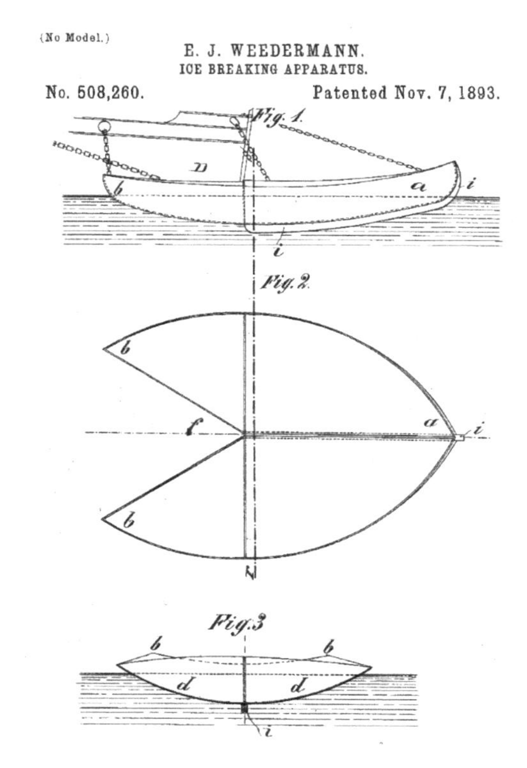

Figure 1. German patent 1893 of icebreaking bow to be placed in front of a ship.

In WINMOS projects, one of the many activities has been the concept study on the next generation of icebreakers. In this context, also removable bow icebreaker has been evaluated, in WINMOS I

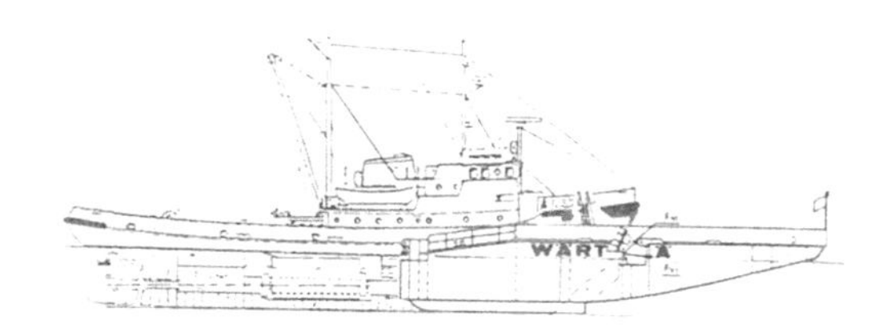

Figure 2. Spoon-shaped removable Wärtsilä bow used with tug protector in Lake Saimaa as an icebreaker.

Baltic icebreaker version and in WINMOS II Lake Saimaa version pilot vessel. The development of the concept was done by ILS Ltd in cooperation with Finnish Transport Agency.

The basic idea in these studies has been to have propulsion also in the removable bow, i.e. to develop a removable bow icebreaker based on the use of a pusher vessel of smaller breadth and lower power than conventional icebreaker and connect it to a removable bow with propulsion. The pusher can be e.g. an existing vessel or a newbuilding optimized for its primary tasks. This way a vessel with wider hull with optimized icebreaker bow shape and higher power is created.

2 FEASIBILITY STUDY OF BALTIC REMOVABLE BOW ICEBREAKER

2.1 *Basic requirements*

In WINMOS I one of the studied next generation icebreaker designs was the removable bow Baltic Icebreaker. As starting point typical Gulf of Finland Icebreaker design requirements were used:

- Minimum speed of 6 knots in 0.8 m level ice ahead to guarantee high enough escort speed
- Total propulsion power about 11 MW (pusher+ removable bow)
- Waterline breadth of removable bow about 24 m
- Fixed coupling of the pusher and removable bow
- The ice strengthening of removable bow and its propulsion to correspond present Baltic icebreakers
- The pusher ice class has to be minimum 1A Super

The selected requirements gave also a good basis for the comparison of the performance and economy of the concept with existing smaller Baltic icebreakers, e.g. Botnica and Tor Viking classes.

As said, the design is based on moderate sized pusher with power less than required for an icebreaker. Thus the removable bow has also own propulsion. If strong enough supply ship is used as a pusher unit, it is possible to have a removable bow without propulsion. Also if smaller and less efficient icebreaker is designed, the pusher power may be enough.

2.2 *Requirements for the pusher*

Based on the general Gulf of Finland icebreaker requirements the main requirements for the pusher are:

- High enough ice class
- Suitable size and draught
- Propellers protected from excessive ice loads
- High bollard pull
- Propulsion power over 4 MW
- Stern shape suitable for backing in ice or possibility to install removable stern section for icebreaking
- Deck arrangement giving possibility to install removable towing notch
- Suitable towing winch
- Accommodation large enough for icebreaker crew
- Preferably high fuel oil capacity

Many ice strengthened vessel types can be used as a pusher including at least:

- Tugs
- Harbor icebreakers
- Supply/offshore ships
- Fairway maintenance vessels
- Environmental protection vessels
- Heavy ice going Coast Guard patrol vessels
- Navy ice going service vessels§
- Ice going research vessels

2.3 *Studied bow propulsion alternatives*

The design is based on DE-propulsion in removable bow either with azimuth thrusters or conventional shafts. The diesel electric propulsion was selected because in all alternatives ice torque in propellers is high and DE-propulsion gives also high flexibility and efficiency in icebreaker use.

Based on about 4 MW propulsion power of the pusher the removable bow should have minimum 6 MW propulsion power.

The studied propulsion alternatives for the removable bow were, see also Figure 3:

- 1*6 MW pulling azimuth thruster at bow
- 2*3 MW pulling azimuth thrusters at bow
- 2*3 MW conventional shafts at bow
- 2*3 MW conventional shafts at stern reamer areas

6 MW propulsion power at the bow means that a quite large power station is needed in the bow with generators and switchboards, propulsion control systems, cooling, FO, LO, fire, ballast, bilge,

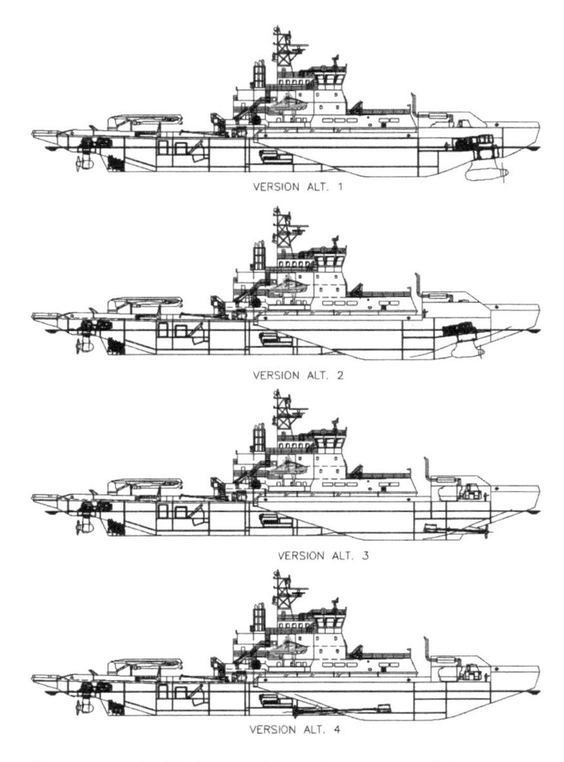

Figure 3. Studied propulsion alternatives of the removable bow.

control etc. systems. Thus, the removable bow is designed with its own diesel generators, machinery and pump rooms.

2.4 *Model test design*

2.4.1 *Selected pusher type*

To give realistic basis for the design M/S Louhi was used as an example pusher. It is a multipurpose icebreaking environmental control and transport vessel built in 2011. It fulfills well the given requirements and thus gives a good basis for the study. The 4 generator DE machinery gives flexibility in icebreaker use. The azimuth thruster propulsion gives maneuvering, backing and ice management power. However the presented design can be easily adjusted according any suitable pusher vessel requirements and even (by rebuilding the aft part of the removable bow) the bow section could be relatively easily changed to suit for another pusher if necessary.

M/S Louhi main characteristics:
Loa 71.4 m
Bwl 14.5 m
Tdesign 5.0 m
DWT 1100 t

- Ice class 1ASuper with ice pressures used are over rule requirements
- Diesel electric azimuth thruster propulsion
- Main generator power, total 7.2 MW
- Propulsion power 2*2.7 MW
- Bollard pull 64 t
- Stern shape designed for backing well in ice

Ice performance:

- 1 m level ice ahead

2.4.2 *Removable bow design*

Generally, the azimuth thruster alternatives give the best maneuvering capability and best performance in heavy ice conditions. However, in this concept the novel propulsion alternative with 2 × 3 MW conventional shafts at reamer area was selected to model tests.

This makes possible to have optimized icebreaking bow shape without limitations caused by propellers in front. Propellers at reamer area give also potentially high turning moment and good flushing effect of the propeller flow at vessel sides. The investment cost of the removable bow with conventional shaft propulsion is as well roughly 15% lower than with azimuth thrusters.

The removable bow hull shape developed is based on modern icebreaker bow shape with moderate frame angles to achieve good channel performance and strongly inclined sides and bottom rise towards sides this way minimizing the "reamer" step between wide bow section and pusher.

The bow section draught was selected as about 6 m based on icebreaking, displacement, space and shaft arrangement requirements.

2.4.3 *General arrangements and main dimensions of the concept*

M/S Louhi with removable bow is shown in Figure 4.

The main dimensions of Louhi, removable bow and combination are given in Table 1.

2.5 *Model tests*

2.5.1 *General*

The ice model test were done in Aker Arctic model basin in Helsinki (Forsén A., 2015a, unpublished).

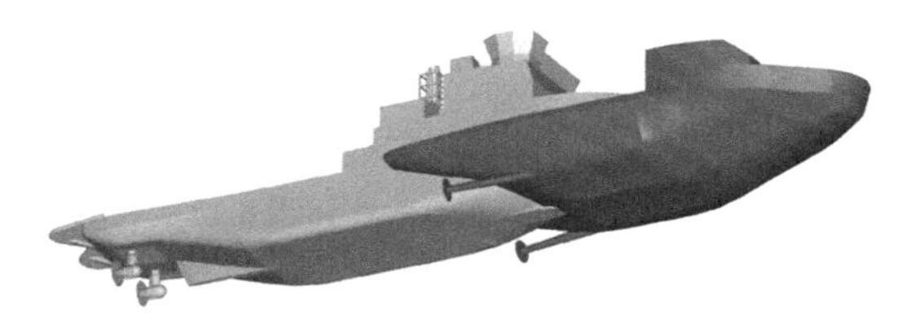

Figure 4. M/S Louhi with removable bow.

Table 1. Main dimensions for the new concept.

	Louhi	Removable bow	Combination
L	71.4 m	56.1 m	95.8 m
Bm	14.5 m	25.1 m	25.1 m
Bwl	14.5 m	24.0 m	24.0 m
T	5.0 m	6.0 m	6.0 m
H	7.0 m	7.9 m	7.9 m

The target of ice model tests was to study the characteristics of removable bow icebreaker concept generally compared to conventional icebreaker designs and to test the novel propulsion arrangement.

Special importance was to test the new operational characteristics to verify how well they work.

In heavy channels and ridges, the bow propellers have direct interaction with ice. Also the suction and flush effect of the bow propeller flows can be used to break heavy channels/ridges. This, together with the wide smoothly shaped reamer bow should make it easier to go through heavy ridges continuously without ramming. Also by turning the pusher steering units from side to side this effect can be further added. In maneuvering the use of the high steering moment, given by bow propellers at sides when driving them unsymmetrically or to opposite directions, together with wider removable bow adds vessels maneuvering performance.

In ice model test, special attention was also paid to how much ice is going to bow propellers in different ice conditions and how it influences the performance. Also of interest was the influence of novel removable bow concept to ice interaction of the pusher propulsion.

Normal open water model tests were done as well. The main target was to measure values needed for ice model test analysis. Open water propulsion test were done also because it is important to estimate the power needed at transit speed for this kind of special design.

2.5.2 *Ice model test results*

Generally it can be said that the reamer hull form with propellers in reamer area worked very well.

Level ice tests ahead were done in 0.6 m and 0.87 m thick ice fields. In level ice tests ahead the ice interaction of bow propellers was quite high, on the other hand less ice is coming to pusher propellers compared to situation where pusher is going in ice alone.

The speed prediction ahead in 0.8 m level ice was 5.7 knots. Botnica and Tor Viking classes have a speed of about 8 knots in the same level ice thickness (Nyman et al., 1999 and Riska et al., 2001). Level ice speed astern in 0.6 m level ice was 6.6 knots, which is almost the same as e.g. Botnica. In level ice astern, there was only quite small ice interaction with the removable bow propellers.

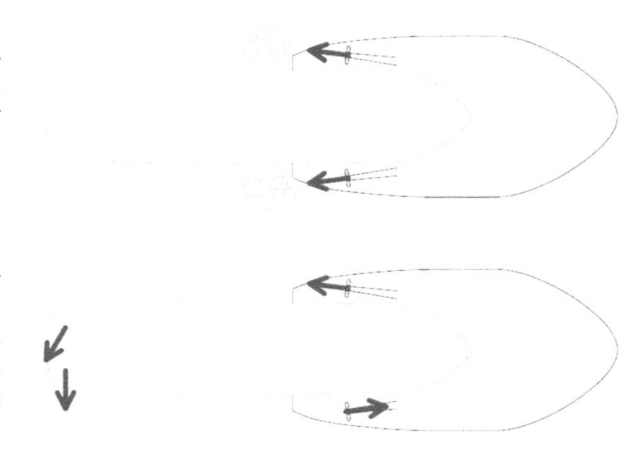

Figure 5. New operational characteristics in ice.

Ridge tests were done in so called Urho ridge (5–8 m consolidated ridge) and in even about 6 m thick consolidated ridges. The ridges represent well big ridges in the planned operation area.

Ridge performance was excellent, meaning that the vessel could go directly through the ridges with initial ramming speed of 8 knots, by turning aft thrusters from side to side when the vessel stopped. Bow propellers direct interaction with ice and the suction and flush effect of the propeller flow broke efficiently ridge and flushed the vessel sides and pushed the ice floes aft. The vessel was able to penetrate also astern the 5.7 m ridge with one attempt.

Channel tests were done in 1.2 m consolidated channels both ahead and astern and also 1.8 m unconsolidated channel ahead. The speeds in 1.2 m channel were 12.8 knots ahead and 11.5 knots astern

Thus, it can be concluded that the speed in channels was high, however the tested channels were light so direct comparison of heavy channel behavior is difficult. Test in box shaped ridge with 6 m even thickness however showed the capability of side propellers to break and flush heavy channel sides.

Maneuvering capability was at the same level as that of multipurpose icebreakers with azimuth thrusters aft and better than conventional icebreakers. The vessel was able to turn in practice on spot in tested 0.6 m thick level ice. The use of the high steering moment (given by bow propellers at sides when driving them unsymmetrically or to opposite directions) together with wider removable bow with inclined sides, worked well making the excellent maneuvering performance possible. Bow propeller flow lifts water onto the ice surface which also breaks ice.

Although channel widening tests were not carried out, it can be considered that because the bow propellers break efficiently heavy channel at bow area sides, the broken ice floes can be pushed aside quite effectively by pusher azimuth units when they are directed outwards.

Concerning the open water test results, it can be mentioned that the max speed of the tested removable bow icebreaker is over 16 knots and the power needed in open water was up to 15 knots, which is at the same level as in Atle/Urho class icebreakers.

2.6 *Economical feasibility*

The investment costs of the removable bow is about 25% of the price of a purpose built icebreaker of same capability making it as an interesting alternative if icebreaking period is limited.

The pusher vessel needs only connection system, control and alarm systems of bow on pusher bridge and relatively small adjustments to able to operate as a pusher in this concept. The removable bow concept has no practical requirements for the hull form and bow shape of a pusher, only the ice class of the vessel must be adequate.

Especially interesting alternative is to use as a pusher existing vessel, which has limited use in wintertime.

3 PILOT VESSEL FOR LAKE SAIMAA

3.1 *General*

In WINMOS II one of the activities is to study, develop and build a pilot removable bow icebreaker with propulsion to Lake Saimaa.

Lake Saimaa is connected to Gulf of Finland by Saimaa channel, which dictates the maximum dimensions of the cargo vessels, which can sail to this large lake area. The Saimaa channel is closed every winter due to difficult ice conditions. Before closing and also in spring when channel is opened icebreaker assistance is needed and e.g. removable bow with a pusher tug shown in the Figure 2 has been used there. For many years, also a purpose built lake icebreaker M/S Arppe was working in Saimaa as icebreaker and also as a pusher tug for cargo barges (Kivimaa & Jalonen, 1995).

Ice conditions in the area are quite heavy. Maximum level ice thickness in Lake Saimaa is about 0.6–0.7 m and 1.5 m thick brash ice conditions are common in channel areas. There are also areas with limited water depth of 5.1 m and width of 34.0 m making progress in brash ice even more difficult. The removable bow concept developed for this purpose is of similar type as the Baltic version. The purpose of the Lake Saimaa removable bow pilot project is to test the concept not only for Lake Saimaa conditions but also to work as 1:2 scale model of a Baltic icebreaker. Thus, the full scale ice tests will include also tests in the Gulf of Finland where also ridges and heavy channels and channel sides etc. can be found. The tug Calypso, which will operate as a pusher, has been already contracted for ten years period by Finnish Transport Agency and the building of the removable bow with propulsion bow is starting in spring 2018, see Figure 6.

3.2 *Pusher tug Calypso and removable bow main data*

The main dimensions of Calypso, removable bow and combination are shown in Table 2.

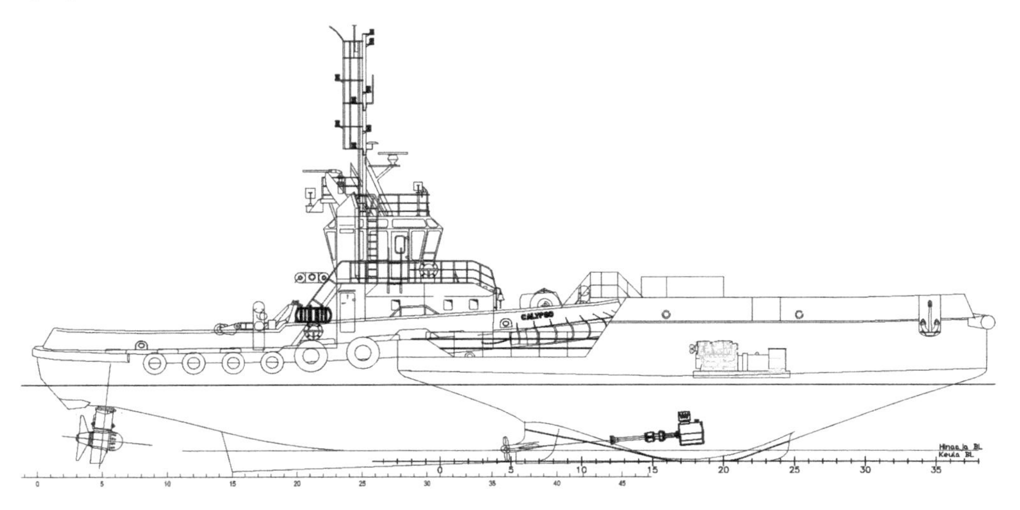

Figure 6. Pusher Calypso with removable bow.

Table 2. Main dimension of the new design for the Lake Saimaa.

	Calypso	Removable bow	Combination
L	25.9 m	25.3 m	40.8 m
Bm	8.95 m	12.6 m	12.6 m
Bwl	8.8 m	12.4 m	12.4 m
Tm	3.0 m	3.4 m	3.4 m
Propulsion power	2 × 700 kW	2 × 600 kW	2600 kW
Ice class	1A	Modified 1A Super	

M/S Calypso has diesel propulsion with two azimuth thrusters aft. For Lake Saimaa use, the propulsion power is limited to 1400 kW and instead of propeller in nozzles the thruster lower parts have been changed to open propeller units.

On the bridge of Calypso bow propulsion and machinery and other control, alarm and monitoring systems will be installed, because the removable bow is unmanned in icebreaker use.

The removable bow has DC-grid diesel electric machinery with two 730 kW diesel generators and also a small auxiliary engine. The electric motors for propulsion are of permanent magnet type with rpm control and high over torque capacity. The bow has also big ballast tanks and own fuel oil tank so that it can be loaded properly and deep enough in all loading conditions.

The removable bow and pusher are connected together at one level with fixed pin at centerline and screw pins at sides at the aft end of reamers. The system can be manually operated, because the connection and opening take place only twice a year.

3.3 *Preliminary model tests*

3.3.1 *General*

Before selection of the pusher tug and final design of the bow, ice model tests were done in Aker Arctic model basin for the preliminary version of the Lake Saimaa removable bow icebreaker to develop and verify the suitability of the concept for Lake Saimaa icebreaking (Forsén A., 2016b unpublished).

The pusher was in these early tests a single screw tug with about 1100 kW propulsion power and the removable bow had two 600 kW conventional shafts with diesel electric drive.

3.3.2 *Preliminary model tests results*

At 70 cm level ice the speed ahead was 3.5 knots. This corresponds to speed of about 4.6 knots in 60 cm level ice. Thus, speed ahead fulfills the speed requirement for class I Saimaa icebreaker which is 4–5 knots. In 40 cm level ice the speed was 7 knots. It was also noted that the influence of power distribution between bow and tug propulsion had only marginal influence on speed. Compared to purpose built lake icebreaker, it can be mentioned, that M/S Arppe has a speed of little under 2 knots in 70 cm ice.

The speed in 1.5 m thick consolidated brash ice was 7.1 knots and correspondingly 6.3 knots in 1.5 m thick consolidated brash ice channel where the water depth was limited to 5.1 m and breadth of channel to 34 m by boards installed in the ice tank. The speed in channel was thus clearly higher than the predicted value of 4.5–5 knots i.e. very good like was already noted in WINMOS I model tests. Also, the turning capability by using the bow propellers thrust to opposite directions was good.

However, there was quite a lot of ice interaction in the bow propellers in level ice. On the other hand, the pusher propellers had only minor ice interaction. The propeller ice interaction is in Lake Saimaa always a problem due to high ice thickness compared to vessel size. For example, lake icebreaker Arppe conventional diesel propulsion with a single screw in nozzle was changed in renovating project of the vessel to diesel electric machinery and open propeller.

Based on the initial model test results, the removable bow has been modified by adding draught, and by developing hull shape and skeg/forefoot arrangements to reduce the ice interaction with bow propellers. Also, because M/S Calypso is quite different from the original tug tested, a full set of model tests both in open water and ice will be done in spring 2018 for the built pilot version of the concept.

4 CONCLUSIONS

The development of removable bow icebreaker concept with propulsion, led to the invention of novel propulsion system for an icebreaker with conventional shafts installed on reamer area of the wider bow.

The concept is economically attractive when the icebreaking period is limited. The price of removable bow is at the level of 25% of the price of a purpose built icebreaker. The pusher can be existing ice going vessel optimized for other tasks than icebreaking. Only small adjustment like connection system and control and alarm systems for bow are needed in the pusher.

In WINMOS I and II projects the concept was developed and model tested both for Gulf of Finland icebreaker and Lake Saimaa versions.

In the ice model tests, the speed in level ice ahead was in Lake Saimaa version higher and in Baltic version lower than the speed of corresponding conventional icebreaker types. Quite high ice interaction with bow propellers was noted in level ice tests ahead. This led to modifications of the bow hull shape, draught and skeg arrangement of the lake Saimaa pilot bow. On the other hand advantageously only minor ice interaction with pusher propellers was noted in ice model tests done.

The concept showed very good performance in ice channels and in ridges. The reamer propellers flush and relief efficiently the ice floes in heavy ice conditions. Thus, the speed in channels was high and also the penetration of a quite heavy ridges was possible without ramming. The high steering moment of bow propellers made it possible to achieve also excellent maneuvering performance even with conventional propulsion with single rudder in the pusher.

Based on the developed concept, a pilot removable bow is built for Finnish Transport Agency for Lake Saimaa icebreaking with tug M/S Calypso contracted to operate as a pusher. Verification of the pilot project will be done in ice tests in Lake Saimaa and also in Gulf of Finland. This way the results can also be scaled to Baltic icebreaker size.

ACKNOWLEDGEMENTS

This work has been enabled by EU-TNT funding through the projects WINMOS I and WINMOS II, which is here gratefully acknowledged.

REFERENCES

Forsén A., 2015a. Ice model tests for a vessel with a self propelling removable icebreaking bow. *Report A-539*. Helsinki. Unpublished.

Forsén A., 2016b. Ice model tests for a vessel with self propelling removable icebreaking bow for Lake Saimaa. *Report A-551*. Helsinki. Unpublished.

Gulling D.L. et al., 1997. Icebreaker attachment. *United States Patent 5660131*.

Johansson, B.M. et al., 1994. Revolutionary icebreaker design. *Icetech '94*. Calgary: O1-O4.

Jones, J., 2008. A history of icebreaking ships. *The Journal of Ocean Technology. Ocean Sovereignity V3*. No. 1:58.

Kivimaa S. & Jalonen R., 1995. Itämerellä liikennöivien pienalusten jäissäkulkuominaisuuksien kartoitus. Espoo: *VTT tiedotteita* 1717: LIITE B 1–3.

Nyman, T. et al., 1999. The ice capability of the multipurpose icebreaker Botnica full scale results. *POAC 99*, Helsinki. 631–643.

Riska K. et al., 2001. Ice performance of the Swedish multi-purpose icebreaker Tor Viking II. *POAC 01*, NRC. Ottawa.

Marine Design XIII – Kujala & Lu (Eds)

Azimuthing propulsor rule development for Finnish-Swedish ice class rules

Ilkka Perälä
VTT Technical Research Center of Finland, Espoo, Finland

Aki Kinnunen
Vibrol Oy, Tuusula, Finland

Lauri Kuuliala
Trafi, Finnish Transport Safety Agency, Helsinki, Finland

ABSTRACT: With increasing number of azimuthing propulsor installed on ships, azimuthing propulsor ice class rule development for Finnish-Swedish Ice Class Rules (FSICR) started in 2010. In the beginning of the development project, suitable load scenarios for azimuthing propulsors were identified and applicable ice load calculation methods were reviewed. The two main load scenarios for the propulsors were considered to be impact to floating ice block and penetration to ice ridge. The impact load scenario drove the development of a new dynamic impact load model for Azimuthing units. This impact model was tested against small-scale controlled tests on sea ice and also validated with selected full-scale measurements. The ice ridge interaction load scenario was studied with state of the art FE-model of ice ridge and ship interaction. The scientific approaches yielded solid physical background for formulating simplified load estimation formulas for the load scenarios. In addition to the main load scenarios, also impact load distributions were investigated.

The whole project lasted from 2010 to 2016, and it had a steering group consisting of classification societies as well as azimuthing propulsor manufacturers. The work was reviewed and guided with the experts in the steering group. The results of the work are being included as a part of new version of Finnish-Swedish ice class rules.

1 INTRODUCTION

1.1 *Historical background*

The azimuthing propulsors begun to appear on the ice going vessels during the 90's. One of the first ones was the Finnish icebreaker Fennica, which has Aquamaster azimuthing propulsors with ducted propellers. Fennica was constructed in 1993.

Other notable ice going vessels with azimuthing propulsors include Icebreaker Nordica (1994), Icebreaker Botnica (1998) and MS Norilskiy Nickel (2006).

With increasing amount of ice going vessels being constructed with azimuthing propulsors, the classification societies begun to address the importance of specific ice class rules for this kind of propulsors. During the late 2000's the classification societies started to develop these rules. Models and calculation approaches were develop at least by DNV (Classification Notes 2010), ABS (Daley, 1999, Daley & Yu, 2009) and Russian Register. Trafi (Finnish Transport Safety Agency) and SMA (Swedish Maritime Administration), who are responsible for the Finnish-Swedish ice class rules (FSICR), wanted to take action on the matter too. VTT, who has been leading partner with Trafi to develop the propulsion machinery rules for the FSICR, was chosen to do the research and development work. The focus was on the thruster body loads. Propeller ice loads had been extensively researched during the '90 s and they were already implemented to the rules.

1.2 *Trafi and VTT – Azirule project*

During 2010 Trafi and VTT begun to develop the Finnish-Swedish ice class rules for azimuthing propulsors. These rules are intended for ships operating in the Baltic Sea. For this project all major classification societies and industrial players were invited to work as a steering group to comment on the progress of the work. Most of the rule development work was carried out by VTT.

In the beginning of the project, all the available load models were reviewed and compared (Tikanmäki, 2010). These included models from ABS, DNV and the first version of the VTT load model. There were some differences in the chosen approaches and thus on the levels of loads. From the beginning, it was clear that the VTT approach was to find realistic loads combined with some safety margin.

Active development and research phase lasted from 2010 to 2015 after which the rules were commented by other parties. The project finally culminated in 2017 when the new FSICR were released. These new rules now include ice loads and design criteria for azimuthing propulsors.

The following chapters describe the research carried out in the Azirule project.

2 LOAD SCENARIOS AND APPLICABLE METHODS

2.1 *Load scenarios*

The first task in the beginning of the rule development work was to define the relevant load scenarios for azimuthing propulsors. The basis for this was to consider what would happen when the ship moves in ice. What kind of loads would the propulsor encounter when the ship is moving ahead or moving astern in ice?

Clearly an impact type scenario was one to be considered. In this scenario an submerged ice block will collide with the propulsor. Depending on the angle and type of the propulsor, the ice block can collide to propeller, propeller hub or thruster housing. All these scenarios were considered important.

Another scenario is the ramming of ice ridge. In this scenario the vessel rams into an ice ridge in a way that the propulsors come into contact with the ridge. See Figure 1 for the considered load scenarios.

The ice block impact scenario differs from the ice ridge ramming scenario in many ways and thus two different methods were chosen to find the suitable loads for these scenarios. These are described in the following chapters.

2.2 *Basic impact load model*

To formulate the load equations and criteria for the rules, an impact model was developed for the ice block impact scenario.

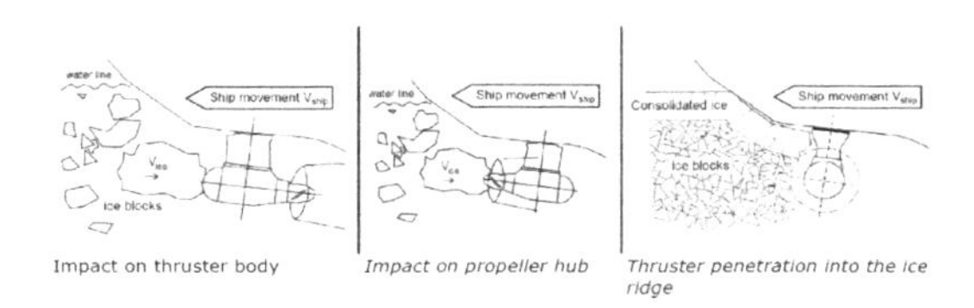

Figure 1. Ice load scenarios for azimuthing propulsors.

This impact model is based on the dynamic behaviour of the thruster. It takes into account the mass of the thruster body, mass of the ship and mass of the ice block. In addition, the ship velocity, thruster structural stiffness, structural damping and ice material parameters are given as input parameters. See Figure 2 for the description of the model.

In this load model the impact is energy limited. The limiting energy is the kinetic energy of the ice block. For example, a ship moving with speed v and colliding with ice block would equal a case where ship with zero speed would be impacted with an ice block with speed v.

The contact force F_c between the thruster body and the ice block is based on indentation pressure-area relationship and is determined by equation 1.

$$F_c = p_0\sqrt{A \times A_0} \tag{1}$$

In equation 1, A is the indentation area which is determined in Figure 4. A_0 is the refer-

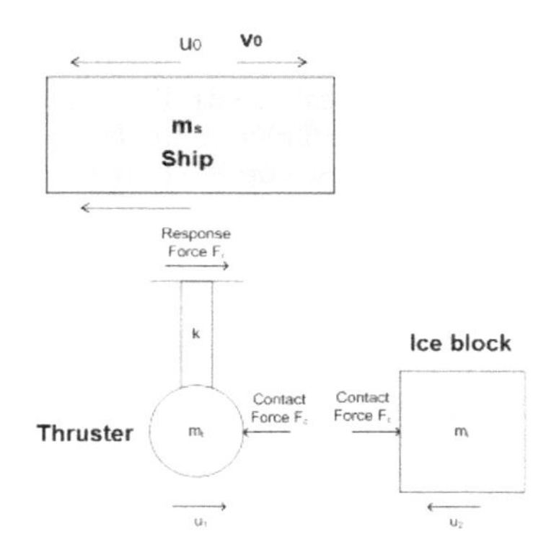

Figure 2. The ice impact model.

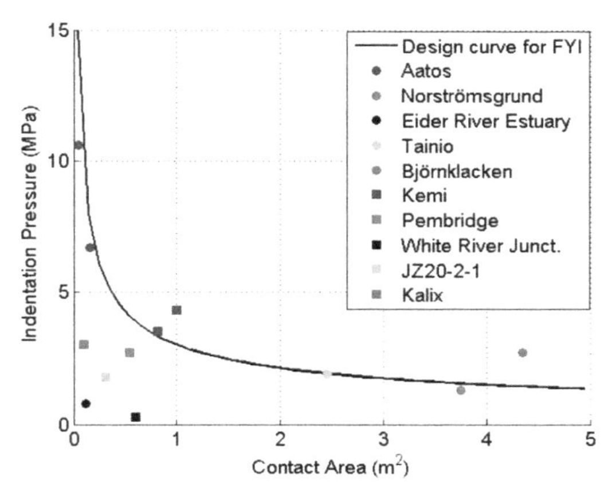

Figure 3. Pressure-area relationship (Tikanmäki, 2010).

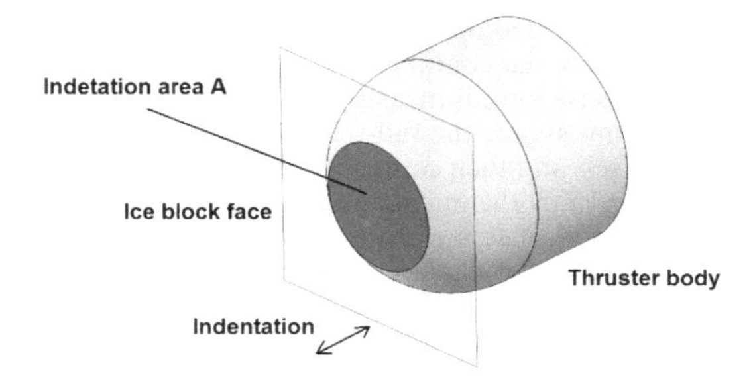

Figure 4. Thruster and ice block impact geometry.

ence area, usually 1 m^2. p_0 is the reference pressure which can be taken as the ice compressive strength. Thruster contact surface is assumed to be spherical. In Figure 3, some of the data used for determining the pressure-area relationship is presented.

The model is solved with difference method in the time domain. The main result is the response force of the thruster body which can be used for dimensioning of the structure. For more detailed description of the model, see references (Kinnunen 2013, 2014, 2016).

2.3 *Ridge penetration FEM studies*

The ship interaction with ice ridge was studied with finite element model simulation. The simulation setup consisted of a rigid ship structure and deformable and destructible model for ice ridge.

The FE model for ice ridge is based on a continuum material model representing the characteristics of consolidated ice ridge. The theoretical background was established by Heinonen (Heinonen, 2004).

The model in this case was set up with rigid ship structure, having initial velocity, mass and buoyancy effects. The ice ridge was modeled with a half-circle surrounded by a boundary elements representing the half-infinite boundary conditions. The simulations were run with ABAQUS Explicit in a VTT simulation workstation(s). The model simulated the interactions on the ship structure surface mesh, from which the net force effects acting on the thruster were extracted for analysis. The simulations were rather time consuming, some lasting several days per run. Figure 5 shows the ship and ice ridge model.

Figure 6 shows force results of ridge penetration study for case where the thruster was turned 90 degrees prior to ridge penetration. It can be seen how the force levels rise with increased vessel speed.

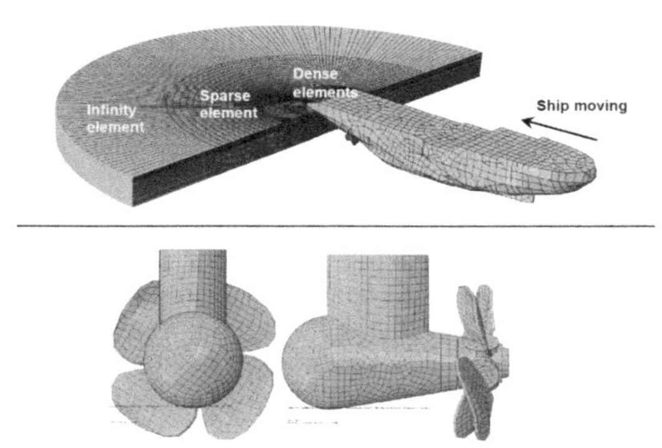

Figure 5. FEM model for the ice ridge penetration study (Kinnunen 2015).

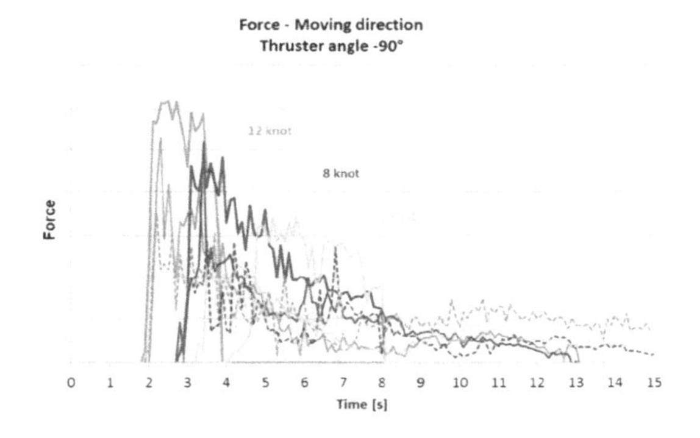

Figure 6. Ice ridge penetration study, results for 90 deg thruster cases (Kinnunen 2015).

3 VALIDATION OF MODELS

With the models developed to simulate the impact and ridge penetration, they had to be validated with tests. For the impact model, a small-scale test setup was build. In addition, some full-scale data was available for the validation.

3.1 *Small-scale tests on natural ice*

The small-scale setup consisted of a pendulum mass with a changeable impact head. The mass was divided into larger mass and a smaller mass to represent mass of the ship and thruster. The idea was to study the dynamic effects, like the vibration of the thruster, during the impact. See Figure 7 for the test setup. The impacted ice blocks were sawn from Baltic sea ice and they were floating in water during the impact tests.

In the tests the ice block and the indentor were both instrumented and thus the contact force and the response force could be measured. The small-scale tests were used to validate the simulation model. See Figure 8. for comparison of simulation and test. F_c and F_r are contact force and response force respectively. Measured values represent the response force.

3.2 *Available full-scale data*

For validation full-scale data from Finnish icebreaker Fennica was availaible for analysis. Fennica azimuthing thrusters had instrumentation and thus the response force of the thruster body was available.

For ridge penetration case there was specific data where the icebreaker was driven into the ice ridge with known speed. The ridge size was determined with measurements. This data was compared to the FEM studies.

The impact model validation with Fennica full-scale data was more difficult. This is because there is no knowledge on the size of the impacting ice block. Also, the contact force is not known. Only the response force is measured. Thus, the approach was to investigate the full-scale data for maximum force levels and then compare these to the response loads given by the impact model. It was concluded that the full-scale measurements were in line with the load levels calculated with the impact model.

The impact load model was also used during full-scale thruster impact tests in laboratory conditions (Perälä, 2017). The tests had scaled down ice blocks which were impacted with the thruster hub cap. The thruster body and ice block were instrumented. Also in these tests the model showed good correlation with measured response force.

Figure 7. Small-scale test setup for impact tests (Kinnunen 2014).

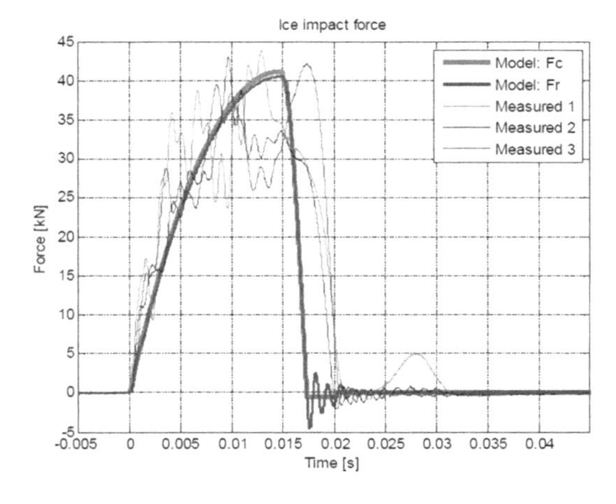

Figure 8. Model scale test results compared to simulations (Kinnunen 2014).

4 IMPLEMENTATION TO RULES

4.1 *Science to simplified rule formulation*

The models developed and validated for the impact load scenario and for the ice ridge interaction scenario were used to simulate several cases with varying input parameters. The parameter study was done to understand the effect of the parameters to the maximum ice load for the thruster.

For the ice impact load scenario, the parameter variation included the impact speed, ice mass and thruster dimension. The time domain simulation model was run with a wide matrix of the input parameters and the maximum contact load output was recorded to correspond the input parameters.

For the ice ridge interaction load scenario, the study was done with ABAQUS explicit, where the ice ridge was modeled, and one ship hull form was used. The variable parameters for the simulations were the ship speed prior to interaction, thruster size, ice ridge size, and thruster orientation. The maximum contact force acting on the thruster was taken from these simulations to represent the load level for the input parameter set used.

Simplified formulations for the load scenarios were drafted, containing the input parameters and the contact load maximum dependency type on the input parameters. These model formulations together with the input- and output values were used with a set of free parameters in nonlinear regression model fit algorithm. The model fitting procedure gives proposals for the model parameters, and also provides estimate on how well the fitted model predicts the output loads based on the input values.

From the nonlinear model fit, the numerical parameter values were truncated to engineering accuracy of one or two decimals. The simplified formula model at this level of finesse was compared with the results from the simulations, with more precise parameter values model, and with experiment and full scale results. Once the accuracy of the simplified formulation was deemed acceptable, it was considered good enough for public review by the steering group and conference publications.

The key idea for the simplified models was to use easily available input parameters, that can be well understood and applied. These parameters now include the thruster dimension, ice dimension and ship speed.

In the new Finnish-Swedish ice class rules, the formulated equations and design criteria can be found in chapter 6.6.5 "Azimuthing main propulsors" (FSICR 2017).

Chapter 6.6.5.2 "Extreme ice impact loads" gives the equation for the impact load calculation. See equation 2 for the exact formula.

$$F_{ti} = C_{DMI} 34.5 R_C^{0.5} \left(m_{ice} v_s^2\right)^{0.333} \quad (2)$$

F_{ti} gives the force acting on the thruster in kN. C_{DMI} is the dynamic magnification factor, R_c is the impacting part sphere radius, m_{ice} is the ice block mass and v_s is the ship speed. Recommended values can be found in the rule text.

Chapter 6.6.5.3 "Extreme ice loads on thruster hull when penetrating an ice ridge" gives the equation for the ridge penetration load calculation. See equation 3 for the exact formula.

$$F_{tr} = 32 V_s^{0.66} H_r^{0.9} A_t^{0.74} \quad (3)$$

F_{tr} gives the force acting on the thruster in kN. A_t is the projected area of the thruster, H_r is the design ridge thickness and V_s is the ship speed. Recommended values can be found in the rule text.

4.2 *Steering group guidance and industry comments*

As mentioned, during the whole process, the steering group was kept informed on the progress, and the methods and results were discussed openly. At this point rule harmonization was not carried out, but all invited classification societies could express their opinion on the proposed rules. This way the rules were kept in line with other ice class rules.

When the rules had been formulated, opinions of manufacturers from industry were also asked. Many adjustments, corrections and clarifications were carried out as aftermath of these comments.

4.3 *Legislative considerations – Trafi*

Trafi was involved in the project for the whole duration as the aim was to amend the Finnish national ice class regulations. Trafi coordinated exchange of information also with the Finnish Transport Agency, which is responsible for the operational part of winter navigation in Finland, as well as the corresponding Swedish agencies: Transportstyrelsen and Swedish Maritime Administration. Similar amendments will also be made to the Swedish national ice class regulations to ensure the consistency of the Finnish-Swedish winter navigation system.

The new rules considering the dimensioning of Azimuthing propulsors are different in nature than previous parts of the FSICR. Only the dimensioning loads and safety factors are given and all technical details are left to the discretion of the Classification Societies and designers. In any case, the FSICR is implemented mainly through the rules of Classification Societies and it was felt that this approach gives a stabile basis for design of ships while providing the safety level that Finnish and Swedish authorities deem necessary for the smooth operation of the winter navigation system in the Baltic Sea. This way, the national rules will not hinder the adoption of new technologies and advances in thruster design will not necessitate rule changes.

5 CONCLUSION

The development of azimuthing propulsor body ice loads for Finnish-Swedish ice class rules was carried out during 2010–2015. The new rules, with added chapter for azimuthing propulsors, were published in 2017 by Trafi.

The need for the new rules came from increasing amount of ice going ships with azimuthing propulsors. With prior experience on propulsion and machinery ice load rule development, VTT carried out most of the studies and development tasks.

The development of azimuthing propulsor ice loads was divided into two parts, ice impact loads and ridge penetration loads. For ice impact loads a model was developed were the impact is solved in time domain and which takes into account the dynamic behavior of the thruster. This model was studied and validated with small-scale tests with natural sea ice. The ice ridge penetration was studied extensively with FEM models. With varying ridge thickness, vessel speed and thruster size, load levels were established. These load levels were validated against full-scale data.

Finally, simplified equations for determining the loads were established for the rules. With the new rules ice loads for azimuthing propulsor body can be calculated with easily available parameters.

REFERENCES

Daley, C. 1999. Energy based ice collision forces. Expanded version of Proceedings of the 15th

International Conference on Port and Ocean Engineering under Arctic Conditions (POAC 99). Espoo, Finland August 23–27, 1999.
Daley, C. and Yu, H. 2009. Assessment of ice loads on stern regions of ice class ships. International Conference on ship and Offshore Technology. Busan, Korea September 28–29, 2009.
DnV Classification Notes No. 51.1 (2010).
Heinonen, Jaakko. 2004. Constitutive modeling of ice rubble in first-year ridge keel. Espoo, VTT Building and Transport. 142 p. VTT Publications; 536 ISBN 951-38-6390-5; 951-38-6391-3 http://www.vtt.fi/inf/pdf/publications/2004/P536.pdf.
Kinnunen, A., Lämsä, V., Jussila, M., Koskinen, P. 2013. Determining of ice loads for azimuthing propulsion units. Proceedings of the Dresdner Maschinenelemente Kolloquium – DMK 2013, 3. and 4. December 2013, Dresden, Germany.
Kinnunen, A., Tikanmäki, M., Koskinen, P., 2014. Ice-structure impact contact load test setup and impact contact load calculation. Proceedings of the 22nd IAHR International Symposium on Ice, Singapore, August 11 to 15, 2014.
Kinnunen, A., Kurkela, J., Juuti, P. 2015. Azimuthing thruster ice load calculation and simplified ice contact load formulation. VTT Research report, VTT-R-00258-15.
Kinnunen, A., Tikanmäki, M., Heinonen, J., 2016. An energy model for ice crushing in ice-structure impact. Proceedings of the 23nd IAHR International Symposium on Ice, Ann Arbor, May 31 to June 3, 2016.
Perälä I., Kinnunen A., Koskinen P., Heinonen J., 2017. Full-scale ice impact to an azimuthing thruster in laboratory conditions. Proceedings of the 24th International Conference on Port and Ocean Engineering under Arctic Conditions June 11–16, 2017, Busan, Korea.
Tikanmäki, M., Heinonen, J., Kinnunen, A., 2010, Comparison of ice load models for azimuthing thruster ice load calculation, VTT Research report VTT-R-10310-10, 25 pp.
Trafi, 2017. Ice class regulations 2017 "Finnish-Swedish Ice Class Rules 2017", Trafi/494131/03.04.01.00/2016.

Marine Design XIII – Kujala & Lu (Eds)
© 2018 Taylor & Francis Group, London, ISBN 978-1-138-34076-3

A method for calculating omega angle for the IACS PC rules

V. Valtonen
Aker Arctic Technology Inc, Helsinki, Finland

ABSTRACT: In design of ice going vessels, the ice load has been traditionally idealized as a horizontal rectangular pressure patch. When the pressure patch is applied on transversally framed side structure, the load is shared by several frames, whereas on longitudinally framed structure, the load is carried by one or two frames. Therefore, the traditional method has been to use one design formula for transverse framing and another one for longitudinal framing. These formulas work very well for parallel midbody region, but for the shape regions in bow and stern, the actual case is neither purely longitudinal nor transverse. Moreover, in some designs, frames are canted instead of transverse or longitudinal. In development of the IACS PC rules, these needs have been taken into account by introducing angle omega, which considers the relative geometry between load patch and framing more accurately than previously used division into transverse and longitudinal framing. However, no way for calculating this angle has been provided in the rules and while approximate angle can be estimated visually or graphically, formulas for calculating the angle would be useful. In this paper, formulas for calculating the omega angle for all typical framing solutions and general hull geometry are presented.

1 INTRODUCTION

The traditional idealization for ice load in design of ice-going vessels has been rectangular pressure patch. Typically, this patch is wide in horizontal direction and narrow in vertical direction, as used in most ice class rules, for example (IACS, 2011), (TraFi, 2010) and (RMRS, 2016). Some studies have further suggested that the load approaches line-like distribution with very narrow vertical extend (Riska, 1991). Therefore, the classical approach has been to use separate formulas for design of transversally and longitudinally framed structures, as the load patch is oriented differently relative to the structure, as illustrated in Figure 1.

As the load is distributed on several frames on transversally framed case and only on one or couple of frames, the individual frames on longitudinally framed structure have to be stronger for equivalent ice load. Equivalent geometrical effect affects individual plate panels as well, leading to greater part of longitudinally framed plate panel being loaded and thus to a thicker plate for longitudinally framed case.

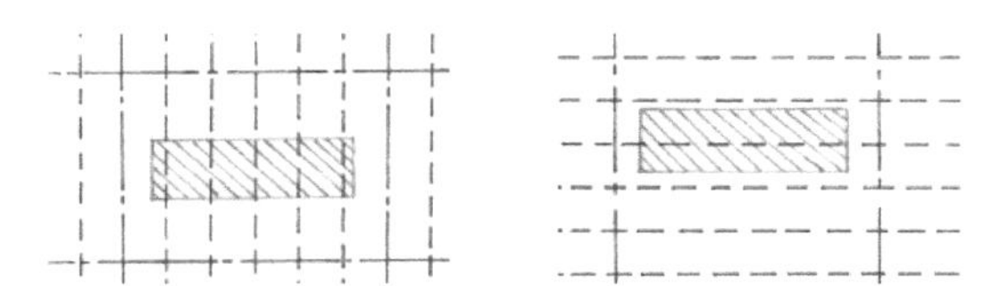

Figure 1. Typical design ice load patch on transversally and longitudinally framed structures.

However, these idealized cases apply only when considering structure where the angle between waterline and the plane of the frame is 90°, i.e. in parallel midbody or pure landing craft bow. For all other shapes, the actual angle between the waterline and frames, noted omega, is not 90° corresponding to the classical transversally framed case nor 0° corresponding to the classical longitudinally framed case, as illustrated in Figure 2.

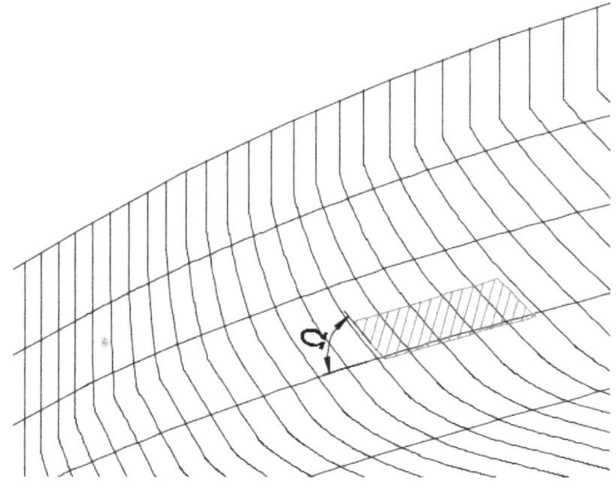

Figure 2. Typical design ice load patch on transversally framed icebreaking bow.

For these regions in bow and stern parts of the vessel, the classical approach has been to use either the formula for transverse or longitudinal case according to the judgement of the designer, as those have been the only tools available. For a very traditional icebreaking bow with relatively small waterline angle of about 20° to 30° and moderate frame angle of about 30° to 50°, this has led to acceptable compromise as the omega angle is around 65° to 75° and transversally framed formula can be still considered reasonable approximation.

For vessels with more modern bow forms with improved icebreaking capability, the hull angles are typically larger, and the approximation may not be any longer sufficiently accurate. Moreover, correct interpretation of the relevant formula to be used in each case has been left up to the consideration of designer and class society. This has led to inconsistent design practices, and, in some cases, incorrect design.

In the IACS PC Rules (IACS, 2011), this issue has been partly solved by introducing the omega angle, which is the angle between frames and waterline when viewed normal to the shell, as shown in Figure 3. The scantling calculation is then done according to transverse case for $\Omega \geq 70°$, longitudinal case for $\Omega \leq 20°$ and with linear interpolation for $20 < \Omega < 70°$. This enables the designer to consider the orientation of the load patch relative to the framing correctly and therefore allows for to more accurate calculation of scantlings.

However, no formulas or tools for calculating this angle have been provided in the rules or public sources, most likely because such tools have not been developed. In principle, it would be possible to determine the angle graphically, but this approach has not been approved by certain class societies. That has led to incorrect formulas being applied in certain cases, leading to incorrect design.

Therefore, accurate and easy to use formulas to calculate the omega angle for typical structures would be needed to carry out the design work in proper manner and to ensure consistent design practices. In this paper, formulas for calculation of

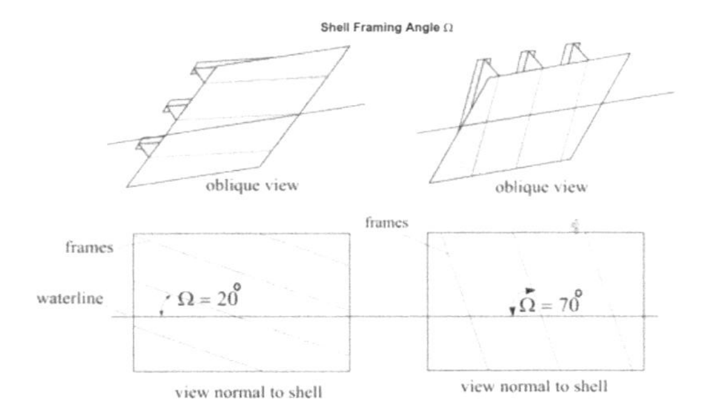

Figure 3. Shell framing angle (IACS, 2011).

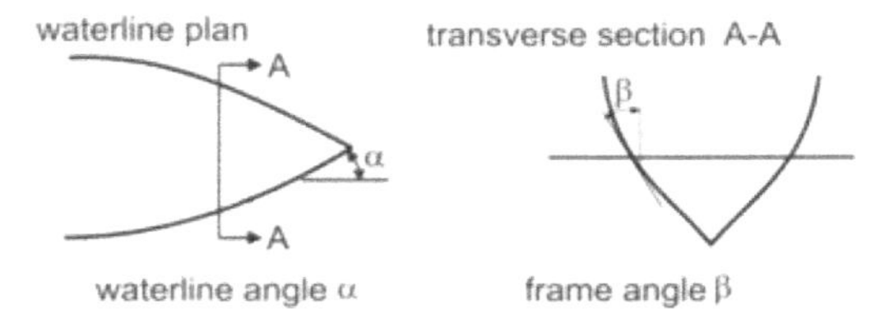

Figure 4. Definition of hull angles (IACS, 2011).

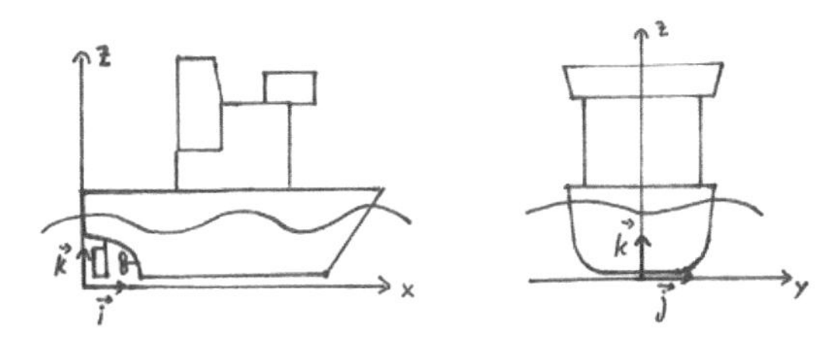

Figure 5. Coordinate system and unit vectors along each coordinate axis.

omega angle are developed and presented, starting from the simple cases of transverse and longitudinal framing systems, and then expanding the same concept for a general case. The notations will be similar to the PC Rules, most importantly the hull angles will be defined in similar fashion, as shown in Figure 4.

Ship coordinate system and unit vectors along these coordinates will be defined as shown in Figure 5.

2 FORMULA FOR TRANSVERSALLY FRAMED STRUCTURE

First, a simple case of transversally framed structure is considered. First, two unit vectors are formed as shown in Figure 6:

$\vec{a} = \cos\alpha\vec{i} - \sin\alpha\vec{j}$, unit vector tangent to the intersection of hull and waterline

$\vec{b} = \sin\beta\vec{j} + \cos\beta\vec{k}$, unit vector tangent to the intersection of hull and frame

The dot product for vectors is defined by

$$\vec{a} \cdot \vec{b} = |\vec{a}||\vec{b}|\cos\Omega \quad (1)$$

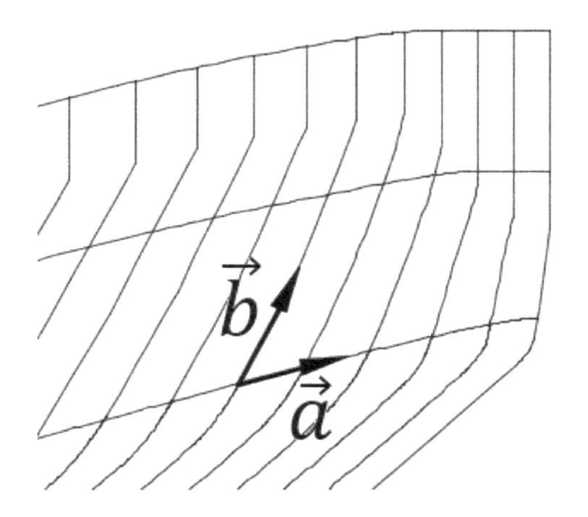

Figure 6. Unit vectors $\vec{a}$ and $\vec{b}$ on transversally framed hull.

This can be rearranged to calculate the angle between two vectors, which for these vectors happens to be the angle omega

$$\cos\Omega = \frac{\vec{a}\cdot\vec{b}}{|\vec{a}||\vec{b}|} \qquad (2)$$

Since the length of unit vector is by definition always 1, this simplifies to

$$\cos\Omega = \frac{-\sin\alpha\sin\beta}{1}, \qquad (3)$$

which can be further arranged to

$$\Omega = \cos^{-1}(-\sin\alpha\sin\beta) = \pi - \cos^{-1}(\sin\alpha\sin\beta) \qquad (4)$$

For the final formula for transverse case, the smaller one of the two possible angles given by the inverse cosine function for the angle between the frame plane and waterline is taken

$$\Omega = \cos^{-1}(\sin\alpha\sin\beta). \qquad (5)$$

3 FORMULA FOR LONGITUDINALLY FRAMED STRUCTURE

Similarly, for the longitudinally framed case with frames laying in the y-plane, the unit vectors are, as shown in Figure 7:

$\vec{a} = \cos\alpha\vec{i} - \sin\alpha\vec{j}$, unit vector tangent to the intersection of hull and waterline

$\vec{b} = \cos\gamma\vec{i} + \sin\gamma\vec{k}$, unit vector tangent to the intersection of hull and frame

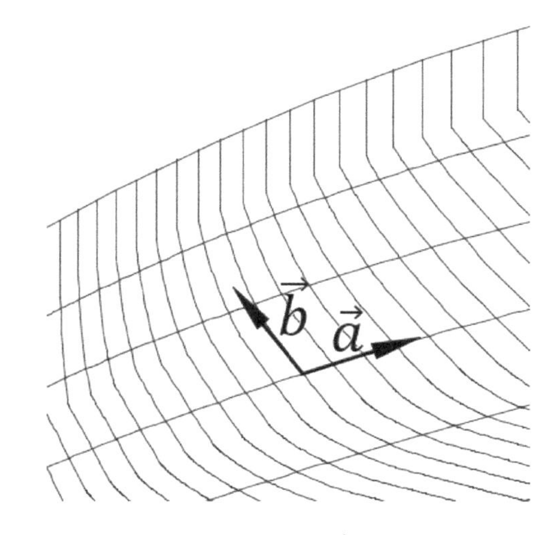

Figure 7. Unit vectors $\vec{a}$ and $\vec{b}$ on longitudinally (y) framed hull.

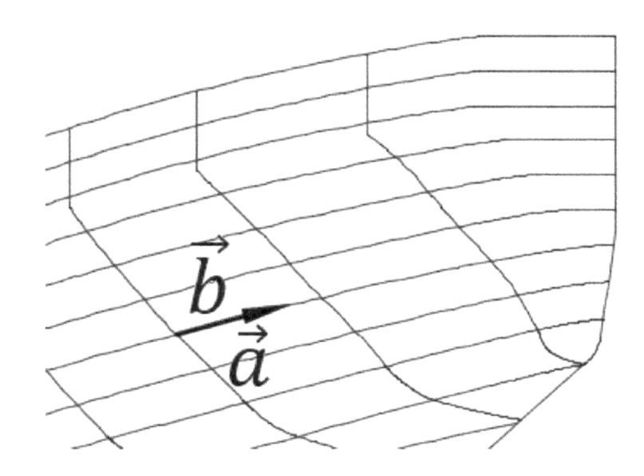

Figure 8. Unit vectors $\vec{a}$ and $\vec{b}$ on longitudinally (z) framed hull.

After similar calculation as was done for the transversally framed case, this yields the formula for longitudinally framed structure

$$\Omega = \cos^{-1}(\cos\alpha\cos\gamma). \qquad (6)$$

It is also noted here for completeness sake that for the case of longitudinal frames laying in the z-plane, $\vec{b} \equiv \vec{a}$ and therefore the angle omega is always zero regardless of the hull shape, as illustrated in Figure 8.

4 FORMULA FOR GENERAL CASE

The same principle can be then used for the general case where frames can lie in any vertical plane with freely chosen angle between the frame plane and athwartship plane as shown in Figure 9, marked by ξ. This formula is suited for both canted and radial framing systems.

Similar to the simple cases presented earlier, the unit vector which is tangent to the intersection of hull and waterline is

$\vec{a} = \cos\alpha\vec{i} - \sin\alpha\vec{j}$, unit vector tangent to the intersection of hull and waterline.

However, now the unit vector $\vec{b}$, tangent to the intersection of hull surface and frame plane has to be formulated in a bit more general way. This vector has to be, by definition, perpendicular to both the normal of the frame plane and the normal of the hull surface. The normal of the frame plane can be defined by

$\vec{c} = \cos\xi\vec{i} - \sin\xi\vec{j}$, unit vector normal to the frame plane.

The normal vector for hull surface can be formed by choosing any two vectors that are tangent to hull surface and by taking the cross product of these vectors, which, by definition, yields a vector that is normal to the hull surface. Vector $\vec{a}$ is tangent to hull surface. In addition, vector $\vec{d}$, which lies in yz-plane and is tangent to hull surface can be defined by

$\vec{d} = \sin\beta\vec{j} + \cos\beta\vec{k}$, unit vector in yz-plane, tangent to hull surface.

Then, by taking the cross product of $\vec{a}$ and $\vec{d}$, the normal vector for the hull surface $\vec{n}$ is formed

$$\vec{n} = \vec{a} \times \vec{d} = -\sin\alpha\cos\beta\vec{i} - \cos\alpha\cos\beta\vec{j} + \cos\alpha\sin\beta\vec{k}. \tag{7}$$

Then, the vector $\vec{b}$ that is tangent to hull surface and frame plane can be formed by taking the cross product of $\vec{c}$ and $\vec{n}$

$$\vec{b} = \vec{c} \times \vec{n} = -\sin\xi\cos\alpha\sin\beta\vec{i} - \cos\xi\cos\alpha\sin\beta\vec{j} - (\cos\xi\cos\alpha\cos\beta + \sin\xi\sin\alpha\cos\beta)\vec{k} \tag{8}$$

This can be simplified slightly to

$$\vec{b} = -\sin\xi\cos\alpha\sin\beta\vec{i} - \cos\xi\cos\alpha\sin\beta\vec{j} - \cos(\xi - \alpha)\cos\beta\vec{k}. \tag{9}$$

Finally, this lets us to calculate the angle between vectors to give the angle Ω

$$\cos\Omega = \frac{\vec{a}\cdot\vec{b}}{|\vec{a}||\vec{b}|}, \tag{10}$$

which can then be simplified to

$$\cos\Omega = \frac{-\sin\xi\cos^2\alpha\sin\beta + \cos\xi\sin\alpha\cos\alpha\sin\beta}{\sqrt{\cos^2\alpha + \sin^2\alpha}} \cdot \frac{1}{\sqrt{\sin^2\xi\cos^2\alpha\sin^2\beta + \cos^2\xi\cos^2\alpha\sin^2\beta + \cos^2(\xi - \alpha)\cos^2\beta}} \tag{11}$$

which can be further arranged to the final formula for general case

$$\Omega = \cos^{-1}\left(\frac{-\sin\xi\cos^2\alpha\sin\beta + \cos\xi\sin\alpha\cos\alpha\sin\beta}{\sqrt{\cos^2\alpha\sin^2\beta + \cos^2(\xi - \alpha)\cos^2\beta}}\right). \tag{12}$$

5 DISCUSSION

To demonstrate the effect of framing orientation on the capacity of the structure, an example case is presented. A vessel with displacement of 20 000 t and ice class PC 4 is used as an example. Frame spacing of 400 mm and steel with yield strength of 355 MPa are assumed, representing rather typical ice-going vessel with relatively high ice class. In Figure 10, the required plate thickness as function

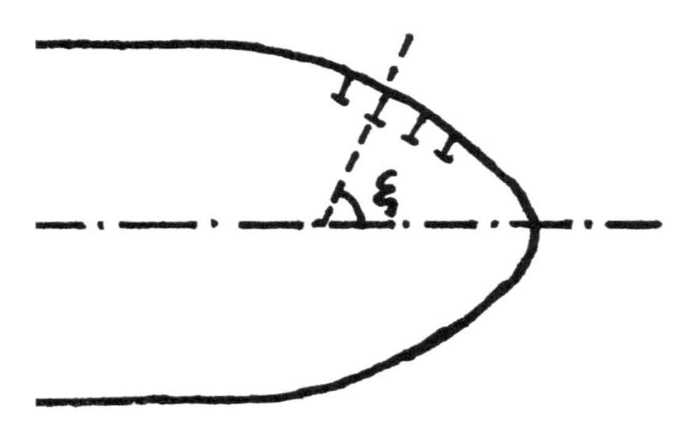

Figure 9. Definition of frame plane angle ξ for canted and radial framing systems.

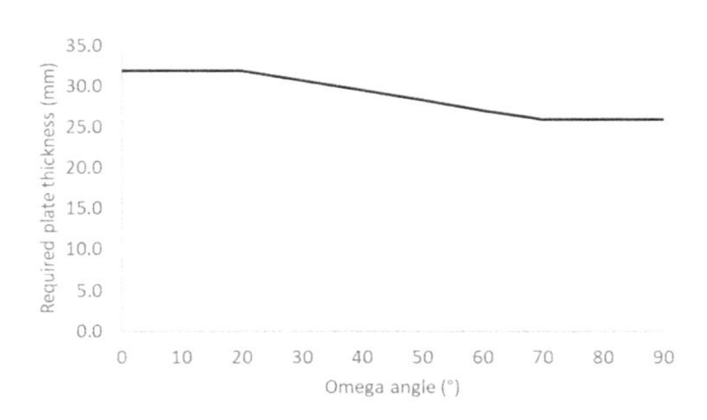

Figure 10. Required plate thickness as function of omega angle.

of omega angle is shown. As can be seen, for lower omega angles, corresponding to longitudinally framed structure, the required thickness is about 23% higher than for high omega angles representing transversally framed structure.

Vice versa, if plate thickness of 28.0 mm is assumed, the corresponding ice pressure is presented in Figure 11. The maximum ice load that longitudinally framed structure can carry is only 63% of the load that the same structure can carry in transverse orientation. For intermediate cases the load carrying capacity is between these two extremes.

Both these figures demonstrate that the omega angle has a significant impact on the design of ice strengthened vessels. If this effect is not correctly accounted for, there is risk of either underdimensioning the structure by a quite significant margin, which will increase risk of damage in service, or of overdimensioning the vessel by equally large margin, which will lead to increased cost and decreased deadweight capacity.

It is expected that considering the actual omega angle instead of simple division to transverse and longitudinal framing will make most difference to vessels with large waterline opening angles at regions which experience large ice loads. Most importantly, longitudinal framing in cylindrical and blunt bows can be properly considered as close to transverse instead of designing it as longitudinal which has led to overdimensioning. Similarly, transversally framed regions with large waterline angle can be calculated according to actual load bearing capability, instead of approximating these as pure transversally framed cases which has led to overestimation of the capacity of the structure and consequently underdimensioning.

While it would then seem at first that the bow and stern parts of most icebreakers, typically built with transverse frames that are due to geometry not actually transverse but instead at some angle to transverse, would be underdimensioned, that is not necessarily the case. Up to date, all ice class rules are to large extent based on measured ice loads, damage records and service experience. Since the background work for these rules has been done with the transverse formulations, it follows that the effect of omega angle for a typical hull shape is indirectly included in the design loads and area factors of rules. Therefore, care should be taken in applying the omega angle, as using it on top of design loads already including the effect would lead the effect to be calculated twice, leading to overdimensioning. To ensure proper and accurate design, the rule background material should be carefully looked through, and if necessary, recalculated considering the effect of omega angle, to ensure correct design loads.

The benefit from doing this would be that the division between load and capacity sides would become clearer, since the geometric effects affecting the load carrying capacity of the structure would be better separated from the design load, improving transparency and clarity of design formulas.

One additional benefit from these formulas is that the ability to calculate omega angle easily gives the designer tool to choose most appropriate framing direction for each hull region.

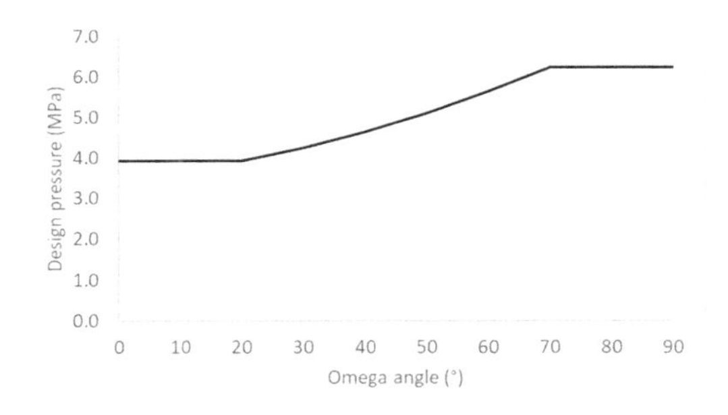

Figure 11. Design pressure corresponding to plate thickness of 28.0 mm.

6 CONCLUSIONS

The introduction of angle omega in PC rules bridges the existing gap between transverse and longitudinal framing in older ice class rules and provides methodology to consider the actual orientation of ice load relative to framing members also in fore and aft parts of the hull as well as canted framing.

However, up to date, no useful formula for calculating this angle has been publicly available and this has made the rules complicated to apply correctly. In many cases, this has led to either estimated omega angle being applied instead of accurate one, or, using either the transverse or longitudinal formula with no regard to effect of omega angle, which has led to inaccuracies in design, and in worst case, incorrect design.

Easy and accurate calculation of omega angle allows more accurate and correct calculation of the actual load bearing capacity of the hull structure when subjected to typical ice load, approximated by a rectangular load patch. Furthermore, ambiguity on the selection of most appropriate formula for each case is removed and design practices can be made consistent. The methods presented in this paper help the industry to apply the PC rules more effectively and in correct way and help in the development of the rules.

REFERENCES

International Association of Classification Societies, 2011. *Requirements Concerning Polar Class, IACS UR I.*

Finnish Transport Safety Agency TraFi, 2010. *Finnish-Swedish Ice Class Rules 2010.*

Russian Maritime Register of Shipping, 2016. *Rules for the Classification and Construction of Sea-Going Ships, Part II Hull.*

Riska, K. 1991. Observations of the Line-like Nature of Ship-Ice Contact. *Proceedings of the 11th International Conference on Port and Ocean Engineering under Arctic Conditions, Vol 2, St. John's, Canada, September 24–28 1991,* pp. 785–811.

Marine Design XIII – Kujala & Lu (Eds)
© 2018 Taylor & Francis Group, London, ISBN 978-1-138-34076-3

Probabilistic analysis of ice and sloping structure interaction based on ISO standard by using Monte-Carlo simulation

Chana Sinsabvarodom, Wei Chai & Bernt J. Leira
Department of Marine Technology, Norwegian University of Science and Technology, Trondheim, Norway

Knut V. Høyland
Department of Civil and Environmental Engineering, Norwegian University of Science and Technology, Trondheim, Norway

Arvid Naess
Centre for Ships and Ocean Structures, Norway
Department of Marine Technology, Norwegian University of Science and Technology, Trondheim, Norway

ABSTRACT: Operation of ships and offshore structures in the Arctic region commonly needs to deal with sea ice. The properties of sea ice are associated with a high degree of uncertainty due to formation of the sea ice under very different conditions within different sea areas. Probabilistic models are frequently introduced in order to cope with this inherent variability of the sea ice and the associated ice loads. In some cases (thin ice, narrow structures and no snow) the thickness and the flexural strength of sea ice are crucial parameters in relation to the interaction between ice and sloping features. For design of Arctic structures, the ISO standard is widely employed by the offshore industry. The objective of the present paper is to investigate the uncertainty associated with the ice forces that are acting on sloping structures based on formulas given in the relevant ISO standard. The Nataf model is applied to study the effect of correlation between the sea ice thickness and its flexural strength. Moreover, varying slope angles both in upward and downward direction are considered in this analysis. Monte-Carlo simulation techniques are applied in order to assess the uncertainty associated with the global force due to the ice structure interaction both in the vertical and the horizontal direction.

1 INTRODUCTION

Ice-structure interaction related to sloping structures is a complex phenomenon, not the least due to the significant variability of the sea ice properties. In the Arctic region, sloping structural features are often presented in connection with offshore structural application such as ships, barges, Gravity-Base Structures (GBS) legs, lighthouses etc. A sloping structural geometry is typically applied in order to reduce the ice load as compared to vertical structures. The slope implies that the failure mechanism of sea ice will turn into the bending mode instead of crushing mode (Bruun P. K. et al., 2006). In connection with ice sheet bending, both compressive and tensile stresses will occur in the ice-thickness direction of the sheet. Inherently, the tensile strength of sea ice is very low. Accordingly, the flexural strength is much lower than the compressive strength, which is also confirmed by field experiments. This implies that the sea ice will fail at a lower load during ice-structure interaction for the flexural mode than for the crushing mode.

Typically, the properties of sea ice are estimated based on experiments both in the field and the laboratory. These properties are associated with aleatory (i.e. random) uncertainties. The ice properties depend on a number of different parameters such as the ice features (e.g. level ice, rafted ice, or rubble field) and the ice conditions (e.g. temperature, salinity, density, porosity, grain size and orientation) (Strub-Klein L., 2017). A probabilistic approach can be applied in order to evaluate the ice-structure interaction based on identification of the most important parameters that govern this interaction. It is found that the ice thickness and flexural strength are dominant quantities with respect to evaluation of the ice-structure interaction effects.

Ranta et al., 2017, studied the failure of level ice pushing against inclined marine structures by performing finite discrete element analysis. The evolution of the ice failure processes is identified. The results from the simulation have been analyzed by means of deterministic and probabilistic methods. The results has shown that the ice thickness has a strong effect on the ice load, especially for thick level ice. In the deterministic analysis, the initial condition of the simulation model has a strong influence on the ice loads. Moreover, for probabilistic analysis, the sample size has a significant effect on the accuracy of the estimated characteristics of the ice load.

Ayobian S. et al., 2016, used Monte Carlo simulation technique to provide the probability analysis of the ice structure interaction for the vertical structures. The keys parameters of sea ice characteristics: Ice thickness, floe diameter and ice crushing strength are defined as random parameters so as to determine the maximum ice load at a designed exceedance probability level, based on the different return periods. However, they focused more on the performance of computer systems. High Performance Computing (HPC) is employed for sea ice load application with Monte Carlo simulation. Consequently, GPUs and Multi-GPUs perform higher potential to reduce the computational time and offer significant speedup over CPU implementations for Monte Carlo simulation.

For design of offshore structures in Arctic region, there are many existing national and international standards for determination of ice loads on sloping structures such as the Canadian CSA S471-04 (2004), the American API RP 2 N (1995), the Russian SNiP code 2.06.04.82* (1995), the Russian VSN code 41.88 (1988) and the Chinese code Q/HSN 30002002 (2002), etc. (Thijssen J. et al., 2014). Among these offshore standards, the ISO standard (ISO 19906) is the most popular, which has been widely using in the offshore industry.

In this work, the objective is to perform the probabilistic analysis of the ice and sloping structure interaction, based on the ISO standard (ISO 19906) in order to investigate the effect of correlation between the thickness and flexural strength of sea ice. The NATAF model (Lui P. & Kiureghian A. D., 1986 and Chai W. et al., 2018) is deployed to generate the transformation of marginal distributions with different correlation coefficients between the sea ice thickness and flexural strength by using Monte Carlo simulation to assess uncertainty associated with the global force arising due to ice interaction with sloping structures with different sloping angles.

2 ENVIRONMENTAL CONDITION

2.1 *Ice thickness*

For ice interacting with sloping structures, the most significant parameters that influence the structural integrity and safety are the ice thickness and the flexural strength of the sea ice. This research focuses on sea ice conditions in the northwestern Barents Sea. During the winter season, the majority of the ice in the Barents Sea has been formed locally between the Svalbard and the Novaya Zemlya areas (Løset, S. et al., 1997 and Strass P. et al., 1997).

The data of ice thickness for the present study is provided by the Norwegian Polar Institute. The data was collected by Helicopter-borne EM (HEM) measurements and by simulation of the level ice thickness.

For some years with lacking data, the ice thickness was simulated by using the Lebedev Freezing Degree Day (FDD) sea-ice growth model (King et al., 2017). The ice thickness data was collected at the location $25°E$ $77°N$ as illustrated in the Figure 1. The maximum annual ice thickness from 1995 to 2015 is shown in Figure 2.

The data of ice thickness in Barents Sea was fitted to various probabilistic models by Normal distribution, Lognormal distribution, Gamma distribution, Gumbel distribution, and Weibull distribution. The Weibull distribution was found to provide the most appropriate fitting of ice thickness. The linear regression of the fitted Weibull distribution has R = 0.9663 as shows in Figure 3. The

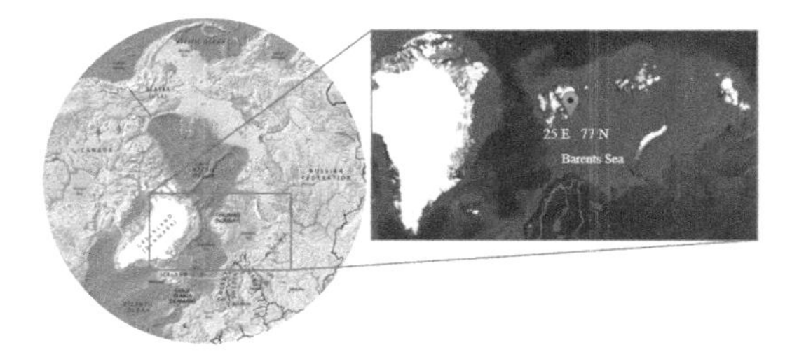

Figure 1. The location of the collected data of ice thickness from HEM measurements (Source: https://www.grida.no).

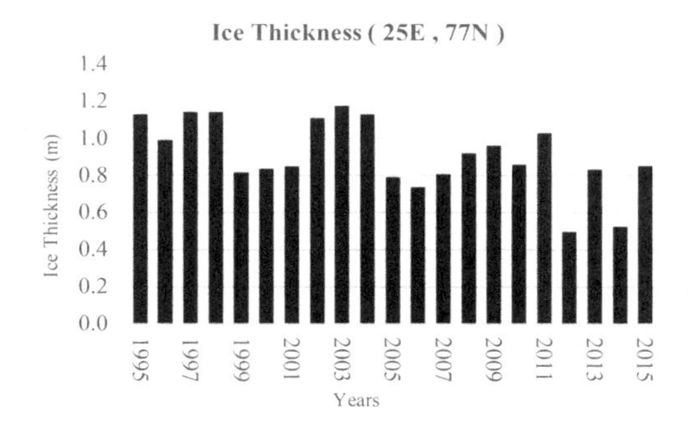

Figure 2. The maximum annual ice thickness from 1995 to 2015.

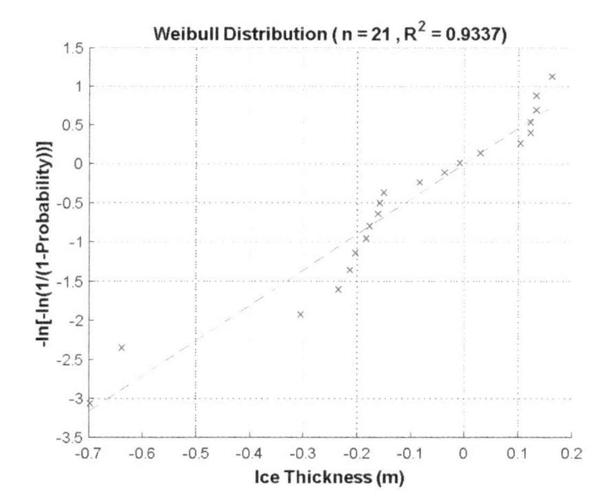

Figure 3. Fitting of Weibull distribution to data of sea ice thickness.

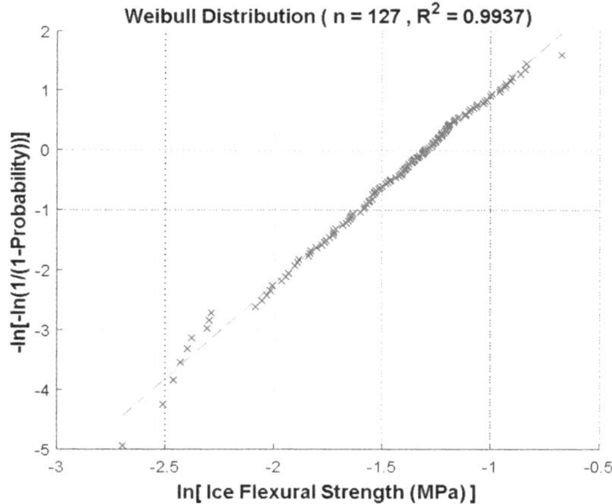

Figure 4. Fitting the data of flexural strength of sea ice by application of the Weibull distribution.

values of scale parameter and shape parameter of Weibull distribtion for ice thickness are 1.00 and 4.521, respectively.

2.2 *Flexural strength of ice*

The flexural strength is in some cases (thin ice, narrow structure and no snow) a key parameter for ice sloping structure interaction. The texture of sea ice is dominated by haphazard alternation of ice layers with different conditions associated with the ice formation. This anisotropy of physical characteristics of sea ice has a significant effect on the random nature of the flexural strength (Karulina M. et al., 2013).

In the last decade, the Arctic and Antarctic Research Institute (AARI) has been conducting extensive expeditions resulting in comprehensive ice research results in the Barents Sea due to the impetus of offshore activities. Therefore, a significant amount of sea ice data has been collected and published both with respect to physical and mechanical characteristic properties.

Typically, there are three types of AARI flexural tests of sea ice, i.e. cantilever beam tests, three points beam tests and small ice plate tests. The cantilever beam test is the most reliable method to estimate the flexural strength because this method has been applied in situ, which allows assessment of ice strength over the whole thickness, hence avoiding relocation of the sample. Disturbance due to relocation can generate inaccuracies associated with the test data (Krupina N. A. et al., 2007). The present study utilizes the data from flexural strength of sea ice from the in situ cantilever beam test in the Barents Sea to fit the Weibull distribution. The linear regression resulting from data fitting gives a coefficient value of R = 0.9937 as shown in Figure 4. The scale parameter and shape parameter of the Weibull distribution for flexural strength are equal to 0.2742 and 3.1655 respectively (Chai W., Leira B. J, 2018).

3 MONTE CARLO SIMULATION

The Monte-Carlo Simulation (MCS) can be applied for analysis of complex problems. In this work, Monte Carlo simulation is applied to generate joint samples of the thickness and flexural strength of sea ice. The objective is to estimate the probabilistic properties of the vertical and horizontal force components due to ice interacting with sloping structures. The random values of the ice thickness x_i and flexural strength x_j are generated from the inverted CDF of the Weibull distribution by drawing uniform random numbers as given by Equations (1) and (2):

$$\text{For the ice thickness: } x_i = F_{thickness}^{-1}\left(x_u\right) \tag{1}$$

$$\text{For the flexural strength: } x_j = F_{flexural}^{-1}\left(x_u\right) \tag{2}$$

In the present analysis the sample size is equal to n = 10,000. The effect of introducing correlation between the ice thickness and flexural strength is also investigated.

In the first case, independence between the basic variables is assumed. Accordingly, the correlation coefficient between the variables is taken as $\rho = 0$, which is given as input to the MCS. The scatter diagram for the case of independent variables is illustrated in Figure 5. The effect of correlation between the two basic variables is subsequently studied by application of the NATAF transformation model.

3.1 *NATAF model*

The NATAF model is used to transform the correlation of marginal probability distributions from the given correlation (Lui P. and Kiureghian A. D.,1986). The assumption of NATAF model is based on transformation of the basic random variables into corresponding Gaussian variables z_i, z_j according to Equations (3) and (4):

$$z_i = \Phi^{-1}\left(F_{thickness}\left(x_i\right)\right) \tag{3}$$

$$z_j = \Phi^{-1}\left(F_{flexural}\left(x_j\right)\right) \tag{4}$$

The correlation coefficient $\rho_{0,ij}$ between the transformed variables z_i and z_j is related to the correlation coefficient ρ_{ij} between the thickness x_i and flexural strength of sea ice x_j. The correlation between x_i and x_j is expressed by Equation (5):

$$\rho_{ij} = \int_{-\infty}^{\infty}\int_{-\infty}^{\infty}\left(\frac{x_i-\mu_i}{\sigma_i}\right)\left(\frac{x_j-\mu_j}{\sigma_j}\right)\varphi_2\left(z_i,z_j,\rho_{0,ij}\right)dz_i dz_j \tag{5}$$

where $E(x_r)=\mu_r, Var(x_r)=\sigma_r^2, r=i, j$; and

$$x_i = F_{thickness}^{-1}\left(\Phi(z_i)\right) \text{ and } x_j = F_{flexural}^{-1}\left(\Phi(z_j)\right) \tag{6}$$

Here, φ_2 is the bivariate standard normal probability density function as given by Equation (7).

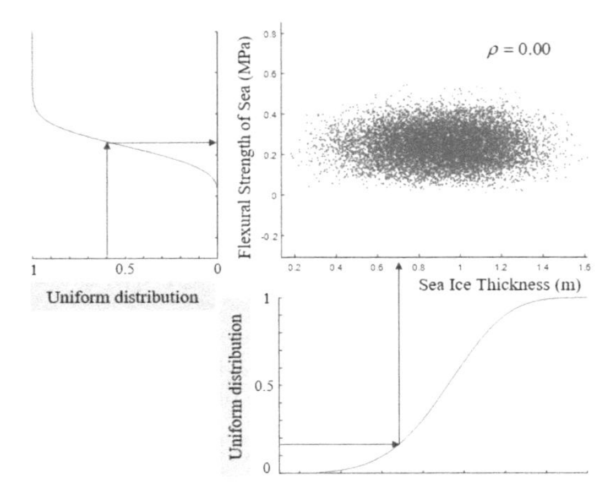

Figure 5. Illustration of the Monte-Carlo simulation for the case with independence between the thickness and flexural strength of sea ice.

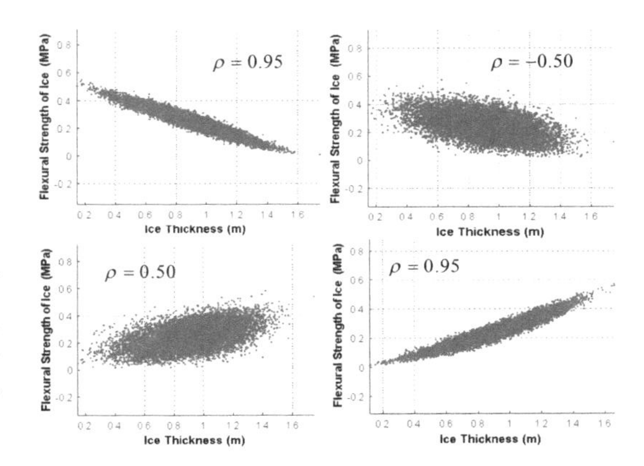

Figure 6. Samples based on the joint probability distribution function of ice thickness and flexural strength of sea ice by application of the NATAF transformation model.

$$\varphi_2\left(z_i,z_j,\rho_{0,ij}\right) = \frac{1}{2\pi\sqrt{1-\rho_{0,ij}^2}} \exp\left[-\frac{z_i^2+z_j^2-2\cdot\rho_{0,ij}\cdot z_i\cdot z_j}{2\cdot\left(1-\rho_{0,ij}^2\right)}\right] \tag{7}$$

Application of the NATAF transformation model is valid under the weak conditions that the cumulative distribution function (CDF) of the ice thickness strictly increases, and that the correlation matrices of **x** and **z** are positive definite. The correlation coefficient can have both positive and negative values.

This analysis employs the values of correlation coefficients, which are varied around the lower bound and upper bound of the limitation of correlation from 95 percent at the negative value to 95 percent of the positive value by dividing into four range at the 50 percent of negative and positive values. The zero correlation at the middle will be set as the benchmark between negative and positive margin. The value of correlation coefficient for this analysis can be written as ρ = [–0.95, –0.5, 0.0, 0.5, 0.95]. The corresponding scatter diagrams for different values of the correlation coefficient are displayed in Figure 6.

4 ICE INTERACTION WITH SLOPING STRUCTURES

4.1 *Ice load on sloping structures*

Typically, the horizontal components of the ice loads acting on sloping structures are smaller than vertical structures. The design of offshore structures in Arctic regions can take advantage of this effect in order to reduce the total loads, which are acting on such structures. It is also found that level ice sheets moving on sloping structures will most likely to fail in a flexural failure mode (Croasdale K. R. et al., 1994 & ISO standard 19906, 2010).

The present study focuses on the formulation of ice interacting with sloping structures according to ISO 19906, which has been widely using for design of offshore structures. The formulation is based on the assumption of flexural failure of level ice sheets, combined with bending of beams on an elastic foundation. The analysis is divided into two slopes categories with upward breaking slope and downward breaking slope as shown in Figures 7 and 8, respectively.

For the downward breaking slope, the weight of the ice fragment or ice rubble in air will replaced by the buoyant ice weight in seawater.

The ice load components in the horizontal and vertical directions can be expressed by Equations (8) and (9):

$$F_H = N\sin\alpha + \mu N\cos\alpha \tag{8}$$

$$F_V = N\cos\alpha - \mu N\sin\alpha \tag{9}$$

The ratio (ξ) of the horizontal and vertical components of the ice load can be written as Equation (10).

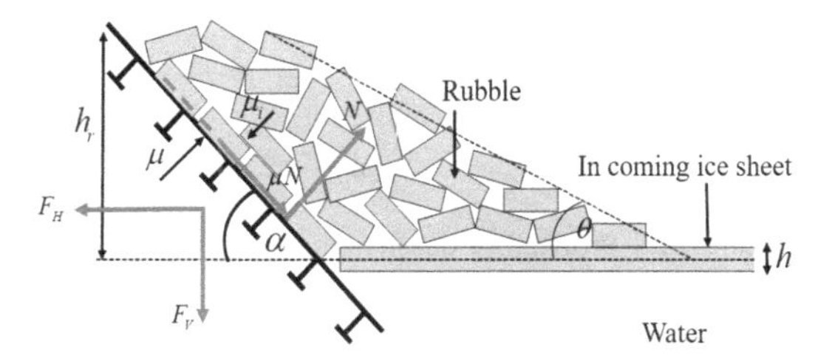

Figure 7. Ice-structure interaction with upward slope.

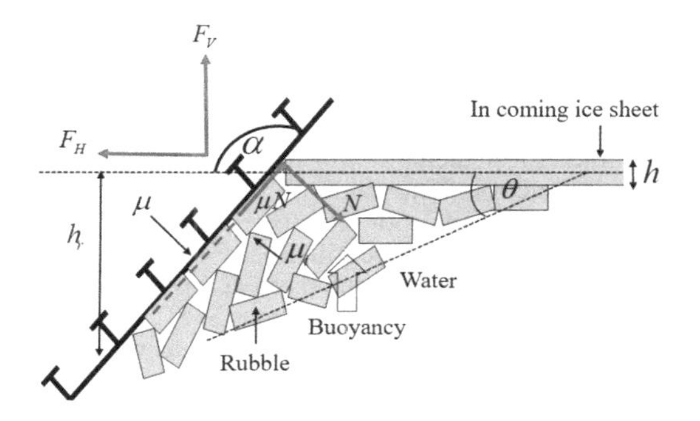

Figure 8. Ice-structure interaction with downward slope.

$$\xi = \frac{F_H}{F_V} = \frac{\sin\alpha + \mu\cos\alpha}{\cos\alpha - \mu\sin\alpha} \tag{10}$$

In the ISO standard, the horizontal force is subdivided into five components which correspond to the breaking load (H_B), the load component required to push the sheet ice through the ice rubble (H_P), the load required in order to push the ice blocks up the slope through the ice rubble (H_R), the load required to lift the ice rubble on top of the advancing ice sheet prior to breaking it (H_L), and the load needed to turn the ice block at the top of the slope (H_T). The horizontal global force of the ice load is accordingly expressed as in the Equation (11):

$$F_H = \frac{H_B + H_P + H_R + H_L + H_T}{1 - \dfrac{H_B}{\sigma_f \cdot l_c \cdot h}} \tag{11}$$

The load component H_B is obtained based on Equation (12):

$$H_B = 0.68 \cdot \xi \cdot \sigma_f \left(\frac{\rho_w \cdot g \cdot h^5}{E}\right)^{0.25} \cdot \left(w + \frac{\pi^2 \cdot L_c}{4}\right) \tag{12}$$

The critical length (L_c) for elastic plate bending is given by Equation (13):

$$L_c = \left[\frac{E \cdot h^3}{12 \cdot \rho_w \cdot g \cdot (1 - v^2)}\right]^{1/4} \tag{13}$$

where E is the elastic modulus, and v is the Poisson ratio.

The load component (H_P) required in order to push the sheet ice through the ice rubble is given in Equation (14):

$$H_P = w \cdot h_r^2 \cdot \mu_i \cdot \rho_i \cdot g \cdot (1-e) \cdot \left(1 - \frac{\tan\theta}{\tan\alpha}\right)^2 \frac{1}{2\tan\theta} \tag{14}$$

where h_r is the rubble height, μ_I is the friction coefficient of ice-to-ice rubble, e is the porosity of the ice rubble, θ is the angle the rubble makes with the horizontal plane. The load component H_R is given by Equation (15):

$$H_R = w \cdot P \frac{1}{\cos\alpha - \mu_{st}\sin\alpha} \tag{15}$$

The parameter (P) is expressed in Equation (16).

$$P = 0.5 \cdot \mu_i (\mu_i + \mu_{st}) \cdot \rho_i \cdot g \cdot (1-e) \cdot h_r^2 \cdot \sin \alpha \cdot \left(\frac{1}{\tan\theta} - \frac{1}{\tan\alpha} \right) \cdot \left(1 - \frac{\tan\theta}{\tan\alpha} \right) + \ldots$$
$$\ldots + 0.5 (\mu_i + \mu_{st}) \cdot \rho_i \cdot g \cdot (1-e) \cdot h_r^2 \cdot \frac{\cos\alpha}{\tan\alpha} \cdot \left(1 - \frac{\tan\theta}{\tan\alpha} \right) + h_r \cdot h \cdot \rho_i \cdot g \cdot \frac{\sin\alpha + \mu_{st} \cos\alpha}{\sin\alpha} \tag{16}$$

The load component (H_L) which is required to lift the ice rubble on top of the advancing ice sheet prior to breaking it is given in Equation (17):

$$H_L = 0.5 \cdot w \cdot h_r^2 \cdot \rho_i \cdot g (1-e) \cdot \xi \cdot \left(\frac{1}{\tan\theta} - \frac{1}{\tan\alpha} \right) \cdot \left(1 - \frac{\tan\theta}{\tan\alpha} \right) + \ldots + 0.5 \cdot w \cdot h_r^2 \cdot \rho_i \cdot g (1-e) \cdot \xi \cdot \tan\phi \cdot \left(1 - \frac{\tan\theta}{\tan\alpha} \right)^2 + \xi \cdot c \cdot w \cdot h_r \cdot \left(1 - \frac{\tan\theta}{\tan\alpha} \right) \tag{17}$$

where c is the cohesion of the ice rubble and ϕ is the friction angle of the ice rubble.

The load H_T which will turn the ice block at the top of the slope is given by Equation (18):

$$H_T = 1.5 \cdot w \cdot h^2 \cdot \rho_i \cdot g \frac{\cos\alpha}{\sin\alpha - \mu_{st} \cos\alpha} \tag{18}$$

The ride-up extent of rubble pile heights increased significantly with the sea ice thickness. The rubble pile height of slope structures is calculated according to the empirical formula from experimental data of Kmei-I lighthouse and Confederation bridge (T. G. Brown and M. Määttänen, 2002 & 2009). The formulation of Kmie-I lighthouse, the rubble pile height is subsequently proportion to sea ice thickness as a linear function as shown in Equation (19), whereas the characteristic formulation of Confederation bridge is obtained by exponential function as given in Equation (20).

$$h_r = 3 + 4 \cdot h \tag{19}$$

$$h_r = 7.64 \cdot h^{0.64} \tag{20}$$

The plotting of the relationship between level ice thickness and height of rubble ice is illustrated in Figure 9.

The probability density function (PDF) of rubble height, which is based on the sea ice thickness, can be calculated by one-to-one transformation as a monotonously increasing function. The example in case of the coloration coefficient ($\rho = 0.00$) of rubble pile high pdf is illustrated in the Figure 10.

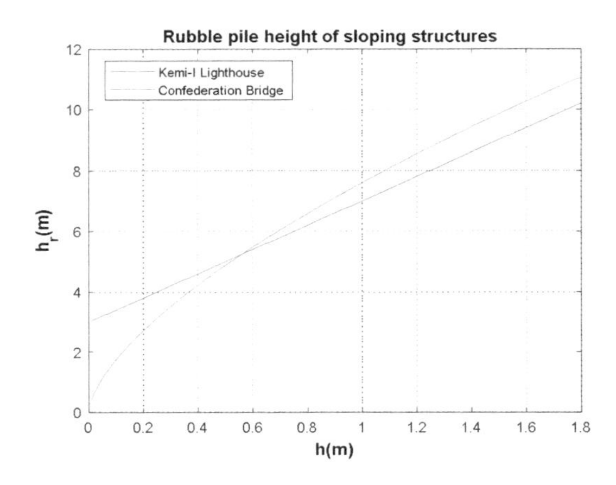

Figure 9. The relationship between level ice thickness and rubble pile height.

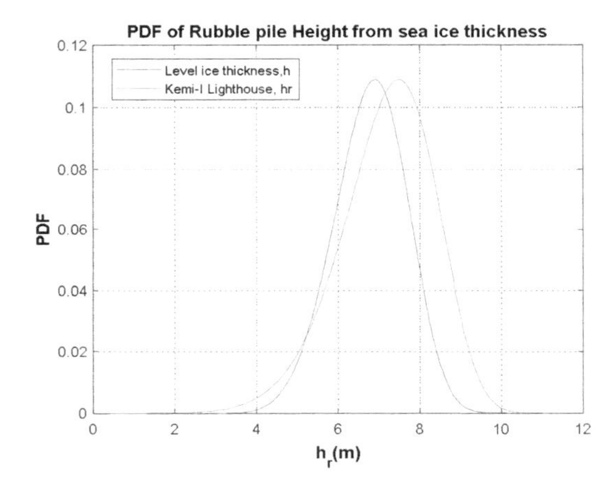

Figure 10. Probability density function of the rubble pile height.

Then the Confederation bridge is selected to apply for this analysis because it provides higher value of rubble pile height, which can be observed from PDF. Therefore, more conservative global forces would be estimated.

Furthermore, the friction coefficient between the sea ice and structure is one of the important parameter for calculation of ice and sloping structure interaction. The value of friction coefficient normally depend upon relative velocity, sea ice temperature and surface roughness. The value friction coefficient in this analysis is approximated according to the contract surface (Saeki H. et at., 1986) between sea ice and uncoated steel materials relating to the average relative velocity of ice drift in Barents Sea as given in Table 1.

The present analysis investigates the uncertainty associated with the forces associated with

Table 1. The input parameters for calculation of ice slop structure interaction.

Data	Value
Elastic Modulus of ice, E_t	567.46 kPa
Ice density, ρ_i	977 kg/m^3
Water density, ρ_w	1025 kg/m^3
Ice Poisson Ratio, v	0.3
The water line diameter, w	1.0 m
The cohesion of the ice rubble, c	5 kPa
The porosity of the ice, e	0.2
The angle of internal friction, ϕ	40 deg.
The coefficient of friction between the ice and the structures, μ	0.07
The coefficient of friction between pieces of ice, μ_i	0.01

Table 2. Variation of slope angles for ice structure interaction.

Up slope (deg.)	Down slope (deg.)
30	120
45	135
60	150

ice interacting with sloping structures according to the formulation given in ISO 19906: 2010. The input parameters associated with this interaction for both upward and downward slopes are summarized in Table 1.

The slope angles have three values both for the case of upward and downward slopes. These angles are listed in Table 2.

5 RESULTS AND DISCUSSION

The samples obtained based on the Monte Carlo simulation are applied for the purpose of fitting distributions. This allows us to investigate the statistical characteristic of the global ice forces, once the joint PDF of the ice thickness and the flexural stress has been determined. The example of fitting global horizontal fore is illustrated in Figure 11.

The mean value, standard deviation, coefficient of variation (COV), and kurtosis of horizontal ice load for the cases of upward and downward slopes are listed in Tables 3 and 4, respectively. The coefficient of variation (COV) is used to indicate the dispersion of the probability distribution of ice loads. The steeper slope angles tend to exhibit slightly higher dispersion of the probability distribution of horizontal ice loads than for milder slopes. Furthermore, the values of the COVs for

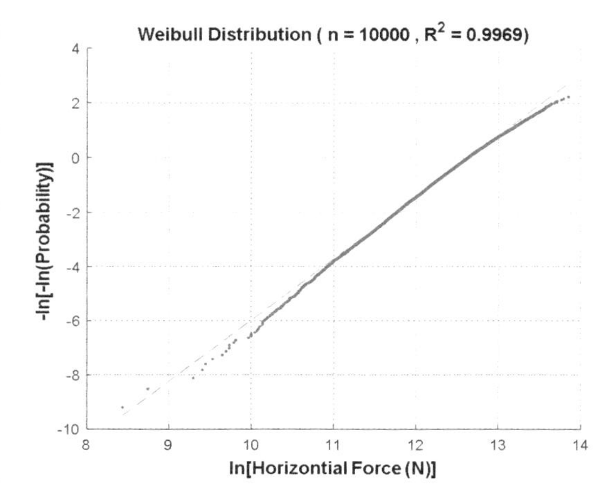

Figure 11. The example of fitting global horizontal loads from Monte-Carlo simulation.

Table 3. Statistical characteristics of the global horizontal force for different upward slopes and different correlation coefficients.

Slope (θ)	ρ	Mean (FH) (N)	Std (FH) (N)	COV%	Kurtosis
30	−0.95	254,447	71,959	28.28	2.84
30	−0.50	265,275	105,062	39.60	3.15
30	0.00	276,309	134,554	48.70	3.49
30	0.50	287,916	160,692	55.81	4.27
30	0.95	298,682	181,390	60.73	4.60
45	−0.95	375,362	98,662	26.28	3.18
45	−0.50	393,666	159,972	40.64	3.24
45	0.00	416,627	215,224	51.66	3.98
45	0.50	437,362	257,780	58.94	4.52
45	0.95	456,542	295,493	64.72	4.98
60	−0.95	689,464	178,215	25.85	3.32
60	−0.50	732,178	307,788	42.04	3.57
60	0.00	773,131	413,942	53.54	4.56
60	0.50	814,646	499,201	61.28	4.36
60	0.95	856,996	577,847	67.43	5.08

the horizontal ice forces are slightly more varying for the steeper slopes.

The correlation coefficient ρ has significant effects to the horizontal ice load. The COV is vary up to 61.66% and 68.06% from the same slope angles for upward and downward slopes, respectively. The values of kurtosis of the probability density functions varies about 38.26% and 38.66% for changing correlation coefficient in upward and downward cases, respectively, and varies about 44.09% and 38.66% for changing slope

Table 4. Statistical characteristics of the global horizontal force for different downward slopes and different correlation coefficients.

Slope (θ)	ρ	Mean (FH) (N)	Std (FH) (N)	COV%	Kurtosis
120	−0.95	482,758	118,552	24.56	3.49
120	−0.50	521,225	244,911	46.99	3.70
120	0.00	563,104	338,055	60.03	4.45
120	0.50	608,328	426,412	70.10	5.41
120	0.95	649,397	494,623	76.17	5.69
135	−0.95	240,166	57,865	24.09	3.53
135	−0.50	258,909	119,908	46.31	3.92
135	0.00	282,149	170,655	60.48	4.72
135	0.50	302,710	210,921	69.68	4.94
135	0.95	322,694	243,401	75.43	5.50
150	−0.95	133,981	32,052	23.92	3.53
150	−0.50	144,951	66,289	45.73	3.44
150	0.00	156,410	92,218	58.96	4.38
150	0.50	168,244	115,304	68.53	5.12
150	0.95	178,209	132,960	74.61	5.67

angles in upward and downward slope direction, respectively.

The Weibull probability density functions (PDF) for upward and downward slopes are displayed in Figures 12 and 13, respectively.

The global horizontal ice loads increase with the slope angle corresponding to physical breaking in terms of the bending failure mechanism. The steeper slopes produce higher bending on the ice sheets, leading to higher horizontal global forces acting on the structure. Comparison of the Weibull cumulative probability distributions for the horizontal ice load for different values of the correlation coefficient ρ and the slope angles α is provided by Figure 14.

For the vertical global forces, the results provide the similar trends as for the horizontal loads. However, the mean values of the vertical loads do not change so much for varying slope angles. The mean values of the vertical loads vary around 19.54% for the interval from 30 to 60 degrees for upward slopes and by 25.6% for the interval from 120 to 150 degrees for downward slopes. The mean value, standard deviation, coefficient of varia-

Figure 12. The Weibull probability density functions for horizontal ice loads in the case of upward slope direction.

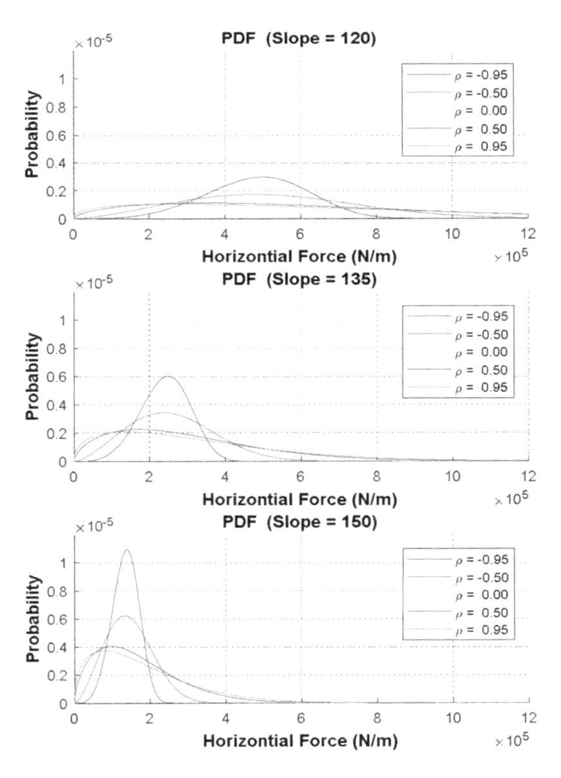

Figure 13. The Weibull probability density functions for horizontal ice loads in the case of downward slope direction.

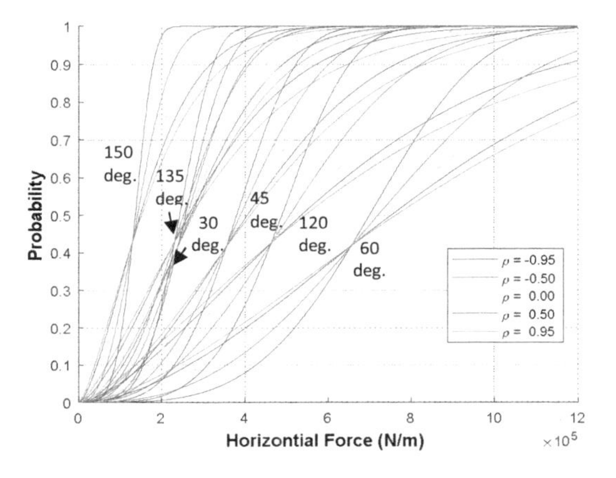

Figure 14. The Weibull cumulative distribution functions for horizontal ice loads in the case of upward and downward slope directions.

Table 5. Statistical characteristics of the global vertical force for different upward slopes and different correlation coefficients.

Slope (θ)	ρ	Mean (FV) (N)	Std (FV) (N)	COV%	Kurtosis
30	–0.95	377,174	106,667	28.28	2.84
30	–0.50	393,224	155,736	39.60	3.15
30	0.00	409,581	199,453	48.70	3.49
30	0.50	426,786	238,198	55.81	4.27
30	0.95	442,745	268,880	60.73	4.60
45	–0.95	326,249	85,753	26.28	3.18
45	–0.50	342,158	139,041	40.64	3.24
45	0.00	362,115	187,064	51.66	3.98
45	0.50	380,137	224,052	58.94	4.52
45	0.95	396,807	256,830	64.72	4.98
60	–0.95	336,212	86,905	25.85	3.32
60	–0.50	357,041	150,091	42.04	3.57
60	0.00	377,011	201,856	53.54	4.56
60	0.50	397,256	243,432	61.28	4.36
60	0.95	417,907	281,783	67.43	5.08

tion (COV), and kurtosis of the vertical forces for upward and downward slopes are listed in Tables 5 and 6, respectively.

In addition, the correlation coefficient has a significant effect on both the vertical and horizontal ice loads. The values of the COV vary between 61.66% and 68.08% a given slope angle both for the upward and downward cases. The kurtosis values vary about 44.09% and 38.66% for the upward and downward slope direction, respectively.

Table 6. Statistical characteristics of the global vertical force for different downward slopes and different correlation coefficients.

Slope (θ)	ρ	Mean (FV) (N)	Std (FV) (N)	COV%	Kurtosis
120	–0.95	235,413	57,811	24.56	3.49
120	–0.50	254,171	119,429	46.99	3.70
120	0.00	274,593	164,850	60.03	4.45
120	0.50	296,647	207,937	70.10	5.41
120	0.95	316,673	241,199	76.17	5.69
135	–0.95	208,742	50,294	24.09	3.53
135	–0.50	225,033	104,219	46.31	3.92
135	0.00	245,233	148,326	60.48	4.72
135	0.50	263,103	183,324	69.68	4.94
135	0.95	280,472	211,554	75.43	5.50
150	–0.95	198,604	47,511	23.92	3.53
150	–0.50	214,865	98,261	45.73	3.44
150	0.00	231,851	136,697	58.96	4.38
150	0.50	249,392	170,918	68.53	5.12
150	0.95	264,164	197,090	74.61	5.67

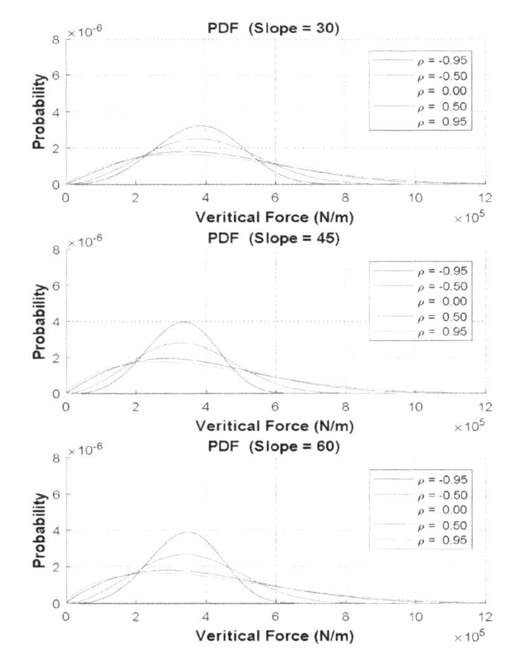

Figure 15. The Weibull probability density functions for vertical ice loads in the case of upward slope direction.

The probability density functions (PDF) of the vertical ice loads for the cases with upward and downward slopes are illustrated in Figures 15 and 16, respectively.

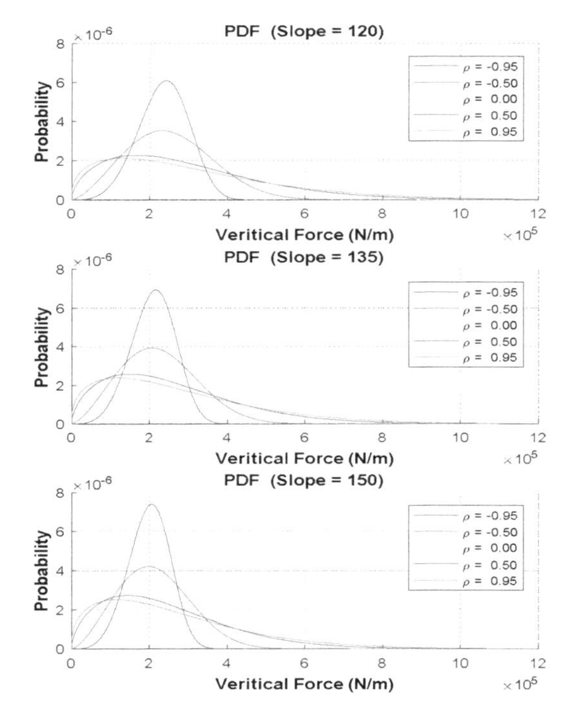

Figure 16. The Weibull probability density functions for vertical ice loads in the case of downward slope direction.

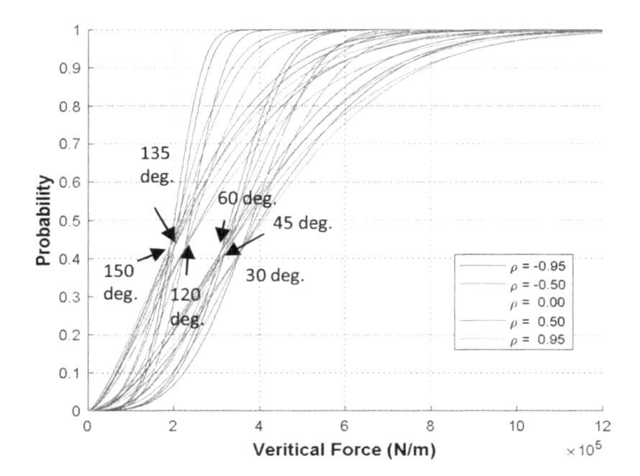

Figure 17. The Weibull cumulative distribution functions for vertical ice loads in the case of upward and downward slope directions.

A comparison of the cumulative probability distribution functions (CDFs) for the vertical ice load for different slopes is provided by Figure 17. The vertical ice loads have a smaller dispersion as compared with the horizontal ice loads. This implies that the inclination angles of the CDF graphs for the vertical ice loads are smaller than those for the horizontal ice loads.

6 CONCLUSIONS AND RECOMMENDATIONS

Probabilistic analysis of ice-slope structure interaction based on the formulation in the ISO standard has been performed by use of Monte-Carlo simulation techniques. The uncertainties associated with the horizontal global forces are associated with higher uncertainty than the vertical global forces. This is reflected by a higher dispersion of the probability distributions, which correspond to the horizontal forces. The inclinations of the cumulative distribution functions (CDF), which correspond to the vertical global forces for ice sloping structure interaction with both negative and positive correlation between the basic variables are steeper than those for the horizontal global ice forces. This also reflects a lower dispersion of the probability distributions for the vertical forces. Similarly, the shapes of the cumulative probability distribution functions (PDF) of the horizontal global ice forces have more flat shapes when compared with those for the vertical global ice forces.

Regarding the effects of slope angle for the ice-structure interaction as expressed by the ISO standard. The flat slope structures imply less uncertainty associated with the global ice forces, which influence by the breaking force component (HB) due to the flexural strength and the reaming global force components due to the sea ice thickness and rubble pile height, as compared to steep slope structures in both the upward and the downward direction. For upslope cases, the uncertainties associated with the global forces for a slope angle of 60 degrees are higher than for slopes with 45 and 30 degrees (60 > 45 > 30), respectively. For downslope cases, the uncertainties associated with the global forces for a slope angle of 120 degrees are higher than for slopes with 135 and 150 degrees (120 > 135 > 150), respectively.

The higher the correlation between ice thickness and flexural strength of sea ice, the higher the uncertainties will be. The results can be explicitly observed from the probability density functions (PDF) of both horizontal and vertical global forces, that when the correlation coefficient increases, they will have more flat shapes, which also imply higher coefficients of variation (COV). This implies that the level of uncertainty also becomes higher. Furthermore, higher values of the correlation coefficient result in higher values of the global forces both in the horizontal and vertical direction. Accordingly, the proper value of the correlation coefficient between ice thickness and flexural strength should be selected with care in order to perform assessment of structural reliability in connection with design of offshore structures in Arctic regions.

ACKNOWLEDGEMENTS

This work is supported by NTNU Oceans Pilot project *Risk, Reliability and Ice Data*, Grants from the *Norwegian Ship-owners Association Fund at NTNU* is acknowledged.

The authors wish to thank Prof. Sveinung Løset for discussion in connection with Sea ice data in the Barents Sea.

REFERENCES

Ayobian S., Alawneh S., and Thijssen J., 2016, "*GPU-Based Monte-Carlo Simulation for a Sea Lce Load Application*". SummerSim-SCSC, Montreal, Quebec, Canada 2016 Society for Modeling & Simulation International (SCS), July 24–27.

Brown T. G. and Määttänen M., 2002, "*Comparison of Kemi-I and Confederation Bridge cone ice load measurement results*", Proceedings of the 16th IAHR International Symposium on Ice, Dunedin, New Zealand, December 2–6.

Brown T. G. and Määttänen M., 2009, "*Comparison of Kemi-I and Confederation Bridge cone ice load measurement results*", Journal of Cold Regions Science and Technology, Vol. 55, Issue 1, January, Page. 3–13.

Bruun P. K. and Gudmestad O. T., 2006, "*A Comparison of Ice Loads from Level Ice And Ice Ridges on Sloping Offshore Structures Calculated in accordance with Different International and National Standards*", 25th International Conference on Offshore Mechanics and Arctic Engineering, Proceedings of OMAE2006–2007, Hamburg, Germany, June 4–9.

Chai W., and Leira B. J., 2018, "Environmental Contours based on Inverse SORM", Submitted, Marine Structures.

Chai W., Leira B. J., and Sinsabvarodom C., 2018, "*Environmental Contours for Design of Ice-capable Vessels*", European Safety and Reliability Conference, ESREL, June 17–21.

Croasdale K. R. and Cammaert A. B., 1994, "*An improve method for the calculation of ice load on sloping structures in first-year ice*", Journal of Hydrotechnical Construction, Vol. 28, No. 3.

HongShuang L., ZhenZhou L., and XiuKai Y., 2008, "*Nataf transformation based point estimate method*", Science in China Press, Springer, September, Vol. 53, no. 17.

ISO 19906, 2010, "*Petroleum and natural gas industries -- Arctic offshore structures*", International Organization for Standardization.

Karulina M., Karulin E., and Marchenko A. 2013," *Field Investigatin of First Year Ice Mechanical Properties In North-West Barents Sea*", Proceedings of the 22nd International Conference on Port and Ocean Engineering under Arctic Conditions, Espoo, Finland, June 9–13.

King J., Spreen G., Gerland S., Haas C., Hendricks S., Kaleschke L., Wang C., 2017, "*Sea-ice thickness from field measurement in the northwestern Barents Sea*", Journal of Geophysical Research: Ocean.,

Krupina N. A. and Kubyshkin N. V., 2007, "*Flexural Strength of Drifting Level First-year Ice in Barents Sea*", International Journal of Offshore and Polar Engineering(ISSN 1053–5381), Vol, 17, No.3, September, pp. 169–175.

Liu P. and Kiureghian A. D., 1986, "*Multivariate distribution models with prescribed marginals and covariances*", Probabilistic Engineering Mechanics, 1986, Vol. 1, No. 2.

Løset S., Shkhinek K., Strass P., Gudmestad O. T., Michalenko E. B. and Kärnä T., 1997, "*Ice Conditions in the Barents and Kara Seas*". Proceedings of the 16th International Conference on Offshore Mechanics and Arctic Engineering, Yokohama, April 13–18, Vol. IV, pp. 173–181.

Ranta J., Polojarvi A., and Tuhkuri J., 2018,"*Ice loads on inclined marine structures—Virtual experiments on ice failure process evolution*", Journal of Marine Structures, 57, 72–86.

Saeki H., Ono T., Nakazawa N., Sakai M., and Tanaka S., 1986, "*The coefficient of friction between sea ice and various materials used in offshore structres*", Journal of Energy Resources Technology, March, Vol. 108/65.

Strass P., Løset S., Shkhinek K.,Gudmestad O. T., Kärnä T. and Michalenko E. B., 1997, "*Metocean Parameters of the Barents and Kara Seas—An Overview*". Proceedings of the 16th International Conference on Offshore Mechanics and Arctic Engineering, Yokohama, April 13–18, Vol. IV, pp. 165–172.

Strub-Klein L., 2017, "*A Statistical Analysis of First-Year Level Ice Uniaxial Compressive Strength in the Svalbard Area*", Journal of Offshore Mechanics and Arctic Engineering-Transactions of the Asme, February, Vol. 139.

Thijssen J., Fuglem M., Richard M. and King T.,2014, "*Implementation of ISO 19906 for probabilistic assessment of global sea ice loads on offshore structures encountering first-year sea ice*" Oceans—St. John's Conference, IEEE, NL, Canada.

Marine Design XIII – Kujala & Lu (Eds)
© 2018 Taylor & Francis Group, London, ISBN 978-1-138-34076-3

Research on the calculation of transient torsional vibration due to ice impact on motor propulsion shafting

Jiang Li
Shanghai Rules and Research Institute, China Classification Society, Shanghai, China

Ruiping Zhou & Pengfei Liao
School of Energy and Power Engineering, Wuhan University of Technology, Wuhan, China

ABSTRACT: Transport by sea gets increasing in Arctic and Antarctic region as climate change progresses. The latest polar class ship rules from the International Association of Classification Societies (IACS) require torsional vibration simulation of ice block impacts on ship's propeller blades. In general, excitation from ice load is considered as transient shock. The simulation should be calculated by the time domain torsional vibration, where transient load is applied. The key techniques and influencing factors such as modeling method, calculation method, excitation, damping, speed drop correction and electric motor characteristic in the calculation of transient torsional vibration on motor propulsion shafting are studied. And the general process of transient torsional vibration calculation for electric motor propulsion shafting under ice load is established. The Newmark-β step-by-step integration method and the MATLAB tool are used to simulate the transient torsional vibration of the ship's propulsion shafting. The steady-state calculation in the time domain shows acceptable agreement with the frequency domain result. The calculation software is developed and validated by motor propulsion example. The research can provide engineering reference for transient torsional vibration calculation due to ice impact on electric motor propulsion shafting.

1 INTRODUCTION

Propulsion shafting design is demanded to ensure reliable and efficient transport solutions with the increase of trades in polar region. Torsional vibration analysis is the vital and indispensable part in reliability evaluation of propulsion shafting.

Generally, under open water conditions, the torsional vibration of propulsion shafting is calculated by frequency domain method only for the steady state. Compared with diesel engine, the torsional vibration response of electric motor propulsion shafting is much smaller under open water conditions, due to smooth excitation output of motor. In ice conditions, the motor propulsion shafting will also produce the vibration response that cannot be ignored, because of the interaction between ice and propeller. The ice load excitation given in International Association of Classification Societies (IACS) is generally regarded as transient ice impact excitation. Therefore, the transient torsional vibration should be simulated by time domain method, and the vibration response results can be used for the fatigue and safety assessment of the shafting.

Electric motor propulsion has better maneuverability and low speed characteristics, and less response time compared with diesel engine. Electric motor propulsion is taken as main propulsion by more and more polar ships.

In this paper, the motor propulsion shafting of Arctic Module Carrier is taken as the research object. The modeling method and key technologies of transient torsional vibration are studied. Based on Newmark-β method, the software of the transient torsional vibration is developed by using MATLAB tool, and the general flow of transient torsional vibration calculation on motor propulsion is built completely.

2 RESEARCH ON CALCULATION METHOD AND KEY TECHNOLOGY

The electric motor propulsion shafting of Arctic Module Carrier as the subject vessel of this paper is given in Table 1.

Table 1. Specification for the electric motor propulsion shafting of Arctic Module Carrier.

Description	Dimension	
Ice class	PC 3	
Electric motor	Max. continuous power	12000 kW
	Max. continuous speed	117 r/min
Propeller	Type	FPP
	Blade No.	4
	Diameter	5.4 m

The subject vessel belongs to direct drive motor propulsion shafting without gear transmission installation. However, in this type of marine propulsion system, the main source of resonance excitation is attributed to propeller-ice interaction.

2.1 *Research on modeling method*

There are two main models in the transient torsional vibration calculation at present: the reduced model and the classical frequency domain model. In order to reduce the time of solving in time domain, some documents recommend reducing the number of inertias for propulsion shafting. Based on the principle of conservation of energy, the motor propulsion shafting can be simplified into two lumped mass-elastic system model which contains only two lumped masses of motor rotor and propeller, and the two-mass is connected by torsional spring. The lumped inertia of motor rotor contains the inertia of the rotor and a part of the shafting, and the lumped inertia of propeller includes the propeller inertia with the water effect and the inertia of the other part of shafting. The basic principle of consistency of the total inertia and lowest natural frequency of the system should be maintained before and after the model simplification.

Although the reduced model can reduce some calculation time, it is difficult to effectively evaluate the response of each shaft section, especially the individual crank throws. The classical frequency domain model takes longer time, the calculation precision is high, and the vibration response of the shaft section is more consistent with the reality.

This paper adopts the classical frequency domain model to carry out the simulation of responses in motor propulsion shafting system, as shown in Figure 1.

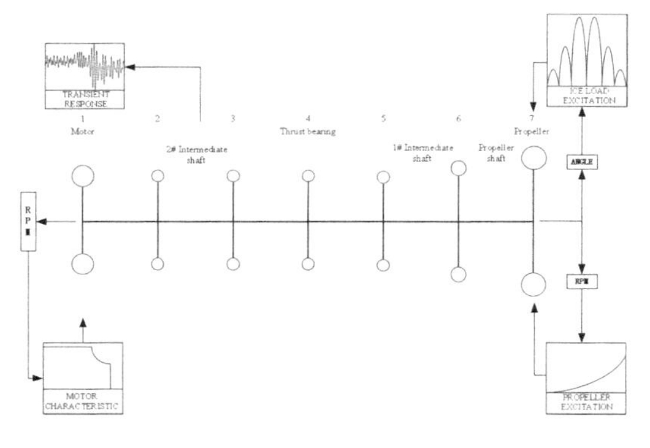

Figure 1. Classical frequency domain torsional vibration system model with excitation.

2.2 *Newmark-β method*

The differential equation of torsional vibration system is as follows:

$$M\ddot{x} + C + K\dot{x} = R \tag{1}$$

where M = inertia; C = damping; K = stiffness matrices; R = excitation torque; x = angular acceleration; x = angular velocity and x = angular displacement.

This paper gets the solution by using the unconditional stability of the Newmark-β method in time domain. The time step is discretized and the vibration equation is transformed into a differential equation of discrete time step. The vibration response values of each inertia point at each discrete time can be calculated.

The Newmark-β method assumes that the acceleration is linearly changed in the time range [t, t + Δt], as follows:

$$\dot{x}_{t+\Delta t} = \dot{x}_t + [(1-\delta)\ddot{x}_t + \delta\ddot{x}_{t+\Delta t}]\Delta t \tag{2}$$

$$x_{t+\Delta t} = x_t + \dot{x}_t\Delta t + [(1/2-\beta)\ddot{x}_t + \beta\ddot{x}_{t+\Delta t}]\Delta t^2 \tag{3}$$

The Newmark-β method should be to meet dynamic equation at every time step:

$$M\ddot{x}_{t+\Delta t} + C\dot{x}_{t+\Delta t} + Kx_{t+\Delta t} = R_{t+\Delta t} \tag{4}$$

According to the formula (2) and formula (3), the following can be obtained:

$$\ddot{x}_{t+\Delta t} = \frac{1}{\beta\Delta t^2}(x_{t+\Delta t} - x_t) - \frac{1}{\beta\Delta t}\dot{x}_t - (\frac{1}{2\beta} - 1)\Delta t\ddot{x}_t \tag{5}$$

$$\dot{x}_{t+\Delta t} = \frac{\delta}{\beta\Delta t}(x_{t+\Delta t} - x_t) + (1 - \frac{\delta}{\beta})\dot{x}_t + (1 - \frac{\delta}{2\beta})\Delta t\ddot{x}_t \tag{6}$$

The equation (5) and (6) substituting (4), the equation of $x_{t+\Delta t}$ can be obtained:

$$\hat{K}x_{t+\Delta t} = \hat{R}_{t+\Delta t} \tag{7}$$

In the equation:

$$\hat{K} = K + \frac{\delta}{\beta\Delta t}C + \frac{1}{\beta\Delta t^2}M \tag{8}$$

$$\hat{R}_{t+\Delta t} = R_{t+\Delta t} + M[\frac{1}{\beta\Delta t^2}x_t + \frac{1}{\beta\Delta t}\dot{x}_t + (\frac{1}{2\beta} - 1)\ddot{x}_t] + C[\frac{\delta}{\beta\Delta t}x_t + (\frac{\delta}{\beta} - 1)\dot{x}_t + (\frac{\delta}{2\beta} - 1)\Delta t\ddot{x}_t] \tag{9}$$

$x_{t+\Delta t}$ can be obtained by solving equation (7), $\dot{x}_{t+\Delta t}$ and $\ddot{x}_{t+\Delta t}$ can be available according to (5) and (6).

The unconditionally stable expression of Newmark-β method is

$$\delta \geq \frac{1}{2}+\gamma, \beta \geq \left(\frac{1}{2}+\delta\right)^2/4 \tag{10}$$

As shown in Figure 2, the torsional vibration stress of the intermediate shaft with the change of propeller angle under the rated speed (117 r/min) is calculated with the Newmark-β method. The larger response of the first half is due to the impact load from 0 to the rated load. After transient excitation impact of the first part, the shafting system operates smoothly in the steady state.

Steady-state analysis is an appropriate way to check the validity of the time domain model. By calculating the time domain response with the Newmark-β method at each specific speed, we can get the curve of shaft torsional vibration stress in the whole speed range. The torsional stress curve in time domain is consistent with the result of frequency domain calculation with the mode superposition method as shown in Figure 3. The figure shows the correctness of Newmark-β method in torsional vibration calculation in time domain, compared with two different calculation method.

2.3 *Ice load excitation*

According to IACS requirements concerning polar class ships, polar classed ships are divided into seven classifications from PC1~PC7. And the propeller ice torque excitation for shaft line transient dynamic analysis (time domain) is defined as a sequence of blade impacts which are of half sine shape and occur at the blade. The torque due to a single blade ice impact as a function of the propeller rotation angle is then defined as:

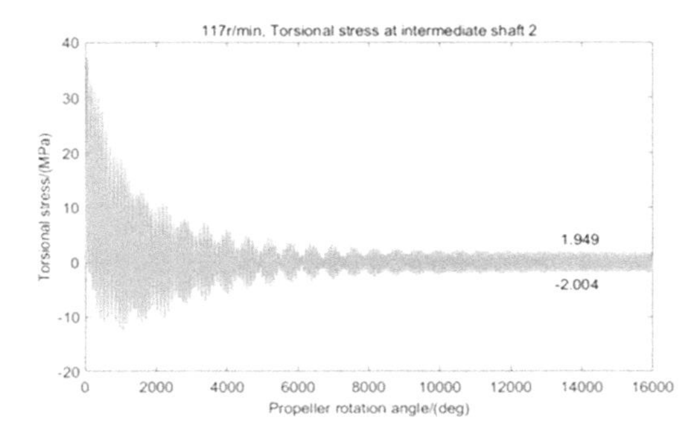

Figure 2. Torsional vibration stress of intermediate shaft 2 in time domain at MCR.

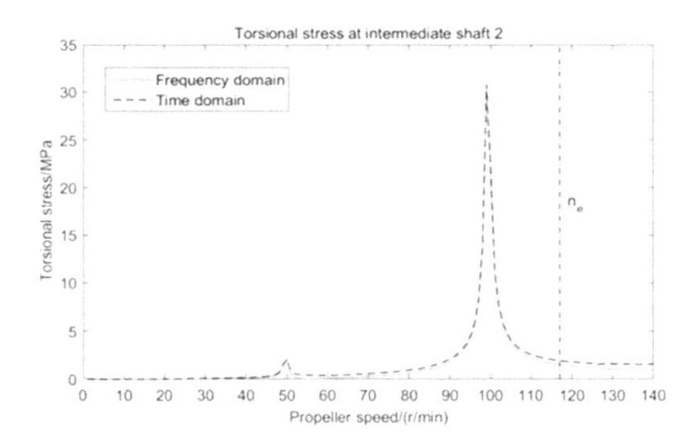

Figure 3. Comparison between time domain and frequency domain of torsional vibration stress of intermediate shaft 2.

Table 2. Ice impact magnification and duration factors for different blade numbers.

Torque excitation	Propeller/ Ice interaction	Cq	α_i(°)			
			Z = 3	Z = 4	Z = 5	Z = 6
Excitation case 1	Single ice block	0.75	90	90	72	60
Excitation case 2	Single ice block	1.0	135	135	135	135
Excitation case 3	Two ice blocks (phase shift 360/ (2·Z) deg.)	0.5	45	45	36	30
Excitation case 4	Single ice block	0.5	45	45	36	30

$$\begin{cases} Q(\phi)=C_q Q_{\max}\sin\left[\phi(180/\alpha_i)\right] & \phi=0\cdots\alpha_i \\ Q(\phi)=0 & \phi=\alpha_i\cdots360° \end{cases} \tag{11}$$

where Φ = rotation angle starting when the first impact occurs; C_q and α_i parameters are given in the table below, α_i = the duration of propeller blade/ice interaction expressed in propeller rotation angle and Q_{max} = maximum torque on the propeller resulting from propeller/ice interaction, kNm.

According to IACS requirements concerning polar class, they are categorized into four different load cases as shown in Table 2 and Figure 4. The torque amplitude is a function of propeller geometry, polar class rating and engine speed. The duration of the ice impact sequence depends on the polar class notation.

The total ice torque is obtained by summing the torque of single blades, taking into account

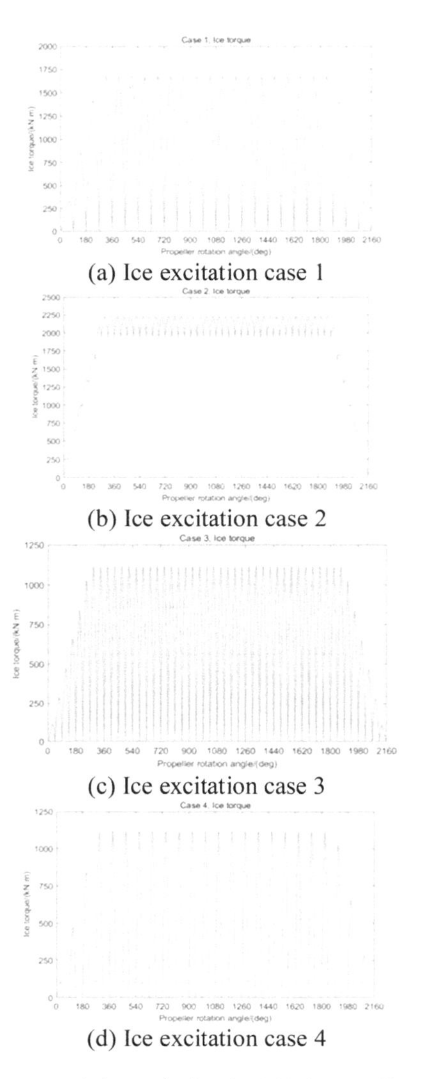

(a) Ice excitation case 1

(b) Ice excitation case 2

(c) Ice excitation case 3

(d) Ice excitation case 4

Figure 4. Relative excitation for 4 blade propeller.

the phase shift 360 deg./Z. At the beginning and at the end of the milling sequence (within calculated duration) linear ramp functions shall be used to increase C_q to its maximum within one propeller revolution and vice versa to decrease it to zero.

The number of propeller revolutions during a milling sequence shall be obtained from the formula:

$$N_Q = 2 \cdot H_{ice} \tag{12}$$

The number of impacts is $Z{\cdot}N_Q$ for blade order excitation. Where H_{ice} = maximum design ice block thickness, m.

$$\begin{cases} Q_{\max} = k_{open} \cdot \left[1 - \dfrac{d}{D}\right] \cdot \left[\dfrac{P_{0.7}}{D}\right]^{0.16} \cdot \left[n \cdot D\right]^{0.17} \cdot D^3 \\ D < D_{\lim it} \\ Q_{\max} = 1.9 \cdot k_{open} \cdot \left[1 - \dfrac{d}{D}\right] \cdot \left[H_{ice}\right]^{1.1} \cdot \left[\dfrac{P_{0.7}}{D}\right]^{0.16} \\ \cdot \left[n \cdot D\right]^{0.17} \cdot D^{1.9} \\ D \geq D_{\lim it} \end{cases} \tag{13}$$

where d = external diameter of propeller hub (at propeller plane), m; D = propeller diameter, m; D_{ilmit} = limit value for propeller diameter, m; n = propeller rotational speed, rev/s; $P_{0.7n}$ = propeller pitch at 0.7R radius at MCR in free running condition, m; k_{open} = 14.7 for PC1-PC5 and 10.9 for PC6-PC7.

The curve of the propeller torque excitation with the propeller angle change is obtained by above calculation method as shown in Figure 4.

2.4 *Torque characteristic of motor*

The following Figure 5 shows the torque characteristic curve of the electric motor. The motor running at rated speed (n_e), if the propeller is impacted by ice load, the motor will have a big drop in the specific speed. In the speed range down process (n_c~n_e), motor operates with constant power, if the speed continues to decline, the motor will run at a constant torque process (0~n_c). The motor has excellent low speed characteristics compared with diesel engine.

2.5 *Damping*

The damping of propeller is considered as main damping in motor propulsion shafting. The typical damping model includes Archer, Prodam and so on. The Archer damping model is used in this

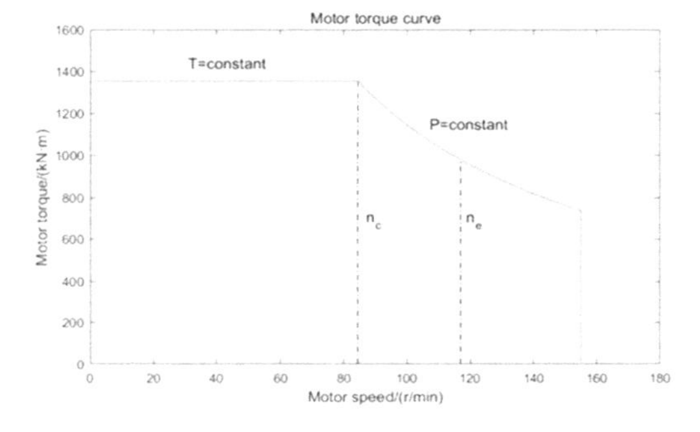

Figure 5. Motor torque characteristic curve.

paper. In the discrete mass-elastic model, the damping of the propeller is mainly expressed in the form of absolute damping. When calculating the transient response under ice load impact, it can include two conditions. One condition is that, output torque of the motor is absorbed by the propeller, that is, the propeller torque in the propeller's hydrodynamic characteristic, which is damping relative to the motor and balanced with the torque output from the motor. And the other condition is that, due to the transient impact of ice load, the shaft speed fluctuation and significant speed drop may occur. The absolute damping torque is caused by the difference between transient speed and average speed.

In the non-steady time domain, the absolute damping cannot be expressed by the damping matrix, but the damping torque is applied as the external excitation force. The mean speed of the lumped mass-elastic system should be subtracted from transient speed to solve the absolute damping torque matrix accurately when simulating the torsional vibration responses at a specific speed. The product of damping torque matrix is the absolute damping coefficient and the torsional speed difference.

$$T_t(t) = C_{abs} \cdot \left[\dot{\phi}(t) - \dot{\phi}_{ave}(t)\right] \tag{14}$$

where C_{abs} = absolute damping coefficient; $\dot{\phi}(t)$ = transient speed and $\phi_{ave}(t)$ = average speed.

2.6 *Motor speed drop analysis*

The motor propulsion shafting will cause speed drop of the motor under the impact of ice load. Adopting the rigid system model (the single mass model), the speed drop of the propulsion shafting under the impact of ice load is calculated by using the following formula.

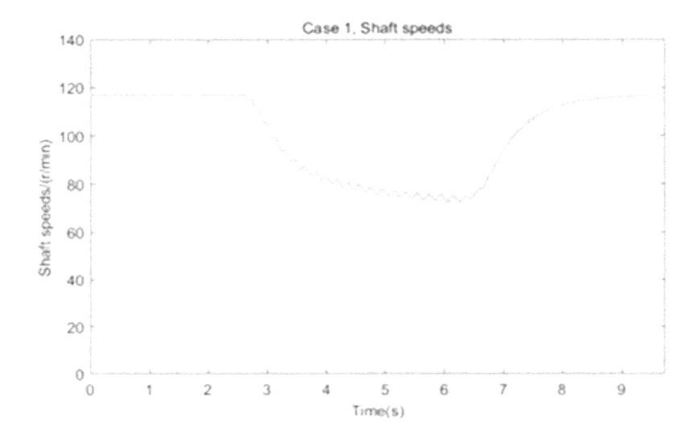

Figure 6. Speed drop curve of shaft system under ice load impact case 1.

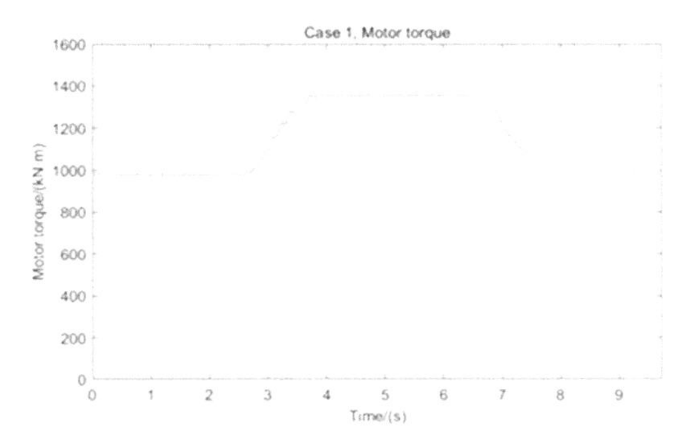

Figure 7. Motor torque curve of shaft system under ice load impact case 1.

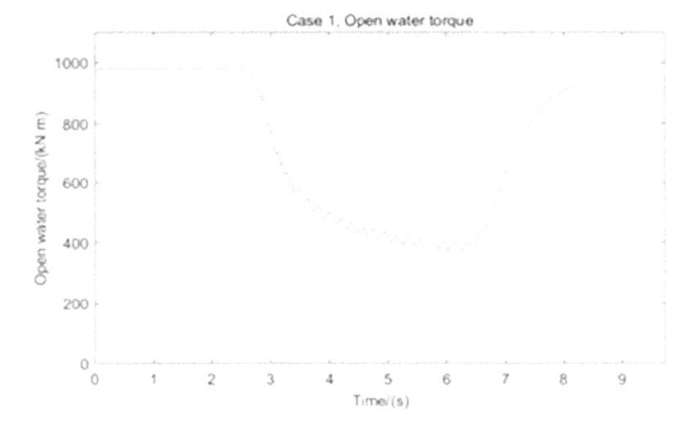

Figure 8. Propeller torque curve of shaft system under ice load impact case 1.

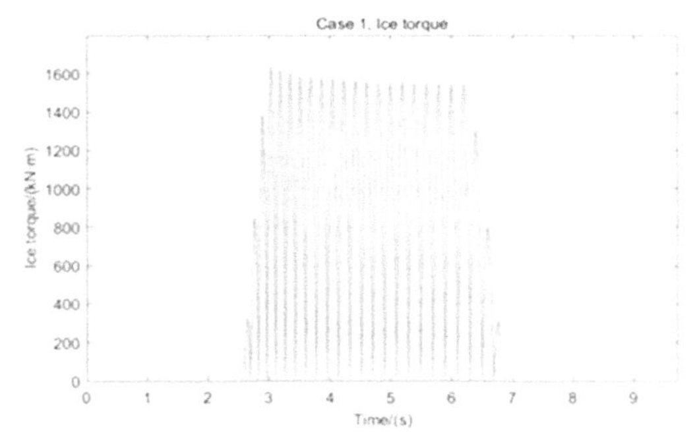

Figure 9. Ice load excitation torque curve of shaft system under ice load impact case 1.

$$\delta n = \delta t \frac{T_{engine} - (T_{prop} + T_{ice})}{2\pi I} \tag{15}$$

where δ_n = speed change in every time step, r/s; δ_t = time step, s; T_{engine} = motor excitation, N·m. In excitation case 1, the speed drop curve of the shafting system can be obtained under ice load

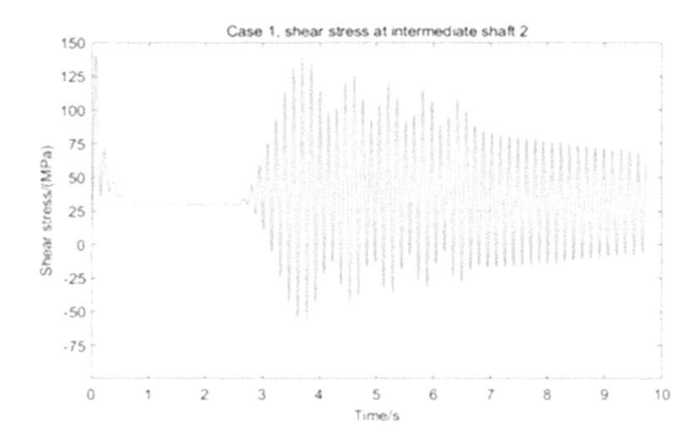

Figure 10. The transient torque stress response curve of the intermediate shaft 2 under case 1.

impact at MCR, as shown in Figure 6. At the same time, torque excitation, propeller torque and ice load torque excitation can be directly obtained by applying to Newmark-β method loading, as shown in Figures 7~ 9. And other three ice load excitation cases could also be applied as above method.

The correctness depends on the motor dynamic characteristic, both MCR speed condition and resonance speed condition.

3 ANALYSIS AND SOFTWARE DEVELOPMENT

3.1 *Response calculation and analysis of influence factors*

The motor damping is negligible compared with the diesel engine. To eliminate the impact of the loading of the motor, the ice load begins to be loaded after 1800 degrees of rotation of the propeller, and the time step for the 1 degree rotation of the propeller is set to solve time. The transient torque stress response curve of the intermediate shaft 2 under case 1, as shown in Figure 10, is obtained.

The paper method has been compared with the solution method of Abaqus CAE6.13–1 finite element software in references for vibration analysis, and it has been concluded that the results of the models agree. Due to ice load impact, the shaft system will have a large speed drop. If the maximum ice torque changes with the speed change, it will affect the final results. Three types of maximum ice torque excitation modes are considered in this paper.

1. The maximum ice torque excitation does not vary with the speed, that is, the maximum ice torque in the speed drop of the shaft is considered as constant value.
2. The maximum ice torque excitation varies according to the speed drop, as shown in the following formula.

$$Q_{\max\_n} = Q_{\max,bollard}\left(\frac{n}{n_{bollard}}\right)^{0.17} \tag{16}$$

where Q_{max_n} = maximum torque in the speed n, kNm; $Q_{max,bollard}$ = maximum torque in bollard condition, kNm and $n_{bollard}$ = the rotational propeller speed in bollard condition, r/s.

3. The maximum ice torque excitation varies according to the propeller open water propulsion characteristics, as shown in the following formula.

$$Q_{\max\_n} = Q_{\max,bollard}\left(\frac{n}{n_{bollard}}\right)^{2} \tag{17}$$

The three maximum ice torque excitation modes of the motor propulsion shafting at MCR speed have little influence on the maximum stress response amplitude in this simulation. And the maximum stress response of the shafting is mainly due to the ice load impact when the propeller has just contacted the ice load.

3.2 *Software development*

The general flow of the transient torsional vibration on motor propulsion shafting by using the classical frequency domain model is as following Figure 11:

The transient torsional vibration calculation software has been developed based on above key technologies, as the following Figure 12.

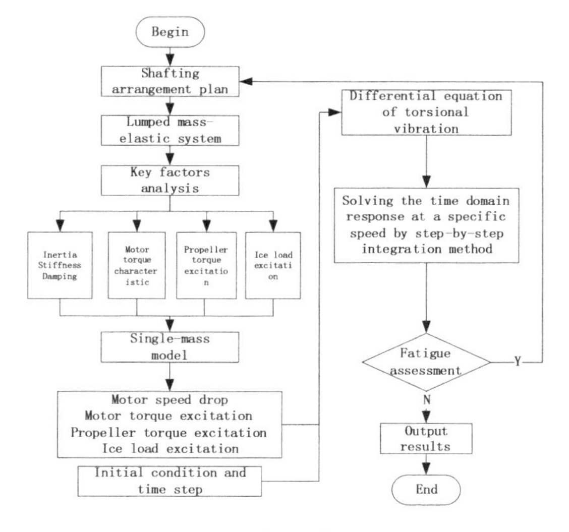

Figure 11. The general flow of the transient torsional vibration on motor propulsion shafting.

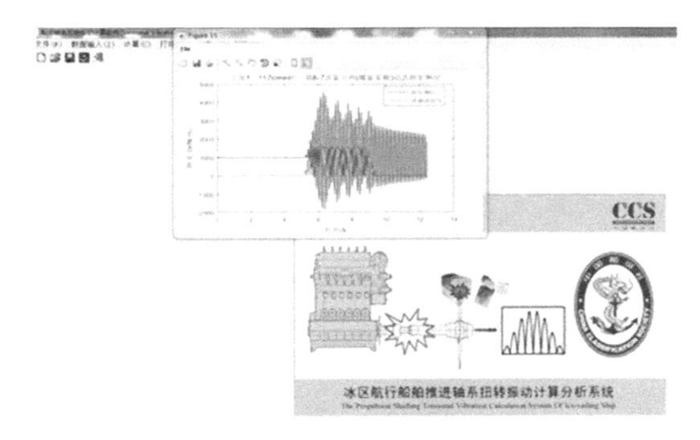

Figure 12. The transient torsional vibration calculation software.

4 CONCLUSIONS

In this paper, the motor propulsion shafting of polar module carrier is taken as the research object. The simulation calculation of transient torsional vibration under the ice impact has been performed. The key technology and influence factors such as modeling method, calculation method, excitation, damping and so on are studied in the calculation of transient torsional vibration of motor propulsion shafting. The general flow of transient torsional vibration calculation of motor propulsion shafting has been put forward. The following conclusions are to be drawn:

1. The classical frequency domain model is recommended in transient torsional vibration simulation on motor propulsion system with the help of MATLAB tool, which improves the accuracy of the calculation, and provides torsional stress/torque on specific shafts.
2. The motor speed will be greatly reduced due to ice load. The influence of the speed drop of the propulsion system must be considered when simulating transient torsional vibration.
3. Based on safety considerations, the design ice torque defined as constant value can be used as the maximum ice load torque for the propulsion system in the full speed range.
4. The absolute damping cannot be expressed by the damping matrix in time domain, but the damping torque is applied as the external excitation force.
5. The simulation correctness due to ice shock depends on the motor dynamic characteristic at specific speed.

REFERENCES

Aker Arctic Technology Inc/ Ikonen T, Valtonen V, 27376–600–313 Rev. D, Propulsion shafting fatigue calculation for Guangzhou Shipyard International[R]. 2014–12–08.

Barro R D, Eom K T, Lee D C. Transient torsional vibration analysis of ice-class propulsion shafting system driven by electric motor[J]. Transactions of the Korean society for noise and vibration engineering. 24(9): 667–674, 2014.

Barro R D, Eom K T, Lee D C. Transient torsional vibration response due to ice impact torque excitation on Marine diesel engine propulsion shafting[J]. Transactions of the Korean society for noise and vibration engineering. 2015,25(5): 321–328.

Batrak Y A, Serdjuchenko A M, Tarasenko A I. Calculation of torsional vibration responses in propulsion shafting system caused by ice impacts[C]//Proceedings of 1st Torsional Vibration Symposium. Austria, Salzburg. 2014.

Brouwer J, Hagesteijn G, Bosman R. Propeller-ice impacts measurements with a six-component blade load sensor[C]// Third International Symposium on Marine Propulsors smp'13, Launceston, Tasmania, Australia, May 2013.

CCS. Rules for classification of sea-going steel ships[S]. Beijing: China Communications Press, 2015.

Dahler G, Stubbs J, Norhamo L. Propulsion in ice-Big ships[M]. Det Norske Veritas, Norway, 2014.

IACS. Requirements concerning polar class[S]. 2016.

Persson S. Ice Impact Simulation for Propulsion Machinery[J]. MTZ industrial, 2015, 5(1): 34–41.

Polic D. The influence of ice loads on the propulsion machinery system[M]. Department of Marine Technology Norwegian University of Science and Technology, 2013.

Yang Hongjun, Che Chidong, Zhang Weijing, et al. Transient torsional vibration anslysis for ice impact of ship propulsion shaft[J]. Journal of ship mechanics, 2015,(1):176–181.

Marine Design XIII – Kujala & Lu (Eds)
 ISBN 978-1-138-34076-3

Simulation model of the Finnish winter navigation system

Morten Lindeberg, Pentti Kujala & Otto-Ville Sormunen
Department of Mechanical Engineering, Aalto University School of Engineering, Espoo, Finland

Markus Karjalainen
Finnish Transport Agency, Helsinki, Finland

Jarkko Toivola
Alfons Håkans LtD, Turku, Finland

ABSTRACT: Increasing maritime traffic, combined with the possible warming of the climate, will affect the demand for icebreaker (IB) assistance in the Baltic Sea. Accurate prediction of the local demand for IB assistance without an appropriate simulation tool is hard because of the number of variables that must be considered. The winter navigation system is fairly complex as the people in charge of the system must consider e.g. the present weather and ice conditions, the capabilities of the merchant ships needing assistance and the number and location of the ships visiting the sea area under consideration. Typically planning of the IB movements is done by the IB captains operating on the studied sea area. This paper describes a simulation tool built around a deterministic, ice-breaker movement based computer model, which simulates the operations of the icebreakers and merchant ships navigating on the varying ice conditions of the studied sea area. Simulation test results, obtained by using real-life historical input-data, indicates that the model is fairly accurate in simulating the winter traffic system of the northern Baltic Sea. The simulation model is applied to study the possible effect of the future EEDI regulations on the ship waiting time and IB demand.

1 INTRODUCTION

Increasing maritime traffic in areas where icebreaker (IB) assistance is needed will naturally also increase the demand for icebreaking assistance. The work load of an IB in its operational area, at a specific time, is strongly dependent on the area specific ice conditions and ship traffic. This leads to large area- and time-specific variations in the demand for icebreaking assistance. Even under constant ice conditions, it is hard to estimate local demand for assistance solely from the estimated increase or decrease in local maritime traffic.

There are a number of earlier work related to the development of the transit simulation models for ships navigating in ice, see e.g. Patey and Riska (1997), Kamesaki et al. (1999), Montewka et al. (2015), Kuuliala et al. (2017) and Bergström et al. (2017). Typically all these models simulate the speed variation of a single ship when it is sailing in varying ice conditions such as level ice, ridged ice and ice channel. In addition the real time AIS data has been used to study e.g. the convoy speed when IBs assist merchant ships, see Goerlandt et al. (2017). Monte Carlo random simulation can also be used to study the uncertainties and variations on the ice conditions and on the calculation methods to evaluate ship speed in various ice conditions (Bergström, 2017).

This paper describes the basic principles of a MATLAB-based simulation tool built around a deterministic icebreaker-movement model. The new approach here is that the simulation model include also the decision principles of icebreakers to determine which ships and when will be assisted. The model also includes the possible assistance and towing principles of merchant ships behind an icebreaker. The tool can be used for predicting local demand for icebreaking assistance under changing ice and traffic conditions. It can also be used to predict how the traffic flow will react to changes in the IB operational areas of the modelled system, i.e. by adding/removing IBs from the system and/or by modifying the boundaries of IB operational areas.

The main idea that this model can be used to optimise the future need of icebreakers so that the total costs of the expensive winter traffic management system can be decreased. The new environment regulations such as the Energy Efficiency Design Index (EEDI) will most probably cause a reduction of the available engine power on ice going ships which can have a remarkable effect on the winter navigation traffic system.

2 BASIC PRINCIPLES OF THE MODEL

2.1 *Constructing the fairway-network for simulation*

The goal is to construct a fairway network that resembles the winter traffic system of interest, see Figure 1. This is accomplished by using two different network building-blocks (BB): (1) a BB resembling a fairway section between two points – i.e. a straight line, and (2) a junction-BB, with three legs going out from the junction point, forming a Y-shape. Leg lengths can be set to zero.

A port (port-node) can be assigned at the end of any BB-leg/section. Usually, majority of the merchant ships enter the simulated network from the same point, which is referred to as the input-node. In this case it is Kvarken at the south end of the Bothnian bay in Figure 1. The input-node differs from a port-node only in the way that a ship cannot spend time (port-time) in the input node, i.e. it functions only as a source and/or a sink. A port-node can also function as a source/sink for a ship.

2.2 *IB operational areas*

As default, one IB is assigned to each BB. Also, as default, the movements of an IB are constrained by the borders (points of connection) of adjacent BBs. This area (or "paths"), formed by the constraining borders, is referred to as the IB operational area. However, an IB operational area can be expanded by assigning the area of adjacent BBs to a single IB. Now the IB is constrained by the borders formed by the adjacent operational areas, and not the individual BB-borders. An IB operational area, no matter how many BBs it spans, always has one single IB operating within it. The total number of icebreakers in the system is indirectly controlled by defining the IB-operational areas in the fairway-network. In addition, there is a possibility of assigning two IBs into one single BB, in which the IBs are programmed to work (assist) optimally in regard to each other (cooperate). However, the operational area of this kind of BB cannot be expanded. Figure 1 below shows an example fairway network.

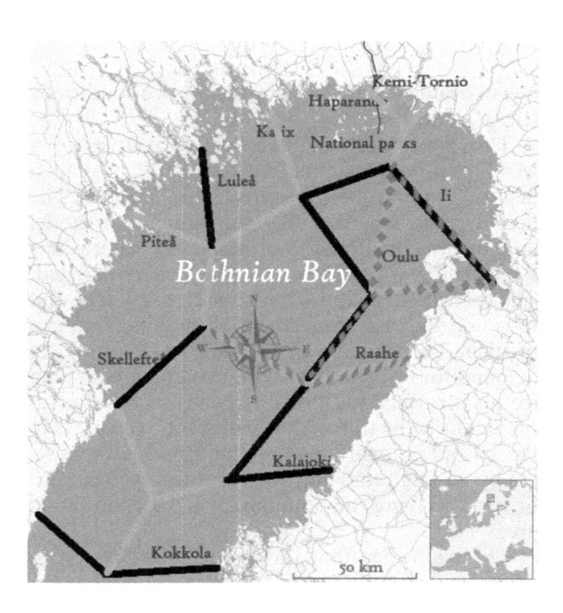

Figure 1. An example fairway-network constructed with the BBs. This network is composed by 16 BBs (10 lines BBs and 6 Y shape BBs). A junction BB is orange in color and a section BB is black. One operational area is explicitly shown: the BBs that comprise the operational area are marked with red stripes.

The example network of Figure 1 could have been constructed in many ways, i.e. by using different leg lengths and/or different BB choices. There is not one right way to construct a network, but in general the junction-BB should be used where ever possible (less BBs). In addition, when constructing a network, the suitable BB choices are largely dependent on the IB operational areas.

2.3 *Model input and output*

For each ship, the time when it enters the network, and from where in the network it enters, and its destination(s) within the network, and the estimated duration(s) of its port visit(s), are all inputted into the program. In addition to this, the program needs ice height – velocity (HV)-curves for the inputted ships and IBs (see 3.1), and an ice-data grid.

The model output consists of all details of all the events that take place in the simulation: individual ship stop-positions and times, individual assistance paths and durations. Possible towing events are also registered: towing start and stop positions and durations, for each towing event. The motor power (60/80/100%) that IBs use during transitions (not assisting) and durations of each transition, are also registered. In addition, the waiting times (waiting for assistance) of individual ships, which are calculated from the above-mentioned information, are also included in the output.

3 ICE CONDITIONS AND SHIP SPEED

Ice thickness and ice-type are the main ice parameters that the model uses. After providing the program with the start and end-coordinate points (long-lat.) of each fairway section, the program automatically reads the ice data from an ice-data grid into the fairways sections.

There is no limit on how often the ice-data can be updated, but a 24 hour period has been used. The resolution of the ice-data is equal to the distance-resolution of the model (adjustable parameter), i.e. the smallest distance that can be used when designing the fairway sections. 1 nautical mile has been used as the resolution.

3.1 *Equivalent ice thickness*

The ice-data provided by the Finnish Meteorological Institute (FMI) has five different ice-types: Level ice, Rafted ice, Slightly Ridged ice, Ridge ice, Heavily Ridged ice, and Brash ice.

By using a model for calculating equivalent ice thickness, the ice thickness of rafted and ridged ice types can be transformed into a corresponding thickness in level ice.

The equivalent ice thickness model was applied to a sample thickness of 0.5 m level ice. For each ice type, the increase in level ice thickness was converted into a relative thickness increase, which then is used for calculating the thickness increase for the corresponding ice type, for any thickness. Rafted ice gave a 7.5% increase, Slightly ridged ice gave +22.5%, Ridged ice gave +60% and Heavily ridged ice gave a 105% thickness increase.

3.2 *HV-curves*

An HV-curve models the ship speed as a function of ice thickness. The ones used by the simulation model are cubic equations. It is possible to have a specific HV-curve for each specific ice type. The equivalent ice thickness model (3.1) was utilized. Hence, the ship speed, in all the different ice types, could be modelled by solely using the HV-curve for level ice. As an exception was the Brash ice type, where the speed was modelled by an HV-curve resembling channel-like conditions.

Every ship-type has its own individual HV-curves, including IBs. The merchant ship HV-curves are assumed to be modelled at 85% engine power, whereas the IBs have three sets of HV-curves, one set for 60% engine power, another for 80% and a third one for 100% engine power. In addition, each merchant ship-type has an HV-curve for the speed under IB assistance, which also uses the equivalent ice thickness.

4 ICEBREAKER MOVEMENTS

4.1 *Basics*

All IB movements in the system strive to minimize the waiting-time that is gained by ships that have stalled in ice (i.e. ships in need of assistance). In addition special emphasis is done also to mimimize the used fuel onboard icebreakers so that icebreakers will use as low engine power as possible to keep the planned time schedule.

Finding the IB movements that would result in the absolute minimum waiting-time of the whole system for the whole simulation period, would be a too difficult problem. Therefore, in this model, an IB's awareness of its surroundings is limited to its operational area, i.e. when striving to minimize the waiting-time, it only considers ships that are, or will be in near future, within its operational area. The timespan of 'near future' is case dependent, but it is at the very least the duration from assistance start (in the last BB prior to entering the operational area) to the time when the ship enters the operational area. In other words; the length of the described time-period is equal to the length of the time-period that is spanned by the moment at which the IB becomes aware of a ship that is soon about to enter the operational area, and by the moment when the ship enters the operational area. Therefore, if the described time-period would be zero, the IB would become aware of a ship at the time the ship enters the operational area.

In situations where a ship doesn't need assistance in the last BB prior to entering the operational area, or when a ship enters the network the first time, the time (before the ship enters the operational area) when the IB becomes aware of such a ship is determined by an adjustable parameter (~8 hours was used).

4.2 *Minimizing the waiting-time*

The basis of the whole simulation model is to have the IBs move and assist in such a way that the total waiting-time is minimized as well as used engine power for as low as possible fuel consumption. An IB movement that results in least gained total waiting-time is considered optimal.

The IB has access to the stop-positions and—times of all the ships that it is aware of. Based on this information, the IB (i.e. the algorithm) finds the assistance-mission that minimizes the waiting-time, see also Figure 2 and Lindeberg et al. (2015).

4.2.1 *Minimization within the BB*

The BB is the foundational structure of the simulation model and the minimizing algorithm operates only within the BB. Therefore, an assistance mission's path cannot cross over to an adjacent BB; an assistance mission's range is within the BB where it is made. This limited operational "range" of the minimizing algorithm radically reduces the size of the minimization problem, and hence enables the use of the brute-force-search method to solve the minimization problem.

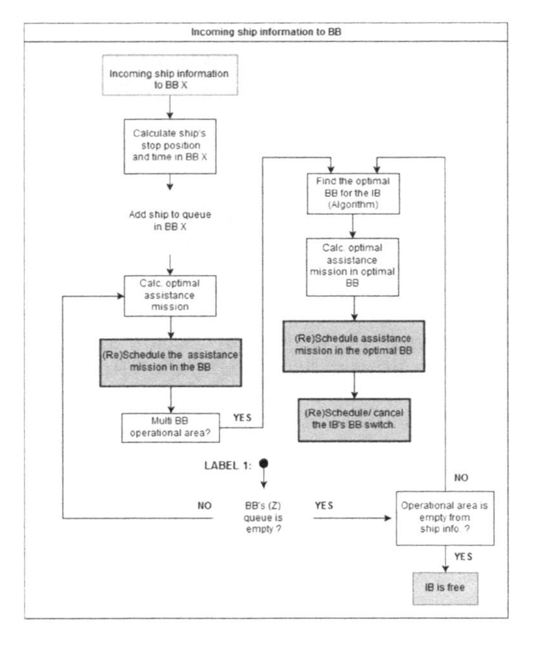

Figure 2a. The flowchart shows the steps that lead to the scheduling of an assistance mission in a BB. (The only purpose of the different shapes and colors is to make it clearer).

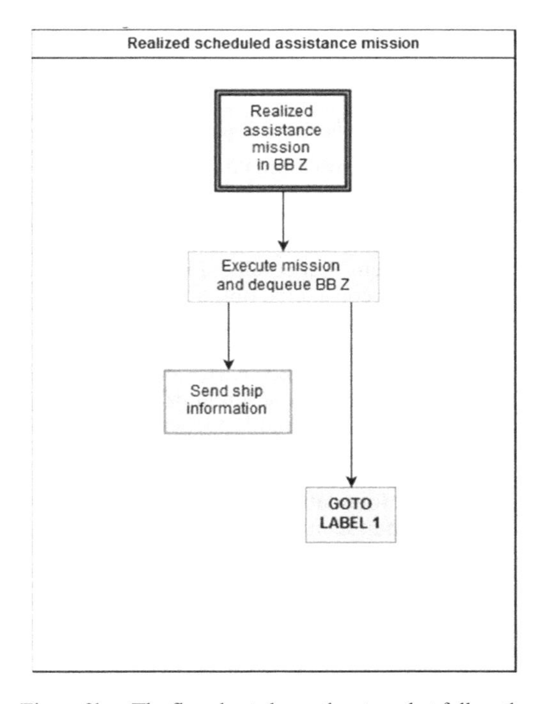

Figure 2b. The flowchart shows the steps that follow the realization of a scheduled assistance mission in a BB.

Applying the brute-force method for this problem means that the algorithm goes through all the possible assistance mission combinations. However, the algorithm doesn't always run though all the possible combinations: in cases where one single assistance mission (A→B) is not enough to serve all ships, brute-force search is only applied for determining the first two assistance mission steps at most (A→B and C→D).

For every assistance mission combination (max. the first two assistance-steps), determined by the brute-force method described above, the minimizing algorithm evaluates the "value" of the new state (IB position, ship stop-positions and – times) in which the BB is in following the completion of the assistance-mission steps. The value of the new state is determined by estimating the gained waiting-time in a case-specific time-period that starts after the completion of the assistance steps. For example, if two ships need assistance at point A, and both are heading to point B, and the IB assists only one of the ships, then the case specific time-period would be from finishing the assistance, to the time when the IB arrives back to point A.

The optimal assistance mission is the one giving the smallest total waiting-time when combining the waiting-time that is gained before finishing the assistance steps, and the waiting-time gained in the case specific time-period, post-assistance.

The reason for using the above described method, opposed to going through all the possible assistance combinations in every situation, is that we are only interested in knowing what the optimal first assistance step (A→B) is.

Theoretically, the used method doesn't guarantee that the found assistance step is optimal, but considering the structure of the model, it is very unlikely that the optimal first mission step is depending on the assistance choices made at the end of a long assistance-step chain to an extent that the used method would not be accurate enough to find the optimal first step.

4.2.2 *IB moving to another BB*

When the IB operational area covers multiple BBs, the IB must know when it's time to move over to another BB, and to which one, if there are more than one alternative.

One simple criterion determines if the IB should move to another BB, and it goes as follows: if the IB can start an assistance mission in another BB earlier than in the current one, then it will switch BB.

If the IB is not in a hurry, it starts moving towards the new BB at the latest possible moment, i.e. just to make it in time for the assistance start in the new BB. Consequently, if the IB receives new information of incoming ships before the start time, it can cancel the scheduled BB move if necessary.

4.2.3 *Algorithm for finding the most suitable BB*

If there is more than one BB that fulfills the BB-switching criterion (3.2.2), the most suitable BB is determined by a developed algorithm that compares all the BB candidates against each other, until there is only one left, the most suitable one. The most suitable candidate is not necessarily the one where the IB could start the assistance mission at the earliest point in time.

As in the minimizing algorithm (3.2.1), gained waiting-time is used as the quality measure in this algorithm as well. When the algorithm compares two BB candidates (marked as X and Y), it produces two quantities, β_X and β_Y:

$$\beta_X = W_a^X + W_a^Y + W_{a \to a+\tau}^Y \tag{1}$$

$$\beta_Y = W_{a+b}^X + W_{a+b}^Y + W_{a+b \to a+\tau}^X, \tag{2}$$

$$\text{if } a + b >= a + \text{, then } W_{a+b \to a+\tau}^X = 0 \tag{3}$$

where a, b and τ are positive real numbers (time). τ is a parameter [hours] ($2 <= \tau <= 4$ should be good. Only $\tau = 2$ tested so far). The notation W_d^Z represents the waiting-time gained in BB Z from the time when the algorithm was executed, till time d. $W_{d \to f}^Z$ represents gained waiting-time in a time-period $d \to f$. The a is the start time of assistance in BB X, which is calculated by the minimizing algorithm (3.2.1), where the IB's arrival time into the BB has been considered. Similarly, $a + b$ is the start time of assistance in BB Y (representing it as a sum is just to make it clear that it is later than a).

The candidate that gives the smaller β-value is the better alternative and the other one is dismissed as a candidate. Equations 1–3 are used until only one candidate remains.

The flowcharts of Figure 2a and 2b show the different steps taken by the program when a BB receives new ship information.

Notice that the box with the text "Multi BB operational area?" only has a "YES" option. This just means that the "NO" option doesn't lead to any new steps. The box marked with "(Algorithm)" represents the BB finding algorithm that was presented in this section (4.2.3). Furthermore, the "LABEL 1" port is a link between the flowchart in Figure 2a and the one shown in Figure 2b below.

From the flowchart in Figure 2b one can see when the ship information is generated (blue box). Notice that the information contained in the blue box of Figure 2a is all originating from the blue box in Figure 2b. Dequeue (green box) refers here to the removal of the assisted ships from the BB's ship queue. Also, notice that the BB in Figure 2b is labeled as "Z" (and as "X" (2a) in the first three boxes). This is just to point out that the BB that generates the ship information (2b) is not the one receiving it (2a).

5 ASSUMPTIONS IN THE MODEL

5.1 *IB's transition movements*

When an IB makes a transition movement, i.e. moving from its present location to the start position of an assistance, or when making the transition to another BB, it has the option of using 60%, 80% or 100% engine power. The IB will always use the lowest possible effect option that enables it to be on time at the destination position. In addition, the start time of a transition move is always as late as possible, while still prioritizing the lower effect option over a higher one. The reason for delaying the start-time as much as possible is to maximize the IB's ship-information before making the assistance decision. The IB's awareness of ships entering its operational area (receiving time of ship-information) was discussed in section 3.1.

5.2 *IB assistance*

A ship will stop and wait for assistance when its speed drops below a limit value (an adjustable parameter in the model of 3 knots is used). A ship that is going to need assistance somewhere within the BB will not pass an already stopped ship, if they both have the same destination within the BB (the stronger ship will stop at the same position as the earlier ship). If the earlier ship is assisted before the stronger ship arrives, then naturally the stronger ship will continue as far as possible.

The assistance speed is determined from the ship specific assistance-HV curve. If the assistance is a convoy, then the assistance speed is determined from the assistance-HV curve that gives the slowest speed. In addition, the assistance speed is always limited by the IB's maximum speed in the current situation.

5.2.1 *Assistance by towing*

A ship is towed if the assistance speed drops below a limit value (here 6 knots, adjustable parameter). The target towing speed is also an adjustable parameter, but the actual towing speed is determined from the IB's HV curves. The IB uses the smallest effect option (60/80/100%) which allows the target towing speed to be reached. In addition, the maximum towing speed is limited by the open water top speed of the ship being towed.

In a convoy situation, ships can follow the towed ship if their speed (form the ship specific assistance-HV curve) doesn't fall below a limit value (adjustable parameter). If the speed drops below the limit, the ship(s) stops.

The idea with having two limit-speed parameters for towing start (one for the initial ship (towed) and the other for the ships following it) is that the following ship(s) can be set to have a smaller limit-speed parameter value than the limit-speed parameter governing the initial towing start.

For the towing to be terminated, the towed ship's assistance-HV curve must give a minimum speed (adjustable parameter) over a specified distance, i.e. the speed must not drop below the minimum speed at any point during the distance. The distance is an adjustable parameter.

6 CHANGES IN THE FAIRWAY-NETWORK DURING A SIMULATION RUN

Changes in the structure of the network can be setup to take place at desired times during the simulation period. This feature becomes especially important when the simulation period is long, as the fairway-network most likely will not remain constant during a longer period. The reason one cannot just do consecutive simulations runs after each change in the network is that all the information (ship and IB positions) contained in the network at the time of the change would be lost.

Everything in the network can be changed, i.e. the whole network can be rebuilt. The most obvious thing to change is probably the borders of the IB operational areas, which could mean adding new IBs into the network, or removing existing ones, or simply changing the operational areas while keeping the IB count constant.

When designing and setting up a simulation that includes a network change, all the fairway sections of the N:th network must be mapped onto the N + 1:th network. In practice, this means one must for each fairway section (a section can any length) in the N:th net assign a corresponding section in the N + 1:th net. When a link like this exist between two different nets, a ship can be transferred to the new net, without losing much or any information. How much information is lost in the transition depends on how accurately the mapping is done, but also on how accurately the mapping can be done: two sections that are geographically at different locations cannot be mapped onto one another without losing information—though the program includes a specific mapping method for handling these kinds of situations.

7 VALIDATION OF THE MODEL

The simulation model was validated by simulating the winter traffic system of the Bay of Bothnia, for the period 15.1–15.2.2010 (32 days). Figure 3 illustrates the ice conditions on that time on that sea area.

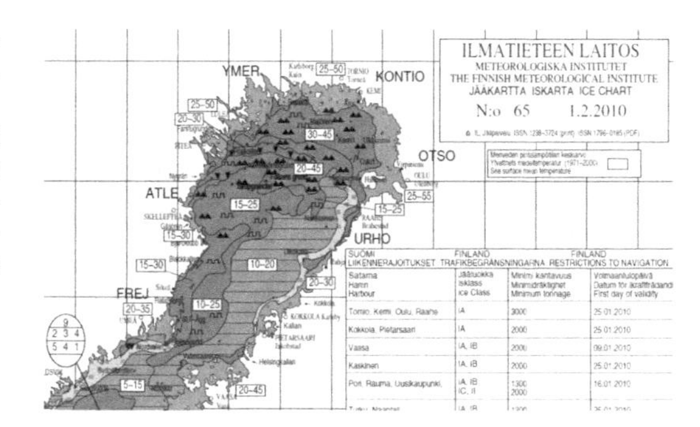

Figure 3. The ice conditions and icebreakers at the Bothnian Bay on February 2010.

The waiting-time that was gained by merchant ships in the area during the simulation period was acquired from data provided by the Finnish Transport Agency.

7.1 *Summary of the ships simulated*

The details of the merchant ships that were sailing through the area, during the time of simulation, were acquired from AIS-data.

For each ship, a timestamp for its first appearance into the simulation domain was approximated from the data, or if possible, acquired directly from the data. In addition, the coordinates associated with each timestamp were collected. Most of the ship traffic entered the simulation domain through the so-called input-node, which was the southernmost point of the simulation domain. A small portion of the ships entered through a port.

For each ship, its destination ports, along with the associated port-visit durations, were all collected from the data. The ships were classified into different ship-types according to their characteristics, with each type having individual HV-curves. A histogram below shows, see Figure 4 the DWT and ice-class of the ship-types that were used in the test for ice class IA and IAS ships as no other ice class was present.

Figure 5 below displays the ice class distribution of the individual ships in the test.

Figure 6 below displays the ice class distribution of the actual ship traffic, and in this context, the term 'ship traffic' refers to a ship's journey from a point x to its destination. There was a total of 408 of such journeys.

If all ships only had one destination within the test area, and they would exit from the destination (port), they would each contribute with two journeys (to port, away from port), i.e. the ice class numbers in Figure 6 would be the same as the ice

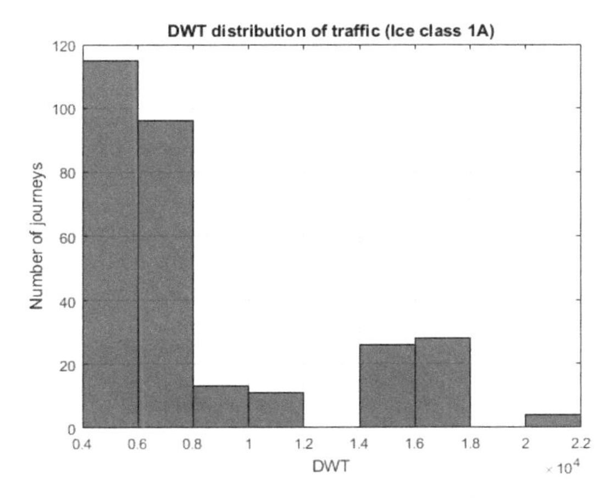

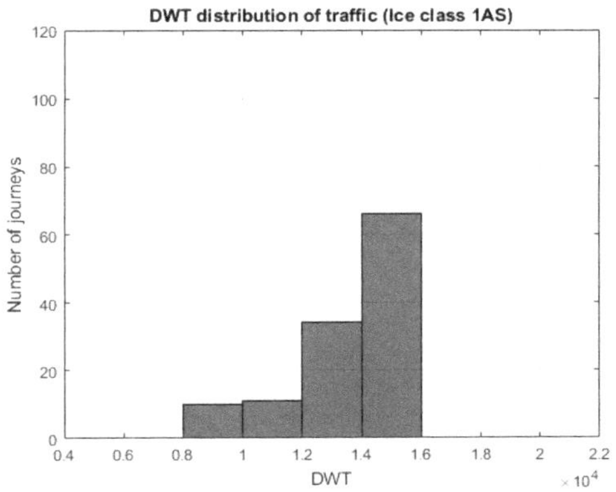

Figure 4. Histogram showing ship size (DWT) and ice classes of the ship-types used in the case study.

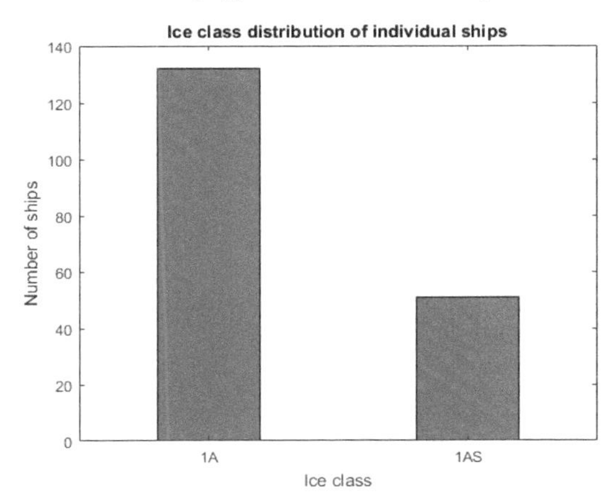

Figure 5. Ice class distribution of the individual ships. There was a total of 183 individual ships.

class numbers in Figure 5, but multiplied by two. By comparing the distributions, most of the ships have had only one destination.

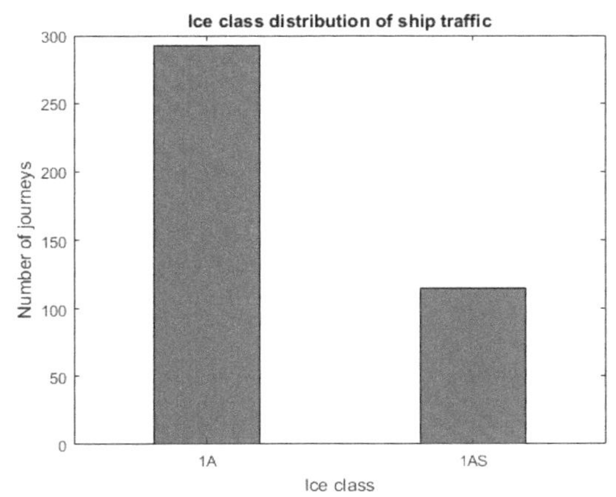

Figure 6. Ice class distribution of the ship traffic.

7.2 *EEDI-ships simulations*

In addition to the simulation run were the ships were classified into ship-types resembling the ship characteristics as much as possible, four additional simulation runs were made where the inputted ships were replaced with Energy Efficiency Design Index (EEDI) compliant ships, see IMO (2011). These ships as a general rule have a lower power-to-weight ratio, which lessens their ice going capabilities. The ships were categorized according to their dwt. into three different EEDI ship-models: 5100-dwt, 15000-dwt and 22000- dwt. Most of the ships fitted into the 15000-dwt category.

The level-ice HV-curves of the EEDI ship-types were linear from the open water-speed till 0 km/h, with 0 km/h speed at 10 cm ice thickness.

In one of the EEDI simulations, all the ships were replaced with EEDI ships. In the other ones, the ships were replaced so that 28%, 51.7%, and 71.8% of the ship traffic would be made by EEDI ships. In this context, the term 'ship traffic' refers to a ship's journey from point x, to its destination. There was a total of 408 ship journeys, so for example in a 50% case, 204 journeys would be made by EEDI ships.

The replaced ships were chosen based on their DWT in the following manner: the 28% case comprised all ships between 14k-18k DWT, 51.7% – all between 4k-6k DWT, and 71.8% case comprised all between 6k-22k DWT.

7.3 *Fairway networks and IB operational areas*

Throughout the whole simulation period, the fairway network(s) comprised four IB operational areas, with one IB in each. The network(s) was designed to resemble the fairway sections that were used during the simulation period in the area. The ports that were included in the simulation area was Kokkola, Raahe, Oulu, Kemi, Tornio, and Luleo (route to Luleo).

During the simulation period, the lengths of the fairways sections changed, as did the borders of the operational areas, and at some point, completely new sections were added to the network. To account for all these changes, the simulation had to be designed with six consecutive networks (five network changes).

7.4 *Results*

The two model parameters which directly relate to reality are the universal "stop-speed" and "limit-speed" parameters. Stop-speed is the speed limit that determines when a ship stops and starts waiting for assistance. Limit-speed is the speed limit that determines when a ship most be towed. The values used for stop-speed and limit-speed were *3kn* and *6 kn*, respectively (with *limit-speed* set to *6 kn,* no towing was needed in any of the three simulation runs in this case study).

The waiting-times that were collected from the data of FTA were labeled under port names in the following manner (in hours): *85* (Kokkola), *293* (Raahe), *308* (Oulu), *154* (Kemi) and *286* (Tornio). This gives a total waiting-time of *1126* hours.

7.4.1 *Non-EEDI simulation run*

The last assistance mission started at 851 h (35.5 days) simulation time and the total waiting-time for the whole simulation was *1209* hours. However, the total waiting-time accumulated within the target period (32 days) was *1145* hours, which is within a 1.7% margin from the value collected from the data. Figure 7 below displays the cumulative waiting-time for the target period.

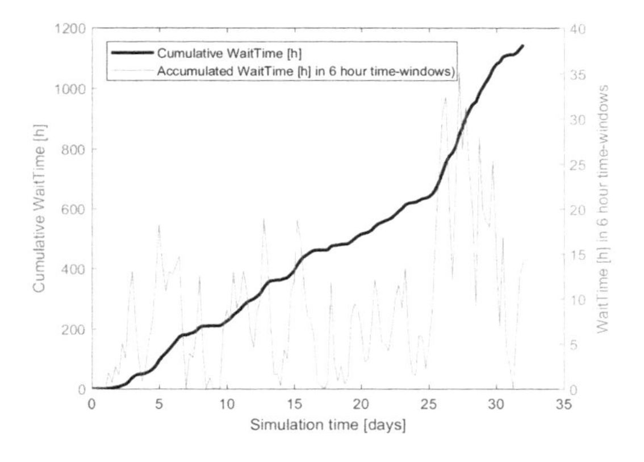

Figure 7. The black line is the cumulative waiting-time [h]. The orange lines (spikes) represent the accumulated waiting—time in 6 hour long time-windows.

Figure 8 below shows the same cumulative waiting-time as in Figure 7, but here the orange spikes represent the number of ships that need assistance at some point within the 6 hour long time-window.

Comparing the orange spiked plots of Figure 7 and Figure 8, it's clear they follow roughly the same pattern. This should be expected, as the number of ships that need assistance is closely related to the gained waiting-time. Note that also ships that doesn't gain any waiting-time (instantly assisted) are included in the spikes of Figure 5.

The model is sensitive to a number of parameters e.g. to the stop-speed parameter: the waiting-time is more than doubled between the 2 to 4 kn range increasing from about 900 hours to 1800 hours. The few downward slopes are due to normal variation in the effectiveness of the assistance missions: the positions of ships in relation to the IBs can easily affect the waiting-time in a positive or negative way. Also, a few vital model parameters should be optimized for every simulation run, and here they are optimized for stop-speed: 3 kn is used. Figure 9 illustrates how sensitive the waiting time is e.g. for this parameter.

7.4.2 *EEDI simulation runs*

The switching from the original ship-types to EEDI-ships, which have significantly lesser ice-going capa-bilities than the original ship-types, had the expected effect on the simulated system: the waiting-times were significantly higher in the EEDI-simulation runs. Figure 10 below shows the total waiting-time as a function of the EEDI-ship percentage.

The waiting-times at the actual data points in Figure 10 are as follows: 0% – 1145 h, 28% – 1786 h, 51.7% – 1987 h, 71.8% – 3677 h, and at 100% – 4040 h. The large increase in waiting-time

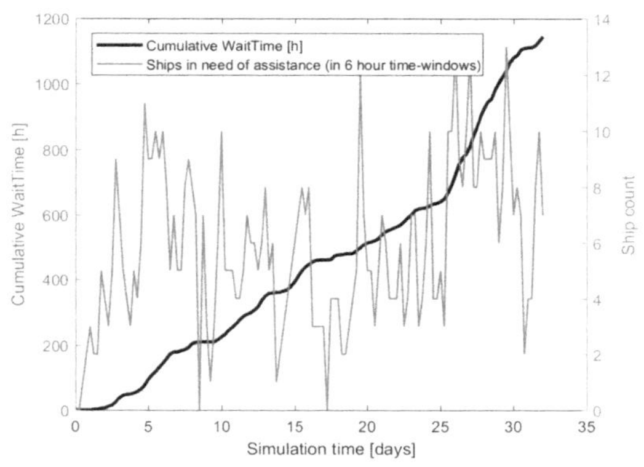

Figure 8. The black line is the cumulative waiting-time [h]. The orange lines (spikes) represent the count of ships that need assistance at some point within the 6 hour long time-window.

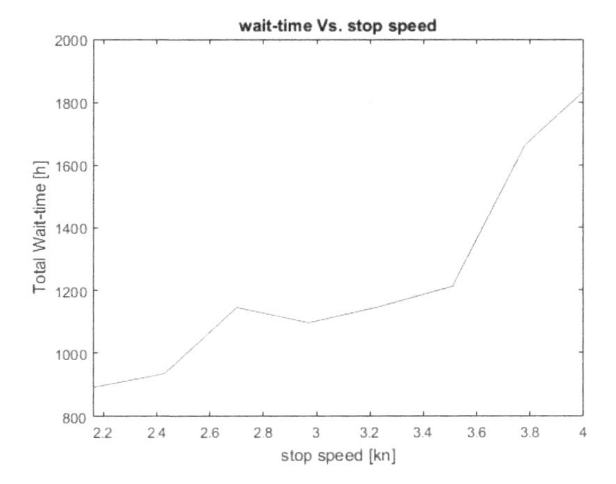

Figure 9. Effect of the used value for the ship stop peed on the waiting time [h].

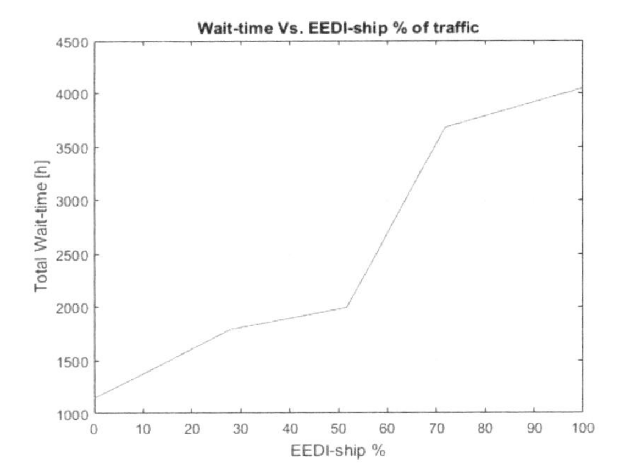

Figure 10. Total waiting-time as a function of the EEDI-ship percentage.

when going from 51.7% to 71.8% could be attributed to the fact that the only ships that are not replaced in the 71.8% case, areships that are modelled by the smallest (4.5 k DWT) ship type, with below average ice going capabilities.

8 CONCLUSIONS

The present model cannot be compared to any other existing models as there aren't any easily comparable models available. However, the results from the conducted validation tests speak for itself. It is highly unlikely that test results from such a comprehensive test (long duration, large area, high traffic volume) would be so close to the actual data, if the model wasn't somewhat accurate. However, when evaluating the results, too much attention shouldn't be put on the attained small margin (1.7%) with the actual data as only total waiting-time was used for validation. Nevertheless, total waiting-time is still the most describing statistic of the system.

The plan is also to apply the model for wider sear areas and longer simulation periods to get the good overall picture of the winter traffic management system with varying extent of the sea ice. When only considering a specific period, there is a risk that the model is adjusted so that it gives the "right" outcome for the considered period, even though it might not be generally accurate. Thus, validation for another periods are important actions in future.

In addition, having accurate ice-conditions (model for equivalent ice-thickness (3.1)) and perhaps even more importantly, having accurate HV-curves, is vital for good model performance.

Further improvements to the simulation model are possible. The main improvement suggestion would concern the points of discontinuity that arise at the connection points between the IB operational areas. These points are probably the main source of unrealistic behavior in the model. Therefore, getting rid of these points, or diminishing their effect, would significantly improve the model. This could be achieved by designing an IB operational area that would be able to include multiple IB operational areas.

ACKNOWLEDGEMENTS

This work has been enabled by EU-TNT funding through the projects WINMOS I and WINMOS II, which is here gratefully acknowledged. In addition the authors would like to thank Mona Zilliacus, Iiro Kokkonen, Mikko Lensu, Tom Mattsson, Tuomas Taivi and Emmi Huvitus in helping to gather the data and help in the practical simulation process.

REFERENCES

Bergström, M., Erikstad. S. and Ehlers, S. 2017. The influence of model fidelity and uncertainties in the conceptual design of arctic maritime transport systems. Ship Technology Research Schiffstechnik pp. 40–64, Volume 64.

Goerlandt, F., Montewka, J., Zhang, W. and Kujala, P. 2017. An analysis of ship escort and convoy operations in ice conditions. Safety Science. (2017). http://dx.doi.org/10.1016/j.ssci.2016.01.004.

International Maritime Organisation (IMO). 2011. MEPC 62/24/Add.1 Annex 19. Available online at: http://www.

imo.org/en/MediaCentre/HotTopics/GHG/Documents/eedi%20amendments%20RESOLUTION%20 MEPC203%2062.pdf#search = EEDI. Accessed: 9.1.2018.

Kamesaki, K., Kishi, S., Yamauchi, Y., 1999. Simulation of NSR shipping based on year-round and seasonal operation scenarios. INSROP Working Paper 164–1999. INSROP.

Kuuliala, L., Kujala, P., Suominen, M. and Montewka, J. 2017. Estimating operability of ships in ridged ice fields. Cold Regions Science and Technology. Volume 135, March 2017, Pages 51–61.

Lindeberg, M. Kujala, P. Toivola, J. Niemelä, H. (2015). Real-time winter traffic simulation tool—based on a deterministic model. Scientific Journals of the Maritime University of Szczecin. 2015, 42 (114), 118–124. ISSN 2392–0378 (Online).

Montewka, J., Goerlandt, F., Kujala, P. and Lensu, M. 2015. Towards probabilistic models for the prediction of a ship performance in dynamic ice. Cold Regions Science and Technology. Volume 112, April 2015, Pages 14–28.

Patey, M., Riska, K., 1999. Simulation of ship transit through ice. INSROP, INSROP Working Paper 155–1999.

Marine Design XIII – Kujala & Lu (Eds)
© 2018 Taylor & Francis Group, London, ISBN 978-1-138-34076-3

Ice management and design philosophy

S. Ruud & R. Skjetne
Sustainable Arctic Marine and Coastal Technology (SAMCoT), Norwegian University of Science and Technology, Trondheim, Norway

ABSTRACT: Ice management (IM) is defined as all activities carried out with the objective of mitigating hazardous situations by reducing or avoiding actions from any kind of ice feature to a protected unit (e.g. a drilling vessel) and includes several types of barriers. IM barriers are ranging from ice observation, ice prediction, ice alerting, ice fighting with icebreakers, and disconnection procedures of the protected unit. The design decisions of the IM barrier systems can be based on qualitative or quantitative performance models. Qualitative descriptions of independent and dependent barriers are first defined and exemplified with qualitative decision criteria. Qualitative concepts for barrier performance of ice prediction are defined and illustrated in event trees. National barrier regulations (e.g. PSA) contain requirements to model quantitatively the barrier performances. Quantification of the IM performance, which are defined by probabilities of barrier functions, is a major challenge due to lack of data and existing uncertainties. Finally, the paper presents a brief plan for demonstration of the performance models in the design phase with experience data collection supporting the safe learning principle.

1 INTRODUCTION

Design decisions for safe and efficient ice management (IM) systems can be based on performance models and decision criteria for the overall or parts of the IM systems. The IM system performance models should be able to cover the following IM system parts (see Fig. 1):

- Ambient ice regime, ice hazards and ice risks to the protected unit operations.
- Operational ice observation, prediction and alerting functions.
- Operational physical IM or ice fighting with e.g. icebreakers.

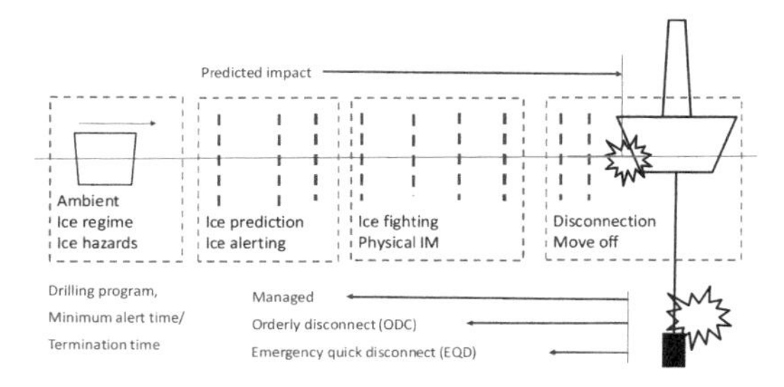

Figure 1. IM performance models cover ice regime models, ice observation and prediction, physical ice management (icebreakers), disconnection and move-off systems for the protected unit. The protected unit will need sufficient time durations for e.g. orderly disconnect or emergency quick disconnect procedures.

- Operational disconnection and move-off procedures and systems for the protected unit. In this paper, a ship-shaped movable exploration offshore unit is chosen as the protected unit.

The Ice Management and Design Philosophy Work Package is a part of the SAMCoT research program. In the beginning of this project, complete methods for combining all the above types of information in qualitative and/or quantitative decision processes were not available. This paper is an attempt to put this together and contribute to the solution of the identified needs. The research and development work are ongoing.

IM design decisions are proposed to be based on a top-down approach of barrier performance descriptions, and the main description types are presented in sections 2–6 as indicated below:

2. The barrier definition and high level definitions of general barrier events are described.
3. Boolean representation of main barrier events. The success of a barrier system is dependent on fulfilment of the barrier internal and external requirements of several barriers.
4. Detailed barrier performance requirements. The detailed combined functional and ambient performance requirements for barrier success are defined.
5. Probabilistic barrier performance. Quantification of the barrier performance is based on the probabilities of successful performance of the Boolean expressions of the barriers.

6. Expected consequences and risk. Consequences associated to the resulting events can be used for estimating residual risk of the barrier system and as basis for design decision-making.

Section 7 describes a plan for a case study on testing and demonstration of the methods given in this paper.

2 IM BARRIERS

The objective of IM is to mitigate hazardous situations by reducing or avoiding actions from any kind of ice (sea ice or glacial ice) at a specific location for planned operations by a protected unit.

The initial premise in this paper is that IM design shall be described in a top-down manner. The ice hazards and risk shall be sufficiently mitigated by the IM barriers. All elements shall be modelled in accordance with the given type of design decision acceptance criteria for the given decision scope. If more detailed information is needed the proposed IM design process shall advice how more detailed information may be incorporated to support an improved decision.

2.1 *Operational and environmental requirements*

The protected unit shall carry out operations within stated environmental limits for given durations.

Two typical operational situations are:

- Long term continuous operational durations (e.g. many years) with fixed environmental limits for e.g. a fixed production unit.
- For short term exploration projects (e.g. drilling 1–3 months) with specific activity durations and variable environmental limits.

Description of the planned activities and operational requirements shall be established in the beginning of an IM design project.

2.2 *Ambient ice regime*

In the beginning of an IM project, available information about the ice regime and the possible hazards for the planned operations shall be compiled.

We assume that the location of the protected unit is at a certain distance from the ice sheet and that possible arrivals of hazardous ice may occur seldom and with a low frequency (e.g., 1–5 times annually). This situation is sometimes denoted as *Workable Arctic Conditions*. The demand for operational IM responses by icebreakers is assumed to occur a few times annually.

The opposite situation is relevant for locations where the ice sheet may continuously surround the protected unit. The demand for IM response may in periods be continuous, and in other periods the ice conditions may be inoperable.

2.3 *Establishing the ice hazards and the ice risks*

Based on the information of the ice regime and the operational ice limitations, an ice hazard and risk analysis shall be prepared and quality assured.

The initial IM design decision shall therefore clarify if the actual risk identified will require additional risk reduction measures (or barriers) in order to comply with the stated acceptance decision criteria.

2.4 *IM performance acceptance criteria types*

The acceptance criteria for IM performance may be defined as requirements to the performance of the overall IM system scope, or at specific subscopes, e.g. ice observation and prediction (Fig. 2). The performance requirements may also be qualitative and quantitative. The main issue, however, is that the overall IM plan shall state the scope and the IM performance acceptance criteria to be complied with for the given scope.

2.5 *The barrier concept*

Operations on the Norwegian Continental Shelf shall be in compliance with the regulations from the Petroleum Safety Authority (PSA 2017). In this paper *§5 Barriers* is used as a reference for defining the barrier concept.

> *'…the responsible party shall select technical, operational and organisational solutions that reduce the likelihood that harm, errors and hazard and accident situations occur. … Barriers shall be established that at all times can a) identify conditions that can lead to failures, hazard and accident situations, b) reduce the possibility of failures, hazard and accident situations occurring and developing, c) limit possible harm and inconveniences… Where more than one barrier is necessary, there shall be sufficient independence between barriers.'*

Figure 3 is an illustration of the main barrier events.

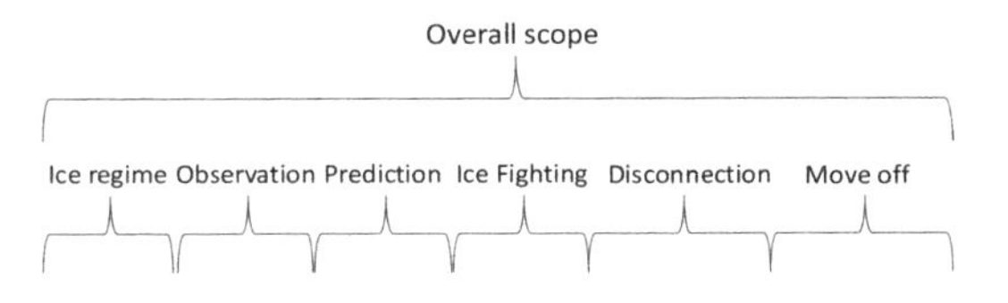

Figure 2. The IM performance acceptance criteria are formulated at the overall barrier system level or at subsystem levels.

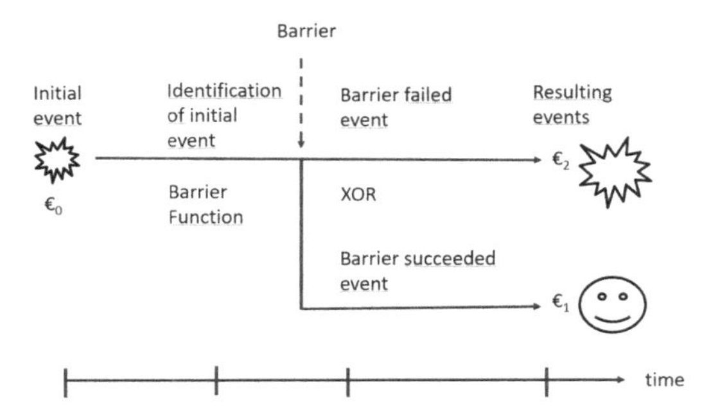

Figure 3. Barrier events in an event tree with one barrier. The barrier shall be able to a) identify conditions or initial events ($€_0$) that can lead to further consequences, b) reduce the possibility of development ($€_1$) of further consequences, c) limit possible harm, and inconveniences ($€_2$). In the figure the exclusive or (XOR) indicates that the barrier will either succeed or fail for one single initial event.

3 BOOLEAN REPRESENTATION OF MAIN BARRIER EVENTS

3.1 *Hazardous ice conditions*

IM is relating to ambient conditions where the conditions in the ambient area may be hazardous to the operation of the protected unit. The ambient conditions may also affect the performance of barriers.

Environmental or ambient parameters concerns:

- wind and current (including depth profiles),
- wave height and spectrum,
- temperature,
- sea ice and iceberg, size, thickness, strength,
- rain, snow, fog,
- visibility, and
- air pressure, polar lows.

Combinations of the environmental parameters must be considered, especially the fact that the ice conditions may take multi-domain characteristics with large spatial and temporal variations.

Examples of ice related events and conditions are:

- drift of ice islands and fragments,
- drift of ice that was land-fast and large areas of pack ice,
- large changing and reversals in ice drift,
- drift of old ice and glacial ice (icebergs) towards the operations sites,
- fast changes of wind speed and direction causing pressure,
- fast changes of currents (outflow from river or ice dam failure), and
- effects of polar lows.

In the context of this paper, occurrence of a hazardous condition at time t is denoted as an initial event $€_0(t)$ and represented as a Boolean variable normally being false, but taking the value true (.T.) when the event occurs (Fig. 3).

$$€_0(t) = .T.$$

During the design phase, the IM designers must apply given acceptance criteria and processes for reduction of the complete set of hazards to be handled in the following IM design process.

3.2 *Barrier function performance*

Traditionally, a safety function or a barrier function is referring to all elements needed for performing risk reduction. A pressure protection system (barrier) may have a sensor (e.g. pressure transmitter), a logic controller, and an actuating device (e.g. valve) which will reduce the pressure (physical risk) when the sensor is measuring an operating pressure above a given limit.

As an IM related example, consider the design of a barrier system with two barriers (A and B) for observing and alerting the protected unit about possible approaching sea ice.

The IM plan typically define 3 zones (Fig. 4) around the protected unit:

1. Zone 1 is secure zone with radius a.
2. Zone 2 is management zone.
3. Zone 3 is monitored zone.

Barrier A is based on a satellite with ice surveillance functionality. In the case that ice is entering Zone 2 from Zone 3, an orderly disconnect (ODC) alert shall be submitted to the protected unit. The protected unit should then initiate procedures for terminating critical activities and start disconnection of systems connected to the sea bed. The performance of barrier A is initially not proved

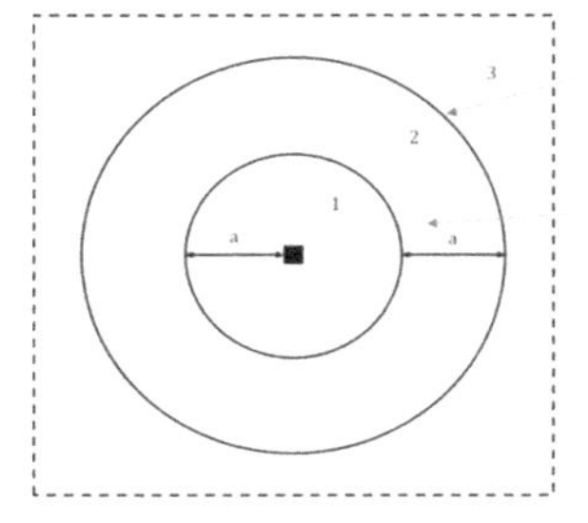

Figure 4. Ice zones for a MODU. 1: secure Zone, 2: management Zone, 3: monitoring Zone (ISO/FDIS 35104). Observation of ice entering Zone 2 shall trigger an ODC alert and this is considered as barrier A. Ice detection system and alerting in Zone 2 is considered as barrier B.

sufficiently high. Hence, according to the safe learning principle, there is a need for extra high contingency (EHC) measures and a barrier B must be included.

Barrier B is based on a supply ship patrolling in Zone 2. The ship has ice detection radars specialized for identification of sea ice, and the crew are also watching for possible ice. In the case that ice floes are observed from the scouting vessel, the protected unit shall immediately be alerted (Fig. 5).

The overall IM system may consist of barriers denoted e.g. A, B, C... which initially are considered independent. This means that there are no common causes leading to simultaneous failures of A and B. Hence, it is assumed that:

- A failure cause leading to failure of barrier A will not cause barrier B to fail in the same manner.
- Failure of B shall not be a consequence of the failure effect of barrier A.

This means that the overall barrier system will alert correctly if either barrier A or barrier B will work correctly (Fig. 5).

The resulting events (€) from the barrier system design in Figure 5 can therefore be expressed by the Boolean equations:

$$€_1 = €_0 \cap F_A$$
$$€_2 = €_0 \cap \neg F_A \cap F_B$$
$$€_3 = €_0 \cap \neg F_A \cap \neg F_B$$

where

F_A = true; barrier A function on demand
F_A = false; barrier A is failing on demand

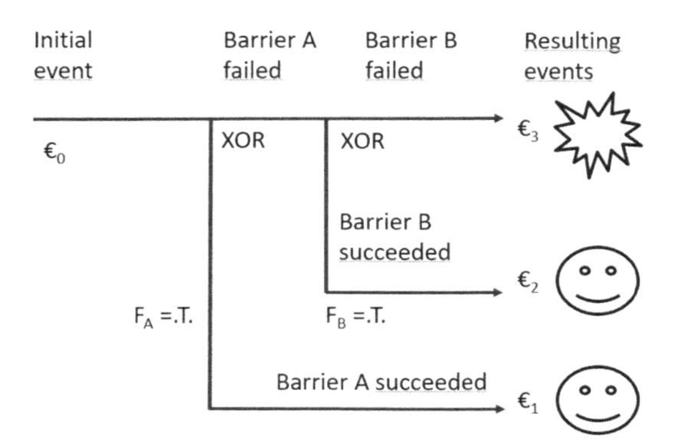

Figure 5. Barrier system event tree representing independent barrier A with function F_A and barrier B with function F_B. The overall barrier system will work as required if either barrier A or barrier B function. The barrier system will fail if both barrier function F_A and barrier function F_B fails.

F_B = true; barrier B function on demand
F_B = false; barrier B is failing on demand
$\cap$: Boolean **AND** operator
$\neg$: Boolean **NOT** operator

The single failure criterion is the traditional type of qualitative functional acceptance criterion used by, e.g. classification societies in prescriptive technical rules for dynamic positioning (DP) system equipment classes 2–3 or for drilling systems (DNVGL-OS-E101, 2018). As the drilling and DP systems are parts of the IM scope, it is natural to start to cross-reference the requirement specifications of these systems to the traditional type of requirements.

In the previous example with barriers A and B, the overall barrier system will comply with the single failure criterion. The reason for compliance is that the two barriers A and B have to fail.

3.3 *Ambient and external conditions*

A technical barrier consist of several subsystems like detection, logic units, and actuating devices. All supporting and utility systems required for the barriers to work are also considered a part of the barrier. The barrier environment and connections to the other systems may also influence the vulnerability of a barrier. In order to organize and visualize these relations, the system boundaries (ISO 14224:2006) should be established. In this context 3 boundaries are described:

1. Barrier boundary, functional parts.
2. External connections and conditions.
3. Environmental and ambient conditions for barriers.

Assume that a Boolean F_A represents the function of barrier A where:

F_A = .T. (true); barrier A function on demand
F_A = .F. (false); barrier A is failing on demand

Also assume that the function of barrier A is dependent on the state of external utilities or ambient conditions, denoted by a Boolean variable X_A where:

X_A = .T.; external/ambient state is workable
X_A = .F.; external/ambient state is not workable

The robustness of a barrier is the ability to function under relevant external conditions at the demand situation (Fig. 6).

The barrier robustness may also be denoted as survivability or vulnerability (Hauge, S. & Øien K. 2016). Robustness and survivability can be related

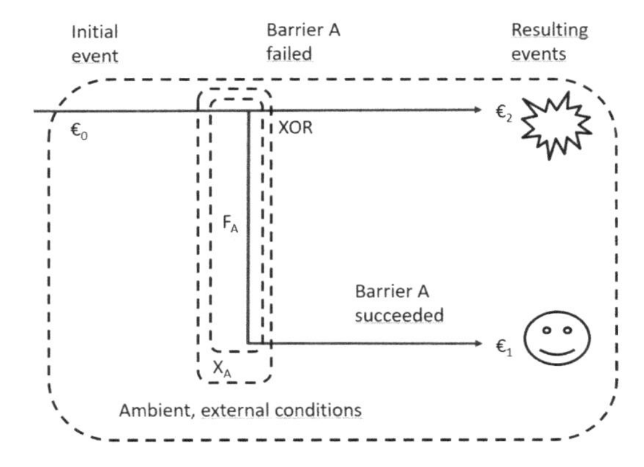

Figure 6. Barrier A will work if the functional part (F_A) is working when demanded and if the external conditions required are complied with (X_A).

to X(t) = .T., and vulnerability can be related to X(t) = .F.

3.4 *Functional and external performance*

The Barrier A Succeed on Demand (BSD_A) if both the function of the barrier and the external conditions are operable. Let $X_A(t)$ and $X_B(t)$ be Boolean variables for external conditions for barriers A and B.

Then

$BSD_A = (F_A \cap X_A)$; barrier A will work
$BSD_B = (F_B \cap X_B)$; barrier B will work

This means that the barrier system of A and B will work (refer to lower branch in Fig. 7) on demand on the initial event ($€_0$) according to:

$$€_1 = €_0 \cap F_A \cap X_A$$

or if (refer to middle branch in Fig. 7):

$$€_2 = €_0 \cap \neg(F_A \cap X_A) \cap (F_B \cap X_B)$$

The barrier system will fail to protect the unit (represented by upper branch in Figure 7 if the following event occurs:

$$€_3 = €_0 \cap \neg(F_A \cap X_A) \cap \neg(F_B \cap X_B)$$

The $€_3$ event represents a hazardous event and the corresponding consequence of the IM barrier system for the protected unit. The analysis of the conditions for event $€_3$ has to start with a screening of the Boolean functions for $€_0$, F_A, X_A, F_B and X_B.

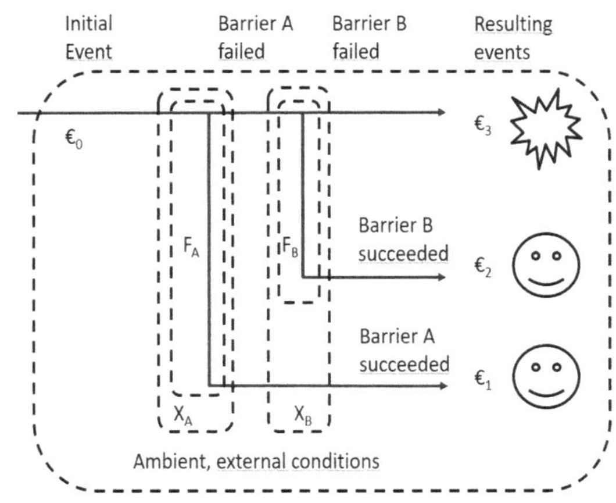

Figure 7. Barrier A will work if the functional part (F_A) is working when demanded and if the external (X_A) conditions required are complied with and similarly for barrier B. The overall barrier system will fail if both barrier A and B fails. Also, if the ambient conditions for both A and B are non-operable, the overall barrier system will fail.

3.5 *Acceptance criteria for loss of protection*

The analysis process of the $€_3$ event shall be based on a stated requirement or acceptance criterion for loss of protection by the given function and ambient conditions of the barrier system.

Acceptance criterion scope: $\{F_A, X_A, F_B, X_B\}$.

Acceptance criterion: No single failure in the acceptance scope shall lead to loss of protection of PU.

The loss of protection condition is given by:

$$€_3(t) = .T.$$

meaning that loss of protection will occur if:

$$€_0 \cap \neg(F_A \cap X_A) \cap \neg(F_B \cap X_B) = T.$$

Therefore, the task is to find possible solutions of this equation.

Ideally the Boolean variables $€_0$, F_A, X_A, F_B and X_B should be independent, and the barrier variables should be true (.T.). In such cases it can be shown that there are no single failure (e.g. X_A = .F.) giving loss of protection such that $€_3$ (t) = .T.

In reality, the independence claim of the above variable set has to be analyzed and justified. In practice, such analysis could be a kind of FMEA or hazard identification in order to identify possible dependencies or common causes that may give $€_3$ (t) = .T.

In the case that the external requirements (e.g. visibility) for barriers A and B are equal:

$X = X_A = X_B = .F.$

then both barriers A and B will be affected simultaneously by the common condition cause X.

By elaborating the Boolean expressions by means of standard Boolean algebra laws for the branches in Figure 7, it may be shown that the event tree may be reorganized as shown in Figure 8. The resulting events (note new numbering of events 3 and 4) will occur given the following conditions:

$$€_1 = €_0 \cap F_A \cap X$$
$$€_2 = €_0 \cap \neg F_A \cap F_B \cap X$$
$$€_3 = €_0 \cap X \cap \neg F_A \cap \neg F_B$$
$$€_4 = €_0 \cap \neg X$$

By inspection of the equations above, event $€_4$ will be the result of $€_0$ and a single non-compliance of the external factor X (e.g. no visibility).

This means that the barrier system design is not complying with the given acceptance criterion which states that the barrier system shall not fail due to a single failure or non-compliance.

Generally, the single failure criterion will require two independent barrier functions in the system. It must be verified that the claimed independent barriers do not contain functional relations which may reduce the number of independent barriers, as was shown in the previous example.

Continuing our example, since barriers A and B are dependent, the barrier B is exchanged with a new barrier C for which the ambient conditions X_C are different than for the A barrier. The cost for the C barrier is significantly higher than the B barrier, but at the decision point it is decided to start operations with this configuration (Fig. 9).

It is also decided to record compliance and non-compliance with respect to ambient conditions related to requirements X_A, X_B and X_C, in the initial operation phase, in order to find the real operational performance of the barriers in the given environment.

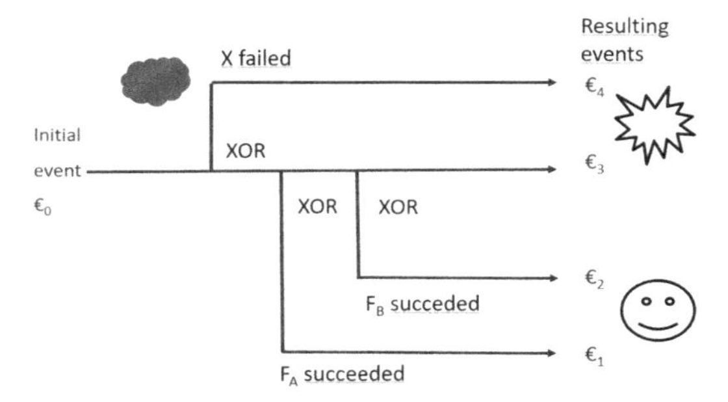

Figure 8. External effect $X = X_A = X_B$ may cause simultaneous failures of barrier A and B. This means that $X_A = X_B$ is a single condition that will not comply with the single failure acceptance criterion, although the functional (F_A, F_B) parts of barrier A and barrier B and X are independent.

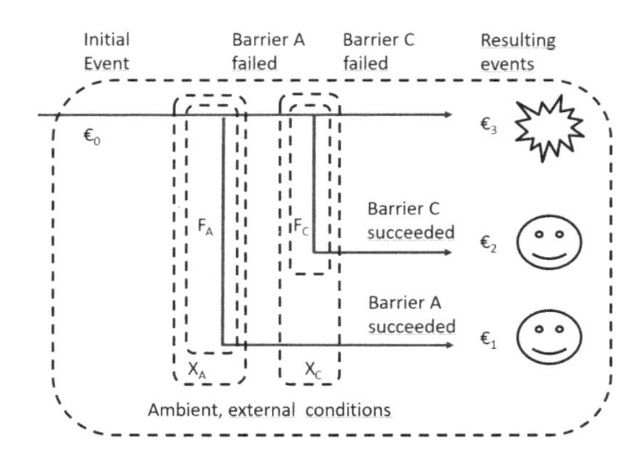

Figure 9. External effects $X_A \neq X_C$ are independent and will not cause simultaneous failures of barriers A and C. The single failure criterion for the barrier system A and C is now complied with.

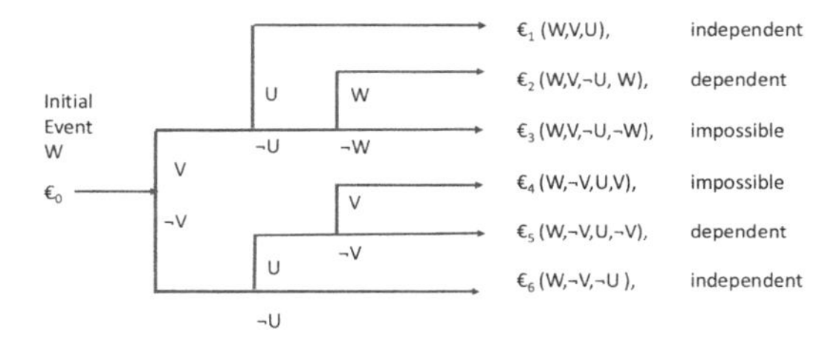

Figure 10. The initial event is dependent on external factor W and the barrier system tree has 5 barriers and 6 resulting events. The barrier functions U, V, W are independent. The resulting events $€_1$ and $€_6$ are based on the states of independent factors and functions. The resulting events $€_3$ and $€_4$ are based on negated dependent factors and might never occur. Resulting events $€_2$ and $€_5$ are based on repeated use of the same barrier function or external condition.

3.6 *Reducible dependent barrier systems*

A barrier system design may consist of e.g. 3 independent functions or conditions (U, V, W) and 6 resulting events (refer to Fig. 10). The barriers may be working with independent functions or external conditions (represented by Boolean variables e.g. U, V, W and functions). These functions and conditions may be combined and influence the outcome of several branches of the designed barrier system, leading to possible dependencies or impossible outcomes.

The general rule for confirming the independence of one barrier branch is that the expressions for the resulting event is such that no barrier functions or external conditions (U,V,W) are used more than once.

Example for resulting event $€_1$ in Figure 10 is based on:

$$€_1 = W \cap V \cap U$$

Here, $€_1$ is consisting of single use of independent functions, and therefore the expression for the resulting event is independent.

Resulting event $€_2$:

$$€_2 = W \cap V \cap \neg U \cap W$$

is the result of 4 barrier branches, and it is referring to W two times making the expression dependent. By the Boolean algebra identity law stating

$$W \cap W = W$$

the dependent expression is reduced to an independent expression for the resulting event $€_2$:

$$€_2 = W \cap V \cap \neg U$$

The assumed resulting events:

$$€_3 = W \cap V \cap \neg U \cap \neg W = 0$$

$$€_4 = W \cap \neg V \cap U \cap V = 0$$

can never occur due to the Boolean algebra law for complementation, stating:

$$W \cap \neg W = 0 \text{ and } \neg V \cap V = 0$$

The resulting event $€_5$:

$$€_5 = W \cap \neg V \cap U \cap \neg V$$

can be reduced to (note that $€_4$ is impossible)

$$€_5 = W \cap \neg V \cap U$$

$€_6$ contains 3 Boolean independent conditions:

$$€_6 = W \cap \neg V \cap \neg U$$

The effects of the dependencies may either improve the overall performance of the barrier system, which could be achieved by e.g. applying operational predictions or forecasts to influence the barrier system to come out with the most favorable resulting events and associated consequences.

On the other hand, common causes may affect the initiating event and several barriers simultaneously, which may degrade the overall performance significantly by influencing the barrier system outcome adversely with less favorable resulting events and consequences.

Possible dependencies between the events in the barrier system may give the IM designer options to avoid events which are reducing the performance of the barrier system, or, on the other hand, to apply possible dependencies (e.g., prediction of the future) to improve the overall performances.

This indicates that special identification and analyses of common causes and dependencies for multiple barriers should be subject to further development of qualitative methods for barrier systems. Reference is made to similar practice where the industry use FMEA as a method for barrier analysis of redundant systems, e.g., for DP systems and redundant propulsion (DNV-RP-D102, 2012).

4 DETAILED BARRIER PERFORMANCE REQUIREMENTS

4.1 *On-demand and continuous performance models*

The on-demand modelling approach is assuming that there will be:

- few ice occurrences per season in the monitored area, and
- sufficient warning time from an ice alert is issued until the ice arrives the protected unit.

An ice occurrence in the monitored area will generate a specific demand on the ice management system, and consequently initiate a sequence of events. An event tree analysis modelling approach is considered in this paper. This on-demand method may be used in the flight mode ice management philosophy. Traditionally, the on-demand mode is generally applied in offshore risk analyses and barrier modelling (IEC 61508, 2010).

In the case where ice presence must be assumed to be continuous, the physical ice management (ice fighting) must be required to operate continuously in periods. This case of ice regime with a possible continuous ice fighting mode is not within the scope of this paper. The continuous ice fighting operation has similarities with requirements to and verification of e.g. maritime propulsion systems, dynamic positioning systems, and heave compensation for drilling systems.

4.2 *Barrier functional types*

IM systems and activities are covering a wide range of types of barriers and events. The outcome of resulting events could be:

1. Ice observations (observed/not observed)
2. Ice predictions (predicted/not predicted)
3. Ice alerting (alerted/not alerted)
4. Ice breakability (breakable/not breakable)
5. Physical ice management (ice broken/not broken)
6. Ice alerting after physical ice management (confirmed ice broken/failed to break ice)
7. Ice arrived in Zone 1.
8. Disconnection (disconnected/connected)
9. Move off (moved off/still at site)

The above events may be modelled as resulting events in barrier event trees, where the actual outcome will be governed by the state of the barrier function when the barriers are demanded.

From the list above, one observes that the nature of the barriers may be classified in at least the following types:

a. Ambient physical events
b. System and functional/failure events
c. Operational ice observations events
d. Operational ice predictions events
e. Operational decisions/commands events

A barrier succeeds on demand (BSD(t)) when the barrier functional requirements are fulfilled at the time t of the demand:

$$BSD_A(t) = \left(F_A(t) \cap X_A(t)\right)$$

which is fulfilled (.T.) when

$F_A(t) = .T.$ and $X_A(t) = .T.$

The detailed conditions for the successful function (F_A) will most often consist of a combination of many parameters.

In the top-down IM design/decision process, the IM designer should initially try to avoid detailed functional modelling of the barriers if a decision process on a high level information is sufficient to reach the acceptance criteria.

But if it is deemed necessary to go into details of a barrier, the detailed functional and ambient variables can be modelled as a kind of 'Safety Requirement Specification' found in IEC 61508 (IEC 61508), which is recommended by PSA for safety systems (PSA 2017). The functional requirements to F_A in the above example is similar to the Safety Requirements Specification (SRS) applied in IEC 61508.

4.3 *Functional and ambient variable requirements*

Detailed functional requirements for a traditional pressure safety function could be expressed by a Boolean safety requirements specification (SRS), according to:

F_A = (Tank pressure above 20 bar) ∩
(Detected overpressure within 2 s) ∩
(Safety valve closed within 3 s)

In this simplified example for F_A, the target reliability (SIL: Safety Integrity Level from IEC 61508) is not included.

The tank external conditions are required to fulfil the following requirements:

X_A = (−30°C < Air temperature < 50°C) ∩
(Incoming fluid viscosity less than Z) ∩
(Incoming fluid not contaminated by sand)

An IM example could be vessel-based ice surveillance functional and external variable requirements:

F_A = (Ice floe size > 10 m) ∩
(Distance to ice floe > 2 km) ∩
(Vessel with detector movements < Z) ∩
(Stable power supplies and communication)

X_A = (Visibility > 2.5 km) ∩
(Wave height < 1.5 m)

4.4 *Boolean operational prediction functionality*

Ice prediction barriers are relating to the functionality to predict correctly if an ice hazard may occur before the ice arrival actually occurs.

The initiating event $€_0$ may be a request for a prediction at t_0 for the future period t_1–t_2. This type of prediction can enable the operational decision maker to start correctly orderly disconnection (ODC) procedures at t_1 in order to be completed at t_2.

The Boolean prediction function F is specified by

$$F = F(X_D(t_1), X_R(t_2))$$

where $X_D(t_1)$ is Boolean variable for the observed environment at t_1, and $X_R(t_2)$ is a Boolean variable for the requested external condition at t_2 (see Fig. 11).

Requirements to observations (deterministic) of the ambient conditions during the lag time period (t_0–t_1):

$X_D(t_1)$ = (Ice floe size < 10 m) ∩
(Optical visibility > 2.5 km) ∩
(Wave height < 1.5 m) ∩
(Sea current < 0.5 m/s)

Note that the optical visibility in the example above is a requirement to the ambient conditions necessary for doing the prediction function F.

Required ambient condition at t_2 to the forecaster at t_1

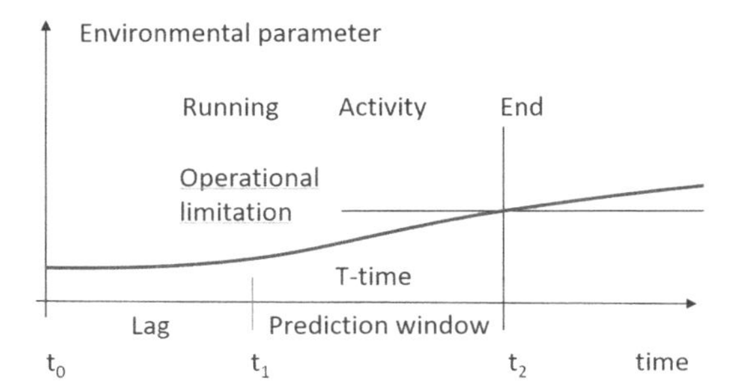

Figure 11. The correct prediction at t_1 in the figure is $€_{44}$ Pessimism Justified. At time t_1 the forecaster is requested to give a prediction for a prediction window starting at t_1 and ending at t_2 in order to start disconnection at t_1 if critical environmental parameters will exceed the stated limitation. The T-time is the required time for termination of the critical operation with environmental limitation.

$$X_R(t_2) = (\text{Ice floe size} < 10\text{ m}) \cap (\text{Wave height} < 1.5\text{ m}) \cap (\text{Sea current} < 0.5\text{ m/s})$$

Observed (deterministic, D) ambient condition at t_2 by protected unit

$$X_D(t_2) = (\text{Ice floe size} < 10\text{ m}) \cap (\text{Wave height} < 1.5\text{ m}) \cap (\text{Sea current} < 0.5\text{ m/s})$$

The prediction functionality is according to requirements if the operable ice condition in $[t_1 - t_2]$ was predicted at time (t_1). There are four resulting events ($€_1$,,$€_2$,, $€_3$, $€_4$) where this is of special interest (DNMI 1979).

Optimism justified, the prediction at t1stated acceptable condition at t_2 (F = true, optimistic) which actually was fulfilled at t_2.

$$€_1 = €_0 \cap F(X_D(t_1), X_R(t_2)) \cap X_D(t_2)$$

Pessimism justified, the prediction at t1stated non-acceptable conditions t_2 (F = false, pessimistic) and this pessimistic situation actually occurred at t_2, or

$$€_4 = €_0 \cap \neg F(X_D(t_1), X_R(t_2)) \cap \neg X_D(t_2)$$

In the example in Figure 11 the Pessimism Justified is the correct prediction type and it was correct to start the orderly disconnect (ODC) procedure at t_1.

In the case that the predictions were not correct, we denote this as failure modes (ISO 14224:2006) of the prediction function.

Too optimistic, the prediction at t1stated acceptable conditions (F = true, optimistic) and this required situation actually did not occur at t_2

$$€_2 = €_0 \cap F(X_D(t_1), X_R(t_2)) \cap \neg X_D(t_2)$$

Too pessimistic, the prediction at t1stated non-acceptable conditions (F = false, pessimistic), but the required situation actually occurred at t_2

$$€_3 = €_0 \cap \neg F(X_D(t_1), X_R(t_2)) \cap X_D(t_2)$$

The prediction function F is not testable or verifiable at the time of prediction or alerting (t_1), but it is observable and testable after the prediction window (t_2) has elapsed. Hence, it is important to log the input and output data for such prediction function in order to get statistics on its performance over time. This can then be used to improve its performance, replace it, or include an additional prediction barrier in a new design decision.

5 PROBABILISTIC BARRIER PERFORMANCE

5.1 *Probabilistic integrity of barrier function*

The integrity of a barrier function may be defined as the ability to function when needed. The integrity of the barrier function F_A may be expressed by the probability of success on demand at $t = t_d$:

$$PSD\ (t_d) = P(F_A(t_d))$$

and probability of failure on demand:

$$PFD\ (t_d) = P(\neg F_A(t_d))$$

Referring back to the example given in Figure 5 with independent barriers A and B, the probabilities of the resulting branch events can be estimated by:

$$€_1 = €_0 \cap F_A$$
$$€_2 = €_0 \cap \neg F_A \cap F_B$$
$$€_3 = €_0 \cap \neg F_A \cap \neg F_B$$
$$P(€_1) = P(€_0) \cdot P(F_A)$$
$$P(€_2) = P(€_0) \cdot P(\neg F_A) \cdot P(F_B)$$
$$P(€_3) = P(€_0) \cdot P(\neg F_A) \cdot P(\neg F_B)$$

5.2 *Probabilistic robustness or survivability*

The robustness (survivability or vulnerability) of the barrier is the ability to function under relevant external conditions at the demand situation.

The probabilistic robustness of the barrier external conditions X may be expressed by the probability that external conditions are complying with the requirements to the environment at $t = t_d$:

$$PSD\ (T_d) = P\ (X(t_d))$$

and probability of failure of robustness on demand:

$$PFD\ (T_d) = P\ (\neg X(t_d))$$

The combined requirements of both functional F_A and external requirements X at demand may be expressed as:

$$BSD_A = (F_A \cap X)$$

and assuming that F_A and X are independent, the probability of success on demand of the barrier A can be expressed as:

$$PSD_A = P(F_A \cap X) = P(F_A) \cdot P(X)$$

Please refer back to example Figure 6. For event trees with common causes (X_A, X_B), we get:

$$€_1 = €_0 \cap (F_A \cap X_A)$$
$$€_2 = €_0 \cap \neg(F_A \cap X_A) \cap (F_B \cap X_B)$$
$$€_3 = €_0 \cap \neg(F_A \cap X_A) \cap \neg(F_B \cap X_B)$$

and assuming $X = X_A = X_B$ the probabilities for the resulting events can be expressed by:

$$P(€_1) = P(€_0 \cap (F_A \cap X))$$
$$P(€_2) = P(€_0 \cap \neg(F_A \cap X) \cap (F_B \cap X))$$
$$P(€_3) = P(€_0 \cap \neg(F_A \cap X) \cap \neg(F_B \cap X))$$

In the case that automatic tools should calculate above estimates, it should be noted that the two lower branches are statistically dependent due to the two occurrences of X in the two branches. Hence, the probabilities for the Boolean expressions for $€_2$ and $€_3$ can not be calculated in the same manner as shown in Section 5.1. The Boolean expressions have to be reduced by application of standard Boolean algebra laws for distributivity, identity and complementation in the same way as shown in Section 3.5, leading to:

$$P(€_1) = P(€_0) \cdot P(F_A) \cdot P(X)$$
$$P(€_2) = P(€_0) \cdot P(\neg F_A) \cdot P(F_B) \cdot P(X)$$
$$P(€_3) = P(€_0) \cdot P(X) \cdot P(\neg F_A A) \cdot P(\neg F_B)$$
$$P(€_4) = P(€_0) \cdot P(\neg X)$$

where we note that the expressions for the 4 resulting events are based on single use of each independent variable F_A, F_B, and X.

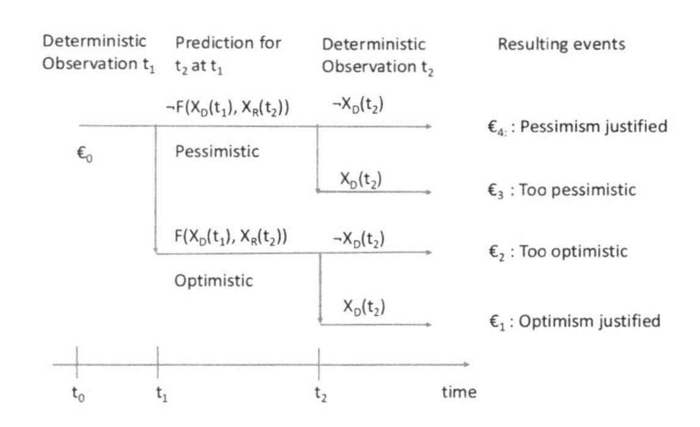

Figure 12. Event tree for representation of 4 resulting events for the prediction function F.

Calculation of the expected performance probabilities have to take into consideration the differences between Boolean and ordinary algebras.

5.3 *Probabilistic operational predictions*

The probability of the Boolean expressions for the resulting events for the 4 types of prediction resulting events are:

Probability of optimism justified:

$$P(€_1) = P(€_0 \cap F(X_D(t_1), X_R(t_2)) \cap X_D(t_2))$$

Probability of pessimism justified:

$$P(€_4) = P(€_0 \cap \neg F(X_D(t_1), X_R(t_2)) \cap \neg X_D(t_2))$$

Probability of too pessimistic:

$$P(€_3) = P(€_0 \cap \neg F(X_D(t_1), X_R(t_2)) \cap X_D(t_2))$$

Probability of too optimistic:

$$P(€_2) = P(€_0 \cap F(X_D(t_1), X_R(t_2)) \cap \neg X_D(t_2))$$

It should be noted that the Boolean expressions may contain dependent elements, and that the effects of the dependencies must be handled in the probability estimates.

Estimates of the prediction probabilities may be produced by comparing time series of environmental data X_D required to be predicted and time series of actual predictions F according to required conditions X_R for a given location.

6 EXPECTED CONSEQUENCE AND RISK

Assume that the consequences £ associated to the resulting events € are $€_1$: $£_1$, $€_2$: $£_2$, $€_3$: $£_3$ (Fig. 13).

The expected consequences or risk of the barrier system may then be defined as:

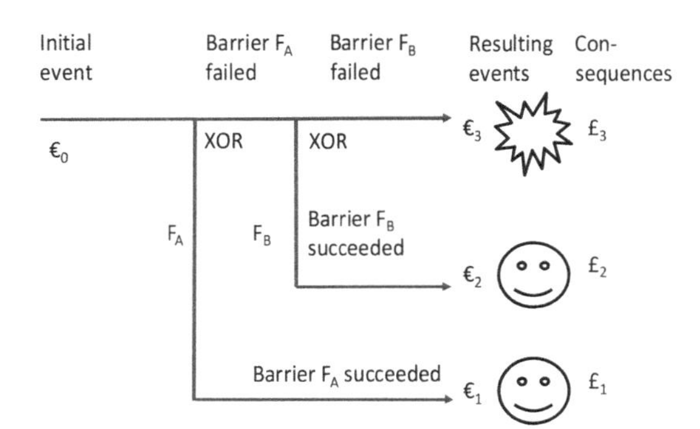

Figure 13. Resulting events $€_1$, $€_2$, and $€_3$ with associated consequences $£_1$, $£_2$, $£_3$.

$$R = P(€_0)\sum_{i=1}^{3} P(€_i)\cdot £_i$$

Expected consequences and risk can be used for estimating residual risk of the barrier system and such estimates can be used as a basis for design decision-making at top-level or at sub-levels according to the stated acceptance criteria.

7 PLANS FOR IM DESIGN CASE STUDY

The descriptions given in previous sections are based on some of the ongoing research and development in the SAMCoT program. A plan for validating the applicability of the IM performance models for design decisions is developed (Fig. 14).

The following topics are considered to be included in the planned study case:

- The ice regime model, assuming an on-demand IM flight response.
- IM design acceptance criteria including acceptable uncertainties with qualitative and quantitative models.
- Establishment and analysis of initial IM barrier system, including extra high contingencies at first design decision in pre-Feed/Feed phase.
- Proposals for improved IM barrier system at design decision after initial operations.
- Identifying the detailed need for collection of operational experience data during the initial operations (Fig. 14).
- Collection of the operational data.
- Conclusion on relaxation of EHC at design decision.

The study case and demonstration is proposed to be presented in 2019.

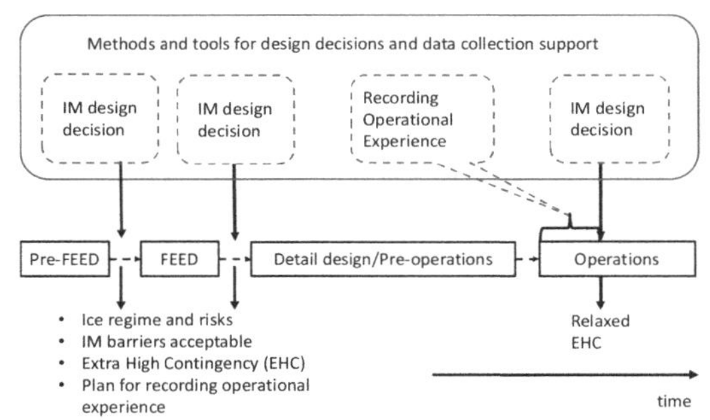

Figure 14. The first IM design decision establishes ice regime and risks, IM barriers system performance, Extra High Contingency compensating for initial uncertainties and a plan for recording operational experience according to stated acceptance criteria. The IM design decision during operations may relax the extra high initial contingencies based on recorded experience.

8 SUMMARY AND CONCLUSIONS

A top down method for modelling ice hazard and barrier performance for IM is presented as ongoing research in SAMCoT. The model is covering the demand mode and the flight mode of ice management. The model is flexible and modular and may be used for qualitative (Boolean) and quantitative (probabilistic) acceptance criteria. The model is flexible with regard to the level of details in the barrier descriptions, and it starts with a Boolean model of barrier performance which may be detailed with analysis of common cause failures and/or probabilistic functional performance. The failure modes for on-demand ice prediction are proposed and defined with probabilities.

The IM model is a natural extension to existing regulations, theories and practices in the O&G industry and the maritime industry. The model is planned to be further developed, tested and validated in study case and demonstrators.

ACKNOWLEDGMENTS

The authors would like to thank the Research Council of Norway (RCN) for financial support through project no. 203471 CRI SAMCoT.

REFERENCES

DNMI 1979, Table with estimates of '*Confidence tag forecast ing Statfjord 1979*'.

DNV-RP-D102: 2012, *Recommended Practice: Failure mode and Effect Analysis (FMEA) of Redundant Systems, 2012*.

DNVGL-OS-E101: 2018, *Drilling facilities, Offshore Standards, 2018*.
Eik, K.J. 2010, *Ice Management in Arctic Offshore Operations and Field Developments, Doctoral Thesis NTNU*.
Haimelin R., Goerlandt F., Kujala P., Veitch B. *Implications of novel risk perspectives for ice management operations*, Cold Regions Science and Technology, Volume 133, January 2017, Pages 82–93.
Hauge, S. & Øien K. 2016, *Guidance for barrier management in the petroleum industry, 2016-09-23, SINTEF A27623*.
IEC 61508–2010, *Functional safety of electrical/electronic/programmable electronic safety-related systems -*.
ISO/FDIS 35104 2017, *Petroleum and natural gas industries—Arctic operations—Ice management*.
070 – *Norwegian Oil and Gas 2004, Application of Iec 61508 and Iec 61511 in the Norwegian Petroleum Industry, Norwegian Oil and Gas Association*.
ISO 14224:2006 – *Collection and exchange of reliability and maintenance data for equipment* Petroleum, petrochemical and natural gas industries.
PSA 2017 *The Management regulations, 2017, Petroleum Safety Authority of Norway (PSA)*.
Ruud, S. & Skjetne, R. 2014. "*Verification and Examination Management of Complex Systems*". *Journal of Modeling, Identification and Control,* 2014, Vol 35, No 4, pp. 333–346.

Marine Design XIII – Kujala & Lu (Eds)
© 2018 Taylor & Francis Group, London, ISBN 978-1-138-34076-3

Towards holistic performance-based conceptual design of Arctic cargo ships

M. Bergström, S. Hirdaris, O.A. Valdez Banda & P. Kujala
Aalto University, Espoo, Finland

G. Thomas & K.-L. Choy
University College London, London, UK

P. Stefenson
Stena Teknik, Gothenburg, Sweden

K. Nordby
The Oslo School of Architecture and Design, Oslo, Norway

Z. Li, J.W. Ringsberg & M. Lundh
Chalmers University of Technology, Gothenburg, Sweden

ABSTRACT: We present an extension of an earlier presented framework for use in the conceptual design of Arctic cargo ships. To enable a holistic approach, the framework integrates a wide range of performance assessment tools and methods namely (a) Discrete Event Simulation (DES) based Monte Carlo simulations, (b) system thinking, and (c) empirical data analysis. The thinking behind the framework is demonstrated by discussing how it could be applied to assess the design impact of four specific technologies: (1) a 'Biomimetic anti-icing coating', (2) a 'Safe Arctic bridge' using augmented reality (AR) technology for improved situation awareness, (3) a 'Arctic voyage planning tool' using Big data analysis, and (4) the use of low flash point fuels' in the Arctic.

1 INTRODUCTION

1.1 *Background*

The design of Arctic ships is challenging. This is because Arctic maritime operating conditions are harsh and demand consideration of the effects of sea ice, icebergs, extreme temperatures, polar lows, and seasonal darkness. Engineering idealisation of these phenomena is challenging as their occurrence is highly stochastic. The performance of ice-going ships is also often dependent on the availability of icebreakers (IBs), making it necessary to extend the design boundaries beyond the individual ship, i.e., to consider a ship as system of systems. (Bergström, et al., 2016a) presents a design framework that deals with these challenges by making use of a combination of system thinking and by treating an Arctic cargo ship as a component of a wider Arctic Maritime Transport System (AMTS). They also present a Discrete Event Simulation (DES) based on the Monte Carlo approach. The later enables simulation of data supporting a well-informed design process. Specifically, their framework allows for:

1. The determination of the required fleet size (number of ships), ship size (ship cargo carrying capacity), and ship speed (specified by the so-called *hv*-curve that determines the speed of a ship as a function of the ice thickness) for a specific transport task.
2. The simulation of scenarios in support of the design process. Examples are: (a) a ship's exposure to various ice conditions, and (b) the temporal distribution between a ship's various operating modes (e.g. operation in open water, independent operation in ice, and IB-assisted operation). Such data may also be used to assess ship (or fleet) fuel consumption, IB tariffs, and level of ice loading.

This framework does not consider accidental events (e.g. collisions, groundings, contacts), specific technologies or 'active' operational measures that might influence how a ship is operated. To this end, this paper outlines an extended version of the design framework presented by (Bergström, et al., 2016a) that addresses these limitations. To open the way toward future research, the paper also discusses how goal based thinking could be applied to assess the design impact of specific technologies. The technologies considered are:

1. A biomimetic anti-icing coating.
2. A 'Safe Arctic bridge' using augmented reality (AR) technology for improved situation awareness.
3. An 'Arctic voyage planning tool' using Big data analytics.
4. The use of Low Flash Point Fuels (LFPFs) such as methanol in the Arctic.

These technologies, together with the design framework, are developed as a part of the EU-funded Horizon 2020 research project SEDNA—Safe maritime operations under extreme conditions: the Arctic case (SEDNA, 2017). The overall aim of the SEDNA project is to develop an innovative and integrated risk-based approach to safe Arctic navigation, ship design and operation.

1.2 *Design regulations for Arctic ships*

Traditionally, the design of Arctic ships has been regulated by main stream IMO statutory instruments (e.g. SOLAS, MARPOL). Certification is supplemented by Flag State ice class rules mitigating Arctic specific risks. To harmonise the regulations, on January 1st, 2017, the International Code for Ships Operating in Polar Waters (Polar Code), was introduced by (IMO, 2015) on all ships operating in the Arctic or Antarctica. This Code incorporates performance based safety by active (operational decision driven) and passive (de-sign orientated) risk control measures. The goal-based approach is ratified via a framework that brings together design requirements and functional requirements (FRs) to meet those. In accordance with Figure 1, compliance with the FRs can be achieved either by meeting a set of prescriptive regulations (traditional regulations that prescribe a specific solution) associated with the FR(s), or by carrying out a safety performance assessment demonstrating that the FR(s) are met. The latter case results in a so-called alternative or equivalent design. In accordance with regulation 4 – "Alternative design and arrangement" of SOLAS Chapter XIV (IMO, 2014d), where alternative or equivalent designs or arrangements are proposed they are to be justified by the following IMO Guidelines:

- "Guidelines for the approval of alternative and equivalents as provided for in various IMO instruments", MSC.1/Circ.1455 (IMO, 2013a).
- "Guidelines on alternative design and arrangements for SOLAS chapters II-1 and III, MSC.1/Circ.1212 (IMO, 2006a).
- "Guidelines on alternative design and arrangements for fire safety", MSC/Circ.1002 (IMO, 2001).

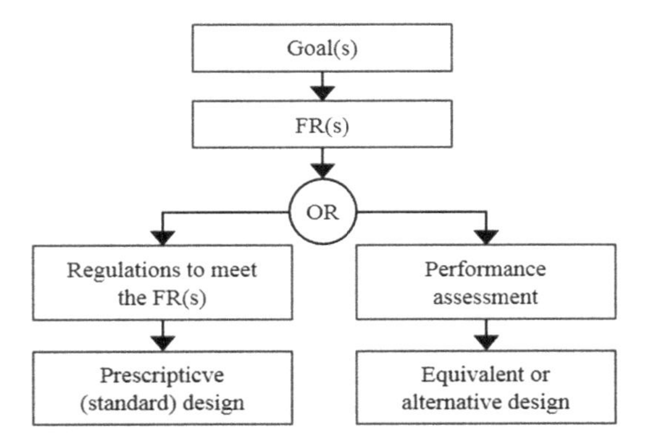

Figure 1. Approval principle of the Polar Code.

A general principle is that any alternative design should be at least as safe as a design determined by prescriptive rules (see Figure 1). It should be noted that alternative solutions are only possible with respect to SOLAS regulations. To date the IMO has not agreed on any similar goal-based approaches with respect to MARPOL regulations or on any environmental risk metrics (Skjong, et al., 2007). The MARPOL code includes a method to quantify a ship's accidental oil outflow performance (IMO, 2006). However, as pointed out by (Papanikolaou, 2009), this metric is not related to any actual environmental risk.

1.3 *Design methods*

We can distinguish between two different types of ship design approaches namely: (a) prescriptive-based design (PBD) in which a design is determined following prescriptive design regulations, and (b) goal-based design (GBD) in which a design is determined following goal-based regulations.

PBD facilitates a time and resource efficient design process by clear-cut and easily applied design rules. However, PBD has two fundamental weaknesses: (a) prescriptive rules might act as design constraints that limit the feasible solution space, and (b) the efficiency of the solution depends on the efficiency of the rules. For instance when applying the Polar Class, the efficiency of the solution might suffer due to the following reasons (Kim & Amdahl, 2015) (LR, 2014):

1. The rules are semi-empirically determined and might therefore be inefficient when applied on new or innovative designs,
2. The rules do not consider the probabilistic nature of ice loading, meaning that a ship operating briefly in some specific ice condition

is assumed to be exposed to the same level of ice-loading as a ship operating extensively in the same ice conditions.
3. The rules are not determined with respect to any clearly communicated goal(s), i.e., in terms of application the rules do not result in any specific known level of performance.

GBD offer solutions to the weaknesses of PBD. First, by applying goal-based regulations, GBD expands the feasible design space, enabling new and innovative solutions (Papanikolaou, 2009). In addition, GBD enhances the 'safety culture' in terms of engineering practice. This is because it challenges the designer to assess design in terms of 'performance' and cost-efficiency. Notwithstanding, the implementation of goal-based regulations implies weaknesses in terms of time and resource allocation, they may lead to risks in terms of technical assurance and may suggest complications in terms of Intellectual Property Rights (IPR) (Papanikolaou, 2009). Recently, (Bergström, et al., 2016b) identified that with reference to application of goal-based rules and the Polar Code, there is a lack of unified performacne measures, criteria and perfromacne assessment tools or methods.

2 DESIGN FRAMEWORK OUTLINE

2.1 *System terminology*

We divide an Arctic Maritime Transport System (AMTS) into a hierarchy of subsystems, where each subsystem serves a specific function towards its overall objectives. Each system is defined in terms of a set of design variables that are precise characteristics of this system, controlled and determined by the designer. Any factor affecting the performance of the system that is controllable by the designer can be turned into a design variable. The uncontrollable conditions under which a system operates are defined in terms of design parameters, including environmental, operational, technical, and financial parameters. Parameters that are stochastic or subject to uncertainty are determined in terms of value distributions (Bergström, et al., 2016b). The feasible design space is limited by design constraints consisting of either physical limit values (bounds) determined in the form space, or of mandatory FRs determined in the function space (Bergström, et al., 2016b). Various types of constraints include operational, technical and regulatory constraints (Bergström, et al., 2016b). The performance of the system is assessed by using a performance assessment model. A performance assessment model is typically subject to some degree of uncertainty, that as per (Bergström, et al., 2016b) may be defined as internal uncertainty.

2.2 *Design process*

The design framework applies a design process as outlined in Figure 2. In the following sections we describe the contents of each step.

2.2.1 *Design context*

The design process starts by determining the design context specifying the primary function of the ship or maritime system being designed as well as its operating conditions, described in terms of a set of design parameters.

2.2.2 *Concepts of operations and preliminary designs*

Based on the design context, a set of preliminary designs, each determined in terms of a set of preliminary design variables, representing various concepts of operations (CONOPS). CONOPS is a performance variable that considers operational strategies and questions how design objectives can be met. A CONOPS for an Arctic ship might for instance include the following (Bergström, et al., 2016b):

1. An Ice Mitigation Strategy (IMS) describing how the AMTS will deal with sea ice (e.g. independent or IB assisted operation).
2. A strategy on how to compose the fleet (e.g. via the use of large or small vessels).
3. A strategy on how to balance the transport demand and capacity under different operation (e.g. ice) conditions (e.g. reserve speed or reserve payload capacity).

2.2.3 *DES-based Monte Carlo simulations*

The operational performance of each preliminary design is assessed probabilistically using DES-Monte Carlo simulation as described by (Bergström, et al., 2016a). Inputs for the DES model may be (a) empirical data, (b) design tools, and (c) models encompassing the impact of specific ship technologies. Applied empirical data might include accident data, full-scale ice load measurements, operating data (e.g. port turn-around times), and financial data (e.g. fuel costs, IB tariffs). Applied design tools include for instance a model for to the calculation of the equivalent ice thickness (Riska, 2010). Models on how to consider specific technologies are still under development in the SEDNA project.

Upon completion of DES data analytics are processed and each design is adjusted and re-simulated until a sufficient level of transport capacity is obtained. In addition to transport capacity, data streams include information on the ship ice exposure (e.g. average distance travelled in various ice conditions per ice-year) and the temporal distribution between various operating modes (e.g.

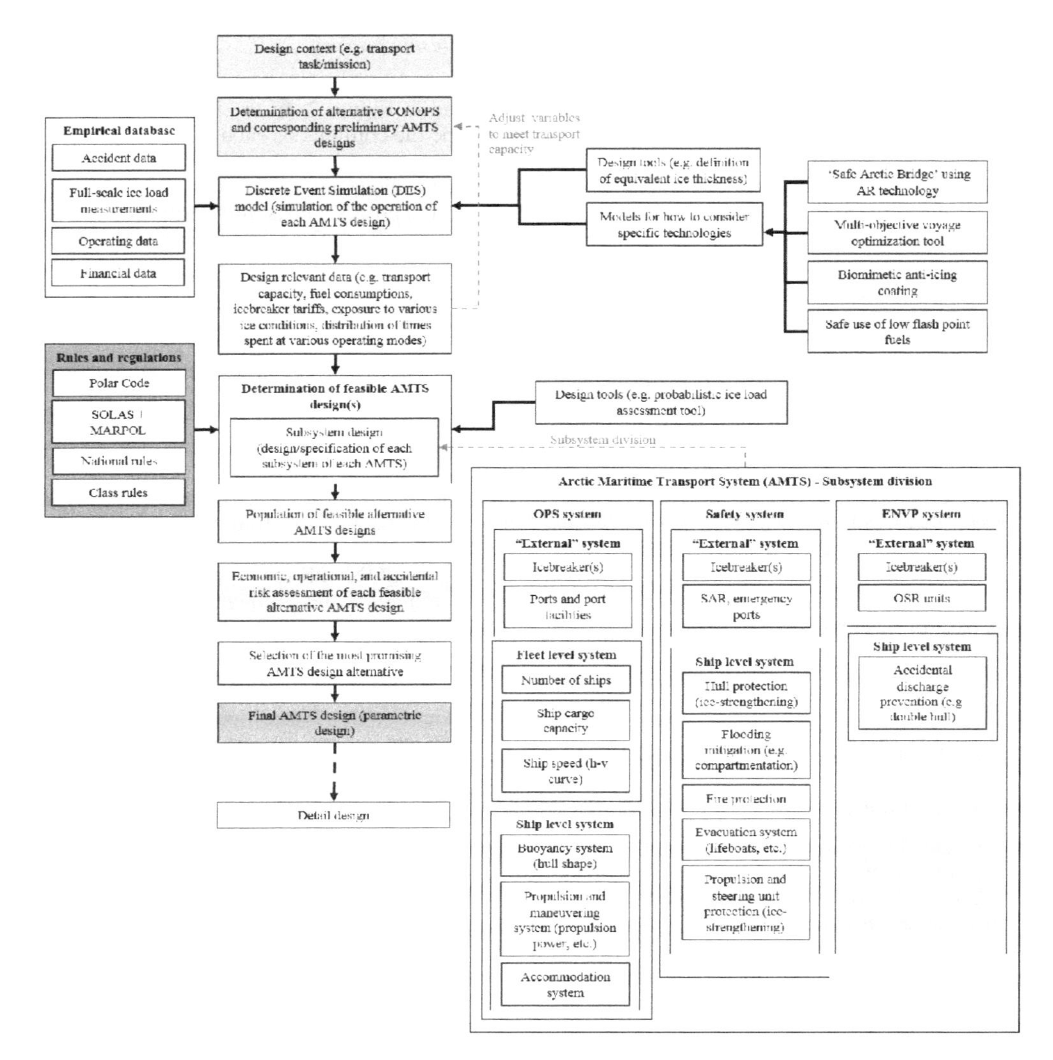

Figure 2. Design process and framework overview.

operation in open water, independent operation in ice, and IB-assisted operation). These data may be used to assess a ship's (or a fleet's) fuel consumption and IB assistance related costs. SEDNA aims to produce models and approaches that also enable the processing of empirical accident data in the simulation. Those may result in risks associated with accidental events (e.g. grounding, collision, contact).

2.2.4 *Design elaboration*

Using the obtained design data, each AMTS design is further elaborated by division into a hierarchy of subsystems. This approach helps to assess the performance of a specific subsystem functions against the original objectives of the AMTS. Based on (Bergström, et al., 2016a), we propose to use the subsystem division presented in Figure 2. Accordingly an AMTS is divided into three main systems namely: (a) an OPS related to transport tasks, (b) a safety system managing safety risks (e.g. risk to human life), and (c) an environmental protection (ENVP) system managing environmental risks (e.g. spill of cargo).

The OPS system consists of: (a) external systems such as IBs and port facilities, (b) fleet-level subsystems determined in terms of the number of cargo ships, the cargo capacity and speed (*hv*-curve) of each ship, and (c) ship-level systems (e.g. the hull buoyancy system, propulsion and manoeuvring systems, accommodation system). Design criteria for the OPS system are determined based on the mission of the AMTS as determined by the owner.

The safety system consists of (a) external systems such as IBs and search and rescue (SAR) resources, and (b) of ship-level systems such as a hull protection system protecting a ship from ice-loading, a flooding mitigation system mitigating the consequences of a potential flooding, a fire protection system preventing and mitigating on-board fires, an evacuation system and a propulsion and steering unit ice-loading protection system that protects the propulsion and steering unit(s) from ice-loading.

The ENVP system consists of (a) external systems such as IBs and Oil Spill Response (OSR) units, and of (b) ship-level systems, such as an accidental discharge prevention system, preventing or limiting accidental discharges of harmful substances such as oil. Design criteria both for the safety and ENVP systems are set by existing rules and regulations (e.g. SOLAS, MARPOL, the Polar Code, national rules, and classification rules).

2.3 *Challenges of subsystem design*

(Bergström, et al., 2016b) provide an overview of design tools for the design of Arctic ships. This work focused on the 'fitness for purpose' of safety and ENVP systems and concluded that there are established approaches in terms of flooding mitigation (SOLAS probabilistic damage stability) fire protection (e.g. fire simulations etc.) and evacuation (e.g. evacuation simulations). On the other hand, for the hull protection systems, they identified that there are no 'well-proven' performance assessment methods or performance based acceptance criteria (Bergström, et al., 2016b). To address this situation, based on (Bergström, et al., 2016b), we propose the validation and implementation of the probabilistic ice load assessment tool presented by (Jordaan, et al., 1993). The tool makes it possible to assess a ship's probabilistic ice loading based on its simulated ice exposure over multiple years of service (an outcome of the DES-based Monte Carlo simulations). As per (Bergström, et al., 2016b) the return period for ice loads should correspond to the plastic limit states of the hull structure. However, this performance measure has not yet been adopted. Since the accuracy of this approach has not yet been validated the SEDNA consortium aims to validate the method by using full-scale data from (ARCDEV, 1998). Other limitations of the method that should be assessed include: (a) the applicability of tools and methods beyond a ship's bow area, and (b) the extension of the theoretical basis toward the calculation of the 'cumulative' effects of ice-loads resulting from exposure to various types of ice conditions (e.g. medium-thick first-year ice plus thick-first-year ice). Until these limitations have been addressed, the level of hull ice strengthening must be determined in accordance with an appropriate IACS Polar Class standard (IACS, 2016a).

Similar to the hull, there are no performance methods, measures or criteria for the propulsion and steering of unit protection systems (Bergström, et al., 2016b). For example, it is believed that to assess the ice-loading acting on a ship's propeller an ice-propeller model is needed. However, as pointed out by (Bergström, et al., 2016b), since existing ice-propeller models are too simplified to provide a sufficient level of accuracy, the validation of any new high-fidelity model would require additional full-scale ice propeller load measurements. Until then, established design rules including those determined by (IACS, 2016a) should be used.

With regards to the design of the ENVP systems, as a first step towards goal-based design, the maritime industry must agree on metrics for environmental risk. Subsequently methods enabling the assessment of those metrics must be developed. This is outside the scope of the SEDNA project, and thus something that will be addressed in the future. Meanwhile, the ENVP system must be designed in accordance with existing prescriptive regulations outlined by MARPOL and other regulations.

2.3.1 *Performance assessment and design selection*

Once each subsystem of each AMTS design is designed in accordance with relevant design criteria, a population of feasible competing AMTS designs is derived. A holistic safety and economic assessment is then carried out for each design. The economic assessment is carried out based on a combination of empirical data (e.g. IB tariffs, fuel prices, time charter costs) and simulated data (e.g. the temporal distribution between a ship's various operating modes). The operational performance of the various AMTS designs is assessed based on simulated data in terms of design robustness (e.g. how sensitive is performance to changes in the operating conditions), and transport reliability (e.g. the likelihood of failing to meet a transport task during a random operating year). Methods for carrying out the safety risk assessment (risk of accidental events) are still under development within the SEDNA project. The intention is to carry out the safety assessment using both empirical accident data, and simulated data, such as data on a ship's exposure to various operating conditions and areas. The applied method could be similar to the Bayesian Network based method presented by (Valdez Banda, et al., 2016).

Based on the outcomes of the economic, operational, and accidental risk assessments, the overall most promising design alternative may be selected. The selected design consisting of a set of design variables (parametric design) may subsequently be used as input for subsequent (detail) design stages.

3 DESCRIPTION OF CONSIDERED TECHNOLOGIES AND DISCUSSION ON THEIR POSSIBLE DESIGN IMPACTS

3.1 *Biomimetic anti-icing coating*

Icing related risks are significant for most Arctic ships. For relatively small ships (e.g. fishing vessels), icing is a serious safety issue as it might complicate ship stability. In larger ships icing may affect the function of deck equipment (e.g. radar antennas and lifeboat release systems), deck mobility. It may also lead to human factor safety concerns (e.g. hazards due to slippery surfaces).

Icing on ships can be managed either by de-icing measures removing already formed ice or by anti-icing measures preventing the formation of ice. Anti-icing measures can be further divided into active and passive anti-icing measures. A commonly applied de-icing measure is to manually remove ice by using hammers, a laborious and potentially dangerous job. The icing rate depends on multiple factors including (a) the air and water temperatures, (b) wind and ship speed, and (c) the angle between a ship's bearings and the wind and wave direction. Whereas active anti-icing measures include adjusting ship speed and bearing (route) (Blackmore & Lozowski, 1994), passive anti-icing measures may include the use of anti-icing coatings.

Anti-icing coating technology is not new. The technology has been applied on aircrafts and wind turbines. However, the fitness for purpose in terms of maritime applications may differ in terms of cost expectations,toughness requirements and ship segment/design or location of the application. In SEDNA project we develop two different types of anti-icing coatings: (a) a low-cost, tough coating for application on the superstructure and other large areas, and (b) a transparent coating for lifeboat surfaces and windows. In the latter case, costs may be of less importance due to the smaller areas to be covered and the higher importance of safety.

Both coatings are biomimetic in nature, because their anti-icing properties is achieved by mimicking a penguin's coat (Choy, 2018).The anti-icing coating developed within SEDNA project could replace traditional anti-icing measures by reducing the speed of a ship or by adjusting her bearing and hence lead to enhanced transport capacity and more cost-efficient design. It could also make manual de-icing redundant, and therefore lead to reduction in terms of crew size and required accommodation capacity. Finally, it could affect the required icing allowance. The later would impact future stability regulations.

3.2 *Safe Arctic bridge*

Navigation in ice-infested waters relates to a number of unique navigation challenges. For example the distance from and movements of nearby ships (e.g. an IB) may change quickly if nearby ships suddenly get stuck in ice. The curvature of an ice channel and local variations in the ice cover (avoidance of large ridges) may also be significant challenges in term of operations. Arctic environmental phenomena such as fog, seasonal darkness, and sun glare present an additional challenge in terms of visibility. Solutions to many of these challenges may emerge by suitable understanding of human behaviour and implementation of principles of ergonomics. In addition we propose applying augmented reality (AR) technology in order to improve arctic bridge systems. AR is a technology that might improve situational awareness and decision-making of a crew by bringing in elements of the virtual world into the real world. Aviation and automobile industries have already successfully applied various implementation of AR technologies in practice in order to support situational awareness (Melzer, 2012). In addition, research within the maritime domain have outlined its potential in supporting the maritime domain (Vasiljević, et al., 2011).

This technology has the potential to improve safety by limiting the need to look away from the outside view and as such significantly reduce operators head down time and directly associate data with the real world setting. In addition, the technology is expected to improve the human-system interface by providing a mechanism to successfully manage large and varied information layers that crews are exposed to (e.g. by an improved possibility to present versatile information at the same time, and by enabling a very large total screen area). The increased versatility in displaying information anywhere in the world increase the potential for effectively integrating new categories of applications that may support users working in the arctic. An example of how AR could be applied in Arctic ship operations is presented in Figure 3.

Facilitating AR for use in the Arctic could provide several operational advantages. For example, the system could be useful on an Arctic ship escorted by an IB in an ice channel, or for operations in areas where icebergs occur. Technology applications could potentially extend the

Figure 3. Example of how AR technology could be applied for an improved situation awareness in Arctic ship operations. ®Amalie Albert.

permitted/safe range of operating conditions for specific manoeuvres (e.g. entering and leaving port, docking), or impact upon on the layout of the bridge/superstructure. In addition, the improved situational awareness provided by the 'Safe Arctic bridge' could help to identify ice features such as growlers and large ice ridges, that may help to minimise ice exposure. This could reduce the required level of ice-strengthening.

3.3 *Arctic voyage planning tool*

When planning an Arctic voyage, the master might have to consider a multitude of factors including ice conditions (e.g. ice thickness, ice ridging, and ice compression), the risk of iceberg collision, the availability of IB support, ship draft and tide level variations, and currents. The choice of a suitable route relates with and impacts upon the operational risk profile of the vessel, voyage time, voyage costs, and emissions. Practically, the operator might wish to minimise either voyage time, ship wear (repair costs), fuel costs, or accidental risk. For example, if the priority is to minimise voyage time, the master will choose the quickest route and such action may require sailing through areas with difficult ice conditions leading to increased risks in terms of ship wear and fuel costs. On the other hand, if priority is to minimise ship wear and accidental risks, the master may chose a detour that minimises ice exposure. The later may result in a longer voyage time. Examples of potential efficiency gains that could be achieved by route optimization are presented by (Markström & Holm, 2013).

To help ship masters achieve their objectives, SEDNA consortium develops an Arctic voyage planning tool (see Figure 4). As such tool could impact upon the chosen ship speed, route, and ice exposure, it may affect the required ship/fleet size or propulsion power specifications for a transport task as well as ice strengthening requirements.

3.4 *Safe use of low flash point fuels*

The Arctic environment is sensitive to emissions and pollution. Using low flash point fuels (e.g. methanol) instead of conventional fuel oil such as Heavy Fuel Oil (HFO) may lead to reduction of exhaust emissions. Low Flash Fuels (LFPFs) dissolve in water, are biodegradable and hence reduce the risk of environmental damage due to accidental spills (Ellis & Tanneberger, 2015). Whereas similar benefits can be obtained by using Liquefied Natural Gas (LNG), LFPFs liquefy at room temperature and in this sense they may provide benefits over LNG in terms of fuel transportation and storage. The main risk in terms of using LFPFs on ships is that their flash point is below the minimum allowed safe flash point for marine fuels as specified by the IMO. This is why long-winded specialist approval is required.

To facilitate the use of LFPFs, the SEDNA project is developing generally acceptable procedures for 'Safe use of low flash point fuels' (LFPFs) in the Arctic.

We believe that a shift from traditional ship fuels (e.g. Heavy Fuel Oil) to LFPFs would influence the implementation and usability of machin-

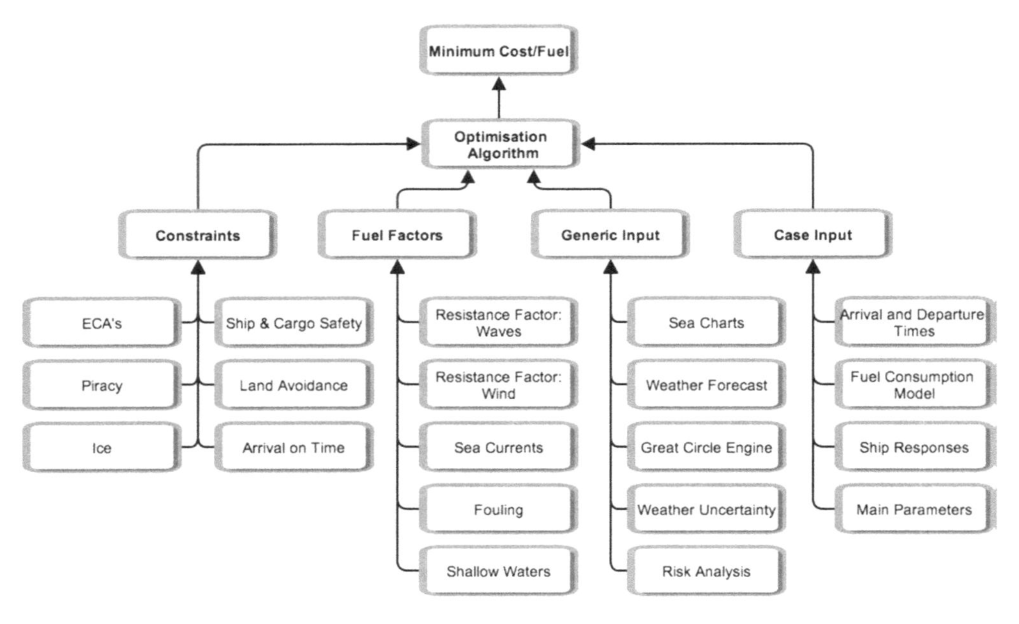

Figure 4. Overall structure of the Arctic voyage planning tool.

ery and fuel systems in the maritime environment. The technology could also prove beneficial within the context of responding to expectations aligned with future environmental regulations at the Arctic Region Emission Control operating Areas (ECAs). Because of its combustion properties and costs, the optimal speed of an LFPF driven ship may differ from that of a ship operating by conventional technology. This may result in a different optimal speed ranges, and challenges with regards to the location of fuel tanks. Yet, LFPF operated machinery would likely require less maintenance (e.g. there would be no need for a scrubber), reducing maintenance costs, as well manning requirements. Combined, this would enable a more cost- and energy-efficient design.

4 DISCUSSION AND CONCLUSIONS

This paper outlines a framework for the holistic performance-based design of Arctic cargo ships. The framework is developed based on the assumption that there is potential for efficiency gains in Arctic shipping by applying such approach. Accordingly, the approach presented utilises systems thinking by treating an Arctic ship as a component of a wider maritime transport system. It is believed that the outlined framework suggests a tangible option for regulatory implementation. Yet, further research is necessary to extend its applicability. Important future re-search work of relevance that will be addressed by SEDNA project include (a) further development of ice-loading assessment methods and validation of ice load assessment tools, (b) development of goal based models that enable performance assessment of accidental events, (c) de-risking of emerging technologies for use by the Polar fleet.

ACKNOWLEDGMENTS

This project has received funding from the European Research Council (ERC) under the European Union's Horizon 2020 research and innovation programme (grant agreement n° 723526).

ABBREVIATIONS

AMTS	Arctic Maritime Transport System
AR	Augmented Reality
CONOPS	Concept of Operations
DES	Discrete Event Simulation
ENVP	Environmental Protection
FR	Functional Requirements
GBD	Goal-Based Design
HFO	Heavy Fuel Oil
IACS	International Association of Classification Societies
IB	Icebreaker
IMO	International Maritime Organization

IMS	Ice Mitigation Strategy
LFPF	Low Flash Point Fuel
LNG	Liquefied Natural Gas
MARPOL	International Convention for the Prevention of Pollution from Ships
OPS	Operations
OSR	Oil Spill Response
SAR	Search and Rescue
SOLAS	International Convention for the Safety of Life at Sea

REFERENCES

ARCDEV, 1998. *Final public report of the ARCDEV project.* [Online] Available at: http://www.transport-research.info/sites/default/files/project/documents/arcdev.pdf

Bergström, M., Erikstad, S.O. & Ehlers, S., 2016a. A Simulation-Based Probabilistic Design Method for Arctic Sea Transport Systems. *Journal of Marine Science and Application,* 15(4), p. 349–369.

Bergström, M., Erikstad, S.O. & Ehlers, S., 2016b. Assessment of the applicability of goal- and risk-based design on arctic sea transport systems. *Ocean Engineering,* Volume 128, pp. 183–198.

Bergström, M., Erikstad, S.O. & Ehlers, S., 2017. The Influence of model fidelity and uncertainties in the conceptual design of Arctic maritime transport systems. *Ship Technology Research - Schiffstechnik,* 64(1), pp. 40–64.

Blackmore, R. & Lozowski, E., 1994. A heuristic freezing spray model of vessel icing. *International Journal of Offshore and Polar Engineering.*

Choy, K.-L., 2018. *University College London, Institute for Materials Discovery.* [Online] Available at: http://www.ucl.ac.uk/institute-for-materials-discovery/research/functional-coatings/anti-icing-coatings [Accessed 20 2 2018].

Ellis, J. & Tanneberger, K., 2015. Study on the use of ethyl and methyl alcohol asalternative fuels in shipping, s.l.: EMSA and SSPA.

IACS, 2016a. *Requirements concerning polar class.* s.l.:International association of classification societies.

IMO, 2000. Formal safety assessment. Decision parameters including risk acceptance criteria. MSC 72/16. London: International Maritime Organization.

IMO, 2001. Guidelines on alternative design and arrangements for fire safety. MSC/Circ.1002. London: International Maritime Organization.

IMO, 2006. Amendments to the Annex of the protocol of 1978 relating to the international convention for the prevention of pollution from ships, 1973. Resolution MEPC.141(54), London: IMO.

IMO, 2006. Amendments to the annex of the protocol of 1978 relating to the International Convention for the Prevention of Pollution from Ships, 1978. Resolution MEPC.141(54), London: IMO.

IMO, 2006a. Guidelines on alternative design and arrangements for SOLAS chapters II-1 and III. MSC.1/Circ.1212. London: International Maritime Organization.

IMO, 2013a. Guidelines for the approval of alternatives and equivalents as provided for in various IMO instruments. London: International Maritime Organization.

IMO, 2014d. Resolution MSC.386(94). Amendments of the international convention for the safety of life at sea, 1974, as amended. London: IMO.

IMO, 2015. International code for ships operating in polar waters (Polar Code). MEPC 68/21/Add.1 Annex 10. London: International Maritime Organization.

Jordaan, I.J., Maes, M.A., Brown, P.W. & Hermans, I.P., 1993. Probabilistic analysis of local ice pressures. *Journal of Offshore Mechanics and Arctic Engineering,* Volume 115, pp. 83–89.

Kim, E. & Amdahl, J., 2015. Discussion of assumptions behind rule-based ice loads due to crushing. *Ocean Engineering,* Volume 119, p. 249–261.

LR, 2014. *Written evidence (ARC0048),* London: Lloyd's Register.

Markström, L. & Holm, H., 2013. Voyage optimization on the shallow waters of the Baltic Sea, s.l.: SSPA.

Melzer, J., 2012. *HMDs as enablers of situation awareness: the OODA loop and sense-making.* s.l., Proceedings of SPIE - The International Society for Optical Engineering. 8383.13-. 10.1117/12.920844.

Papanikolaou, A., 2009. *Risk-Based Ship Design - Methods, Tools and Applications.* 1 ed. Berlin Heidelberg: Springer.

Riska, K., 2010. Ship-ice interaction in ship design: theory and practice (Course Material), Trondheim: NTNU.

Rolls-Royce, 2016. *Autonomous ships - The nexy step.* [Online] Available at: http://www.rolls-royce.com/~/media/Files/R/Rolls-Royce/documents/customers/marine/ship-intel/rr-ship-intel-aawa-8pg.pdf

SEDNA, 2017. Safe maritime operations under extreme conditions: the Arctic case (SEDNA). [Online] Available at: https://sedna-project.eu/

Skjong, R., Vanem, E. & Øyvind, E., 2007. *Risk Evaluation Criteria,* s.l.: SAFEDOR.

Valdez Banda, O. et al., 2016. Risk management model of winter navigation operations. *Journal of Marine Pollution Bulletin,* 108(1–2), p. 242–262.

Vasiljević, A., Borović, B. & Vukić, Z., 2011. Augmented reality in marine applications. *Brodogradnja: Teorija i praksa brodogradnje i pomorske tehnike,* 62(2), pp. 136–142.

Marine Design XIII – Kujala & Lu (Eds)
© 2018 Taylor & Francis Group, London, ISBN 978-1-138-34076-3

Comparison of vessel theoretical ice speeds against AIS data in the Baltic Sea

O.-V. Sormunen
Department of Mechanical Engineering, Aalto University School of Engineering, Espoo, Finland

R. Berglund
VTT Technical Research Centre of Finland, Espoo

M. Lensu
Finnish Meteorological Institute, Helsinki, Finland

L. Kuuliala
Finnish Transport Safety Agency, Helsinki, Finland

F. Li, M. Bergström & P. Kujala
Department of Mechanical Engineering, Aalto University School of Engineering, Espoo, Finland

ABSTRACT: The northern Baltic Sea freezes every year, complicating and slowing down ship traffic due to added resistance and from at times having to wait for ice breaker assistance in severe ice conditions. This interferes with optimization of commercial vessel logistics. On-time deliveries are important for industry for countries such as Finland as the industry is mainly shipping its exports by sea. Thus knowledge of how well winter traffic sailing times can be forecasted is paramount for proper logistical planning and cost-effectiveness: better predictability of ship sailing times leads to cost savings due to lower delivery time buffers and overall better utilization of transport capacity. While vessel speed in ice can be theoretically predicted, another question is how accurately this can be done given the limited availability and resolution of e.g. ice data. For this purpose the performance of an ice breaker and a commercial vessel are analyzed and compared along with discussion on how their design affects their ice-going predictability.

1 INTRODUCTION

1.1 *Ship resistance in ice*

The resistance of a ship in ice is caused by breaking ice by crushing and bending as well as displacing the broken ice and ice rubble present in channels, rubble fields and ridges. The relative contribution of the different resistance components varies according to the encountered ice conditions and the hull-form of the ship. Especially the bow form has a big influence on ice resistance and the bow forms optimized for ice breaking performance differ significantly from the bow forms of open water ships. Due to stricter emission regulations such as the IMO (2011) EEDI-regulations, ice-classed merchant vessels generally have a worse ice-going capability than many older designs. The design point for the machine power requirement for the Finnish-Swedish ice classes is 5 knots speed in a new ice channel and this can be attained with traditional open water bow forms. However, some ships with ice class IA Super – such as double-acting tankers – have previously been able to operate also in off-design conditions such as breaking out of a channel or breaking their own channel at times. It remains to be seen how the optimizing of ice classed ships for open water performance affects the overall winter traffic system once older ships with superior ice-going capabilities are decommissioned.

The ice conditions encountered by merchant vessels in the Baltic are level ice, ice channels, rubble fields and ridged ice. Normal merchant vessels are not expected to operate independently in rubble fields or ridged ice but rather be assisted by ice breakers along official routes called dirways. Ice channels are formed by ice breakers and merchant vessel traffic. Channels in the land-fast ice zone near the coastal area can remain open for the whole winter and become progressively more difficult to navigate as new ice is formed on the surface and broken by passing vessels to contribute to the channel thickness. In the dynamic ice field at open sea, channels are often closed by compression in the ice field.

Outside the land-fast ice zone, compression in the ice field produces also ridged ice and rubble fields. Ridging typically starts as rafting. When the stresses in the ice sheet caused by friction and roughness of the ice cover become sufficiently large, ice fails in buckling and bending. Ridging occurs until the driving forces can no longer cause additional ridge forming or until the ice sheet fails elsewhere resulting in a new ridging event. (Tuhkuri et al., 1999).

Ridges consist of an underwater part called the keel and an above ice surface part known as the sail. The sail and keel are composed of ice blocks that may be frozen together. In mature ridges there is a consolidated layer near the waterline, where water filling the space between the ice blocks has frozen. The consolidated layer is inhomogenous and has a varying thickness. The resistance in ridged ice depends mostly on the size of the keel and the thickness and mechanical properties of the consolidated layer.

Sail and keel masses are in hydrostatic balance over long length scales but significant local deviations from balance may occur. The deviations from hydrostatic balance are compensated by deformations and stresses in the surrounding ice field. (Boven and Topham, 1995; Kankaanpää, 1989; Tin and Jeffries, 2003) Due to ridging usually beginning as rafting, there can be significant lateral displacements between ridge sails and keels. (Melling et al., 1993).

Lindqvist (1989) proposed a semi-empirical level ice resistance formula, that is still used at least as a basis of methods to estimate resistance of ships in ice. Lindqvist's formula was validated with ships that have traditional ice breaking hull-forms. The assumptions of Lindqvists formula are not necessarily valid for hull forms of merchant vessels optimized for open water service. Riska et al. (1998) studied merchant vessels in the Baltic and presented resistance formulas for level ice, ice channels and ridged ice. The channel ice formula was modified by Juva and Riska (2003) and included in the national ice class regulations in Finland and Sweden.

Methods for predicting the ridge resistance of ships have been presented by Keinonen (1979), Mellor (1980), Malmberg (1983), Abdelnour et al. (1991) and Riska et al. (1998). Most of the ridge resistance formulations are based on soil mechanics and the resistance predictions are sensitive to assumptions concerning the material properties of ridge rubble. See for instance Kuuliala et al. (2017) for further details.

A numerical method to predict the resistance and performance of ships in level ice was presented by Wang (2001). Su (2011) and Tan et al. (2013) used the method in semi-empirical simulation models of ship's performance in ice. These models are also sensitive to assumptions and parameter values, see for example Kuuliala (2015).

Transit simulation models in various ice conditions have been presented by La Prairie et al. (1995), Patey and Riska (1997), Kamesaki et al. (1999), Montewka et al. (2015), Goerlandt et al. (2017), Kuuliala et al. (2017) and Bergström et al. (2017).

Transit simulations pose several sources of uncertainty. The available data concerning ice conditions has typically a low resolution and different ice conditions such as leads and channels cannot be identified and taken into account. Also, other considerations than the ice conditions influence the speed of ships. For ships not in scheduled traffic, there may not be an incentive to maintain maximum possible speeds if for instance berthing space, loading or offloading services are not available. Furthermore, the waiting times for ice-breaker escort may have an effect on the speed that merchant vessels use. Depending on the chartering model, the ship owner may have an incentive to reduce speeds and conserve fuel and not push for a fast delivery.

Mechanical properties of ice are variable and not known in an operational setting. Even the thickness of level ice or deformed ice cannot be accurately measured over large areas. The spatial distribution of ice ridge keels affects the operability of ships significantly but it cannot be observed from the surface as there is no one to one correspondence between sails and keels. These factors combine to make the problem of predicting the performance of ships in ice essentially stochastic. Within the resolution of available data and models of the sea ice cover, there can be no strict correspondence between the attainable speed of a ship and the ice conditions.

Bergström et al. (2017) investigated some of the uncertainties related to commercial winter traffic in the Baltic sea. Among others, they was found that the closer the ice thickness is to a ship's maximum ice breaking capacity, the more sensitivity and scatter there are in the results with respect to estimating ship speed based on h-v curves using various assumptions regarding ice thickness. This is because when a ship gets closer to its maximum ice-going capability, any uncertainty in the assumed H_{eq} value results in a larger relative uncertainty in the estimated ship speed, see Figure 1.

1.2 *Objectives of the study*

The objective of this study is to assess the accuracy of theoretical approaches for estimating the real-world speed of ships operating in natural ice To this end, recorded real world speeds of two different ships (an icebreaker and a commercial vessel) operating in various ice conditions are compared with theoretical

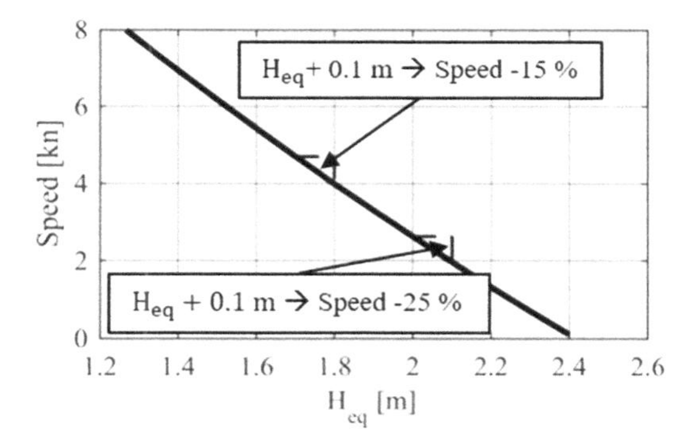

Figure 1. Variations in the equivalent ice thickness (H_{eq}) vs. relative variations in ship speed (Bergström 2017).

speed estimates for the same ships and ice conditions. In the following we describe the applied methodology and data. In this paper the analysis is done based on limited case studies utilizing data at the accuracy level that is readily and realistically available for larger sea areas. The aim is not to provide be-all answers but to make an initial study and discussion.

2 DATA AND METHODOLOGY

This paper focuses on data from February–March 2010. The data consists of AIS positional data filtered by VTT, ice condition data from FMI and theoretical ice going capacity of vessels using h-v curves found in literature.

2.1 *Winter 2010*

The onset of freezing was in the beginning of December and the Bay of Bothnia was ice covered on 7th January and the Gulf of Finland before the end of the month. The winter was stormy and with multiple deformation phases with ridging in the northern part of the basins and opening in the southern parts. Difficult compressive conditions were encountered especially in the Gulf of Finland. The sea of Bothnia became fully ice covered after mid-February and the ice cover extended to the Northern Baltic proper in March. Measured by the ice extent the season was somewhat more severe than average from the point of view of ships the difficult conditions prevailed to the end of March in the Gulf of Finland and to the end of April in the Bay of Bothnia.

2.2 *AIS data*

The AIS data is from 1st of February to 31st March 2010. The data was obtained from the terrestrial AIS network maintained by the Finnish Traffic Agency. VTT has archived AIS data covering the Baltic Sea since many years as part of the monitoring and maintenance activities related to the icebreaker information system IBNet. The data contains AIS position report messages (number 1 and 3) with an unfiltered temporal resolution, which means a position report every 10 seconds if the speed of the vessel is up to 14 knots, every 6 seconds if the speed is up to 23 knots and every 2 seconds if the speed is higher than 23 knots. Due to occasional reception errors, not all messages have been recorded, resulting in occasional longer intervals, but sporadic missing packets do not have an impact on the usability of the data. However, when the vessel is too far from the nearest base station, AIS data is not obtained, which may cause lack of validation data for some areas (e.g. middle of the Sea of Bothnia). The archived AIS data has all the values in the AIS messages plus an additional timestamp field indicating the time of data reception in the server that has stored the data in a file. This is necessary as the data itself only contains the number of seconds (0–59) of the position recording at the vessel AIS transponder. The fields used in this study were the Time, MMSI, Lat and Long. Ships were selected for further studies that had AIS tracks overlapping with available icebreaker dirway point and ice thickness data.

The method consists of calculating the actual vessel velocities from the AIS data and comparing it to the predicted sailing speed based on the theoretical ice velocity curves under the current ice conditions in the area. The ice conditions are updated once per day while the AIS data sampling frequency is typically around a couple of observations per minute.

2.3 *Ice data*

Ice data is extracted from gridded Finnish Meteorological Institute's (FMI) ice charts. In the gridded charts the regions are sets of grid cells, and the 1 nautical mile grid resolution renders the graphical information practically without loss. The data is presented both as gridded that express ridging in terms of a numeral in similar manner to Table 1 as well as graphical charts the ridging is indicated qualitatively, see Figures 3 and 5. The numerals are ice analyst's estimates based on the deformation history, SAR images and reports from icebreakers. The FMI ice data is filtered into relevant tables, defined between the dirway-points, see Table 1. These waypoints define official routes for icebreakers where commercial vessels can expect ice breaker assistance to navigate through the ice. The FMI ice data classifies ice as

Table 1. Snapshot of FMI ice data for the ice breaker route on 17.3.2010 between the dirway points. Empty fields indicates that the data is missing.

Concentration [%]	Mean thickness [cm]	Min thickness	Max thickness	Type	Lat N	Lon E
–	–	–	–	–	65.283	23.3
98	40	30	60	4	65.1	23.75
98	40	30	60	4	–	–
98	40	25	50	3	64.867	23.783
98	40	25	50	3	–	–
98	40	30	60	4	64.67	24.033

0 Level ice
1. Rafted ice
2. Slightly ridged ice
3. Ridge ice
4. Heavily ridged ice
5. Windrow (brash barrier)

and provides the following variables:

1. Ice Concentration (%)
2. Ice Thickness (cm)
3. Min Thickness (cm)
4. Max Thickness (cm)
5. Ice Type (0–5)

For the purposes of the equivalent ice thickness calculations, FMI's ice types 1–4 are considered level ice with added ice due to various degrees of ridging, see equation 1. For these ice conditions an equivalent ice thickness is calculated as follows: For level ice (type 0) with ice concentrations close to 100% the ice thickness is simply read from the FMI ice data mean thickness column.

T_I = Ice thickness [m]

When there are ridges present, additional ice thickness is added using the following assumptions. Ridged ice (equivalent) thickness in [m] is

$$T_{RE} = \frac{3\,C_R\,T_R^2}{2000} + T_I \tag{1}$$

T_R = ridge thickness [m], assumed to be 5 m.
C_R = Ridge density (number of ridges per 1000 m).

Ridge density is not specified in the FMI data but can be roughly estimated (Lensu 2016) as [raft ed,slight,ridged,heavily] = [1 3 8 14] ridges/km.

2.4 *H -v curves*

The h-v curves describe vessel speed (v) in ice as a function of ice equivalent thickness (h). They are obtained theoretically through the balance between ice resistance and propeller net thrust. According to the benchmark study by Kämäräinen (1993), the level ice resistance formula by Lindqvist (1989) generally gives a better predition over other investigated methods. Therefore, in this paper, we use Lindqvist's formula to calculate the resistance in level ice. Riska's (1997) formula is the base of the channel ice resistance calculation in the Finnish-Swedish ice class rules (TRAFI 2010), thus it is applied here due to its wide use. Propeller net thrust refers to the available thrust to overcome ice resistance. According to Riska et al. (1997), the net thrust can be approximated by

$$T_{net} = T_b\left(1 - \frac{1}{3}\frac{v}{v_{ow}} - \frac{2}{3}\left(\frac{v}{v_{ow}}\right)^2\right) \tag{2}$$

where T_b is the bollard pull as a function of propeller received power and propeller diameter; v is the ship speed; v_{ow} is the open water speed under certain power P. The two investigated ships in this paper are installed with fixed pitch propellers. Assume the icebreaker uses 100% of the full propeller power and the commercial ship uses 85% when transiting on the sea, the open water speed under transit power is estimated according to Li (2016) by

$$v_{ow} = \left(\frac{P}{P_m}\right)^{\frac{1}{3}} v_m \tag{3}$$

where P_m and v_m are the maximum propeller received power and maximum open water speed. $\frac{P}{P_m}$ then equals 80% and 85% for the icebreaker and the commercial ship.

The ice resistance is theoretically calculated by the formulas of Lindqvist (1989) and Riska et al. (1997) for level ice and the formula of Riska et al. (1997) for channel ice. Equating the net thrust with ice resistance gives the obtainable speed under given ice thickness. The h-v curves of the investigated ships are plotted in Figure 1.

In order to facilitate calculation, we make polynomial regression fit for each h-v curve. The regressions are expressed in a uniform way as

$$v = a_0h^3 + a_1h^2 + a_2h + a_3 \tag{4}$$

Table 2. Coefficients for different h-v curves represented as a 3rd degree polynomial function.

	Ice breaker			1 AS ro-ro		
	Lindqvist, level ice	Riska, level ice	Riska, channel	Lindqvist, level ice	Riska, level ice	Riska, channel
a_0	0.0044	–0.0044	–0.0146	–3.7702	0.3693	–1.8943
a_1	0.9189	0.5777	–0.0444	14.3866	0.4107	3.5430
a_2	–6.2631	–5.0875	–1.4079	–20.3701	–7.9147	–6.9509
a_3	10.1115	10.0295	9.9892	9.6007	9.5208	9.5341

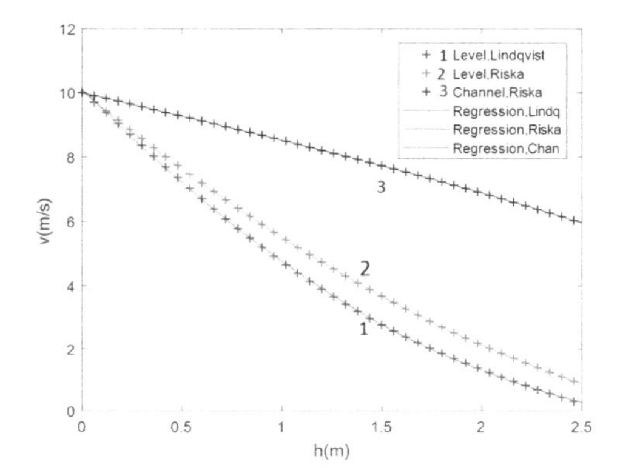

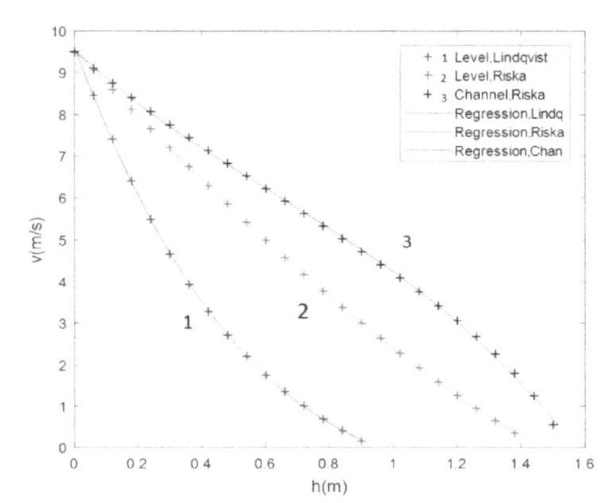

Figure 2. H-v curves for level ice and channel conditions for icebreaker (top) and 1 A super ice class ro-ro vessel (bottom).

The coefficients are summarized in Table 1. The regression lines are also plotted in Figure 2. As shown in the figure, the regression fits the calculated h-v curves very well.

It should be pointed out that the ice resistance formulas applied in this study are semi-empirical, and therefore unreliable when applied on ships that significantly differ from those based on which they were determined. Reliable ice resistance estimates can be obtained by model testing, or by existing numerical methods (von Bock und Polach and Ehlers, 2015) but this level of detail is not possible for analyzing complete regional ship traffic systems.

3 VELOCITY COMPARISON

In this chapter the actual velocity of various vessels sailing under different conditions are compared to the theoretical velocities to closer investigate how well theoretical velocities correspond to actual velocities. This is done for 2 vessels: one ice breaker and one Finnish Ice Class IAS ro-ro vessel. The candidates were chose based on available and overlapping data from AIS, ice data as well as vessel details for h-v curve estimation.

3.1 *Icebreaker 17.3.2010*

The icebreaker has $l = 90$ m, $b = 23.4$ m and $P_m = 15$ MW. It is sailing in the Gulf of Bothnia between an area south of Kalix in Sweden to Raahe in Finland, see Figure 3. The mean (level) ice thickness is constant throughout the dirway at 0.40 m in the FMI ice data with ice type being FMI type 3 (ridged) or 4 (heavily ridged), which corresponds to an added additional 0.3 / 0.525 m ice thickness divided into 80 segments in the sampled data. Two runs—one in the night from North-West between 2.00–5.59 and one in the morning from South-East between 9.00–12.59 are analyzed.

From the AIS data the real vessel velocity is calculated for each data point at time t as the difference in distance to the last observation at time t-1 divided by the time difference:

$$Vt = \frac{D_t - D_{t-1}}{T_t - T_{t-1}} \tag{5}$$

The calculated velocities are illustrated in the histograms below along with the mean velocity of the vessel in the current timeframe, with and without taking away observations where the icebreaker was standing still.

In Figure 4, it can be seen that in reality the ice breaker's velocity varies during the run. Using the

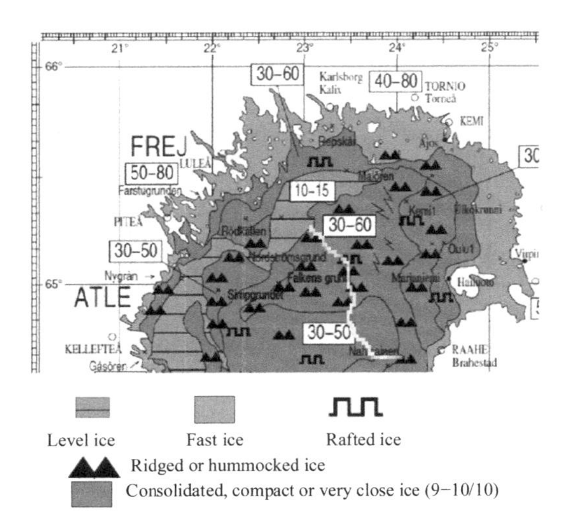

Figure 3. Prevailing ice thicknesses [cm] from FMI data in Northern Bay of Bothnia and the IB back-and-forth route on March 17, 2010.

available ice data we have only one mean ice thickness (0.4 m) throughout the dirway with additional 0.3 or 0.525 m added ice thickness along the way depending on the location's specified FMI ridge type, see Table 1 and equation 1. This would be possible to make more detailed by including minimum and maximum ice thicknesses from Table 1 but the result will not be as varied as the histograms in Figure 4. As such, we cannot realistically predict velocities for a small time frame: More detailed data should be collected—or mean values over longer time periods should be used. From a logistics planning point of view this might not be a concern; the most important estimate is the exact travel time between two ports. As the ice conditions are known only roughly, the comparison between actual (AIS) and theoretical (h-v) speeds is done using several different assumptions, in Table 3.

The different ice assumptions are the (simple) mean ice thickness ($H = 0.4$ m) as well as the equivalent ice thickness (H_{eq}) where to H the added ice due to ridging is included. Calculations are done for both ice assumptions using the vessel ice channel h-v curve as well level ice h-v curve (i.e. no ice channel present), see Figure 2. The uppermost quadrant describes the theoretical average velocity for the route in Figure 3 (which are more or less equal for 2:00–5:59 and 9:00–12:59 as the ice breaker operates in a loop). The 2 lower quadrants describe the% difference between the AIS mean speed (from Figure 4) and the theoretical speeds as taken from the upper quadrant. As can be seen in the Table 3, the theoretical mean velocity for the stretch changes drastically with the assumptions

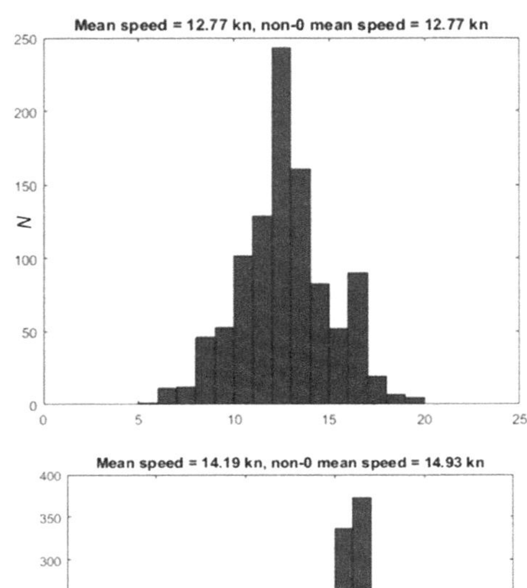

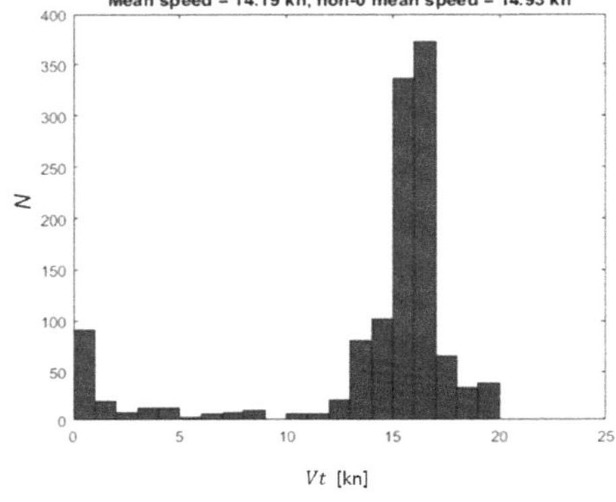

Figure 4. Histogram of AIS velocities for the icebreaker during 2.00–5.59 (top) with mean speed of 12.77 knots and 9.00–12.59 (bottom) with a mean speed of 14.19 (14.93) knots on 17.3.2010.

Table 3. Mean velocity comparison to the non-zero AIS mean speed for ice breaker using various assumptions.

	H-v curve	Speed [kn]
Ice	Level ice	Ice channel
H	13.848	15.918
H_{eq}	8.845	13.216
2:00–5:59	H-v curve	Difference
Ice	Level ice	Channel
H	8.44%	24.65%
H_{eq}	–30.74%	3.49%
9:00–12:59	H-v curve	Difference
Ice	Level ice	Channel
H	–7.25%	6.62%
H_{eq}	–40.76%	–11.48%

between an absolute value of 3.49–49.33% difference between the observed AIS and the theoretical h-v curve velocities.

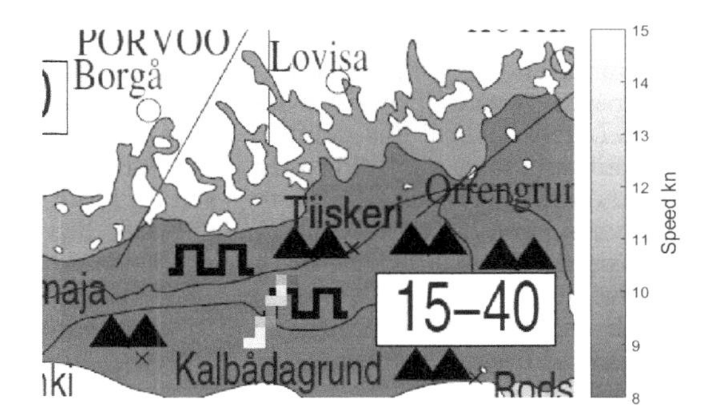

Figure 5. Ro-ro AIS track color-coded by velocity 0.00–0.59 on 2.3.2010.

The problem is that with the available data it was not possible to discern exactly which conditions should be prevailing: The ice data does not tell us whether there is a channel (and how consolidated the ice is in such channel). Not enough detailed AIS data was available for this paper to see exactly which vessels had previously passed and when, though in the AIS data the ice breaker was going back and forth on the analyzed day. Thus channel ice conditions can be argued to the most realistic assumption with the ice breaker. This assumption also gives the most accurate results, though the exact conditions of the channel could also not be determined such as if the ice thickness should be considered to be something else than the ice mean thickness due to ice consolidation in channel. Furthermore adding to the uncertainty, the number of discrete ice data points along the route is relatively small (with some missing values) meaning extrapolations from other observations have to be made, see Table 1. The ice sample points in Table 1 also do not fully match the vessel route as the vessels make some deviations from the appointed ice breaker route at the captain's discretion.

The numbers here can be considered quite good as we do not know the engine setting of the ice breaker and we are currently just assuming full power.

3.2 *IAS ro-ro cargo ship 2.3.2010*

The second comparison is done for a commercial vessel of the highest Finnish-Swedish ice class 1 A Super with ~11 500 GT, $l = \sim160$ m, $b = \sim20$ m and $P_m = 12.6$ MW. The timeframe is 0.00–0.59 on 2.3.2010. The vessel is sailing from South-West to North-East in an area south of the town of Loviisa in the Gulf of Finland, see Figure 5. The ice conditions vary between ridged to heavily ridged ice (3–4) with the level ice component thickness between 8–30 cm and an ice concentration of 90–95%, which in these calculations is assumed to be equivalent to full ice cover.

According to our AIS-data, during the passage the ro-ro vessel is not escorted by any icebreaker assisting in the area. As such using the vessel's own h-v curves is valid. The ro-ro's h-v curves for the level ice vs channel are significantly different than for the ice breaker: the relative difference of level ice vs channel ice speed increases faster than for the ice breaker, see Figure 2. The AIS-data based velocities for the time period are given in Figure 6.

Again, a wide range of velocities is observed as shown in Figure 6. The calculated mean velocities utilizing various h-v curves and ice thickness assumptions for the ro-ro are shown in Table 3.

The mean velocity differences between the observed ones from AIS and the theoretical ones are between 21–71%, which is more than for the ice breaker but the conditions are also more uncertain as there is a wider range of ice thicknesess and the vessel is sailing less in a straight path between the dirway points than the ice breaker and the time period is shorter. The ro-ro is also more sensitive to changes in ice thickness (see Figure 2) and we have only limited information available regarding the ice conditions. These compound to the added uncertainty of reliable speed estimation for the ro-ro.

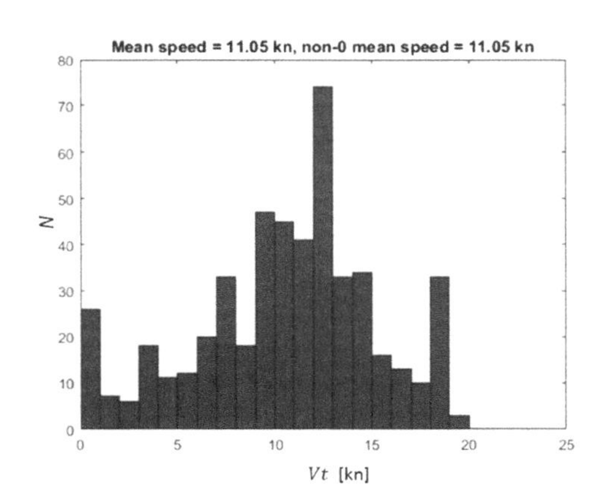

Figure 6. Ro-ro velocity histogram for 0.00–0.59 on 2.3.2010, mean speed is 11.05 knots.

Table 4. Mean velocity comparison to the non-zero AIS mean speed for the ro-ro using various assumptions.

0.00–0.59	H-v curve	Speed [kn]
Ice	Level ice	Ice channel
H	13.76	18.25
Heq	3.21	13.34
0.00–0.59	H-v curve	Difference
Ice	Level ice	Channel
H	24.55%	65.12%
H_{eq}	–70.94%	20.75%

4 RESULT, DISCUSSION AND CONCLUSIONS

In this paper we compared the theoretical and practical performance of one ice breaker and one commercial vessel. Depending on the assumptions and the situation the theoretical mean velocity difference between the real velocity from AIS data and the theoretical h-v curve based velocity found to be 3.5–71%, see Tables 3–4. In this paper several simplifications were made; we do not know the exact ice conditions and what kind of channel conditions potentially are present as the ice data is relatively rough in sampling frequency and resolution. This paper, however does contribute by highlighting one of the most critical shortcomings in predicting vessel ice speed based on readily available big data: The effect of ice channel versus no channel is the most significant one in this study. The ship engine settings are also not known in our data. Collecting better and more comprehensive data of the former is proposed as future research. As the sample size in this paper is very limited, expanding the number of vessels, ice conditions and time frames is also proposed as well; the results presented here should be taken more as initial pointers than robust conclusions based on large sample sizes. Furthermore, the assumptions about the added ice thickness (i.e. equivalent ice thickness) due to ridges are rough estimates, see Bergström (2017), Bailey et al. (2015) and Tuhkuri et al. (1999) for more.

In this paper the ice breaker was much less sensitive to the assumptions regarding the ice type/h-v curve under the analyzed conditions as the velocity % differences are lower in Table 3 (icebreaker) than in Table 4 (ro-ro). As can be seen in Figure 2, the discrepancy between the level ice and channel h-v curves is bigger for the commercial vessel than for the ice breaker: The slope of the h-v curve is steeper for the commercial vessel than for the ice breaker, meaning that an uncertainty of +/–10 cm in the ice conditions affects the operability of the commercial vessel more than the ice breaker. Conversly, given the same ice conditions but decreasing vessel ice performance e.g. due to more stringent emission regulations not only increases travel time and ice breaker assistance demand but also makes it more difficult to predict vessel ice speed as more vessels will more often be at ice conditions close to their maximum, where the uncertainty is the largest on the h-v curves. Due to more ice-optimized design the ice breaker is able to maintain a better speed in hard ice conditions. This comes with a trafe-off with respect to open water handling in conditions such as waves due to the more spoon shaped bow. Thus the specific focus for conventional Finnish ice breaker design on maximizing the ice going capability gives additional robustness in terms of being able to predict the speed in ice under uncertain ice conditions. Commercial vessels with good ice-going capabilities on the other hand have more capital, maintenance, insurance and running costs in open water traffic compared to pure open water designs, see Solakivi et al. (2017). Overall this means that any ice going capability is a trade-off; in order to optimize this trade-off the exact sailing conditions and the resulting sailing speeds must be known and modelled. This paper—to a limited degree—shows what is currently known with respect to the ice conditions and the theoretical ice gong speed estimations along with what should be investigated further.

4.1 *Conclusions*

The findings of this study indicate that there is a significant uncertainty in estimating a ship's real-word speed in ice based on its h-v curve and the prevailing ice conditions. This result is relevant for future studies and assessment of the performance of ice-going ships and winter navigation systems, among others. Reasons for deviations between theoretical speed estimations, and actual speeds are that theoretical speed estimations might not consider a number of factors such as (1) a ship might not use its full MCR rating, (2) a ship might have a captain who can avoid local areas with difficult ice conditions, (3) there might be an old ice channel available, etc. Thus, a typical theoretical (h-v curve based) speed estimate is oversimplified, and that more factors (see above) need to be considered for a realistic estimate.

The results might be considered good as we don't know e.g. power settings nor if the velocity had to be reduced due to traffic, navigational challenges or to avoid coming to port too early etc. as the results came as close as 3.5% with certain assumptions. Though it must be noted that the assumptions—especially regarding if there is a channel—heavily influence the results. This should be taken into account in logistical planning and full-scale modelling of winter navigation in a sea area with more buffer time and/or added transport capacity. In order to fully model and predict the winter, other factors than commercial vessel h-v curves need to be taken into account namely the smaller scale ice conditions and other traffic. The last affects when and how ice breakers are available as well as affects the condition of potential ice channels that the vessels sail through. Bergström et al. (2017) noted that the uncertainty of ship travel time was particularly sensitive to the definition of ice breaker assistance waiting time and the equivalent ice thickness definition.

ACKNOWLEDGEMENTS

The authors would like to thank Morten Lindeberg, Iiro Kokkonen, Tom Mattsson, Mona Zilliacus, Emmi Huvitus, Jarkko Toivola, Markus

Karjalainen, Helena Niemelä, Liangliang Lu and Jani Romanoff. This article is written as a part of the EU's TEN-T WINMOS (Winter Navigation Motorways of the Sea) I-II projects and the Finnish Transport Agency sponsored SimWinNS (Simulation of the Finnish-Swedish Winter Navigation System) project. The contributions made by the fifth author are supported by the research project Kara-Arctic Monitoring and Operation Planning Platform (KAMON).

REFERENCES

Abdelnour, R., Comfort, G., Peirce, T., 1991. Single pass ridge penetration model. Proceedings of the 11th International Conference on Port and Ocean Engineering under Arctic Conditions. vol. 2. pp. 600–622.

Bailey, E., Taylor, R., Croasdale, K., 2015. Mechanics of ice rubble over multiple scales. OMAE2015.

Bergström, M. 2017. A simulation-based design method for Arctic maritime transport systems. PhD thesis, Norwegian University of Science and Technology, Trondheim, Norway.

Bergström, M., Erikstad. S. and Ehlers, S. 2017. The influence of model fidelity and uncertainties in the conceptual design of arctic maritime transport systems. Ship Technology Research Schiffstechnik, 64, 40–64.

Bowen, R., Topham, D., 1995. A study of the morphology of a discontinuous section of a first year arctic pressure ridge. Cold Reg. Sci. Technol. 24 (1), 83–100.

Finnish Transport Safety Agency (TRAFI) 2010. Maritime safety regulation. Ice class regulations and the application thereof, Finnish Transport Safety Agency, Helsinki, Finland.

Goerlandt, F., Montewka, J., Zhang, W. and Kujala, P. 2017. An analysis of ship escort and convoy operations in ice conditions. Safety Science. http://dx.doi.org/10.1016/j.ssci.2016.01.004.

International Maritime Organisation (IMO). 2011. MEPC 62/24/Add.1 Annex 19. Available online at: http://www.imo.org/en/MediaCentre/HotTopics/GHG/Documents/eedi%20amendments%20RESOLUTION%20MEPC203%2062.pdf#search=EEDI. Accessed: 9.1.2018.

Juva, M and Riska, K. 2002. On the Power Requirement in the Finnish-Swedish Ice Class Rules. Research report No 53 of the Winter Navigation Research Board.

Kamesaki, K., Kishi, S., Yamauchi, Y., 1999. Simulation of NSR shipping based on year-round and seasonal operation scenarios. INSROP Working Paper 164-1999. INSROP.

Kankaanpää, P., 1989. Structure of first year pressure ridges in the Baltic Sea. Finnish Institute of Marine Research.

Keinonen, A., 1979. An Analytical Method for Calculating the Pure Ridge Resistance Encountered by Ships in First Year Ridges. Helsinki University of Technology. (Ph.D. thesis).

Kuuliala, L., Kujala, P., Suominen, M. and Montewka, J. 2017. Estimating operability of ships in ridged ice fields. Cold Regions Science and Technology.135, March 2017, 51–61.

Kämäräinen, J. (1993). Evaluation of ship ice resistance calculation methods. A thesis for the degree of licentiate of technology, Helsinki University of Technology.

Lensu, M. 2016. Personal communication by e-mail on 21.12.2016.

Lensu, M. and Kokkonen, I. 2016. Exploring ship performance and traffic system through integrated data sources. Inventory of ice performance for Baltic IA super traffic 2007–2016. STORMWINDS Deliverable D4.1. EU BONUS STORMWINDS, 52.

Li, F. 2016. Evaluation of a semi-empirical numerical simulation method for level ice breaking. M.Sc. Thesis, Aalto University, Espoo, Finland.

Lindqvist G. 1989. A straightforward method for calculation of ice resistance of ships. Proceedings of 10th International Conference on Port and Ocean Engineering under Arctic Conditions (POAC),12.-16.6.1989, Luleå, Sweden, Luleå University of Technology, pp. 722–735. Available at http://www.poac.com/Papers/POAC89_V2_all.pdf. Accessed: 22.9.2017.

Malmberg, S., 1983. Om fartygs fastkling i is (of ship's becoming beset in ice). M.Sc. Thesis, Helsinki University of Technology, Espoo, Finland.

Melling, H., Topham, D., Riedel, D., 1993. Topography of the upper and lower surfaces of 10 hectares of deformed sea ice. Cold Reg. Sci. Technol. 21, 349–369.

Mellor, M., 1980. Ship resistance in thick brash ice. Cold Reg. Sci. Technol. 3, 305–321.

Montewka, J., Goerlandt, F., Kujala, P. and Lensu, M. 2015. Towards probabilistic models for the prediction of a ship performance in dynamic ice. Cold Regions Science and Technology. Volume 112, April 2015, Pages 14–28.

Patey, M., Riska, K., 1999. Simulation of ship transit through ice. INSROP, INSROP Working Paper 155–1999.

Riska K, Wilhelmson M, Englund K, Leiviskä T. 1997. Performance of merchant vessels in ice in the Baltic. Research Report No. 52, Winter Navigation Research Board, Helsinki, Finland.

Solakivi, T., Kiiski, T. and Ojala, L. 2017. On the cost of ice: estimating the premium of Ice Class container vessels. Marit Econ Logist. DOI 10.1057/s41278–017–0077–5.

Su, B. 2011. Numerical predictions of global and local ice loads on ships. Phd thesis, Norwegian University of Science and Technology, Trondheim, Norway.

Tan, X., Su, B., Riska, K., & Moan, T. A six-degrees-of-freedom numerical model for level ice-ship interaction. *Cold Regions Science and Technology,* 2013, 92, 1–16.

Tin, T., Jeffries, M., 2003. Morphology of deformed first-year sea ice features in the southern ocean. Cold Reg. Sci. Technol. 36 (1–3), 141–163.

Tuhkuri, J., Lensu, M., Saarinen, S., 1999. Laboratory and field studies on the mechanics of ice ridge formation. Proc. POAC 1999.

von Bock und Polach, R., Ehlers, S., 2015. On the scalability of model-scale ice experiments. Trans. ASME J. Offshore Mech. Arct. Eng. 137.

Wang, S. 2001.A dynamic model for breaking pattern of level ice conical structures. Phd thesis, Helsinki University of Technology, Espoo, Finland.

Autonomous ships

Marine Design XIII – Kujala & Lu (Eds)
© 2018 Taylor & Francis Group, London, ISBN 978-1-138-34076-3

The need for systematic and systemic safety management for autonomous vessels

O.A. Valdez Banda, P. Kujala, F. Goerlandt & M. Bergström
Department of Mechanical Engineering (Marine Technology), Research Group on Maritime Risk and Safety, Aalto University, Aalto, Finland

M. Ahola
Department of Mechanical Engineering (Marine Technology), Research Group on Maritime Risk and Safety, Aalto University, Aalto, Finland
Aalto University, Department of Design, Design and Architecture, School of Arts, Helsinki, Finland

P.H.A.J.M. van Gelder
Faculty of Technology, Policy and Management, Safety and Security Science Group, TU Delft, Delft, the Netherlands

S. Sonninen
Finnish pilots (Finnpilot Pilotage), Helsinki, Finland

ABSTRACT: The date when the first fully autonomous vessel starts operations seems to be close. The readiness and development of technology for creating and combining the needed operational components of an autonomous ship provide strong evidence to believe that the maritime transport industry is prompt to witnessing this date. However, technology is not the only element that has to be set in order to initiate the operations of these new smart sea vessels. Other crucial elements have to be systematically and systemically integrated into the initial design phase of an autonomous vessel. In this study, a general review of such elements is presented, making the representation of a process for building the initial safety management strategy in an early design phase of an autonomous vessel. This process particularly focuses on identifying determined safety controls for ensuring and managing the safety of the autonomous vessels.

1 INTRODUCTION

Today, autonomous vessels have become a topic of high level of importance in the maritime transport industry (Levander 2016). Significant efforts are currently made by ship manufacturers and technology developers for making an efficient integration of the components needed to have an autonomous vessel ready for operation (Tevainen 2017). However, the practical implementation of the final constructed vessel is subject also to the efficient integration of several other components integrated into the entire maritime autonomous system (Rødseth & Burmeister, 2012).

This study presents a general description of the components that need to be aligned to the operational readiness of a safe autonomous vessel. The aim is to make a systematic and systemic representation of a process that defines safety controls that need to be integrated into the design of autonomous vessels. The systematic characteristic refers to the need for a methodological approach to define the initial safety management strategy of the autonomous vessels. The systemic characteristic refers to make this approach efficient to cover the different elements that need to be included in the design an autonomous vessel and its entire autonomous system (Carr, 1996). This approach demands covering new issues that are not commonly included and reviewed in a traditional ship design process.

For this, the study offers the description of several elements that need to be considered in the evaluation of the challenges that autonomous vessels need to overcome before operating. The focus is on detecting those safety management elements that are essential to consider since an early design phase (Poel and Robaey, 2017). The aim is to present a systematic and systemic safety management strategy which can ensure the correct functionality of an autonomous vessel.

2 THE CURRENT SAFETY CONTEXT FOR AUTONOMOUS VESSELS

2.1 *Safety research on autonomous vessels*

The research of autonomous shipping has considerably been increased in the last decade. This research has identified initial safety challenges that autonomous vessels need to overcome before they begin operations. Studies have accurately detected concrete safety challenge for the execution of operations and prevention of accidents in the context of an autonomous vessel (Wróbel et al. 2016; Wróbel et al. 2017). These studies make an estimation of the possible effect on the prevention and response to common maritime accidents such as collisions, groundings, fire on-board, structural failure among others.

Other analyses focus on the review of the safety risks for a particular type of autonomous vessel. These reviews a semi-defined context of operation of an autonomous vessel (Burmeister et al. 2014a, 2014b; Porathe et al. 2014; Rødseth and Burmeister, 2015). These have provided a defined systematic approach which gradually ensures a certain degree of autonomy in the navigational operation of a cargo vessel.

Moreover, there are studies which particularly focus on the analysis of the roles of personnel involved in the management of safety of the current maritime operations, and its potential integration and transformation into an autonomous maritime system (Ahvenjärvi 2016; Man et al. 2015; Wahlström et al. 2015). The studies explicitly identify some challenges for transferring the current practices of safety management from a manned vessel to an autonomous one.

Finally, there are also studies presenting an initial analysis of the challenges to be faced in the legal implementation of autonomous shipping (Hogg and Ghosh 2016; Van Hooydonk 2014). These present a description of the complexity for determining a safety regulatory framework which can cover all elements of the autonomous maritime system.

2.2 *Safety of autonomous vessels: manufacturer and operator perspective*

Autonomous vessels will be a product subject to the efficient implementation of a service (Burmeister et al. 2014b). Manufacturers aim to provide autonomous vessels which are efficient and safe. However, the operation of such vessels depends on the role of many other stakeholders that share the responsibility for ensuring its appropriate functioning with a demanded or expected level of safety.

The main objective of manufacturers is to make an efficient integration of each required component for building an autonomous vessel ready for operations. However, the readiness of the vessel can only represent the first green light needed to begin operations. The vessel will still be dependent on the readiness of the other elements of the entire autonomous system.

So, what are then the operational components of such a system? Initially, autonomous vessels will be dependent on connectivity, including the alignment between the components of artificial intelligence and satellite services. Other services need to be also integrated to ensure the communication between vessels interacting in a common maritime environment. This represents a huge challenge as the communication carried out by merchant vessel is very rich and active in the exchange of information with the maritime traffic and with other maritime stakeholders such as VTS operators and Pilots.

Additionally, regulations have to be created, adopted and efficiently implemented for the operations of autonomous vessels. This requires to efficiently integrating a regulatory framework for construction, equipment, arrangements, operations, and the new aspects needed in the services linked to the vessel. This demands the involvement of other critical stakeholders of the autonomous system.

These challenges seem to be known and currently reviewed by manufacturers and operators. Therefore, in the path towards a fully autonomous vessel, certain categories of autonomy have been set (Burmeister et al. 2014b). These categories attempt to represent a gradual process for reaching the fully autonomous operation of a vessel in the future. However, the actual delegation of responsibilities and tasks for the entire stakeholders of the system has not yet been officially specified.

2.3 *Safety of autonomous vessels: the perspective from some other stakeholders*

When ensuring safety in autonomous vessels, other stakeholders become also relevant in the process. There is a long list of organizations which are essential in the management of the operations of autonomous vessels. This list includes among others:

- Marine equipment manufacturers
- Suppliers
- Designers
- Ship repairs and offshore yards
- Port and port operators
- Financers and insurances
- Maritime Authorities
- Pilots
- VTS
- SAR services
- Classification societies

- Marine trainers
- Unions
- General public
- Etc.

Thus, what about the safety perspective of those? Marine equipment manufacturers have a positive view towards developing more advanced equipment for ensuring the safety of an autonomous vessel. For them, autonomous vessels represent the opportunity to transform the maritime transport in a more proactive industry (Kretschmann et al. 2017). This is aligned with the aim of making a clear step towards proactive maritime safety management, an objective constantly pursued by the International Maritime Organization (IMO) (Schröder-Hinrichs et al. 2013).

Ship designers represent another group which has the opportunity to apply innovative designs in shipbuilding. The shipbuilding process based on digital design, operation and maintenance is pointed out as the alternative to be further developed in autonomous vessels (Kim et al., 2002). Digital shipbuilding is the process of performing the whole shipbuilding process from conceptual design to operation and maintenance in computer model and simulation. It processes and implements the whole manufacturing of the ship using an integrated database.

In port operations, the progress of technology has provided new solutions for making the operations at port more efficient and smoothly. Automation of port and ship operations has brought simplification and improvements of cargo process (Bahnes et al. 2016). However, as the effect of autonomous ships on port operations is hard to estimate at this stage, the relation between the implemented changes in port operations, with the incorporation of autonomous vessels, and the cost-effectiveness of those will determine the actual posture of this group towards autonomous vessels.

Maritime authorities are clear about the perspective and expectative of safety for autonomous vessels. Autonomous vessels will be subject to same safety demands for manned ships (Sonninen, 2017). This group also expects a gradual but slow progress in the update of the maritime legislation towards autonomous maritime systems (Van Hooydonk, 2014).

Marine trainers represent a key player in the role of autonomous vessels. The expectation with the introduction of autonomous vessels is that the current jobs of mariners cannot be completely eliminated by automation, but many of these will be redeployed (Frey & Osborne, 2013). With this in mind, marine trainers have to plan the potential reallocation of the personnel into other new demanded tasks in autonomous maritime systems. This has created the introduction of new educational programs focused in autonomous maritime operations (NOVIA, 2017).

Financers and insurance companies remain skeptical about their position towards supporting the concept of autonomous vessels (Macauley 2017). However, similar trends have been previously shown in other transport domains such as cars and air drones (Colonna 2012; North 2014). Eventually, the functioning of these is becoming more reliable and these are gradually receiving support from this type of organizations.

Unions have a stronger posture against autonomous vessels. They have recently released public critics claiming that the plan for having autonomous vessels in operations is unrealistic and lacks understanding about the actual needs of maritime navigational operations (Järvinen 2017).

In the perception and expected demands by the general public, safety constitutes one of the basic human needs (Van Rijswijk et al. 2016). Fear, panic, risk and correspondingly safety are highly related to the environment. Fear influences the experience of the environment as much as the environment influences our experiences of fear (Koskela and Pain, 2000). Therefore, the safety perception of people towards autonomous vessels varies depending on the believes and personality of the people. In a recent analysis made by (Kruskopf, 2017), people expressed trust in the implementation of artificial intelligence to operate an autonomous ferry. The perception and expectation of the general public play an essential role in the operations of an autonomous vessel. For autonomous vessels, the demands and expectations will be high, gaining trust will take time, but losing it can occur fast.

3 SYSTEMATIC AND SYSTEMIC SAFETY MANAGEMENT FOR AUTONOMOUS VESSELS

This section offers the general structure of a framework to manage safety since an early design phase of autonomous vessels. It introduces a general approach which supports the initial phase of the vessel design. The aim is to provide an approach that can support the role of the ship manufacturer and operator with an extended view from other safety stakeholders that will be involved the operation of the vessel.

3.1 *Definition of the mission and context of operation*

A crucial initial step for determining the basis of the structure of a systematic and systemic safety management for an autonomous vessel is to define the actual mission and the operational context of the vessel. This enables a clear focus for the identification of the actual safety risks that the safety strategy aims to mitigate.

This represents the initial step in the design of a ship concept. The difference for autonomous vessels is that it needs to incorporate more aspects that go beyond the traditional approach used for designing a vessel between ship manufacturer and ship operator. This refers to the initial design of the new services linked to the functioning of the vessel, including elements such as the remote monitoring and control center, sensors systems, and satellite services.

3.2 *The use and value of the information extracted from navigation and technology experts, and system users*

Designing an autonomous vessel and the services linked to it is a demanding task that includes designing the elements of a system which does not exist in the current operation mode of the maritime transport. Therefore, the value of the information coming from the experience from a multidisciplinary group of experts is essential. This points out the need for establishing initial analyses which can combine the view of experts in navigational operations (masters, chief officers, mariners, etc.) and experts in the different new technology domains (artificial intelligence, software designers, sensor developers, experts in automation, etc.).

In addition, the perspective and expectation of the system users have to be considered in the design of the systematic safety management. The users refer to the stakeholders of the systems who can be an operator located in a remote monitoring and control center or the passengers on-board autonomous vessels. This is just a simplified representation of two members of a long list of system designers and users that have to be clearly identified in the initial design phase. The analysis of this group is essential to define the expectations in the user experience expectations.

3.3 *Initial identification of the accidents and hazards to be prevented and controlled*

With the two previous steps defined, the next is the definition of the accidents and the identification of hazards that this initial systematic safety management strategy will focus on. This refers to the detection of the main accidents and the identification of hazards leading to those accidents that may result in damages and injuries, and which may have a severe impact in the operation of the autonomous vessel and the entire autonomous system.

3.4 *A preliminary hazard analysis*

This preliminary hazard analysis provides a detailed description of the identified hazards, including their potential effect on different components of the vessels and the autonomous system. This description represents a justification of relevancy of the hazard, including the initial estimation of its severity and its consequences. This aims at providing the initial estimation of the potential affectations that the hazard may have. This estimation includes an initial assessment of the difficulty and cost of their implementation. This estimation represents the establishment of the actions who act as the basis for defining a mitigation strategy.

3.5 *Definition of safety controls*

This task demands the review and prioritization of the mitigations actions that will be further developed as the safety controls of the initial safety management strategy. The focus is on defining if these actions have a significant effect on the mitigation of the hazard. The purpose is to define the objective and relevant safety controls that represent the basis of the initial safety management strategy of the autonomous vessel.

3.6 *The representation of the initial systematic safety management strategy*

The safety controls systematically represent the structure of the initial safety management strategy. This serves as the basis to define the actual safety controls and processes that need to be planned in the earliest concept design phase. The aim is to anticipate those roles, tasks, and responsibilities that need to be delegated in the implementation of the safety management strategy. This supports the identification of new potential roles and demands that are needed in the functioning of a new autonomous ship concept.

Figure 1 presents the process introduced in previous sections. It describes the steps needed for the definition of the safety control elements of the initial safety management strategy. In the figure, the steps are represented as processes of a flowchart. These have an attached reference number which specifies the sections where the steps are described. The flowchart includes the influence of the system stakeholders mentioned in Section 2. This attempts to remark the importance of considering the view and expertise of all relevant stakeholders.

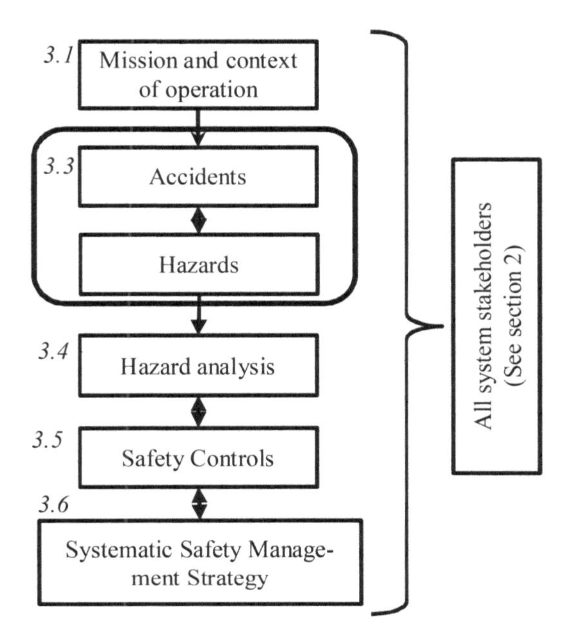

Figure 1. Process to define the initial structure of the safety management strategy of a autonomous vessel.

4 COMPARISSON WITH EXISTING APPROACHES

4.1 *Managing safety in current ship design*

The assessment of the risks and the goal-based design in traditional ship design approaches such as the ship design spiral is focused on the safety regulations for the current maritime traffic operations and the safety specifications defined mainly by the customer (Papanikolaou 2014). This systematic approach is essential to ensure the safety of the vessel under design. For supporting this, several safety analysis techniques are applied to evaluate different elements in the concept design, preliminary design, contract design and detailed design of the ship, including:

- Quantitative and qualitative safety analysis
- Cause and effect relationship
- Hazard analysis
- Hazard and Operability
- Fault tree analysis
- Event tree analysis
- Markov Chains
- FMECA

These techniques support the ship design phases and produce the most advanced ship projects (Molland, 2011). However, these techniques are commonly implemented in the analysis of the safety in the technical aspects of the vessel. Although this is highly important, this approach is commonly limited to the demands of the regulations implemented to assess the safety of the current ship fleet, and the demands and specifications defined by the customer. The customer specifications require the input from technical expertise across many disciplines to design a ship for the current operation modes. In the case of autonomous shipping, these specifications need to cover the view of the stakeholders of a bigger system.

4.2 *Aspects lacking in current approaches*

Although systematic approaches such as the ship design spiral are essential to produce efficient designs which are transformed into advance ship constructions, this systematic approach needs to be strengthened in the design of autonomous ships and autonomous maritime systems.

The implemented techniques for hazard analysis, risk and safety assessments, and safety management in traditional ship design are not commonly focused on providing an actual systemic way to model accidents and define how these can be prevented. This issue must not be transferred in the analysis of potential accidents in autonomous shipping. In the analysis of the hazards and risk for autonomous vessels, the lack of information of this new operational context demands the development of analyses that go beyond component failures and the review of past accidents. Analyses focusing on the definition of the cause of not yet existing failures is a risky approach. Autonomous vessel and maritime systems demand to create understanding about the functioning of the entire system and the safety risks on the functioning of each element of the system. This requires the incorporation of multiple safety viewpoints and interpretations coming from a number of safety stakeholders which is bigger than the number used in traditional ship design.

5 DISCUSSION

5.1 *The general safety needs of autonomous vessels*

True systematic and systemic safety management seems to be required for ensuring the functioning of autonomous vessels. Due to an increased interaction between all system stakeholders, the management of safety needs to be better coordinated and efficiently implemented. The implementation of this type of approach will result in a more proactive and transparent safety management among the different stakeholders of the autonomous vessel.

This approach has to be also suitable for increasing the competitiveness of the maritime transport industry. It has to provide input information for the elaboration of new business models which can consider safety as part of their competitive advantage. The approach must clarify of the roles of some safety organizations in the maritime industry such as SAR services and VTS centers which have not a clear position in the management safety for autonomous vessels.

5.2 *The proposed process for structuring the safety management strategy of an autonomous vessel*

The process aims at producing detailed information which guides the initial design process of an autonomous vessel and the other crucial services linked to it. The process focuses on supporting the design and management of complex systems and maintaining it functional during its complete operational life (Blanchard 2004). The process aims to initiate the design of safety in the earliest conceptual design phase for engineering a more efficient and safer system (Leveson 2011).

The process structure represents a systematic approach that is capable of analyzing accidents and hazards detected in exercises developed with the system stakeholders, including designers, builders and different type of system users. This open a broad view that integrates not only the perspective of experts of ship safety design and operation but also the expected perspective by the system users. This is essential as the safety perception in the system should be influenced since the initial design phase (Ahola, 2017).

The process focuses on formulating safety controls to prevent and/or respond to accidents and hazards. These controls represent the basis of the foundations of an initial safety management strategy. With this approach, information rich in content and quality can be transferred to the ship designer, manufacturer, and operator.

The definition of the safety controls and specification of the safety management strategy supports the development of a system structure where safety management responsibilities and tasks can be systematically delegated to the system stakeholders.

5.3 *Future work*

The process presented in this study aims at defining the initial safety management strategy for two concept ferries currently under analysis. The results of this implementation will be published during 2018.

Another essential aspect to for further research is the development of a new regulatory framework for autonomous vessels. This is a crucial task for making the operations of autonomous vessels a reality. The parallel development of this framework and the approaches to formulate the initial safety management strategy of an autonomous vessel is essential.

6 CONCLUSIONS

This study proposes a process to define safety controls that need to be integrated into the initial safety management strategy of autonomous vessels. The process remarks the need for designing a safety management strategy which focuses on the function of the autonomous vessel and the services linked to its autonomous maritime system.

The process presents a simple structure which starts with the definition of the mission and operational context of the vessel, continues with the specification of the accidents that the strategy aims to prevent or respond to. Then, a hazard identification and hazard analysis are executed. Based on the hazard analysis, certain safety controls are analyzed and consistently defined. Finally, the functioning of the safety controls is defined in a systematic and systematic representation of the initial safety management strategy.

The process represents a guide for obtaining itemized input information for the initial design phase. The characteristics of the proposed process enable its simple integration into systematic approaches used for ship design (e.g. ship design spiral). The process is structured to include input information coming from the safety views of the stakeholders of the entire autonomous maritime system. The aim is to design safety and the safety management strategy in the earliest design phase of the autonomous vessel.

ACKNOWLEDGEMENTS

The work presented in this study is part of the research project "Smart City Ferries" (ÄLYVESI) and the Design for Value (D4 Value) program. ÄLYVESI is funded by the European Regional Development Fund (ERDF). Additional financiers are Finnish Transport Safety Agency and the cities of Helsinki and Espoo. The D4 Value program is partially funded by Tekes.

RFERENCES

Ahola, M., 2017. Tracing Passenger Safety Perception for Cruise Ship Design. Aalto Doctoral Disseration, Aalto University. ISBN: 978-952-60-7241-8.

Ahvenjärvi, S. 2016. "The Human Element and Autonomous Ships." TransNav : International Journal on Marine Navigation and Safety of Sea Transportation Vol. 10 nr 3.

Bahnes, N., Kechar, B., Haffaf, H., 2016. Cooperation between Intelligent Autonomous Vehicles to enhance container terminal operations. J. Innov. Digit. Ecosyst., Special issue on Pattern Analysis and Intelligent Systems – With revised selected papers of the PAIS conference 3, 22–29.

Blanchard, B.S., 2004. System Engineering Management. John Wiley & Sons.

Burmeister, H-C., Bruhn W., Rødseth, O.J., and Porathe T. 2014a. "Can Unmanned Ships Improve Navigational Safety?" In Chalmers Publication Library (CPL). http://publications.lib.chalmers.se/publication/198207-can-unmanned-ships-improve-navigational-safety.

Burmeister, H-C., Bruhn W., Rødseth, O.J., and Porathe T. 2014b. "Autonomous Unmanned Merchant Vessel and Its Contribution towards the E-Navigation Implementation: The MUNIN Perspective." International Journal of E-Navigation and Maritime Economy 1 (December): 1–13.

Carr, A.A., 1996. Distinguishing Systemic from Systematic. TechTrends 41, 16–20.

Colonna, K., 2012. Autonomous Cars and Tort Liability. Case West. Reserve J. Law Technol. Internet 4, 81.

Frey, C. and Osborne, M., 2013. The Future of Employment: How Susceptible Are Jobs to Computerisation? Oxford Martin School, University of Oxford.

Hogg, T., and Ghosh, S. 2016. "Autonomous Merchant Vessels: Examination of Factors That Impact the Effective Implementation of Unmanned Ships." Australian Journal of Maritime & Ocean Affairs 8 (3): 206–22.

Järvinen, J. 2017. Merimies-Unioni tyrmää miehittämättömät laivat utopiaksi – laiva ei ole lennokki. YLE Uutiset, October 2017.

Kim, H., Lee, J.-K., Park, J.-H., Park, B.-J., Jang, D.-S., 2002. Applying digital manufacturing technology to ship production and the maritime environment. Integr. Manuf. Syst. 13, 295–305.

Koskela, H., Pain, R., 2000. Revisiting fear and place: women's fear of attack and the built environment. Geoforum 31, 269–280.

Kretschmann, L., Burmeister, H.-C., Jahn, C., 2017. Analyzing the economic benefit of unmanned autonomous ships: An exploratory cost-comparison between an autonomous and a conventional bulk carrier. Res. Transp. Bus. Manag.

Kruskopf, J. 2017. Archipelago as a Recreational Service. Master Thesis, Service Design, Aalto University 2017.

Levander O., 2016. Ship intelligence – a new era in shipping. Smart Sh. Technol., London. RINA; 2016, p. 25–32.

Leveson, N., 2011. Engineering a Safer World: Systems Thinking Applied to Safety. MIT Press.

Macauley, M. Insurance Implications of Autnomous Shipping. Global Trade. February 2017.

Man, Y., Lundh, M., Porathe, T. and MacKinnon, S. 2015. "From Desk to Field - Human Factor Issues in Remote Monitoring and Controlling of Autonomous Unmanned Vessels." Procedia Manufacturing, 6th International Conference on Applied Human Factors and Ergonomics (AHFE 2015) and the Affiliated Conferences, AHFE 2015, 3 (January): 2674–81.

Molland, A.F., 2011. The Maritime Engineering Reference Book: A Guide to Ship Design, Construction and Operation. Elsevier.

North, D., 2014. Private Drones: Regulations and Insurance. Loyola Consum. Law Rev. 27, 334.

NOVIA University of Applied Science, 2017. Master of Engineering - Autonomous Maritime Operations. Web Link

Papanikolaou, A., 2014. Ship Design: Methodologies of Preliminary Design. Springer

Poel, I. van de, Robaey, Z., 2017. Safe-by-Design: from Safety to Responsibility. NanoEthics 11, 297–306.

Porathe, T., Prison, J. and Man, Y. 2014. "Situation Awareness in Remote Control Centres for Unmanned Ships." In Chalmers Publication Library (CPL), 93

Rødseth, O.J., and Burmeister, H.C. 2015. "Risk Assessment for an Unmanned Merchant Ship." TransNav, International Journal on Marine Navigation and Safety Od Sea Transportation 9 (3).

Rødseth, Ø.J. & Burmeister, H.-C., 2012. Developments toward the unmanned ship. Hamburg, Germany, German Institute of Navigation.

Schröder-Hinrichs, J.-U., Hollnagel, E., Baldauf, M., Hofmann, S., Kataria, A., 2013. Maritime human factors and IMO policy. Marit. Policy Manag. 40, 243–260.

Sonninen S., 2017. Preparing for change in the maritime services in context of autonomous shipping: A case study focusing on Finnish maritime cluster stakeholders. Master Thesis, Department of Applied Mechanics (Marine Technology), School of Engineering, Aalto University.

Teivainen A., 2017. Rolls-Royce to set up R&D centre in Turku, Finland. Helsinki Times. March 2017.

Van Hooydonk, Er. 2014. "The Law of Unmanned Merchant Shipping - an Exploration" 20 (3):403–23.

Van Rijswijk, L., Rooks, G., Haans, A., 2016. Safety in the eye of the beholder: individual susceptibility to safety-related characteristics of nocturnal urban scenes. J. Environ. Psychol. 45, 103e115.

Wahlström, M., Hakulinen J., Karvonen, H. and Lindborg, I. 2015. "Human Factors Challenges in Unmanned Ship Operations – Insights from Other Domains." Procedia Manufacturing, 6th International Conference on Applied Human Factors and Ergonomics (AHFE 2015) and the Affiliated Conferences, AHFE 2015, 3 (January):1038–45.

Wróbel, K., Krata, P., Montewka, J. and Hinz, T. 2016. "Towards the Development of a Risk Model for Unmanned Vessels Design and Operations." TransNav : International Journal on Marine Navigation and Safety of Sea Transportation Vol. 10 nr 2.

Wróbel, K., Montewka J., and Kujala, P. 2017. "Towards the Assessment of Potential Impact of Unmanned Vessels on Maritime Transportation Safety." Reliability Engineering & System Safety 165 (September): 155–69.

Marine Design XIII – Kujala & Lu (Eds)
© 2018 Taylor & Francis Group, London, ISBN 978-1-138-34076-3

Do we know enough about the concept of unmanned ship?

Risto Jalonen
Finland

Eetu Heikkilä & Mikael Wahlström
VTT Technical Research Centre of Finland Ltd., Finland

ABSTRACT: The main objective of this paper is to widen and deepen the scope of discussion related to the concept of autonomous and tele-operated ships. It is not quite clear, whether all relevant arguments and issues within this field are already identified, discussed and well understood. The impacts of such new technology on all stakeholders must be taken into account, in the phases of design, building, operation, maintenance, repair and scrapping. Driving forces, motivations and goals in the background, as well as all identified hazards, risks, policies and limitations need to be brought into open discussion. In the maritime sector the pace of propagation of some new technological developments may sometimes be felt to be too slow by the developers in their race to the market. However, the ability of the regulators to communicate with all stakeholders must be confirmed. The gradual, slow process of growth of a common and sufficient understanding of the proposed change of technology and of its all relevant features is in crucial role. Thinking slow is sometimes more beneficial than haste, if a justifiable, reasonable, sound and safe future of the society and its transportations must be confirmed. A set of controversies and problems related to the concept of unmanned ship are presented here and discussed with the hope of further discussion and contribution to a safer future.

1 INTRODUCTION

The concept (idea) of autonomous, tele-operated and unmanned ships has been brought into publicity in various media during the past few years. Nice artwork related to them may have already taken some root in our brains, when repeatedly brought into our eyes. Some results of a few, rather recently finalized research and development projects related to these concept candidates of new ships have been published, but other papers, discussing additional pros and cons of the concept (autonomous ship, tele-operated ship and/or unmanned ship) will slowly grow in numbers, too.

To our knowledge, real merchant ships applying new technology to the absolute theoretical maximum (in an unmanned ship, with no humans onboard) have not been taken into use, in operational use, in maritime traffic, yet. Automation, digitalization and artificial intelligence plus machine learning are some of the keywords in many discussions, but do we really understand what they mean to us, to others and to the logistics and to the market, if applied in ships? It may not be so easy to know, so should we ask someone who may know? Whose answers can we really trust?

We, as researchers, must try to be objective, even if we cannot confess us being under or fully aware of all the public and private (shareholder and organisational) pressures acting towards taking the futuristic novelties into use as quickly as possible. Some, even rather seemingly small political changes and redirection of the goals and policies may have big effects on the structure, operation and performance of the maritime system and the pace and direction of its development. However, responsible persons should confirm that the cargo transportation will not get disturbed too much or get totally stopped due to some tempting, although queer and perhaps even risky ideas covered by an alluring cover.

Although many of our issues and topics may have discussed elsewhere, we do not aim to present any full list of the publications dealing with them. This paper is an attempt to supplement the work published in the incomplete report by Jalonen et al (2017) by an additional effort to widen and deepen the scope of discussion and the field of arguments related to the concept of autonomous and tele-operated ships, and especially into many challenges that may arise with their use. More holistic and clear views on these concepts, and what they actually would mean, need to be created. However, concurrently, we want to go into a few specific topics. We also try to find out and reveal at least some important wicked questions that may not have asked or answered before.

It is not quite clear, what relevant problems, topics, arguments and issues related to the impacts of the new technology, and especially the consequences on all stakeholders are already known, well understood, thoroughly discussed and properly managed. Driving forces, motivations and goals in the background, as well as hazards, risks, policies and limitations need to be identified and brought into daylight and into open discussion to make it possible.

The ability of the regulators to recognize all stakeholders needs, to facilitate open communication with all of them and the gradual growth of a common and sufficient understanding of the proposed change of technology and of its all relevant features and challenges are in crucial role. A justifiable, reasonable, sound and safe future of the society must be confirmed. The maritime system is an important sector and context to support a well-balanced development of the whole, global society. A set of challenges, controversies and problems related to the concept of unmanned ship are presented here and discussed with the hope of further open discussions and with the aim of an additional contribution to a safe future in the maritime context.

It is known that the topic and context discussed here are hot. Therefore, it is assumed that a full and thorough list of relevant references would be out-of-date in a very short time. Due to this reason, deep apologies are made, in case of all significant, but unintentional omissions of references.

2 CONTEXT

In the maritime sector the pace of propagation of some new technological developments may sometimes be felt to be rather slow. However, it may sometimes be backed by commercial contest and hasty race to the market, but, quite opposite trends may also be recognized. More importantly, some of the more slowly matured inventions in the maritime world have been real improvements, although not always straight from the first steps of their development. Regardless of the time used, big efforts required and price paid for them, some of the innovations applied in marine technology (like oars, sails, a caravelle, rudder, propeller, steel, welding, tanker, radio, radar, container ship etc.) have dramatically changed the world. Many of the new innovations have been socially constructed in a gradual, stepwise pace with time, see e.g. Bijker et al (1987).

An autonomous ship is an idea that may have its roots in the world of ideas like the idea of a tele-operated ship. Whether these ideas are feasible and good or not have been studied in some rather recent research projects, see e.g. MUNIN (Rødseth et al, 2015) and AAWA (Jokioinen, 2016), with main emphasis on the technical-economical aspects and deskwork. It is not well known to what extent similar or even deeper and wider studies on the same topics have been made earlier as a full, cumulative gathering of all available knowledge.

A full synthesis of all research in the scope of unmanned vessels is not necessary here. An examination of factors that impact the effective implementation of unmanned ships has been carried out e.g. by Hogg, T., Ghosh, S. (2016), maybe by some others, too.

This study is limited to merchant vessels only. With this paper it is hoped to add some important issues on the list of challenges that need to be resolved. As our knowledge of many structural and operational matters and chosen solutions related to autonomous and tele-operated ships is limited the aim of this paper is to concentrate on many important challenges and problems that need to be highlighted. Swarms of unmanned ships, see e.g. Zhang and Furusho (2017) are not discussed in this paper.

A socio-technical approach, see e.g. Perrow (1984), has been selected for this paper. It will follow to some extent a multi-disciplinary approach presented in (Le Coze et al, 2017), with some additional issues and features included. One big problem in techno-economical assessment is found to be that it may not have space enough for some additional dimensions that may be more important for many stakeholders, outside the scope of those who get most of the benefit from the new technology. Therefore, in order to avoid omission of any important issue, multi-disciplinarity needs to be emphasized.

The scope of things to be studied in relation to unmanned ships is vast, if we can let us freely think about all direct and indirect effects and consequences of unmanned shipping at several hierarchical levels. It must be confessed that it may be even a mission impossible. Thus, the main emphasis in the next chapters is laid on just a few topics:

- alternative practices in the assessment of new technology, socio-technical and multi-disciplinary approach,
- seaworthiness of the ships and good seamanship, as two focal parameters of the structural and operational topics in shipping, that may be endangered, and on a more specific topic,
- artificial intelligence, as a new alternative solution, some uses of which in the shipping world may create tools to change it a lot, in case it may be supported by the decision-makers.

Although the selection of the topics for this discussion is here limited to these three examples, it is believed that more topics can be discussed elsewhere in the future. In order to bring some uniformity in the text the following definitions are applied throughout the text:

An *autonomous ship* is a ship that "has systems based on automated software that may advice human operators, or at the highest level of autonomy replace human decision making and action. Autonomous ships may have crew onboard, e.g. for service tasks or inspections or for occasional manual control." See, Bertram (2016).

Unmanned ships have no crew members onboard, "but may be under autonomous or remote control. Most visions of unmanned shipping involve a mix of autonomous systems and remote control by humans." See, Bertram (2016). "Unmanned ships can be either remote controlled or autonomous. In practice they will have to be both." See, Levander (2016).

3 SOCIO-TECHNICAL AND SOME OTHER CHALLENGES IN AUTONOMOUS SHIPPING

3.1 *Techno-economic versus multidisciplinary assessment of technology*

It is rather easy for some engineers to limit an assessment of new technology in only some limited technical, technological and in economical issues, especially if the budget is limited. "Is it possible?" and "Will it be feasible?", may be the first questions. "Will it be safe?", is one of the most important issue, too. The whole idea should be forgotten, if the unmanned ship is not economical or if it is unsafe. Due to the fact that the operating environment, the maritime system around the new technological innovation will be for a long time a socio-technical system, positive influences and support by method(s) ensuring an all-round view on the potential problems of safety are highly recommended for this purpose (Jalonen et al., 2017).

The socio-technical approach, see e.g. Perrow (1984), has widened our eyes to possibilities to consider, not only many important technical details, but also wider aspects. This leads us to systemic thinking with the important effects of operational and organizational factors shaping the system design. Problems in any of the three named areas (technology, economy, safety) above can very easily affect others, too. It would be narrow-minded to restrict us to discuss about and try to solve any more or less wicked questions found with a limited number of representatives of the technocracy sector only.

The maritime system includes many channels for interaction and feedback loops between actors, some of which are not always well known. A lot of expert knowledge is distributed among the actors in the maritime system. Why not to utilize it when trying to set safe limits for newcomers of much different design and when shaping new innovations to make them able to conform themselves in the system without any major problems for its performance?

Visions of new developments in the field of automation and their applications in ships may be identified as first messengers of a new technological and operational era, a future that can be built, shaped and restricted by us, at least to some extent, to resemble more a Utopia than a Dystopia. Both extremes are possible, although maybe not so good for any stakeholder or for the society. The development may also turn out to be faster or slower, but the society must be able to manage the risks.

Emergent new technology may include many hazards, however, and even some disruptive effects. These features set high demands on the social responsibility of the developers to cover all important aspects in their assessments of impacts on safety and security. So, obligations are high for meticulous and over-arching work before the last practical test cases and applications are thoroughly scrutinized, results assessed and understood, before the first commercial solution can be taken into serious use with a conditional approval only.

One of the many challenges identified in autonomous ships is to assess the pace of the development. If the technological development seems to be faster than the design and construction of all necessary and feasible safety controls and improvements, more efforts should be put on getting them developed at the same pace. This process of Safety Qualification Process for Autonomous Ship Concept Demonstration is described elsewhere in more detail, see e.g. Heikkilä (2016). At high speed of development the safety assessments of technology will often assess technology that is already out-of-date.

One of the crucial questions regarding the new technology is the cost. It should be so that a system of Shore Control Stations and autonomous ships, with highly reliable new technical equipment, will be much cheaper than the existing system of onboard seafarers. Cost calculations, which predict that the cost profiles are favorable for unmanned ships, have been done in the MUNIN (2015) project. However, the calculations apply hypothetical ship concepts, business models and forecasts. Overall, we do not have fully reliable cost data. With such high risk technology the cost of a full and continuous connectivity, linked to reliable sensor and other autonomy-enabling technology equipment would probably be high, as it must be of extremely good quality. The cost of the needed new technology involved is probably high, too, even if the widening of the market might mean some cost reduction. The contradiction between safety and cost still remains.

The need for multidisciplinarity in technological assessments become very clear from e.g. a recent report (see, Le Coze et al, 2017). The following related scientific and engineering disciplines were considered to be important enough to be shown in (Ibid.):

- Natural sciences,
- Mathematics,
- Engineering,
- Ergonomics,
- Human Factors and Cognitive Engineering,
- Naturalistic Decision Making
- Work,
- Organisational Psychology and Social Psychology
- Sociology,
- Management,
- Law and Political Sciences

The referred report (Ibid.) was a product of a project related to process safety, so an abridged list of more widely applicable topics, issues and aspects to be considered and discussed as outlined in it is given below:

- Hazard, threat, accident, act of terror
- Probability, failure, reliability & risk
- Safety barriers & defence in depth (hardware, software)
- Redundancy
- Human–machine/computer interaction
- Human error
- Situation awareness
- Expertise
- Sensemaking
- Teamwork
- Resilience
- Whistle blowing
- Groupthink
- Mindfulness
- Learning
- High reliability organisations
- Safety management systems
- Safety culture (& climate)
- Migration, drift, normalisation of deviance & organisations at the limits
- Risk & robust regulation regimes
- Regulatory capture
- Safety and risk as socially constructed

As can be easily seen, the topics on the above list, obtained from (Ibid.) give already a rather fruitful base to raise several questions with regard to autonomous ships, with their design (as a system) and the operation of the system (or its parts), too. It is presumed that experienced safety assessors on any field can easily continue from it. As an example, they could ask from the proponents of a new technological innovation:

"Is there sufficient evidence regarding the operability and reliability of all equipment alone and also as part of the sub-systems, integrated sub-systems and the whole system, in all imaginable environmental conditions?", or

"Do you really think that the new idea and its use can fulfil the society's requirements regarding sensemaking?", or

"How do you think you can get operators with good expertise on navigation of big ships in the future, if all ships would be unmanned in the future?"

Some of the questions above may already have been discussed somewhere. If not, they need to be discussed in open forums.

The autonomous ships have to be both safe and economically viable. Business cases, use scenarios, safety assessments and technological design solutions all depend one from another. The development of commercial autonomous shipping could be described with the chicken and the egg causality dilemma: there is not a clear leading element in the change. Given these features, it is not certain if viable concepts for autonomous shipping can be developed in research programs with separate work packages formulated based on scientific disciplines, that is, with different sections considering technology development, liability issues, business cases and user-studies and so worth.

All of the above implies that in managing the development of autonomous shipping, it is important to take in to use methods that foster cross-disciplinary discussions. Face-to-face meetings should be prevalent while secrecy should be avoided. Models for the development work could be derived, for instance, from software development: scrum sprint involves close collaboration towards a specific development aim in a clearly defined time-window. Open data and idea sharing akin to the Wikipedia project could be beneficial.

In the authors' own experience, creation of developmental suggestions based on safety-analysis is difficult without understanding of the features of the to-be-developed ship.

3.2 *Seaworthiness and good seamanship in autonomous operation*

Seaworthiness is related to a number of issues onboard, but the main idea is believed to be in short as follows: a seaworthy ship is fit for the voyage it is planned to do. This means that its structures, equipment, loading, supplies and manning should all be in such condition and level that a safe voyage, especially at the beginning of the voyage can be expected. However, although the work to improve safety at sea has been successful during the past 50 or 100 years, it should not be totally forgotten that there is still some risk involved in seafaring and operations at sea, even in spite of the seaworthiness of the ships. The contents and meaning of the word seaworthy may have some slight variations depending on the type and text of regulations, insurance rules, agreements etc. It has some relation to the environmental issues, too. In any case, safe manning has been one important cornerstone of seaworthiness and safe shipping for long.

During the past 200 years it has been possible to reduce the manning levels of ships considerably with the many steps of technological development. In many tasks the number of seamen has been reduced drastically when the physical work related to many tasks, once carried out by men, could be carried out by new machines. Autopilot was invented roughly 100 years ago, but more control tasks that could be handled by automation were introduced almost during each decennium since then. The spreading of automation in ships has turned out to be a success story, so far.

Today, or let's say in 2016–2017, a general cargo or container feeder ship with a gross tonnage below 4000 may be operated along short se shipping routes with a safe manning of seven to nine (7–9), whereas a modern ultra large container ship being able to bring hundredfold amount of containers across the oceans may be operated with only about fourfold manning of twenty-eight (28).

In many modern ships of today it is allowed to leave the engine room unmanned for considerably long times, i.e. several hours per day. However, according to our knowledge and memory, on the navigational bridge deck it has never been allowed for many good reasons. Most of the latter issues belong under the broad titles: navigational duties and seamanship.

Good seamanship is believed to be something that is difficult for many of us to describe prescriptively in words. The deeds and way of thinking related to good seamanship may vary from small and tiny issues, habits and practices to bigger and more laborious processes, but sometimes, very rarely, however, they may extend even to heroic deeds. Operational ship safety, with respect to good seamanship, is believed to be largely built on everyday deeds onboard, but it won't grow from nothing without training.

One group of big challenges in the development of autonomous or tele-operated ships is related to the implementation of good seamanship. Is it possible to change all items and aspects of maritime humanware into hardware and software onboard? It is not well and widely known, how this important part of safe shipping, with all its relevant nuances could be built in unmanned vessels, their equipment and preplanned routines. Is it, or will it ever be possible to implement them in algorithms? How do unmanned ships participate in maritime rescue operations and what kind of shelter they may offer e.g. for the victims of marine accidents? Will it be possible or impossible, and in case of the latter, can such a moral choice be accepted?

3.3 *Artificial intelligence in the maritime setting*

The increasing autonomy is likely to require application of artificial intelligence (AI) technologies. These technologies will most probably be used in implementing some of the fundamental functions in autonomous operation. The use of AI technologies potentially provides efficient tools especially for various image processing and object recognition tasks needed when developing marine situational awareness systems, as well as in increasingly important condition monitoring tasks (see e.g. Li & Jiang, 2017). In some visions, AI can also provide methods for adaptive or learning capabilities in the actual navigational decision-making.

While the use of AI technologies can be seen as justifiable from a purely technological point of view, there are many unsolved issues regarding their implementation and performance in the actual maritime environment. When developing an AI-enabled system for autonomous shipping, the design and validation will require new approaches throughout the entire product development process. The lack of prescriptive standards for such systems further complicates the process (Danks & London, 2017).

The sufficiency and reliability of data poses one of the most significant challenges. AI applications generally need to be developed and optimized using a large amount of data. The acquisition of the data produces challenges in the maritime environment. It is difficult to collect sufficient data by simply monitoring existing ships, as a sufficient amount of significant events may not be accumulated during the data collection period. Additionally, the validity and relevance of the collected data needs to be considered carefully. Furthermore, the performance of the actual learning process needs to be considered: how to validate the learning results, and how to set the acceptance criteria for the results of machine learning system's training and performance. This is likely to increase the significance of simulator-assisted development and testing throughout the development process.

To address machine learning-related issues in development of AI-enabled systems, Falcini & Lami (2017), for example, present a data-driven approach for developing road transport AI systems. A similar approach, tailored for the development of marine systems, would be beneficial.

AI transparency is another largely unexplored theme in the maritime context. By AI transparency, we mean the ability of a withstander to understand the decision-making rationale of the AI system. Currently, many neural network based solutions, which can be applied to object detection and perhaps for anticipation of the movements of the surrounding traffic at the open sea, could be described as "black boxes" that process the input data and produce predictive outputs without even the system designer being able to understand how the system achieved its results. Assumedly, given the complexity of a many-phased deep neural network, the only easily understandable aspect of information processing through this method could be percentual probability of the given predictions. Lately, various initiatives have been started to improve the awareness regarding black box AI systems. Technological solutions for increasing the transparency, however, can often be seen as disadvantageous from the system performance point of view, making AI transparency a central theme for further research. Transparency is especially important in remote operation scenarios, where the operator needs sufficient understanding of the decision-making logic. Similarly, in possible problem and accident scenarios, the rationale behind decisions needs to be traceable.

Cyber security issues, see e.g. reports: DNV-GL (2015), Lloyd's Register (2016), and the one by Felten (2016), add further concerns also for the design of AI solutions. Even with a robust cyber security implementation in place that prevents direct, software-based way to attack the AI system, the system might be vulnerable to new kinds of attacks. AI systems in object recognition tasks, for example, could be confused by adversarial placement of fake targets to produce false positive detections. This is a serious threat for safe shipping.

Additional important aspect that is related to the planned AI solutions onboard is the sufficiency of the brain-power resources of the regulators available to carry out their control tasks with them. Pure self-control may not be the best option onboard in the ships of local, regional or international traffic.

4 DISCUSSION

A huge number of questions can be raised in relation to autonomous, unmanned and tele-operated ships. They can be allocated in many ways, under several multidisciplinary topics and be presented in relation to various parts of the maritime system and on several hierarchical levels, too. Some of the questions may include e.g.:

- Is there any sense at all, or a lot of it, in the concepts?
- What are the most relevant arguments for and against these concepts and why?
- Do we know enough about the new technology and the background of the arguments above? Is it based on reliable evidence?
- Will IMO, the states and the maritime society accept the new types of ships?
- What are the benefits and the drawbacks of the concept for different stakeholders?
- Have all stakeholders been identified and included in discussions?
- Are all stakeholders and stakeholder groups unanimous?
- What are the arguments in case of each group?
- Could a real autonomous ship operate in full scale without problems in harsh environmental conditions?
- Will it be safe enough? (As safe or more safe as a conventional ship?
- If it is not safe, what does it mean?
- What should be included in the assessment of consequences? (Direct consequences and indirect ones?)
- If it turns out that it is not safe, who is responsible for harms, accidents and damage?
- If it is safe, does it apply to all more or less similar ships? Is type approval OK?
- If yes, for how long period of time? (one journey, one year, the whole lifetime or more?)
- Capability of automation to detect vessels and floating objects on route and to steer clear off them, if necessary?
- Capability of automation to avoid collisions in case of encounters of multiple ships?
- Capability of automation to navigate safely on coastal fairways, and in ports?
- Reductions on preventive and corrective maintenance that are currently largely carried out during voyages?
- Capability to handle emergencies, such as firefighting or failure recovery and repairs at sea?
- Errors and malfunctions in software (including updates)?
- Disturbances, malfunctions and vulnerabilities in data communication connections?
- Undue trust on the capability and flawlessness of ICT systems?

It would be utmost important to get mariners with a wealth of experience of sea service, onboard different ship types in various parts of the world, to participate in structured discussions, where the issues and topics are identified, challenges and ideas as possible candidates for their solutions are handled. Co-operation would surely be needed.

From a security point of view concerns have been raised as to the higher vulnerability of envisaged unmanned ships to hijacking or piracy with the purpose of stealing the cargo or "kidnapping" the vessel for ransom. Similar to the raised concerns regarding cyber security of ICT systems in general, potential vulnerability of unmanned ships to cyber-attacks by different adversaries, allowing them to illegally manipulate or exploit the attacked system, have been especially underlined.

Contrary to the negative safety and security effects raising concerns, claims have also been made for the higher safety levels of ships with higher levels of automation and operation autonomy. Such claims have been reasoned e.g. based on high involvement of human error in accidents at sea in the past, and the high crew fatality rate when compared to other industries observed currently. Both of these issues could be hypothesised to be reduced by increased ship autonomy by reducing the human involvement in direct control of ships, and by reducing the size of the crew on-board and exposed to hazards of the hostile sea environment. However, will then the number of technical/technological and organisational errors increase?

While addressed initially in a few studies, it appears that the impacts of unmanned merchant ships on maritime safety have not been studied comprehensively, yet. Furthermore, there is no experience available on such ships and their safety in everyday use as far as we know. Therefore it is of high importance in any new development projects that the safety risks and many other aspects are systematically addressed from the very beginning, and that a common knowledge base on safety implications is established and systematically built up, without forgetting applicable experience from other applications in the past.

Unfortunately, based on the limited sources of information in use it is not possible to give a short and clear answer to the question: Are unmanned ships safe? However, it is believed, that some better questions may also be found.

As it has been shown in the previous chapters some of the questions have a large and complicated background. Therefore, it is not possible to condense all results into short answers to simple looking question, being in fact more or less "wicked problems".

Potential hazards, risks and problems in relation to concept of unmanned and tele-operated ships are numerous. So far we haven't seen an unmanned ship based on artificial intelligence that could make correct decisions in unforeseen situations, too. This is believed to be a serious drawback, fundamentally related to all AI-based control systems.

Instead of trying to answer to all imaginable questions it is believed that this explorative study helps designers to recognise and understand many new aspects, limitations, mechanisms, and the social and technological viewpoints related to the topics mentioned in the text above. Thus, when making assessments of the feasibility of the concept of unmanned ship, the effects of the new technology in relation to the following topics should be analyzed:

- Economy,
- Engineering (efficiency),
- Ecology,
- Ergonomics,
- Employment,
- Safety,
- Sosiology, and
- Ethics

Wide, open and trustworthy co-operation with all relevant stakeholders of the maritime society and even beyond needs to be started in the flag states and at least as a regionally cooperative effort between them. No relevant stakeholder should be neglected.

Safety of the unmanned ships has been discussed in some publications (e.g. Rødseth and Burmeister, 2015; Levander, 2016; Wróbel et al, 2016), so have some technically related issues and topics been reported before. However, some of the topics of the list above seem to have been fully neglected, e.g. ethics and some sociological questions and societal risk. Some parts of these issues have been discussed already by Berkeley (1952). From the economical point of view and also linked to the feasibility of the concept one should not forget the importance of the orderliness and reliability of the transport chain.

A good old advice in relation to any well operating system is to **not** change it in any way. At the moment, however, it is believed, that nobody on any ship, on the sea, on shore or in heaven knows how many aspects of the old maritime system with orderly manned ships should remain or need to be transformed into adaptations to the new system with unmanned ships. It may lead us to an endless series of changes with trials and errors, the previous series (with manned ships) we may just have managed to overcome during the past two hundred years.

So far we have mostly seen the arguments of the proponents of the unmanned and autonomous ships. Is it due to the wickedness of the questions or due to mere lack of information, why we seem to be lacking a lively discussion about all these topics? Why is it so easy for the advocates of the new technology to bypass or belittle the positive impact on safety of the current maritime legislation? Why is it so easy for many of us to forget how many hundreds of years it has taken to develop the maritime safety to its current level? Let us hope that important issues have not been totally forgotten. Human element, human factors and human errors will not disappear with unmanned ships. They will just get transformed as discussed e.g. by Ahvenjärvi (2016).

If the pace of development is too fast and hasty, some very important issues may not get the consideration that they would deserve. Such issues can be learned e.g. from the Manhattan project, see Jungk (1958) or Rhodes (1986). Technology should not be developed fully without the control of the society. If the previous example raised too dystopic thoughts in the readers mind, the story of the "Flying Enterprise" and captain K. Carstens, see Kofod-Hansen (1952), can be recommended to attain a better balance again.

Social responsibility of each individual, including politicians as well as authorities, directors, managers, engineers and researchers, should be emphasized and valued higher than ever due to the threats of social problems connected to unemployment and uncertainty related to new technology.

5 CONCLUSIONS AND RECOMMENDATIONS

Regarding AI applications in autonomous shipping, several uncertain areas were identified. Selected key issues, with recommendations how to address these issues, are listed below:

- Design and validation process: Further research is needed on how to integrate the AI development as a part of a product development process, and how to evaluate sufficiency, reliability and relevance of the data in use.
- AI transparency: Guidelines should be developed for AI transparency in the maritime context.
- AI security: Awareness should be increased regarding the new ways of attacking AI systems, for example by using real-world objects.

The challenges in relation to unmanned and tele-operated ships are numerous. So far we haven't seen an unmanned ship that could make correct decisions in unforeseen situations. However, with a spirit and dedication, similar to that, what has brought men first, hundreds of years ago, over the seas and oceans, it is believed that at least some of the problems involved in the unforeseen technology can be solved. On the other hand, as the majority of the area on the planet earth is covered by seas and oceans, it would be strange to change it to a playground of unmanned or tele-operated ships as long as able-bodied professionals can be found. Do we really want that the mankind would lose the experience it has achieved and gathered as a professional heritage since the days of first brave sailors?

There are many challenges in getting the new technology to be safe straight from the beginning of its use. Therefore, many carefully-thought, widely and deeply considered precautionary measures need to be found out. The hazards involved need to be identified beforehand. If not, safe management of improvement design and application need to be applied. Still, some uncertainty in case of unidentified hazards and risk will always remain.

Instead of fully unmanned or tele-operated ships we prefer recommending the use of AI as separate support tool. Let the master mariners, officers and mariners onboard be well-trained. In case the level of automation is increased, it should be confirmed that the users are sufficiently well-trained to be able to use properly and to understand the new technology (and implications thereof) they may get onboard. It is hard to believe that improving the efficiency, economy, safety etc. is not possible in manned ships with improved technology.

The choices to be made, i.e. the available alternatives, when making decisions concerning policies related to unmanned and tele-operated ships need to be discussed thoroughly. It should not be forgotten that these choices are not just about technology and economy, they are ethical and moral choices, too, at the same time, although some proponents of the new technology may not think about the latter.

The decision-makers should let the professionals do their work, and let them use time as needed, without unbearable time pressures, or political pressures, to be able to understand what they are actually doing. Otherwise they may not be able to explain the politicians what is safe (and what might not be, or what are the uncertain issues that could be shed some light on), and why, to guarantee a safer future via good decisions. This will require more discussion between all stakeholders, thus enabling a path to sound, transparent, and understandable, more widely negotiated and more widely accepted policies (with sufficient budgets of money and time for precedent research, education and training). Multi-disciplinary research is required, as well as responsible, good quality development work, with creative thinking about a variety of alternative solutions and halts or best possible improvements in a well-controlled way, with sufficiently long testing phase.

We do not believe in a pure, inevitable process of techno-determinism without any real alternatives or choices. A real option to halt, regret and make a change in the direction, if felt necessary, should always be preserved available for free men and women, whether onboard or on shore. Men and women should be trained to use and to improve the capabilities and use of their own brains, too. This and continuous search for the truth will finally ensure that we are approaching the correct questions.

At the moment we may assume that the society is **not yet ready** for technological systems of unmanned ships. There seems to still be a huge area of uncertain issues related to safe possibilities to apply them in the near future. If we will get all the answers to our present, forgotten and future questions and get almost all the involved problems solved, we are soon facing the last, final question: Do we and does the society really want machines and robots to replace real men and women from work?

Research is one of the best options to improve our knowledge and understanding of new ideas, new observations and to decrease uncertainty. Without free and autonomous research, and experiments, it will be difficult to get sufficient knowledge and understanding about the hypothetical concept of unmanned ship in the maritime system.

ACKNOWLEDGEMENTS

We would like to acknowledge our previous colleague, Mr. Risto Tuominen (M.Sc.), now retired, for his dedicated mental support to us during the early thinking processes in relation to this paper. We present our acknowledgement to our unidentified reviewers, too, and hope that this paper will take into account most of their good and valuable comments.

REFERENCES

Ahvenjärvi, S. (2016) *The Human Element and Autonomous Ships*. the International Journal on Maritime Navigation and Safety of Sea Transportation, Volume 10, Number 3, September 2016, DOI: 10.12716/1001.10.03.18

Berkeley, E.C. (1952) *Giant Brains or Machines That Think*. John Wiley & Sons, Inc., New York, 270 p.

Bertram, V. (2016) *Autonomous Ship Technology – Smart for Sure, Unmanned maybe*. The Royal Institution of Naval architects, Smart Ship Technology, International Conference proceedings, 26–27 January, London, UK, pp. 5–12.

Danks D., London A. 2017. *Regulating autonomous systems: Beyond standards*. IEEE Intelligent Systems. Vol. 32:1. IEEE.

DNV-GL (2015) Digitale Sårbarheter Maritim Sek-tor – Lysneutvalget. Rapport nr. 2015–0569, Rev. 1.

Falcini F., Lami G. 2017. *Deep Learning in Automotive: Challenges and Opportunities*. Communications in Computer and Information Science. Vol 770. Springer.

AAWA (2016) Jalonen, R., Tuominen, R., Wahlström, M.: *Safety and security in autonomous ships* - pp. 56–73, in Jokioinen, E. Position papers. Rolls-Royce Marine. http://www.rolls-royce.com/~/media/Files/R/Rolls-Royce/documents/customers/marine/ship-intel/aawa-whitepaper–210616.pdf.

Bijker, W.E., Hughes, T.P., Pinch, T. eds. (1987) The Social Construction of Technological Systems: New Directions in the sociology and history of technology. Cambridge: MIT Press.

Felten, E. (2016) Preparing for the Future of Artificial Intelligence, White House Office of Science and Technology Policy.

Heikkilä, E., Tuominen, R., Tiusanen, R., Montewka, J., Kujala, P. (2017) *Safety Qualification Process for an Autonomous Ship Prototype – a Goal-based Safety Case Approach*. Marine Navigation: Proceedings of the 12th International Conference on Marine Navigation and Safety of Sea Transportation (TransNav 2017).

Hogg, T., Ghosh, S. (2016) *Autonomous merchant vessels: examination of factors that impact the effective implementation of unmanned ships*. Australian Journal of Maritime & Ocean Affairs, Vol. 8, 2016, Issue 3, pp. 206–222. Published online: 06 Sep 2016. https://doi.org/10.1080/18366503.2016.1229244.

Jalonen, R., Tuominen, R., Wahlström, M. (2017) *Safety of Unmanned Ships*. Safe Shipping with Autonomous and Remote Controlled Ships Aalto University publication series, SCIENCE + TECHNOLOGY 5/2017.

Jokioinen, E. Position papers. Rolls-Royce Marine. http://www.rolls-royce.com/~/media/Files/R/Rolls-Royce/documents/customers/marine/ship-intel/aawa-whitepaper-210616.pdf.

Jungk, R. (1958) Brighter than a Thousand Suns: A Personal History of the Atomic Scientists, New York: Harcourt Brace, 1958.

Kofod-Hansen, M. (1952) Kapteeni Carlsen, "Flying Enterprisen" sankari, (translated by Olli Nuorto to Finnish language from the original book: Captajn Carlsen, "Flying Enterprise"), Gummerus, 216 p.

Le Coze, J-C, Pettersen, K, Engen, O.A., Morsut, C., Skotnes, R., Ylönen, M., Heikkilä, J., Merlele-Coze, I. (2017) *Sociotechnical systems theory and the regulation of safety in high-risk industries* - White paper. VTT TECHNOLOGY 293, 44 p., http://urn.fi/ URN: ISBN: 978-951-38-8522-9

Levander, O. (2016) *Ship intelligence – a new era in shipping*. The Royal Institution of Naval architects, Smart Ship Technology, Internation-al Conference proceedings, 26–27 January, London, UK, pp. 25–32.

Li, X., Jiang, H. 2017. *Artificial Intelligence Technology and Engineering Applications*. Applied Computational Electromagnetics Society Journal. Vol. 32:5.

Lloyd's Register. 2016. Cyber-enabled ships – a Lloyd's Register Guidance Note

Perrow, C., 1984. *Normal accidents*: living with high-risk technologies. Basic Books, 386 p. ISBN 0-465-05142-1

Rhodes, R. 1986. The Making of the Atomic Bomb. A Touchstone Book, Simon & Schuster, New York, 886 p., ISBN 0-671-65719-4.

Rødseth, Ø.J. and Burmeister, H.-C. (2015) *Risk Assessment for an Unmanned Merchant Ship*. TRANS-NAV, the International Journal on Marine Navigation and Safety of Sea Transportation, Vol. 9, Nr. 3, September 2015.

Williams, D.L., de Kerbrech, R.P. 1982. *Damned by Destiny*. Teredo Books LTD.

Wróbel, K., P. Krata, P., Montewka, J., Hinz, T. (2016) *Towards the Development of a Risk Model for Unmanned Vessels Design and Operations*. International Journal on Marine Navigation and Safety of Sea Transportation, Vol. 10, No. 2, June 2016, pp. 267–274.

Zhang, R.L., Furusho, M. (2017) *Conversion Timing of Seafarer's Decision-making for Unmanned Ship Navigation*. International Journal on Marine Navigation and Safety of Sea Transportation, Vol. 11 No. 3, September 2017, pp. 463–468.

Marine Design XIII – Kujala & Lu (Eds)
© 2018 Taylor & Francis Group, London, ISBN 978-1-138-34076-3

Towards autonomous shipping: Operational challenges of unmanned short sea cargo vessels

C. Kooij, M. Loonstijn, R.G. Hekkenberg & K. Visser
Delft University of Technology, Delft, The Netherlands

ABSTRACT: Autonomous shipping is considered a highly promising development in the shipping industry. Technologically it is believed to be possible, but there are several challenges that still need to be solved before autonomous shipping can become a reality. In this paper the main challenges for autonomous shipping are identified. This is done by taking the main functions of the ship and combining them with the crew tasks. That way it is possible to identify where problems occur when the crew is no longer on board. Combining the challenges found with prior research on autonomous shipping leads to three elements into which little to no research has been conducted. However, they are important for the operation of an autonomous ship. The elements that need to be further investigated are; navigation and situational awareness, ship-to-ship communication, physical intervention of the crew on the ship and its systems, and maintenance and repair.

1 INTRODUCTION

With the technology for autonomous vehicles evolving rapidly, the shipping industry has also started to research its potential. There are several projects completed and in the works which all come to the same conclusion: the technological building blocks to build an autonomously navigating ship are available. However turning these building blocks into actual systems and control algorithms still requires significant development. At this point there are several small vessel that can operate autonomously or via remote control, but there are no functioning cargo vessels yet (Israel Aerospace Industries, no date; Naveltechnology.com, no date).

This paper looks at the direct problems that arise from the removal of all crewmembers from a conventional short sea cargo vessel. This allows for the identification of the biggest challenges that lie ahead. Finally these challenges are compared to information and articles about research that has already been done to find the remaining challenges. These results are the starting point for further research.

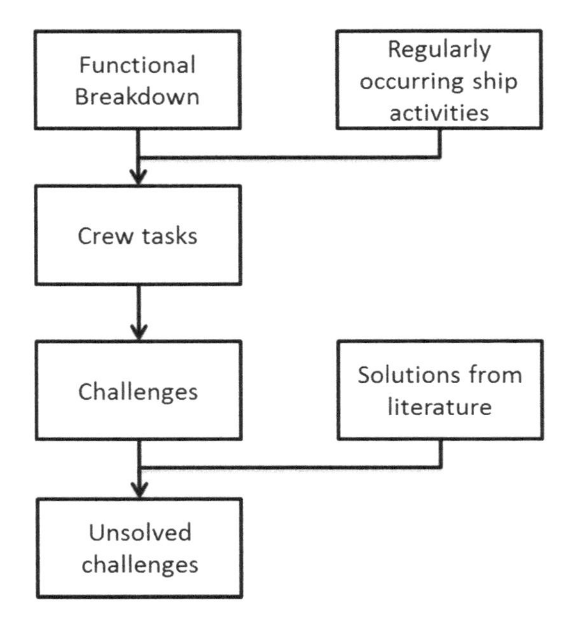

Figure 1. Graphical overview of the method followed.

2 METHOD

This section discusses the method by which the unsolved challenges have been found. Figure 1 gives a graphical representation of the method followed.

The first step in identifying the aforementioned challenges is making a functional breakdown of the ship. In the functional breakdown, every function of the ship and crew which is relevant to fulfil the ship's mission is identified. This is done by analyzing what functions need to be fulfilled in order to accomplish certain operations such as 'move'. This analysis is repeated until basic functions, i.e. functions that can be split further and can be performed by a single system, can be found (Viola *et al.*, 2012). For example: the function 'communicate' can be split into two functions 'to allow for external communication' and 'to allow for internal communication'. The function 'to allow for external communication' can then be split into the functions 'communicate with other ships' and 'communicate with the shore'.

In his book *Practical Ship Design* (Watson, 1998), Watson states that in war ship design there are three main functions (in order of importance): float, move and fight. But if 'fight' is identified as 'perform the ships main mission', this method can be applied to any ship type.

During the course of making the functional breakdown, it was found that in addition to these three functions, two additional functions play an important role in the operation of the ship. The first is communication. This is a part of almost every function the ship can perform, from communication between two ships to the communication for a sensor to a computer. To better identify the different communication types, communication is added as a main function.

It was also found that the three main functions identified by Watson do not cover what happens in case of failure. This is an interesting area for autonomous ships, since the crew is now the first line of defense in many cases. Therefore it is important to know how failures are prevented or currently solved, what systems exist to prevent or mitigate these failures and what role the crew plays. For this reason this function is also added as a main function.

To summarize: the main functions of the ship identified in this paper are:

- Float
- Move
- Communicate
- Perform mission
- Prevent and mitigate failure

The next step is to identify which role the crew plays in the operation of the ship. A ship is a complicated system of which the operational profile differs depending on the situation and the environment. To get a good view of the crew involvement their tasks and actions are identified in six key activities that take place during a ship's journey. These activities are ordered chronologically to ensure that every aspect of the journey is covered. Additionally, the processes that take place during specific parts of the journey in each department are documented, although they differ between ships and companies. These six activities are:

- Preparing for the mission
- Quayside operations (e.g. loading and unloading, bunkering and taking on stores)
- Preparing for arrival and departure
- Near and in port sailing (arrival and departure)
- Normal sailing in open sea
- Emergency situations

For each of the activities mentioned, the relevant functions from the functional breakdown are identified. The functional breakdown allows for an identification of which functions are performed by systems and which functions require human interference or a human to perform the function completely. The human involvement is determined from handbooks, observations in practice and expert interviews. Together these elements give a good overview of the level of human involvement. As this paper focusses on the general operational aspects of shipping where challenges can arise, no further details are required within the functions or the crew tasks.

From there, it is possible to identify what would happen if the crew is no longer on board. This is determined by expert interviews.

The final step in this paper is to combine the identified problems with the challenges and solutions found in literature. That way the challenges that have not been identified or solved in previous research can be found, thus showing where further research is required.

3 FUNCTIONAL BREAKDOWN AND CREW TASKS

The first step is to identify which functions a short sea cargo vessel must have, to perform its mission. Next, the crew tasks for each of the identified key activities are identified.

3.1 *Functional breakdown*

The developed functional breakdown consists of the five main functions of the ship. Each of these five functions in broken down into smaller functions until a function can be defined that can be performed by a single system. A graphical representation of the functional breakdown, showing the first two levels, can be found in Figure 2.

3.1.1 *Float*

The function 'float' covers sub functions such as have structural integrity and be watertight. While these functions are not performed by a system per se, they are an important element in the ship's functions.

3.1.2 *Move*

The function 'move' not only covers the physical movement of the ship, but also the planning of this movement: navigation.

The physical moving of the ship consists of functions covering the movement of the ship in

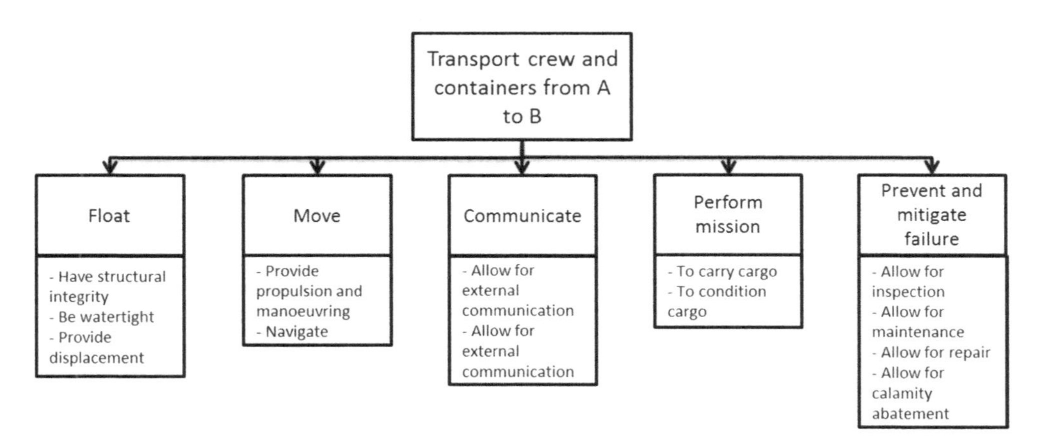

Figure 2. Functional breakdown showing the first two levels of the identified functions.

all directions, for example, proving lateral thrust at slow speeds, which can be performed by the bowthruster.

The navigational part of the 'move' function mostly consist of planning the route and being able to follow it without any problems. This includes functions such as 'to display navigational data' which can be performed by several systems alone or a combination of them.

3.1.3 *Communicate*

There are several forms of communication that take place during a mission. There is internal communication between different crewmembers, external communication with the shore (e.g. the vessel owner, port authorities, and terminals) and communication between ships. Additionally the ship is able to send out several different emergency signals.

Internally the most used methods of communication are verbal communication between crew and data transfer between systems. Finally the systems also communicate with the crew, for example by showing data on a screen or giving alarms.

In external communication verbal communication and data transfer are also important, additionally communication happens via lights and sound signals. Each of these manners of communication takes place in both ship-to-ship and ship-to-shore communication.

3.1.4 *Perform mission*

This function includes all elements that are required to move cargo from a to be. Therefore it not only includes the ability to load and unload cargo, but also all elements that are required for safe and comfortable passage of the crew, such as providing HVAC and electricity.

3.1.5 *Prevent and mitigate failure*

The final main function of the ship is the ability to solves problems if something breaks down as well as maintenance to decrease the possibility of a failure. This is a very broad function as it covers many potential problems that can occur on board of a ship. In general it covers the ability of each system and space to be inspected and be repaired if required as well as calamity abatement for every possible calamity.

3.2 *Crew tasks*

For this paper a only a general overview of the crew tasks is required. A general understanding of what the crew does is enough to determine if problems will arise when the crew is no longer available. Further details can always be added if they are required. Commonly, the crew of a ship is subdivided into four departments:

- The bridge crew
- The deck crew
- The engine room crew
- The catering crew

The bridge crew is responsible for the general operation of the ship. This means that they steer the ship, determine the precise route and make decisions to deviate from this plan, if necessary. Since the bridge crew steers the ship, they are also responsible for navigational safety and situational awareness of the ship.

The deck crew is responsible for planned maintenance and general upkeep of the deck equipment. Additionally they operate the equipment and handle the lines during mooring and anchoring operations.

The engine room crew spends most of their time on inspection and planned maintenance to keep the equipment in good condition.

Table 1. Crew involvement for functions during different operational activities. (The checkmarks show crew involvement).

	Preparing for journey	Quayside operations	Preparing for arrival and departure	Near and in port sailing	Normal sailing in open sea	Emergency situations
Float	X	X	X	X	X	X
Move	✓	✓	✓	✓	✓	✓
Communicate	✓	✓	✓	✓	✓	✓
Perform mission	✓	✓	✓	✓	✓	✓
Prevent and mitigate failure	✓	✓	✓	✓	✓	✓

Finally, the catering crew is responsible for the food on board of the ship

For each of the six identified key activities, the crew tasks were identified. Table 1 shows whether crew is required for the performance of a specific function during different operational activities.

3.2.1 *Preparing for the journey*

The preparation for the journey is done by the bridge officers prior to departure. The whole route is planned and elements such as traffic conditions and weather reports are reviewed (International Chamber of Shipping, 2016). The exact list of steps that are taken during the preparation of the trip depends on the company, the ship and the mission it will execute.

During the trip preparation, the ship also undergoes basic maintenance and repairs if required.

3.2.2 *Quayside operations*

The most important part of the quayside operations is the loading or unloading of the ship. In most cases, the cargo is handled by the port crew, leaving the ship's crew free to perform other tasks. Labor organizations even advise against working on a ship where loading and unloading operations fall under the crews authority (ITF, 2015). Some other operations during the loading and unloading, such as ballasting, are already fully automated.

Not only mission-specific items (e.g. cargo) are loaded onto the ship, the ship also takes in bunkers and stores. According to industry experts this task is performed by the crew from the responsible department. For example, a member from the engine room crew would oversee the bunkering of the fuel.

3.2.3 *Preparing for arrival and departure*

The final preparations for arrival and departure are listed on a checklist, to ensure that no tasks are forgotten. However, these procedures are not standardized, they differ from ship to ship. In general the following elements can be expected to be listed on the checklist (International Chamber of Shipping, 2016):

- Check main engines and rudder control
- Check navigational equipment
- Check cargo and cargo handling systems
- Check communication systems

Each of the departments has their own checklists and a required number of people required for the tasks. Generally, this is a higher number than is required for normal sailing to assist in case of an incident.

3.2.4 *Near and in port sailing*

Near port sailing differs from normal sailing in several key points. The first is the amount of traffic, different ships with different functions are moving around the port. This does not only include cargo vessels but also smaller vessels such as yachts and pleasure crafts. This means that the communication density with other ships and shore is much higher. The second area where the two differ is the fact that the water depth is limited and the domain in which the ship can sail safely becomes more confined. This means that some procedures on board change. For example, the watertight doors are closed, as the chance of grounding increases near shore.

These elements cause more crew to be involved in the operation of this ship during near port sailing than is required for normal sailing. In the engine room the number of crew members on standby increases, allowing for quick intervention should the situation call for it. The additional crewmembers have no specific additional tasks, they are mainly there to decrease the consequences should something go wrong.

On the bridge, there are two main tasks for the crew, communicating and traffic identification and processing. Communication is required between the ship and both port control and other ships.

Additionally, contact between the different areas of the ship and between the ship and its pilots might be required. In addition to the communication, the crew is using both equipment and their own visual capabilities to determine if the selected route is clear of traffic and how to respond if it is not. Experts agree that in busy waters the navigational equipment, such as the radar, AIS and electronic charts, is not sufficient to identify all other traffic. This means that human intervention is required.

The deck crew is responsible for the maintenance and operation of the deck equipment, such as mooring winches and lines and anchors. During arrival and departure the deck crew is on deck to operate the equipment and afterwards secure it and prepare it for later use.

3.2.5 *Normal sailing*

During normal sailing the crew works on their normal duties. In the engine department this means inspection, planned maintenance and, in some rare cases, unplanned repairs.

The deck crew also performs planned maintenance on the equipment on the deck and also on the hull and superstructure of the ship. Additionally they monitor the cargo; should some of the cargo shift or become loose they are tasked with re-securing the lashings if possible.

The bridge department navigates the ship. There is a continuous watch on the bridge as well as a crewmember tasked with steering the ship. Additionally each of the officers have specific responsibilities such as keeping the navigational charts up to date or periodically checking the safety equipment.

3.2.6 *Emergency situations*

There are many emergency situations that could occur during sailing. This means that it is also very difficult to detail what the crew does during those situations. Each situation is different and has its own procedures. However, in most situations, the crew is heavily involved, for example by fighting the fire or by navigating the ship to a safer location.

4 IDENTIFICATION OF CHALLENGES

The crew tasks are combined with the functional breakdown of the ship. This gives an overview of the challenges that arise on an unmanned vessel during the different phases of the mission. The largest challenges for each of the main functions are discussed here. A summary of the findings can be found in Table 2.

4.1 *Float*

The functions that are listed under float are all ship functions that can be performed without human intervention. The prevention and mitigation of failures with respect to these functions of the ship are discussed in paragraph 4.5.

4.2 *Move*

The function move can be divided into two parts, the first is the actual propulsion of the ship, the second the navigation of the ship. In the following paragraphs, the challenges for each part are discussed.

4.2.1 *Provide propulsion and maneuvering*

Most of the propulsion and maneuvering equipment requires very little interaction with the crew after starting up. The crew only steps in to maintain the equipment and monitor its state.

4.2.2 *Navigation and situational awareness*

The second part of the main function move consists of navigation. A large part of the navigation is already automated. For example; Meteo Group offers a service where weather data is used to find the optimal route for a ship, sending a route to the ship that only needs to be followed (MeteoGroup, no date).

Table 2. Operational challenges found the defined activities. (The checkmarks show that there is a challenge to be found).

	Preparing for journey	Quayside operations	Preparing for arrival and departure	Near and in port sailing	Normal sailing in open sea	Emergency situations
Float	X	X	X	X	X	X
Move – Provide propulsion	X	X	X	X	X	X
Move – Navigation and situational awareness	X	X	✓	✓	✓	✓
Communicate – external	X	X	✓	✓	✓	✓
Communicate – internal	X	X	X	X	X	X
Perform mission[1]	X	✓	✓	✓	✓	✓
Prevent and mitigate failures[2]	✓	✓	✓	✓	✓	✓

[1]Due to physical interaction with the system.
[2]Due to physical interaction with the system and current maintenance strategies.

It is the situational awareness where the crew plays an important role. Much of the navigational and steering equipment can help with the following of a route, for example by using the autopilot function, which can not only go in a straight line but can also use waypoints. However, the crew is still required for the observation of the environment. While radar and AIS do provide a good view of what obstacles are around the ship, these systems are not infallible. AIS updates in intervals and radar does not pick up all ships in the area, especially smaller ones like recreational crafts. In this case a human serving as lookout is relied upon to identify potential hazards that the ship might encounter. Experts feel that at this point in time it is not possible to sail the ship blindly on the navigational instruments, the crew is required.

4.3 *Communicate*

As mentioned before, the ship's communication can be divided into two parts, internal and external communication. Both parts are discussed separately.

4.3.1 *Internal communication*

As stated above the internal communication consists of either verbal communication or data transfer. The verbal communication is no longer relevant if there are no crewmembers left on board. The data communication between systems remains in operation as long as no problems occur, repairing the systems will pose a problem, which is discussed further in paragraph 4.5.

The communication of the systems with the crew are also no longer required onboard. However, these systems could be required in a shore control station, which is discussed in paragraph 4.3.3.

4.3.2 *Ship-to-shore communication*

When the ship is sailing, the crew keeps in regular contact with the owner or main office, updating them on their location and any situations that might have occurred. On a crewed ship, this contact takes place approximately once a day, depending on company policy. In case of an unmanned ship the number of updates as well as the amount of information might increase, to ensure that the ship remains safe.

In the AWAA and MUNIN projects it is suggested that the ships are monitored or controlled from a shore control system (Jokionen, 2015; MacKinnon, Man and Baldauf, 2015). This would mean that a lot of information would have to be sent from shore to the ship via satellite communication, which is currently expensive. The MUNIN project has estimated that communicating all the data collected by sensors on the ship to the shore by satellite, would cost approximately 150.000 US dollars a month (Porathe, Prison and Man, 2014).

4.3.3 *Ship-to-ship communication*

The other important form of communication is between ships. This contact is both active, by the crew on one ship talking to the crew on another ship by VHF, and passive, for example by using navigation lights or sound signals.

The active communication is expected to be a challenge, especially between a manned and an unmanned vessel, as humans and computers have very different ways of communicating. While speech recognition is getting increasingly better, mistakes are still made and the technology does not allow for a full conversation.

The passive communication from the autonomous ship is relatively simple, the ship shall be equipped with navigational lights and horn to produce the required signals at the required time. However, interpreting the signals that other ships and structures are giving might pose a larger challenge. For example, registering a sound signal might pose a problem for a system, especially when it is important to determine the direction.

Taking these elements into account it is clear that communication could pose a major challenge in autonomous shipping.

4.4 *Perform the mission*

As mentioned, the mission 'transport of cargo' requires very little human intervention outside of the observation of cargo and intervention if there is a problem with the cargo. Both these tasks require a physical interaction between the crew and the ship which is very difficult to recreate on an unmanned ship at this point in time.

In addition to transporting the cargo, one of the ships functions is also to transport the crew, with all of the provisions required to do so. However, with no crew, these provisions are no longer required.

4.5 *Prevent and mitigate failures*

In case of a failure or an emergency the crew has a large role to play. Once again a large part of this is the crew interacting with the ship and its systems, something that might by difficult to do on an unmanned ship.

The state of the equipment is monitored closely. This is mostly done by sensors, which inform the crew if something is amiss. Should the sensor indicate a problem someone is send to investigate and solve the problem. The ability of the crew to physically interact with the ship and its systems is one of the big challenges to solve.

Experts also expect planned maintenance to be a challenge. While a lot of the equipment is self-regulating (e.g. switching to a second filter if the first does not function anymore), a lot of planned maintenance still occurs during normal sailing. This would mean that the entire maintenance plan for the ships would have to be restructured.

4.6 *Summary of the challenges found*

To summarize; the following primary challenges arise with the removal of the crew from the ship;

- Situational awareness
- External communication
- Physical interaction of the crew with the systems and the ship
- Maintenance and repair

5 PROMISSING SOLUTIONS

The next step is to find the promising solutions within literature that provide solutions for the challenges identified above. The effects of the found solutions on the challenges from section 4 can be found in Table 3.

5.1 *Autonomous navigation and situational awareness*

As concluded in the previous section, the crew plays an important role in the identification of hazards and in the positioning of the ship. Fully autonomously navigating ships are not yet a possibility. Most projects propose the use of some form of a shore control station. In this station an operator has the ability to directly control the ship and steer it away from danger if necessary. This does mean that enough information has to be send to the shore control station for the operator to see what is happening to the ship.

In the AAWA project's white paper (Poikonen *et al.*, 2017) an analysis is performed on the different sensor types that are available to provide the ship with situational awareness. In that paper it is stated that a wide variety of sensors is required to

Table 3. The operational challenges that are not solved in previous research. (The symbol explanation can be found below).

	Preparing for journey	Quayside operations	Preparing for arrival and departure	Near and in port sailing	Normal sailing in open sea	Emergency situations
Move – Provide propulsion	–	–	–	–	–	–
Move – Navigation and situational awareness	–	–	/ AAWA: Sensors and remote control require further research	/ AAWA: Sensors and remote control require further research	/ AAWA: Sensors and remote control require further research	/ AAWA: Sensors and remote control require further research
Communicate – External	–	–	X Ship-to-ship communication not investigated	X Ship-to-ship communication not investigated	/ Ship-to-ship communication and data transfer needs further investigation	/ Ship-to-ship communication and data transfer needs further investigation
Perform mission	–	✓ Port crew	✓ AAWA and Lloyds: Use of sensors and cameras	X Physical interaction has not been researched	X Physical interaction has not been researched	X Physical interaction has not been researched
Prevent and mitigate failures	✓ MAN: CBM	✓ MAN: CBM	X Physical interaction has not been researched	X Physical interaction has not been researched	X Physical interaction has not been researched	X Physical interaction has not been researched

–: no challenges found.
✓: Challenges solved in research.
/: Partially solved in research, further research is required.
X: No solutions found.

get a complete view of the landscape around the ship. A combination of high frequency radar, infrared cameras and normal cameras, as well as the standard navigational equipment such as AIS and ECDIS should give a good overview of the world around the ship. Additionally it is suggested that the ship is outfitted with sound sensors to allow the ship to capture sound signals that are produced by other ships as a means of communication.

Lloyds register goes even further in their booklet *Global Marine Technology Trends 2030* (Lloyds Register, QinetiQ and University of Southhampton, 2017), focused on autonomous systems. They suggest the use of a Global Positioning System (GPS), an Inertial Navigation System (INS), optical and infrared cameras, radar, Lidar, high-resolution sonar, microphones and wind and pressure systems.

Both Lloyds Register and the AAWA project conclude that with sensors, the ship could have a situational awareness that is at least as good as the situational awareness of a human crew, they do however, both expect a challenge in the communication of this information.

The level of detail in the situational awareness is determined by the phase of the journey in which the ship is operating. During the journey preparation and quayside operations it is not necessary to know exactly what is going on around the ship. The ship might only need a watch to ensure that it remains moored. During the arrival and departure preparation and the near and in port sailing the level of detail needs to be very high due to the difficult navigational situation and the presence of other traffic. On the other hand the communication of the relevant data will be easier due to the proximity to land where the coverage of different communication methods is much better than out at sea.

During normal sailing the reverse is true. It might be possible to decrease the level of detail in the situational awareness due to the lower number of possible obstacles however, it will be more difficult to send the required data to shore.

Autonomous navigation and situational awareness are considered partly solved. The largest challenge that remain is the sensor types that are required for a complete view of the area around the ship.

5.2 *External communication*

5.2.1 *Ship-to-ship communication*

Communication is an element that is discussed in many different autonomous shipping projects (Rødseth *et al.*, 2013; Poikonen *et al.*, 2017). However, these mostly covers the communication between the ship and the shore control station as discussed above. There are no solutions given for ship to ship communication.

The ship to ship communication is mostly important during near and in port sailing due to the high traffic density.

5.2.2 *Ship-to-shore communication*

With regards to the communication between the ship and shore the biggest challenge lies within the data transfer. To cut down on the cost of this data transfer several solutions have been suggested, especially for the transfer of the camera images. In the AWAA white paper (Poikonen *et al.*, 2017) it is suggested to use traditional methods of decreasing the file size; such as a lower frame rate as well as efficient compression of the data. Part of the data could be processed on board, for example by using horizon detection to remove the irrelevant parts of the image before it is send to an observer in a shore control station. It is stated that this should give a remote observer enough information in non-critical situations. As mentioned above this solution is more relevant during normal sailing when the ship is far from shore with limited communication options.

For critical situations, such as near and in port navigation and emergencies, a shore control station with direct control is required (MacKinnon, Man and Baldauf, 2015).

Neither forms of communication can be considered a solved area within autonomous shipping. There has been no research into ship-to-ship communication and ship to shore communication is too expensive to be feasible.

5.3 *Physical interaction of the crew with the system and the ship*

There are many occasions where the crew physically interacts with the systems or the ship. For some problems, a specific solution has been found. For example, research has been done into automatic mooring, either by a magnetic connection or by vacuum suction cup (Port of Rotterdam, 2014; Cavotec, 2017). These systems means that no deck crew is required to assist with the mooring operations. However, these systems are currently only available on certain locations and for certain ship types.

For inspection work, drones are suggested as a good alternative to human inspectors. DNV GL has used drones to inspected the tanks on ships (DNV GL, 2015). This method could perhaps be expanded towards inspections of other areas and at different moments.

However, these solutions are stand alone and solve only a part of the interaction challenge. There are other elements that do not have a solution yet. For example, when the anchor is used, a human operator is used to determine how much anchor cable is fed into the water. This cannot be done by remote observation due to the speed of the release and the limited view due to airborne rust particles.

Additionally the integration of all these solutions could also lead to additional challenges. Therefore this subject is considered partly solved.

5.4 *Maintenance and repair*

There are no clear solutions to the challenges in maintenance and repair that are currently applied in shipping. There are some promising techniques used in system maintenance that could also work on board of unmanned ships, and specifically unmanned machinery spaces. Modern propulsion plants use Condition Based Maintenance (CBM). This means that systems are used to monitor the health of the propulsion system and send a signal to the engine room crew in case of degradation of performance.

This system is slowly being integrated into the Marine Diesel engine market (Oskam, 2014; Basurko and Uriondo, 2015). The systems can help provide information that could allow for a completely unmanned machinery space, for example by sending the information not to the engine room crew but to a monitoring system on shore.

Full CBM systems already exist in the shipping industry (Rolsted, no date), however they are monitored by a full crew which is on board of the ship.

A wider adoption of CBM systems can improve the technology, increasing, for example, the reliability of both the system and the machinery space it monitors. However, the problem of fixing a broken system due to an unforeseen event still remains. CBM only allows for better planning of planned maintenance, which will have to be done while the ship is in port.

In the MUNIN project a different solution is suggested. They feel it might be beneficial to switch from a diesel engine to a less complicated system that is easier to maintain. They propose LNG as a solution. Additionally it is proposed to increase the redundancy of the system by having two main engines (Rødseth *et al.*, 2013). However, this solution does carry a very high cost with it.

6 DISCUSSION: UNSOLVED CHALLENGES

With the analysis of the challenges the solutions in literature, a few challenges remain that have no clear solution at this point in time. These challenges are discussed in this part of the paper.

6.1 *Navigation and situational awareness*

At this point, the solutions suggested for autonomous or remote navigation and situational awareness are not a feasible option. They are simply too costly to implement on a large scale. Further research as well as a steady advancement in imaging and data transfer capabilities should make this technology more widely available and useable in the future.

6.2 *Ship-to-ship communication*

While communication between the ship and the shore, mostly in the way of data transfer, is a topic that is frequently discussed and researched the communication between two ships is not addressed. There are some intricacies not easily solved, such as the communication via sound signals or spoken communication.

In case of the shore control station, communication is possible between a manned ship and an operator on shore, or between two operators on shore. This way clear communication between two humans is still possible.

Later on in the process the communication between two autonomous ships should also pose no further challenges, as once again the ships speak the same language. The largest challenge will be when both autonomous ships and manned ships interact with each other.

6.3 *Physical interaction of the crew with the systems and the ship*

It has become clear that a large part of the identified challenges are related to the physical interaction of the crew with the ship. While there are some isolated solutions, such as the automatic mooring systems mentioned above, there is no good complete solution.

This element comes into play in a lot of different areas, from standard everyday tasks to emergencies. Because it is so diverse, finding an all-encompassing solution seems unlikely. Each problem has to be looked at individually to find which solution is the most suitable.

6.4 *Maintenance and repair*

As mentioned there is very little solved about how maintenance will look on an autonomous ship. The maintenance strategy mentioned above provide a partial solution but it does not solve all the problems. Additionally, the repairs that have to be executed in case of problems are not covered in any research.

In the MUNIN project (Rødseth *et al.*, 2013) it is suggested that redundancy is a good solution to decrease the potential for failures. However, it does mean that there is a large additional cost in the building of the ship. Additionally this solution is difficult for conversions.

They also suggest using a simpler propulsion system, as a diesel engine has a lot of different part that can break. While this is a valid point, their solution, LNG, comes with its own challenges and elements that can fail. It is also less well known, making predicting the probability of failure more difficult.

7 CONCLUSIONS

Based on the performed analysis there are four elements that are barely researched or only researched in part:

- Navigation and situational awareness
- Ship-to-ship communication
- Physical interaction of the crew with the systems and the ship
- Maintenance and repair

These four elements will be a starting point for further research as they are critical elements in the operation of the ship that need to be solved before autonomous shipping can become a reality.

ACKNOWLEDGEMENTS

This paper would not have been possible without the input of several people with significant experience on board of ships. We would like to thank Harmen van der Ende of the Maritiem Insituut Willem Barentsz, Gaby Steentjes of MARIN, and the crew of the training ship the Delfshaven, who were kind enough to share their experience and ideas on our analysis.

REFERENCES

Basurko, O.C. and Uriondo, Z. (2015) 'Condition-based maintenance for medium speed diesel engines used in vessels in operation', *Applied Thermal Engineering*. Elsevier Ltd, 80(x), pp. 404–412. doi: 10.1016/j.applthermaleng.2015.01.075.

Cavotec (2017) *MoorMasterTM Automated Mooring systems*.

DNV GL (2015) *Surveys without scaffolds – DNV GL conducts drone inspection of ship tanks – DNV GL*. Available at: https://www.dnvgl.com/news/surveys-without-scaffolds-dnv-gl-conducts-drone-inspection-of-ship-tanks-51628 (Accessed: 8 December 2017).

International Chamber of Shipping (2016) *Bridge procedure Guide, Release 9.1.3*.

Israel Aerospace Industries (no date) *Katana USV System*. Available at: http://www.iai.co.il/2013/36781-46402-en/BusinessAreas_NavalSystems.aspx (Accessed: 2 January 2018).

ITF (2015) 'ITF UNIFORM "TCC" COLLECTIVE AGREEMENT'. Available at: http://www.itfseafarers.org/files/seealsodocs/33560/itfuniformtcccba20122014.pdf.

Jokionen, E. (2015) 'Introduction'. AAWA.

Lloyds Register, QinetiQ and University of Southhampton (2017) *Global Marine Technology Trends 2030 – Autonomous Systems*.

MacKinnon, S., Man, Y. and Baldauf, M. (2015) 'D8. 8 : Final Report : Shore Control Centre'.

MeteoGroup (no date) *RouteGuard*.

Naveltechnology.com (no date) *Fleet Class Common Unmanned Surface Vessel (CUSV) – Naval Technology*. Available at: https://www.naval-technology.com/projects/fleet-class-common-unmanned-surface-vessel-cusv/ (Accessed: 2 January 2018).

Oskam, G.J.. (2014) *Optimizing diesel engine condition monitoring*. Tu Delft. Available at: https://repository.tudelft.nl/islandora/object/uuid:037a6a9d-8655-429d-a361-bb5d9caf9c7f.

Poikonen, J. *et al.* (2017) 'Technologies for Marine Situational Awareness and Autonomous Navigation.pdf'. AAWA.

Porathe, T., Prison, J. and Man, Y. (2014) 'Situation awareness in remote control centres for unmanned ships', *Human Factors in Ship Design & Operation*, (February), pp. 1–9.

Port of Rotterdam (2014) *Are magnets taking mooring into the future?*

Rødseth, O.J. *et al.* (2013) 'Communication architecture for an unmanned merchant ship', *OCEANS 2013 MTS/IEEE Bergen: The Challenges of the Northern Dimension*, (314286). doi: 10.1109/OCEANS-Bergen.2013.6608075.

Rolsted, H. (no date) 'On board diagnostic, a web-based engine information management system', pp. 1–32. Available at: http://marine.man.eu/docs/librariesprovider6/marketing-publications/chalmers-universary-of-technolgy-2016/09-time-between-overhaul-condition-monitoring.pdf?sfvrsn=4.

Viola, N. *et al.* (2012) 'Functional Analysis in Systems Engineering: Methodology and Applications', in *Systems Engineering- Practice and Theory*, pp. 71–96.

Watson, D.G.M. (1998) *Practical Ship Design*. Elsevier.

Marine Design XIII – Kujala & Lu (Eds)

Towards the unmanned ship code

M. Bergström, S. Hirdaris, O.A. Valdez Banda, P. Kujala & O.-V. Sormunen
Aalto University, Marine Technology, Espoo, Finland

A. Lappalainen
Rolls-Royce, Turku, Finland

ABSTRACT: Maritime operations are disrupted by smart innovative technologies enabling an ever-higher level of on-board automation. Recently, developments reached a point where unmanned remotely controlled ships are thought to be in principle technically feasible. However, for unmanned ships to deliver on their promise for safer, cleaner and resource-efficient transport, new regulations are essential. This paper paves the way toward a performance driven regulatory framework for unmanned ships along the lines of the recently introduced IMO Polar Code. The work presented is thought to be useful supplement to the existing regulations of conventionally manned ships.

1 INTRODUCTION

In recent years, the maritime industry witnessed dramatic technology developments resulting in an ever-increasing level of on-board automation. Today this trend reached a break—even point where completely unmanned, remotely controlled, ships are thought to be in principle technically feasible. It is thought that unmanned ships may enable safer, cost-efficient and environmentally friendly maritime transport. However, the origins of existing maritime rules and regulations come from an era before the introduction of such disruptive technologies. To enable the design and operation of unmanned ships from a design for safety and overall regulatory perspectives, several performance driven regulatory challenges have to be addressed. Along these lines, this work suggests the introduction of a new regulatory framework for unmanned ships, namely the 'Unmanned Ship Code' (USC). Our proposal takes under consideration the recently introduced IMO code on the safety for ships operating in polar waters (Polar Code). This means that USC is fundamentally performance driven, goal-based and supplements existing conventional regulations.

The paper is structured as follows: Firstly, an overview of the existing maritime regulatory framework is presented and regulatory challenges of unmanned ships are highlighted; Secondly, the proposed USC is outlined and discussed. The study is limited to regulatory issues concerning ship design and operation. Legal issues, such as liability, are only briefly touched upon.

2 MARITIME REGULATIONS

2.1 *Overview of the regulatory framework*

The design and operation of ships are regulated by a mixture of international, European Union (EU), national, and Classification specific rules and regulations. The implementation of mandatory international regulations is facilitated by the International Maritime Organization (IMO). Nevertheless, enforcement of these regulations depends upon the individual IMO member states, acting both as flag administrations and port/coastal states (IMO, 2016c).

While individual IMO members are obliged to make international IMO conventions part of their own national law, they also have the right to determine additional regulations (and bans) on top of the IMO requirements (UN, 1982). This implies that, within their own internal waters, they could deviate from international conventions, unless they have accepted limitations to that right in other international agreements.

2.2 *Traditional maritime regulations*

Traditional maritime regulations consist of rules prescribing the required means, i.e., the required solutions to achieve safety in design and operations. Generally, these regulations are empirically determined based on in service experience.

The benefit of prescriptive Rules is that they are purpose specific in terms of application. Verification against compliance helps to assure design or operation objectives that consist the backbone of associated insurance requirements. However, because prescriptive rules refer to specific solutions, they imply design constraints and do not encourage innovation (Papanikolaou, 2009). According to (Papanikolaou, 2009) the rules are often reactive. They are determined in response to individual catastrophic events and based on older (parent) designs. Therefore, they may not be efficient in terms of assurance or cost efficiency when innovation poses unforeseen challenges.

Today technology advances (e.g. the ever-increasing knowledge and computer speed) often suggest improved possibilities for cost-efficient and theoretical or semi-empirical assessments. This improved ability to assess the performance of a ship, irrespective to the afore mentioned weaknesses of prescriptive rules, initiated a trend toward performance-based standards. This trend is expected to lead to the formation of goal based design rules and operations that may fundamentally change how ships are assured (IMO, 2015b).

2.3 *Polar Code*

The Polar Code (IMO, 2015), is the first international regulatory framework addressing performance driven arctic shipping risks. It comprises of a set of regulations that interlink with various IMO statutory instruments, such as the International Convention for the Safety of Life at Sea (SOLAS), the International Convention for the Prevention of Pollution from Ships (MARPOL) and other binding IMO instruments that account for Arctic specific conditions and hazards (e.g. ice, remoteness and severe weather conditions). In this sense, the Polar Code de-risks polar ship design, operations, construction, equipment, training, and pollution prevention.

The Polar Code is fundamentally performance-based. This is because it helps to determine mandatory provisions in terms of goals, functional requirements FR(s), and regulations that meet those. Whereas regulations are generally prescriptive, the objective of the Polar Code is not to enforce a specific solution, but to ensure that the applied solution meets the FRs and design goals. Thus, a ship can be approved either as a prescriptive design or as an equivalent design (see Figure 1). A prescriptive design assurance process should meet all requirements associate with the FR(s) as stipulated in prescriptive rules. On the other hand, an equivalent design is approved in accordance with Regulation 4 of SOLAS Chapter XIV. This regulation (IMO, 2014), states that '*any solution may deviate from the prescriptive requirements determined by the Polar Code, on the condition that the alternative design meets the intent of the goal and functional requirements concerned and provide an equivalent level of safety as the prescriptive design*'. Accordingly, to

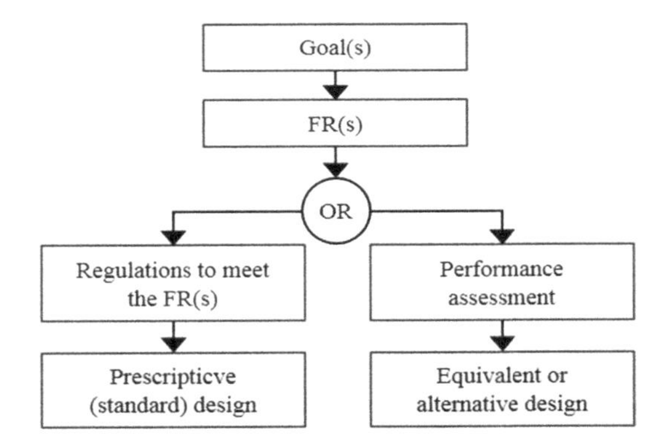

Figure 1. Approval principle in accordance to the Polar Code.

prove equivalency, a design should be analysed, evaluated, and approved on the basis of:

- Guidelines for the approval of alternative and equivalents as per various IMO instruments, MSC.1/Circ.1455 (IMO, 2013a).
- Guidelines on alternative design and arrangements as per SOLAS chapters II-1 and III, MSC.1/Circ.1212 (IMO, 2006a).
- Guidelines on alternative design and arrangements for fire safety, MSC/Circ.1002 (IMO, 2001).

A ship approved in accordance with the Polar Code will be issued a Polar Ship Certificate that classifies her as:

- Category A, when allowed to operate in at least medium thick (> 0.7 m) first-year ice.
- Category B, when allowed to operate in at least thin (≤ 0.7 m) first-year ice.
- Category C, when allowed to operate in ice conditions less severe than those included in Category A-B.

The ice certificate also clarifies in detail operational limits (e.g. minimum temperature, worst ice conditions, etc.).

2.4 *The regulatory challenges of unmanned ships*

Studies carried out by research projects MUNIN (Maritime Unmanned Navigation through Intelligence in Networks) and AAWA (Advanced Autonomous Waterborne Application Initiative) highlighted key legal and regulatory challenges for unmanned ships. This is because existing international maritime conventions do not consider autonomous ships and do not provide a path toward their regulatory approval (MUNIN, 2015b) (AAWA, 2016). Both projects concluded that future regulatory challenges will relate with obligatory crew functions in general and the role of the shipmaster in specific. Currently, the following regulations directly refer to the shipmaster and crew functions (AAWA, 2016) (MUNIN, 2015b) (Raw & Craney, 2017):

- COLREGs (Convention on the International Regulations for Preventing Collisions at Sea), Rule 5: A ship must always maintain a proper lookout by sight and hearing as well as by all available means appropriate in the prevailing circumstances and conditions to make a full appraisal of the situation and of the risk of collision.
- STCW (International convention on Standards of Training, Certification and Watchkeeping for Seafarer), Ch. VIII, Reg. VIII/2: Officers in charge of the navigational watch must be physically present on the navigating bridge or in a directly associated location at all times.
- SOLAS, Reg. 24: The on-board track control system (autopilot) must enable an immediate switch from automatic to manual control. In addition, the functionality of the manual steering must be tested frequently.
- SOLAS, Reg. 33: The master of a ship is required to assist persons in distress at sea.

3 THE 'UNMANNED SHIP CODE' (USC)

3.1 *Main characteristics*

To address specific hazards related to unmanned ships, the proposed Unmanned Ship Code (USC), similarly to the Polar Code, is determined as a supplement to existing rules and regulations. However, USC cannot simply add provisions on the top of those for conventional ships, but must also replace existing regulations; especially those that define obligatory safe crew functions on-board a vessel (see Sec. 2.4).

Given the general limitations of prescriptive rules (see Sec. 2.2), USC is largely performance driven and goal-based. It aims to enable unmanned ship operations and to ensure that unmanned ship operations are at least as safe as (preferably safer than) conventional ship operations. In this sense, USC determines ship system specific sub-goals and related FRs. However, because of general challenges related to the qualification and quantification of safety performance metrics (Bergström, et al., 2016b), USC, similarly to the Polar Code, would rely upon new and established industry standards (e.g. regarding integrated sensor systems) as well as 'common interpretations' of existing rules for manned ships.

A performance-based regulatory approach requires the definition of goals and FRs for various ship functions. Hence, it motivates systems-driven thinking, i.e. it promotes the thinking that a ship is part of the maritime transport eco-system. The ship system may also be divided into a hierarchy of subsystems that perform a specific function that supports her overall function or mission (Bergström, et al., 2016b). This notion drives the design process in a way that considers both functional requirements and limitations of subsystems (Bergström, et al., 2016a) and enables design optimisation early on (Bergström, et al., 2016a).

3.2 *The 'Unmanned Ship Certificate'*

USC suggests that a ship is issued an "Unmanned Ship Certificate". This, similarly to the Polar Ship Certificate would be awarded on the merit of ship-specific operational limitations and constraints determined along the lines of autonomous operations envelope. For example, permitted conditions could be specified in terms of the type of fairway (e.g. open water, narrow fairway), traffic (e.g. vessel density) or the prevailing weather conditions (e.g. wind speed). Whenever a ship encounters a situation exceeding the operational limitations set by the certificate, she would either need to operate at a lower level of autonomy. This implies increased involvement by the Remote Operating Centre (ROC) in charge of the ship. As last resort, each safety critical system would require an integrated 'fail-to-safe' function.

The level of autonomy could be specified in accordance with Table 1. The general approach

Table 1. Autonomy Levels (AL) as determined by Loyd's register (Marine Electronics & Communications, 2016).

Level	Description
AL0	No autonomous functions. All operations are manual
AL1	On-ship decision support. Data will be available to crew
AL2	Off-ship decision support. Shore monitoring
AL3	Active human-in-the-loop. Semi-autonomous ship. Crew can intervene.
AL4	Human-on-the-loop. Ship operates autonomously with human supervision.
AL5	Fully autonomous ship. There is a means of human control.
AL6	Fully autonomous ship that has no need for any human intervention.

Table 2. System specification.

System function	Specification
Unmanned look-out	The goal of this system is to ensure proper lookout at all times. The FRs of the system must be at least equivalent to the function of an on-board human look-out as per COLREGs Rule 5, SOLAS Reg. 24, and the STCW. FRs should be determined both with regards to visual detection of objects ('sight') and the detection of sounds ('hearing'). For remote control, it is enough that the system transfers sufficient visual and sound data to the ROC. For higher level of autonomy, the system should also enable the identification of objects and sounds. Especially the ability to detect and identify small objects (e.g. persons, liferafts) in water in harsh conditions (e.g. high seas, darkness, rain/snow fall) may be difficult yet critical for safety (Rødseth & Burmeister, 2015). The performance of the system could be tested and validated using 'test tracks', along which an unmanned lookout system's ability to detect various objects could be measures. Once well-proven solutions have been identified, these could be described and prescribed in terms of a set of technical standards (e.g. on the required technical characteristics of integrated sensor systems).
Unmanned collision avoidance	The goal of the collision avoidance system is to avoid collisions between the ship and other objects. To this end, the FRs of the system should be determined so that the functioning of the system replaces the collision avoidance actions normally taken by the crew (e.g. the 'Officer of the Watch'). To this end, the system must be able to plan and execute proper collision avoidance manoeuvres. The system depends on input from, and is thus strongly integrated with, the lookout system. It should be noted that the system is only needed for levels of autonomy higher than AL3 as defined by Table 1.
Unmanned Search and Rescue (SAR)	The goal of the SAR system is to provide sufficient assistance to persons in distress at sea. A manned ship, encountering persons in distress, could take measures including: (1) contact and inform SAR services, (2) lower life rafts/lifeboats, (3) recover persons from the water or life rafts /lifeboats, (4) provide shelter and first aid to recovered persons. Out of these measures sufficient input from the lookout system, an unmanned ship could be expected to provide measure 1–2. We assume that measures related with decision-making under uncertain and complex circumstances would always be carried by remote control, i.e., at autonomous AL3 (see Table 1).
Unmanned Cybersecurity	The goal of cybersecurity system is to provide sufficient and holistic protection against all types of cyber hazards. Cyber security is already an issue for conventional ships and the IMO recently issued guidelines on maritime cyber risk management (IMO, 2017). However, cyber security risks are obviously more significant for unmanned ships. Cybersecurity could also be ensured by prescribing a specific cybersecurity standard, combined with appropriate FRs. Such standard should account for "penetration tests" during which human testers try to exploit all the vulnerabilities of a system (Bayer, et al., 2016).
Unmanned Technical reliability	The goal of technical reliability and maintenance system is to ensure sufficient technical reliability of an unmanned ship. Without an on-board crew able to deal with possible on-board technical failures, any technical failure – however minor – may lead to serious consequences (AAWA, 2016). Thus, the risk of accidents caused by technical failures may be considered as one of the most significant challenges for unmanned/autonomous ships (MUNIN, 2015a). FRs that may help to ensure sufficient technical reliability include: (1) condition monitoring devices that enable addressing technical issues before the occurrence of failures; (2) inbuilt technical resilience systems that may limit the consequences of technical failures; (3) backup systems; and (4) fail-to-safe functions incorporated into safety critical systems as last resort (MUNIN, 2015b) (Rødseth & Burmeister, 2015).
Unmanned fire protection	The goal of the fire protection system is to ensure a sufficient level of fire protection. Existing fire protection regulations mainly aim to protect sea-going personnel. On an unmanned ship, the priority of fire protection system would be to protect the ship and her cargo. Accordingly, efficient automated fire extinguishing methods, such as the use of CO_2, should be applied (MUNIN, 2015b).
Unmanned physical security	The unmanned physical security system is used to ensure sufficient security for the ship and her cargo. An important FR of the system is to prevent unauthorised persons from gaining access to critical ship systems and cargo. On a manned ship the shipmaster is responsible for the security of the cargo (and passengers) and armed security personnel might be deployed to fight piracy. An advantage of an unmanned ship is that physical security can be achieved partially by other means as those used on a manned ship. For instance, because there is no need for a traditional deck and superstructure, the access to a ship and her cargo can be made significantly more difficult by passive means such as locks and other protective physical barriers. If such passive measures are insufficient, additional active measures such as remotely controllable anti-piracy 'weapons' (e.g. long range acoustic devices, anti-piracy laser and water cannons) can be installed (Kantharia, 2017c).

to reduce the level of autonomy in challenging situations should not be based on the premise that human involvement is always safer. Instead, operational limits for various levels of autonomy should be determined so that they maximise safety, considering the strengths and weaknesses of technical systems and human operators.

From a legal point of view, we believe there is a significant difference between AL 3-4, implying 'remotely-controlled' operations, and AL 5-6 implying 'fully autonomous' operations. Remotely-controlled operations imply ships with crews that have been relocated to a ROC. In such situations, the crew would still oversee operations and be responsible for the ship, albeit not from on-board the ship. On the other hand, fully autonomous operations imply ships that, at least periodically, do not have any crew. As a result, in case of an accident, caused by a weakness or failure in any autonomous system, there could be a complicated situation in terms of liability (i.e. it might be unclear who is legally responsible for the accident, the system provider or the operator). For this reason alone, it is believed that fully autonomous ships may not be feasible any time soon. Notwithstanding, similarly to manned ships that often operate on autopilot, remotely-controlled ships may periodically be allowed to operate at a higher level of autonomy (Marine Electronics & Communications, 2016).

3.3 *Specific unmanned ship functions*

Alike the Polar Code, USC supplements existing regulations where it is necessary to consider risks specific to unmanned ships (e.g. a heightened risk of cyber-attacks and technical failures), that in turn require the removal of regulatory barriers (see Sec. 2.4). To this end, it addressees the following ship and crew functions:

- Look-out
- Collision avoidance
- Search and Rescue (SAR)
- Cybersecurity
- Technical reliability
- Fire protection
- Physical security

Following the principle of system thinking, each of these functions is to be provided by a specific system. Exemplary functions are described in Table 2.

Continuous autonomous operations (AL4 or higher) would, in addition to the functions listed in Table 2, require an operational and strategic decision-making function for:

- Safe weather routing;
- Planning of complex manoeuvres such as entering/leaving port;
- Determination of safe speed margins that consider prevailing operational conditions (e.g. visibility, traffic density, manoeuvrability, wind, currents, potential navigational hazards and draught).

According to (MUNIN, 2015b), state-of-the-art technology is not able to help with the completion of such tasks.

4 DISCUSSION AND CONCLUSIONS

This paper paves the way toward USC. It addresses existing regulatory barriers for unmanned ship operations. The thinking behind this proposal is based on the recently enforced goal-based Polar Code. Our proposal inter-connects existing rules and regulations for conventional manned ships and suggests to consider unmanned operations along the lines of technical standards. To enable autonomy without compromising safety, USC proposes to translate safety critical crew functions into technical FRs that at least ensure the basis of equivalent level of safety.

Because it is practically not feasible to quantify all types of safety performance we advocate, similarly to the Polar Code, a 'hybrid' regulatory approach i.e. an approach that that is well aligned to multiple paths to approval would present the best way forward. In this way, once there are established technical standard solutions that are known to meet the FRs, approval can be achieved by applying such standards. Also, with regards to performance assessment over time there could be multiple acceptable alternative performance assessment methods to demonstrate that a given FR has been met.

Today the "human factor" is the reason behind most (75–96%) marine accidents (Rothblum, 2000). Because the societal acceptance is higher for accidents caused by humans than for accidents caused by machines, applied technical solutions that enable remote and periodically autonomous control should be robust enough to ensure minimal risk. This is why performance assessment may require validation against real-world data and testing of unmanned ship operations. Limited test operations are already underway or planned at least both in Norway and Finland (The Maritime Executive, 2016) (World Maritime News, 2017). For this reason, the next steps in USC development will entail the definition of goals and FRs for the safe operation and assessment of unmanned ship systems.

ACKNOWLEDGEMENTS

This paper was produced within the frame of the DIMECC project D4VALUE (Design for Value—Value driven ecosystem for digitally disrupting supply chain).

ABBREVIATIONS

AAWA	Advanced autonomous waterborne Application Initiative
AL	Autonomy Level
COLREGs	Convention on the International Regulations for Preventing Collisions at Sea
D4VALUE	Design for Value – Value driven eco-system for digitally disrupting supply chain
EU	European Union
FR	Functional Requirements
IMO	International Maritime Organization
MARPOL	International Convention for the Prevention of Pollution from Ships
MUNIN	Maritime Unmanned Navigation through Intelligence in Networks
Polar Code	International Code for Ships Operating in Polar Waters
ROC	Remote Operating Centre
SAR	Search and Rescue
SOLAS	International Convention for the Safety of Life at Sea
STCW	International convention on Standards of Training, Certification and Watch-keeping for Seafarer
UN	United Nations
USC	Unmanned Ship Code

REFERENCES

AAWA, 2016. *Remote and autonomous ships—The next steps,* Espoo: AAWA—Advanced Autonomous Waterborne Applications Initiative.

Bayer, S., Enderle, T., Dennis-Kengo, O. & Wolf, M., 2016. Automotive Security Testing—The Digital. *Energy Consumption and Autonomous Driving,* pp. 13–22.

Bergström, M., Erikstad, S.O. & Ehlers, S., 2016a. A Simulation-Based Probabilistic Design Method for Arctic Sea Transport Systems. *Journal of Marine Science and Application,* 15(4), p. 349–369.

Bergström, M., Erikstad, S.O. & Ehlers, S., 2016b. Assessment of the applicability of goal – and risk-based design on arctic sea transport systems. *Ocean Engineering,* Volume 128, pp. 183–198.

IMO, 2001. Guidelines on alternative design and arrangements for fire safety. MSC/Circ.1002. London: International Maritime Organization.

IMO, 2006a. Guidelines on alternative design and arrangements for SOLAS chapters II-1 and III. MSC.1/Circ.1212. London: International Maritime Organization.

IMO, 2013a. Guidelines for the approval of alternatives and equivalents as provided for in various IMO instruments. London: International Maritime Organization.

IMO, 2014. Amendments to the international convention for the safety of life at sea, 1974, AS AMENDED. RESOLUTION MSC.386(94). Annex 7. s.l.: International Maritime Organization.

IMO, 2015b. International goal-based ship construction standards for bulk carriers and oil tankers, London: International Maritime Organization.

IMO, 2015. International code for ships operating in polar waters (Polar Code). MEPC 68/21/Add.1 Annex 10. London: International Maritime Organization.

IMO, 2016c. *Implementation, Control and Coordination.* [Online] Available at: http://www.imo.org/en/OurWork/MSAS/Pages/ImplementationOfIMOInstruments.aspx [Accessed 1 December 2016].

IMO, 2017. Guidelines on maritime cyber risk management. MSC-FAL.1/Circ.3., London: International Maritime Organization.

Kantharia, R., 2017c. *Marine Insight. 18 Anti-Piracy Weapons for Ships to Fight Pirates..* [Online] Available at: https://www.marineinsight.com/marine-piracy-marine/18-anti-piracy-weapons-for-ships-to-fight-pirates/ [Accessed 15 12 2017].

Marine Electronics & Communications, 2016. *Creating class procedures for autonomous shipping.* [Online] Available at: http://www.marinemec.com/news/view,creating-class-procedures-for-autonomous-shipping_44157.htm [Accessed 20 2 2018].

MUNIN, 2015a. *D9.2: Qualitative assessment,* s.l.: MUNIN—Maritime Unmanned Navigation through Intelligence in Networks.

MUNIN, 2015b. *D9.3: Quantitative assessment,* s.l.: MUNIN—Maritime Unmanned Navigation through Intelligence in Networks.

Papanikolaou, A., 2009. *Risk-Based Ship Design—Methods, Tools and Applications.* 1 ed. Berlin Heidelberg: Springer.

Raw, J. & Craney, P., 2017. *Ghost ships.* [Online] Available at: http://www.kennedyslaw.com/article/ghost-ships/ [Accessed 10 5 2017].

Rødseth, Ø.J. & Burmeister, H.-C., 2015. Risk assessment for an unmanned merchant ship. *TransNav—The International Journal on Marine Navigation and Safety of Sea Transportation,* 9(3).

Rothblum, A., 2000. Human Error and Marine Safety. Presented at the Maritime Human Factors Conference 2000. Linthicum, s.n.

Rylander, R. & Yemao, M., 2016. Autonomous safety on vessels—an international overview and trends within the transport sector, s.l.: Lighthouse—Swedish maritime competence centre.

The Maritime Executive, 2016. *Norway Readies for Autonomous Ship Testing.* [Online] Available at: http://www.maritime-executive.com/article/norway-readies-for-autonomous-ship-testing[Accessed 10 5 2017].

UN, 1982. United Nations Convention on the Law of the Sea. New York: United Nations.

World Maritime News, 2017. *First Test Area for Autonomous Ships Opened in Finland..* [Online] Available at: https://worldmaritimenews.com/archives/227275/first-test-area-for-autonomous-ships-opened-in-finland/ [Accessed 14 December 2017].

Marine Design XIII – Kujala & Lu (Eds)
© 2018 Taylor & Francis Group, London, ISBN 978-1-138-34076-3

Autonomous ship design method using marine traffic simulator considering autonomy levels

Kazuo Hiekata, Taiga Mitsuyuki & Kodai Ito
University of Tokyo, Tokyo, Japan

ABSTRACT: These days, autonomous ships or remote ships are being developed all over the world. However, development of autonomous ship is at the beginning and in which sea area ship can be automated, and what level of autonomy is needed, should be discussed. Based on systems approach, this study proposes the method to discuss the feasible and implementable specification of autonomous ships. In case study, this study shows the superior designs of autonomous ship from the viewpoint of cost and safety, using developed marine traffic simulator which inputs are autonomy levels of avoidance and lookout. As a result of using proposed method, some Pareto designs of autonomous ship were derived, and proposed method was confirmed to work.

1 INTRODUCTION

1.1 *Background*

These days, many companies and projects, like Rolls-Royce, DNV-GL and MUNIN are developing autonomous or remote ship. Autonomous or remote ships may deliver following advantages.

- Solve shortage of ship crew
- Mitigate work load on board
- Avoid human errors

Especially, in Japanese domestic vessels, shortage and aging of ship crew is serious problem for long time, and the problem is getting worse. More than 40 percent out of 20,258 domestic ship crew in Japan as of October are over 55 years old.

Also, the high turnover rate of ship crew due to poor work environment is a problem. Retention rate of ship crew is roughly 70%, and 30% quits on the way. Even if the salary is high, it is due to work style (3 months on board, 1 month off) and harsh working environment. It is pointed out by interviews with persons involved in the domestic shipping that living environment · meal problem · poor connectivity of internet are main problems.

An idea of autonomous ship was invented to deal with such problems. It is stated that autonomous ship will contribute to some sustainability (Burmeister 2014, Rødseth 2012).

- Economic sustainability: Reduce operating costs (such as personnel expenses)
- Ecological sustainability: Reduce fuel consumption, for example by reducing equipment related to human life
- Social sustainability: Improve safety by avoiding involvement of fatigued crew members in ship maneuvering, and reduce the burden on ship crew by eliminating long voyages by remote systems

1.2 *The importance of developing concept*

Denmark, Estonia, Finland, Japan, the Netherlands, Norway, the Republic of Korea, the United Kingdom and the United States submitted "Maritime Autonomous Surface Ships Proposal for a regulatory scoping exercise" in IMO MSC98 held on June, 2017. In those documents, following matters were agreed and decided.

- To start investigation of laws about autonomous ship
- To start investigation of definition of autonomous ship and autonomy levels
- To identify IMO regulations which preclude unmanned operation, which would have no application to unmanned operation, and which do not preclude unmanned operation

As inferred from above, development of autonomous ship is at the beginning. In order to design autonomous ship, in which sea area ship can be automated (or remote controlled), and what level of autonomy is needed, should be discussed. However, this concept design is difficult because wide range of things should be considered like stakeholders and technology levels. Currently, these kinds of concept design are carried out empirically by teamwork, and framework to design concept is needed.

1.3 *Objective*

In this research, we propose design process for consideration of complex stakeholders and maritime logistics and feasible adoptions of emerging technologies for ship automation by employing systems approach. Specifically, we focus on evaluation of safety in navigation of autonomous ship using a marine traffic simulator.

2 PROPOSED METHOD

2.1 *Overview of proposed method*

The problem of designing the autonomous ship is complex issue. In order to solve this problem, variety of things should be considered. For example, the changes for automation may have unpredictable impacts on other stakeholders in maritime industry. These side effect may decrease the expected benefit. These kinds of problems are caused in sociotechnical systems, which is combination of technology and social system.

This study demonstrates design process aligned with systems approach. Systems approach is a set of methods of system thinking. Crawley defines system thinking as, "*System thinking is, quite simply, thinking about a question, circumstance, or problem explicitly as a system—a set of interrelated entities. System thinking is not thinking systematically*" (Crawley 2016). In this study, the design process consists of stakeholder value network (SVN), Object Process Methodology (OPM) and Morphological Matrix (MM).

Figure 1 shows the overview of proposed process. Proposed method has 5 stages. Simulator is required to predict the system performance in stage 4.

1. Analyze stakeholders and select metrics
2. Analyze conventional ship by OPM
3. Make Morphological Matrix
4. Calculate metrics value
5. Calculate Pareto solutions

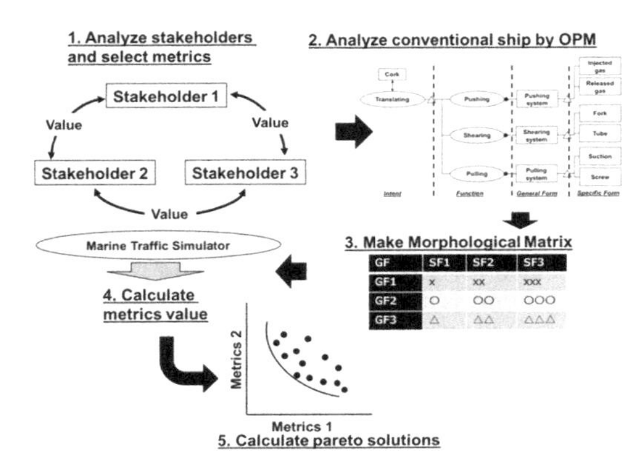

Figure 1. Overview of proposed method.

2.2 *Analyze stakeholders and select metrics*

The 1st stage has 2 steps, stakeholder analysis and metrics selection. Metrics for evaluation of needs are selected in this stage. In metrics selection, we adopt the notion of "ility" which is used in systems engineering. ility is "*requirements of systems, such as flexibility or maintainability, often ending in the suffix ility; properties of systems that are not necessarily part of the fundamental set of functions or constraints and sometimes not in the requirements*" (De Weck 2011).

The 1st step is to analyze stakeholders concerning autonomous ship using stakeholder value network (SVN). Autonomous ship project involves many stakeholders. Each stakeholder has different needs, and some needs has commonality, but some does not. Some needs can be realized by autonomous ship, but some are not and even worse can be disturbed. These needs are recognized implicitly and empirically by each stakeholder, and sharing needs are sometimes omitted. So sometimes contradiction can occur.

Therefore, we try to visualize stakeholders and value flow (needs) around stakeholders by using SVN. Stakeholders and needs are clarified by visualizing. Then, we define the decision maker and primary beneficially in this study. By defining decision maker, design area of autonomous ship is restricted within control area of decision maker. By defining primary beneficially, what needs can be accomplished by autonomous ship projects is clarified.

The 2nd step is to define the metrics to major fulfillment of needs. Cost and ility are ordinary in tradeoff relations. In this study, we choose a ility and some metrics related to that ility and analyze tradeoff space between cost and metrics.

2.3 *Analyze conventional ship by OPM*

The 2nd stage is to analyze conventional ship by conceptual modeling language, Object Process Methodology (OPM). Output of this stage is OPM modeling of conventional ship. OPM can model things in the form of object and process. Rectangle is "Object", oval is "Process", white triangle is "Generalization-Specialization", bidirectional arrows is "Effect link", and black dot is "Agent link". We adopt Crawley's writing style (Crawley 2016).

Figure 2 shows the OPM example of "pump" system (Crawley 2016). According to Crawley, processes and objects can be divided into Intent, Function,

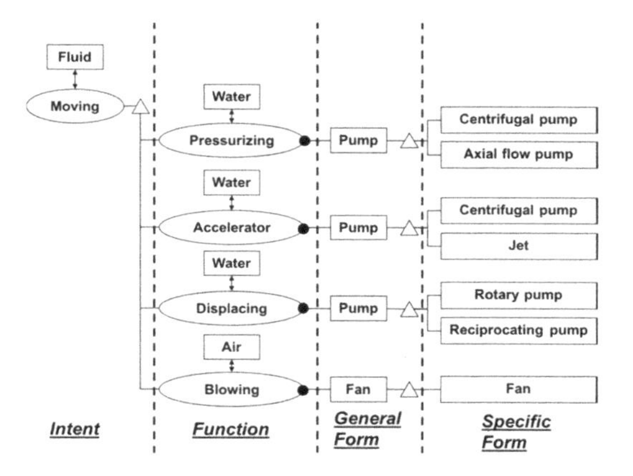

Figure 2. OPM style example (Crawley 2016).

General Form, and Specific Form. Intent is what the system wants to achieve, always present and solution-neutral function. Function is what the system does. Form is what the system is, and can be divided into General Form and Specific Form. "Pump" system's intent is "moving fluid", and this can be realized by selecting one function like "Pressurizing", "Accelerator", "Displacing", and "Blowing". Each Function has its agent (General Form) like "Pump", and "Fun". General Form can be selected by choosing Specific Form like "Centrifugal pump".

2.4 *Make morphological matrix*

The 3rd stage is to list the architectural decisions using Morphological Matrix (MM) which can represent decisions in matrix form. Output of this stage is MM of autonomous ship design options. Making MM looks easy, however its difficulty is mentioned (Ritchey 2003). So, this study utilizes OPM modeling which is output of "2.3 Analyze conventional ship by OPM". We set the problem of selecting Specific Form as General Form using MM like Table 1. Each row represents General Form, and each column represents Specific Form. Autonomous ship design options are represented as selecting specific form in each row. For example, Table 1 has $3^2 = 9$ design options.

2.5 *Calculate metrics value*

The 4th stage is calculating metrics value. Output of this stage is metrics value of each metric. Metrics were chosen in "1. Analyze stakeholders and select metrics". There are some ways to calculate metrics value. Setting the metrics value in each cell of MM and summing or multiplying them is one way. We can also calculate metrics value by using simulator.

Table 1. Example of MM (GF = General Form, SF = Specific Form).

GF	SF1	SF2	SF3
GF1	x	xx	xx
GF2	o	oo	ooo

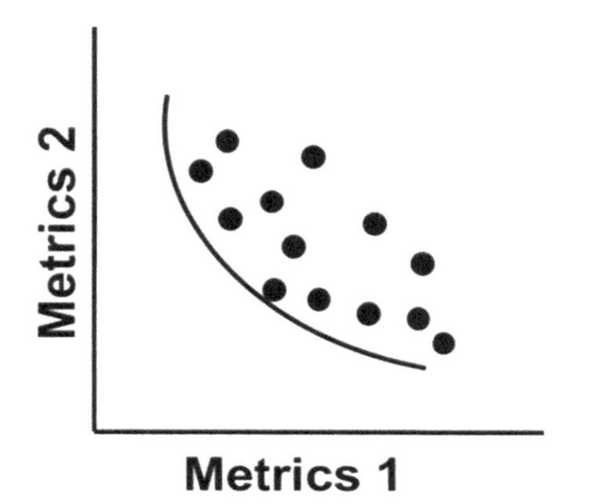

Figure 3. Pareto solutions plotting image (When 2 metrics are chosen).

2.6 *Calculate Pareto solutions*

The 5th stage is plotting each design options according to metrics value and calculating Pareto solutions. Output of this stage is Pareto solutions.

3 CASE STUDY

This case study follows the 5 stages of proposed method. The objective of case study is to apply proposed method to real case and validate proposed method can work. Target ship of this case study is SR108 container ship.

3.1 *Analyze stakeholders and select metrics*

The 1st step is to analyze stakeholders. We mapped stakeholders in maritime industry using SVN by modifying work of Hiekata (Hiekata 2017) (Figure 4). This SVN is made from discussion among maritime researchers, employers of shipping company, shipbuiliding company, and classification society. Shipping company is set as decision maker, and cargo client is set as primary beneficially. Shipping company gives shipping service to cargo client and cargo client returns freight charge. So, the need of the primary beneficially is receiving good shipping service.

The 2nd step is to define the metrics to major fulfillment of needs. De weck listed ilities of system, for example, flexibility, agility, safety, maintainability, etc... (De Weck 2011). In Japan, Ministry of Land, Infrastructure, Transport and Tourism recruited funding projects called i-shipping in June, 2017 (Table 2). 7 projects are selected and they are related with safety, maintainability, reliability, etc..., and all projects are related with safety.

In this study, because it is assumed that safety is serious issue when autonomous ship is realized, we chose safety as ility and collision number as metrics. Trade-space analysis between safety and cost will be conducted in following chapters.

3.2 *Analyze conventional ship by OPM*

Figure 5 shows OPM modeling of conventional container ship. The intent is shipping and shipping affects cargo and ship. Shipping can be divided into following functions, routing, lookout, maneuvering, routing for avoiding hazard, loading cargo, maintaining, networking, accommodating, providing power, berthing/unberthing, IT security, communicating, and life supporting. In order to divide intent into functions, documents (Brown 1998) and interview with worker in maritime industry are referred. Each function has general forms which accomplish functions. In order to choose which specific form as general form, MM will be made in next chapter.

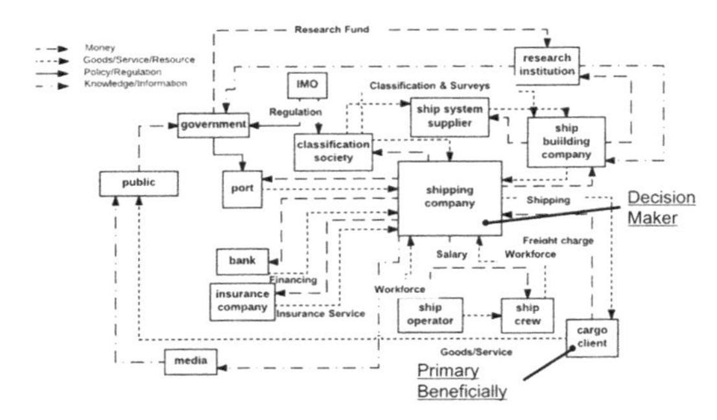

Figure 4. SVN of marine industry.

3.3 *Make morphological matrix*

Table 3 shows MM which represents options of specific form. By selecting one specific form in each general form, one autonomous ship design option can be obtained. In this report, decisions of Lookout system (ID = 2, 3), Rudder controller (4, 5), Collision avoidance routing system (8, 9)

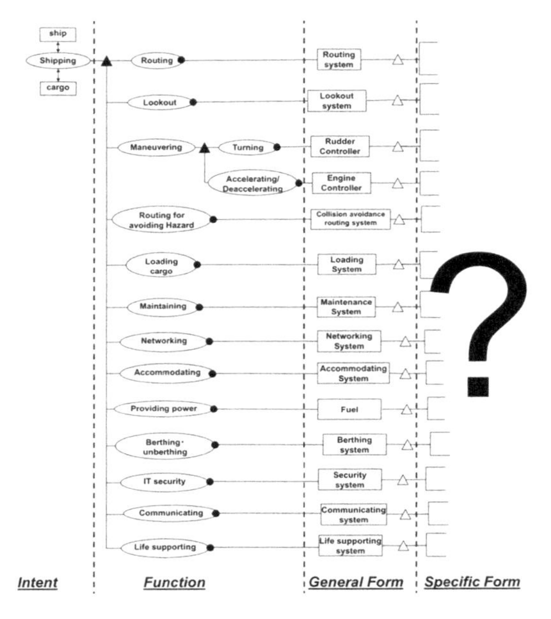

Figure 5. OPM of cargo ship.

Table 2. Projects selected in i-shipping.

Field	Title	Ility
Operation support by maneuvering simulator	Collision risk assessment and autonomous maneuvering	safety, operability, flexibility
	Auto observation and sending system of marine weather	operability, safety
	Improvement of analysis accuracy by auto correcting hull characteristic model	accuracy, operability, safety
	LNG safety transportation support using inter-ship communication	accuracy, safety
Safe design by hull monitoring	Hull structure health monitoring on large containerships	maintainability, reliability, safety
Preventive maintenance of marine equipment and systems	Prevention of engine accidents by utilizing big data	reliability, safety
	IoT conversion of deck machine for cargo ship/bulk cargo ship	reliability, safety

Table 3. MM of autonomous ship (OS: Open sea, CA: Congested Area).

ID	General Form		SF1	SF2	SF3
1	Routing system		Human onboard	Weather routing	
2	Lookout system	OS	Human onboard & Radar & AIS	Human remote & Radar & AIS	No human & Radar & AIS
3		CA	Human onboard & Radar & AIS	Human onboard & Radar & AIS	No human & Radar & AIS
4	Rudder controller	OS	Human onboard	Human remote	Automated
5		CA	Human onboard	Human remote	Automated
6	Engine controller	OS	Human onboard	Human remote	Automated
7		CA	Human onboard	Human remote	Automated
8	Collision avoidance routing system	OS	Human onboard	TCPA·DCPA·Fuzzy	Deep learning
9		CA	Human onboard	TCPA·DCPA·Fuzzy	Deep learning
10	Loading system		Human onboard	Human remote	Automated
11	Maintenance system		Human onboard	Human remote	Automated
12	Networking system		L band	Ka band	Ka & L band
13	Accommodating system		Hotel	Small hotel	No hotel system
14	Fuel		C heavy oil	LNG	Electricity
15	Berthing system		Human onboard	Expert system	Deep learning
16	Security system		High	Mid	Low
17	Communicating system		SCC	Whistle & Radiotelephone	
18	Life supporting system		None	Yes	

Table 4. Restriction among each decision.

Scope	Equation
ID2, 3, 4, 5, 13	(ID13 = = SF1 && ID2,3,4,5 = = SF1)|| (ID13 = = SF3 && ID2,3,4,5 = = SF3)

and Accommodating system (13) will be considered. Collision avoidance system in this study is fixed to TCPA·DCPA·Fuzzy.

However, these decisions are not independent, and there are some restrictions. For example, in accommodating system, hotel system is needed unless onboarding crew is unnecessary. These constraints are written down in a syntax which Simmons advocated (Simmons 2008) (Table 4).

3.4 *Calculate metrics value*

As noted in 3.1, metrics values of cost and safety are calculated according to each autonomous design options from MM. There is restriction between ID2, 3, 4, 5 and 13, and collision avoidance routing system is fixed. So, from MM, there are $3^4 = 81$ autonomous ship design options which we should evaluate.

3.4.1 *Cost*

To evaluate the cost of automation, shipping cash-flow model of Stopford is used (Stopford 2009) (Figure 6). Stopford divides cash flow into revenue and costs, which are consisted from capital costs, operating costs and voyage costs. In this study, ship revenue of autonomous ship is assumed to be the same as conventional ship. We will explore which costs will change because of automation.

Kretschmann considered cost changes of autonomous ship (Kretschmann 2017). Table 5 shows cost changes in operating costs, voyage costs and capital costs. In this study, ID2, 3, 4, 5 in MM affects operating costs and autonomous ship technology in capital costs. ID13 affects capital costs.

From the researches of Stopford and Kretschmann, variation of cash-flow from current conventional ship is calculated according to equation (1)

$$\begin{aligned}&\Delta \textit{Cash flow of design} = \\ &\Delta \textit{Autonomous ship technology cost} + \\ &\Delta \textit{Crew wages} + \\ &\Delta \textit{Hotel system development cost}\end{aligned} \quad (1)$$

The values of equation (1) is set as is shown in Tables 6, 7. The values of Table 6 is set by referring to documents (Kakuta 2007), and the values of Table 7 is from Kretschmann's research (Kretschmann 2017). The cost of autonomous ship design option is calculated by summing values of each cells. For example, when specific form1 is selected from ID2 to 5, and specific form2 is selected in ID13, Δ*Cash flow of design* will be $0 + 0 + 0 + 0 - 175 = -175$.

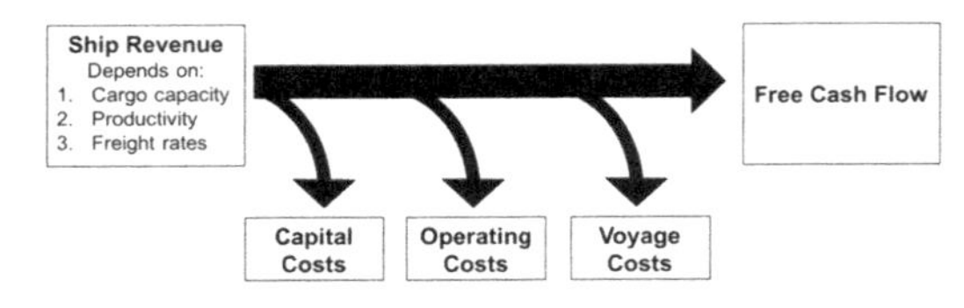

Figure 6. Revenue model (Stopford 2009).

Table 5. Cost changes (Kretschmann 2017) Minus (–) represents a reduction of costs; plus (+) an increase.

Operating costs	Voyage costs	Capital costs
Crew wages (–)	Air resistance (–)	Deckhouse (–)
Crew related costs (–)	Light ship weight (–)	Hotel system (–)
Shore control center (+)	Hotel system (–)	Redundant technical systems (+)
Maintenance crews (+)	Boarding crew for port calls (+)	Autonomous ship technology (+)

Table 6. Value of Δ *Autonomous ship technology cost* and Δ *Crew wages* (10 KUSD).

ID	GF		SF1	SF2	SF3
2	Lookout system	OS	+0 +0	+10 –60	+20 –120
3		CS	+0 +0	+20 –90	+40 –180
4	Rudder controller	OS	+0 +0	+5 –30	+10 –60
5		CA	+0 +0	+30 –30	+20 –60

Table 7. Value of Δ *Hotel system development cost* (10 KUSD).

ID	GF	SF1	SF2	SF3
13	Accommodating system	0	–175	–350

3.4.2 *Safety*

As safety metrics, we select collision numbers calculated by marine traffic simulator which applied to 2 areas, Opens Sea (OS) and Congested Area (CA). The detail of marine traffic simulator is written in Appendix.

3.4.2.1 Basic simulator settings

Table 8 shows the basic settings of simulator. Maneuverability parameter K and T are referred from values of SR108 container ship whose length is 175 m, deadweight is 24801 ton, and draft is 9.5 m.

Table 8. Settings of simulator.

	OS: Open sea	CA: Congested srea
Time step	2 [sec]	
Ship length	175 [m]	
Ship speed	5 [m/s] = 9.7 [knot]	
Area	Square (5 km × 5 km)	Tokyo Bay
Route	Crossing of 2 ship	Keihin port · Chiba port · Isogo · Nakanose route · Uragasuidou route
$\boldsymbol{D_w}$	2000 [m]	
$\boldsymbol{R}$	2000 [m]	
$\boldsymbol{Ds}$	1500 [m]	
minTCPA	720 [sec]	
minDCPA	1200 [m]	
K	0.155 [sec^{-1}]	
T	80.5 [sec]	

3.4.2.2 How to consider autonomy levels

To calculate collision numbers depending MM, simulator is able to calculate different output according to each autonomous ship design.

In this study, new parameters "Lookout miss rate: p" and "Uncertainty of rudder controller: Z" are given to the developed simulator. "Lookout miss rate: p" is probability that TCPA and DCPA won't be calculated. Normally, TCPA and DCPA are calculated every time step. Without updating TCPA and DCPA of ship, some ship could miss other ship and there could be possibility of collision. "Uncertainty of rudder controller: Z" is the parameter that gives randomness—max $\pm Z$ degree—to the course of ship.

"Lookout miss rate: p" depends on the lookout system (ID2, 3) and "Uncertainty of rudder controller: Z" depends on the rudder controller (ID4, 5).

3.4.2.3 Parameter settings and result

Parameters about lookout system and rudder controller are set as shown in Table 9. Each simulation is calculated changing parameter value according to Table 9. When total number of created agents reaches 1000, simulation ends. Table 10 shows the results.

3.5 *Calculate Pareto solutions*

Figure 7 shows plotting of all 81 autonomous ship design options derived from MM. Vertical axis is collision numbers calculated by simulator and horizontal axis is variation of cash flow. Autonomous ship design options located in left bottom are considered better design options. From 81 autonomous ship design options, 8 Pareto

Table 9. Parameter settings of simulator.

ID	SF1	SF2	SF3
2: p (OS)	0.0	0.3	0.5
3: p (CA)	0.0	0.3	0.5
4: Z (OS)	0.0	2.0	5.0
5: Z (CA)	0.0	2.0	5.0

Table 10. Collision number Left: OS/Right: CA.

ID2	ID4	Times	ID3	ID5	Times
SF1	SF1	0	SF1	SF1	1
	SF2	0		SF2	21
	SF3	0		SF3	70
SF2	SF1	0	SF2	SF1	3
	SF2	0		SF2	63
	SF3	1		SF3	53
SF3	SF1	0	SF3	SF1	3
	SF2	0		SF2	68
	SF3	2		SF3	58

Table 11. Breakdown of Pareto solutions (M = MUNIN).

	ID2: Lookout system (OS)	ID3: Lookout system (CA)	ID4: Rudder controller (OS)	ID5: Rudder controller (CA)	Collision number [Times]
A	SF3: No human	SF3: No human	SF3: Automated	SF3: Automated	60
B	SF3: No human	SF1: Human	SF3: No human	SF2: Remote	23
C	SF3: No human	SF3: No human	SF3: Automated	SF1: Human	5
D	SF3: No human	SF1: Human	SF2: Remote	SF1: Human	1
E	SF1: Human	SF1: Human	SF1: Human	SF1: Human	1
M	SF2: Remote	SF1: Human	SF2: Remote	SF1: Human	1

solutions are derived. Table 11 shows breakdown of Pareto solutions, A, C, D, E, design of MUNIN and B in reference.

4 DISCUSSION

4.1 *Which Pareto solutions should be selected?*

Selection of autonomous ship design options depends on utility function of decision maker. However, as an example, let's consider the problem of selecting one Pareto solution out of the above

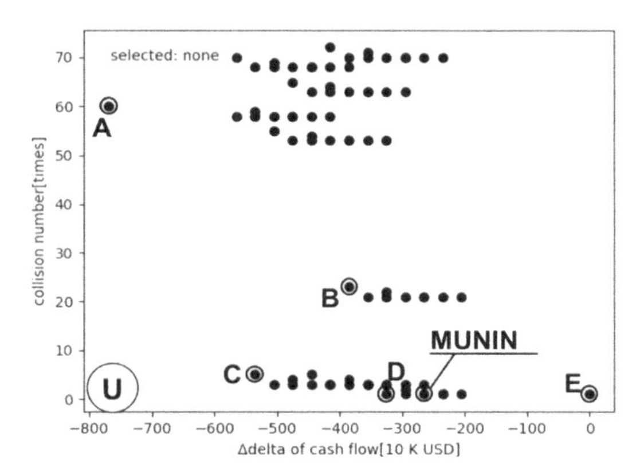

Figure 7. Plot of autonomous ship designs.

A to E. Collision is serious problem, so collision number should be 0 or no more than 1. According to this notion, D and E should be selected. Compared D and E from the view point of cash flow, D should be selected.

4.2 *Accuracy of parameter settings*

There are some discussions about parameter settings. Since the result depends on the parameter setting of the simulator, it is hard to say that the result is completely correct. This is due to the difficulty of modeling the uncertainty of the human system and the uncertainty of the incomplete future autonomous system. However, we think that the framework to design the whole concept works if the correct parameter is set.

4.3 *Comparison with MUNIN*

We plotted concept of MUNIN in Figure 7. MUNIN's concept is on the Pareto frontier. So, in our simulation, concept of MUNIN is concluded reasonable. Also, this can suggest that our result of simulation is somewhat right qualitatively.

5 CONCLUSION

In this research, we proposed design process for consideration of complex stakeholders and maritime logistics and feasible adoptions of emerging technologies for ship automation by employing systems approach. Specifically, we focused on evaluation of safety in navigation of autonomous ship using a marine traffic simulator. As a result, some Pareto solutions are derived and though there are some discussions, proposed method was confirmed to work.

As a future work, all decisions about MM will be considered. Also, we will compare autonomous ship design focusing on other stakeholders, like between shipbuilding company and shipping company.

REFERENCES

Brown, A. 1998, Reengineering the Naval Ship Concept Design Process, *Research to Reality in Ship Systems Engineering Symposium, ASNE, September.*

Burmeister, H.-C., Bruhn, W.C. & Rødseth, Ø.J. 2014. Can unmanned ships improve navigational safety? *Proceedings of the transport research arena 2014*, Paris, France Grove.

Crawley, E., Cameron, B. & Selva, D. 2016. System Architecture, Boston, the United States of America, Pearson.

De Weck, O., Roos, D. & Magee, C. 2011, Engineering Systems, *Cambridge, the United States of America, The MIT Press.*

Hasegawa, K., & Kouzuki, A. 1987, Automatic Collision Avoidance System for Ships Using Fuzzy Control (in Japanese), *Journal of the Kansai Society of Naval Architects, Japan, Vol. 205, pp. 1–10.*

Hiekata, K., Mitsuyuki, T., Ueno, R., Wada, R., & Moser, B. 2017, A Study on Decision Support Methodology for Evaluating IoT Technologies Using Systems Approach. *Proc. of the 18th International Conference on Computer Applications in Shipbuilding, Vol. 2, pp. 7–15.*

Iwasaki, H. & Hara, K. 1986, A Fuzzy Reasoning Model to Decide the Collision Avoidance Action (in Japanese), *The Journal of Japan Institute of Navigation, Vol. 75, pp. 69–77.*

Kakuta, R., Yamato, H. & Ando, H. 2007. Evaluation Method of Bridge Team Performance by Using of Simulation (in Japanese), *Journal of the Japan Society of Naval Architects and Ocean Engineers, Vol. 5, pp. 47–55.*

Kretschmann, L., Burmeister, H.-C. & Jahn, C. 2017, Analyzing the economic benefit of unmanned autonomous ships: An exploratory cost-comparison between an autonomous and a conventional bulk carrier, *Research in Transportation Business & Management, (in Press).*

Nomoto, K. 1964, Ship Maneuverability (in Japanese), *Text of the 1st Symposioum on the Ship Maneuverability, the Society of Naval Architects of Japan.*

Ritchey T. 2003. Modelling Complex Socio-Technical Systems Using Morphological Analysis. *Swedish Parliamentary IT Commission, Stockholm, December.*

Rødseth, Ø & Burmeister, H.-C. 2012. Developments towards the unmanned ship. *Proceedings of International Symposium Information Ships—ISIS 2012*, Hamburg, Germany, August.

Simmons, W. 2008, A Framework for Decision Support in Systems Architecting, *Doctoral thesis, Department of Aeronautics and Astronautics, Massachusetts Institute of Technology.*

Stopford, M. 2009. Maritime economics. *London, United Kingdom, Routledge.*

APPENDIX (DEVELOPED MARINE TRAFFIC SIMULATOR)

Overview of marine traffic simulator

We programmed multi-agent based marine traffic simulator. Marine traffic simulator is used in 2.5. Simulator calculates metrics value by changing the output. For example, output (metric) will be collision numbers if selected ility is safety, or will be distance traveled if selected ility is economy. Simulator is programmed by agent based modeling language NetLogo. Figure 8 shows developed marine traffic simulator which area is Tokyo bay.

Simulation flow

Figure 9 shows the simulation flow. First, the simulator starts. Then, agents (ships) are created. Next,

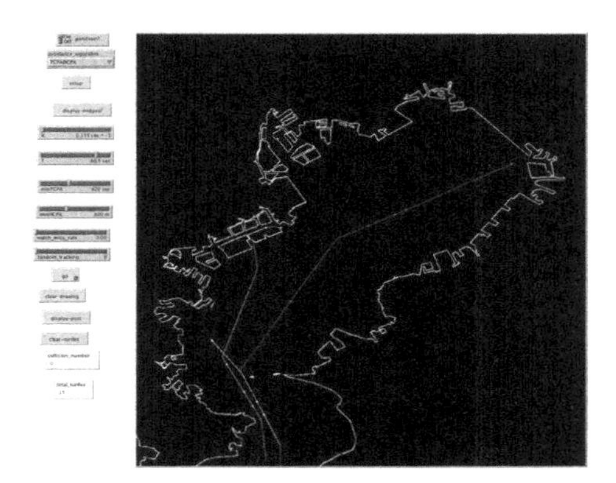

Figure 8. Capture image of simulator (Tokyo bay area).

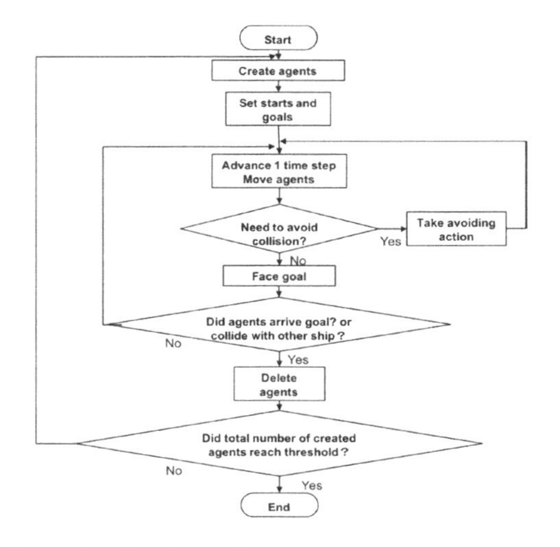

Figure 9. Simulation flow.

simulator sets starts and goals to each agent. Next, simulator advances 1 time step and moves agents. Agents proceeds straightly to the directions which it faces. Next, each agent checks the need to avoid collision. This avoiding rule is written in "Avoiding rule and avoiding action" in appendix. If the agent has to avoid collision, it takes avoiding action. This action is also noted in "Avoiding rule and avoiding action" in appendix. If no, it faces goal. Next, if agents arrive goal or collide with other ship, agents are deleted. Lastly, if the total number of created agents reaches the threshold, simulation ends.

Basic navigation rule

Each ship navigates in a route. Route is constructed of sum of waypoints. Ship navigates straightly between waypoints. The timing of changing from the waypoint currently set as the target to the next waypoint is assumed to be when the distance to the target waypoint is within D_w m.

Avoiding rule and avoiding action

In this study, 2 types of avoiding cases are considered, one is with static objects like coastline, the other is with other ship. With static objects, ship turns right 10 degree when there is obstacle in D_s m.

With other ship, TCPA (Time to Closest Point of Approach) and DCPA (Distance of Closest Point of Approach) are used. When there are multiple ships in the simulation area, TCPA · DCPA of all the pair of ships should be calculated. However, it takes a lot of time, so in this study, TCPA · DCPA of owned ship and other ship within R m are calculated. TCPA · DCPA are calculated from these equations.

$$\text{DCPA} = \frac{D\left|V_0 sin\alpha + V_t sin\beta\right|}{\sqrt{V_0^2 + V_t^2 + 2V_0V_t\cos(\alpha-\beta)}}[m] \tag{2}$$

$$\text{TCPA} = \frac{D\left(V_0 cos\alpha + V_t cos\beta\right)}{\sqrt{V_0^2 + V_t^2 + 2V_0V_t\cos(\alpha-\beta)}}[sec] \tag{3}$$

V_0: Owned ship speed [m/sec].
V_t: Target ship speed [m/sec].
D : Distance between 2 ships [m].
α : Target ship azimuth from owned ship [rad].
β : Owned ship azimuth from target ship [rad].

Now, we can calculate TCPA · DCPA, however it does not tell us what TCPA · DCPA are dangerous and when we should take avoiding action. Iwasaki et al (Iwasaki 1986) defines this threshold

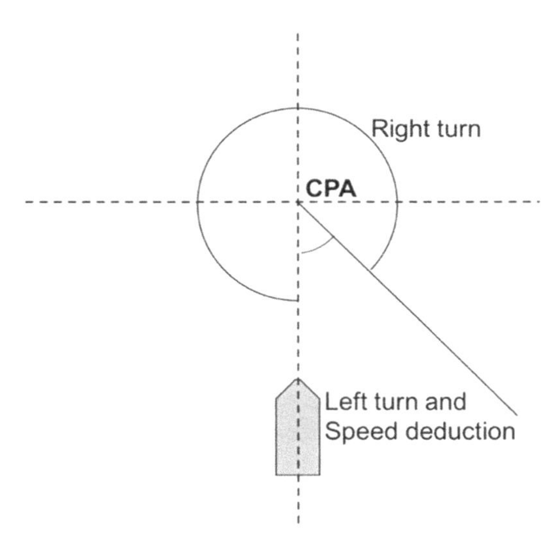

Figure 10. Changing course rule of ship.

using fuzzy rule. In this study, we use partly their method. When TCPA · DCPA becomes NM (Negative Medium: Little time left to avoid), owned ship changes course for 10 degree. This TCPA and DCPA is called *minTCPA* and *minDCPA* in this study. Turning direction is decided from azimuth in CPA (Closest Point of Approach) (Hasegawa 1987). When target ship is in 135°~ 180°(Travel direction is set as 0° and right is positive), owned ship turns left and deduces speed. When target ship is in other direction, owned ship turns right.

This avoiding decision is made for every time step. When TCPA · DCPA goes beyond threshold, ship stops avoiding action and faces goal.

Dynamics of ship

First order linear KT model (Nomoto 1964) is used for dynamics of ship for simulator. r is angular velocity, δ is rudder angle, K, T is maneuverability index. Each ship defines δ in response to environment around them, and r is calculated according to equation (4).

$$T\dot{r} + r = K\delta \tag{4}$$

Marine Design XIII – Kujala & Lu (Eds)
 ISBN 978-1-138-34076-3

Toward the use of big data in smart ships

D.G. Belanger, M. Furth, K. Jansen & L. Reichard
Stevens Institute of Technology, Hoboken, New Jersey, USA

ABSTRACT: The evolution towards "smart" vehicles, including ships, is well under way. It is driven by the promise of many valuable applications, but made possible, in large part, by the revolutions in technologies such as Big Data, Machine Learning, and Artificial Intelligence. We are a few decades into the production application of Big Data in industries such as Telecom, Finance, and Media; and nearly a decade into advanced development in areas such as Healthcare, Autonomous Vehicles, and Smart Grid. Shipping involves some unique challenges in networking, analytics, and implementation, but some of the fundamental tasks will be similar. We focus on the issue of Big Data, and on the essential role of understanding the data that exists, the data that will be needed, and, through careful metadata creation and management, the fitness of that data in terms of properties such as: quality, latency, structure, volume/velocity, provenance, and others.

1 INTRODUCTION

The 21st century will be the century of Internet of Things, Smart Technology and Big Data. This development is being pioneered by Telecom, Financial, Media, Automotive, Manufacturing, and Healthcare Providers, many of whom have been involved in this journey for over a decade. It is a path the shipping industry must take, however, the journey towards smart ships will be incremental. Technical knowhow, regulatory aspects and the attitude of ship operators and crew must be taken into account. In some respects, the journey will not be easy. Shipping is an old industry built on several thousand years of experience, the industry can, in many aspects be said to be conservative. Furthermore, ships frequently operate in some of the most remote, hostile, and least connected parts of the world. In another sense, it can be significantly accelerated by learning from other participants.

Big Data, combined with the machine learning and artificial intelligence (AI) engines to use the data, is an essential component of the evolution to Smart Technologies. It refers to the amount, latency (time it takes to store or access data), and structure of the data available to be analyzed. It has already been responsible for allowing many industries, e.g. Telecom, Finance, Media, Retail, and Transportation, to create applications that have fundamentally changed operations in those industries. It will do the same for the Shipping Industry, as has been the case in other industries, by allowing large amounts of timely data to uncover applications previously too costly, too hard, or just impossible. Big Data will shift the analysis to looking for patterns in the data that cannot be found without processing massive amounts of data. This type of analysis will make it possible to detect events much earlier than existing techniques, allowing for an increase in performance, an optimization of maintenance, and an enhancement of design technology.

This project focuses on the first steps in this journey. The foundation of any data analysis is the structure, availability, timeliness, and quality of the data. Essential to this is the governance of data, and, in particular, the creation and management of high quality metadata. This controls the syntax, semantics, quality control, and access to the data, enabling the data integration and, in some cases, sharing, that is essential to production use of information flowing from the data. Ships are already generating a large amount of data ranging from noon reports, data collated by the Voyage Data Recorder (VDR) to data generated by the engine. In this paper, metadata in the marine environment is discussed with the aim of providing guidelines for an effective metadata protocol, taking advantage of knowledge leveraged from other industries and international consortiums, while focusing on the unique challenges of the marine industry.

The use of big data can facilitate a shift from time based maintenance to condition based maintenance. By analyzing the data currently collected on-board, and employing advanced algorithms and machine learning, this data can be used for better condition based scheduling of maintenance. Current sources of data can be used to build and test predictive, and ultimately proscriptive, maintenance algorithms. From there, it can be determined which data sources are the best predictors of failure, wear, or breakdown. This can also inform the optimal use of advanced sensors, both in terms of value of their data and of cost. With better metadata and connectivity, alerts and alarms can be generated in real time, improving management of vessels while at sea.

The remainder of this paper is divided into four sections. Section 2 is an introduction to essential concepts in Big Data, with particular emphasis on Metadata and Metadata Management as it applies to the data thrown off by the operations of ships. Section 3 focuses on specific data, and its related metadata, that exists on ships today, and that might be required on ships as we move toward smart ships. Sections 4 and 5 consider the implications of this data, now and in the future, with a focus on data strategies to be implemented at the design phase in order to move towards smart predictive maintenance and ship architecture and design.

2 BACKGROUND ON BIG DATA AND METADATA MANAGEMENT

2.1 *Big data*

The journey to "smart" technology in fields that involve Cyber Physical Systems (CPS) is one in which data can be used to automate many of the activities currently requiring skilled humans, and inform activities which are beyond human ability in terms of volume and/or velocity/latency. The technologies required nearly always include: sensing, movement, storage, and analysis/visualization of the data relating to the underlying system, in our case a vessel. Each of these is a significant challenge in itself. On most current vessels there are many sensors, but probably far fewer, and perhaps less sophisticated, than will be needed for smart operations. Large ships are a challenging environment for the networking required to move data to computers for integration, analysis, and control (i. e. action). This is due not only to the materials that make up a ship, but their size and complexity, and the sometimes hostile and distant locations in which they operate. Depending on the analytics deployed for various operations, the data may be required to be sent to central locations, or be analyzed on board (edge/fog analysis (Bonomi, 2012)).

Whereas in some smart applications such as smart cities, smart grid, smart networks, all of which have been extensively studied, the greatest challenge is deployment of adequate sensors to gather the data, on ships the greatest challenge may be moving the data to the right place within the required latency constraints. This is likely to require careful attention to the movement and collection of data, and given the likely application of "Big" Data, careful understanding of the properties and uses in the complex data ecosystem. For example, if automated systems are used to control the steering and speed of a ship, including in congested and/or dangerous areas, they will require a huge amount of data from many sources, e.g. radar, GPS, cameras, and will require sufficient bandwidth in associated networks to deliver and analyze that data in subseconds. An analog would be the approach of an autonomic automobile to a congested intersection. Most of the required analysis will need to be done onboard, but some may also be done remotely.

By now, the underlying concepts of Big Data are well known, starting from the definitions and going through to the many applications using networking, machine learning, artificial intelligence, and the related technologies. The classic definition is data that has one or more of these three characteristics: Volume, Velocity, Variety. Another frequently used definition is "data that exceeds the capability of commercial tools" (Manyika, 2012). For the purposes of this paper we will use a somewhat expanded definition which includes not only the properties of the data, and the properties of the technology needed to manage and manipulate it, but the properties and types of applications enabled by Big Data. These include applications which have been infeasible using existing techniques.

As we consider the evolution of Big Data Techniques in industries that have been involved in these technologies for well over a decade, and, in addition, have made substantial progress in "smart" technology, for example: Electric Power (Smart Grid), Telecommunications (Intelligent Networks) (Kalmanek, 2009), and Finance (Algorithmic Trading), we see that such applications are seldom only supported by a single "big" source of data, but include a large variety of data sources, with very different structural properties, integrated together. These data sources often exemplify one or more of the "V"'s, but just as often have, for example, very different sizes, update cycles, and structure. There are several application types that are often associated with big data, and which, with traditional data, have proven; too costly, too difficult, or simply impossible. These applications types include those in Table 1.

Within and between vessels, applications will often be associated with real time discovery of rare events such as structural weakening, possible collisions, or imminent machinery breakdowns, using sensors, including visual and audio sensors, to uncover issues much sooner than otherwise possible, and enable deep automation of current processes. They will often be associated with complementary technologies such as machine learning, artificial intelligence, and visualization. Equally, they often depend on effective application of networking technology and the use of Edge and/or Fog Computing.

Table 1. Big data applications.

Change	Classical	Big data
Granularity	Transactional or Aggregate	Atomic, Personalized
Rare Events	Sample Based	Population, Signatures, Classification
Broad Search	Indexed Retrieval, Flat Files	Structured and Unstructured data
Relationships	Customized	Graphs and (social) Networks
Analytics at Scale	Sample	Population
Personalized Traits	Survey based	Behavioral
Process Control	Silo'ed	End to End, Feedback Control
Recommendations, Learning	Rules	ML, Collaborative, Relationship based
Transparency	Within Silo	Across Processes
Prediction	Limited, Aggregate	Real time, Specific
Control	Customized, Rule Based	Learning Based

All of these techniques are embedded in "smart" vehicle technology (Keertikumar, 2015), and will certainly be true of smart vessel technology.

This project focuses on the first steps in this journey. The foundation of any data analysis is the structure, availability, timeliness, and quality of the data. Essential to this is the governance of data, and, in particular, the creation and management of high quality metadata. Metadata controls the syntax, semantics, quality, timing, and access to the data, enabling the data integration and, in some cases, the sharing that is essential to production use of information flowing from the data. Ships are already generating a large amount of data including, for example, noon reports, data collated by the Voyage Data Recorder (VDR), and data generated by the engine room. In this paper, metadata in the marine environment is discussed with the aim of providing guidelines for an effective metadata protocol, taking advantage of knowledge leveraged from other industries and international consortiums, while focusing on the unique challenges of the marine industry.

2.2 *Metadata management*

Given the size, complexity, and diversity of the data that is currently available, or may become available, on commercial ships, it will be essential to create a map of data, including not only the names and definitions of various data fields, but their semantics, relationships, timing, quality, and other properties. This activity is typically thought of as a subset of data governance known as Metadata Management, and is the subject of various standardization efforts as well as a common practice in any serious data management activity (ISO, 2014).

Metadata is data that describes other data. It includes schemas for structured data, usually in relational data structures. It also includes logs of activity, and quality, integrity, timing, semantics, type (i.e. alphanumeric, audio, text, image, and video), access, security/privacy risk, and many other traits of the

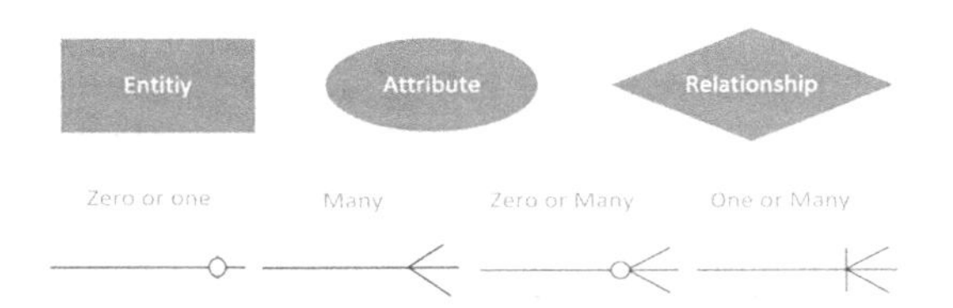

Figure 1. Basic entity relationship shapes and types.

data. It is essential to effective management of the data asset, but also to the effective use of the data in applications. As a simple example, it is often the case that large, high velocity datasets must be integrated with much smaller datasets most of which are updated by transaction and only occasionally. These activities must be effectively synchronized, hence update timing is essential to correct use of the data. A specific example is the analysis, and setting off alerts/alarms when necessary, on the health or security of a network, including that of a ships network. Today many of these networks use Internet Protocol (IP), hence health analysis will include very large flows of IP Header or Session Data. In addition, however, there are databases which evolve on completely different timescales. For example, the owners of communicating IP Addresses (e. g. for White or Black Lists), or expected Ports for various communication types (e.g. Port 80 for http web traffic), or if using Network Address Translation (NAT) to increase IP Addresses for security or network management.

Understanding the available data through a careful mapping of metadata will be essential to creating smart applications for vessels, and the start of that activity is the purpose of the remainder of this paper. It will guide better understanding of the data that is available for applications; the networking, data management, and analytic/visualization technologies that are appropriate for the data; and a better understanding of the data resources that should be added. Various standards have been created or are evolving for this activity (nist.gov, 2017).

In aid of this activity, there are several tools that support the management of metadata. In succeeding sections, we give some specific examples of them as an introduction to the mapping of a vessel's data resource now, and as anticipated. The most traditional of these tools is Entity Relationship Diagrams (Chen, 1976), originally developed for design of relational databases, but now evolved to cover most types of data structures. ER Diagrams are meant to make explicit the relationships between various attributes, entities, and datasets (Figure 1). For example, the shapes in Figure 1 are sufficient to express simple elements in the design of a relational database (entities, attributes, relationships), and the nature of some of the relationships. They are essential to ensure quality related to referential integrity, that is to ensure that elements which are functionally related to each other change according to that relationship. This is fundamental in relational data structures, known as normalization, but at a more abstract level, essential to the entire data fabric. Of course, there are now a large number of such data and relationship types, and on-line tools to help draw them (Lucidchart.com, 2018).

Other helpful tools include several forms of visualization which will be seen in succeeding sections specifying such items as data latency, data quality, risk, volume/velocity, and others. We have left out from this discussion the collection of more advanced tools associated with the semantics of the data, for example semantic webs. This is essential, but the subject of a further paper. A reference to this area is (Lee, 2001).

With the various tools now available for description of data, we expect that control of the data asset aboard a ship, or class of ships, can be understood in detail, managed optimally including across heterogeneous suppliers, and organized to support the evolution to smart ships. The succeeding sections in this paper are a start in details of that journey.

3 SHIP METADATA

3.1 *Currently available data*

Shipping produces much data on a daily basis, data that could be very valuable for the design process and maintenance of ships when captured and analyzed accordingly. Figure 2 shows a basic schedule of how data produced onboard is being used. Examples of data currently available include daily noon reports and VDR data.

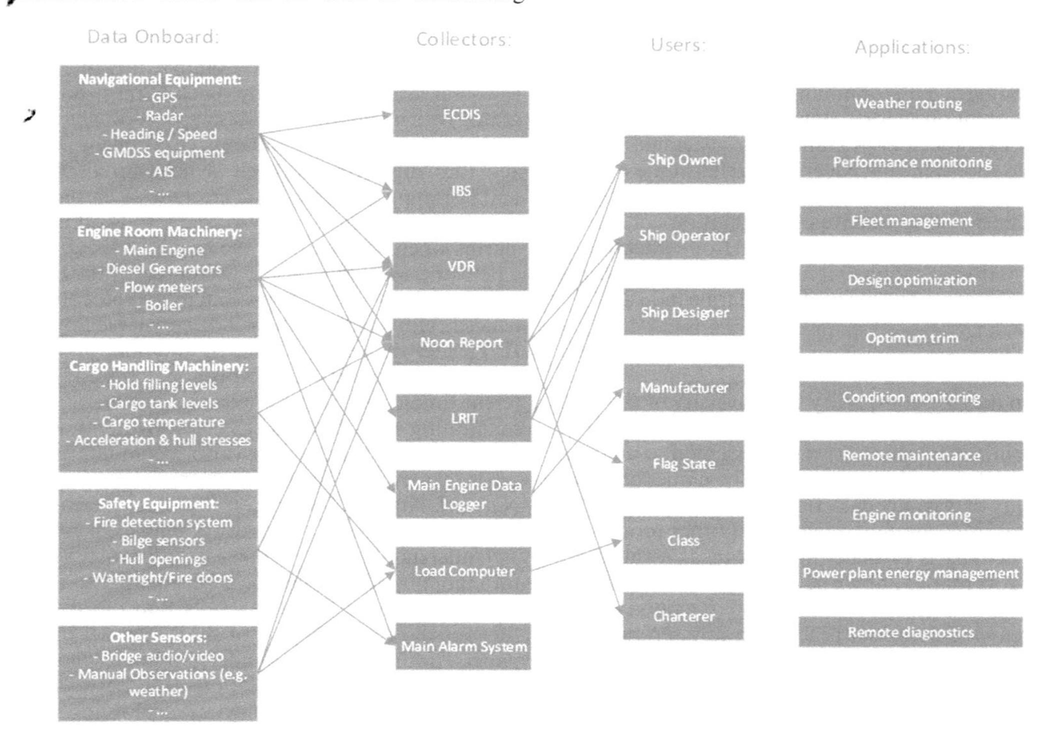

Figure 2. The role different data sources play for design, operation, maintenance, rule compliance and their required latency today.

3.1.1 *Noon reports*

A noon report is a collection of ship and voyage data that is sent to the shipping company, ship owner and charterer on a daily basis. This is usually done around noon, which dates back to the time when celestial navigation was used and noon was the only time seafarers could accurately determine their position. If the report cannot be sent at noon, it will be saved and sent as soon as ship to shore connection is restored.

The report provides the vessel's position and other relevant data to assess the health and performance of the ship based on its speed, course and environmental forces including weather conditions. The noon report generally contains at least the following:

- Name or Call sign of the ship
- Voyage number
- Date and time
- Position of the ship
- Average speed since last submitted report
- Propeller Slip
- Average RPM
- Wind direction and force
- Sea and swell condition
- Distance to destination
- Estimated Time of Arrival
- Remaining on board (R.O.B)
    - o Fresh Water
    - o Fuel Oil (HSFO & LSFO)
    - o Low Sulphur Marine Gas Oil (LSMGO)
    - o Lube oil for ME
    - o Lube oil for Generator
    - o Hydraulic oil

Some information in the report is specific to vessel type such as information about the cargo. For example, for a ship with liquid cargo the cargo density is of great importance and will be included in the noon report. However, this does not apply to general cargo ships.

There are various ways to generate the report. Bigger shipping companies usually have an in-house program that gathers most data automatically and the use of a data collecting program means that only some information need to be manually added by the captain, such as the weather condition, stoppages and special remarks. Smaller companies on the other hand often generate their noon reports by filling in a PDF, excel form or a text file which is then sent to shore by means of email.

The metadata differs between different shipping companies. Even something as basic as the vessel's name and identification are often placed in varying locations within the noon reports. Apart from the specific order of the information, there are discrepancies in various notations.

Firstly, the notation for the date used by Americans (mm/dd/yyyy) versus the notation used by the rest of the world (dd/mm/yyyy). Another example of this aspect is the diverse decimal separator being used in different countries, while some use a comma, others use a dot. Accuracy in the amount of decimal numbers is also an inconsistency, so is the use of either a percentage or a percentile. Similarly, the GPS position (latitude & longitude) is currently being sent in a variety of formats: degrees or decimals, N/S and E/W placed in front or behind the numbers. Wind and sea direction (090° or East), wind force (Beaufort scale or m/s) and sea height (meter or feet) are some further examples of other unstandardized unit notations. A very elementary thing like the ship's time can be written in numerous formats (Greenwich time, local time, 0700 PM or 1900).

Finally, the noon reports contain numerical data as well as written information. Examples of written information are the Captain's name, departure/arrival port, weather conditions, stoppages and general remarks.

3.1.2 *VDR*

The Voyage Data Recorder (VDR) is required onboard ships and, like a black box on an airplane, continuously records vital information related to the operation of a vessel and can be used for reconstruction of the voyage details and vital information during an accident investigation (IMO, 2016). The data is collected in a recording medium consisting of three parts: the fixed recording medium, float-free recording medium and long-term recording medium. The VDR is equipped with the software program needed to retrieved the stored data for playback. An external laptop can be connected and the data can be copied onto a portable storage device. Some ships are currently sending this data to external companies that will, for example, analyze it for bridge communication improvement options.

The data collected by the VDR includes:

- Date and time
- Ship's position
- Speed
- Heading
- Bridge audio
- Communications audio
- Radar
- Electronic Chart Display and Information System (ECDIS)
- Echo sounder
- Main alarms
- Rudder order and response
- Engine and thruster order and response
- Hull openings status

- Watertight and fire door status
- Accelerations and hull stresses (when the ship is fitted with hull stress and response monitoring equipment)
- Wind speed and direction
- Automatic Identification System (AIS)
- Rolling motion (if an electronic inclinometer is installed on ship)
- Configuration data
- Electronic logbook (when fitted)

3.2 *Structured vs unstructured data*

Most data listed in Figure 2 falls into the category structured data. That is, information with a high degree of organization. Text files for example, displayed in titled columns and rows. As most data will be a numerical value with a time stamp, it is categorized as structured data. However, it would be beneficial if the format of each individual software or system would be converted into a standardized format.

Examples of unstructured data currently produced on board include bridge audio and or video that is collected by the VDR. Also, recorded VHF radio communications fall into this category. The VDR also stores images of the ECDIS and radar display so that in case of an investigation after an emergency the information available to the officer of the watch can be retrieved. Future possibilities include video and or audio used for health monitoring. For example, drones can be used to inspect the ships structural status, even in spaces inaccessible for humans. A broken bearing could be detected by audio before any other sensors would have noticed this defect.

3.3 *On-board data collector*

The VDR and noon report are two examples of on board data collectors. However, there is much more data produced on board a ship, all by their individual software or system. For example, the main engine data logger collects values of the engines' pressures, temperatures, RPM, etc. This data is send to the manufactures who uses it for remote monitoring and diagnostics in addition to optimizing future design. Another data collector is the integrated bridge system (IBS) that collects all relevant navigational information and displays it on bridge so is can be used by the offices of the watch for direct decision making. This data is currently not sent to shore for analysis or monitoring, however, this could provide valuable insight on ship performance.

Instead of having all separate collectors, it is desirable to create one 'master collector' that retrieves all data from on board equipment. As all systems and software use different formats, a converter would be needed to change it into a standardized format. Communications between on board equipment and master database will be done by means of the ships LAN connection. From the master collector the data will be send to shore, where it can be distributed to the different users. Of course, not every user needs the whole data set, but the standardized format will make separating the data more convenient. Possible applications range from navigational support, remote maintenance and condition monitoring.

3.4 *Connectivity*

Sending data from the ship to shore can be done by physically saving the data to a portable usb-drive and posting it when the vessel is in port, by using a terrestrial mobile system when in reach of land or, through satellite communication when at sea.

The first option is used when the data is large and the information is not time sensitive. An example is data collected by the VDR that is to be sent onwards for analysis by an external company. The terrestrial mobile systems can be used when in reach of land, for example 3G coverage ranges about 10 nm.

Some information, like the noon reports, needs to be sent on a more regular basis. At sea, the internet access is provided through satellite communication. There are different systems that can provide a ship with an internet connection. Both Inmarsat-C and Iridium are Mobile Satellite Systems (MSS) that operate on L-band and can provide a few hundred kbps (marinesatellitesystems.com, 2014, Lag, 2015). Inmarsat-C can be part of the Global Maritime Distress and Safety System (GMDSS) equipment of the ship, which makes it the dominant player using L-band. Iridium has the advantage that it provides coverage in the Polar Regions, other systems only cover up to approximately 70° latitude. Due to limited available spectrum on L-band, data packages must be limited in size and prices are relatively high. Very Small Aperture Terminal (VSAT) is another system that operates on C-, Ku- and Ka-bands, where there is a large spectrum available. This makes the data rates higher and costs lower than for L-band systems. Data speeds provided by a VSAT are typically up to 6 Mbit/s. Compared to Inmarsat-C and Iridium, the throughputs of VSAT are ample. The cost for a VSAT subscription depends on the size of the satellite dish, coverage and bandwidth. A VSAT antenna needs an accurate and reliable stabilization and tracking mechanism, which causes high installation costs.

There are several vendors working on developing new systems, Inmarsat Global Xpress and Iridium Next have both launched new satellites in the past few years also using Ka-band but still having L-band as a backup. Another trend is the use of communication brokers. A communication broker is capable of using several communication types and selects which one to use depending on availability, urgency, priority, data size, cost and bandwidth.

Altogether, a proper satellite connection is costly and the owner of a ship should be willing to invest in the installation and usage of bandwidth. This is an important factor to keep in mind when considering transferring more data from ship to shore. However, it should be noted that most ocean-going vessels are fitted with a satellite system of some sort. Coastal vessels on the other hand will depend on terrestrial mobile systems.

3.5 *Moving towards smart ships*

Latency is the delay in delivery of a signal across a network, and can include required activities such as database access and analytics. For example, if you are using a search engine on the web, latency would be the time between your clicking "search", and the time the results are returned to you. Search engines target searches to have latency in the very few seconds, with standard deviations in the 10 s of milleseconds. Of course, in Cyber Physical Applications, such as shipping, there are some activities that have hard time constraints. Often these are related to automatic control of machinery, including power and steering. In smart ships, increasing amounts of control will have no human intervention. It is therefore essential that the timing of everything in the data to control pipeline is well understood. The role different data sources plays and may play in the future to the ship design process is shown in Figure 3 and Figure 4 respectably.

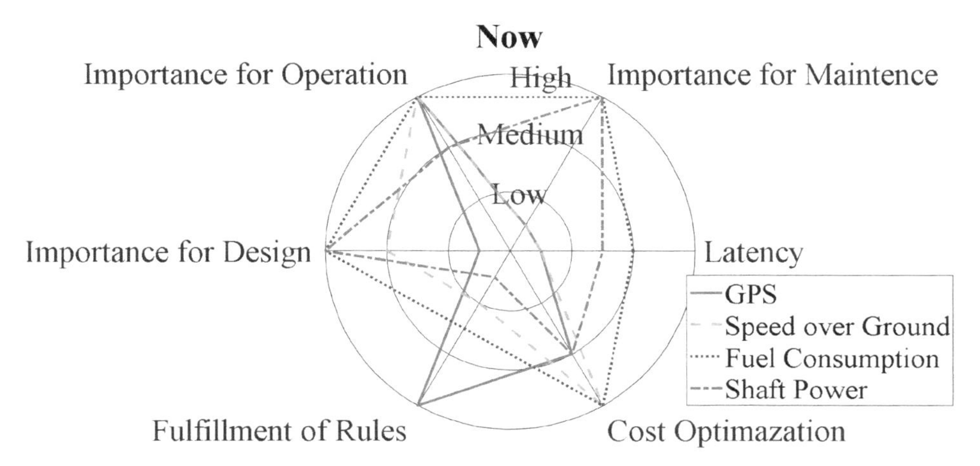

Figure 3. The role different data sources play for design, operation, maintenance, rule compliance and their required latency today.

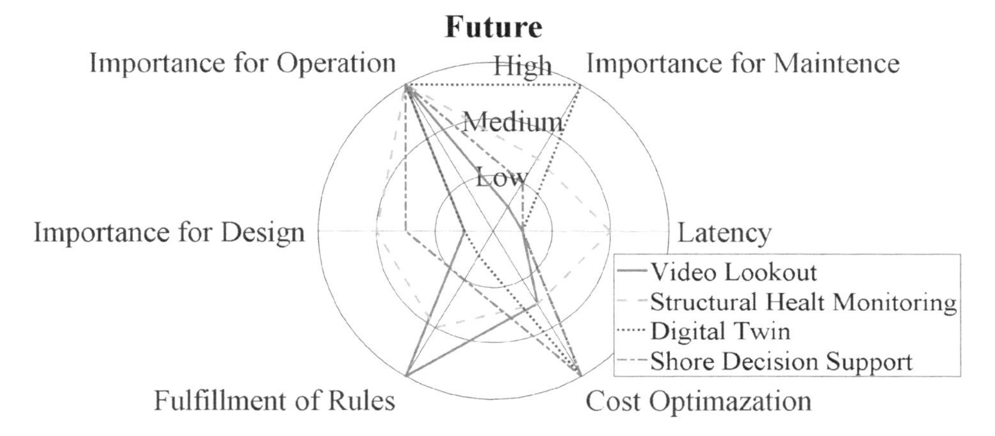

Figure 4. The role different data sources may play for design, operation, maintenance, rule compliance and their required latency in the future.

3.6 *Entity relationship*

The diagram in Figure 5 describes a very simplified version of an Entity Relationship Diagram for a set of machinery in which a set of 3 devices, described in the process in Figure 6, each have sensors to measure the temperature, pressure, and time of the fluid which moves between them. The fluid travels from the Sea Water Pump through the Engine Auxiliary Cooling Header then through the Generator Auxiliary Cooling Header. This diagram is

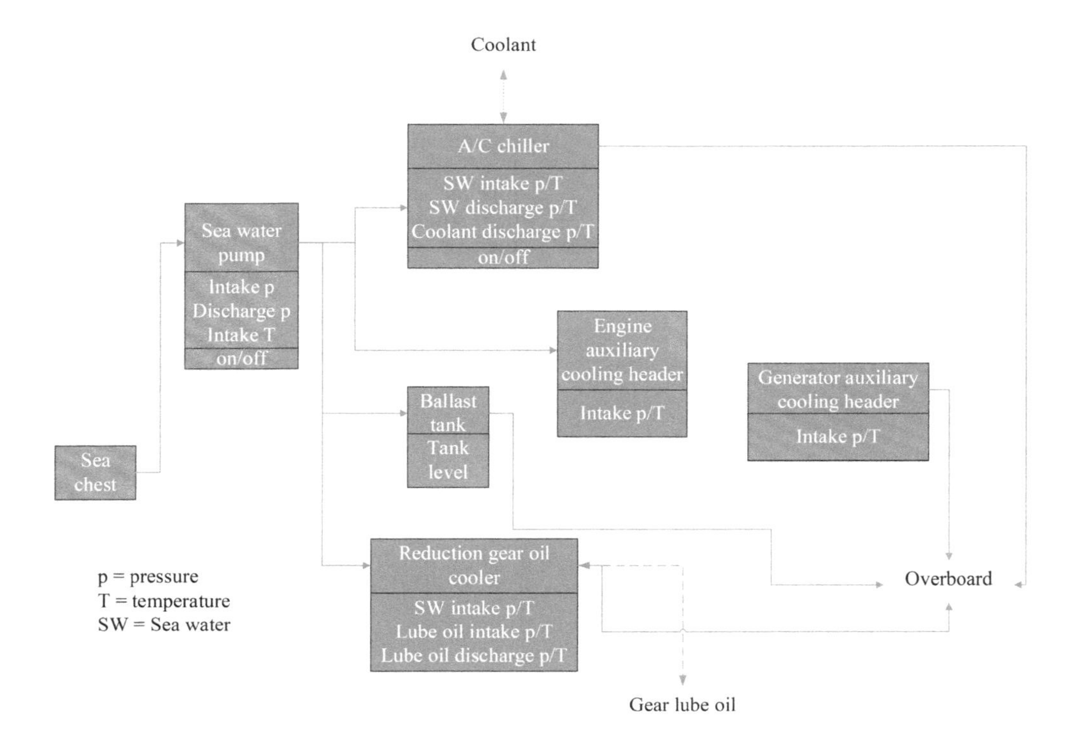

Figure 5. Diagram of sea water flow.

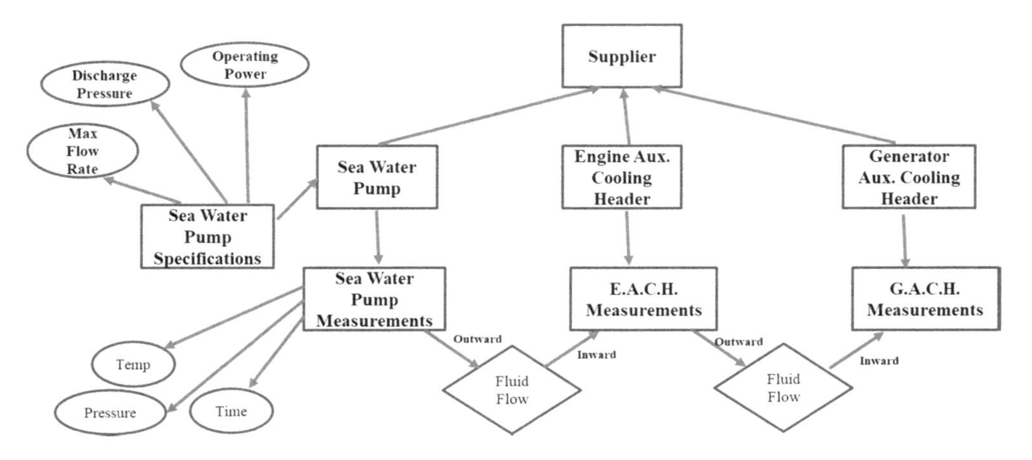

Figure 6. Entity diagram of sea water pump system.

only a representation of the data related to each of the elements in this process, and a very simplified description.

Rectangles represent entities. In this case the devices, the device suppliers, and device specifications. Ellipses represent attributes of the entity connected to them. That is the fields which contain values recorded by each piece of machinery. Finally, diamonds represent relationships between entities, with the nature of the relationship on the edges.

As one can see, the map of the data related to even a small section of the ship can be complex. This mapping does not include other attributes of the data such as timing, semantics, and other traits described in Section 2.

4 PREDICTIVE MAINTENANCE

4.1 *Introduction*

Maintenance operations evolve over decades as operators and designers become more familiar with their specific class of equipment. Three main types of maintenance persist today, run-to failure, scheduled and preventative. A fourth type, predictive, is making steps towards mainstream use. The goal of predictive maintenance is to remove invasive actions and unnecessary repairs from the maintenance process. Progress is slow, but fully achieving the goal hinges on effective data collection and the subsequent use of that data.

4.2 *Current methods of predictive maintenance*

Predictive maintenance is time sensitive. In the case of rotating machinery, a failure can occur rapidly, within a matter of seconds; or components can fatigue and degrade from prolonged operating cycles. For this reason, methods and required data for assessing failure modes varies. As the word predictive implies, the data must aid the methods in detecting these impending modes. The limitations of current predicative practices will be assessed as a precursor to the requirements for enhanced maintenance. The focus is shipboard machinery as it is often the most difficult to evaluate and the costliest to maintain.

4.2.1 *Operating data*

Equipment operating data is one of the only ways to identify a rapidly degrading machine. Operating data can be retrieved manually from local gauges or recorded digitally through control systems and data acquisition units. This data constructs a vague operating profile for the machine as it only centers around required operating and control parameters. Common types of data are; temperature, pressure, voltage, rotational velocity. Collection intervals, on average, range from fractions of a second to hourly. System alarms and control actions are continually modified and added based on the data collected from previous failures or advised by manufacturers. Predictive alarm indicators include widening parameter differentials, cycling and out of specification operation. Alarms can trigger a variety of actions, most common are shutting down equipment and de-rating operation.

What the operating data and related alarm set points have in common is latency. The data can only be effectively assessed to determine when a piece of equipment is in a failing state or recently experienced a failure, not before. This is easily depicted through a simple, best case scenario, example: detecting an overheating bearing. Individual bearing temperatures across an engine are sampled close to once per second. An alarm will be triggered if the temperature differential between two bearings reaches a critical point, as determined by operating data from a previous failure. This alarm indicates that one bearing is not operating like the others. Upon immediate engine shutdown the usual result is the hotter of the two bearings has suffered a minor failure. The bearing will require replacement and the engine will need to be thoroughly inspected for associated damage. This is a costly time consuming process. The reality of this scenario is it could be avoided by greatly reducing the latency of the incoming temperature data. Further, there are preceding signals to the failure before the bearing temperature rises to a critical point. These signals may not be inherent to the bearing; they could be present in surrounding components or the bearing lubricant. This highlights the need to broaden the scope of data and shows that generic operating data and sample rates are not sufficient for detecting these signals.

4.2.2 *Fluid monitoring*

For rotating machinery such as diesel engines, generators and turbines, fluid quality monitoring is widely practiced for predicting long-term fatigue failures. This involves taking a small sample of lubricant or coolant and evaluating its current state versus the previous samples' state. Samples are taken in large time intervals, 50–100 operating hours or greater. A primary focus is on foreign material inclusion, excessive copper, fuel oil, or carbon. For example, increasing carbon content indicates a piston ring or cylinder liner could be degrading. While fluid sampling is effective it has limitations. If a vessel does not have onboard fluid testing capabilities the sample must be analyzed on shore; obtaining data can take weeks. Establishing equipment baseline requires, at minimum, 10 data sets. The volume of data is minimal, a table listing foreign material content and some other fluid properties. Fluid sampling is not definitive. If a negative trend is identified further, invasive, investigation is necessary; requiring significant equipment downtime. With long time gaps between sampling and report generation fluid monitoring might not detect a degraded component until it has already failed. As shown by operating data monitoring, latency is an important factor.

Operating data assessment and fluid quality monitoring are just a few types of none invasive predictive maintenance practices. But they are among the most widespread. In addition, there are few alternatives to these methods. With advancements in ship design, data analysis, and sensor technology these processes do not need to be replaced, they just need to be enhanced.

4.3 *Future of predictive maintenance*

As expressed in the previous section prevailing methods of predictive maintenance suffer from a common issue, data latency. Reducing latency increases the number of options for enhanced predictive methods. There is another factor, the type of data collected must broaden. Broadening the scope of data is introducing different types of sensors beyond the traditional thermocouple or pressure transducer; it can go as fair as non-relational data.

Attaining high sample rates, well into millisecond region, is a reality. It is done in the aerospace industry today. Real-time, in flight, gas turbine engine monitoring is now common practice. Those same data acquisition methods can easily be implemented in a maritime application. With exponential advancements in data storage, from traditional warehousing to cloud based, the assumption that all data gathered can be stored effectively will hold.

4.3.1 *Operating data*

As mentioned previously operating data can be very sparse, it is often limited to simple control parameters. A prime example, a centrifugal pump; temperature and pressure data about the fluid is recorded but nothing about the actual pump. This means operators and maintainers must interpolate pump health from its intake and discharge pressure. Placing vibration and temperature sensors on the impeller shaft bearing cases is an effective prognostic tool. Curtis Wright has developed sensors that detects contact shock and friction events on ultrasonic wavelengths. They have achieved great results on ball bearing supported pumps (Famos, 2017). If the rotating assembly of a pump is degrading and out of balance it will manifest in the bearing load profile. In contrast, if a bearing is failing it will emit irregular loading. High sample rates will depict the difference between the two scenarios. An almost constant data flow from the sensors will differentiate the harmonic frequency loading of the unbalanced rotating assembly from the sporadic nature of bearing chatter in a failing bearing. The greater the data latency the greater the difficulty in delineating between the two. To further support the bearing data, flowmeters can be added to monitor additional flow characteristics in conjunction with the existing temperature and pressure data. This enhanced data can be used to build a more robust operating profile for understanding the pump components and their fluid interaction. Non-constant intake flow will affect the bearing loading as well.

Having a robust operating profile and enhanced, diverse data can alleviate the uncertainty and error that arises when a skilled operator must use limited data to generate failure assumptions. The goal of predictive maintenance is to eliminate assumptions that lead to invasive actions and unnecessary maintenance by better targeting the failure source. In order to achieve this, the enhanced data should be put to further use. Popular machine learning applications require low latency existing operating data to be used as training sets. A common application is classification models (Accorsi, 2017). In order to establish a model each class, or plausible failure, must be trained in. The more continuous and correlated the data is the more accurate the model will be. Continuous meaning low latency. Gaps between data points could lead to poor training of the model. The intended result is for the machine learning model to ingest near real-time operating data and determine if a component is degrading. For that to happen the model must be fed data of equal or lower latency.

In conjunction with popular machine learning algorithms many machinery designers and operators are utilizing digital twins. A digital twin is a computer rendered 3-dimensional model and associated simulations that mimic operating scenarios of the machine (Ansys.com, 2017). These simulations are targeted to find weak components and operating thresholds. The twin can become very specialized by honing in on specific conditions surrounding the machine (General Electric, 2016). An example would be specifying that a centrifugal pump moves only seawater between 50 and 72°F, is constant service and suffers from cycling head pressures between 40 and 70psi. This twin can determine a multitude of concentrated predictions about component health. The actual machine must be configured with the proper sensors to detect the conditions that the twin is producing. The more varied and abundant the sensor array the more effective the digital twin is.

4.3.2 *Fluid monitoring*

With advancements in the analysis of operating data through machine learning and digital twins, less emphasis can be placed on fluid monitoring. Nevertheless, there are a few instances where fluid monitoring can be a part of the machine learning model. Companies are developing sensors that continuously monitor fluid properties; the unit housing is mounted to the pipe and sensor sits in the fluid stream. The data feed will simply be another source of operating data acquired at the same sample rate as the others. Constant fluid quality data will add another angle of diversity to any machine learning algorithm for any diesel engine, generator or gas turbine.

4.4 *Predictive maintenance conclusion*

This assessment of predictive maintenance has strengthened the continuing themes; reduce data latency as much as possible and diversify the types of data collected. Through this a variety of predictive tools become useful. There is no doubt that the future of predictive maintenance relies on vast amounts of data. However, it is yet to be seen how ship operators, ship designers and their equipment suppliers willpivot. It will need to be a team effort so that machinery capable of producing low latency diverse data is placed in a ship that can fully utilize it.

5 CONCLUSION AND FUTURE DIRECTIONS

Smart technology is all around us, ranging from phones to fridges, smart cars are technically available, yet shipping is lagging. Shipping too will have to move in this direction and the journey has indeed started as shown by the many resent smart ship initiatives. The experience for the shipping industry will largely differ from that of other industries that were early adaptors. The two main reasons for this are that the wheel has already been invented; Big Data, Machine Learning and Artificial Intelligence have already been widely adopted in other industries. The second reason is that ships travel in some of the remotest areas of our planet; the high seas, where connectivity is less reliable and more expensive.

Today, the data analytics capability varies profoundly between operators and is thus hard to give an overall picture that accurately describes all operators. The analysis in this paper is aimed at parts of the industry that are yet to start or are in the process of starting their data analytics operations. There is a wide range of data available on-board, some of which is sent ashore for further analysis, and some analyzed onboard in more time sensitive applications. This paper covers some of these data sources and their importance to maintenance. Maintenance, and especially the downtime associated with failure, are costly to the industry. Moving to predictive maintenance from time based maintenance will facilitate cost savings, as maintenance will be performed when needed as determined by data analytics rather than being based on a time schedule. Moving towards predictive maintenance will also allow for a better determination of imminent failure and therefore reduce failure and downtime. The data analytics process that enables predictive maintenance will also allow for performance optimization.

It is often hard for humans to realize how unstructured much data are. However, even smart computers struggle with discrepancies in the data. Thus, metadata protocols are of fundamental importance in order to facilitate the journey towards smart ships. This is something on which the industry needs to collaborate; setting standards spanning operators and equipment manufacturers.

Ships are going to become moving local area network with hulls, much as autonomous cars are quickly becoming. This means that ability to collect data must be taken into consideration during the design phase. Ships must be designed for easy transfer of data on-board. This means not only existing path ways such as from the engine room to the bridge, but from the structure itself to shore.

Change will come with associated risks. For example, relying too heavily on shore support may cause the crew to be less knowledgeable and less prepared for emergencies. Further, there are risks related to the data itself. For example, it is important to determine the risks associated with data quality. For example, what happens when the data is missing, or in error; how does one mitigate these conditions; and how severe are the consequences? Significant risk is also related to latency, as more decisions are made by computers they will need both better and faster information, often supporting actions in real time. This is something that will require a much more robust connection and collection systems.

In short, Big Data, Machine Learning and Artificial Intelligence will dramatically change the marine landscape, and significantly improve shipping, as we know it. This process will not be easy or pain free but much can be learned from other industries. One of the lowest hanging fruits, where great cost savings can be made, will be moving from time based maintenance to predictive maintenance. An important step in this direction will be the careful description of the actual data, its properties and provenance, and where possible, the standardization of this description, i.e. metadata. Merging the application data needs, and the detailed metadata, informed by opportunities for useful analytics based on the data is a direction of future work.

REFERENCES

Accorsi, R., Manzini, R., Pascarella, P., Patella, M. & Sassi, S. 2017. Data mining and machine learning for condition-based maintenance. *Procedia manufacturing* (volume: 11): 1153–1161.

Ansys.com, 2017. *Delivering a digital twin*. [online] Available at: http://www.ansys.com/about-ansys/advantage-magazine/volume-xi-issue-2–2017/delivering-a-digital-twin [Accessed Nov. 2017].

Bonomi, F., Milito, R., Zhu, J. & Addepalli, S. 2012. Fog Computing and Its Role in the Internet of Things. *Proc. 1st Edition MCC Workshop Mobile Cloud Comput.*, pp. 13–16, Helsinki, Finland.

Chen, P., 1976. The Entity-Relationship Model-Toward a Unified View of Data. *ACM Transactions on Database Systems*. Vol. 1, No. I.

Famos.scientech.us, 2017. *StressWave Systems*. [online] Available at: http://famos.scientech.us/StressWave.html [Accessed Dec. 2017].

General Electric, 2016. *GE: Digital Twin: Analytic Engine for the Digital Power Plant*. General Electric Company.

Hong, HJ., Fan, CL., Lin, YC. & Hsu, CH. 2016. Optimizing Cloud-Based Video Crowd sensing. *IEEE Internet of Things Journal*, (Volume: 3, Issue: 3).

IMO, 2016. Performance standards for shipborne radiocommunications and navigational equipment, 2016 Edition. International Maritime Organization.

ISO, 2014. ISO/IEC Standard 11179 [online] Available at: metadata-standards.org/11179/ [Accessed Dec. 2017].

Kalmanek, C.R., Ge, I., Lee, S., Lund, C., Pei, D., Seidel, J., van der Merwe, J. & Ates, J. 2009. Darkstar: Using exploratory data mining to raise the bar on network reliability and performance. *7th International Workshop on Design of Reliable Communication Networks*, Washington DC, USA.

Keertikumar, M., Shubham,M. & Banakar,R. 2015. Evolution of Smart Vehicles in IoT, An overview. *International Conference on Green Computing and Internet of Things* (ICGCIoT), Delhi, India.

Lag, S., 2015. Ship Connectivity. DNV GL strategic research and innovation position paper. DNV-GL.

Lee, B., Hendler, J., Lassila, O. 2001. The Semantic Web. *Scientific American*, Vol 284, Issue 5, (May).

Lucidchart.com, 2017, *Diagrams Done Right*. [online] Available at: *https://www.lucidchart.com/*, [Accessed Jan. 2018].

Manyika, J., Chui, M., Brown, B., Bughin, J., Dobbs, R., Roxburgh, C. & Hung Byer, A. 2012. *Big Data: The Next Frontier for Innovation, Competition, and Productivity*. McKinsey Global Institute.

Marinesatellitesystems.com, 2014. [online] Available at: http://marinesatellitesystems.com/index.php?page_id=809 [accessed Oct. 2017].

Nist.gov, 2017. *IEEE Big Data*. [online] Available at: https://bigdatawg.nist.gov/bdmm2017.html [Accessed Dec. 2017].

rd-alliance.org, 2017. *Data Fabric IG*. [online] Available at: https://www.rd-alliance.org/group/data-fabric-ig.html [Accessed Dec. 2017].

Marine Design XIII – Kujala & Lu (Eds)
© 2018 Taylor & Francis Group, London, ISBN 978-1-138-34076-3

Simulations of autonomous ship collision avoidance system for design and evaluation

J. Martio, K. Happonen & H. Karvonen
VTT Technical Research Centre of Finland Ltd., Finland

ABSTRACT: The main features of our simulation platform aimed for autonomous marine navigation research are described in this paper. The platform includes a Collision Avoidance System featuring two subsystems: a decision-making protocol and an autopilot. The decision-making subsystem navigates the vessel according to the collision avoidance regulations (COLREGs), whereas the autopilot commands propulsors and control de, vices. The utilized autopilot features three modes: track-, heading- and docking modes. During an evasive action, the autopilot mode is changed from the track mode to heading mode. However, the location of the ship related to the track and track limits is monitored in the heading mode also. For harbor maneuvering, the function of the track mode is extended to slow speeds. The system has been implemented to a ship handling simulator environment and selected simulation scenario with object crossings is presented in this paper. The developed system can be used for the design and evaluation of autonomous ship systems. More specifically, we suggest the following application fields for the system: 1) development of autonomous navigation systems, 2) human factors studies with potential users, 3) verification and validation activities, and 4) scenario tests before real implementations on autonomous vessels. Finally, the general requirements and boundary conditions related to autonomous navigation systems are also discussed in the paper.

1 INTRODUCTION

Autonomous navigation system involves various subsystems, such as route planning or optimization and Collision Avoidance System (CAS), for instance. In this work, a research platform aimed for development and testing purposes of autonomous navigation systems is presented. The first implemented version includes a CAS utilizing a decision-making system and a wide-speed-range autopilot. At this stage, the work has concentrated mostly on the development of the autopilot.

Fundamentally, the CAS follows the Convention on the International Regulations for Preventing Collisions at Sea (COLREGs) maintained by the IMO.

There are several ways to categorize the decision-making protocols of Collision Avoidance Systems; one possibility is to divide them to reactive and proactive systems. The current decision-making protocol is reactive

VTT's medium-size ship handling simulator is utilized as the platform for the implementation. In this environment, various traffic scenarios in realistic conditions can be evaluated. All normal environmental conditions including wind, current, waves and banks can be taken into account.

1.1 *Previous work*

Several scientific publications have been published concerning autonomous navigation systems, including various Collision Avoidance Systems. The proactive decision-making protocols are quite often based on fuzzy logic, as for example in Perera et al. (2015) where the decision-making module consisted of a fuzzy logic-based decision-making process that generates parallel collision avoidance decisions with respect to each ship that is under collision course with the own vessel. Fuzzy logic-based systems, which are formulated for human-type of thinking, facilitate a human-friendly environment during the decision-making process. Other approaches have been also developed, for example based on the velocity obstacles method by Kuwata et al (2014). Moreover, the safety aspects including validation and qualification of the autonomous navigation systems has been discussed in various papers, such as Heikkilä et al (2017) and Wróbel, K et al (2017).

2 AUTONOMOUS NAVIGATION TEST PLATFORM

2.1 *Simulation environment*

VTT's ship handling simulator is a typical middle-size bridge simulator including the following models and features:

- ECDIS, radar simulator, control devices for rudders, propulsors and joysticks

- Hydrodynamical ship model including all essential environmental effects, also six-degree-of-freedom model and uniform ice
- Tug and escort tug simulation capabilities
- Traffic simulation
- Virtual world
- Fast time/real time simulations

VTT's simulator is designed as a research simulator, meaning that the main components are modular and can be easily modified.

2.2 *Autonomous collision avoidance system*

The Collision Avoidance System comprises of three subsystems: Situation Awareness (SA), decision-making and autopilot (Fig. 1). The SA system creates an assessment of the instantaneous traffic situation and environmental conditions using sensors and sensor fusion. The decision-making system utilizes the evaluation of current situation provided by the SA, Based on this information, it navigates the vessel according to COLREGs and assessment of the current situation. The decision-making commands the autopilot to steer the vessel to a desired location, that is, the autopilot controls the propulsors and control devices.

For the time being, in this work, the SA system is assumed to be perfect, meaning that the simulated traffic vessels and environmental conditions are provided to the decision-making system directly.

2.2.1 *Decision-making protocol*

A reactive decision-making system has been implemented at the first phase focusing on the crossing encounter situations. The decision-making protocol follows the identified objects and, if necessary, performs evasive actions according to COLREGs. The required actions and decisions are made according to the distance and relative bearing angles between the objects. In this specific case, the simulated cases involve vessels approaching both from starboard and port side. If the incoming ship approaches to a certain distance—this specific limit is a function of ship's velocity—and the relative bearing angle between the ships does not change, the decision-making protocol changes its state to evasive mode as described in Fig. 2.

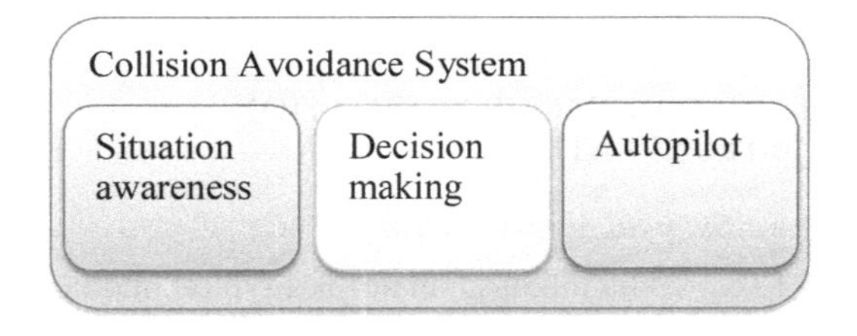

Figure 1. Generic view of collision avoidance subsystems.

2.2.2 *Autopilot*

A wide-speed-range autopilot has been implemented to VTT's ship handling simulator in order to find out operational weather limits for ships. Due to large number of simulations needed, so-called fast-time simulations have been utilized for the majority of the situations. The autopilot features the usual heading mode, that is, a PID-control, which steers the vessel with the given heading angle and speed. At the track mode, the ship is controlled by the track autopilot, which is trying to keep the ship on the given track using desired speed. Track-mode uses both the ship heading and cross track error to navigate of the vessel.

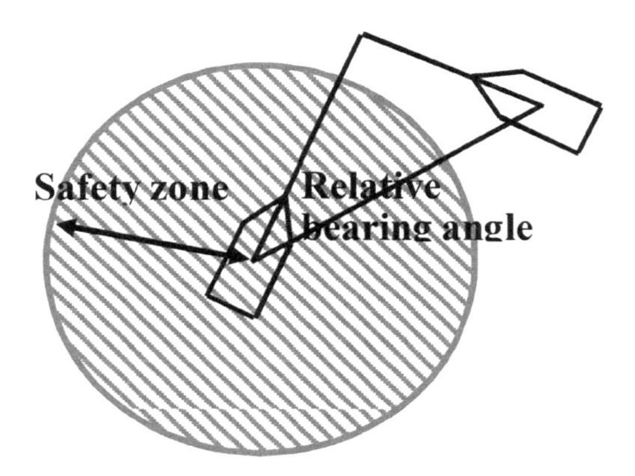

Figure 2. Track and track limits with waypoints.

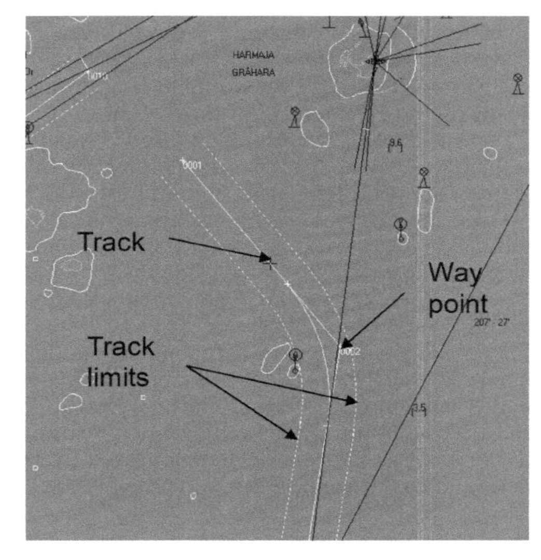

Figure 3. Track and track limits with waypoints.

For the harbor maneuvering, the track autopilot operation has been extended to slow speed and even astern speed range by combining joystick and autopilot control. At this specific docking mode, the autopilot gives orders to the joystick mode, which gives orders to all propulsion and steering units, including lateral thrusters. The docking mode includes also a crabbing mode option in order to maneuver the vessel to dock. Furthermore, the docking mode utilizes a predictor, which simplifies the control of the ship during the harbor maneuvers.

Regarding the development of CAS, additional features have been implemented to the autopilot so that the autopilot is able to conduct evasive actions. In the normal mode, the autopilot uses track mode. However, if the decision-making module commands an evasive action, the autopilot changes it's mode from track mode to heading mode. In this particular case, the heading mode utilizes the track information, that is, the distance to track limits and track are evaluated (see Fig. 3). If the vessel exceeds the track limits, the CAS system mode is changed to alert mode. This approach enables the ship to remain inside the safe zone—the track limits has to be designed so that the ship is able to maneuver inside these specific limits. If only the track mode would be applied without the heading mode, the position of waypoints should be re-located or 'virtual waypoints' should be utilized.

3 APPLIED SIMULATION SCENARIO

A simulation scenario with own ship and six traffic vessels has been applied for the test and demo case. The vessel's own track is shown in Fig. 4 and the initial traffic condition is presented in Fig. 5.

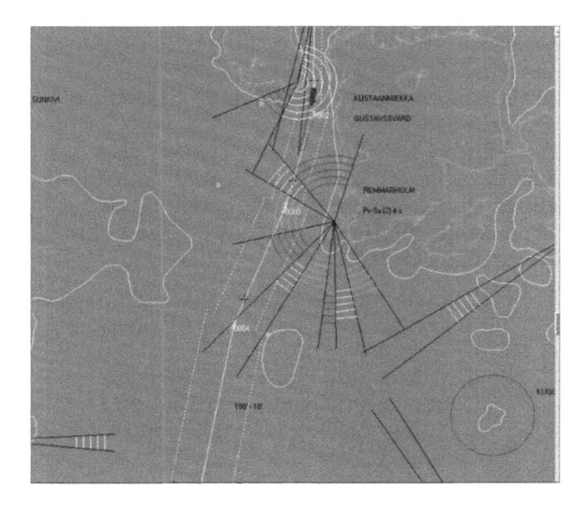

Figure 4. Vessel's own track.

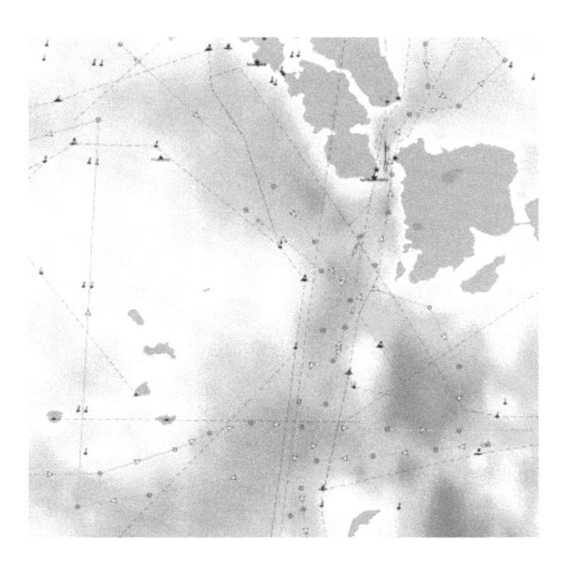

Figure 5. Generic view of the simulation scenario with the traffic vessels and their tracks.

At the end of the simulation, the vessel performs a docking by using the crabbing mode. Simulation are conducted with a generic ice-going vessel equipped with two azimuth devices and a bow thruster.

4 RESULTS

An example of evasive action condition is presented in Fig. 6. At the first picture, the 'own ship' has detected the traffic vessel and a potential collision situation. The decision-making system has commanded an evasive action, that is, the autopilot track mode has been changed to heading mode. At the second picture, the vessel has changed the heading angle and it is avoiding the traffic vessel. In the last picture, the vessel returns to the original track, the autopilot has changed back from the heading mode to the track mode.

In Fig. 7, the location and orientation of the 'own ship' related to the track during the evasive action is shown. The autopilot is in heading mode, and still utilizing the track information (distance to track and track limits).

During the crabbing mode, the propulsors are set to desired angles in order to control together with the bow thruster the translational and rotational motion of the vessel. An example of crabbing configuration is shown in Fig. 8 and the motion is shown in Fig. 9.

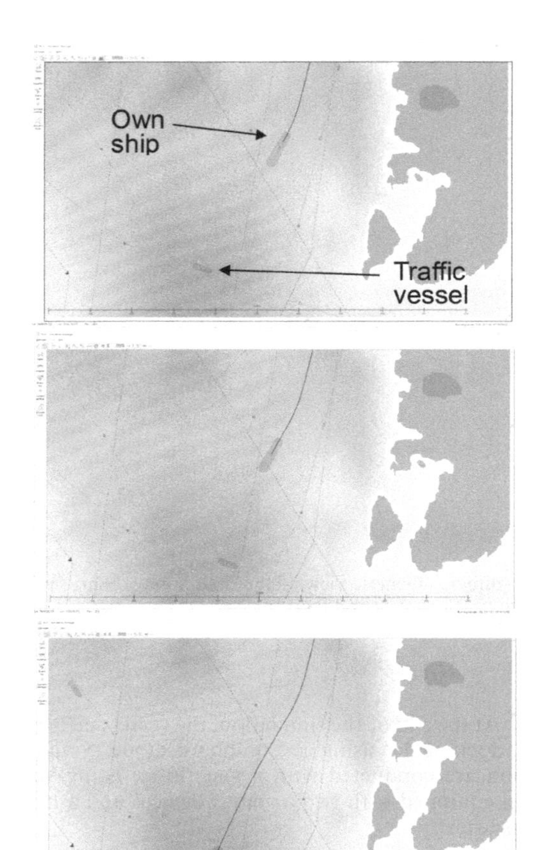

Figure 6. Snapshots of evasive action with three simulation time steps.

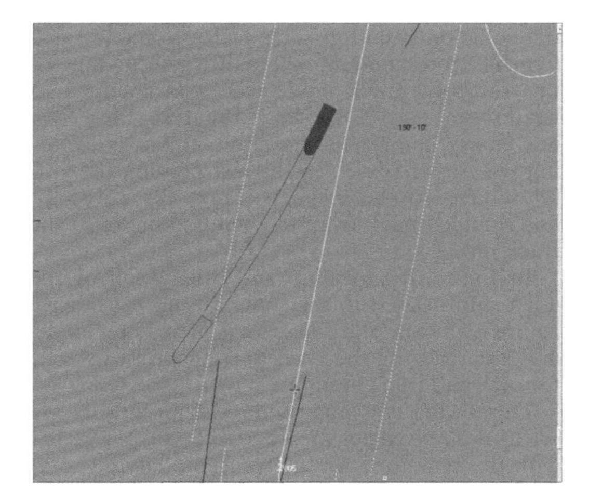

Figure 7. 'Own ship' performing an evasive action with heading mode.

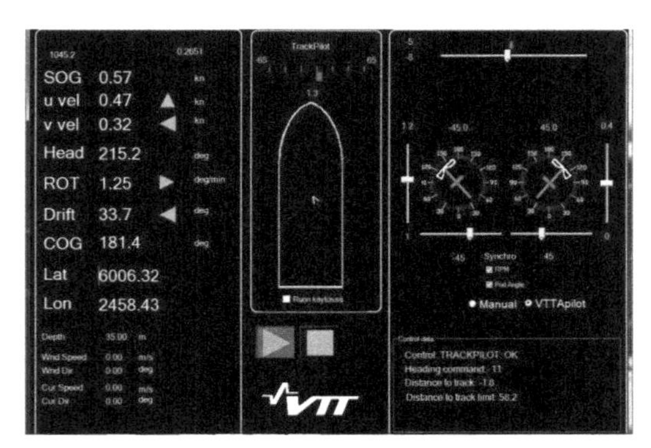

Figure 8. Conning display; the rotated azimuth devices are shown at right.

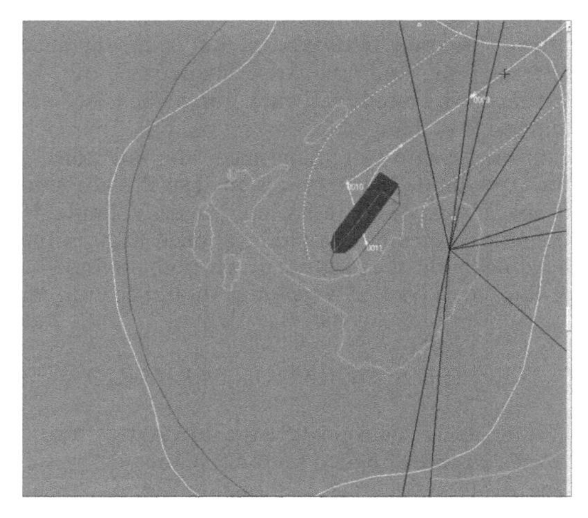

Figure 9. Ship motions in crabbing mode utilizing the prediction display.

5 UTILIZATION OF THE DEVELOPED SYSTEM

The developed system can be used both for the design and evaluation of autonomous ship systems. More specifically, we suggest the following applications for the system:

1. development of autonomous navigation systems and other systems related to autonomous ships,
2. human factors studies with potential users,
3. verification and validation activities, and
4. scenario tests before real implementations on autonomous vessels

Next, each of these application areas will be gone through in detail.

First of all, autonomous navigation systems include route planning, obstacle detection and classification, and Collision Avoidance functionalities. Currently, these functionalities are typically operator-assisted and monitored. However, particularly path following abilities and obstacle detection and collision avoidance in specific cases can be done rather autonomously. In the future, an autonomous navigation system needs to function autonomously and adapt its performance based on the detection and classification of obstacles. Consequently, avoiding collisions and path planning needs to be done autonomously with minimum human monitoring and intervention. Therefore, these functionalities will require robust algorithms in order to function correctly. For example, artificial intelligence, machine learning, and deep learning approaches become essential in this functioning.

In addition, other technical aspects than autonomous navigation systems of autonomous vessels, such as situational awareness (SA) systems, connectivity issues, engine and other systems' remote monitoring, and remote operating center (ROC) designs can be studied and developed further by the use of the simulator environment. Related to the latter, it is comparatively easy to integrate also various virtual and augmented reality visualizations to the simulator environment's visualizations.

Second, the simulator can be used for human factors studies with potential real users. For example, prototype remote control user interfaces can be studied with users to gain feedback. In addition, the effects of communication network delays can be simulated to find thresholds for too long delays for users. In this way, requirements for communication delays in different operational situations can be determined. Finally, the appropriate level of automation in different operational situations can be experimented with various automation mode conditions and potential real users. The change of modes can be easily done on the fly in the development simulator environment during experiments.

Third, verification and validation activities (V&V) aim at an independent assessment of the safety and functioning of the proposed solutions. The main goals of V&V activities are the following (adapted from O'Hara et al., 2002 [NUREG-0700, Rev. 2]):

- to determine the fulfilment of design specifications (verification)
- to determine that the developed system adequately supports safe operation (validation)
- to provide evidence of the adequacy of the system for its purpose (validation)

To reach these goals, V&V utilizes evidence-based argumentation in order to create a demonstration of an autonomous system's safety and present it in an understandable way, for example, for authorities and other stakeholders. In addition to this safety demonstration, V&V activities also support stakeholder learning and iterative design of the system by pointing out safety-relevant deficiencies and inconsistencies with the defined requirements.

The second objective, showing that the designed systems support safe performance, is fulfilled by the validation activities. In these activities, it can be demonstrated that the systems can meet reliability and performance criteria regarding safety. This demonstration is based on both quantitative and qualitative analysis of gathered data from different tests. The final assessments providing the acceptance or rejection/recommendations of specific improvements of the solution can be conducted based on this data. Therefore, the results of the validation tests provide combined evidence of the safety and functionality of the system.

This approach includes constructing a Safety Case (e.g., Bishop & Bloomfield, 2000 and Kelly, 1998) from the acquired data and results of the validation tests. The objective of the Safety Case is to gather the safety evidence gathered in validation activities and to present a coherent argument of safety. To produce the safety evidence in addition to data gained from the simulator environment, also different complementary approaches are possible, such as analysis of documented earlier experiences, analytical/numerical/experimental methods, and model/HIL/field testing.

Finally, scenario tests before real implementations on autonomous vessels can be conducted. This saves considerable amount of resources and benefits safety as some possible safety-critical problems can already be identified in the simulator environment. Our developed simulator can be used to build simulation scenarios to test, for example, how well COLREGs are followed and how proactive/reactive the autonomous navigation system is. Also, operational scenarios in difficult environmental conditions (e.g., harsh weather), or with degraded ship systems (e.g., deficient propulsion) or uncertain situational awareness (e.g., sensor failures) can be tested.

6 CONCLUSIONS

A Collision Avoidance System has been implemented to the shiphandling simulator. At the first stage, the decision-making protocol is reactive. The development has been concentrated mostly on the autopilot, that is, the implemented wide-speed-range autopilot utilizes three modes: track-, heading- and docking-modes for the autonomous maneuvers. During the evasive actions the autopilot mode is changed to heading-mode, in which the location related to track is still monitored. A simulation scenario including crossings of vessels was utilized to demonstrate the features of current implementation.

As a conclusion, autonomous vessels need to be at least as safe as corresponding conventional vessels. Therefore, there cannot be increased risk in own operation or additional risk to others in operation. The technology developer is responsible for demonstrating the safety of an autonomous vessel. We see that simulator testing is an important method in producing the safety evidence. We also see that this testing requires comprehensive risk assessment and human factors expertise combined with simulator testing facilities for cost-effective development.

REFERENCES

Bishop, Peter, and Robin Bloomfield. *A methodology for safety case development*. Safety and Reliability. Vol. 20. No. 1. Taylor & Francis, 2000.

Heikkilä, E. (2016) *Safety Qualification Process for autonomous ship concept demonstration.* Master's thesis, Aalto University.

Heikkilä, E., Tuominen, R., Tiusanen, R., Montewka, J., Kujala, P., (2017). *Safety Qualification Process for an Autonomous Ship Prototype – a Goal-based Safety Case Approach, in: Marine Navigatison.* CRC Press, pp. 365–370. doi:doi:10.1201/9781315099132–63.

Kelly, T. (1998) *Arguing Safety—A Systematic Approach to Managing Safety Cases.* Ph.D. Thesis, University of York.

Kuwata, Yoshiaki et al, *Safe Maritime Autonomous Navigation With COLREGS, Using Velocity Obstacles*; IEEE JOURNAL OF OCEANIC ENGINEERING, VOL. 39, NO. 1, JANUARY 2014.

O'Hara, J.M., Brown, W.S., Lewis, P.M., & Persensky, J.J. (2002). *Human-system interface design review guidelines.* NUREG-0700, rev, 2.

Perera, L et al, *Experimental Evaluations on Ship Autonomous Navigation and Collision Avoidance by Intelligent Guidance,* IEEE Journal of Oceanic Engineering, VOL. 40, NO. 2, April 2015.

Wróbel, K., Montewka, J., Kujala, P., 2017. *Towards the assessment of potential impact of unmanned vessels on maritime transportation safety*. Reliab. Eng. Syst. Saf. 165. doi:10.1016/j.ress.2017.03.029.

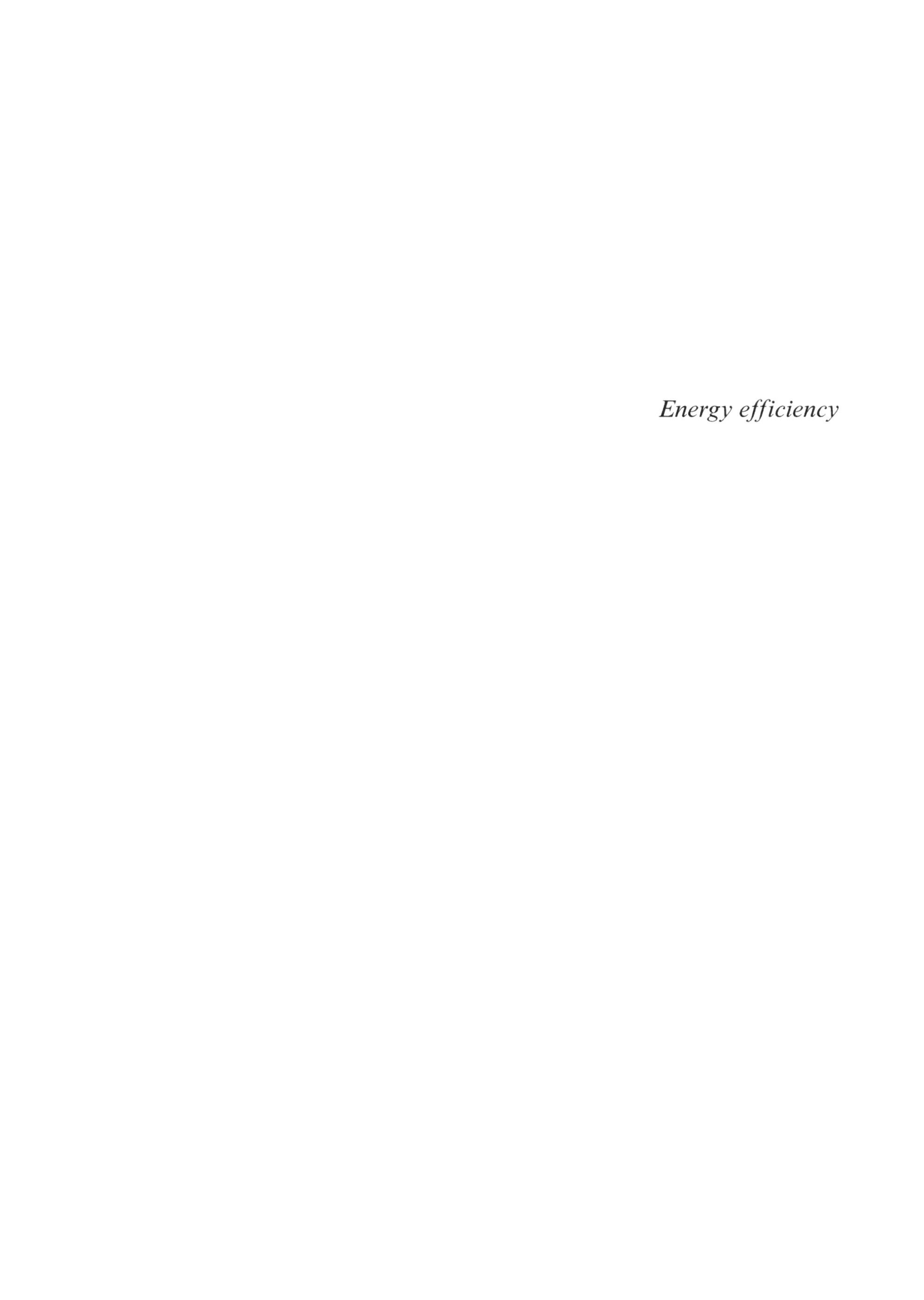

Energy efficiency

Marine Design XIII – Kujala & Lu (Eds)
© 2018 Taylor & Francis Group, London, ISBN 978-1-138-34076-3

Feedback to design power requirements from statistical methods applied to onboard measurements

T. Manderbacka & M. Haranen
NAPA Ltd., Helsinki, Finland

ABSTRACT: Definition of the ship main dimensions at the contract design phase plays an important role on the profitability of the ship project. Cargo carrying capacity and design speed are fixed at this stage. Power requirement is then estimated and an allowance to weather margin and hull fouling is reserved. Clean hull resistance in calm sea conditions can be generally estimated at high accuracy. However, the sea margin is generally taken as 10–20% of the clean hull resistance. Onboard measurements, with the information gathered on similar ships provide important insights into actual power requirements. The ship may be operated on different loading conditions (draft and trim), speed profile, and weather conditions than what was expected at the contractual design stage. Statistical methods applied to onboard measurements can capture the dependencies of the ship power consumption from different factors, and a decomposition into weather, shallow water, and hull-fouling effects, of the required power can be presented. In this paper, a statistical method based on regression model and application to the onboard measurements is presented. The statistical method enables constant monitoring of the power requirement, and the development of hull condition. Regression model can make use of the entire data set, instead of limiting the analysis on the variation of fixed set of parameters.

1 INTRODUCTION

The cost of main engine represents around 16% of the total cost of a merchant ship according to Stopford (2009). The cost of the engine and required installations for ship propulsion increase with increased engine power. The Energy Efficiency Design Index (EEDI) (IMO Resolution MEPC 245(66) 2014) also set limitations to installed power aiming to reduce Green House Gas (GHS) emissions. Then again, in terms of safety and the ability of the ship to maintain its required speed, it must have sufficiently installed propulsion power (Papanikolaou et al. 2016). The hull condition also deteriorates over time. The hull fouling increases with the growth of algae and barnacles, which can have significant effects on the resistance of the ship. Moreover, the condition of hull plating gets worse over time contributing to the increased resistance. Thus, the ship need to be equipped with sufficient engine power to be able to overcome these effects over its lifetime, but at the same time limiting the construction costs without installing more engine power than needed. At the phase of estimating the commercial feasibility of a ship project and at the contractual design phase the question of power requirement is essential, among the definition of the main dimensions of the ship. The clean hull resistance in calm water conditions can be quite accurately estimated by numerical methods and with experiments. However, the added resistance in waves is more complicated to estimate. At small to moderate wave conditions, the added resistance is mainly caused by the reflection of waves, and according to Valanto and YongPyo (2017), power requirement at smaller waves is also increased by lower propeller efficiency due to increased resistance. Whereas at higher waves the added resistance is affected by ship motions, mainly pitching and heave motions. These motions can be estimated by numerical methods (Kalske and Manderbacka, 2017), but the estimation of the added resistance of pitching and heaving ship is more challenging as shown in the benchmark by SHOPERA (2016) project. Moreover, for computational estimation of the required power, the navigational conditions of a ship need to be defined. For this purpose, the historical, statistical data, scatter diagrams, of wave conditions at different sea areas at are generally applied. However, the actual navigation routes do not cover the entire areas of the scatter diagrams and the weather routing is also often applied during operations.

Analysis of the onboard data offers and alternative approach to the conventional one described above. In this way providing information of the actual realized power consumption and its origins: wave, wind, and shallow water effects, and hull fouling. In this paper, we describe a statistical method to assess the power consumption based on the onboard measurements. We apply an integrated onboard data acquisition, monitoring and analysis system implemented in NAPA software to study the power consumption. Statistical methods for estimation of ship's resistance and requirement of propulsion power have a long history. The most common method, still widely used, for estimation of calm water hull resistance was published by Holtrop (1984). Vesterinen (2012) applies statistical regression model to estimate the effect of hull fouling on the power consumption. Petersen et al. (2012) investigate artificial neural networks and Gaussian process approach to estimate ship propulsion efficiency. Perera et al. (2015) applied statistical analysis to study trim effect on consumption. Eide (2015) estimated the sea margin based on noon-reports and concluded that the margin is speed dependent. Mao et al. (2016) establish statistical methods to predict ship speed for given power and sea environment.

2 METHOD

Conventional approaches for the ship performance evaluation have limitations. Separate sea trials are expensive and usually available for the limited number of combinations of the loading conditions. Towing tank tests are not always available and CFD results can be unreliable. The method described in the ISO 19030 standard (ISO, 2017) has limitations, as it is based on the physical modeling and data correction approach. The main problems of such approach are well summarized in Logan (2011). At this moment, the ISO standard approach takes into account only the effect of wind for correction of the propulsion power measured values. In addition, the approach produces the reliable results only for restricted operational profiles, for example when ship is constantly moving on fixed engine rpm and speed. The development leading to the current statistical method has its origins in the more conventional methods (Kariranta 2010) of defining hull performance, which were then improved by statistical approach (Vesterinen 2012; Haranen et al. 2016)

In this paper, we describe the statistical modeling approach, which is based on fitting of a non-linear regression model to the high-density measurement data. The measurement data is collected onboard the vessel during the normal operations from numerous sensors. We assume the ship measurement data to be the data from a continuous randomized experiment in which numerous combinations of different variables are tested. The main target of the model is to estimate the expectation of the dependent variables given the independent variables. This means that after we create the model we can calculate the average values of the dependent variables for any fixed combination of values for the independent variables. In data-driven science and data mining the process of model creation is often called the model training, thus we will use this concept in the further text. For example, the ship measurement data is used to train the statistical model, which help us to understand the relation between variables. When we obtain more data from the ship, we can re-train the existent model and obtain a new one.

What variables we are using in the model? There are independent and dependent variables when we are talking about the regression model. In the case of the ship performance, the independent variables are variables, which influence the ship performance. We separate them into controllable and uncontrollable variables. Controllable variables are variables that can be affected by a ship crew, such as engine shaft revolutions, rudder angle, draft and trim. Uncontrollable or environmental variables are, for example, the weather and swell conditions, shallow water, and seawater temperature. In our modeling method, those variables are the main factors, which define the current ship performance, which in turn can be described by a ship speed through water and propulsion power. Thus, the ship speed through water and the propulsion power are the model dependent variables.

We assume that the dependences between all above described independent and dependent variables remain constant with one exception. This exception is the hull condition. Even if the ship would sail with the same engine rpm, draft and trim, upon the same wind, swell the speed through water and the propulsion power would not be the same because of the hull fouling. Assumingly the hull fouling will increase the propulsion power consumption and decrease the speed through water of the ship even if all other variables remain the same. Thus, the effect of the hull fouling needs to be included into the model. In proposed method, the hull fouling is treated as an effect of the time by adding the time variable into the set of the independent variables. The general idea of the modeling is to retrain the model each time we receive a new portion of the data from the ship and assess the marginal effect (i.e. effect of one variable, while keeping the others constant) of all independent variables on the model response.

After the model is created the fouling effect can be assessed as a conditional expectation of the propulsion power and speed through water for different time moments assuming all other independent variables remain constant and fixed at the certain values. Usually, this is done for the zero weather, no swell and deep-water conditions.

To describe the ship performance each dependent variable is modeled by a Generalized Additive

Model (GAM), see (Hastie 1990), which general form can be expressed as follows:

$$g(Y) = \beta_0 + s_1(x_1) + s_2(x_2) + \ldots + s_n(x_n) \qquad (1)$$

where, $g(Y)$ is some transformation of the output variable, x_i are the additive predictors and $s_i()$ are the smoothing functions. We assume that $g(.)$ is an identity function in our method.

To control the model flexibility and avoid an overfitting the most terms of the additive model are set to be polynomial functions. For example, we assume that the relationship between the engine rpm and propulsion power consumption or the speed through water can be described by a polynomial function having degree three at maximum. The term, that is responsible for the time effect, is a smoothing spline. The flexibility of this term is controlled by regularization or penalizing the spline curvature to avoid overfitting. Typically, regularization can be done through cross-validation technique, see (Hastie 2003). The main idea of the technique is to minimize the Mean Squared Error (MSE) of the model prediction over the testing data. The data set is divided into training and testing data sets. Then only the training data set is used for the model training and the MSE is calculated over the testing data set. This is a greedy approach in which, we re-train the model with different smoothing parameters in order to find one, which minimizes the MSE.

The main idea of the smoothing parameter selection is demonstrated in Figure 1. In the figure, the development of the MSE is shown for the model, which is trained with different smoothing parameters and for the data with different number of records (n). The minimum for the MSE is dependent on the size of the data set; this defines the correspondent optimum value for the degrees of freedom.

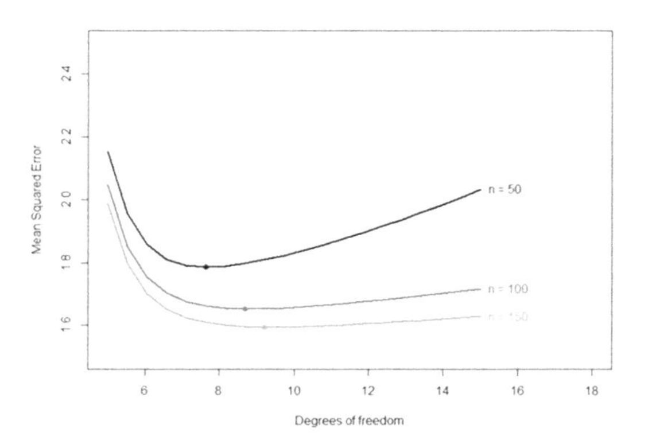

Figure 1. Finding the optimum degrees of freedom for the smoothing spline.

Before the model training, the ship data is not heavily filtered. In contrary, we want to have as many observations as possible. The main idea of the modeling is to find how different variables affect the consumption of propulsion power and speed through water, thus the high variability of the independent variables is critical. However, some specific data is filtered out. For example, clear sensor malfunctions must be recognized and filtered out from the data set before the modeling. In addition, the ship performance is assessed during normal operations when ship is moving with a constant engine rpm and without heavy maneuverings. Acceleration and deceleration periods are filtered out.

The procedure for the modeling of the ship data is straightforward. The models for the propulsion power and speed through water are trained every time when a new portion of the data is received from the ship. After the model is built, we can assess the marginal effects of any independent variables, visualize and interpret them.

3 RESULTS

In this paper, we apply the presented statistical method to the onboard data of Very Large Crude Carriers (VLCC). These two tankers have navigated mainly on routes between Asia and Americas crossing Southern Atlantic and Indian Oceans and rounding Cape of Good Hope. The histories of locations of the ships are presented in Figure 2. The data over a four years period, from the beginning of year 2013 until the end of 2017 is used for the analysis and modeling.

In Figure 3, the marginal effects of the wind speed and direction are visualized. The marginal effects of the input variables are calculated by the model, while keeping all the other input variables constant. The propulsion power increase is estimated for different wind speed and direction angles. When estimating the wind effect, we keep all other independent variables constant and fix their values to the most typical values, which have

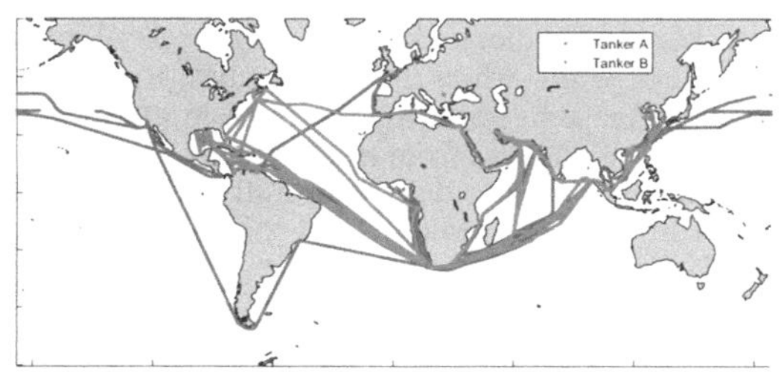

Figure 2. Navigated routes of two tankers during period between 2013 and 2017.

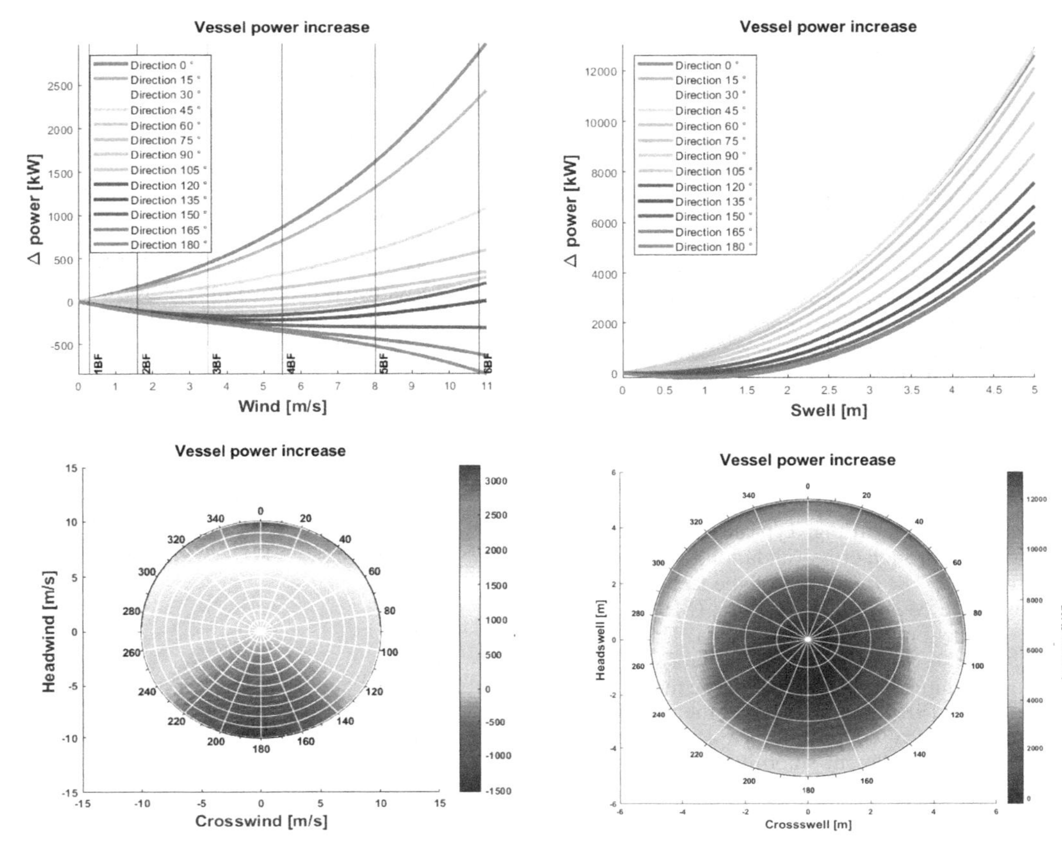

Figure 3. Wind effect to the power requirement Curves for different headings, 0 deg is head wind, and polar plot up to 11 m/s wind speed.

Figure 4. Wave effect to the power requirement. Curves for different headings, 0 deg is head waves, and polar plot up to 5 m wave height.

been measured during the ship operations. In real navigation, wind incurs power increase by affecting to other variables, such as rudder angle. These are taken into account in the model; however, the rudder angle effect or changes in other variables caused by the wind are not shown in these figures. The effect of waves is estimated similarly and viewed in Figure 4. The power increase due to wind is mainly dictated by the magnitude of relative head wind, while the waves cause nearly the same increase in power requirement in a bow sector up to 30 degrees from head waves.

The effect of the decreased hull condition can be estimated and visualized in the same manner. We keep all variables in the model constant, except the time and simulate how the propulsion power consumption is changing at the certain speed through water. The calculation is usually done for zero weather, no swell and deep-water conditions. The hull index represents the change of the

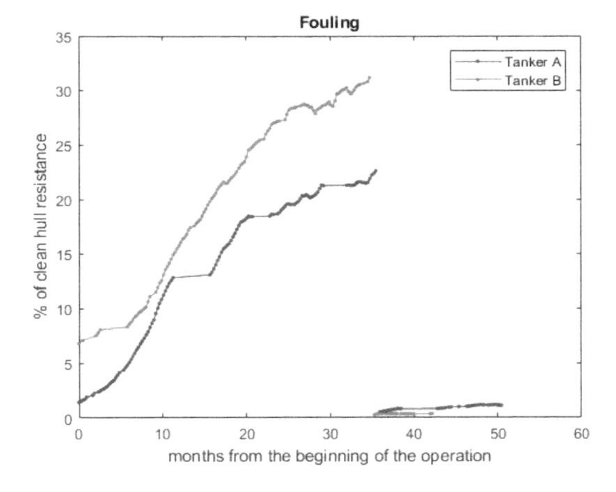

Figure 5. Change in percent of the hull condition during the inspected period. The vertical line denotes the dry dock.

consumption of propulsion power in percent and the baseline is selected to be the lowest power consumption over the data collection period.

In Figure 5, the effect of the fouling is visualized. In these cases, we can see that the hull condition was constantly decreased since the beginning of data collection. In this case, the ships were experiencing problems with the hull coating. After the three years of operation, the hull condition was decreased so that at the same typical speed the ship was consuming 30% more propulsion power. Within 36 months of operations, the ships were taken to the dry dock. After the maintenance and repainting, the hull condition was improved to the original level.

3.1 *Power usage decomposition*

The statistical model, which we have trained with the ship sensor data, can be successfully used to decompose the power usage onboard the ship. This could be a very valuable information for the ship owner or operator because it shows how much propulsion power or fuel is consumed due to different factors, such as weather, swell, shallow water or decreased hull condition.

The main idea of the power decomposition is based on simulation of the optimal power consumption for the ideal conditions and real conditions. We can use the statistical model to simulate, first, the power consumption for the ideal condition. This means that the power consumption is calculated for the clean hull condition, zero weather, no swell and deep water. Then, taking in account the fact that the model is additive, we can simulate the power consumption by setting the input variables, one at a time, to the measured values and calculate the difference from the power consumption for ideal conditions. The calculated difference is the addition power consumed due to certain variable. In Figure 6, the monthly power usage for the weather is show for two tankers. Resistance caused by the weather effects, wind and waves, is in average less than 10% of the clean hull resistance in calm water.

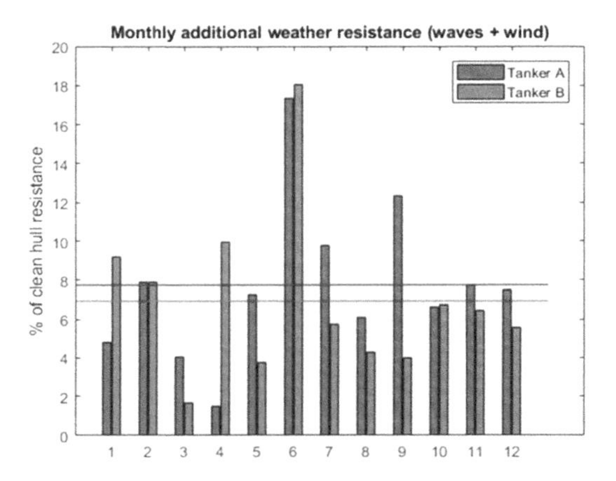

Figure 6. Monthly increase in power usage due to weather conditions, waves and wind additional resistance for two tankers.

4 CONCLUSIONS

We presented a statistical approach to estimate and decompose the power consumption into different sources, weather and hull condition, and applied the method to onboard measurements of two tankers. We showed that with the applied method, the weather effects can be distinguished from the hull fouling. And vice versa, the deterioration of hull condition, fouling, can be distinguished from other factors affecting the power consumption. This is particularly important when planning the hull maintenance. The weather effects were divided into monthly distributions, in this way we showed how the consumption varies seasonally, for the example vessels navigating mainly in the southern hemisphere the weather effects were the most remarkable in June. In average, the weather effects were less than 10% of the clean hull resistance for the example ships on their navigational area. Hull fouling effects were more severe, causing up to 30% of increase into power consumption before the maintenance was carried out. The usual values of sea margin up to 20% would be suitable for these ships in these navigation routes, if similar fouling rate should be expected. However, in light of the weather effects even smaller sea margins could be used, if the hull fouling rate is expected to be more moderate than in the example cases, and the maintenance is carried out in such a way that the hull fouling is kept in moderate level. We believe that, the presented statistical method, able to decompose the power requirement into different factors, can be of important benefit at the contractual, and initial design phase, providing essential, actual operational data based information.

REFERENCES

Bialystocki, N., Konovessis, D., 2016. On the estimation of ship's fuel consumption and speed curve: A statistical approach, Journal of Ocean Engineering and Science, Volume 1, Issue 2. https://doi.org/10.1016/j.joes.2016.02.001.

Eide, E. 2015. Calculation of Service and Sea Margins. Master Thesis. Trondheim: NTNU.

Haranen, M., Pakkanen, P. Kariranta, R., Salo, J., 2016. White, grey and black-box modeling in ship performance evaluation, 1st Hull Performance & Insight Conference, Pavone, pp.115–127.

Hastie, T., Tibshirani, R., Friedman, J., 2003. The Elements of Statistical Learning, Springer.
Hastie, T.J., Tibshirani, R.J. 1990. Generalized Additive Models, Chapman & Hall/CRC, ISBN 978-0-412-34390-2.
Holtrop. J. 1984. A statistical re-analysis of resistance and propulsion data. International Shipbuilding Progress, 31(363):272–276.
IMO. MEPC 245(66). 2014 Guidelines on the Method of Calculation of the Attained Energy Efficiency Design Index (EEDI) for New Ships. 2014.
ISO/CD 19030–1, Ship and marine technology—Measurements of changes in hull and propeller performance, Part 1: General Principles (2017).
ISO/CD 19030–2, Ship and marine technology—Measurements of changes in hull and propeller performance, Part 2: Default method (2017).
Kalske, S., Manderbacka, T. 2017. Development of a new practical ship motion calculation method with forward speed. Proceedings of the International Ocean and Polar Engineering Conference (ISOPE 2017). San Francisco, CA, USA.
Kariranta, R.J., 2010. Analysis of the Technical Performance of a Ship in Service. Master's Thesis. Espoo: Aalto.
Logan, K.P. (2011), Using a Ship's Propeller for Hull Condition Monitoring, ASNE Intelligent Ships Symp. IX.
Mao, W., Rychlik, I., Wallin, J., Storhaug, G. 2016. Statistical models for the speed prediction of a container ship. Ocean Engineering, 126(152–162). https://doi.org/10.1016/j.oceaneng.2016.08.033.
Papanikolaou, A., Zaraphonitis, G., Bitner-Gregersen, E., Shigunov, V., El Moctar, O., Guedes Soares, C., Reddy, D.N., Sprenger, F. 2016. Energy Efficient Safe SHip Operation (SHOPERA), Transportation Research Procedia, Volume 14, Pages 820–829, ISSN 2352–1465, https://doi.org/10.1016/j.trpro.2016.05.030.
Perera, L.P., Mo, B, Kristjánsson, L.A. 2015. Identification of Optimal Trim Configurations to improve Energy Efficiency in Ships, IFAC-PapersOnLine, Volume 48, Issue 16, Pages 267–272, ISSN 2405–8963, https://doi.org/10.1016/j.ifacol.2015.10.291.
Petersen, J.P., Jacobsen, D.J., Winther, O.J. 2012 Marine Science and Technology Volume 17, Issue 30, https://doi.org/10.1007/s00773-011-0151-0.
SHOPERA. 2016. SHOPERA Energy Efficient Safe Ship Operation EU FP-7 Project.
Stopford, M. 2009. Maritime Economics. 3rd Edition. Routledge, Taylor and Francis Group, London and New York.
Valanto, P., Yong Pyo. 2017. Wave Added Resistance and Propulsive Performance of a Cruise Ship in Waves. Proceedings of the International Ocean and Polar Engineering Conference (ISOPE 2017). San Francisco, CA, USA.
Vesterinen, H. 2012. Statistical Regression Models for Ship Performance Analysis. Master's Thesis. Espoo: Aalto.

Marine Design XIII – Kujala & Lu (Eds)

Reducing GHG emissions in shipping—measures and options

Elizabeth Lindstad
Sintef Ocean AS (MARINTEK), Trondheim, Norway

Torstein Ingebrigtsen Bø
Sintef Ocean AS (MARINTEK), Trondheim, Norway
Norwegian University of Science and Technology (NTNU), Trondheim, Norway

Gunnar S. Eskeland
Norwegian School of Economics (NHH) and SNF and CenSES, Bergen, Norway

ABSTRACT: CO_2 emissions from maritime transport represent around 3% of total anthropogenic CO_2 emissions. These emissions are assumed to increase by 50% to 250% up to 2050 in business-as-usual scenarios with a tripling of world trade, while climate target of 1.5° – 2°C requires 50–85% reductions across all economic sectors. The maritime sector thus faces demanding challenges to reduce its emissions. Previous studies (Buhaug et al. 2009; Lindstad, 2013; Bouman et al. 2017) have indicated that by combination of design and operational measures based on today's technologies, emissions can be reduced by 75% up to 2050. This study examines the main reduction measures identified in previous studies and investigates to what degree the measures are implemented and used by the industry. Moreover, we assess how current policies encourage or discourage the implementation of the main reduction measures, and point towards important areas of policy realignment.

Keywords: Maritime transport; Shipping and the environment; Greenhouse gases; Abatement options; Emission reductions IMO

1 INTRODUCTION

From the first days of human civilization, sea transport has dominated trades between cities, nations, regions, and continents. Together with telecommunication, trade liberalization and international standardization, transport—maritime in particular—has enabled the process we call globalization (Kumar and Hoffman, 2002), entailing productivity gains from specialization and comparative advantages. World trade in the form we know today started around 1850 as global communication developed with steam engines allowing vessels to move without wind, steel hulls enabling larger ships, screw propellers making ships more seaworthy and deep-sea cables allowing traders and ship owners to communicate across the world (Stopford, 2009).

Products are increasingly being manufactured in one part of the world, transported to another country, further refined, and then redistributed to their final country of consumption.

Figure 1 illustrates the strong globalisation of the world from 1970 up to 2012 with all monetary figures adjusted to 2010 levels. First, the growth in sea transport measured in tons transported and ton- miles (freight work) broadly has followed annual global GDP growth of 3%. Second, growth in the *value* of international trade is twice the annual growth in tons moved, i.e. 6%. This means that movement of high-value items has increased more than movement of low value cargo such as iron-ore, crude-oil, coal and grain. Third, freight work measured in ton-miles increased as much as tons moved, which implies that average freight distance has been constant. Fourth, fuel consumption in maritime transport has increased less than the freight work, i.e. by 2% per year, which implies 1% annual reduction in fuel use and CO_2 emissions per ton nm.

The environmental consequences of increased international trade and transport have become important because of the current climate challenge (Rodrigue et al., 2016). With a business-as-usual (BAU) scenario with continuous transport growth, i.e. around 3% and 1% efficiency improvement, both annually (as seen from 1970) future emissions might increase by 150% – 250% up to 2050 (Buhaug et al., 2009; Lindstad 2013).

Figure 2 based on Smith et al. (2014) shows shipping emissions up to 2050 for the 16 different scenarios developed by the Third IMO GHG

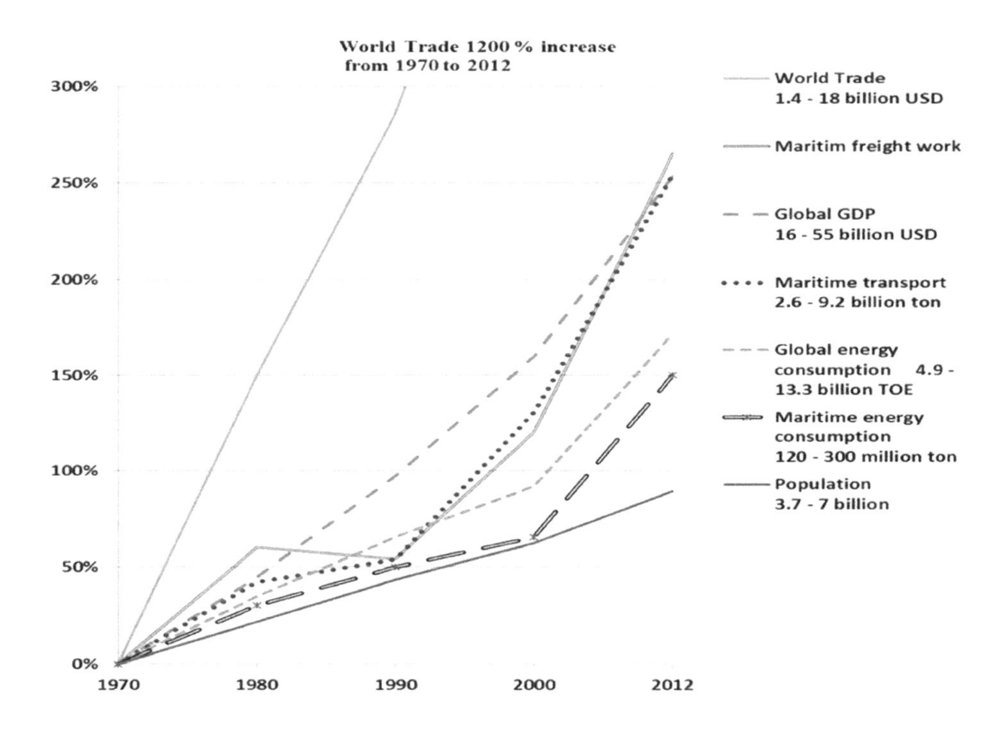

Figure 1. Global development 1970–2012. Sources: Lindstad (2013); Eskeland and Lindstad (2016).

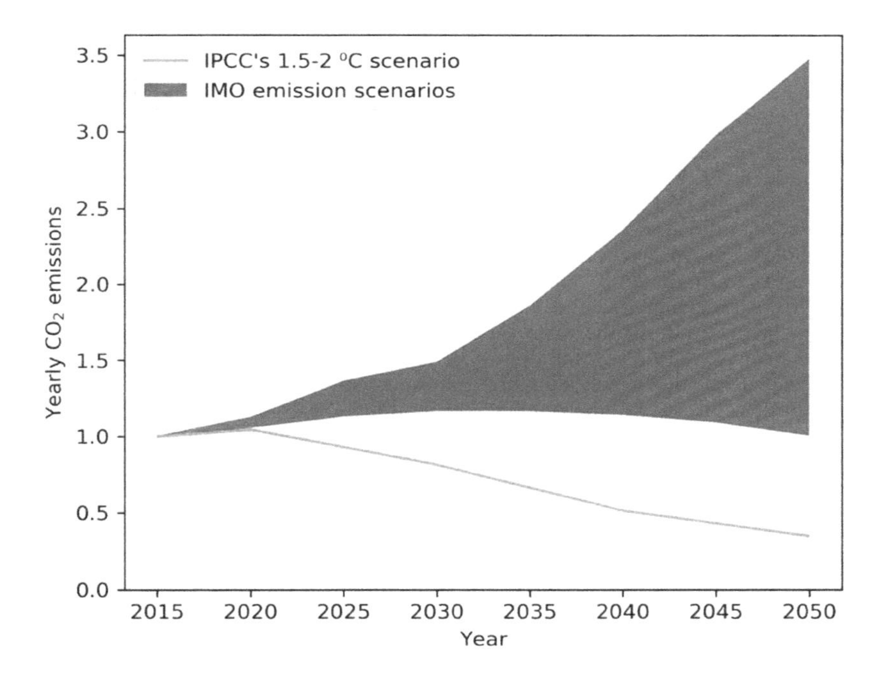

Figure 2. Scenarios for global shipping emission. Source: Smith et al. (2014) and IPCC (2013).

study. These scenarios contain various growth and technology assumptions, however none of them indicates a decrease in emissions up to 2050. In best case (for climate mitigation), emissions will stabilize and in worst case they will increase by 250%. These emission growth prospects are opposite to what is required to reach a climate targets by 2100. Global GHG emissions must decrease to net zero and even further to negative values across all sectors by the second half of this century, as indicated by the slope of the 1.5–2-degree scenario IPCC (2013). Nevertheless, it is a controversial issue how the annual greenhouse gas reductions shall be taken across sectors. Given a scenario where all sectors accept the same percentage reductions, the total shipping emissions in 2050 may be no more than 15% – 50% of current levels.

In unit terms to reach the 1.5–2 degrees target, the CO_2 emissions under a business as usual scenario should be reduced from 20–25 gram of CO_2 per ton nautical mile in 2007 to 4 gram or less of CO_2 per ton-nautical mile in 2050 (Lindstad 2013), i.e. an 80–85% reduction.

2 MODEL DESCRIPTION

The model enables a full evaluation of fuel consumption, costs and emissions as functions of vessel operation, abatement options and fuel prices; see Lindstad et al. (2011, 2014, 2015a, 2015b, 2017)

A vessel's fuel consumption is given by Equation (1)

$$F = \sum_{i=1}^{n} \left(\frac{D_i}{v_i} \cdot P_i^{dmvs} \cdot K_{fp} + T_{lwd} \cdot P_{lwd} \cdot K_{fp} \right) \quad (1)$$

During a voyage, the sea conditions will vary and we divide each voyage into sailing sections, with a distance D_i, speed v_i and power P_i^{dmvs} as a function of vessel design d, speed v, total weight carried m and sea conditions s. K_{fp} is fuel per produced kWh as a function of engine load, T_{lwd} is time spent in port loading, discharging, and waiting and P_{lwd} is average power used in port.

The cost per freight unit transported, i.e. per ton-mile is given by Equation (2):

$$C = \frac{1}{D \cdot M} \cdot \left(\left(\sum_{i=1}^{n} \frac{D_i}{v_i} + T_{lwd} \right) \cdot (TC + C_{abatement}) + F \cdot C_{Fuel} \right) \quad (2)$$

The first factor transforms cost to cost per ton-mile: M is the weight of the cargo and D is distance sailed with cargo. Large bulkers and tankers typically sail one way fully loaded and return empty in ballast, while liner vessels usually tend to be neither empty nor completely full. Total days per voyage is given by sailing days $\sum_{i=1}^{n} \frac{D_i}{v_i}$ and days in port T_{lwd}. The vessel's daily cost is given by its operational and financial costs plus the cost for the abatement option. Fuel cost is a function of consumed fuel F and the fuel price C_{Fuel}.

Emissions, E per pollutant, comprises fuel and freight work as expressed by Equation (3):

$$\varepsilon = \left(\frac{F}{D_C \cdot M \cdot N_C} \right) \cdot K_{ep} \quad (3)$$

K_{ep} is the emission factor for each exhaust gas as a function of power and fuel type. D_c is the distance of the cargo voyage, M is the weight of the cargo and N_c is the annual number of cargo voyages. SOx and CO_2 are always strictly proportional to fuel consumption by fuel type, while the other pollutants increase relative to fuel consumption when engine operates at high or low power.

Metrics that weight emitted greenhouse gases according to their global warming potential (GWP), and report them in terms of "*CO_2 equivalents*" have become a standard (Shine, 2009). Equation (4) gives total GWP impact per energy unit produced and ton transported.

$$GWP_t = \sum_{i=1}^{n} \varepsilon_e \cdot GWP_{et} \quad (4)$$

were, ε_e represents emissions per exhaust gas i and GWP_{et} is the GWP factor for each pollutant within the given time frame, i.e. usually, 20 or 100 years consistent with Houghton et al. (1990).

3 MITIGATION MEASURES IN THE LITERATURE

Maritime emission and reduction measures are commonly divided into two main categories: technical and operational (Psaraftis, 2016). Technical measures focus for example on energy savings through more energy efficient designs, improved propulsion and power system, and alternative or cleaner fuels. Some technical measures can be applied as retrofit measures, while others will practically and economically be considered only when building new vessels. Operational measures aim at reducing emissions during operations both for existing and newbuilt vessels.

Bouman et al. 2017 identified twenty-two (22) types of measures for which sufficient, reliable and comparable data are available in the peer-reviewed literature. Figure 3 shows the CO_2 reduction

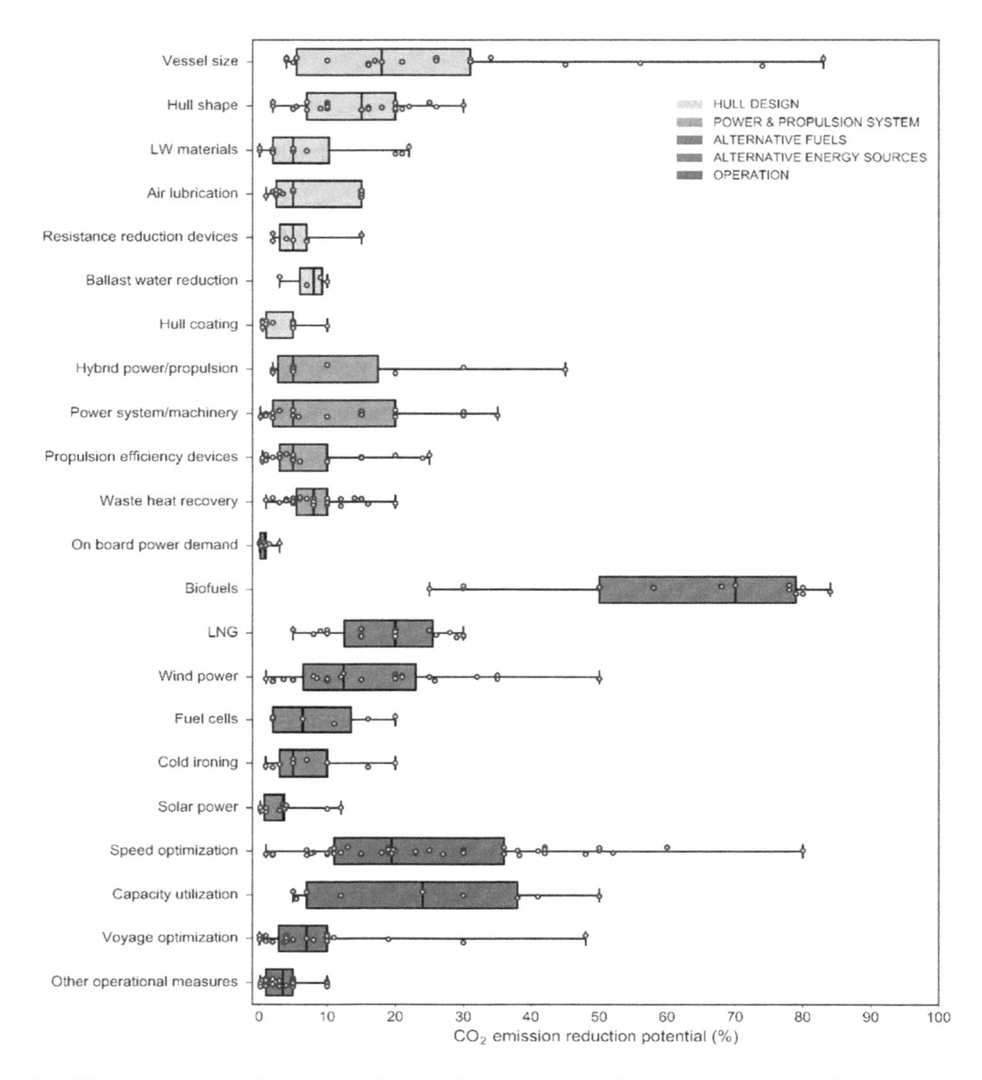

Figure 3. CO_2 emission reduction potential from individual measures. Source: Bouman et al 2017.

potential for each of the 22 measures. For each, a solid bar indicates the typical reduction potential area, i.e. from 1st to 3rd quartile of the dataset, and a thin line indicates the whole spread. In addition, the data points are shown by a small circle. Moreover, the study grouped the measures in five main categories: *hull design, power and propulsion, alternative fuels, alternative energy sources,* and *operations*.

From Figure 3 we observe a large range in emission reduction potential per measure reported by the individual studies. Some of the variability can be explained by differences in assumptions and benchmarks across the selected studies, but it also indicates large uncertainty as to the reported reduction potentials.

If all options depicted in Figure 3 could be combined, which is a highly hypothetical exercise, the emission reductions would be over 99% based on 3rd quartile values, 96% based on the median, and 82% based on 1st quartile values. A more likely feasible combination would be: Vessel size; Hull shape; Ballast water reduction; Hull coating; Hybrid power/propulsion; Propulsion efficiency devices; Speed optimization; Weather routing and Trim/Draft optimization. Assuming relatively large independence between the individual measures, combining these options can lead to emission reductions of 80% based on 3rd quartile values, 59% based on the median, and 34% based on 1st quartile values.

4 MITIGATION MEASURES IN THE SHIPPING INDUSTRY

Here we discuss to what extent the reduction measures from the literature are used by the industry. The first observation is that each of the 22 measures are used on at least one vessel. Second, the operational measures, represented by the four blue bars at the bottom of Figure 3, are relevant both for existing and new vessels. These are really the core of running a shipping business to make a profit, and are thus partly or fully employed across the industry. Third, economies of scale through larger vessels and operational speed reductions are the only ones for which we have seen large scale utilization, even though these simple measures also have potential to deliver a lot more.

First larger vessels. The key observation is that when the ship's cargo-carrying capacity is doubled, the required power and fuel use typically increases by about two thirds, so fuel consumption per freight unit is reduced. The vessel's building cost increases with about half of the increase in cargo capacity, and also costs of crew, maintenance and management rise less than proportionally with cargo capacity. Table 1 (Lindstad et al. 2015c) shows how average vessel size has increased from 2007 to 2015 based on IHS Markit data (www.ihs.com) and ISL data (ISL 2014). From 2007 to 2015, the average cargo vessel size has increased from 22 500 to 31 500 ton. Moreover, it can be observed that the its's the average size which has increased most. The explanation is the large increase in number of dry bulkers, which has an average size more than twice the average for the fleet.

Second, reducing operational speeds, The explanation is that the power output required for propulsion is a function of the speed to the power of three and beyond (Silverleaf and Dawson, 1966). This implies that when a ship reduces its speed, the power required and therefore the fuel consumed per transported unit is considerably reduced (Corbett et al., 2009; Psaraftis and Kontovas, 2010; Lindstad et al., 2011, Psaraftis and Kontovas, 2013). Table 2 show the development of average vessel size, design and operational speeds per vessel type from 2007 to 2012 (Smith et al., 2014; Lindstad et al., 2015c).

5 HOW LEGISLATION ENCOURAGE OR DIS-ENCOURAGE GHG REDUCTION

Presently, the policy objectives behind regulations for NO_X and SO_X relate to human health and ecosystems. CO_2 regulation, in contrast such as the Energy Efficiency Design Index (EEDI), is

Table 1. Average vessel increase 2007–2015.

	Average vessel size (dwt)			
Vessel type	2007	2012	2015	Change
Dry Bulk	52 500	68 600	69 300	32%
General Cargo	4 600	5 300	6 200	35%
Container	34 200	41 600	44 300	30%
Reefer	5 400	5 700	6 000	11%
RoRo & Vehicle	7 200	7 600	8 900	24%
Crude oil tank	178 700	183 500	185 800	4%
Product tank	9 800	10 700	10 700	9%
Chemical tank	15 800	18 000	19 000	20%
LNG & LPG	22 800	27 600	29 000	27%
RoPax	1 400	1 600	1 800	29%
Average	22 500	30 800	31 500	40%

Table 2. Design and operational speeds 2007–2012.

	Average vessel size (dwt)		Design speed		Operational speed	
Vessel type	2007	2012	2007	2012	2007	2012
Dry Bulk	52 500	68 600	14.1	14.8	12.2	11.5
General Cargo	4 600	5 300	12.1	12.5	10.0	9.3
Container	34 200	41 600	20.3	21.3	16.3	14.6
Reefer	5 400	5 700	16.2	16.2	16.2	13.4
RoRo & Vehicle	7 200	7 600	16.3	16.3	15.0	15.0
Crude oil tank	178 700	183 500	15.5	15.7	13.8	11.9
Product tank	9 800	10 700	12.3	12.4	10.6	9.4
Chemical tank	15 800	18 000	13.4	13.6	12.1	11.1
LNG & LPG	22 800	27 600	14.9	15.6	13.1	12.9
RoPax	1 400	1 600	17.9	16.6	13.8	10.7
Average	22 500	30 800	14.1	14.6	12.0	11.1

motivated by the need to reduce global warming. Other exhaust gases in shipping internationally are unregulated. Separate regulations for each exhaust gas exists, despite the fact that the emissions are interlinked both through the reductions measures and through their environmental impacts. One example is that the present approach to NO_X reductions through technical standards neglects that the reductions tends to come at the cost of higher fuel consumption (Lindstad et al., 2015b), and thus CO_2 emissions. Similarly, stricter SO_X rules tend to raise fuel consumption on a well to propeller basis, i.e. either when refineries remove sulphur from heavy fuel oils (HFO), or in scrubber operation and increased speeds at sea due to the higher capex with onboard abatement options (Lindstad et al., 2017).

As CO_2 is regulated through the energy efficiency design index (EEDI), the policy aims to address directly the ratio between the CO_2 emitted and the freight work. The EEDI verification—as a new vessel is built—takes place under assumptions of i) design speed; ii) design loads, and iii) still water conditions. This are important abstractions from real-life conditions; calm sea is the exception in shipping and—even at calm water—vessels generally operate at speeds different from design speed, depending inter alia on fuel prices and market conditions.

A major challenge if the emission reductions envisaged for CO_2 shall be achieved through EEDI will be to identify EEDI compliant solutions which are energy- and emission-efficient for power outtake under realistic operational conditions, from lying

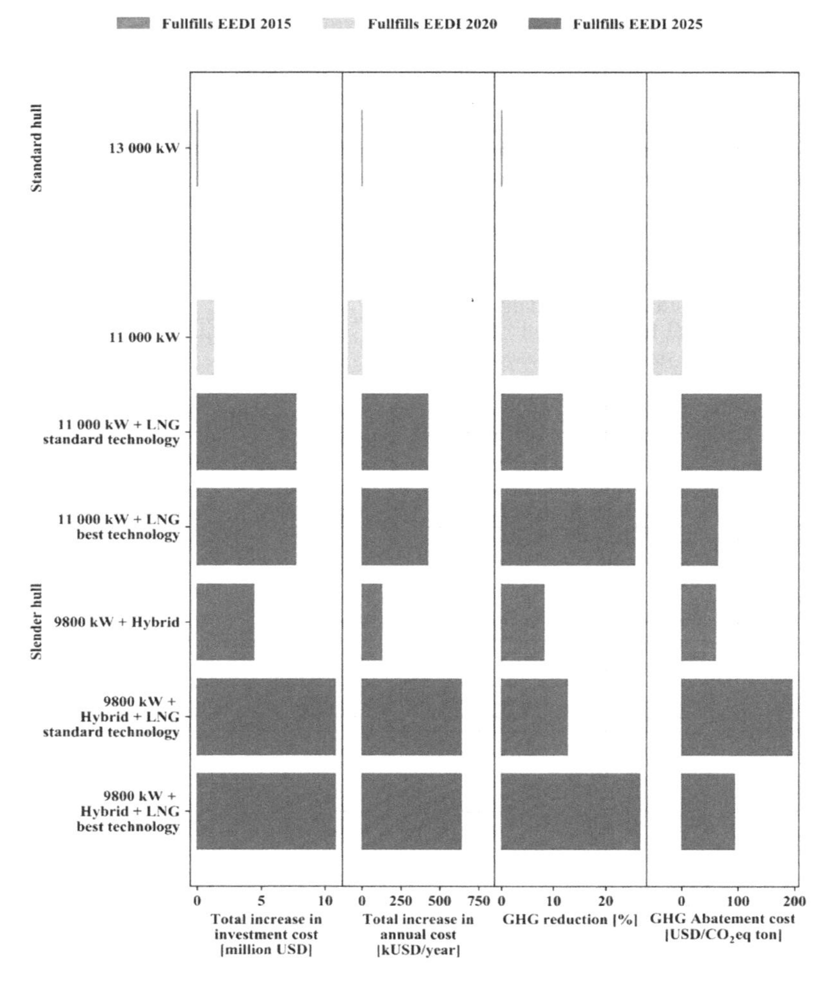

Figure 4. Investment cost, yearly cost, CO_2-eq emission and abatement cost per ton of CO_2 for an Aframax crude oil tanker. Based on Lindstad and Bø (2018).

idle at berth in port to realistic combinations of sea states, including ensuring that the vessels have the required power in critical situations in high sea states.

In Lindstad and Bø (2018) the effect of different abatement measures are investigated as EEDI compliance methods. An Aframax tanker (110 000dwt) is used as the case study vessel, and the measures evaluated are slender hull design, LNG, and hybrid propulsion. Tankers typically sail one way fully loaded and return empty in ballast, which gives an average capacity utilization of 50% for a roundtrip voyage. The vessels spend around 200 days at sea, 100 days loading, discharging, leaving or entering ports or in speed-restricted zones such as estuaries and canals, and the remaining 65 days idle in port or waiting at anchor.

One of the results from this study is shown in Figure 4. Here the first column shows the increase in newbuilding cost for the alternative abatement measures compared with the baseline vessel. The second column shows increased yearly cost, combining amortized CAPEX from column 1 with operational costs such as fuel expenditures. The third column shows the reduction in Greenhouse gas (GHG) emissions expressed as CO_2-equivalent on an annual basis. The fourth column shows GHG abatement cost per ton of GHG reduced (CO_2 eq.).

The red colour is used for today's standard design which has a main engine of 13 000 kW. This vessel fulfils the EEDI requirements for 2015, i.e. 10% reduction compared to 2013. From 2020, 20% reduction is required and from 2025, 30% reduction is required compared to 2013.

Main observations from Figure 4 are the large spread in cost and GHG emissions. For the cost the annual cost increases ranging from less than zero with the 11 000 kW slender design in 2020 up to 0.8 MUSD per year for the most expensive options in 2025. For the GHG emission the reductions range from 5%, i.e. when applying only the standard LNG technology to 25–27% when combining a slender hull form with best LNG technology and a hybrid power setup.

However, shipping lines are in business to make a profit, which suggests that their ranking will be based on cost minimizing, i.e. the slender vessel with a 11 000 kW engine to meet the 2020 EEDI standard, and the slender vessel with the 9 800 KW engine and a hybrid power plant to meet the 2025 EEDI standard. Consequently, the real reductions of GHG emissions through the EEDI scheme might be less than half of what is indicated by the test.

6 CONCLUSIONS

It is an important intervention point when a vessel is being commissioned, to influence its emissions through its lifetime in a cost-effective fashion. These interventions will then be based on assumptions about how the vessel is used.

For international shipping, apart from technologies (drivetrain, fuel, hybrid, hull), very important factors in future emissions will be the vessel size and speed, and it is thus problematic that policies such as EEDI will embody: i) unrealistic operative assumptions (like still water, and speed); ii) more generous EEDI limits for smaller vessels, thus to some extent failing to incentivize sufficiently a further move towards larger vessels; iii) speed limitations through power limitations, thus rendering vessels poorly equipped for power in situations of need and/or resulting in vessels operating at full power in normal conditions; iv) a slowdown in newbuildings resulting in a slowdown in modernization of the fleet, to some extent resulting in existing vessels being active longer and used more intensively.

ACKNOWLEDGMENT

This study has been financially supported by the Norwegian Research Council project (Norges Forskningsråd) 237917. *The SFI Smart Maritime*

REFERENCES

Bouman, E., A., Lindstad, E., Rialland, A. I, Strømman, A., H., 2017 State-of-the-Art technologies, measures, and potential for reducing GHG emissions from shipping – A Review. Transportation Research Part D 52 (2017) 408–421.

Buhaug, Ø.; Corbett, J.J.; Endresen, Ø.; Eyring, V.; Faber, J.; Hanayama, S.; Lee, D.S.; Lee, D.; Lindstad, H.; Markowska, A.Z.; Mjelde, A.; Nelissen, D.; Nilsen, J.; Pålsson, C.; Winebrake, J.J.; Wu, W.-Q.; Yoshida, K., 2009. Second IMO GHG study 2009. International Maritime Organization, London, UK, April.

Corbett, J, J. Wang, H, Winebrake, J, J. 2009. The effectiveness and cost of speed reductions on emissions from international shipping. Transportation Research D, 14, 593–598.

Eskeland, G., S., Lindstad, H., E. 2016 Environmental Taxation of Transport. The International Journal on Green Growth and Development 2:2 (2016) Page 51–86. ISSN 2393-9567.

Kumar, S. Hoffman, J., 2002, Globalisation the Maritime Nexus. In Grammenos, C.T. (editor): The Handbook of Maritime Economics and business. Page 35–64, LLP ISBN 1-84311-195-0.

Houghton, J.T., G.J. Jenkins, and J.J. Ephraums (eds.), 1990. Climate Change. The IPCC Scientific Assessment. Cambridge University Press, Cambridge, UK and New York, NY, USA.

IPCC 2013. Fifth assessment report of the Intergovernmental panel on climate change www.ippc.ch.

ISL 2014, Vessel fleet database and New-buildings up to 2017.

Lindstad, H. 2013. Strategies and measures for reducing maritime CO_2 emissions, Doctoral thesis PhD. Norwegian University of Science and Technology – Department of Marine Technology. ISBN 978-82-461-4516-6 (printed).

Lindstad, H. Asbjørnslett, B., E., Strømman, A., H., 2011. Reductions in greenhouse gas emissions and cost by shipping at lower speed. Energy Policy 39: 3456–3464.

Lindstad, E., Bø. T. (2018). Potential power setups, fuels and hull designs capable of satisfying future EEDI requirements. Under Journal review.

Lindstad, H., Eskeland. G., S., 2015a Low-carbon maritime transport: How speed, size and slenderness amount to substantial capital energy substitution. Transportation Research Part D, 41, pp. 244-256.

Lindstad, H., E. Eskeland. G., Psaraftis, H., Sandaas, I., Strømman, A., H., 2015b. Maritime Shipping and Emissions: A three-layered, damage-based approach. Ocean Engineering, 110 (2015) 94–101.

Lindstad, H., Verbeek, R., Blok, M., Zyl. S., Hübscher, A., Kramer, H., Purwanto, J., Ivanova,O. 2015c. GHG emission reduction potential of EUrelated maritime transport and on its impacts. European Commission: CLIMA.B.3/ETU/2013/0015.

Lindstad, H., Steen, S., Sandaas, I. 2014. Assessment of profit, cost, and emissions for slender bulk vessel designs. Transportation Research Part D 29: 32–39.

Lindstad H.E. Rehn C.F., Eskeland, G.S. 2017. Sulphur Abatement Globally in Maritime Shipping Transportation Research Part D 57 (2017) 303–313.

Psaraftis, H, N. Kontovas, C, A. 2013. Speed model for energy-efficient maritime transportation: a taxonomy and survey. Transp. Res. Part C, 26 (2013), pp. 331–351.

Psaraftis, H.N. Kontovas, C.A. 2010. Balancing the economic and environmental performance of maritime transport. Transportation Research Part D 15 (2010). Page 458–462.

Psaraftis, H.N., 2016. Green Maritime Transportation: Market-based Measures, In: Psaraftis, H.N. (Ed.), Green Transportation Logistics. Springer International Publishing, pp. 267–297.

Rodrigue, J.-P., Comtois, C., Slack, B. 2016. The Geography of Transport Systems, 4th Edition–Routledge.

Shine, K., 2009: The global warming potential: the need for an interdisciplinary retrial. Climate Change, 96, page 467–472.

Silverleaf, A., Dawson, J., 1966. *Hydrodynamic design of merchant ships for high speed operation*. Summer meeting in Germany 12th – 16th of June, 1966. The Schiffbau-technische Gescaft E.V, The institute of engineers and shipbuilders in Scotland, The North-East Coast Institution of Engineers and shipbuilders, The Royal institution of naval architects.

Smith et al. (2014) The Third IMO GHG Study.www.imo.org.

Stopford, M. 2009. Maritime Economics 3rd edition. 2009. Oxon: Routledge.

Marine Design XIII – Kujala & Lu (Eds)
© 2018 Taylor & Francis Group, London, ISBN 978-1-138-34076-3

Alternative fuels for shipping: A study on the evaluation of interdependent options for mutual stakeholders

Shinnosuke Wanaka & Kazuo Hiekata
Graduate School of Frontier Science, The University of Tokyo, Japan

Taiga Mitsuyuki
Graduate School of Engineering, The University of Tokyo, Japan

ABSTRACT: The objective of this research is to evaluate mutual stakeholders' interdependent options for introducing alternative fuels for shipping on the basis of SoS technique and simulation, which is an agent-based discrete event simulation of maritime transportation. The evaluation is carried out for not only each stakeholder's lifecycle properties but also the overall system's properties and clarifies tradespace of their actions. It helps stakeholders understand their system's phenomena and make a decision. In the case study, the introduction of LNG for crude oil tanker in a route between Japan and Persian Gulf is assumed, actions of ship operating company and terminal operator are modeled, and the impact of actions are evaluated by the developed simulation. On the basis of the result, it demonstrates that we can take a look at system's behavior by the simulation and discuss strategy of ship operator with some insight from the simulation result.

1 INTRODUCTION

It is highly complicated to answer the question that which fuels we should select as alternative fuels which is for corresponding to the new and more sever emission control regulation, MARPOL Annex VI (International Maritime Organization, 2009). From 2020, the emission of NOx, SOx and volatile organic compounds (VOC) are regulated in not only emission control areas (ECAs) but also the other regular areas. All the current vessels won't be operated any more without any changes, such as change of fuel type, introducing devices for emission reduction, and so on. The industry has to make a decision for the question, but the decisions are distributed to multiple stakeholders, such as ship owner, shipbuilding company, terminal operator, and so on, and their decisions are interdependent. Ship owner can't introduce LNG-fueled ship for transportation without introducing LNG bunkering facilities by terminal operator. Also, the decisions are highly affected by fuels' prices which nobody can foresee and are uncertain in the future. So, this problem is an interdependent decision making problem of mutual stakeholders considering future uncertainties.

The difficulties of such an interdependent decision making problem of mutual stakeholders are the complexity and uncertainty. Transportation system by vessels is large system of systems (SoS) (Maier, 1998), in which various kinds of forms and functions are interrelated and generate complexity. Also, the system of ship transportation is highly affected by future uncertainties including market and fuel price situations. The other stakeholders' decision makings and actions are also uncertain. These complexity and uncertainty make it difficult for stakeholders to make a valuable decision.

The objective of this research is to evaluate mutual stakeholders' interdependent options for introducing alternative fuels for shipping on the basis of SoS techniques and simulation which is an agent-based discrete event simulation of maritime transportation. SoS technique is utilized for mitigation of the system's complexity, and provides a holistic viewpoint over stakeholders. By modeling and simulation, the system's phenomena are evaluated. The evaluation is carried out for not only each stakeholder's lifecycle properties but also the overall system's properties and clarifies tradespace of their actions. It helps stakeholders understand their system's phenomena and make a decision.

In the case study, the introduction of LNG for crude oil tanker in a route between Japan and Persian Gulf is assumed, actions of ship operating company and terminal operator are modeled, and the impact of actions are evaluated by the developed simulation. On the basis of the result, it demonstrates that system's behavior can be observed by the simulation and strategy of ship operator with some insight from the simulation result can be discussed.

The rest of this paper is organized as follows. First, we discuss some of the existing works related to our research. Next a result of system's context analysis of introducing alternative fuels for shipping is given based on the SoS technique. And then, we explain the full detail of the developed simulation of ship transportation, which is for evaluation of stakeholders' action comes from the system context analysis. Lastly, case study based on the analysis and simulation is conducted and the result is discussed. Section 7 concludes this paper with a brief summary and mention of future work.

2 RELATED WORK

In this section, we review 2 kinds of papers, one is about existing analysis related to MARPOL ANNEX VI regulation, the other is about SoS techniques.

The compliance of the regulation for reducing gas emission is one of the most important topic for maritime industry. Researchers have been very active and a bunch of papers can be found in the field. Burel et al. (2013) estimate the impact of installation of LNG fueled ship from the perspective of operating cost efficiency and emission gas reduction. Their work reveals that LNG can be one of the solutions for the compliance of the regulation and, in particular, it demonstrates that there are some possibilities for handy size tanker to improve the efficiency. Brynolf et al. (2014) focus on alternative fuels and carry out environmental assessment of LNG, liquefied biogas, methanol and bio-methanol by using life cycle assessment (LCA). Schinas & Stefanakos (2014) state that the bibliography is extensive, but only a few academic references discuss operational pragmatism. They discuss the limitation of the financial assessment of technologies enhancing compliance with the SOx regulations, and propose the approach based on AHP and ANP for decisions of selecting technologies related to the compliance of the regulation. And also, Schinas & Butler (2016) study the feasibility of LNG as marine fuel by their method to evaluate the commercial incentives and assess policy that encourages the use of alternative technologies. In summary, there are researches about the reduction of gas emission for ship transportation from not only social but also technical aspect. The important point is that this topic and ship transportation system itself can be recognized as sociotechnical system and need to analyze by both social and technical perspective. And all of the works don't consider the topic as multiple stakeholders' problem.

For solving problems, design new architecture of sociotechnical system of systems (Davis et al. 2014), systems thinking (Frank, 2000) and the methods of engineering systems are studied over many years. The methods are widely adoptable for a bunch of domains, for example project of space development airplane and so on. Cameron et al. (2008) applied the stakeholder approach based on the system thinking to NASA's Vision for Space Exploration. Their work reveals the value delivery mechanism among stakeholders of the project. Doufene et al. (2013) proposed an architecture framework in order to design complex system with a holistic approach, and Góngora et al. (2013) proposed a method to search optimal solutions on the basis of the framework. They applied system of systems methods to frame a problem, which is the problem for introducing electric vehicles to a city, with multiple stakeholders' viewpoint, and applied game theory to search architectural equilibrium of the problem. They stated that understanding stakeholders' expectation, mainly from a strategic view point is not easy and it is important and helpful in the defining the system which is the target of design work. System of systems technique provides stakeholders' viewpoint to our design process and enables us to define a frame of the problem appropriately.

The novelty of this paper is that we treat the decision-making problem of alternative fuels for shipping as mutual stakeholders' problem, and apply the system of systems technique to analyze the system context and frame the problem. On the basis of the result of system context analysis, both overall system's lifecycle properties and stakeholders' lifecycle properties can be defined, and their interdependent actions can be evaluated and discussed. Our focus is on not only the impact of the introduction of alternative fuels or gas emission processing devices but also how is the interrelationship between stakeholders' options and the decision making.

3 SYSTEM CONTEXT ANALYSIS

In this section, we conduct system context analysis of the case, introducing alternative fuels for shipping, by using SoS techniques, in particular, stakeholder value network (SVN) (Cameron, 2011), and object process methodology (OPM) (Dori, 2002). These techniques provide a holistic viewpoint among multiple stakeholders and help us clarify what the system is and what the important properties are for stakeholders.

3.1 *Stakeholder analysis*

Stakeholder value network is a network diagram to map multiple stakeholders and show the relationship between the stakeholders and value flows. Boxes represent stakeholders and arrows represent value exchanged between stakeholders. The value can be defined as benefit at cost delivered from a stakeholder to another stakeholder. The aim of SVN is to clarify the list of stakeholders, their needs and value loop in the system. SVN around introducing alternative fuels for shipping is shown in Figure 1.

We identified 6 stakeholders in the system. The roles and definitions are shown in Table 1. Primary beneficiary of this system is cargo owner and the need is transportation of cargo. However, for this project, design of less emission transportation by vessels, the need comes from regulator. So, it is needed to take cargo owner and regulator into account when considering this project.

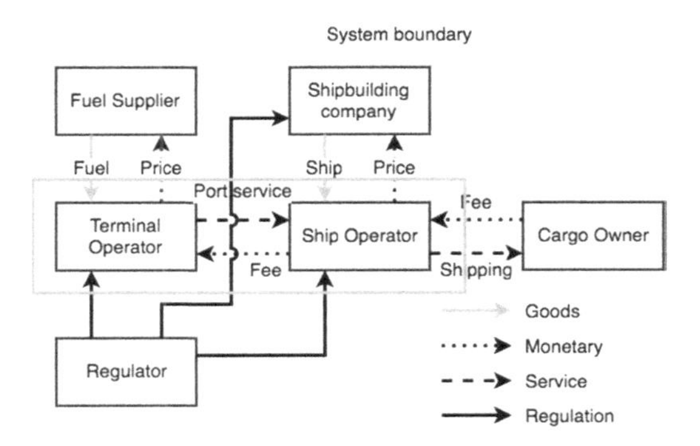

Figure 1. Stakeholder value network for making decisions about alternative fuels for shipping.

Table 1. Definition and role of stakeholders in the SVN.

Stakeholder	Definition & Role
Ship operator	Companies which own and operate vessels. Role: Decision maker.
Cargo owner	Companies which pay shipping fee to ship operator for shipping service. Role: Beneficial stakeholder
Terminal operator	Companies which operate storage and bunkering facilities at port. Role: Decision maker.
Fuel supplier	Companies which supply terminal operators with fuel. Role: Beneficial stakeholder.
Shipbuilding company	Companies which supply ship owners with ships or retro-fit system. Role: Beneficial stakeholder.
Regulator	Organizations which indicate long term energy plan and support the construction of infrastructure. Role: Problem stakeholder

For setting up the system boundary which defines what is inside our system, we classified the role of stakeholders, beneficial stakeholders, problem stakeholders and decision makers. The definition of beneficial stakeholder and problem stakeholder come from (Crawley, 2015). Beneficial stakeholder is the people who receive valued outputs and also provide valued inputs, and Problem stakeholder is the people who provide something to the others and need nothing from them. Additionally, we add decision maker as a role, who is a beneficial stakeholder and has ability to take an action for changing the system. The result of classification is shown in Table 1. On the basis of this classification, we set up the system boundary as shown in Figure 1. It means the behavior of fuel supplier, regulator, shipbuilding company and cargo owner are treated as external context and the analysis mentioned after is focused on the impact by decision makings of terminal operator and ship operator.

3.2 *Ilities: System lifecycle properties*

The ilities are properties of engineering systems that often demonstrate and determine value after a system is put into initial use. (de Weck, 2012) The ilities usually have relationship of trade-off and construct tradespace based on the stakeholders' decisions. In order to analyze the actions of decision makers mentioned above, we need to define not only system's overall ilities but also each stakeholder's ilities. Each stakeholder will select the action on the basis of his or her ilities, but the result will emerge as the system's overall ilities. Table 2 shows our definition of ilities.

The definition of system overall ilities derive from needs of regulator and cargo owner, outside the boundary. Environmental cleanliness is needed and constrained by regulator, and transported cargo amount is how the system fulfill the need of primary beneficiary, cargo owner. The ilities of ship operator are defined as operating cost and transported cargo amount and the ilities of terminal operator are defined as operating cost and

Table 2. Definition of the ilities for the overall system and the decision makers.

	Ilities
Overall system	Environmental cleanliness Transported cargo amount
Ship operator	Operating cost (Fuel efficiency) Transported cargo amount
Terminal operator	Operating cost Handled cargo amount.

handled cargo amount. Their purpose is to provide service to others with keeping operating costs as low as possible. Safety can be another important factor for the system and every stakeholder. However, their decisions will be made after the safety is guaranteed somehow. This is the reason we treat safety as not ilities but constraints.

3.3 *Architectural decisions of stakeholders*

In this subsection, options of the decision makers mentioned before are enumerated and listed up. For the enumeration, "intent-function-form" template (Crawley, 2015) by OPM is used. OPM is a system modeling language and "intent-function-from" template derive concept from solution-neutral function intent. Figure 2 shows the diagram by OPM. The intent of this project is to reduce gas emission from ship transportation. This intent will make an impact directly to the ship transportation's property, environmental cleanliness. There are 2 functions to realize this intent, changing type of fuel itself or introducing additional devices which are for processing the emission gas from vessels. As a result of breaking down of system's intent, general forms for implementing the functions are obtained. It means that decision makers have several options about these general forms.

On the basis of the result of analysis by OPM, an list of architectural decisions is obtained as shown in Table 3.

Ship operator is able to select a configuration for vessels including engine, fuel, and fuel tank and also, select a configuration about how to operate those vessels including fleet portfolio and their routes. As for the route, starting port and destination are fixed by the shipping demand, but they can change intermediate ports for bunkering. Terminal operator has options about type of bunkering facility, type of bunkering method and the portfolio. For example, they can select if they put a facility for LNG, HFO or the others, this is the selection of type of bunkering facility. Also, they can select to take shore to ship bunkering, ship to ship bunkering, or truck to ship bunkering. This is the selection of bunkering method.

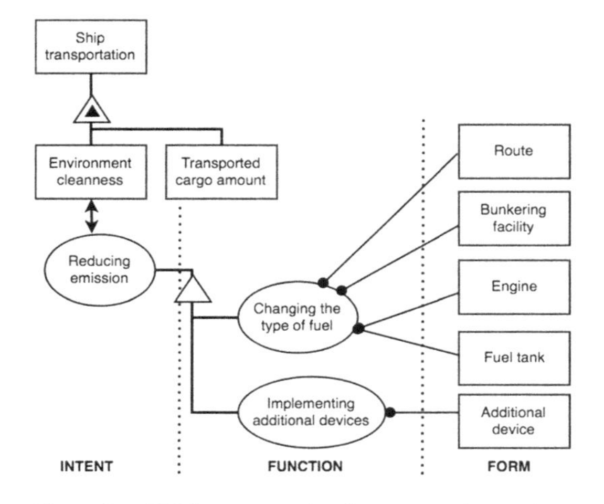

Figure 2. OPM representation for enumerating options of decision makers by "intent-function-form" template.

Table 3. The list of architectural decisions for each stakeholder.

Stakeholder	Decision item
Ship operator	Type of fuel, Type of engine Type of emission processing device, Type of fuel tank, Routing Fleet portfolio
Terminal operator	Type of bunkering facility Type of bunkering method Bunkering facility portfolio

4 SIMULATION OF SHIP TRANSPORTATION

In the previous section, we conduct system context analysis and obtain the list of decision makers, their ilities, and their decisions. For the evaluation of their decisions, we need to estimate how their decisions make an impact to the current transportation system. For the evaluation, simulation approach is applied in this paper. Agent base discrete event simulation for representing ship transportation system is developed, which is shown in this section. There are 2 reasons to take the agent base approach. One is that complex system's phenomena can be represented by agents' simple rule and attribute. And the other is that not only holistic system's properties but also detailed system's behavior can be observed from the result of simulation.

4.1 *Agent base modeling*

In the developed simulator, there are multiple agents and demands between them. Overview of this modeling is shown in Figure 3. Agents have their brief decision rules and demands assign tasks to the agents. Agents select their actions from the assigned demands according to their decision rules and take the selected actions.

Agents have 4 functions, "observe", "select", "checkNextEvent" and "update". For each iteration of the simulation, agents observe their status and demanded task list, select their next tasks, and estimate their next event. And then, they update their status until the next event. This is the core logic of the developed simulation on the basis of agent base modeling.

4.2 *Type of agents and demands*

Table 4 shows the type of agents and demands in the simulation. There are 5 types of agents and 5 types of demands. Cargo owner makes a shipping demand to ship operator. According to the shipping demand, ship operator makes a transportation demand to their ships and berthing demand to terminal operator. Terminal operator makes a bunkering demand and loading demand to owned facilities considering the vessel's fuel type and cargo type. After the distribution of the demands, ships and facilities take actions in accordance with the demanded tasks.

All kinds of demand have start and end time as the properties. Shipping and transportation demand has cargo type, cargo amount, starting port and destination as the other properties. Berthing demand is split into bunkering demand and loading demand, and bunkering demand has the type and amount of fuel for bunkering, loading demand has the type and amount of cargo for loading or unloading.

For the simulation, it is needed to represent outside of the system boundary because the outside system affects the system in the boundary, for example, fuel price, marker price and so on. It is called external environment in this paper.

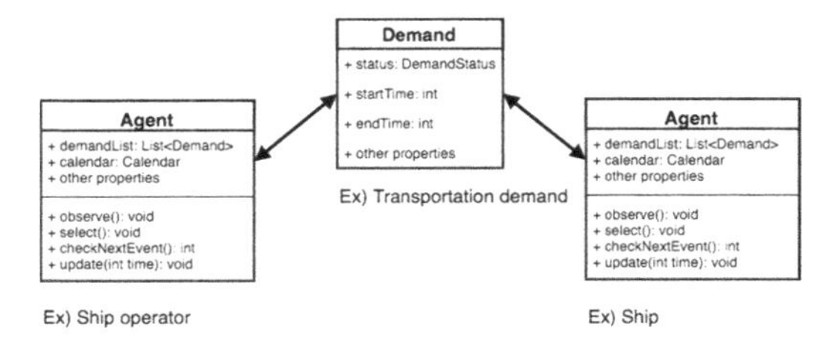

Figure 3. UML diagram for representing agent base modeling framework in the developed simulation.

Table 4. Type of agents and demands.

Agent	Demand	To
Cargo owner	Shipping demand	Ship operator
Ship operator	Transportation demand	Ship
	Berthing demand	Terminal operator
Terminal operator	Bunkering demand	Port facility
	Loading demand	Port facility
Ship	N/A	N/A
Port facility	N/A	N/A

5 CASE STUDY

This section provides the case study on the basis of the result of system context analysis and the simulation. In this case study, we narrow down the target of evaluation and introduce some assumptions, for example, cargo type, route of transportation, stakeholders' business scale and so on, and evaluate that how the stakeholders' decisions are interdependent and make an impact systemically.

5.1 *Assumptions*

In this case study, some factors are assumed as external environment. First of all, we consider only oil tanker (VLCC) as ship type and a route between Yokohama and Persian Gulf as target route for narrowing down the target of analysis and simplifying the problem. Table 5 shows all of the assumptions for fuels the VLCC and port facilities. Heavy fuel oil (HFO), liquefied natural gas (LNG), and marine gas oil (MGO) are assumed as alternative fuels. Those prices are not constant and uncertain in the future, but for this case study, we pick up 1 scenario from many forecasted scenarios (Danish Maritime Authority, 2012). The value of emission rates for NOx, SOx, and CO_2 are from the reference of (Winnes & Fridell, 2009).

Table 5. Assumptions of fuels, VLCC and ports.

Fuel properties			
Fuel type	HFO	LNG	MGO
Price [$/ton]	620	570	1040
Emission rate [ton/ton]			
NOx	0.0045	0.0009	0.0045
SOx	0.002	0	0.0006
CO_2	3.1	2.75	3.2
Ship properties			
Item	Unit	Value	
Ship type	-----	VLCC	
Average speed	knot	15	
Deadweight (Cargo)	ton	300,000	
Fuel oil tank capacity	m^3	5,000	
Fuel oil consumption rate	ton/mile	0.20	
Port properties			
Item	Unit	Value	
Cargo on/off loading speed	ton/h	3125	
Bunkering speed	ton/h	104	

All of the ships that ship operator has are the same type of ships, and all of the facilities in the ports has the same performance for loading and bunkering.

5.2 *Decisions of stakeholders*

We assume 1 ship operator and 2 terminal operators as the decision makers in the case. Ship operator has options for their ships' configuration and fleet portfolio. Terminal operators have options of their bunkering facilities' ability and the portfolio. In the first situation, the ship operator has 20 HFO fueled vessels and each port has 5 facilities for oil loading and HFO bunkering.

The ship operator's decisions are how many scrubbers are installed, how many vessels are retrofitted with MGO fueled engine system, and how many vessels are exchanged with LNG fueled vessels. If the scrubber is installed for a HFO fueled ship, the SOx emission is 100% reduced and the NOx emission is 80% reduced. In exchange for the advantage for eco-friendliness, the cargo space is 20% reduced because space is needed for the installation. The installation cost is 7.2E+06[$]. The retrofitting for MGO fueled engine system doesn't need initial cost, and just change the fuel type from HFO to MGO. The exchange between HFO fueled ship and LNG fueled ship needs 1.8E+08[$].

The terminal operators' decisions are how many MGO and LNG bunkering facilities are installed in the ports. When those bunkering facilities are not installed, MGO and LNG fueled vessels cannot be operated in the route. As the initial situation, it is assumed that each port has 5 HFO bunkering and loading facilities which are enough facilities for 20 HFO fueled ships. The terminal operators can make a decision they will introduce additional bunkering facilities for alternative fuels. The cost of introduction of LNG bunkering is 1.0E+07 per 1 facility, and the cost of introduction of MGO bunkering facility is 0 per 1 facility.

The summary of the decisions is shown in Table 6 whose format is like Morphological Matrix (MM) (Zwicky, 1967). One option can be selected from each row and the combination of selected option represents one decision making of ship and terminal operators.

5.3 *Evaluation result*

According to the Table 6, 240 cases are evaluated by using the simulation mentioned previously. The time step of the simulation is 1 hour and the simulation duration is 1 year. The amount of emitted NOx, SOx and CO_2 and the amount of transported cargo are calculated as system's overall properties, and fuel efficiency is calculated as the ship operator's property. Operating cost for the ports is calculated for terminal operators' property. The amount of transported cargo is calculated by Equation 1, and the fuel efficiency is calculated by Equation 2. M_{cargo} is the amount of transported cargo by 1 ship, $D_{transported}$ is the transported distance of 1 ship, and FCC is the cost for the fuel. For this kind of decision making, net present value (NPV) analysis is usually applied, and initial cost is compared with the operating cost. But, in this research, environmental performance and transportation capacity are also compared with the initial cost. For the comparison by such multiple criteria, NPV analysis using discount rate is not applied.

Table 6. Decision matrix for ship and terminal operator.

Ship operator					
Fuel type	HFO		LNG		MGO
Scrubber	Yes		No		
Fleet portfolio					
HFO with scrubber	0	5	10	15	20
LNG	0	5	10	15	20
MGO	0	5	10	15	20
Terminal operator					
Additional bunkering facility portfolio					
LNG bunkering	1	3	5		
MGO bunkering	1	3	5		

$$A_{cargo} = \sum_{ship} M_{cargo} \cdot D_{transported} \tag{1}$$

$$E_{fuel} = \sum_{ship} \frac{FCC}{M_{cargo} \cdot D_{transported}} \tag{2}$$

As a result of the evaluation, 82 pareto optimal configurations are obtained. Transporting 80% of cargo amount transported by 20 HFO vessels is considered as the minimum requirement, and 58 configurations are picked up. The result from the viewpoint of the overall system is shown in Figures 4 and 5.

Figure 4 shows a tradespace which explains the relationship between CO_2 gas emission and the amount of transported cargo amount. On the pareto front, we obtain only the configuration in which ship operator and terminal operators have the same strategy for selecting fuel type. In the case ship operator highly invests to LNG fueled ships, the pareto optimal decision for terminal operator is to highly invest for LNG bunkering facilities. In the case terminal operator moderately invests for alternative fuel, the pareto optimal decision of ship

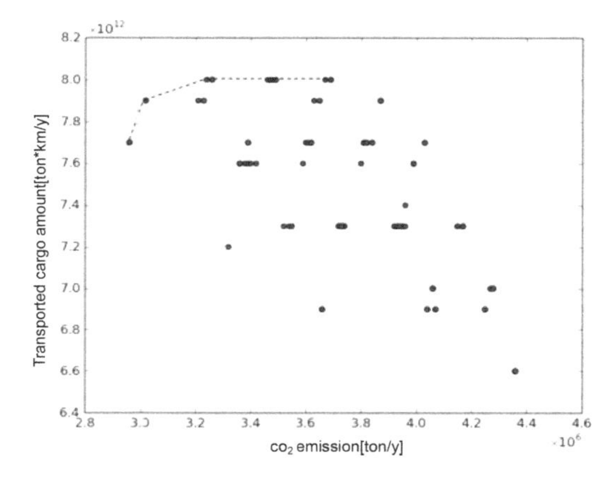

Figure 4. Tradespace by the amount of CO_2 gas emission and transported cargo amount. Less gas emission and more transported cargo amount are better for the system, so the upper left side of the graph is the utopia point.

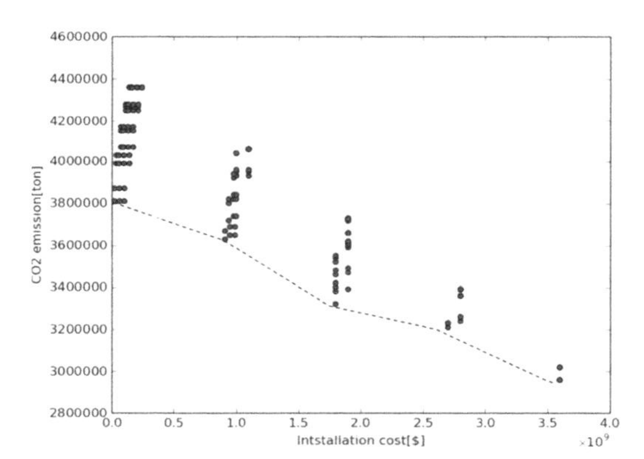

Figure 5. Tradespace by the amount of CO_2 gas emission and the installation cost. Less gas emission and cost are better for the system, so the lower left side of the graph is the utopia point.

operator is to invest moderately to the alternative fueled ships. Basically, the shared and same strategy make a pareto optimal impact for gas emission and transported cargo amount.

Figure 5 shows a tradespace which explains the relationship between the installation cost and CO_2 gas emission. The installation cost includes all of the cost for ship construction and retrofitting and port facilities installation. The figure shows that there is a trade-off relationship between the CO_2 gas emission and the installation cost, and it means that if it is needed to obtain better environmental performance, it is needed to invest more for changing the transportation system.

By Figures 4 and 5, it is possible to observe the behavior of the overall system. For evaluating the interdependent stakeholders' action, some characteristic cases are picked up and summarized in Table 7.

6 DISCUSSION

6.1 *Strategy of ship operator*

Table 7 shows the characteristics of the options. LNG fueled transportation system has the best environmental performance, but also it needs high initial cost and it is highly affected by the uncertainties of fuel price and other stakeholders' actions. MGO fueled transportation system has the worst fuel efficiency, however it hardly needs initial cost and that will lead to the situation that it is easier to collaborate with the other stakeholders. It means the system is not affected by uncertainty derived from the others' action so much. And, the advantage of HFO fueled transportation system with scrubber is that it doesn't need high initial cost and doesn't have risks for now because it can be realized by just improvement of the current system. However, it just fulfills the minimum requirement of environmental performance.

When focusing on the ship operator's decision making, their interests are the fuel efficiency and

Table 7. Evaluation result of specific characteristic cases.

Ship portfolio			Port facility portfolio			Gas emission			Cargo amount	Fuel efficiency	Initial cost	
H*	M*	L*	H	M	L	NOx	SOx	CO_2			Ship	Port
20	0	0	5	0	0	6.34 E+03	0	4.36E+06	6.56E+12	1.22E-04	1.44E+08	0
0	20	0	5	1	0	3.15E+03	4.20E+02	2.24E+06	2.49E+12	1.62E-01	0	0
0	20	0	5	5	0	5.45E+03	7.26E+02	3.87E+06	7.93E+12	9.34E-02	0	2.00E+07
0	0	20	5	0	1	5.72E+02	0	1.75E+06	2.49E+12	1.29E-04	3.54E+09	2.00E+07
0	0	20	5	0	5	9.87E+02	0	3.02E+06	7.93E+12	7.48E-05	3.54E+09	1.00E+08

*H: HFO (Heavy fuel oil), M: MGO (Marine gas oil), L: LNG (Liquefied natural gas).

the amount of transported cargo. Any type of fuel can realize enough amount of the cargo transportation when enough infrastructures exist. So, when they are able to share their policy and build a cluster, they can choose any type of fuel with focusing on the fuel efficiency because the consensus can mitigate the uncertainty of other stakeholders' action. If it is impossible, the ship operator needs to mix the options and perform risk hedge. A way of the risk hedge is using MGO because it doesn't need initial cost and must be robust for the future uncertainty. The characteristic of MGO, the worst fuel efficiency means that the MGO fueled transportation system doesn't have scalability, and so it cannot be main fuel. However, it also has the robustness for the future uncertainty and it can be an effective ingredient in the context of risk mitigation. The problem is in focusing on the fuel efficiency because there are a bunch of estimations for LNG fuel price in the future, and it will directly make an impact for the fuel efficiency. Fuel price naturally has future uncertainty, but the change of ships' fuel is a big shift from oil to gas and it is not possible to foresee that the shift affects the price positively or negatively. As a result of the situation, the ship operator cannot make a valuable decision until the behavior of the LNG price becomes clear or the ship operator has a confidence of the behavior. Beyond the ship operator's local perspective and when considering the sustainability of the transportation system itself, environmental sustainability, LNG is the best solution because the other fuels have disadvantage about NOx and CO_2 emission. This perspective is out of ship operator's interest. However, the sustainability can be a requirement by pressure from regulators. In fact, a regulation for LNG carriers about CO_2 emission by EEDI is coming from IMO (Attah & Bucknall, 2015).

The result of case study reveals that the system's behavior and options' characteristics. Ship operators can make a decision with those characteristics and their expectation about fuel price and regulation of gas emission.

6.2 *Method to analyze mutual stakeholders' interdependent options*

The evaluation procedure in this paper is following.

1. Stakeholder analysis by SVN.
2. Definition of ilities for the system and stakeholders.
3. Enumeration of stakeholders' options.
4. Simulation based on the result of 1–3.
5. Visualization of the simulation result and observation of the system's behavior.

This paper just shows one example applied the evaluation procedure. However, this procedure is not specific for the problem, selecting alternative fuels for shipping, and can be applied for another decision-making problem which is about interdependent decisions by mutual stakeholders.

It means that the procedure can be design template for that kind of problems and this paper shows the applicability by one concrete example.

6.3 *Limitation*

In this research, the situation that only one shipping company exists in the world is assumed. The situation one shipping company to one terminal operator is very ideal for considering the decisions of terminal operator because the terminal operator has other customers and the risk will be hedge and income will be more increased than the result of this study. Next direction is to become more realistic and get insight for the actual decision making of the terminal operators.

7 CONCLUSION

This paper evaluated mutual stakeholders' interdependent options for introducing alternative fuels for shipping on the basis of SoS techniques and simulation which is an agent-based discrete event simulation of maritime transportation. SoS technique is utilized for mitigation of the system's complexity, and provides a holistic viewpoint across stakeholders. By modeling and simulation, the system's phenomena were evaluated. The evaluation was carried out for not only each stakeholder's lifecycle properties but also the overall system's properties which and clarifies tradespace of their actions. It reveals the relationship between a bunch of diversified ilities, such as initial cost, fuel cost, environmental performance, and the transportation capacity.

In the case study, the introduction of LNG for crude oil tanker in a route between Japan and Persian Gulf was assumed, actions of ship operating company and terminal operator were modeled, and the impact of actions were evaluated by the developed simulation. The result reveals the characteristics of the options taken from each stakeholder. On the basis of the result and defined stakeholders' interest, it demonstrated that strategy of ship operator with some insight from the simulation result could be discussed.

The procedure we applied to evaluate the selecting alternative fuel for shipping is not specific about the problem and can be applied for the same kind of problems, mutual stakeholders' interdependent decision making problems. One of our future directions is that we will apply the procedure to some different problems, normalize and refine the procedure and propose a general approach for the mutual stakeholders' interdependent decision making problems.

REFERENCES

Attah, E.E., & Bucknall, R. 2015. An analysis of the energy efficiency of LNG ships powering options using the EEDI. *Ocean Engineering* 110: 62–74.

Brynolf, S., Fridell, E., & Andersson, K. 2014. Environmental assessment of marine fuels: liquefied natural gas, liquefied biogas, methanol and bio-methanol. *Journal of cleaner production* 74: 86–95.

Burel, F., Taccani, R., & Zuliani, N. 2013. Improving sustainability of maritime transport through utilization of Liquefied Natural Gas (LNG) for propulsion. *Energy*, 57: 412–420.

Cameron, B.G., Crawley, E.F., Loureiro, G., & Rebentisch, E.S. 2008. Value flow mapping: Using networks to inform stakeholder analysis. *Acta Astronautica*, 62(4): 324–333.

Cameron, B.G., Seher, T., & Crawley, E. 2011. Goals for space exploration based on stakeholder value network considerations. *Acta Astronautica* 68(11): 2088–2097.

Crawley, E., Cameron, B. & Selva, D. 2015. System architecture: strategy and product development for complex systems. Boston, Pearson.

Danish Maritime Authority. 2012. North european LNG infrastructure project. A feasibility study for an LNG filling station infrastructure and test of recommendations: 20.

Davis, M.C., Challenger, R., Jayewardene, D.N., & Clegg, C.W. 2014. Advancing socio-technical systems thinking: A call for bravery. *Applied Ergonomics* 45(2): 171–180.

Dori, D. 2002. *Object process methodology – a holistic systems paradigm*. Berlin: Springer Science & Business Media.

Doufene, A., Chalé-Góngora, H.G., & Krob, D. 2013. Complex systems architecture framework: Extension to multi-objective optimization. *Complex Systems Design & Management*: 105–123.

de Weck, O.L., Ross, A.M., & Rhodes, D.H. 2012. Investigating relationship and semantic sets amongst system lifecycle properties (ilities). *Proceedings of the 3rd International Symposium on Engineering Systems CESUN 2012*.

Frank, M. 2000. Engineering systems thinking and systems thinking. *Systems Engineering* 3(3): 163–168.

Góngora, HGC., Doufene, A., & Krob, D. 2013. Sharing the Total Cost of Ownership of Electric Vehicles: A Study on the Application of Game Theory, *INCOSE International Symposium* 23 (1): 988–1005.

International Maritime Organization. 2009. Amendments to the annex of the protocol of 1997 to amend the international convention for the prevention of pollution from ships 1973, as modified by the protocol of 1978, relating thereto (Revised MARPOL Annex VI). *Technical Report MPEC* 176(58).

Maier, M.W. 1998. Architecting principles for systems-of systems. *Systems Engineering* 1(4): 267–284.

Schinas, O. & Stefanakos, Ch. H. 2014. Selecting technologies towards compliance with MARPOL Annex VI: The perspective of operators. *Transportation Research Part D: Transport and Environment* 28: 28–40.

Schinas, O., & Butler, M. 2016. Feasibility and commercial considerations of LNG-fueled ships. *Ocean Engineering*, 122: 84–96.

Winnes, H., & Fridell, E. 2009. Particle emissions from ships: dependence on fuel type. *Journal of the Air & Waste Management Association* 59(12): 1391–1398.

Zwicky, F. 1967. The morphological approach to discovery, invention, research and construction. *New Methods of Thought and Procedure*: 273–297.

Marine Design XIII – Kujala & Lu (Eds)

On the design of plug-in hybrid fuel cell and lithium battery propulsion systems for coastal ships

P. Wu & R.W.G. Bucknall
Marine Research Group, Department of Mechanical Engineering, University College London, London, UK

ABSTRACT: A plug-in hybrid propulsion system comprising of a Proton Exchange Membrane Fuel Cell (PEMFC) and lithium battery capable of being recharged in port offers a promising low carbon propulsion system for small coastal ships, e.g. small container ships, tankers and ferries, which typically sail over short routes at modest speeds. PEMFC operate at high efficiency and emit no harmful emissions, but their poor transient performance necessitates the need for an energy storage system such as a lithium battery. A shore-to-ship electrical connection is needed to recharge the lithium battery from the grid so as to improve the propulsion system performance both environmentally and economically. Production of both H_2 and grid electricity have a carbon footprint. In this paper a two-layer optimisation based methodology is used for the design of plug-in hybrid fuel cell and lithium battery propulsion systems for coastal ships. Results from a case study suggest that the design of hybrid PEMFC and battery propulsion systems should be influenced by the 'well-to-propeller' carbon footprint.

Keywords: Plug-in hybrid fuel cell; energy storage system; hydrogen; ferry; propulsion system design

1 INTRODUCTION

1.1 *Fuel cells*

Fuel cells offer the desirable combination of high efficiency and environmentally benign operation (Sharaf & Orhan 2014). Among the main fuel cell types, the low-temperature Proton Exchange Membrane Fuel Cell (PEMFC) and high-temperature fuel cells (e.g. solid oxide fuel cell and molten carbonate fuel cell) offer the most promising power sources for future marine propulsion applications (Luckose et al. 2009). However, the economic feasibility of fuel cells is currently compromised by their high cost, poor transient performance, poor reliability, availability of alternative fuel supplies e.g. H_2 and associated fuel bunkering facilities (de-Troya et al. 2016).

High-temperature fuel cells offer higher efficiency when compared to the PEMFC (van Biert et al. 2016). A higher operating temperature makes it possible to recover heat from the exhaust gas so as to improve overall thermal efficiency, e.g. a combined cycle plant. Importantly high-temperature fuel cells can use a range of fuel types including natural gas. However, the main challenges of high-temperature fuel cells in marine applications are their low overall power to volume density, long start-up times, limited cycling times and transient performance (Welaya et al. 2011).

PEMFC have been successfully applied to a range of propulsion applications, e.g. road vehicles, submarines and inland water boats (Sasank et al. 2016, Pei & Chen 2014, Han et al. 2012). PEMFC offer improved power to volume density but their efficiency is lower than high-temperature fuel cells and they can only operate on H_2 (van Biert et al. 2016). Unlike natural gas, H_2 does not exist naturally on earth so is produced using various means including electrolysis and reformation of hydrocarbon fuels. Therefore, H_2 through life Global Warming Potential (GWP), production cost, bunkering and onboard storage will all influence the feasibility of using PEMFC in ships.

For coastal ships operating on short routes at modest speeds then PEMFC with their better power to volume ratio appear to be more suitable. The PEMFC is well developed and its price is falling (DOE 2015). PEMFC with lithium batteries will provide acceptable transient performance. The low volumetric energy density of the hydrogen fuel suggests efficient operation is required to minimise onboard storage facilities. The production of H_2 has a carbon footprint as does grid electricity production. This paper explores these factors for a low carbon propulsion system.

NOMENCLATURE

Acronyms

AC	Alternating current
DC	Direct current
EMS	Energy management strategy
ESS	Energy storage system
GWP	Global warming potential
HHV	High heating value
MOO	Multi-objective optimisation
NGSR	Natural gas steam reforming
PEMFC	Proton exchange membrane fuel cell
PIHFCB	Plug-in hybrid fuel cell and battery
SOC	State of charge

Roman symbols

C	Lithium battery C-rate
c_{eq}	Equality constraint
F_1	Multi-objective optimisation 1st objective function
F_2	Multi-objective optimisation 2nd objective function
f	Single-objective optimisation objective function
g_{fc}	Fuel cell specific hydrogen consumption function
J	Time step number when the ship calls at the port
K	Total time step number
L_{fc}	Fuel cell lifetime, h
L_{ESS}	Lithium battery lifetime, h
M_1	Multi-objective optimisation 1st constraint function
M_2	Multi-objective optimisation 2nd constraint function
P_{ESS}	ESS power, kW
P_{shore}	Shore power, kW
P_l	Load power, kW
P_{fc}^R	Fuel cell rated power, kW
P_{fc}	Fuel cell power, kW
P_{ESS}^R	Lithium battery rated power, kW
p_h	H_2 price, $/kg
p_{fc}	PEMFC price, $/kW
p_{ESS}	Lithium battery ESS price, $/kWh
p_e	Shore electricity price, $/kWh
R_{fc}	Fuel cell power ramp up/down limit
SOC_i	Lithium battery state of charge at *i*-th time step
T	Voyage time, h
Δt	Time step length, h
V_D	Equivalent diesel system total volume, m^3
V_{df}	Diesel fuel tank volume, m^3
W_D	Equivalent diesel system total weight, t
w_h	H_{2We} specific GWP, kg CO_2/kg
w_e	Electricity specific GWP, kg CO_2/kWh
x	Decision vector
$x_{1,2,\ldots,K}$	PEMFC stack per unit power output
$x_{K+1,K+2,\ldots,2K}$	Lithium battery C-rate

Greek symbols

ρ_{fc}^v	PEMFC stack volumetric power density, kW/m^3
ρ_{ESS}^v	ESS stack volumetric energy density, kWh/m^3
ρ_t^v	H_2 tank volumetric energy density, m^3/kg H_2
ρ_{dg}^v	Diesel engine volumetric power density, kW/t
ρ_{fc}^g	PEMFC stack gravimetric power density, kW/t
ρ_{ESS}^g	ESS gravimetric energy density, kWh/t
ρ_t^g	H_2 tank gravimetric energy density, kg/kg H_2
η_1	Uni-directional converter efficiency
η_2	Bi-directional converter efficiency
η_b	Lithium battery efficiency

1.2 *Energy storage systems*

Energy Storage Systems (ESS) such as lithium batteries have already been adopted for use in commercial ship applications, often in a configuration of hybridisation with the diesel engine (Luo et al. 2015). When hybrid configurations are used, they can potentially achieve 15% annual fuel saving depending on operational profile, e.g. the Viking Lady offshore supply vessel (Stefanatos et al. 2015). When only a battery is used alone for propulsion, e.g. the Norled Ampere battery powered ferry, then the low volumetric energy density of the batteries restricts both speed and range.

For a marine Plug-in Hybrid Fuel Cell Battery (PIHFCB) propulsion system, the ESS provides transient capability and greater plant efficiency. Furthermore, when a shore charging facility is available, integration of ESS can further improve the overall energy efficiency through direct utilisation of clean grid electricity e.g. electrolysis of water generating H_2 rather than reformation of hydrocarbon fuels. For a PIHFCB propulsion system, lithium batteries are preferable over other main ESS types for better energy density (Hannan et al. 2017).

1.3 *Design methodology*

Using fuel cells and batteries together the overall GWP can potentially be very low or even zero when renewable energy is utilised for electricity generation and hydrogen (Hansen & Wendt 2015). There are some research studies, e.g. Bassam et al. (2016) and Mashayekh et al. (2012) who have looked into the optimisation of hybrid ship propulsion systems. These works only focus on cost optimisation without considering the overall environmental performance of the propulsion plant i.e. well-to-propeller.

When multiple power sources are integrated into one propulsion system, two problems need to be resolved: 1) How to size the different energy and power sources to achieve an optimised well-to-propeller emission performance; 2) How to manage the different power sources to maintain high overall efficiency.

Since this paper is considering a PIHFCB propulsion system suitable for coastal ships which typically sail on short routes at modest speeds the analysis needs to consider GWP emissions and operating costs. The propulsion system design methodology consists two layers of optimisation:

1. An external layer applies a controlled elitist Multi-Objective Optimisation (MOO) scheme using evolutionary algorithms to optimise environmental and economic performance thereby overcoming the constraints on the propulsion plant design such as volumetric and gravimetric limits of the ESS and H_2 (Deb 2001).
2. The inner layer optimisation scheme utilises dynamic programming to generate most optimal Energy Management Strategy (EMS) for multiple power sources knowing the powering requirements and the operating profile.

2 PLUG-IN HYBRID PEMFC AND ESS PROPULSION SYSTEM

2.1 *Basic concept of operation*

There are different operating modes for coastal ships which need to be considered independently:

When the ship is at sea, both the PEMFC stack and lithium batteries work concurrently to power the ship propulsion and its service loads. The ESS has two functions: 1) Levelling the PEMFC stack loads to achieve the best overall efficiency; 2) Utilising the stored clean grid power to achieve the best overall environmental performance.

When the ship is manoeuvring, then the battery should supplement the fuel cell set at a lower output. The battery will charge or discharge as needed to reduce transients to the fuel cell but also to maintain high overall efficiency.

When the ship is in port, shore power is available to charge the ESS and to power the ship's services, i.e. cold ironing.

Due to the high volumetric demands of H_2 fuel it is assumed that the ship bunkers H_2 fuel for each voyage, i.e. every time it calls at the port.

2.2 *System layout*

Figure 1 presents the PIHFCB propulsion system layout. DC power distribution architecture is preferred since the power out from both PEMFC stack, and ESS is DC electrical power (Zahedi et al. 2014).

2.3 *Propulsion system dynamics*

According to energy conservation principle, the relationship between the PEMFC stack output power P_{fc}, battery power P_{ESS}, shore power P_{shore} and the lumped system power demand P_l can be determined as:

$$P_{fc}\eta_1 + P_{ESS}\eta_2\eta_b + P_{shore}\eta_1 - P_l = 0 \quad (1)$$

where η_1, η_2 and η_b are uni-directional, bi-directional converter efficiency and lithium battery ESS

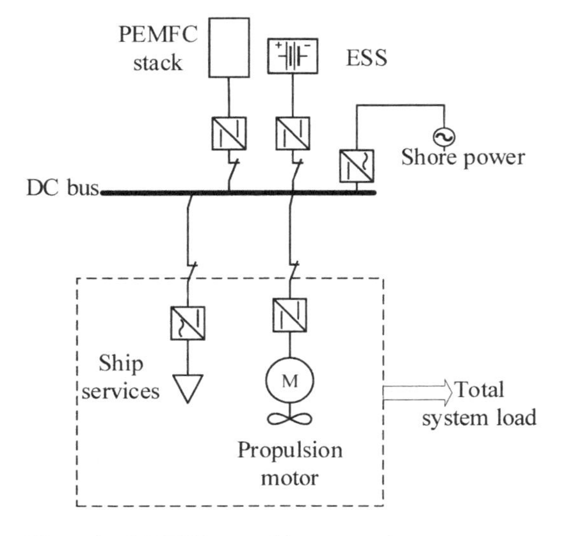

Figure 1. PIHFCB propulsion system layout.

efficiency respectively; and $P_{shore} = 0$ when ship is sailing, $P_{shore} \geq 0$ when ship is at port. $P_{ESS} > 0$ when lithium battery ESS discharges, and $P_{ESS} < 0$ while ESS charges.

2.4 *Proton exchange membrane fuel cell*

The PEMFC stack model is developed and calibrated using the methodology and data from (Larminie et al. 2003), (Tremblay & Dessaint 2009) and (Li et al. 2009). The PEMFC model is simplified to represent per unit power versus specific H_2 consumption based on the 141.8 MJ/kg H_2 High Heating Value (HHV) as presented in Figure 2 (Koroneos et al. 2004). The PEMFC stack specific H_2 consumption is given by:

$$SHC = g_{fC}(x) \tag{2}$$

where SHC is the specific H_2 consumption and is a function g_{fC} of the PEMFC stack per unit power x, $0 \leq x \leq 1$. For the rated PEMFC stack power of P_{fc}^{R}, the power output from the PEMFC stack is:

$$P_{fc} = P_{fc}^{R} x \tag{3}$$

2.5 *Energy storage system*

As lithium battery features high efficiency for charging and discharging, the round-trip efficiency of ESS charging/discharging is assumed as $\eta_b =$ 0.98 within allowed State of Charge (SOC) range, e.g. $0.2 \leq \boldsymbol{SOC} \leq 1$ (Ovrum & Bergh 2015). The SOC range is set to avoid excessive degradation due to over-discharge. Note that, the initial SOC is one. At time step t, SOC is calculated by:

$$SOC = 1 - \int_{0}^{t} C(t)\,dt \tag{4}$$

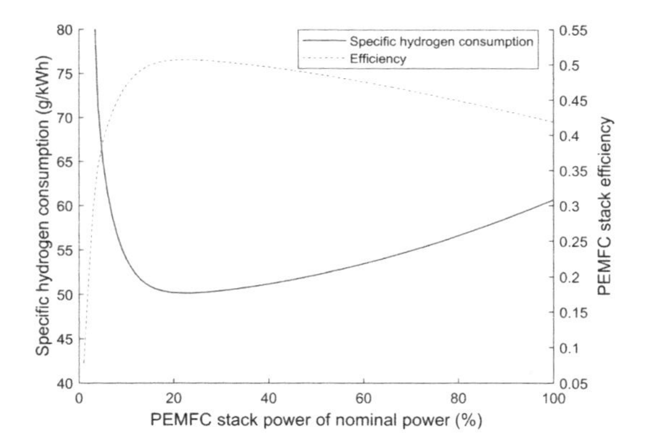

Figure 2. PEMFC stack specific H_2 consumption and efficiency.

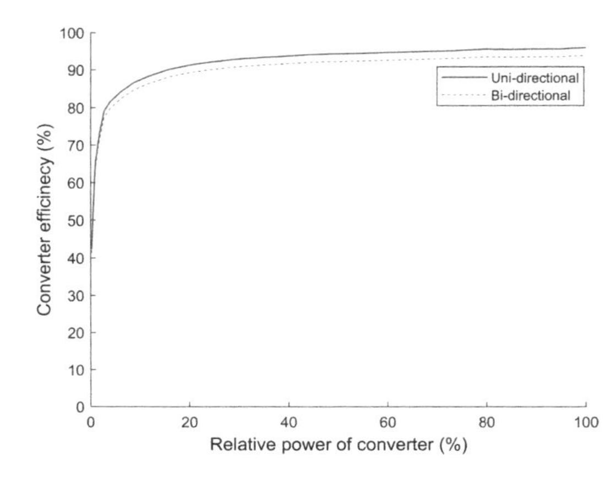

Figure 3. Power electronics characteristics.

where $C(t)$ is ESS chrage/discharge C-rate at time step t. And $P_{ESS} = C(t) P_{ESS}^{R}$, where P_{ESS}^{R} is ESS power when C-rate is 1.

2.6 *Power converters*

Figure 3 shows the power converter efficiency characteristics used in this study (Martel et al. 2015). The uni-directional efficiency is slightly higher than that of a bi-directional one.

3 HYBRID SYSTEM DESIGN METHODOLOGY

Ship power and propulsion systems are customised for individual ships to provide efficient and reliable operation. The design of hybrid propulsion systems comprising multiple power sources should be optimised for the specific operational requirements and scenarios to exploit merits and avoid drawbacks of each type of power sources effectively. The electricity and alternative fuels (e.g. H_2) characteristics can vary from place to place. Also, novel power technologies such as fuel cells and batteries are typically limited by high production costs, limited lifetime and power/energy density for marine propulsion systems. These factors need to be considered for propulsion system designs.

3.1 *Methodology overview*

The proposed design methodology includes two layers of optimisation schemes as presented in Figure 4. The external MOO scheme searches predefined ranges to find optimum PEMFC stack rated power, ESS capacity and shore charging power. The power and propulsion solutions need to meet both volumetric and gravimetric constraints on the propulsion plant. The inner optimisation scheme

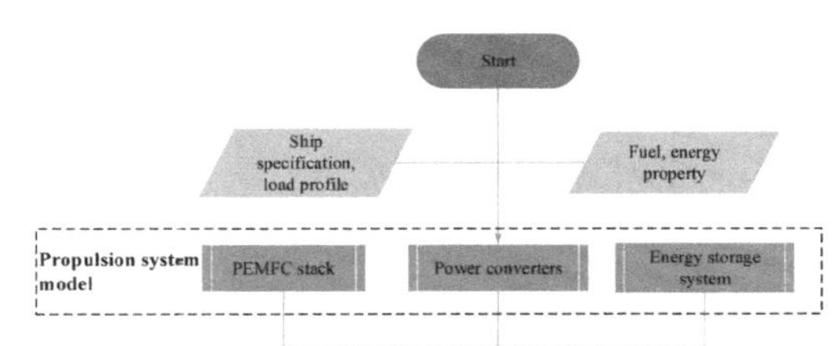

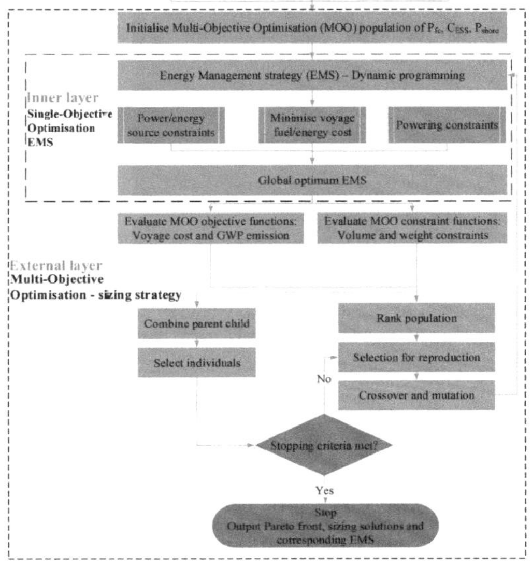

Figure 4. Hybrid system design methodology.

based on dynamic programming determines the EMS for each combination of power sources generated by the external layer. The EMS minimises the voyage fuel costs satisfying the powering demands and power sources constraints.

3.2 *Multi-objective genetic algorithm—sizing optimisation*

The MOO solutions, in the form of Pareto fronts, allow the decision makers to make informed decisions by seeing a set of acceptable trade-off optimal solutions (Ngatchou et al. 2005). In this case, the trade-offs are between equivalent voyage GWP and average voyage cost. The former includes the equivalent CO_2 emission throughout the lifecycle of H_2 and electricity. The average voyage cost consists of H_2 cost, electricity cost and PEMFC and ESS degradation costs.

3.2.1 *Decision variables*

The decision variables of the external optimisation layer are rated PEMFC stack power P_{fc}^R, ESS capacity C_{ESS}^R and shore charing power P_{shore}^R. The searching range of the three variables are set considering the maximum power and total energy demands in operating profile as following:

$$P_{fc}^{min} \le P_{fc}^R \le P_{fc}^{max} \tag{5}$$

$$C_{ESS}^{min} \le C_{ESS}^R \le C_{ESS}^{max} \tag{6}$$

$$P_{shore}^{min} \le P_{shore}^R \le P_{shore}^{max} \tag{7}$$

3.2.2 *Objective functions*

The first objective function of MOO is the average voyage cost, which includes H_2 fuel and electricity cost for one voyage, battery and PEMFC stack degradation costs for one voyage:

$$\begin{aligned} F_1 = & \sum_{i=1}^{K} g_{fc}(x_i)\, x_i P_{fc}^R \Delta t p_h \\ & + \sum_{i=J}^{K} P_{shore}^R \Delta t (K-J)\, p_e \\ & + \left(\frac{p_{fc} P_{fc}^R}{L_{fc}} + \frac{p_{ESS} C_{ESS}^R}{L_{ESS}} \right) \end{aligned} \tag{8}$$

where p_h is the H_2 price in \$/kg, p_{fc} is the PEMFC stack price in \$/kW, p_{ESS} is the battery price in \$/kWh, K is the total time step number, i is i-th time step, J is the time step number when the ship calls at the port, T is the voyage time, $\Delta t = T / K$ is time step length, L_{fc} and L_{ESS} are fuel cell and battery lifetime respectively.

The second objective function of MOO is the GWP emission for one voyage, which is the sum of H_2 fuel GWP and shore electricity GWP in equivalent kg CO_2:

$$F_2 = \sum_{i=1}^{K} g_{fc}(x_i)\, x_i P_{fc}^R \Delta t w_h + P_{shore}^R \Delta t (K-J)\, w_e \tag{9}$$

where w_h and w_e are H_2 and electricity specific GWP respectively.

3.2.3 *Constraints*

The first constraint function limits the hybrid propulsion system volume does not exceed the equivalent diesel-mechanical system volume. The difference between the hybrid propulsion system and the diesel-mechanical plant is:

$$\begin{aligned} M_1 = & P_{fc}^R \rho_{fc}^v + C_{ESS}^R \rho_{ESS}^v \\ & + \sum_{i=1}^{K} g_{fc}(x_i)\, x_i P_{fc}^R \Delta t \rho_t^v - V_D \le 0 \end{aligned} \tag{10}$$

where $\rho_{fc}^v, \rho_{ESS}^v$ and ρ_t^v are volumetric density of PEMFC stack, ESS and H_2 tank (contains H_2 for one voyage) respectively, V_D is the equivalent diesel system total volume and $V_D = P_{dg}\rho_{dg}^v + V_{df}, \rho_{dg}^v$ is the diesel engine volumetric power density, and V_{df} is the diesel fuel tank volume. It is assumed the original case ship refuels diesel once a week in the subsequent analysis.

The second constraint function limits the hybrid propulsion system total weight does not exceed the equivalent diesel-mechanical system weight:

$$M_2 = P_{fc}^R \rho_{fc}^g + C_{ESS}^R \rho_{ESS}^g + \sum_{i=1}^{K} g_{fc}(x_i) x_i P_{fc}^R \Delta t(\rho_t^g + 1) - W_D \leq 0 \quad (11)$$

where ρ_{fc}^g, ρ_{ESS}^g, and ρ_t^g are the gravimetric density of PEMFC stack, battery and H_2 tank respectively, W_D is the diesel based system total weight including the diesel engine, gearbox and fuel weight.

3.3 *Dynamic programming—Energy Management Strategy solving*

The inner optimisation scheme applies dynamic programming based on Bellman's optimality principle to find the most optimal EMS with load profile is known before solving (Bellman 2013). The dynamic programming approach can be used to find the optimal EMS which can be used as a benchmark to evaluate the effectiveness of on-line real-time EMS (Song et al. 2014). The EMS solution for each power and energy source combination is passed to external MOO to evaluate the objective and constraint functions. The objective function values of MOO are infinite if no EMS solution exists.

3.3.1 *Decision variables*

The decision variables represent specific loading conditions for PEMFC stack and ESS. The shore connection delivers rated power whenever the ship is in port. The decision vector is:

$$x = [x_1, x_2, \ldots, x_K \mid x_{K+1}, x_{K+2}, \ldots, x_{2K}] \quad (12)$$

where $x_1, x_2, \ldots, x_K$ are per unit power of the PEMFC stack, and $x_{K+1}, x_{K+2}, \ldots, x_{2K}$ are the C-rate of the ESS from 1_{st} to K_{th} (final step of one voyage) time step.

3.3.2 *Objective functions*

The objective function of the inner optimisation scheme is the voyage total fuel and electricity cost:

$$f = \sum_{i=1}^{K} g_{fc}(x_i) x_i P_{fc}^R p_h + P_{shore}^R \Delta t (K - J) p_e \quad (13)$$

3.3.3 *Constraints*

3.3.3.1 Powering

For each time step, the power provided by all the power and energy sources should equal to the sum of load demand and system losses, therefore rewrite Eq. (1) to discrete form:

$$c_{eq,i} = x_i P_{fc}^R \eta_1(x_i) + x_{i+K} P_{ESS}^R \eta_2 (x_{i+K}) \eta_b + P_{shore} \eta_1(1) - P_{l,i} = 0 \quad (14)$$

3.3.3.2 ESS state of charge

The battery SOC needs to be within a range to avoid over-charge or over-discharge, therefore:

$$SOC_{min} \leq SOC_i \leq SOC_{max} \quad (15)$$

moreover:

$$SOC(i) = 1 - \sum_{1}^{i} x_{K+i} dt \quad (16)$$

3.3.3.3 Fuel cell power ramp up/down rates

Compared to diesel engines, PEMFC stack is weak in transient loads. Therefore, the fuel cell power change between two adjacent time steps should satisfy:

$$|x_i - x_{i-1}| \leq R_{fc} \quad (17)$$

where R_{fc} is the fuel cell power ramp up/down limit.

4 CASE STUDY

4.1 *Case ship specification*

In the case study, the proposed methodology is applied to design the PIHFCB propulsion system considering both environmental and economic performance for case ship which sails on short routes. The vessel specification is shown in Table 1 (Traffic 2015).

4.2 *Case ship operating profile*

For system level design and optimisation, the load profile of the case ship is modelled as a lumped power profile including both propulsion and service loads as shown in Figure 5 (Mashayekh et al. 2012). The load ramps up to a high value in the

Table 1. Case ship specification (Traffic 2015).

Vessel type	Ro-ro/passenger ship
Gross tonnage	3,193 tons
Deadweight	572 tons
Length overall	87 m
Breadth extreme	17 m
Designed speed	12 knots
Installed engine power	2,148 kW

first 10 minutes and fluctuates to follow a sinusoidal wave to mock the power demand variations. Then the ship power ramps down (90–100 mins) to the port where shore connection charges the battery if necessary. Shore power is available to charge from 100 to 140 mins. This system load profile is converted into a discrete time series and repeats for each voyage.

Table 2 presents the price and specific GWP of H_2 generated via three typical approaches (Acar & Dincer 2014). The three types of H_2 were analysed to investigate the impacts from H_2 properties to the design of propulsion system. The electricity price is assumed to be $0.12/kWh, and its GWP is 0.289 kg/kWh (Eurostat 2017).

4.3 *Case study parameters*

Table 3 describes the parameters applied in the case study. The PEMFC stack and battery properties, the prices are all for system level, i.e. including the ancillary devices. It worth mentioning that the results are sensitive to the parameters.

4.4 *Sizing results*

4.4.1 *Pareto fronts*

Figure 6 shows the Pareto fronts for H_2 produced from the three sources mentioned in Table 2. For the case

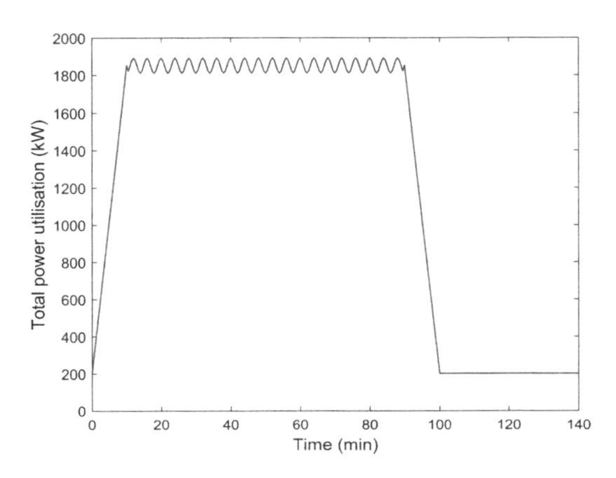

Figure 5. Case ship lumped load profile.

Table 2. H_2 characteristics (Acar & Dincer 2014).

H_2 generation method	Price ($/kg)	GWP (kg CO_2/kg)
Nuclear Cu-Cl	1.7	1.6
Wind	7.2	1.3
Natural gas steam reforming (NGSR)	1.5	7.5

Table 3. Case study parameters.

Parameters	Value	Reference
Annual operating days	300 days	(Traffic 2015)
Daily voyage number	6	(Traffic 2015)
Fuel cell price	$1200/kWh	(Isa et al. 2016)
Fuel cell lifetime	3 years (or 10,800 h)	(Ballard 2017)
Battery price	$800/kWh	(Ovrum & Bergh 2015)
Battery lifetime	3 years	(Stroe et al. 2015)
Shore electricity price	$0.12/kWh	(Eurostat 2017)
Shore electricity GWP	0.289 kg CO_2/kWh	(Eurostat 2017)
PEMFC volumetric specific power	128.2 kW/m^3	(Ballard 2017)
PEMFC gravimetric specific power	200.0 kW/t	(Ballard 2017)
ESS volumetric specific energy	91.8 kWh/m^3	(Corvus 2017)
ESS gravimetric specific energy	80.6 kWh/t	(Corvus 2017)
Battery maximum C-rate	6.0	(Corvus 2017)
Diesel engine volumetric specific power	43.9 kW/m^3	(Wartsila 2016)
Diesel engine with gearbox specific power	54.8 kW/t	(Wartsila 2016)
Marine gas oil price	$0.64/kg	(BunkerIndex 2017)
H_2 tank volume	0.17 m^3/kg H_2	(Choi et al. 2016)
H_2 tank weight	28.5 kg/kg H_2	(Choi et al. 2016)

of H_2 generated via Nuclear Cu-Cl, as both the H_2 specific GWP and price low, the Pareto front points only distribute in a small region of shore power. H_2 generated using wind power features for the lowest GWP, but the highest price can achieve best emission performance but leads to high voyage costs. The Pareto front of NGSR H_2 can contribute the lowest average cost, but also the highest GWP. In general, Nuclear Cu-Cl generated H_2 excels the other two.

4.4.2 *Optimal sizing*

Figure 7 presents the detailed Pareto front solutions including the information of PEMFC stack rated power, battery capacity and rated shore power. The optimal shore power distributed between a narrow region from 180 to 185 kW, which is because both the H_2 specific GWP and price are low amongst the three H_2 production methods. Furthermore, increasing the average voyage cost cannot further improve emission performance effectively. ESS mainly functions as an energy buffer to optimise PEMFC stack loading to achieve higher efficiency.

For the wind power generated H_2 case, as presented in Figure 8, the optimal solutions scatter in more substantial space. The combinations with high shore charging power and larger ESS capacity correspond to better emission performance (Figure 8b), but worse economic feasibility (Figure 8a). Such trends match the wind power generated H_2 property—high price, but low specific GWP.

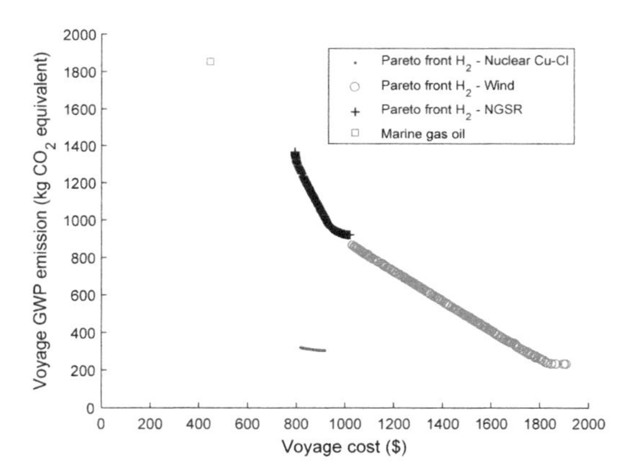

Figure 6. Pareto fronts.

In contrast, Figure 9 shows the trade-off between economic and environmental performances for the H_2 generated via NGSR (high specific GWP and low price). Higher PEMFC stack power leads to lower running cost (Figure 9a) but higher GWP (Figure 9b).

Figure 10 presents the most optimal EMS for H_2 generated by Nuclear Cu-Cl case: the ESS capacity is 692 kWh, PEMFC stack power is 1823 kW, and shore power is 182 kW. The battery starts to provide most of the power demands at the beginning while the PEMFC stack increases the power output gradually and takes over most of the load. The battery tackles most of the power transients during cursing. It is interesting that when the ship is at the port, the fuel cell stack still delivers power to the system, which is mainly due to the H_2 generated by Nuclear Cu-Cl is cheap.

Figure 11 compares the voyage cost and GWP emission breakdown between diesel-mechanical

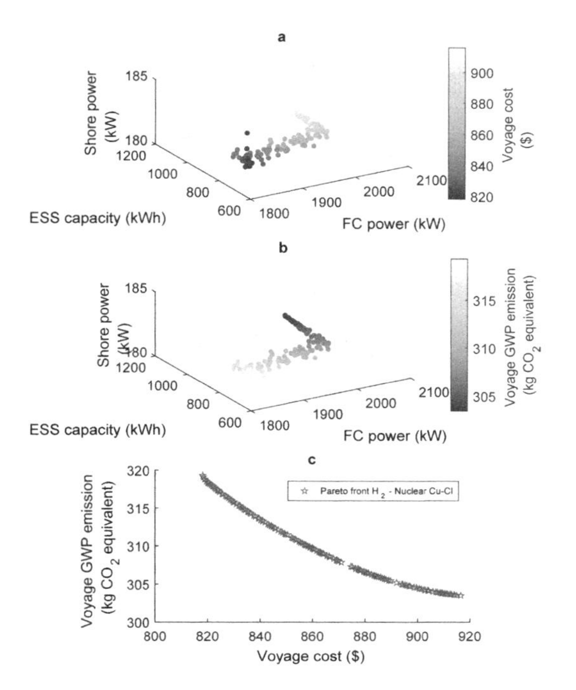

Figure 7. Power source sizing results of H_2 generated via Nuclear Cu-Cl method: (a) average voyage cost, (b) voyage GWP and (c) Pareto front.

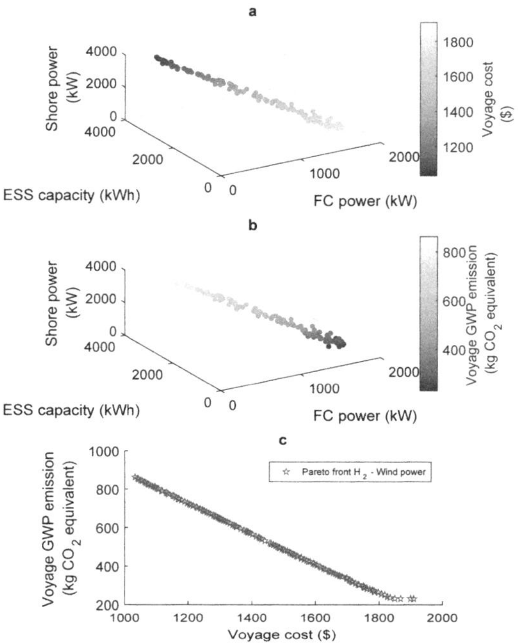

Figure 8. Power source sizing results of H_2 generated via wind power: (a) average voyage cost, (b) voyage GWP and (c) Pareto front.

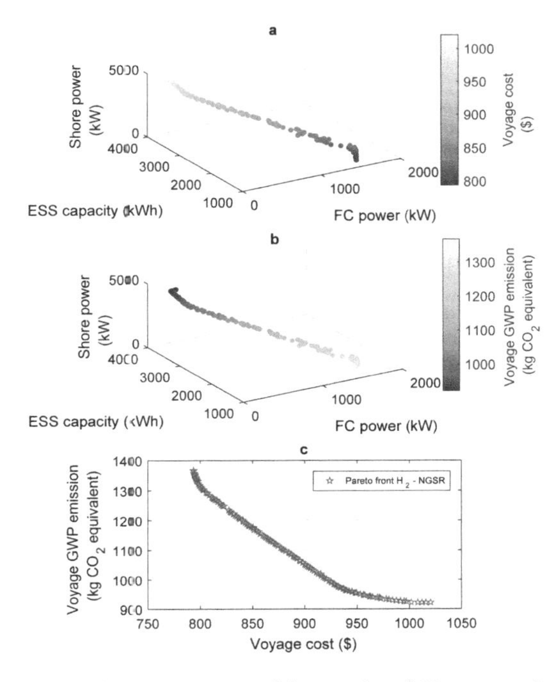

Figure 9. Power source sizing results of H_2 generated via NGSR: (a) average voyage cost, (b) voyage GWP and (c) Pareto front.

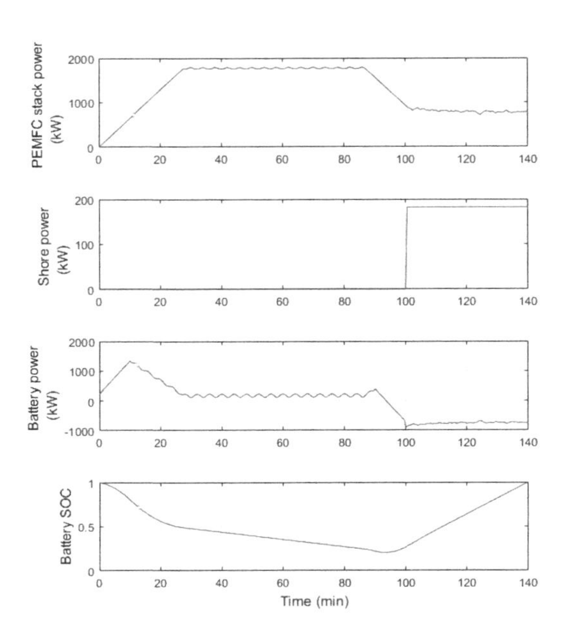

Figure 10. EMS of Nuclear Cu-Cl generated H_2 sample case: ESS capacity – 692 kWh, PEMFC stack power – 1823 kW and shore power – 182 kW.

plant operating on marine gas oil and the alternative PIHFCB propulsion system (the scenario discussed in Figure 10). The average voyage cost of the hybrid system is approximately 70% higher than the diesel-mechanical system. Nevertheless, about 60% of the hybrid system voyage cost is from battery and PEMFC stack degradation. The fuel cell and battery technologies have been evolving rapidly in the past decade, which can potentially cut down the cost significantly (Sharaf & Orhan 2014, Nykvist & Nilsson 2015).

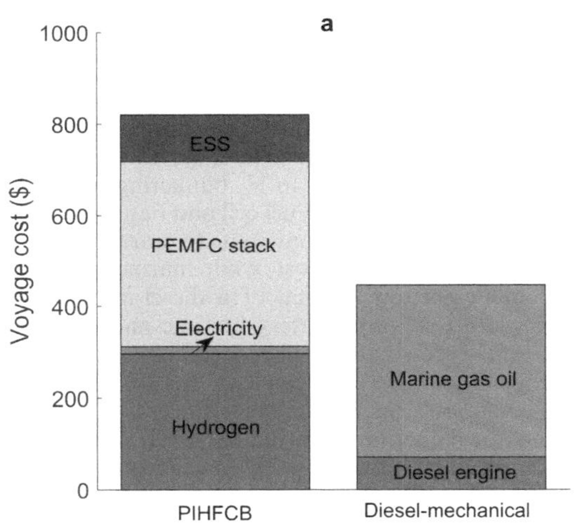

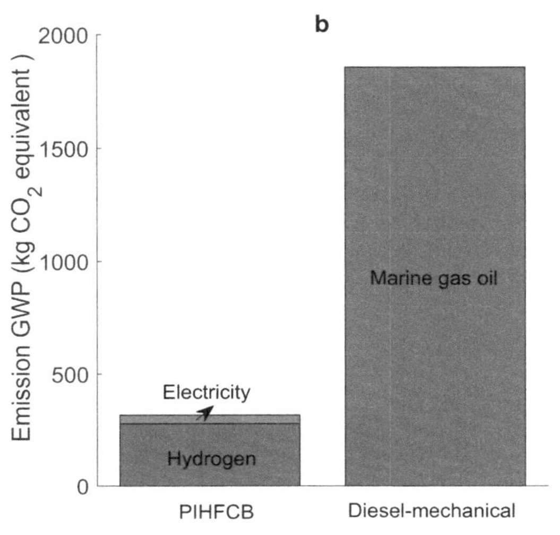

Figure 11. Voyage cost (a) and voyage GWP (b) breakdown comparison of Nuclear Cu-Cl generated H_2 sample case: ESS capacity – 692 kWh, PEMFC stack power – 1823 kW and shore power – 182 kW vs baseline diesel-mechanical system.

5 CONCLUSIONS

This paper presents a PIHFCB design methodology for coastal ships sail on short routes and have accessibility to H_2 bunkering and battery charging facilities. The two-layer optimisation methodology has been shown to generate optimal sizing solutions with an energy management strategy for each design point. Instead of providing a single design point, the solution space provides the decision makers with a better view of the trade-offs between overall emission reduction and commercial feasibility.

The case study results show that electricity and H_2 characteristics have a significant influence on the design of hybrid PEMFC and battery propulsion system. The volumetric and gravimetric impacts from H_2 fuel, PEMFC stack and battery can be mitigated for coastal ships sail on short routes with easy access to H_2 bunkering and battery charging facilities. Fuel cell and battery degradation can potentially contribute to more than 50% of the average voyage cost, while marine gas oil is the main portion for that of a diesel-mechanical plant. Fuel cell and battery lifetime and durability are expected to be improved to be commercially competitive with conventional diesel engine based propulsion plants. Nevertheless, the GWP emission reduction from the PIHFCB propulsion system can be more than 25%, even using H_2 produced from NGSR.

As the degradation of both PEMFC and battery could potentially impact the average voyage cost significantly, more detailed PEMFC and battery degradation models are expected to be included in future work.

ACKNOWLEDGEMENTS

The first author would like to thank the China Scholarship Council (CSC) and University College London for supporting his studies at University College London, UK.

REFERENCES

Acar, C. & Dincer, I. 2014. Comparative assessment of hydrogen production methods from renewable and non-renewable sources. *International Journal of Hydrogen Energy,* 39(1), 1–12.

Ballard, 2017. *Key Advantages of Ballard's Fuel Cell Stack Products* [online]. Available from: http://www.ballard.com/fuel-cell-solutions/fuel-cell-power-products/fuel-cell-stacks [Accessed 10th/Oct 2017].

Bassam, A.M., Phillips, A.B., Turnock, S.R. & Wilson, P.A., Sizing optimization of a fuel cell/battery hybrid system for a domestic ferry using a whole ship system simulator. In: *2016 International Conference on Electrical Systems for Aircraft, Railway, Ship Propulsion and Road Vehicles & International Transportation Electrification Conference (ESARS-ITEC)*, 2–4 Nov. 2016 2016, 1–6.

Bellman, R., 2013. *Dynamic programming.* Courier Corporation.

BunkerIndex, 2017. *Index Summary* [online]. Available from: http://www.bunkerindex.com/ [Accessed 28th/10 2015].

Choi, C.H., Yu, S., Han, I.-S., Kho, B.-K., Kang, D.-G., Lee, H.Y., ... Kim, M. 2016. Development and demonstration of PEM fuel-cell-battery hybrid system for propulsion of tourist boat. *International Journal of Hydrogen Energy,* 41(5), 3591–3599.

Corvus, E., 2017. *Orca Energy Specifications* [online]. Available from: http://corvusenergy.com/technology-specifications/ [Accessed 10th/Oct 2017].

de-Troya, J.J., Álvarez, C., Fernández-Garrido, C. & Carral, L. 2016. Analysing the possibilities of using fuel cells in ships. *International Journal of Hydrogen Energy,* 41(4), 2853–2866.

Deb, K., 2001. Multi-objective optimization using evolutionary algorithms. John Wiley & Sons.

DOE, 2015. *Fuel cell technologies office accomplishments and progress* [online]. Available from: http://energy.gov/eere/fuelcells/fuel-cell-technologies-office-accomplishments-and-progress [Accessed 10th/Sep 2015].

Eurostat, 2017. *Electricity price statistics* [online]. Available from: http://ec.europa.eu/eurostat/statistics-explained/index.php/Electricity_price_statistics [Accessed 10th/Dec 2017].

Han, J., Charpentier, J.F. & Tianhao, T., State of the art of fuel cells for ship applications. In: *Industrial Electronics (ISIE), 2012 IEEE International Symposium on*, 28–31 May 2012 2012, 1456–1461.

Hannan, M.A., Hoque, M.M., Mohamed, A. & Ayob, A. 2017. Review of energy storage systems for electric vehicle applications: Issues and challenges. *Renewable and Sustainable Energy Reviews,* 69, 771–789.

Hansen, J.F. & Wendt, F. 2015. History and State of the Art in Commercial Electric Ship Propulsion, Integrated Power Systems, and Future Trends. *Proceedings of the IEEE,* 103(12), 2229–2242.

Isa, N.M., Das, H.S., Tan, C.W., Yatim, A.H.M. & Lau, K.Y. 2016. A techno-economic assessment of a combined heat and power photovoltaic/fuel cell/battery energy system in Malaysia hospital. *Energy,* 112, 75–90.

Koroneos, C., Dompros, A., Roumbas, G. & Moussiopoulos, N. 2004. Life cycle assessment of hydrogen fuel production processes. *International Journal of Hydrogen Energy,* 29(14), 1443–1450.

Larminie, J., Dicks, A. & McDonald, M.S., 2003. *Fuel cell systems explained.* J. Wiley Chichester, UK.

Li, X., Xu, L., Hua, J., Lin, X., Li, J. & Ouyang, M. 2009. Power management strategy for vehicular-applied hybrid fuel cell/battery power system. *Journal of Power Sources,* 191(2), 542–549.

Luckose, L., Hess, H.L. & Johnson, B.K., Fuel cell propulsion system for marine applications. In: *2009 IEEE Electric Ship Technologies Symposium*, 20–22 April 2009 2009, 574–580.

Luo, X., Wang, J., Dooner, M. & Clarke, J. 2015. Overview of current development in electrical energy

storage technologies and the application potential in power system operation. *Applied Energy,* 137, 511–536.
Martel, F., Kelouwani, S., Dubé, Y. & Agbossou, K. 2015. Optimal economy-based battery degradation management dynamics for fuel-cell plug-in hybrid electric vehicles. *Journal of Power Sources,* 274, 367–381.
Mashayekh, S., Wang, Z., Qi, L., Lindtjorn, J. & Myklebust, T.A., Optimum sizing of energy storage for an electric ferry ship. In: *2012 IEEE Power and Energy Society General Meeting*, 22–26 July 2012 2012, 1–8.
Ngatchou, P., Zarei, A. & El-Sharkawi, A., Pareto multi objective optimization. In: Intelligent systems application to power systems, 2005. Proceedings of the 13th international conference on, 2005, 84–91.
Nykvist, B. & Nilsson, M. 2015. Rapidly falling costs of battery packs for electric vehicles. *Nature Clim. Change,* 5(4), 329–332.
Ovrum, E & Bergh, T.F. 2015. Modelling lithium-ion battery hybrid ship crane operation. *Applied Energy,* 152, 162–172.
Pei, P. & Chen, H. 2014. Main factors affecting the lifetime of Proton Exchange Membrane fuel cells in vehicle applications: A review. *Applied Energy,* 125, 60–75.
Sasank, B.V., Rajalakshmi, N. & Dhathathreyan, K.S. 2016. Performance analysis of polymer electrolyte membrane (PEM) fuel cell stack operated under marine environmental conditions. *Journal of Marine Science and Technology*, 1–8.
Sharaf, O.Z. & Orhan, M.F. 2014. An overview of fuel cell technology: Fundamentals and applications. *Renewable and Sustainable Energy Reviews,* 32, 810–853.
Song, Z., Hofmann, H., Li, J., Hou, J., Han, X. & Ouyang, M. 2014. Energy management strategies comparison for electric vehicles with hybrid energy storage system. *Applied Energy,* 134, 321–331.
Stefanatos, I.C., Dimopoulos, G.G., Kakalis, N.M.P., Vartdal, B. & Ovrum, E., Modelling and simulation of hybrid-electric propulsion systems : the Viking Lady case. In: *12th International Marine Design Conference*, 2015 Tokyo, 161–178.
Stroe, A.I., Swierczynski, M., Stroe, D.I. & Teodorescu, R., Performance model for high-power lithium titanate oxide batteries based on extended characterization tests. In: *2015 IEEE Energy Conversion Congress and Exposition (ECCE)*, 20–24 Sept. 2015 2015, 6191–6198.
Traffic, M., 2015. *Marine Traffic* [online]. Available from: http://www.marinetraffic.com/ [Accessed 10th/Jan 2016].
Tremblay, O. & Dessaint, L.-A., A generic fuel cell model for the simulation of fuel cell vehicles. In: *Vehicle Power and Propulsion Conference, 2009. VPPC'09. IEEE*, 2009, 1722–1729.
van Biert, L., Godjevac, M., Visser, K. & Aravind, P.V. 2016. A review of fuel cell systems for maritime applications. *Journal of Power Sources,* 327(Supplement C), 345–364.
Wartsila, 2016. *Wartsila 20 Project Guide* [online]. Available from: http://cdn.wartsila.com/docs/default-source/product-files/engines/ms-engine/w%C3%A4rtsil%C3%A4-20-product-guide.pdf?sfvrsn = 2 [Accessed 11th/Oct 2017].
Welaya, Y.M.A., El Gohary, M.M. & Ammar, N.R. 2011. A comparison between fuel cells and other alternatives for marine electric power generation. *International Journal of Naval Architecture and Ocean Engineering,* 3(2), 141–149.
Zahedi, B., Norum, L.E. & Ludvigsen, K.B. 2014. Optimized efficiency of all-electric ships by dc hybrid power systems. *Journal of Power Sources,* 255, 341–354.

Marine Design XIII – Kujala & Lu (Eds)

Estimation of fuel consumption using discrete-event simulation—a validation study

E. Sandvik, B.E. Asbjørnslett & S. Steen
Department of Marine Technology, Norwegian University of Science and Technology, Trondheim, Norway

T.A.V. Johnsen
SINTEF Ocean, Trondheim, Norway

ABSTRACT: In this paper, we investigate the validity of quasi-static discrete-event simulation for estimation of fuel consumption and assessment of energy effective ship designs. Stricter emission regulations for ships and developments in computer science have sparked an interest in virtual testing and simulation approaches to enhance our understanding of vessel performance early in the design process. Our methodology uses discrete-event simulation and historical weather data to replicate the operational conditions, and quasi-static calculations to estimate wave and wind added resistance on the ship hull. The validity of this approach is tested in a case study using full-scale measurements from a deep-sea vessel. Results show that we are able to recreate the voyage in a manner that show several similarities to the case vessel measurements. Speed policies that better replicate real operation and fuel curves that take the engine state into account is recommended in order to improve fuel consumption estimates.

1 INTRODUCTION

Fuel efficiency of ships has become increasingly important in recent years due to higher fuel costs and stricter emission regulations. This has sparked an interest in the industry and academia towards establishing new and improved design methodologies to enhance the designer's understanding of design performance. Required power and fuel consumption is commonly estimated using static numerical and empirical tools in the early design stage. The impact of weather is approximated using a sea margin, relying on experience and statistics. Virtual testing and benchmarking schemes has been developed in recent years for marine applications, enabling the designer to factor in the characteristics of the area of operation. The IDEAS project worked on developing a simulation-based benchmarking tool for evaluation of ship designs (Fathi et al. 2013). VISTA (virtual sea trial) was developed for assessing operability of complex marine operations during design (Erikstad et al. 2015). ViProMa (virtual prototyping of maritime systems and operations) presents an open virtual prototyping tool based on distributed co-simulation (Skjong et al. 2017).

The majority of Norwegian ship builders and design companies has for many years had vessels for the offshore oil and gas industry as their core activity and source of income. Due to the rapid drop in oil price in 2014, the demand for such vessels plummeted, resulting in near empty order books and challenging financial times. Many of the companies that experienced this downturn looked to new industries and segments for projects. Cruise/expedition vessels, fishing/aquaculture vessels, ferries and offshore wind support vessels are some of the segments now replacing offshore oil and gas vessels in Norwegian ship yards. This poses challenges for designers, evolving and adapting to a new set of requirements and considerations to assess during design and engineering. In this context, a methodology for rapid testing and exploration of design performance is advantageous due to the often limited knowledge base applicable across vessel type.

Ship owners and operators focus on the operation and economy of vessels. Their studies often require a more detailed operational scenario, specifying a vessel schedule for cargo pick-ups and drop-offs as part of a logistics system. Facilitating detailed descriptions of such scenarios results in a flexible platform capable of providing knowledge also during a vessels lifetime.

In this paper, we investigate the validity of quasi-static discrete-event simulation models for estimation of required propulsion power and fuel consumption. A case study is presented where we attempt to replicate the voyage of a ship in transit between China and the United States, using the simulation based GYMIR workbench (see chapter 3 for further description) developed in the research project SFI Smart Maritime (SFI Smart Maritime 2015). The case vessel is a general cargo carrier outfitted with an onboard performance monitoring system. Logged data from the voyage is used for comparison towards the simulation results. Discrete-event simulation and hindcast weather data is applied to replicate the vessel sailing conditions along the route. Quasi-static calculations are applied to estimate the required propulsion power and fuel consumption, taking calm water and added resistance due to wind and waves into account. Our focus is to compare and evaluate the simulation-based results towards the performance monitoring system measurements, emphasizing challenges and potential sources of error in view of the uncovered differences.

2 CASE VESSEL DATA

2.1 *Vessel particulars*

The particulars for the case ship studied in this paper are given in Table 1.

2.2 *Data acquisition*

The vessel is outfitted with a real-time monitoring system which stores operational data with a sampling period of 15 minutes. A list of the applied parameters are given in Table 2.

Table 1. Case vessel particulars.

Length meters	Beam meters	Draft meters	Gross tons tons
204	32.3	13	37 000

Table 2. List of parameters applied in case study and respective measuring techniques.

Parameter	Measuring technique
Shaft torque, RPM and power	Optical sensors
Fuel consumption	Fuel line flowmeter
Speed through water	Doppler sonar
Speed over ground	GPS
Position	GPS
Wind	Anemometer

Table 3. Route legs and key information.

Leg	Speed knots	Duration hours	Distance nautical miles	Draft meters	Trim meters
1	15.8	72.4	1144.1	8.5	0.7
2	15.5	133.6	2071.1	8.5	0.7
3	16.0	80.8	1292.2	8.5	0.7

2.3 *Route*

The vessel route is from Qingdao to Seattle, covering a distance of 4,514 nautical miles over the North Pacific. The voyage was conducted during fall of 2016. The route is simulated as three successive legs as shown in Table 3.

Vessel speed is varied between routes in order to minimize the spatial distance between the real and simulated vessel in time. Loading condition is kept constant for all legs according to the case vessel.

3 METHODOLOGY

3.1 *Simulation workbench*

GYMIR applies discrete-event simulation to replicate the vessel voyage. Methodology flowchart is shown in Figure 2. The simulation is set to follow the vessel route as shown in Figure 1, updating weather condition according to the current position and time.

A speed policy must be set in order to determine the simulator actions towards maintaining speed in weather. For this study, a constant speed policy was applied. In this setting the powering calculation function determines the propulsion power required to maintain the specified speed in the current weather condition. A new event is triggered either by an update in position, defined by the route waypoints, or an update in weather condition between two successive waypoints.

3.2 *Quasi-static estimates*

The discrete event simulation in GYMIR is using static calculations of speed and power for each event. The average added resistance is calculated for the given speed and sea state, and the power required to reach the specified speed with the increased resistance is calculated by interpolation of calm water propulsion characteristics. The

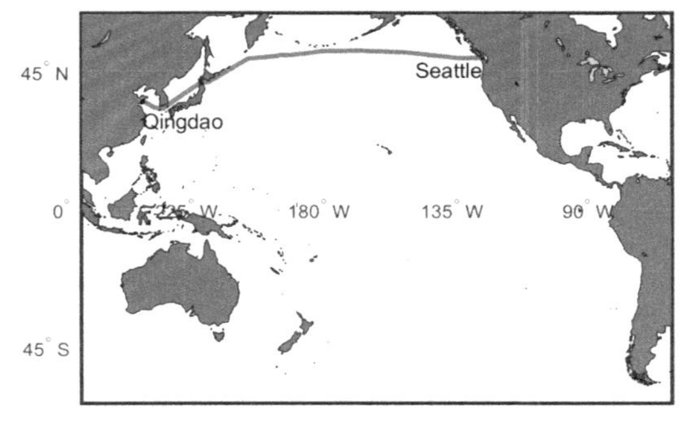

Figure 1. Vessel route map.

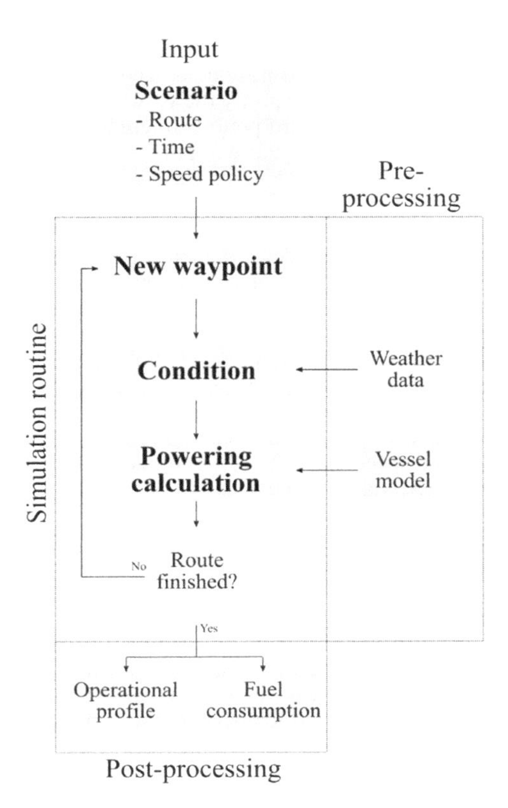

Figure 2. Methodology flow chart.

increase of added resistance is found by using quadratic added resistance transfer functions computed by the pressure integration method of Faltinsen (Faltinsen et al. 1980) and extended Gerritsma & Beukelman (Loukakis & Sclavounos 1978), implemented in the linear frequency-domain strip-theory seakeeping program ShipX Veres.

3.2.1 *Resistance*
The total vessel resistance, R_T, in our model consists of the calm water resistance R_{T0} and added resistance due to waves R_{AW} and wind R_{AA}.

$$R_T = R_{T0} + R_{AW} + R_{AA} \quad (1)$$

The calm water resistance term is weather independent, defined only by loading condition and vessel speed through water. The added resistance terms are highly influenced by weather conditions and relative direction.

3.2.2 *Calm water resistance*
Calm water resistance curves are taken from towing test results of the vessel hull. Admiralty coefficient C_{adx}, based on the towing power P_E, is applied to account for differences in loading condition.

$$C_{adx} = \frac{\nabla^{\frac{2}{3}} \cdot V_s^3}{P_E} = \frac{\nabla^{\frac{2}{3}} \cdot V_s^2}{R_t} \quad (2)$$

The resistance curve is calibrated using the case vessel measurements by comparing the power requirement in a series of calm sea states.

3.2.3 *Added resistance due to waves*
Added resistance due to waves is caused by two separate physical phenomena; wave reflection and vessel motion induced wave generation. Two approaches are applied for calculation of short-term added resistance coefficients due to waves. The Gerritsma & Beukelman method is derived from radiated energy consideration and strip-theory (ST) approximation for head seas. Loukakis and Sclavounos generalized this approach to cover oblique waves. The second approach is the pressure integration method (PI) where the pressure is integrated along the intersection between the ship hull and water surface and over the average position of the wetted ship hull. Both methods are combined with the asymptotic formula for low wavelengths to account for wave reflection.

3.2.4 *Added resistance due to wind*
Vessel area above water is approximated using general arrangements drawings. The influence of cargo cranes is neglected. Drag coefficients for the specific vessel type is applied using the ShipX database, based on (Brix 1993).

3.3 *Propulsion*

The vessel is a single screw vessel outfitted with a Mewis duct. Open water and propulsion test reports for the case vessel are used for propulsion power calculations.

3.4 *Weather data*

Hindcast data from the recently released ECMWF ERA5 catalogue is used to replicate the environmental conditions. It combines models and observations which allow high temporal and spatial resolution.

3.5 *Fuel consumption*

The last step in the presented methodology is estimation of fuel consumption. This is done using shop test results from the main engine onboard the vessel.

4 SOURCES OF ERROR

Each step in the simulation process, i.e. from the occurrence of weather to calculation of fuel consumption, is a potential source of error and deviations from the measured data. These errors are caused by physical phenomena not taken into account in the simulation, or models that are not able to capture and sufficiently describe the system state and underlying factors. This section provides an overview of the most prominent sources of error and their influence on fuel consumption estimates.

4.1 *Hydrodynamics*

Replicating vessel performance in a sea way presents great modelling challenges. For propulsion system fuel consumption, our main concern is related to the resistance and propulsion efficiency.

4.1.1 *Added resistance*

Added resistance of ships in a seaway is notoriously difficult to estimate due to the complex interaction process between the waves and the ship hull. The two applied methods, strip-theory and pressure integration, have limited accuracy in certain scenarios due to their underlying assumptions. Strip-theory is known to give conservative estimates of added resistance (Fathi & Hoff 2017), implying a conservative estimate of required power. Coefficients for following and stern quartering seas are found to be considerably larger for strip-theory than the equivalent pressure integration coefficients for the case study vessel. The pressure integration approach requires an accurate description of the flow surrounding the hull. Added resistance coefficients are calculated using 15-degree increments for relative wave heading, interpolating for intermediate headings during simulation.

4.1.2 *Propulsion coefficients*

As a ship advances through waves, the wake field is influenced by the motion of the hull and the incident wave induced particle velocity. Significant changes in propulsion performance can therefore be expected (Taskar et al. 2016). These effects are not captured by the presented methodology due to the assumption of steady-state propeller inflow conditions for a given ship speed described by the calm water test results.

4.1.3 *Steering losses*

Manoeuvring causes an increase of resistance due to rudder drag and hull angle of attack relative to the direction of travel. The hydrodynamic model applied in the simulation does not account for these effects, and the contribution to required power is hard to estimate, as there are no information regarding rudder use and relative angle of attack in the monitoring data. The monitoring system does however specify the vessel state/mode, allowing us to disregard periods where the vessel is in manoeuvring mode close to port.

4.2 *Weather data*

Even though the hindcast database relies on state of the art meteorological services, we can not say with certainty that the weather events occurring in the simulation coincide perfectly with the real voyage weather. In addition, the six-hour time resolution of the weather data can cause significant errors if conditions are rapidly changing in time and space. Figure 3d shows a comparison of relative wind speed calculated from hindcast data and measured data from the vessel.

4.3 *Vessel and machinery condition*

Vessel and machinery condition will certainly affect vessel performance. Marine growth, roughness and substrate fouling on the hull can cause a significant increase in resistance. Propeller fouling causes an increase in required power due to reduced efficiency. Machinery maintenance affect the fuel consumption and required power. As mentioned in section 3.2.1, the calm water resistance curve is calibrated using the measured propulsion power. Hence, the error in power estimation due to hull condition is minimized.

4.4 *Quasi-static estimation*

Quasi-static estimation of added resistance due to waves assumes a characteristic steady-state value to be present for the duration of the sea state. Added resistance is in reality a dynamic process, with significant transient loading in most commonly observed sea states. However, for the applied propulsion power, which is of more interest here, we do not expect the same rapid variation since the captain usually do not vary engine power settings between individual resistance peaks.

5 RESULTS

Simulation results and performance monitoring data are compared in Figures 3–5. A comparison of the simulation and real voyage is presented in Table 4.

Vertical lines are used in Figure 3d to indicate updates in weather data, occurring at six-hour intervals in the hindcast data applied in the simulation. Between these lines the simulation routine updates weather as a consequence of changes in vessel posi-

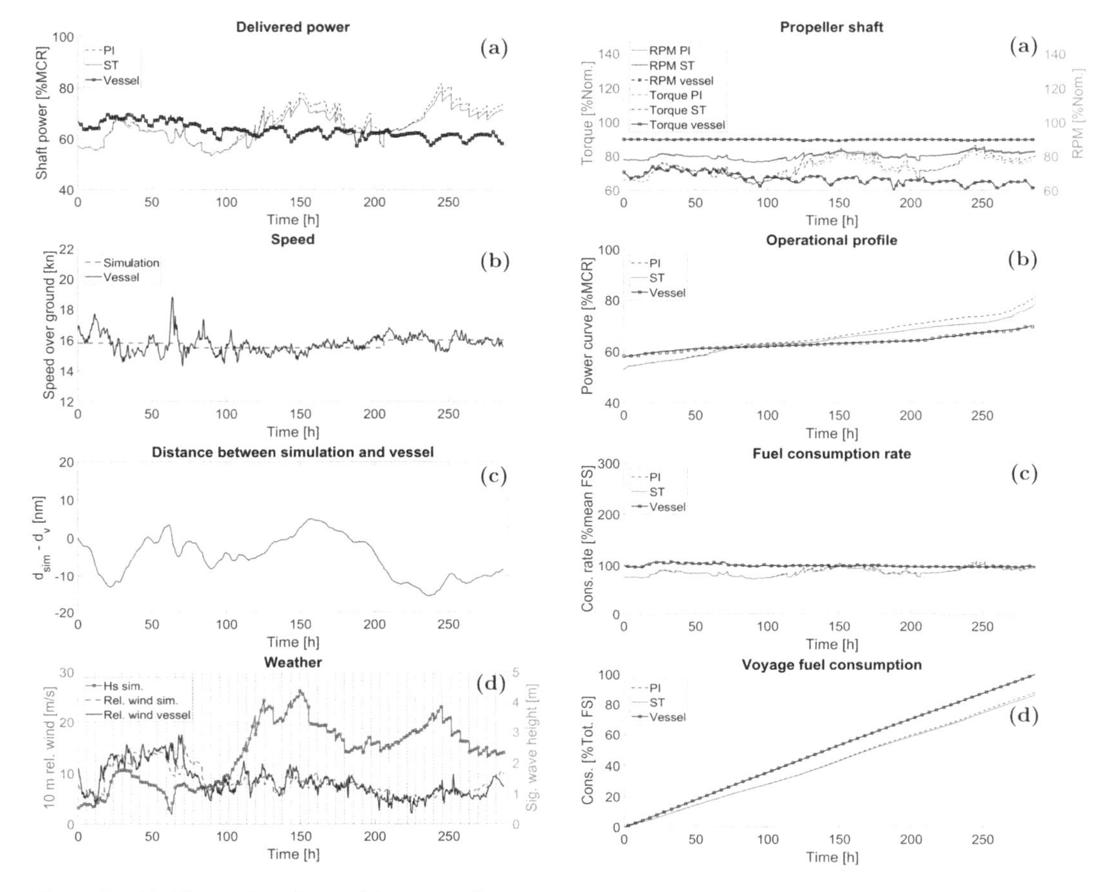

Figure 3. (a): Time series of propulsion power for case vessel and simulation for pressure integration (PI) and strip-theory (ST) calculations of wave added resistance. (b): Speed over ground for case vessel and simulation. (c): Distance between simulation and real vessel position. (d): Relative wind speed measured onboard case vessel. Significant wave height and relative wind speed from hindcast data applied in simulation.

Figure 4. (a): Time series of propeller shaft characteristics from case vessel and simulation. (b): Operational profile taken as the sorted power output in figure 3(a). (c): Fuel consumption rate normalized using the mean consumption rate of the case vessel measurements. (d): Total fuel consumption normalized using the total voyage consumption for the case vessel. Each plot show results for pressure integration (PI) and strip-theory (ST) calculations of wave added resistance.

tion, disregarding temporal changes. This causes a saw tooth-pattern clearly visible in most simulation variables. The most prominent cases occurs between time instant 100–130 and 230–280. In the first period, the simulated vessel is approaching an area with harsh weather, as seen in Figure 2. The harsh weather is moving in the same direction as the vessel but remains fixed for six hours in the simulation. This causes a steep increase in significant wave height as the vessel moves between updates in weather, and a sudden drop once the weather and storm location is updated. Between time instants 230–280, a storm is passing in front of the vessel at an angle relative to the direction of travel. We observe the same pattern here since the vessel sails for six hours into the storm between each weather update. As the weather is updated, the significant wave height is reduced as the storm moves further away to the side of the vessel.

The simulation propeller shaft torque and rpm characteristics differs from that measured onboard the case vessel, as seen in Figure 4a. Lower and frequently varying RPM is applied in the simulation, as apposed to the measurements where the RPM is kept seemingly constant at 90% nominal rating. The case vessel has for the majority of the time a lower torque and a higher RPM than the simulation, indicating a lighter propeller load.

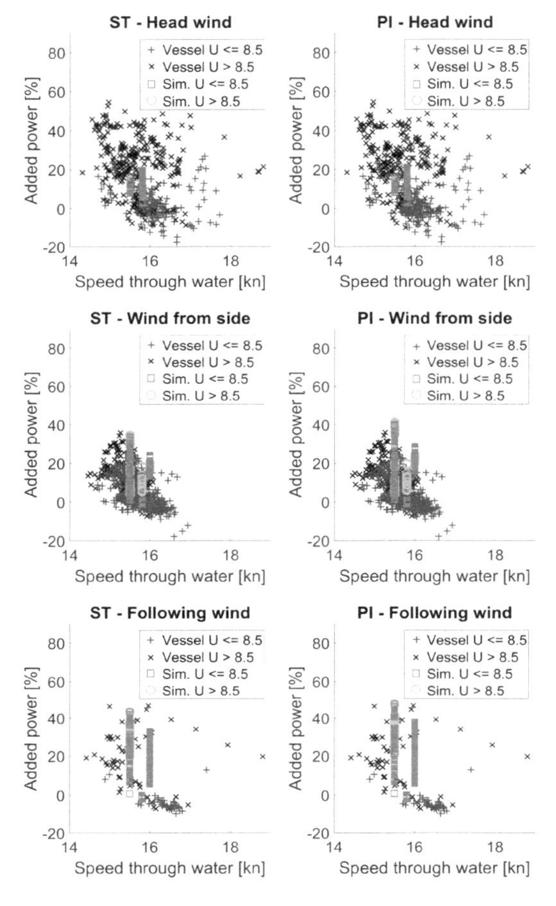

Figure 5. Added power relative to the calm water estimates from the vessel model sorted according to relative wind direction and wave added resistance calculation method. Results using strip theory (ST) in the left column and pressure integration (PI) in the right column. Each plot show cases for relative wind speed U above and below 8.5 m/s.

Table 4. Voyage comparison.

	Duration	Average speed	Distance	Consumption
	Hours	Knots	Nautical miles	% Case vessel total
Vessel	286.8	15.73	4514	100
GYMIR	287.0	15.71	4510	PI: 88.4 ST: 86.7

A clear difference between the simulation and real operational profile is that the simulator applies a wider range of power, both higher and lower than the case vessel. The logged data indicate a control setting of constant RPM for the real vessel while constant speed is used in the simulation. Higher rating occurs since the simulator opts for an increase in power rather than accepting a speed loss in harsh weather.

Lower rating occurs in the calm waters at the start of the route, where the case vessel measurements show an increase in speed and the simulator holds a constant speed of 15.8 knots.

The fuel consumption results indicate that the use of the engine shop test curve in the simulations is too optimistic, as shown in figure 4c and 4d. Even for the high engine loads towards the end of the route, the simulation results indicate only a marginally higher fuel consumption than the case vessel measurements. Overall, the fuel consumption in the simulation is approximately 12% lower than the measured data.

Figure 5 show the added power relative to calm water power calculated using the steps outlined in section 3.2.1 and 3.3. The results are sorted in terms of relative wind direction due to the lack of information regarding wave condition in the case vessel monitoring system. The results indicate a reasonable level of agreement between simulation and case vessel. Pressure integration gives a higher power requirement than strip theory, shown in Figure 3a to be most prominent for harsh sea states. The case vessel data have higher counts of negative added power than the simulation results for higher speeds. Since the resistance curve was calibrated such that the calm water power in the vessel model coincide with the case vessel power at approximately 16 knots, this indicate that the model is conservative in its prediction of power requirement at higher speeds.

6 DISCUSSION

Attempting to replicate the case vessel voyage, using what must be described as low-fidelity models, is clearly ambitious. Vessels moving in a seaway are subject to hydrodynamic effects notoriously difficult to estimate, affecting both resistance and propulsion characteristics. However, even though our models are built on significant simplifications, we do observe several similarities between the simulation results and measured data.

First of all, to have a basis for comparison, it is imperative to have similar weather conditions in our simulation as for the case vessel. Since the onboard performance measurement system is limited to wind measurements, we are limited to comparison between relative wind speed time series for verification. Figure 3d indicate that the wind measured on the vessel deck and the combination of historical data and calculation of relative wind speed are closely matched. We do however observe a distinct saw-tooth pattern in our weather time series as a consequence of the six-hour time discretization, which makes us question the validity of the intermediate results. These patterns are most prominent where the weather changes rapidly in time and space, i.e. along storm edges. For this reason, we recommend careful use of weather data with poor time resolution, especially in combination with scenarios where the probability of storms is high.

The quality of the simulation result is heavily dependent on the vessel model. In this case study, we have applied experimental data for both the resistance and propulsion characteristics. It can therefore be argued that this model has the best possible foundation for estimation of required power. This was done to factor out the differences in calm water, allowing us to test the quality of the added power estimates due to weather. The added power is applied to maintain speed when resistance is increased due to wind and waves. When comparing the operational profiles in Figure 4b, it is apparent that the range of applied power is significantly wider than the case vessel measurements. This is partly due to the assumed constant speed policy applied in the simulation. The propulsion machinery onboard vessels are not regulated according to a set speed, but rather constant RPM. This causes an involuntary speed loss in the presence of harsh weather, but limits engine wear and fuel consumption. For instances of harsh weather in the historical weather data, we also observe a reduction in measured case vessel power, suggesting a voluntary speed loss. Both voluntary and involuntary speed loss are disregarded in the presented simulation routine, causing higher maximum power in the operational profile from the simulation.

The added resistance estimates are based on pre-processed coefficients calculated in ShipX using pressure integration and the Gerritsma & Beukelman method. These methods are known to overestimate the added resistance and resulting increase in required power. Our comparison with the case vessel measurements is based on calculation of added power relative to the calm water power estimates of the vessel model. Figure 5 indicate that the added power applied in the simulation is comparable to the measured data. However, our application of calm water propulsion characteristics is likely to reduce the required propulsion power relative to that of the real vessel. In addition, our simulation routine does not factor in the influence of current, affecting speed trough water. The difference between strip theory and pressure integration for the added resistance due to waves is most prominent in the rougher sea states. For the calm sea states early in the simulation, the difference in resulting propulsion power is found to be negligible. In more rough sea states the pressure integration method result in higher estimates.

Our fuel consumption estimates are found to be highly optimistic. Even though the simulator opts for an increase in power when subjected to harsh weather conditions rather than accepting a speed loss, the resulting estimated fuel consumption is lower than the case vessel measurements. Figure 6 shows the difference in added consumption between the measured data and the shop test fuel curve used in simulation. The shop test fuel curve is created based on tests performed in an engine laboratory, where the engine loads are reasonably static and the inlet and surrounding air temperatures are lower than for operating conditions. In addition, we lack information regarding engine service and maintenance history. The explanation for the pattern in Figure 6 may therefore be the exposure to dynamic engine loads, higher air temperatures and reduced performance due to engine condition.

Overall, the most prominent causes of differences in required power and fuel consumption is the speed policy and fuel consumption curve. It is clear that further description of the weather-speed relationship must be included to get simulation behaviour more similar to realistic operations. As an alternative, requiring constant RPM seems to be an option according to the case vessel measurements presented here. Furthermore, especially with regards to long-term simulations, models that describe hull degradation and operational machinery performance should be included to achieve more realistic fuel consumption and emission estimates.

Further research, with a larger test program of routes, seasons, loading conditions and ship types, is needed to conclude on the validity of the methodology. The focus in this paper has been on the

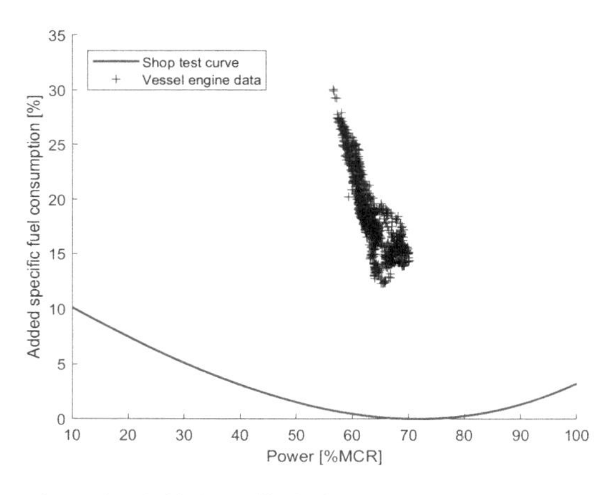

Figure 6. Added specific fuel consumption for shop test curve and case vessel measurements normalized using shop test curve minima.

power requirements related to propulsion for a general cargo carrier, with emphasis on the impact of wind and waves. For more complex ship types, such as cruise vessels, the power requirement and fuel consumption must be evaluated while considering hotel and equipment power loads.

7 CONCLUSION

In this paper we have investigated the validity of quasi-static discrete-event estimation of operational profile and fuel consumption in early design. A case study was performed where we replicated the voyage of a general cargo carrier across the North Pacific. Our results indicate that even though the applied hydrodynamic models and quasi-static calculations are based on significant simplifications, the methodology provides knowledge regarding the vessel's performance in realistic operating conditions. Recommended improvements include vessel speed-policy, hull degradation and fuel curves that take the operational engine performance into account. However, further research including more routes, ship types and seasons are required to provide a general conclusion.

ACKNOWLEDGEMENT

The authors are grateful for the financial support from the Research Council of Norway through the Centre for Research based Innovation (SFI) Smart Maritime project number 237917. In addition, we are thankful for the comments our anonymous reviewers provided towards improving this article.

REFERENCES

Brix, J., 1993. Manoeuvring Technical Manual. *Seehafen Verlag*.

Erikstad, S.O. et al., 2015. VISTA (Virtual sea trial by simulating complex marine operations): Assessing vessel operability at the design stage. In *12th International Marine Design Conference 2015*.

Faltinsen, O.M. et al., 1980. Prediction of resistance and propulsion of a ship in a seaway. In *13th Symposium on Naval Hydrodynamics*. Tokyo.

Fathi, D.E. et al., 2013. Integrated Decision Support Approach for Ship Design. In *OCEANS - Bergen, 2013 MTS/IEEE*.

Fathi, D.E. & Hoff, J.R., 2017. ShipX Vessel Responses (VERES) Theory Manual.

Loukakis, T.A. & Sclavounos, P.D., 1978. Some extensions of the classsical approach to strip theory of ship motions, including the calculation of mean added forces and moments. *Journal of Ship Research*, 22.

SFI Smart Maritime, 2015. Norwegian Centre for improved energy efficiency and reduced harmful emissions. Available at: http://www.smartmaritime.no.

Skjong, S. et al., 2017. Virtual prototyping of maritime systems and operations: applications of distributed co-simulations. *Journal of Marine Science and Technology*.

Taskar, B. et al., 2016. The effect of waves on engine-propeller dynamics and propulsion performance of ships. *Ocean Engineering*, 122.

Marine Design XIII – Kujala & Lu (Eds)

Voyage performance of ship fitted with Flettner rotor

Osman Turan, Tong Cui, Benjamin Howett & Sandy Day
Department of Naval Architecture, Ocean and Marine Engineering, University of Strathclyde, UK

ABSTRACT: In order to tackle global warming, the maritime sector is trying alternative power sources to reduce the carbon emissions from ships. Wind assist technologies are considered as renewable alterative power sources and ships designed with such technologies may provide the important step toward decarburization. Among many effective measures, Flettner rotor, as one of the wind assist technologies, is increasingly becoming a topic of interest within the shipping industry. However, the performance of a ship fitted with Flettner Rotor is very much dependent on the wind magnitude and direction, which are random by nature. This paper aims to make an in-depth discussion on the benefits from Flettner rotor in terms of energy efficiency. Firstly, a ship weather routing system is briefly introduced which is very helpful to assess the true performance of ships fitted with Flettner rotor. Next, many case studies are carried out based on a generic ship with and without rotors. Different combinations of several traditional routes including both outward and return directions, different departure time throughout the whole operation year and average slow, medium and fast ship speeds are taken into account as shipping conditions for comparison. Finally, the benefit obtained through Flettner rotor technology – in various shipping routes, departure times and ship speeds—is presented and a framework for assessing the performance of wind assist technology is proposed.

Keywords: Voyage Optimisation, Weather Routing, Flettner Rotor, Energy Efficiency

1 INTRODUCTION

Nowadays, in order to counter fuel price rising and tackle global warming, shipping industry makes great efforts to find measures to increase fuel efficiency and reduce carbon emissions from ships. It is well known that compare to traditional fossil fuel, wind energy is extremely clean, free and abundant. Therefore, as a tendency of low carbon economy development, shipping with wind assist technologies is attracting more interest from shipping industry.

This research focuses on one of effective wind assist technologies – Flettner rotor. A Flettner rotor is a smooth cylinder with disc end plates which is spun along its long axis and, as air passes at right angles across it, the Magnus effect causes an aerodynamic force to be generated in the third dimension (Seifert, 2012). A rotor ship is a type of ship designed to use the Magnus effect for propulsion (Betz, 1925; Nuttall et al, 2016). The ship is propelled, at least in part, by large vertical rotors (Wikipedia-Rotor ship). German engineer Anton Flettner was the first to build a ship which attempted to tap this force for propulsion in 1920s, so Flettner rotors are also named after their inventor. Rizzo (1925) discussed the fundamental principles of the Flettner rotor ship in the light of Kutta-Joukowski theory. Actually, this technology is treated indifferently since it was first deployed. But these years, with the development of technology and proposition of "low carbon shipping" concept, people are turning their attention back to rotors as an immense potential measure (Howett et al, 2015). They made many achievements on Flettner rotor technology. In 2008, the German wind-turbine manufacturer Enercon launched a new rotor ship, E-Ship 1 (as shown in Fig. 1) and claimed that it can save up to 25% fuel compared to same sized conventional vessels

Figure 1. "E Ship 1" from Germany.

after 170,000 sea miles in 2013 (Enercon, 2013). In 2014, Norsepower Company developed a "Norsepower Rotor sail solution" which is a modernized version of the Flettner rotor (Norsepower, 2014). They installed two Norsepower Rotos on a RoRo vessel "M/V Estraden" in 2014 and later 2015, and announced that this technology has potential for fuel savings of up to 20% for vessels with multiple, large rotors traveling on favourable wind routes (Norsepower, 2016). In 2016, Viking Line also considers the rotor concept for their next planned new building. They showed a new 63,000 GT vessel with large Flettner rotors which could help the ship to save up to 15% fuel consumption (Viking Line, 2016). Besides, Traut et al (2012) assessed the wind power performance of a bulk carrier fitted with three Flettner rotors on the route from Brazil to UK and suggested possible fuel savings of 16%. They also (2014) researched the average power contribution from the Flettner rotor on the analysed routes ranges from 193 kW to 373 kW. When three Flettner rotors are fitted on a 5500dwt cargo carrier, they could provide more than half of the required main engine power under slow-steaming condition. De Marco (2016) analysed the Flettner rotor performance with the method of unsteady Reynolds averaged Navier-Stokes simulations and also presented the applicability of such device for marine applications.

This paper aims to make a primary discussion on the benefits from Flettner rotor in terms of energy efficiency shipping. Firstly, a voyage optimization tool with weather routing function is introduced briefly as the methodology for assessing the performance of ships fitted with Flettner rotor. Next, a generic ship is taken as the vessel model for case studies. For both ship with and without rotors, many comparative case studies are carried out in various shipping routes, departure times and ship speeds. Finally, the benefit obtained through Flettner rotor technology is analysed based on these case studies and a framework for assessing the performance of wind assist technology is proposed.

2 METHODOLOGY

The voyage optimization model presented in this paper is not only for single fixed route optimisation but also for global optimisation which also takes the direction optimisation into consideration. According to the requirement from stakeholders, the only input items are the ship model, departure time and ETA, positions of departure and destination, and then the model will determine an optimum route towards minimum fuel consumption after complicated calculation.

The workflow of the model is shown in Fig. 2. It can be seen the model generally contains five modules, and each of them will be introduced briefly as below:

2.1 *Module 1: Ship performance calculation*

Ship performance prediction is the basis of the whole model, which decides whether the ship operation has high quality or not. In this research, in order to obtain ship performance very quickly and conveniently, a Ship Performance Profile File, is generated from WASPP (Wind Assisted Ship Performance Prediction) (Howett, B., 2016). "This file is to compile performance related attributes of an individual ship for a whole range of environmental and operational conditions in a single file, allowing data to be pre-calculated for later use in time intensive applications" (Howett, B., 2016). Howett made a concise explanation of the Ship Performance Profile File in his paper.

One simple file contains these dimensions: speed, significant wave height, relative wave angle, true wind speed, true wind angle and an output attribute: brake power. In WASPP, the brake power values under all potential environmental conditions have been pre-calculated by using modified Kwon's method (Lu, R. et al, 2015). All of these attributes are stored in a database. When the values of five dimensions are given, through the 5-D interpolation, their corresponding brake power can be obtained. So that the fuel consumption for a certain route can be easily determined by:

$$FC = P_B \cdot sfoc \cdot t \tag{1}$$

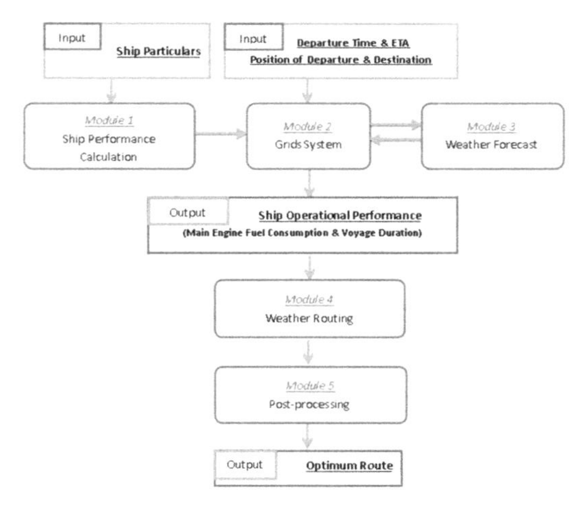

Figure 2. Workflow of voyage optimisation model.

where, *FC* is fuel consumption, *PB* is brake power obtained by ship performance profile, *sfoc* (g/kWh) is specific fuel oil consumption of the engine, *t* is ship navigation duration, which can be easily obtained when the length of the route and actual ship speed are known.

2.2 *Module 2: Grids system*

Grids system means the waypoints distribution strategy in the shipping area. The ship will travel along the potential route formed by these waypoints. In this grids system, as presented in Fig. 3, great circle route (The shortest distance between two points on a sphere) is taken as the reference route, which is divided into several equal stages with certain numbers of nodes. Through every point, a straight line can be drawn perpendicular to its tangent line around the circle. Next, certain numbers of nodes can be distributed along this vertical line, including upper and lower parts of the great circle. The ship in one stage can travel to any waypoint in next stage. Here, considering the larger course deviations are not feasible and would be unrealistic for an optimum route, a limit is set for the ship in one waypoint, which can only go to nearest five waypoints in next stage. For some special shipping tasks, such as the ship going to several fixed ports during the whole navigation, the grids system can be automatically divided into several stages based on the same principle. Besides, for most real situations, the ship always travels around some islands or avoid the prohibited military zone, as land avoidance function is also developed.

2.3 *Module 3: Weather forecast*

This model currently takes only winds, waves into account at the moment. Waves and winds data are both downloaded from ECMWF (European Centre for Medium-Range Weather Forecasts) website. The data is constructed in gridded binary (GRIB) data form, including 10 meter U wind component, 10 meter V wind component, mean wave direction, mean wave period and significant height of combined wind waves and swell, and they all update every 6 hours. These weather data are converted to Beaufort Number and relative weather direction according to ship heading for ship performance calculation. The weather forecast files are linked with three common parameters: time, latitude and longitude. With given time, latitude and longitude, the weather forecast module provides corresponding sea conditions to grids system.

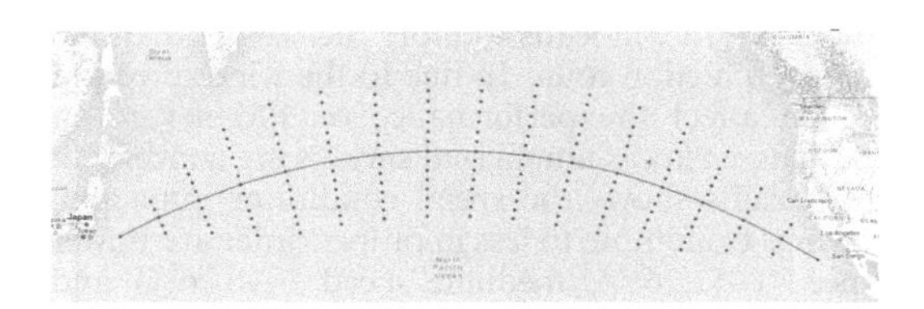

Figure 3. A typical grids system.

2.4 *Module 4: Weather routing*

Weather routing module is the core of the whole voyage optimization model. There are many different types of methods developed for ship weather routing (Simonsen et al, 2015; Walther et al, 2016). In this module, the global searching optimization strategy is introduced.

As mentioned in Module 2, the ship will travel from the departure to destination stage by stage (Fig. 4). For the ship in any waypoint, the system reads the weather data at that point in accordance with local longitude, latitude and departure time firstly. Next, a random speed is assigned to a travel direction. This speed ranges from minimum speed to maximum speed with an interval speed. These speed options are set by operator according to the ship and shipping situation. Having the weather data and ship speed, ship performance database generated in module 1, the fuel consumption on this arc is calculated for this particular speed and direction. Then the navigation information of this arc (fuel consumption, ship speed, navigation duration, local time and coordinates of local departure point) is stored at its arrival point. After that, the ship will travel to next waypoint based on all information saved at this arrival point.

The whole calculation starts from the departure port. Under this simulation principal as described above, the fuel consumption modelling will be continuously utilised for all stages between departure port and destination. This process covers all the speed and direction options. In the end, the total fuel consumption and voyage duration of the different potential route with different speed set will

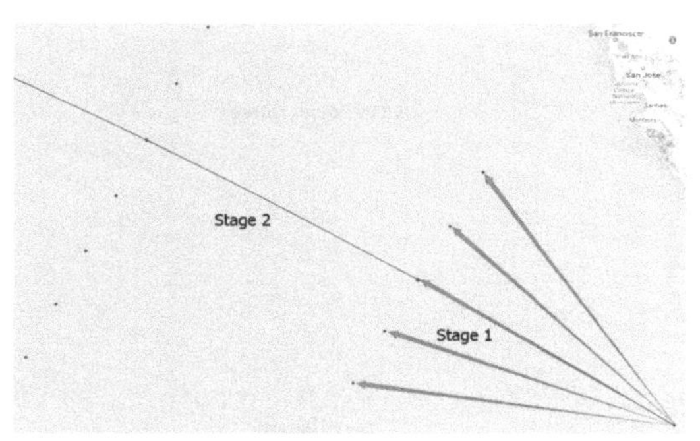

Figure 4. Ship routing stages.

be stored in the destination node. The recorded information will be used to select the minimum fuel consumption route at given ETA, which is regarded as the global searching optimization. During the optimisation process, several smart algorithms are added to the weather routing module, which can reduce the calculation time considerably and make the whole calculation run faster to obtain the results for the required ETA.

2.5 *Module 5: Post-processing*

After the calculation in weather routing module is finished, there will be many results stored in final destination waypoint. This module will select the minimum fuel consumption route under required ETA for stakeholders. According to the information stored in every waypoint on this route, the speed, duration, fuel consumption and weather data etc. of every stage will be extracted in detail. Users can see and handle them very easily.

3 CASE STUDIES AND DISCUSSIONS

A generic motorised KCS container ship fitted with and without 936 m^2 Flettner rotor is taken as the ship model in order to perform the energy saving performance of the ship with Flettner Rotor compared to the fully motorised baseline ship. This KCS hull form is 230 meters long, 32.2 meters wide, 19 meters deep and has 52030 m^3 displacement. It is fitted with a five-blade fixed pitch propeller with a diameter of 7.9 meters (Simman 2008 website, accessed 2017). The speed-power curve for pure motorised KCS ship in calm water is shown in Fig. 5. The main aim of carrying out a systematic calculations is to calculate the performance of the wind assisted ship and determine the correct way of assessing the performance of the ship fitted with Flettner rotor.

3.1 *Case Study 1*

Following parameters are investigated in details:

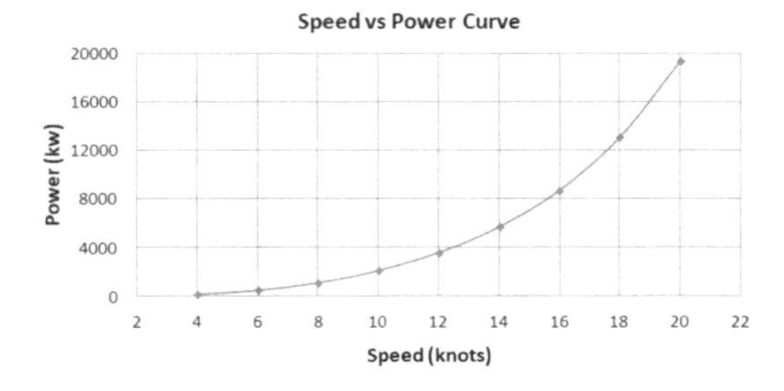

Figure 5. The speed-power curve for pure motorised KCS ship in calm water.

Effect of average ship speeds: 6 knots, 10 knots and 14 knots, represent slower, medium and faster speed respectively.

Effects of shipping season: departure time is on 05:00 am of JAN-05-2014 (Winter), APR-05-2014 (Spring), JUL-05-2014 (Summer), OCT-05-2014 (Autumn).

Effects of geographical location of routes and the travel direction: Five traditional routes over the world including both outward and return directions are investigated. They cover various common oceans and all the travel directions (North-South and West-East). They are:

Route 1: from Chiba in Japan to Los Angeles in USA
Route 2: from Rio de Janeiro in Brazil to Cape Town in South Africa
Route 3: from Puerto la Cruz in Venezuela to Gibraltar
Route 4: from Brisbane in Australia to Yokosuka in Japan
Route 5: from Banda Aceh at Malacca to Durban in South Africa

All of these routes are labeled with letter “a” and “b”, which means the outward direction and return direction respectively.

All the results are shown as below, the percentage means how much benefits can be obtained with Flettner Rotor technology in terms of fuel consumption saving.

As can be seen from Table 1, for both route 1a and 1b under slower speed and medium speed, the most fuel consumption saving happens in January while the least saving happens in July. It illustrates that for shipping area 1, if the ship speed is not fast, winter will be the most favourable season for winds assisted shipping while summer only generates least benefit in terms of fuel saving. The same situation also happens at route 1a under faster speed, but in July, it shows the ship not only has not fuel saving, but actually needs more fuel for wind assisted shipping, which means the summer season is totally not a good time for the ship travel in area of route 1 under fast speed. For route 1b under faster speed, when the ship sails in summer, it still needs consume more fuel if fitted with rotors, but the most favourable season for that situation becomes autumn (October) instead of winter (January). For all three speed conditions, the route 1a will leads to more fuel saving than the return direction route 1b due to the wind direction which affect the performance of Flettner rotor. Besides, as far as annualized savings are concerned, the results shows the speed conditions impacting savings from more to less in proper order are slower speed (–51.26%), medium speed (–16.76%) and faster speed (–5.32%) in shipping area 1. Thus, the slower speed is, the more fuel consumption savings will be obtained.

Table 1. Fuel consumption saving as % for Route 1 (Chiba in Japan – Los Angeles in USA).

Route/Speed (knots)	JAN	APR	JUL	OCT	Average	Combined Annual Average
1a/6	−83.38	−56.15	−17.32	−76.08	−58.57	−51.26
1b/6	−67.64	−47.47	−20.20	−39.17	−43.45	
1a/10	−33.16	−28.31	−1.64	−25.85	−22.27	−16.76
1b/10	−15.71	−12.70	−5.53	−11.22	−11.26	
1a/14	−14.18	−7.61	0.26	−12.28	−8.46	−5.32
1b/14	−2.13	−0.98	0.78	−6.43	−2.20	

Table 2. Fuel consumption saving as % for Route 2 (Rio de Janeiro in Brazil – Cape Town in South Africa).

Route/Speed (knots)	JAN	APR	JUL	OCT	Average	Combined Annual Average
2a/6	−16.93	−14.71	−47.76	−30.99	−27.71	−25.5
2b/6	−12.87	−19.96	−35.08	−24.3	−23.22	
2a/10	−1.14	−4.93	−17.43	−0.82	−6.13	−6.28
2b/10	−5.58	−8.66	−6.57	−4.97	−6.44	
2a/14	3.63	−1.22	2.58	4.26	2.32	0.28
2b/14	−1.66	−1.87	−1.05	−2.41	−1.75	

Table 3. Fuel consumption saving as % for Route 3 (Puerto la Cruz in Venezuela – Gibraltar).

Route/Speed (knots)	JAN	APR	JUL	OCT	Average	Combined Annual Average
3a/6	−10.97	−6.98	0.93	−21.73	−9.59	−14.83
3b/6	−33.64	−19.23	−13.32	−13.70	−19.91	
3a/10	−7.31	2.15	4.00	−5.16	−1.58	−3.18
3b/10	−9.75	−3.89	−3.47	−2.00	−4.76	
3a/14	1.98	3.26	4.79	2.57	3.15	0.8
3b/14	−1.60	−3.18	−1.05	−0.38	−1.55	

As can be seen from Table 2, in area 2, when the ship sails with slower speed, it can achieve relative huge fuel saving (−25.5%) throughout the whole year; when the ship sails with medium speed, it can still achieve good fuel saving (−6.28%); but when the ship speed increases to 14 knots, the situation reverses as the ship consumes more fuel consumption (average 0.28%) in that area. The reason is the wind resistance generated by the Flettner rotor structure consumes more fuels which exceed the savings it generated. This indicates clearly that higher the ship speed lower the savings generated by the Flettner rotor. For route 2a with slower speed, the best season for wind assisted shipping is July, which can lead to 47.76% of saving in fuel consumption, while the worst season is April, but it can still save 14.71% fuel. For the return direction under this speed, the best season is still July, while the worst season turns to January, but it still obtains 12.87% fuel savings. The fuel savings of both directions are almost same at average value of around 25% throughout the whole year. For route 2a with medium speed, the most favourable season is still July, which can lead to 14.43% of saving in fuel consumption, while January and October becomes two worst season as their fuel savings are both around only 1%. For the return direction with medium speed, the fuel saving caused by Flettner rotors does not change a lot during the whole year at an average value of −6.44%. Both directions have almost same average fuel savings throughout the whole year. However, for faster speed route 2a, besides April, is not suitable for shipping with Flettner rotors throughout the year, as it consumes more fuel compared to fully motorized baseline ship. For the route 2b, the situation becomes a little better for faster speed as it can manage only average 1.75% fuel savings.

As can be seen from Table 3, shipping area 3 has the same trend as shipping area 3 that the slower speed will generate more fuel saving than faster speed for the whole year. For route 3a under slower speed, it is clear that shipping in October obtains most benefits from the Flettner rotor while shipping in July consume more fuels unexpectedly. For the return direction under this speed, all seasons can lead to relative huge fuel consumption savings. From

Table 4. Fuel consumption saving as % for Route 4 (Brisbane in Australia – Yokosuka in Japan).

Route/Speed (knots)	JAN	APR	JUL	OCT	Average	Combined Annual Average
4a/6	−26.73	−20.61	−11.51	−9.31	−16.92	−20.23
4b/6	−45.07	−19.09	−13.39	−15.51	−23.55	
4a/10	−3.53	−6.97	−2.59	−3.34	−4.10	−4.48
4b/10	−6.96	−9.73	1.43	−3.85	−4.86	
4a/14	−2.68	−0.90	1.21	0.44	−0.48	−0.4
4b/14	−0.15	−2.71	0.07	1.50	−0.32	

Table 5. Fuel consumption saving as % for Route 5 (Banda Aceh at Malacca – Durban in South Africa).

Route/Speed (knots)	JAN	APR	JUL	OCT	Average	Combined Annual Average
5a/6	−10.55	−14.02	−40.26	−20.16	−21.23	−19.99
5b/6	−5.02	−26.67	−31.77	−11.70	−18.82	
5a/10	−1.56	−0.40	−6.60	−6.78	−3.84	−4.78
5b/10	0.75	−11.45	−10.64	−1.40	−5.71	
5a/14	0.02	−0.94	−0.80	−1.57	−0.82	−0.47
5b/14	2.17	−3.59	−1.96	2.90	−0.12	

the aspect of the whole year, the backward direction route can generate 19.91% fuel savings which even double the savings from outward direction route (9.59%). For route 3a under medium speed, both January and October are appropriate seasons for shipping with the Flettner rotors while April and July are not. For the return direction route (3b) with medium speed, the January is still the best season and all of other seasons the ship generates lower savings, and it can obtain a little more fuel saving than the outward direction. For route 3a with faster speed, the whole year is totally unsuitable for using Flettner rotor technology. For route 3b with faster speed, almost all year around the ship can obtain only little fuel saving (−1.55%), which is a bit better than the outward direction-route 3a.

Route 4 has different characteristic from all other shipping areas as the sailing direction of this route is almost between south and north. As can be seen from Table 4, although the sailing directions are totally different, the operation of slower speed can still lead to more fuel savings than faster speed almost all year around. For route 4a with slower speed, January and April are two good seasons in which the sailing ship can get more than 20% fuel savings and July and October followed with around 10% savings. For route 4b with slower speed, January becomes the only season receive huge benefits while other three seasons also have good performance (from 13.3% to 19.09%). The return direction route obtains more fuel savings than outward direction route at slower speed. For route 4a with medium speed, the average 4.10% fuel saving can be achieved almost whole year. For route 4b with medium speed, April will be the best season to gain more benefit and July should be avoided. For route 4a and 4b with faster speed, there is still very little benefit in all seasons with an average saving of 0.4%, so it can be considered sailing with faster speed in route 4 with Flettner rotors is meaningless. Whether the speed is medium or faster, the two opposite directions under same speed have almost same average fuel savings throughout the whole year.

As can be seen from Table 5, the trend follows the same principle that operating at slower speed in route 5 can generate more fuel saving than faster speed. For both route 5a and 5b at slower speed, the most fuel consumption saving happens in July while the least saving happens in January. For route 5a at medium speed, July and October will be the most appropriate shipping season while for the return direction, the best shipping seasons become April and July. When the ship has faster speed, the benefit from Flettner rotor is very little for both directions. And for all year around, average fuel savings at the two opposite directions under same speed do not have much difference.

To show more intuitive results, three figures are produced to compare the savings as well as the combined global average for all the routes combined as shown in Fig. 6, Fig. 7 and Fig. 8.

From Fig. 6, Fig. 7 and Fig. 8, it can be clearly seen that, wherever the ship is sailing in, the operation of slower speed always generate more benefits using the Flettner rotor technology than the faster speed. Furthermore, sailing along route 1 can save more fuels than other routes, which can illustrate that the west-east lines in Pacific Ocean will be the best shipping area with Flettner rotor technology.

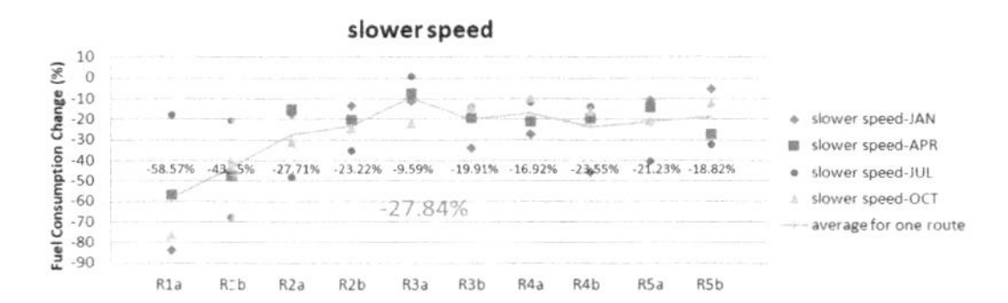

Figure 6. Fuel consumption saving percentage under slower speed.

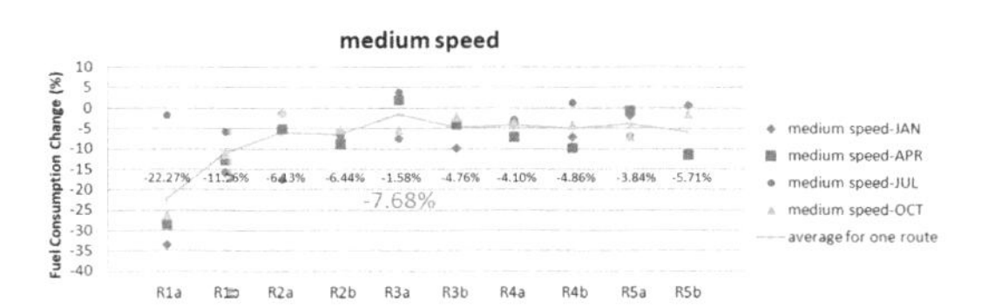

Figure 7. Fuel consumption saving percentage under medium speed.

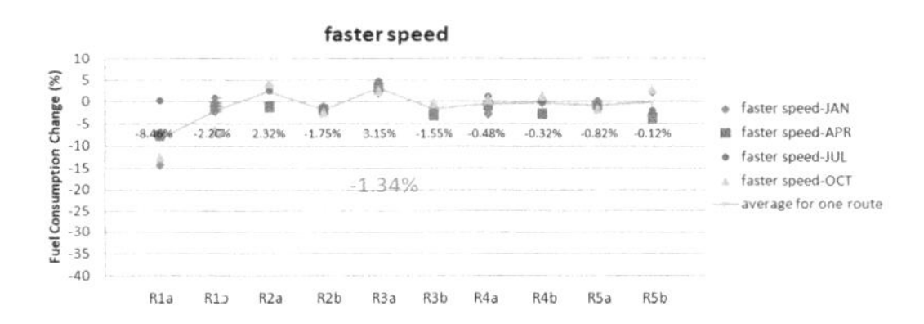

Figure 8. Fuel consumption saving percentage under faster speed.

In order to better understanding the ship sailing situation, the case route 1a with the departure time of January 5th is taken as the example to show the relevant data in detail. The voyage optimization model divided this route into 16 stages. The ship speed and relevant wind conditions (wind speed and relative wind angle) recorded in every stage of route 1a departure at January 5th are extracted in Fig. 9, Fig. 10 and Fig. 11.

These figures show that the ship speed will change with the local and instant wind conditions as the ship passing. The ship fitted with Flettner rotor is more willing to chase the sea conditions of higher wind speed in the downwind direction for greater fuel efficiency. It is already known more fuel savings can be obtained from Flettner rotor technology when the ship operates at slow or medium speed. Through these three figures, it can be also clearly seen that the situations of slower and medium speed always corresponds to the fact that the ship speed is slower than the wind speed. So One conclusion can be drawn that, when the ship operates at the speed slower than wind speed, more benefits in terms of energy efficiency can be obtained from Flettner rotor technology.

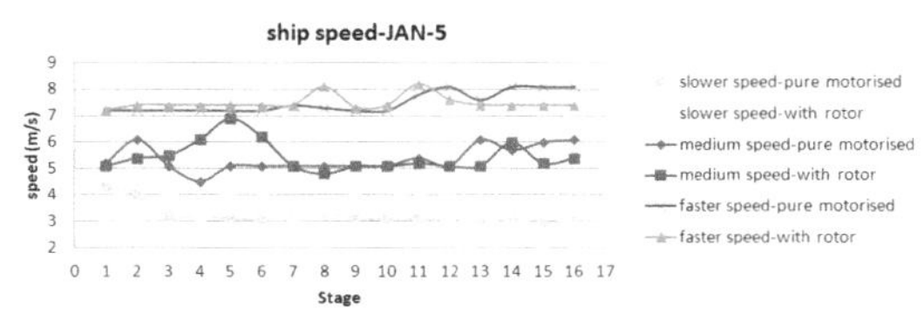

Figure 9. Ship speed at each stage of route 1a departure at JAN-5.

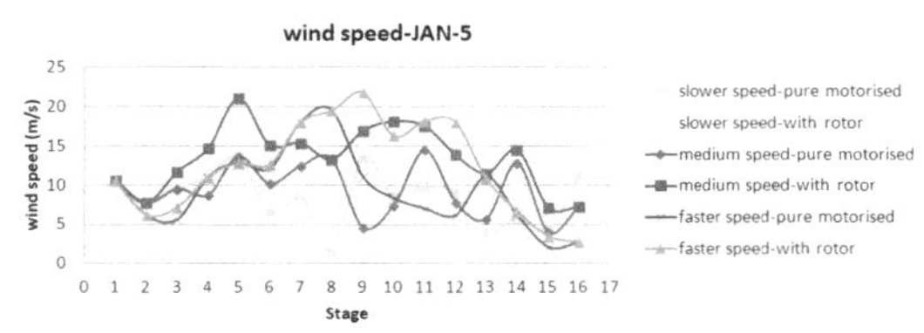

Figure 10. Wind speed at each stage of route 1a departure at JAN-5.

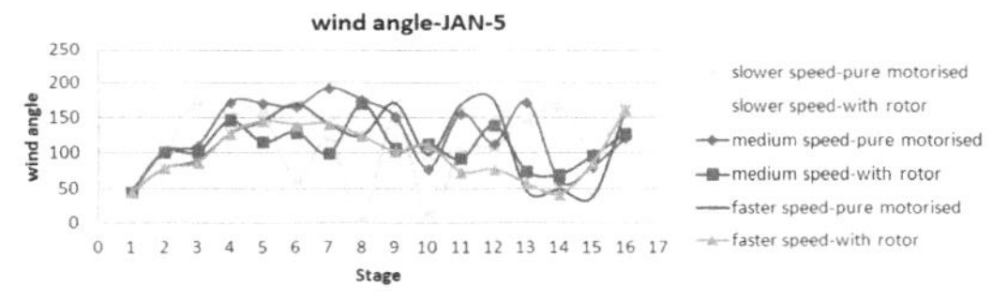

Figure 11. Relative wind speed at each stage of route 1a departure at JAN-5.

At last, for the purpose of routes visualization, and also due to limited space, only route 1a departure at different seasons with different conditions are shown as below (Fig. 12):

3.2 *Case Study 2*

This case study focuses on Route 1 (Chiba in Japan – Los Angeles in USA). Two opposite sailing directions (outward and return) and three different ship speed (slower, medium and faster) are still investigated, by taking into account different departure times. Four different departure times in most favourable month, January, and four departure times in least favourable month, July, for Flettner rotor are investigated. Eight departure dates selected are: JAN-05-2014, JAN-12-2014, JAN-19-2014, JAN-26-2014 and JUL-05-2014, JUL-12-2014, JUL-19-2014, JUL-26-2014.

After performing the calculations, all the results are presented in Fig. 13 and Fig.14.

As can be seen from Fig. 13, in January, route 1a under slower speed still obtains most fuel savings of average 74.40% and return direction under slower speed takes the second place with the value of average 67.16%. Route 1a and 1b under medium speed take the middle places with the value of

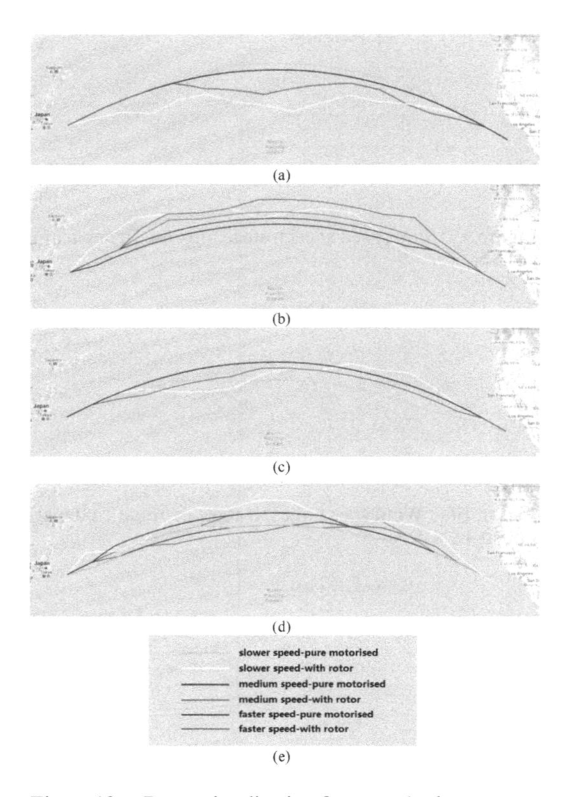

Figure 12. Route visualization for route 1a departure at (a) January; (b) April; (c) July; (d) October; (e) Legends.

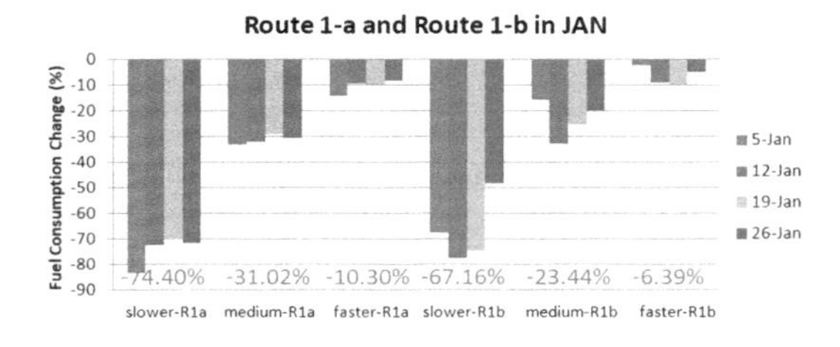

Figure 13. Fuel consumption saving percentage in January.

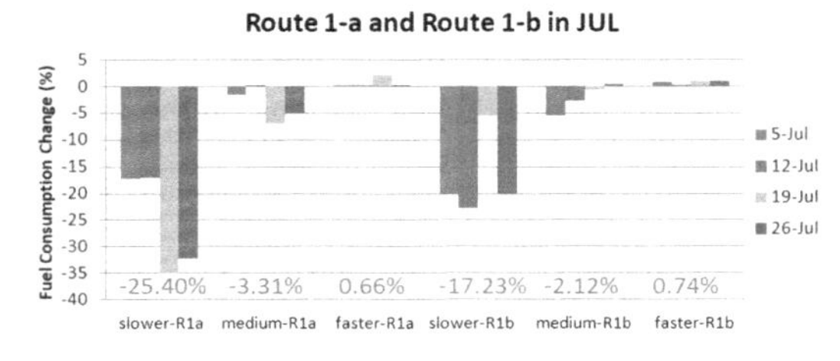

Figure 14 Fuel consumption saving percentage in July.

average 31.02% and 23.44% while route 1a and 1b under faster speed gain least benefits which are average 10.30% and 6.39% respectively. That illustrates January is really suitable for sailing with Flettner rotors in shipping area 1.

As can be seen from Fig. 14, in July, both directions of route 1 under slower speed can get good benefits from Flettner rotors with the savings 25.40% and 17.23% respectively. However, besides them, all other sailing conditions lead to very little fuel savings. Route 1a and 1b under medium speed obtain very small benefits which are only 3.31% and 2.12%. When the benefits are examined for the faster speed, all 8 departures in both directions create additional fuel consumption in July. This means the weather in July is only appropriate for ships fitted with Flettner rotors sailing under slower speed. Thus, relative higher speed are not good choices for ship operation in this area..

4 CONCLUSION

This paper presents a methodology on how detailed performance analysis of a ship fitted with Flettner rotors should be performed. Firstly, a voyage optimization model with weather routing function is introduced to assess the ship's fuel consumption performance. Based on this model, many case studies are carried out for a motorized ship with and without Flettner rotors. According to results, it can be seen that the ship fuel consumption performance varies with different sailing conditions, various shipping routes, departure times and ship speeds. One principle extracted from the results is that, without considering other conditions, the ship sailing at relatively slower speed can always generate more fuel saving benefits than faster speed. What is more, when the ship operates at the speed slower than wind speed, more benefits in terms of energy efficiency can be obtained from Flettner rotor technology. Besides, the west-east lines in Pacific Ocean can be regarded as the best shipping routes with Flettner rotor technology among all the case studies. This paper can offer some reference value for shipping industry for the research on the application of Flettner rotors to shipping.

The main conclusion is that, performance of ship fitted with wind assist technologies should be calculated based on annualized fuel savings rather than focusing on individual voyages. This is due to the fact that, the performance of the ship with wind assist technologies heavily depends on the wind magnitude and directions, which are random by nature and the time of sailing. Therefore, many calculations should be carried out in different dates, times and directions to determine the average annualised performance of the ship with flettner rotor in a given route. Due to the wind magnitude and direction not all routes are feasible for ships with wind assist technologies. Therefore, the tools presented here will support the decision makers to select the best route for a given wind assist technology or select the best technology for a given route.

At last, based on all above analysis, together with some personal thoughts, a framework is proposed for assessing performance of wind assist technology which is shown as below:

STEP 1: Identify the operating area (Route(s));
STEP 2: Determine the operating profile of the ship including min and max operating speed range;
STEP 3: Check whether average annual wind speed is greater than the ship speed. Wind assist technologies highly sensitive to speed and the wind direction;
STEP 4: Calculate the annualised average saving by simulating a number of voyages at different times of the year including return legs of the journeys;
STEP 5: If the gain is acceptable please utilise weather routing approaches to extract maximum benefits—especially at lower average speeds;
STEP 6: Perform similar analysis for different speeds around the ideal operating speed to weigh the gains and losses;
STEP 7: Results should be provided in terms of net fuel saving

Same steps can be repeated for different wind assist technologies as different wind assist technologies perform differently depending on the technology selected.

ACKNOWLEDGEMENTS

This research is part of project: Shipping in Changing Climates (EPSRC Grant no. EP/K039253/1). The author would like to express sincere thanks for the support from UK Research Council and University of Strathclyde.

REFERENCES

Betz, A. 1925. The "Magnus effect"—the Principle of the Flettner Rotor. Technical Memorandums, National Advisory Committee for Aeronautics, January, 1925.

De Marco, A.& Mancini, S. & Pensa, C. & Calise, G. & De Luca, F. 2016. Flettner Rotor Concept for Marine Applications: A Systematic Study. International Journal of Rotating Machinery, Vol.2016, 2016, 1+12.

Enercon, 2013. Rotor sail ship "E-Ship 1" saves up to 25% fuel.

Howett, B. & Lu, R. & Turan, O. & and Day, Sandy. 2015. The Use of Wind Assist Technology on Two Contrasting Route Case Studies.Shipping in Changing Climates Conference 2015, Glasgow, UK.

Howett, B. & Turan, O. & Day, S. 2016. WASPP: WIND ASSISTED SHIP PERFORMANCE PREDICTION. Shipping in Changing Climates Conference 2016, Newcastle, UK.

KCS container ship geometry, http://www.simman2008.dk/KCS/kcs_geometry.htm

Lu, R. & Turan, O. & Boulougouris, E. & Banks, C. & Incecik, A. 2015, A semi-empirical ship operational performance prediction model for voyage optimization towards energy efficient shipping, Ocean Engineering Vol 110, 2015, 18–28

"Norsepower".www.norsepower.com

Nuttall, P. & Kaitu'u, J. 2016. The Magnus Effect and the Flettner Rotor:Potential Application for Future Oceanic Shipping. The Journal of Pacific Studies, Vol.36 (2), 2016, 161–182.

Rizzo, F. 1925, The Flettner Rotor Ship in the Light of the Kutta-Joukowski Theory and of Experimental Results, Technical Notes, National Advisory Committee for Aeronautics, October, 1925.

"Rotor Sail Solution", www.norsepower.comSeifert, J. 2012. A Review of the Magnus Effect in Aeronautics", Progress in Aerospace Sciences Vol. 55, 2012, 17–45.

Simonsen, M.H. & Larsson, E. & Mao, W. & Ringsberg, J.W. 2015. State-of-the-art within Ship Weather Routing. Proceedings of the ASME 34th International Conference on Ocean, Offshore and Arctic Engineering, St. John's, Newfoundland, Canada, OMAE.

Traut, M. & Larkin, A. & Gilbert, P. & Mander, S. & Stansby, P. & Walsh, C. & Wood, R. 2014. Low C for the High Seas Flettner Rotor Power Contribution on a Route Brazil to UK, Low Carbon Shipping 2012 Conference, Newcastle, 2012.

Traut, M. & Gilbert, P. & Walsh, C. & Bows, A. & Filippone, A. & Stansby, P. & Wood, R. 2014. Propulsive power contribution of a kite and a Flettner rotor on selected shipping routes. Journal of Applied Energy, Vol. 113, 2014, 362–372.

Viking Line. https://www.sales.vikingline.com/

Walther, L. & Rizvanolli, A. & Wendebourg, M. & Jahn, Carlos. 2016, Modeling and Optimization Algorithms in Ship Weather Routing, International Journal of e-Navigation and Maritime Economy 4,2016, 31–45.

Wikipedia-Rotor ship.

Marine Design XIII – Kujala & Lu (Eds)

Time based ship added resistance prediction model for biofouling

D. Uzun, R. Ozyurt, Y.K. Demirel & O. Turan
University of Strathclyde (UoS), Glasgow, UK

ABSTRACT: The selection of antifouling coating is a significant process that affects fuel consumption of ships and therefore it is important to select the most effective coating system. Since each ship has a particular route and operational characteristics, appropriate antifouling paint should be selected based on the ship's operating profiles to get the maximum efficiency from the coating. The paper presents a time-based biofouling prediction model that uses ship route, ship speed, time at sea, time in port and data from paint performance tests. The model basically predicts the increase in roughness due to biofouling. The model converts the roughness into frictional resistance coefficients, which are stored in a database for a range of ships from 10 m to 300 m.The developed model was used to predict added resistance of a real ship which is coated with an antifouling paint. The predicted added resistance values were presented for 1-year operation time.

1 INTRODUCTION

The Shipping industry is one of the industries based on profit; meaning that acquisition is the most considerable criterion when it comes to decide whether a ship design is successful or not. There are many obligatory and third-party costs that ship owners and designers have to deal with such as administration, ports, fuel, operating, insurance, repair and maintenance etc. However, building well-designed correctly equipped ships can help reduce the fuel consumption and emissions including CO_2.

There are many studies indicating importance of fuel consumption, which is the most substantial cost and by itself constitutes 30%–40% of total ship expenses (Notteboom and Vernimmen, 2009). As Milne (1990) mentioned in his study, approximately 184 million tonnes of fuel are consumed by the world fleet. A recent study indicated that it's between 140 and 289 million tonnes per year (Endresen et al., 2004, Corbett and Koehler, 2004). For these reasons even, any small change in fuel consumption can make a significant contribution on ship's profit as well as environmental friendliness.

Selection of antifouling coating process plays an important role that drastically affects fuel consumption. However, selection of the most effective system is still uncertain (Swain et al., 2007). Since routes and operating profiles are specific to each ship, antifouling paint should be selected accordingly to obtain the maximum efficiency from the coating.

As mentioned paint decision is a key procedure in terms of profit but surprisingly the way that hull coating is selected is in general is not very scientific without full awareness such that it`s selection is usually made using information provided by the manufacturer, by word of mouth, and negotiations with sales personnel (Swain et al., 2007).

To the best of the authors' knowledge, no model exists, which takes into account the effect of the role of different coatings on fuel consumption by simulating real ship operations. As Swain et al., (2007) stated—the lack of knowledge situation leads to the potential for incorrect coating selection resulting in penalties in speed, performance and fuel consumption as well as the requirement for unscheduled hull cleaning and dry-docking.

As previously shown biofouling causes added resistance (Demirel et al., 2017), speed reductions and increases in GHG emissions (Townsin, 2003). For this reason, today's ships are generally equipped with monitoring devices to record fuel consumption on a daily basis. Measured data is sent to a computerised analysis of ship performance tool. This tool uses a mathematical model to calculate the added resistance due to biofouling by eliminating transmission system performance, weather conditions, ballast/laden conditions, speed etc. (Munk, 2006). This kind of framework can be for all intents and purposes accommodating on cost-benefit choice for coating determination, optimal dry-docking intervals or timing for surface treatment.

The developed model in this paper is a time-based biofouling prediction model that uses ship route, ship speed, time at sea, time in port and performance tests of paint. In other words, the model is designed as a decision-support tool for the antifouling paint selection of any type of ship cruising on any sea in the world. It also provides a specified profitable effect of monitoring system such as cost-benefit comparison of distinctive sorts of antifouling paints.

In Section 2, antifouling paint performance index (API) was explained. Then structure of API vs. time model and regression analyses on static paint tests were presented while conversion of API to roughness height was performed.

In Section 3, details of model and required data and preparations were provided. Also added frictional resistance database, developed neural network on this database used in this study as assistant parts of model, were introduced.

In Section 4, the developed model was tested on a real case. The result in increase on effective horse power was compared with a report giving increases on effective horse power written by a company which use computerised analysis ship performance tool.

In Section 5, discussions were mentioned pointing out the novelty of the developed model together with potential utilisation of the model in various areas such as, maintenance scheduling and paint decision support tool were stated.

2 CORRELATION BETWEEN TIME AND FOULING

Fouling growth is a complex biological phenomenon, authors with given resources are not in a position to find a physical basis for a model of its progression. So, what it can be developed instead is empirical model as is commonly done with biological systems. In order to come up with empirical models, results of immersion tests done over an extended length of time in diverse geographic locations were used. Flat plates coated with a biocide based, tin-free, self-polishing anti-fouling paint from a paint company were kept submerged in Mediterranean Sea and in a place in Equatorial region and were occasionally lifted above water-surface to measure the level of fouling accumulated on the plates. The results were noted against the observation dates, and this dataset was used to build regression models of fouling progression. The source data-set for this model is essentially a time-stamped series of ASTM fouling rating called API, for each immersed sample. This rating used in this study ranges between 0 (representing zero fouling) and 100 (representing extreme calcareous fouling covering the entire plate).

The following regression candidate was found appropriate for fitting API as a function of exposure time:

$$API(t) = a * e^{\left[-\left(\frac{t-t0}{\tau}\right)^2\right]} \tag{1}$$

where t is the immersed time as measured in days. a is the upper limit for biofouling rating and it changes respect to the type of biofouling.

The free parameters of the model are t0 and τ. The model has the structure of a delayed-onset of exponential growth. The API value is 0 to start with, and after a time it gently increases using an exponential growing process with a time constant of τ.

Following are the values of the free parameters as fitted on the immersion experiment. It should be noted that static immersion paint dataset is a collection of static test which conducted throughout 1 to 2 years periods with many number of different test panels immersed at various seasonal times. Due to temperature differences between seasons in a year, test panel which is immersed in summer and the test panel which is immersed in winter show different performance profile. Therefore, data points that stayed out of the fitting line were removed by taking data point clusters into account to keep the fit on the general trend.

Figure 2, 3, 4, 5, 6 and 7 show model curves corresponding to these models plotted along with the tidied data points.

From a micro-modelling perspective, the fouling process would be a complex stochastic process dependent on local flora/fauna and physical conditions. However, a model can be created based on sea surface temperature (SST), which is widely accepted as the principal determiner of fouling level (Yebra et al., 2004) (Cao et al., 2011). As the authors of this study have two data points, as far as sea-temperature is concerned—Equatorial and Mediterranean. From the model parameters it can be seen that the onset time t0 varies wildly across sea-temperature data-points, and hence it can be assumed to be a strong function of the coating's efficacy.

Mean sea water surface temperature (SST) of the region can be can be predicted by using Figure 1.

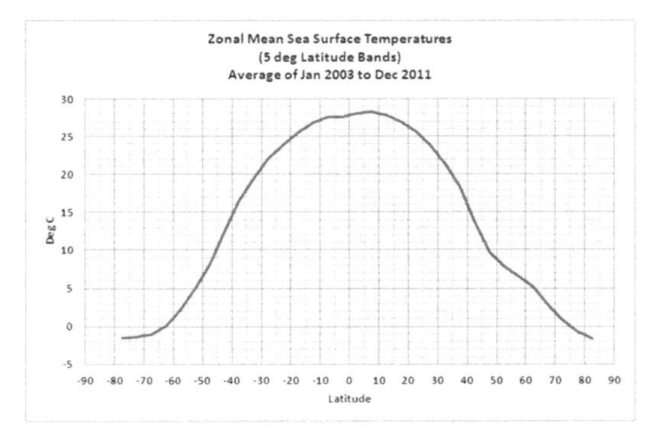

Figure 1. Mean sea water surface temperature for 5-degree latitude bands.

Minus values on the x axis of Figure 1 represent the latitudes on north hemisphere while positive values represent the latitudes on south hemisphere. Mean sea water surface temperature of any region in world can approximately be obtained through Figure 1. Sea surface temperature difference between the base regions' and the region in question can be calculated by the equations shown below.

$$\begin{aligned}\Delta T_E &= 30° - T°(x) \\ \Delta T_M &= T°(x) - 20°\end{aligned} \tag{2}$$

where T indicates the sea water surface temperature (SST) and subscripts E indicates the Equator whereas subscript M indicates the Mediterranean.

$$\begin{aligned} t0(x) &= \frac{t0_E * \Delta T_M + t0_M * \Delta T_E}{10} \\ \tau(x) &= \frac{(\tau_E * \Delta T_M + \tau_M * \Delta T_E)}{10}\end{aligned} \tag{3}$$

Lastly, new coefficients can be calculated as explained in Equation 5 to be used in API calculation for any region in question.

2.1 *Regression analyses of API values*

Calculated API values over test durations were utilized in graphs for each type of fouling separately.

These types of fouling are presented namely as;

- Type C as tube worms, barnacles, hydroids, molluscs, tunicates and algae (longer than 5 mm)
- Type B as algae (shorter than 5 mm) and weeds
- Type A as diverse group of microorganisms forming biofilm layers. (slimes, marine bacteria, diatoms)

A Gaussian type equation, shown at Equation 1, was fitted on data to be able to predict API values for any desired time. Both regression coefficients and statistical values were presented on graphs as well.

2.1.1 *Paint efficiency in an equatorial region*

Paint efficiency in an Equatorial region was evaluated through regression analyses for each fouling type regarding changes in API values according to increasing time. It can be clearly seen from figures biofouling rapidly grows in an Equatorial region when compared with the growth in a Mediterranean region. It is important to note that this growth matches expectations since Equatorial region is the most suitable region in terms of biofouling growth.

2.1.1.1 Regression analysis for Type C (Animals and algae larger than 5 mm)

API values for Type-C fouling in Equatorial region were shown in Figure 2 for more than 700 days. A

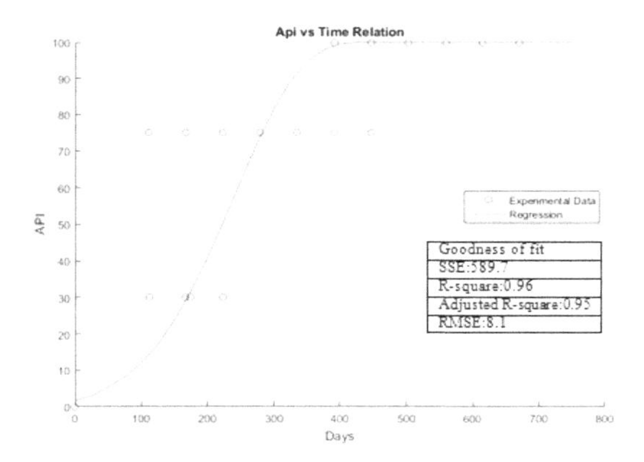

Figure 2. Regression analysis for animals and algae larger than 5 mm.

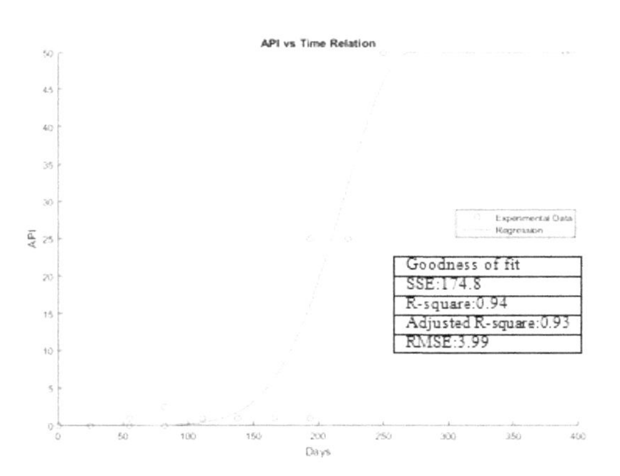

Figure 3. Regression analysis for algae shorter than 5 mm.

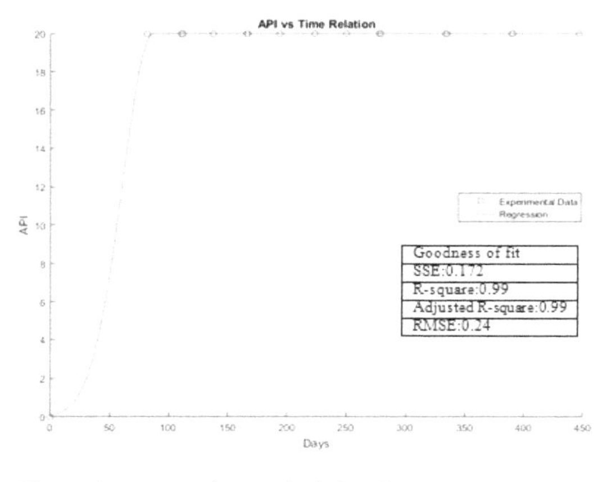

Figure 4. Regression analysis for slime.

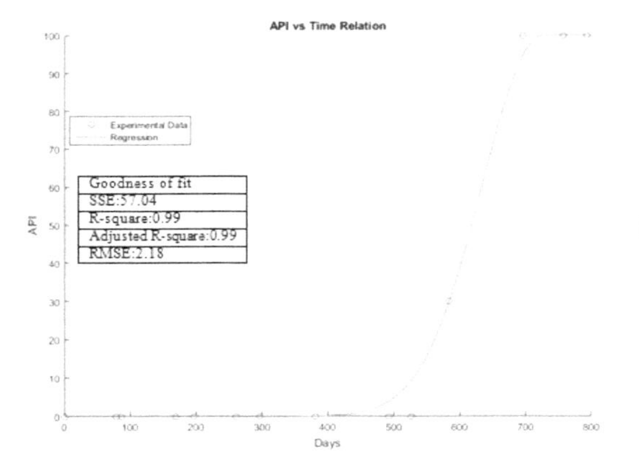

Figure 5. Regression analysis for animals and algae larger than 5 mm.

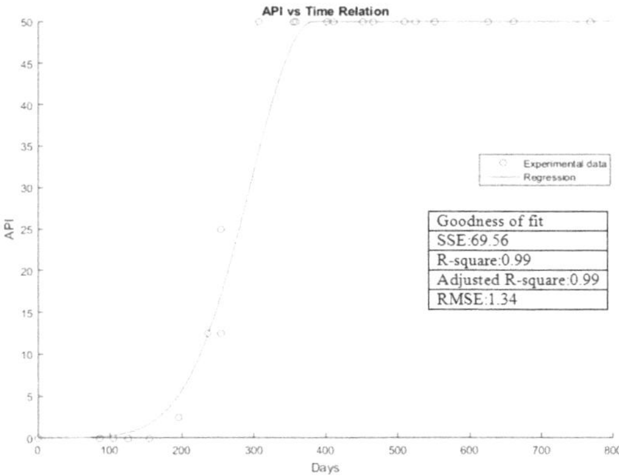

Figure 6. Regression analysis for algae shorter than 5 mm.

regression curve was fitted on this data by using Gaussian type equation. Statistical values for regression curve were also presented in tables on the graph.

2.1.1.2 Regression analysis for Type B (Algae shorter than 5 mm)

API values for Type-B fouling in Equatorial region were shown in Figure 3 for more than 400 days. A regression curve was fitted on this data by using Gaussian type equation. Statistical values for regression curve were also presented in table on the graph.

2.1.1.3 Regression analysis for Type A (Slime)

API values for Type-A fouling in Equatorial Region were shown in Figure 4 for more than 450 days. A regression curve was fitted on this data by using Gaussian type equation Statistical values for regression curve were also presented in table on the graph.

2.1.2 *Paint efficiency in a mediterranean region*

Paint efficiency in a Mediterranean Region was also evaluated through regression analyses for each type of fouling regarding changes in API values according to increasing time. It can be clearly seen from figures biofouling growth is relatively slow in Mediterranean Region when it is compared with the growth in an Equatorial Region. The reasons of comparatively slow biofouling growth in Mediterranean Region are obviously hydro-geographical features of the region.

2.1.2.1 Regression analysis for Type C (Animals and algae larger than 5 mm)

API values for Type-C fouling in Mediterranean region were shown in Figure 5 for more than 800 days. A regression curve was fitted on this data by using Gaussian type. Statistical values for regression curve were also presented in table on the graph.

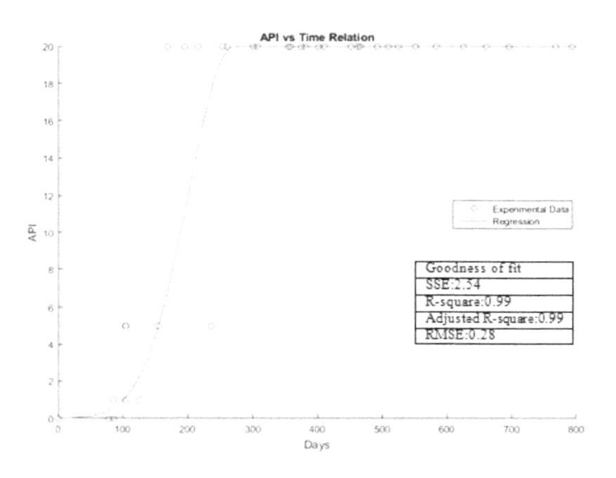

Figure 7. Regression analysis for slime.

2.1.2.2 Regression analysis for Type B (Algae shorter than 5 mm)

API values for Type-B fouling in Mediterranean Region were shown in Figure 6 for more than 800 days. A regression curve was fitted on this data by using Gaussian type equation. Statistical values for regression curve were also presented in table on the graph.

2.1.2.3 Regression analysis for Type A (Slime)

API values for Type-A fouling in Mediterranean Region were shown in Figure 7 for more than 800 days. A regression curve was fitted on this data by using Gaussian type equation. Statistical values for regression curve were also presented in table on the graph.

2.2 *Conversion of antifouling paint efficiency index to equivalent sand roughness height*

Schultz (2007) presented the equivalent sand roughness for a range of coating fouling conditions together with the NSTM (Naval Ships' Technical Manual) rating based on his extensive experiments in his previous studies (Demirel 2015). To the best of this authors' knowledge it is the only study providing this type of conversion between fouling rating and the equivalent sand roughness heights. Thus, using the similarity between these two ratings, API ratings calculated in this study were converted to equivalent sand roughness height to be able to calculate added frictional resistance. Related calculations can be seen step 6 in Section 3.2.

3 MODEL AND PROCESS ORDER

3.1 *Required data and preparation for the code*

3.1.1 *Time*

Operation duration was separated into two sections as idle time which refers to stationary ship including loading, unloading and cruise time means moving ship.

3.1.2 *Sea water surface temperatures and latitudes*

The model focuses on sea water surface temperature (SST) as the most important hydro graphic features. Latitude degrees come directly from the noon reports of the ship.

Latitudes then are separated into three groups regarding their SST values as seen in Figure 8. Tropical and subtropical regions were taken as one group since their SST values were similar to each other.

$$\textit{Latitude Groups} = \begin{cases} \textit{Tropical and Subtropical} & \rightarrow \quad 0 \leq \textit{latitude degree} \leq 40 \\ \textit{Temperate} & \rightarrow \quad 40 < \textit{latitude degree} \leq 60 \\ \textit{Polar Region} & \rightarrow \quad 40 < \textit{latitude degree} \leq 60 \end{cases}$$

3.1.3 *Ship length and speed*

Ship length and ship speed is vital parameters which have important effects on ship frictional resistance thus they have to be assigned into the model.

To sum up only the required inputs for model are idle times (time in stationary position-anchored), latitude degrees of the region where ship is anchored, speeds of ship, length of ship and regression coefficients coming from paint tests.

3.1.4 *Neural network system on added frictional database*

3.1.4.1 Added frictional resistance database

A numerical in-house code, which was developed by Demirel (2015), was used with the aim of creating an extensive database for frictional added resistance of ships. The inputs of the mentioned code are roughness height, roughness functions and corresponding roughness Reynolds numbers and desired ship lengths. The code basically applies Granville scale-up procedure to find the effect of measured roughness condition on full-scale ships. Therefore, it calculates the frictional resistance of ships for a given roughness on their hull for several ship speeds. (depending on the roughness functions).

Having approximately 7000 predictions for ship lengths from 10 m to 300 m with 10 m increments by using the equivalent sand roughness height in Table 2, an added resistance database was created. Thus, in case of providing roughness heights, by using this database added frictional resistance can be predicted for different ships of any desired lengths and speeds.

3.1.4.2 Neural network system on database

A simple neural network system was created using the added frictional resistance database to be able

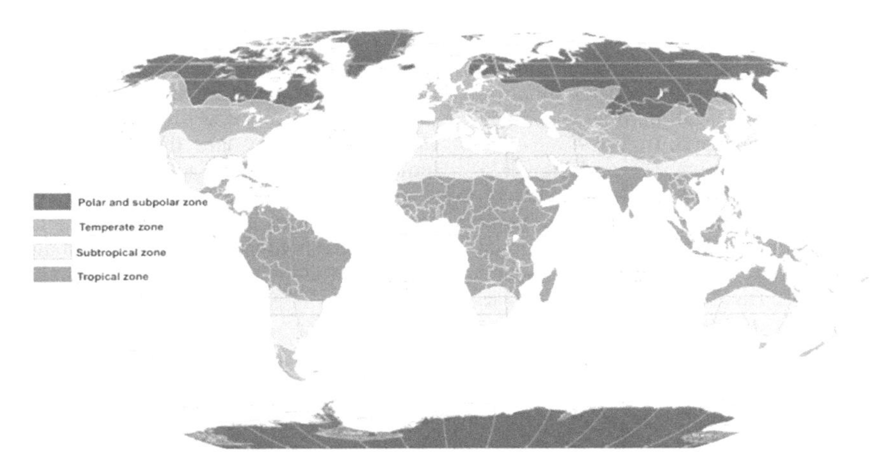

Figure 8. Climate zones.

Table 2. Correlation between NSTM and API ratings for different fouling conditions and the values of equivalent sand roughness (k_s) Except API ratings other values are adapted from (Schultz, 2007).

Description of condition	NSTM rating	API rating	k_s (μm)
Hydraulically smooth surface	0	0	0
Typical as applied AF coating	0	0	30
Deteriorated coating or light slime	10–20	10–20	100
Heavy slime	30	30	300
Small calcareous fouling or weed	40–60	40–60	1000
Medium calcareous fouling	70–80	70–80	3000
Heavy calcareous fouling	90–100	90–100	10000

Table 3. Required inputs for the model.

Idle Times	[DD](days)
Latitude degrees	[°]
Speed of Ship	[m/s]
Length of Ship	[m]
Paint Regression Coefficients (t0, τ, a)	[-]

to make predictions for any ships in different fouling condition. Data was formed of 6817 × 3 input and 6817 × 1 output data in which Input data contains length of ship, speed of ship, roughness height whereas output data contains ΔC_F (change in frictional resistance).

Levenberg-Marquardt method was selected as training algorithm and at the 24th iteration ideal results were obtained.

In neural network guided user interface, as usual, 70% of samples were used for training whereas 15% of samples for validation and 15% of samples for testing. The standard network is used for function fitting which is a two-layer feedforward network, with a sigmoid transfer function in the hidden layer and a linear transfer function in the output layer. The default number of hidden neurons was set to 10 as seen in Figure 9.

It can be seen from the Figure 10 and Figure 11 training process was validated at 24th epoch with 10^{-8} mean standard error and 0.99 R values. Thus, it shows that prediction on changes in fractional resistance will be precise enough to use this neural network for the added frictional resistance database.

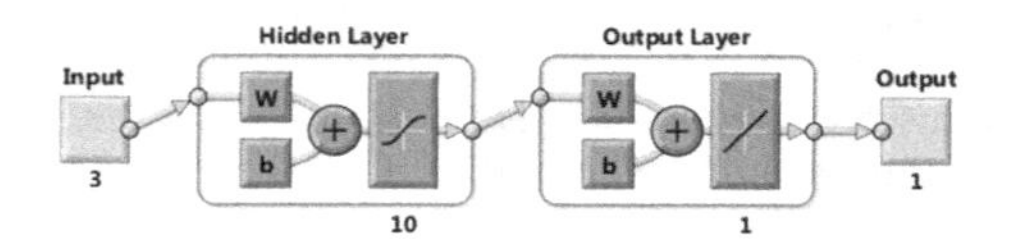

Figure 9. Schematic of the neural network.

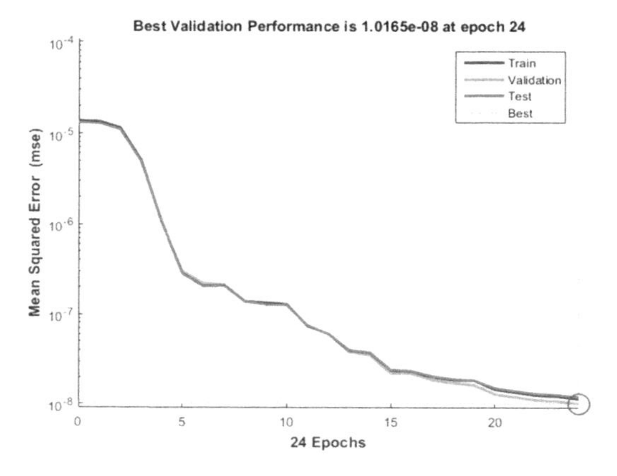

Figure 10. Neural network training performance.

	Samples	MSE	R
Training:	4771	1.14535e-8	9.91675e-1
Validation:	1023	1.01653e-8	9.92568e-1
Testing:	1023	1.19707e-8	9.91769e-1

Figure 11. Results of neural network training.

3.2 *Process steps of the code*

1. *Categorisation of Idle Times*

First step is to categorise idle times according to the latitude groups. This basically creates three fouling zones regarding latitudes degrees and idle times in each of these three zones. Tropical and subtropical regions are the most suitable zones (equatorial region) in terms of fouling growth since SST temperature is relatively higher than the other regions. Essentially, this separation aims to distinguish time for different climate zones in order to evaluate effect of time on fouling growth accurately.

$$\text{Idle times Groups} = \begin{cases} \text{Idle times in Tropical and Subtropical Region} \rightarrow 0 \le \text{latitude degree} \le 40 \\ \text{Idle times in Temperate Region} \rightarrow 40 < \text{latitude degree} \le 60 \\ \text{Idle times in Polar Region} \rightarrow 60 < \text{latitude degree} \le 90 \end{cases}$$

2. *Paint Selection*

With the aim of stating efficiencies of paints in the model, paint regression constants obtained from regression analyses are identified as inputs. In this step paint is selected and its constants are automatically assigned for the following steps.

3. *Determination of SST's*

An equation was fitted for the Figure 1 and SSTs for other regions are determined by using this equation.

$$SST(\,^{\circ}\mathrm{C}) = 14 + 15 * \frac{\cos(\,\textit{latitude degrees})}{28.64} \tag{4}$$

4. *Extrapolation of Paint Coefficients for Other Regions*

In order to have a biofouling growth model, it has to cover all world seas wherever ships cruise. Since the author does not have such an extensive experimental data, the existing data must be extrapolated for other regions to predict paint efficiency in other regions. Since this model was created based on the data obtained in the Mediterranean Region (SST 20°C) and Equatorial Region (SST 30°C). By using linear regression between these temperatures, paint coefficients were derived for other regions. It is important to note that the model can be improved in case of having more experimental paint data in order to determine paint efficiency accurately for other regions. In this step firstly, SST value of the region is predicted via Equation 6 and then paint coefficients are derived by using the equations 4 and 5 respectively for the region in question.

5. *Calculation of Antifouling Paint Efficiency Ratings (APIs)*

Equation 3 was formed based on general formula of Gaussian equation. For each region where the ship has been anchored the 3 types fouling growth models are created using both derived paint coefficients and the total time that has been spent by ship at ports. It's important to note that the time here is a cumulative value which is equal to the sum of times in ports. The sample equations simulating 3 types fouling growths are created for a Mediterranean Region, as can be seen below. These formulas were created as examples. The model produces these equations at every step for each place where ship is anchored.

- Animal and Long Algae Fouling growth in a Mediterranean Region

$$API(t) = 100 * e^{\left[-\left(\frac{t-726.4}{129.7}\right)^2\right]} \tag{5}$$

- Short Algae Fouling Growth in a Mediterranean Region

$$API(t) = 50 * e^{\left[-\left(\frac{t-383.5}{124.4}\right)^2\right]} \tag{6}$$

- Slime Fouling Growth in a Mediterranean Region

$$API(t) = 20 * e^{\left[-\left(\frac{t-271.9}{99.31}\right)^2\right]} \tag{7}$$

As clearly seen from the equations, regression coefficients are assigned into the equations and t will be defined by the model through using the idle times in ports. An important point here is that the model separates the idle times into three groups as explained in the idle times categorization step. After making groups in idle times, the model calculates the fouling ratings for the increasing idle times without considering the group of idle times. However, if the result rating was lower than the rating calculated at previous step, code simply continues with the higher API rating. The reason of this process is to eliminate cleaning effect when especially ships move to the colder regions after spending a certain time in warmer waters.

6. *Calculation of Equivalent Sand Roughness Heights for calculated API ratings*

Each fouling rating has to be converted to k_s values since the neural network works on k_s values.

Therefore, an equation needs to be fitted between NSTM and k_s as shown as below.

$$k_s = f(\,API\ rating) \tag{8}$$

NSTM and API ratings are similar in terms of evaluating fouled surface as seen from the Table 2. Whilst NSTM ratings accept value of 0 as clean surface, API ratings in this study also accept 0 as clean surface. A function can be created by using Table 2 in order to find equivalent sand roughness of any API rating that would be calculated by the developed model.

$$x = API \tag{9}$$

$$k_s = \begin{cases} 119.7937 x^{2.8169518} e^{\left(0.2047757x - 0.0003197x^2 - 2.5814013\sqrt{x}\right)}, & x > 2 \\ 30 & x \le 2 \end{cases} \tag{10}$$

If there is no fouling on the ship surface or x less than or equal to 2 equivalent roughness heights are set to 30 as newly applied SPC type antifouling paint surface roughness height are approximately 30 μm. A regression equation was created based on the values in Table 2 to find corresponding k_s values for any calculated API values. If x is greater than 2 k_s is calculated with the equation shown above.

7. *Calculation of Added Frictional Resistance and Increase in Effective Power (P_E)*

Calculated k_s values at previous step with ship length and ship speed were taken as inputs for the developed neural network. Frictional resistance increases for each k_s was calculated in 12 months operation period for each operation day. Then the increase on the effective power of a ship, P_E, due to the effect of fouling can be expressed by:

$$\%\Delta P_E = \frac{C_{T_R} - C_{T_S}}{C_{T_S}} \times 100 = \frac{\Delta C_F}{C_{T_S}} \times 100 \tag{11}$$

4 IMPLEMENTATION OF MODEL ON REAL SHIP (CASE STUDY)

160,000 DWT Crude Oil Tanker was selected as a case study for the developed model. Ship's route and ship dimensions were taken as inputs and model was run for 12 months period. Ship's all operation in that time was in tropical and subtropical regions which are the most suitable regions for biofouling. Furthermore, calculated increase on effective horse power of ship was compared with a report written by a company which uses commercial software (tool) to monitor the ship performance. It works basically taking the data from any kind of monitoring and data recording system and running them in a tool with also making several corrections such as wind, waves, sea current, fuel oil quality, water temperature and etc.

As clearly seen in Figure 12, there is a good agreement between the present results and the actual results provided by the company over a period of approximately 12 months. The increase in the effective power of the ship due to biofouling at cruise speed were predicted to be 26.6% whereas according to the calculations based on the actual operation data the increase on the effective power was 25% for the same ship over a same operation period.

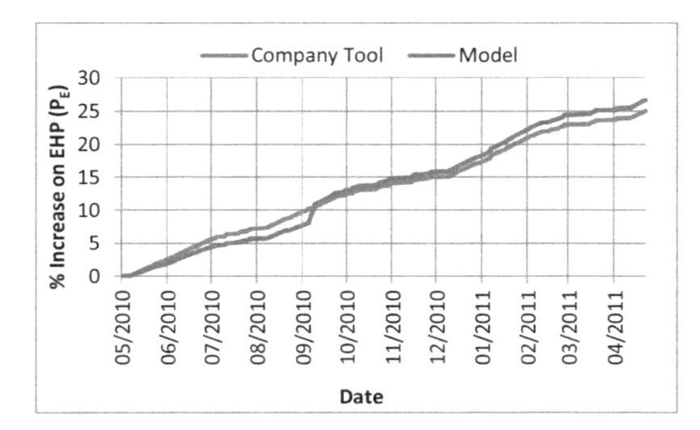

Figure 12. Percentage of increase on P_E for ship respect to time.

5 DISCUSSIONS

To the best of this author's knowledge, no specific numerical model exists to predict the biofouling effect on ship frictional resistance therefore the developed model is novel. This model also can be used as a decision-making tool which determines the suitable coating type for particular types of ships. It may also be used to decide the best maintenance and/or hull cleaning activities as well as maintenance intervals.

NOMENCLATURE

GHG	Greenhouse gases
API	Antifouling paint performance index
SST	Sea water surface temperature
NSTM	Naval ships technical manual
DWT	Dead weight Tonnage
P_E	Effective horse power
k_s	Equivalent sand roughness
t0, τ, a	Paint regression coefficients
ΔC_F	Change in frictional resistance
C_{T_R}	Total resistance coefficient in rough condition
C_{T_S}	Total resistance coefficient in smooth condition

REFERENCES

ASTM-D6990-05 2011. Standard Practice for Evaluating Biofouling Resistance and Physical Performance of Marine Coating Systems.

Cao, S., Wang, J.D., Chen, H.S., & Chen, D.R. (2011). Progress of marine biofouling and antifouling technologies. Chinese Science Bulletin. http://doi.org/10.1007/s11434-010-4158-4.

Corbett, J.J. & Koehler, H.W. 2004. Considering alternative input parameters in an activity-based ship fuel consumption and emissions model: Reply to comment by Øyvind Endresen et al. on "Updated emissions from ocean shipping". Journal of Geophysical Research: Atmospheres, 109, n/a-n/a.

Demirel, Y.K. 2015. Modelling the roughness effects of marine coatings and biofouling on ship frictional resistance [PhD thesis]. Glasgow: University of Strathclyde.

Demirel, Y.K., Uzun, D., Zhang, Y., Fang, H.-C., Day, A.H. & Turan, O. 2017. Effect of barnacle fouling on ship resistance and powering. Biofouling, 33, 819–834.

Endresen, Ø., Sørgård, E., Bakke, J. & Isaksen, I.S. A. 2004. Substantiation of a lower estimate for the

bunker inventory: Comment on "Updated emissions from ocean shipping" by James J. Corbett and Horst W. Koehler. Journal of Geophysical Research: Atmospheres, 109, n/a-n/a.
Mcentee, W. 1916. Variation of Frictional Resistance of Shjps With Condition of Wetted Surface. Journal of the American Society for Naval Engineers, 28, 311–314.
Munk, T. 2006 Fuel consumption through managing hull resistance. Copenhagen: Motorship Propulsion Conference.
Milne, A. 1990. Roughness and drag from the marine paint chemist's viewpoint. Marine Roughness and Drag Workshop, London
Naval Ships' Technical Manual. 2006. Chapter 081 – waterborne underwater hull cleaning of Navy ships. Publication # S9086-CQ-STM-010/CH-081 Revision 5. Washington (DC): Naval Sea Systems Command.
Notteboom, T.E. & Vernimmen, B. 2009. The effect of high fuel costs on liner service configuration in container shipping. Journal of Transport Geography, 17, 325–337.
Schultz, M.P. (2007). Effects of coating roughness and biofouling on ship resistance and powering. Biofouling. http://doi.org/10.1080/08927010701461974
Swain, G.W., Kovach, B., Touzot, A., Casse, F. & Kavanagh, C.J. 2007. Measuring the Performance of Today's Antifouling Coatings. Journal of Ship Production, 23, 164–170.
TOWNSIN, R.L. 2003. The Ship Hull Fouling Penalty. Biofouling, 19, 9–15.
Yebra, D.M., Kiil, S., & Dam-Johansen, K. (2004). Antifouling technology—past, present and future steps towards efficient and environmentally friendly antifouling coatings. Progress in Organic Coatings, 50(2), 75–104. http://doi.org/10.1016/J.PORGCOAT.2003.06.001.

Hull form design

Marine Design XIII – Kujala & Lu (Eds)
© 2018 Taylor & Francis Group, London, ISBN 978-1-138-34076-3

Utilizing process automation and intelligent design space exploration for simulation driven ship design

E.A. Arens, G. Amine-Eddine & C. Abbott
Siemens PLM, UK

G. Bastide
Siemens PLM, France

T.-H. Stachowski
Digitread AS, Norway

ABSTRACT: In order to increase market share, companies need to design more energy efficient and cleaner vessels to meet new environmental regulations and increased societal expectations. Discovering better designs, faster for such complex engineering systems requires easy-to-use CAE simulation solutions with fast turnaround time. But even more so, the engineering process itself should allow automated design space exploration to drive innovation in the form of better performing vessels and offshore platforms that are guaranteed to meet current and future regulatory requirements. This paper will demonstrate state of the art technology for simulation driven design of a typical offshore supply vessel. Firstly industry trends driving the need for innovation and automated design space exploration are reviewed. Next the methodology for setting up an automated simulation driven workflow is explained, including an overview of the parametric CAD preparation process, the CFD simulation approach for evaluating vessel performance, and the intelligent search technology employed to determine optimal designs. Finally, results of a multi-objective design space exploration study are presented.

1 INTRODUCTION

1.1 *Trends driving innovation in today's marine market*

International Maritime Organisation (IMO) regulations surrounding emissions and fuel efficiency amplified by the entry into force of the Paris agreement on Climate change are playing an increasingly pivotal role in the direction of ship design.

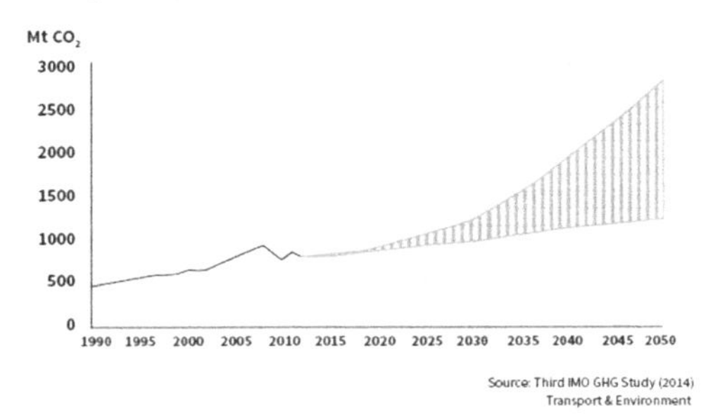

Figure 1. Range of expected increase in GHG emissions from shipping (Transport & Environment).

The range in expected growth of CO_2 from shipping until 2050 is shown in Figure 1 (Transport & Environment). The wide range indicates the level of uncertainty in the projection, however, the trend is always upwards. As ratified by Paris agreement, this trend must be reversed for shipping to become carbon neutral by the year 2040 if the rise in global temperatures is to be kept below 1.5C (see Figure 2, Transport & Environment).

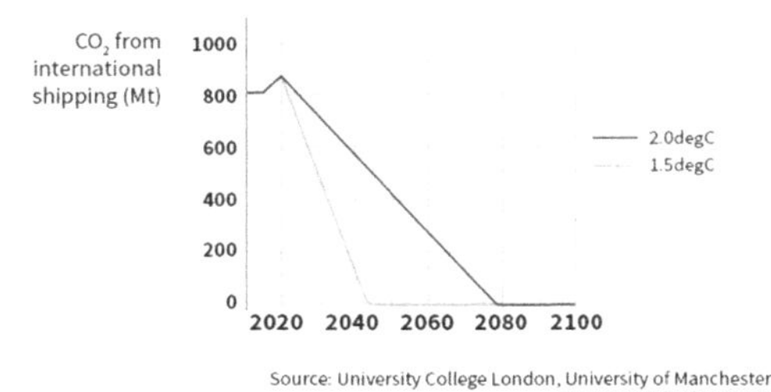

Figure 2. Trend which are necessary to meet the 1.5/2 degrees objective (Transport & Environment).

In addition to the regulatory side, the rapid growth of production capacity in the market since the start of this century is leading to an extremely tough and competitive environment. This has already led to significant consolidation in the market with, many shipyards going out of business. This emphasizes the need to be able to adapt quickly to new market demands and offer innovative and compliant products with ever increasing pace.

1.2 *The problem*

In a highly competitive market where regulations become increasingly stricter, adoption of the right technology and innovation differentiates those organization that thrive from those that fail. Whether it is the requirement to upgrade existing technologies, increase R&D budgets, or look at ways to reduce costs throughout the design process without compromising quality, each one of these factors has a major impact on the organizations bottom line.

In the typical scenario when lots of organizations have to react quickly on tenders, they rely on their domain knowledge and previous designs, modifying them accordingly to meet new requirements. This approach remains standard across marine engineering firms, and it's clear that companies have already realized good value out of using engineering simulation to validate designs, to troubleshoot challenging design flaws, and even to predict the effects of certain proposed design changes. The bottleneck however remains in the time that it takes to provide all this information within a given project timetable.

A major reason for this has to do with the nature of Computer Aided Engineering (CAE) simulation and Computer Aided Design (CAD) tools engineers use throughout project lifecycles. Typically, tools used in the process are a set of disconnected, domain-specific applications. Meaning that each type of drawing, such as general arrangement, pipe routing or mechanical system is done in different software packages. This can create complex user workflows, create redundant work and impede the data flow and transfer of designs between simulation domains.

As a consequence, the simulation process, as it exists today can be challenging, inefficient and too time consuming to truly achieve the level of innovation and pace that market demands. Typically, simulation is done on only a handful of design concepts to simply verify they will meet performance requirements.

Companies may struggle to use simulation knowledge at a higher level as a means to provide input into design decisions and the process itself, may struggle simply because by the time a simulation is complete, the design has already iterated a few more times, rendering the simulation obsolete.

This also means good design suggestions that could improve product performance go unnoticed, or problems become noticed far too late in the development process to be implemented without significant cost to the shipbuilder.

1.3 *Related work in automated CFD (Computational Fluid Dynamics) and design space exploration technology for hull design*

Several proposals to improve engineering processes within ship design and make it more holistic have been published in recent years, see for example Papanikolaou (2009). There is also an ongoing EU-funded project involving a number of partners in an effort to develop framework standards for linking various design tools to perform holistic ship optimization (HOLISHIP). Some major limitations observed in these approaches are:

- The hull design space exploration is disconnected from the overall vessel design and drawing tools. This will create redundant work and optimized hull design might not satisfy all requirements.
- To allow for sufficient design iterations and reasonable turnaround time, CFD is often done using inviscid methods or as a number of standalone pre-defined runs to create Response Surface Model data sets.
- Disparate tools are used that require significant scripting effort and IT expertise to work together.
- Knowledge of optimization algorithms is often required.

These difficulties are acknowledged by Papanikolaou (2009), which ends with the statement: "*It should be noted, however, that the implementation of the required relevant optimisation procedures needs still to be developed for a long list of practical cases by experienced software programmers supported by ship designers (ideally by naval architects), which will be a demanding R&D task for the decade(s) to come*".

It is because of these difficulties that many organizations are still using traditional approaches to ship design.

2 THE MODERN APPROACH

This paper demonstrates an approach whereby innovation is driven directly by the simulation processes, themselves. By using modern technology a more efficient and cost-effective vessel can be designed.

The key to avoid out-of-sync data and really incorporate the simulation process in vessel design is to make use of a Master Model as presented by Stachowski & Kjeilen (2017): building a 3D general arrangement with a linked 2D drawing, aesthetic views and different models suited for analysis and simulation (Figure 3). The Master Model holds both solid and surface representations which are available for different use cases and decision making.

In this study, the number of details in the Master Model are limited for incorporating a CFD simulation, but the approach itself can be truly holistic.

The Master Model is created within the NX for shipbuilding CAD tool which enables a holistic design approach from initial concept to production. Parameterization techniques can be utilized, where dimension/parameter updates in the model result in automated regeneration of CAD to reflect the updated parameter values. By parameterizing the design instead of building the model as a traditional "one-off" approach, geometry modifications in the model can be updated automatically and driven with ease to evaluate potentially infinite design possibilities. General Arrangement drawings can be exported automatically, as well as all CFD simulation input data. For CFD, Simcenter STAR-CCM+ is used, enabling a virtual prototype of our vessel to be tested in a simulated full-scale towing tank environment. Finally HEEDS MDO (hereafter referred to as HEEDS) is used to automate the entire process of changing variables in the parametrized CAD, updating the virtual prototype in our CFD simulation, executing the analysis, and intelligently searching for better designs.

Utilizing this approach allows designers and engineers to focus more on gaining insight and discovery, rather than the laborious efforts associated with the traditional manual approaches to ship design. Now, better designs can be found, faster!

2.1 *Case description*

For the purpose of this paper, a Typical Offshore Supply Vessel (OSV) is used as a case study example. The D-OSV is designed by Digitread AS (See Figure 4, Stachowski & Kjeilen (2017)). Offshore vessels are built in small production runs, are highly customized to complex requirements, and are therefore expected to benefit most commercially using the advanced design approach demonstrated here.

In this study we want to achieve both a competitive and efficient vessel design by finding the best tradeoff between capital expenditure (CAPEX) and operational expenditure (OPEX).

In this paper we will not look at all contributions of CAPEX and OPEX but rather look at the main difference with respect to our baseline design (our initial design).

For the CAPEX we will monitor the change of additional steel used for the lower hull and the cost associated to this, so a ΔCAPEX.

For the OPEX we will look at the vessel resistance (and thus only the fuel consumption related to voyage). We will not consider any other power requirements like hotel loads since we assume these contributions will not change significantly with the change in hull design.

For the OPEX it is important to take the whole operating envelope of the vessel into account. Each significant operating condition requires a separate simulation. Based on the occurrence of each operating condition the fuel consumption can be calculated and related to the OPEX. The design approach presented can take a large number of simulations into account (linked of course to hardware resources that are available for the simulations), however for our offshore supply vessel example we will take only one operating condition into account. This design condition is a sailing condition at 13 knots at a design displacement of 15000 mt.

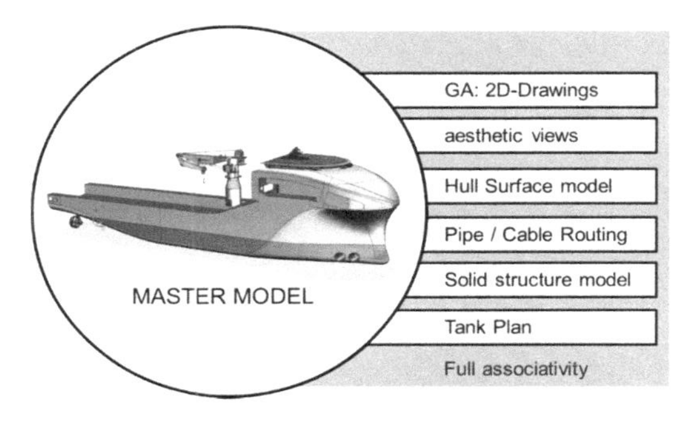

Figure 3. Master Model. One CAD model with full associative links to drawings, views and different models suited for analysis and simulation.

Figure 4. D-OSV (Courtesy of Digitread AS).

3 DESIGN APPROACH SETUP

3.1 *Parametric CAD*

Using HEEDS to perform a CAD Robustness study on design variables allowed us to identify and highlight conflicts of certain variable combination choices that led to CAD rebuild failures. These problems were significantly reduced in the baseline model by refactoring CAD tree features and removing redundant operations. The final success rate of CAD robustness was 75%, an adequate regeneration rate to ensure robust coverage of the design landscape, specified by the bounds in our variables.

Changing the vessel parameters (length, width, etc.) will propagate throughout the design assembly and update all associative relationships automatically including the general arrangement. The 3D General Arrangement assembly built in NX has an associative relation to the hull volume as input to the compartmentation process. As a consequence there is no need to redraw this manually with every new design change. Of course the designer is free to modify such layout at a later stage of development, however this immediate insight can be leveraged very early on in the vessel's development lifecycle.

Table 1 lists the hull design variables modified in this study, along with their ranges. In practice, there is no restriction on the number of variables that can be included in such a design exploration study, but in this study the number required was eight, to capture the intended geometric variation. The first two parameters control the overall size of the hull and the additional six parameters control the shape of the hull. The bilge radius in the aft is larger than the main bilge radius and is setup as a function of the main bilge radius. It is therefore not considered a design variable, but a so-called dependent variable; it gets calculated for a given design based on the other inputs. The propeller clearance is a constraint to ensure specified azimuth thrusters will fit.

Table 1. Hull design parameters.

Parameter			Allowable range Min	Max
Length between perpendiculars	L_{pp}	[m]	80	120
Beam	B	[m]	20	24
Bilge radius	R_b	[m]	1.5	4
Bow shape	T_{cwl}	[m]	6	8
Bow tightness	β_{bow}	[deg]	45	75
Run entry fwd	L_{fwd}	[m]	40	80
Run entry fwd shoulder	$L_{fwd}2$	[m]	5	50
Run entry aft	L_{aft}	[m]	20	70
Bilge radius aft	$R_{b,aft}$	[m]	Rb+1.8	
Propeller clearance	Z_{aft}	[m]	6.5	

Figure 5 shows the influence of the beam (B), length between perpendiculars (L_{pp}) and the run entry forward (L_{fwd}) parameters on the hull design.

Figure 6 below shows how the changes depicted in Figure 5 will automatically update the general arrangement. This can be seen for the lower deck where the engine room is situated: the engine room stays within the parallel middle body of the hull to ensure it fits. This can especially be seen in the bottom right drawing where the maximum run entry

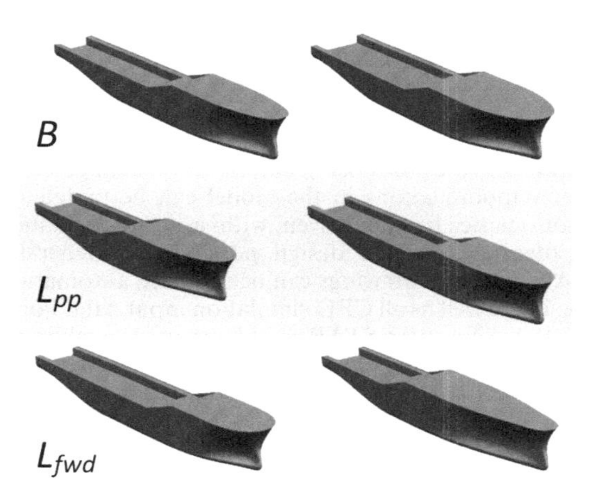

Figure 5. Main hull design parameters.

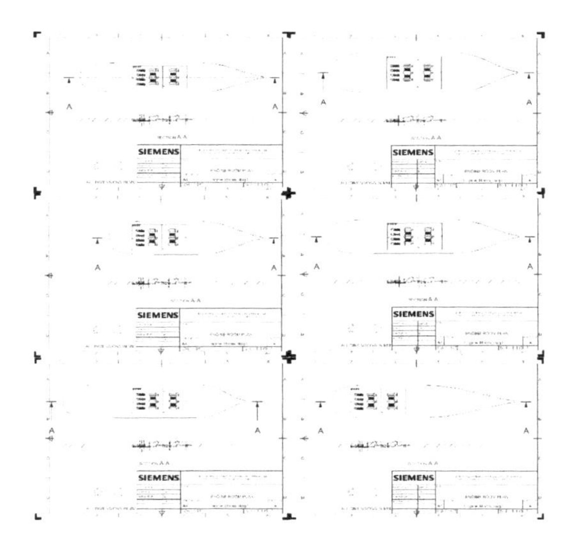

Figure 6. Master model approach ensures general arrangement updates with design changes. Here the deck with engine room is shown.

Table 2. Constraints.

Parameter	Allowable range Min	Max
Deck space [m^2]	800	n/a
GM – maximum expected VCG [m]	0.1	n/a
Freeboard – main deck height [m]	0.25	n/a
Trim [deg]	–3	3

forward results in a sharp and narrow bow shape and therefore pushes the engine room all the way to the aft of the vessel.

In order to find a realistic design the parameter ranges as defined in Table 1 are used combined with a constraint on deck space. For supplying goods to and from a platform, we specify a minimum required deck space of 800 m^2. The constraints are listed in Table 2.

3.2 *Weight assessment*

For the weight assessment a large amount of detailed information is required on hundreds of parts to get an accurate estimate. This can all be done in NX. However for this study we only look at the amount of steel used in the lower hull to calculate the ΔCAPEX, therefore we use a simpler approach of a simple Python journal executed form the top level part assembly. We measure the surface areas of the hull, all decks and directly compute the weight based on the different plate thicknesses and a steel density of 8 g/cm^3.

The ΔCAPEX is based purely on the weight change. A steel price of 500$/mt is assumed.

$$\Delta CAPEX = \Delta M \cdot 500$$

3.3 *Hydrostatics and stability assessment*

Based on the design displacement of the baseline design, and the change in weight, vessel draft is determined for each design iteration. It is assumed that the center of gravity is at deck level and always above the center of buoyancy. Simple hydrostatic calculations are performed in Simcenter STAR-CCM+ producing the longitudinal center of gravity and draft, as well as the freeboard and small angle stability. These are used as constraints in the design study to ensure a feasible design.

While the hydrostatics in this study are determined using Simcenter STAR-CCM+ in principle this can also be substituted with a dedicated stability software allowing for more advanced calculation of center of gravity and stability by including tank fillings and locations of the various loads from the loading condition.

In order to find a realistic design we also add two more constraints on the results of the hydrostatics and stability assessment. For the freeboard we specify that the distance between the main deck and maximum draught should be larger than 0.25 m. For the stability we specify a GM larger than 0. Since we use a simplified weight assessment in this we cannot calculate an accurate VCG. Instead it is assumed that the maximum expected VCG will be at main deck height. The constraints are listed in Table 2.

3.4 *Resistance simulations*

The vessel resistance is made up of both a hydrodynamic and an aerodynamic contribution. Although the method is not in any way restricted from taking into account all details of the hull and superstructure, a more computationally efficient approach is used in this study.

The hull hydrodynamic resistance will make up the largest part of the total resistance and will be most sensitive to design changes. So for each design evaluation the resistance of the clean hull without small appendages is computed (see Figure 7).

For the hydrodynamic resistance Simcenter STAR-CCM+ is used to solve the Reynolds-Average Navier-Stokes Equation (RANS). The flow is assumed incompressible and for modelling turbulence the standard k-epsilon model is used with a two-layer all y+ wall treatment. For two-phase flow (air and water) the volume of fluid (VOF) method is used. An implicit scheme is used for solving the time discretization.

Since the hull is symmetric, only half the computational domain is considered to reduce the computational expense. The computational domain can be seen in Figure 8. To avoid wave reflections from domain boundaries a method called 'wave forcing' is activated across specified adjacent boundary zones. This allows a relatively

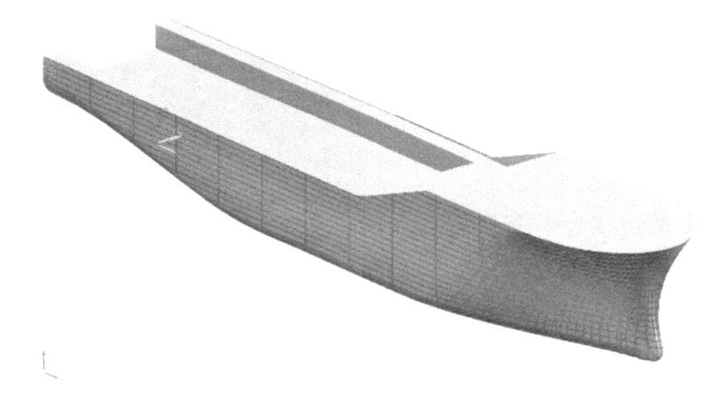

Figure 7. Geometry used in CFD simulation.

small domain to be used enabling further reductions in computational cost.

A trimmed mesh is used in combination with prism layers to capture the boundary layer near the walls. The mesh can be seen in Figure 9 and a more detailed view of the prism layers can be seen in Figure 10.

To speed up the simulation a model called DFBI Equilibrium is used. This model helps to quickly achieve a quasi-steady-state equilibrium position of the vessel that is subject to fluid forces. The sink and trim are not updated every time-step according to the equations of motions, but instead the vessel moves in a stepwise manner. The motion itself is performed using mesh morphing,

Another method applied to speed up the simulation is a multi-grid method. With this approach the simulation starts with a very coarse mesh and is run for a fixed number of time-steps. Next a new finer mesh is created and the previous results are interpolated on the new mesh. Note that the vessel is kept in the moved position and the mesh is realigned with the free surface. Based on the previous resistance results a convergence criteria is defined for this new mesh level. The simulation continues until the convergence criteria is satisfied. This procedure uses a total of five different mesh levels. The multi-grid method is not only faster, but since the morphed mesh is remeshed with every step, the final cell quality is better which results in better overall solution convergence.

Figure 8. Computational domain.

Cell Count: 1.72615e+06

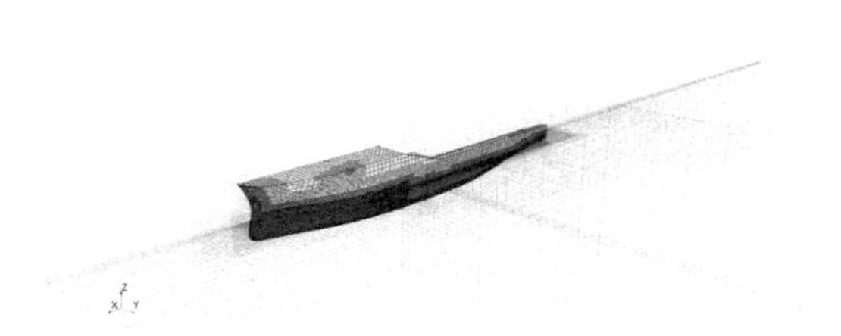

Figure 9. Volume mesh showing refinement areas to capture the free surface and kelvin wake.

A constraint is specified for the dynamic trim to be between -3 and +3 degrees. Vessel designs which do not meet this criteria are flagged as unfeasible. The constraint is listed in Table 2.

The aerodynamic resistance is determined in a separate simulation, only for the baseline design. Discrete contributions are derived for the superstructure, antenna, hull and crane (see Figure 11). For the superstructure a coefficient is derived by dividing the result by the width of the vessel to estimate the aerodynamic resistance of each design iteration (note in this study only the width of superstructure will change). Again, because the hydrodynamics make up the largest part of the total resistance this is believed to be a fair assumption.

For the aerodynamic part Simcenter STAR-CCM+ is again used to solve the Reynolds-Average Navier-Stokes Equation (RANS). The flow is assumed to be incompressible. For modelling tur-

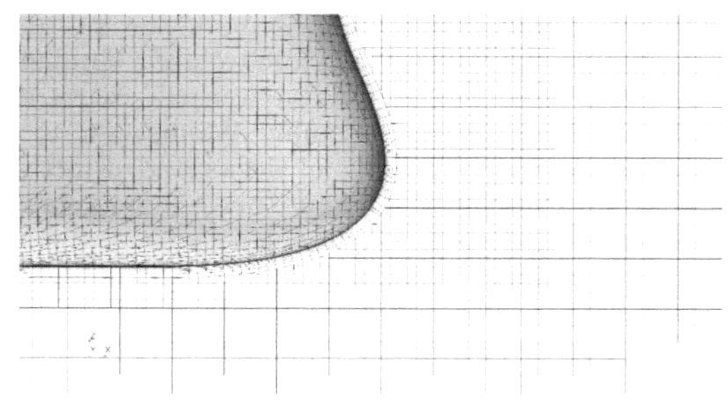

Figure 10. Mesh near hull with prismatic layers to capture the boundary layer.

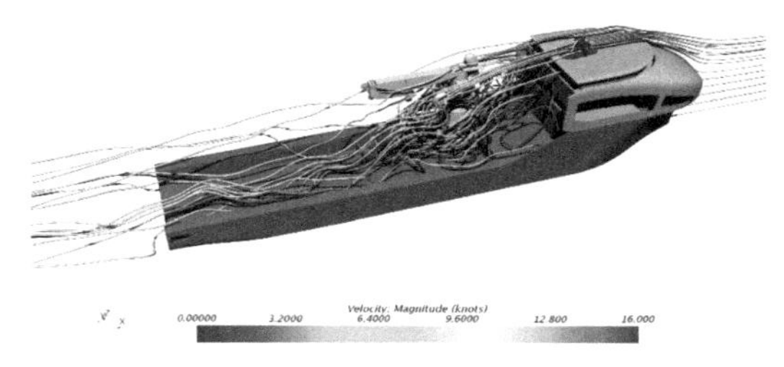

Figure 11. Aerodynamic simulation showing streamlines. Separate coefficients are derived for superstructure (blue), crane (yellow), antenna-mast (grey) and hull (red).

bulence the k-omega SST turbulence model is used combined with a two-layer all y+ wall treatment. The implicit scheme is used for solving the time discretization.

The OPEX in this study only includes fuel costs and is derived from the resistance. First the total resistance is determined by summing the hydrodynamic resistance and the aerodynamic resistance (derived from coefficients).

$$R_{total} = R_{Hydrodynamic} + R_{crane} + R_{mast} + B \cdot c_{superstructure}$$

Next the power required for sailing is determined by multiplying R_{total} with the vessel speed and taking into account propulsive efficiency (65%), gearbox and shaft losses (95%) and a sea margin (85%).

$$P_{Engine,sailing} = v_{vessel} \cdot R_{Total} / (0.65 \cdot 0.85 \cdot 0.95)$$

Note that because we are only interested in the OPEX due to the improved hull design we do not consider any other power requirements like hotel load, crane operations, dynamic positioning (DP) operations, etc.

For this study we assume a specific fuel consumption of 200 g/kwh. It is also assumed that the OSV is sailing 4% of the time during its 20 years design life time. The average fuel price throughout its lifetime is assumed to be $550/mt. The OPEX for sailing throughout its complete design lifetime is:

$$OPEX = 0.55 \cdot 0.2 \cdot P_{Engine,sailing} \cdot (0.04 \cdot 20 \cdot 365 \cdot 24)$$

Note that the assumptions above to derive CAPEX and OPEX are by no means detailed, but since they are directly derived from the change in resistance and weight, and the relative comparison of different designs will not change, they will not influence the goal of this study to find a better design.

3.5 *Design space exploration*

In this study HEEDS is used to automate the entire engineering process: update of geometry in CAD, transfer of geometry through automated meshing for CFD, distributed execution on parallel hardware resources, post-processing and analysis of simulation data, and subsequent calculation of CAPEX & OPEX. To efficiently search the design space, HEEDS uses a proprietary hybrid and adaptive search method called SHERPA.

During a parametric optimization study, SHERPA uses multiple search methods simultaneously, both local and global in a uniquely blended manner. Evolving knowledge gained about the nature of the design space is used to determine when and to what extent each approach contributes to the search. In essence, SHERPA efficiently learns about the given design space and adapts itself to effectively search it regardless of the class of engineer problem. SHERPA is a direct optimization algorithm in which all function evaluations are performed using the virtual prototype itself, as opposed to a response surface model approximation.

The SHERPA method can also perform a Pareto Optimization by handling multiple objectives independently of one another, to provide not only the best design, but a best set of design solutions, all of which are optimal in some sense for one of the objectives. (Chase et. al 2009). SHERPA is used specifically in this study to determine the best set of OSV design solutions, while assessing the trade-off between minimizing both CAPEX and OPEX, whilst simultaneously satisfying all constraints and design requirements in our problem statement.

A graphical representation of the workflow and the outputs is shown in Figure 12. The CFD simulation will only be executed if the constraints requirements for deck space, GM and Freeboard are satisfied (see Table 2). Otherwise the CFD simulation will be skipped and the next design evaluation will commence.

In this study, 300 design evaluations were chosen as the allowed budget for the design space exploration. This was chosen based upon the desired target solution of less than two weeks, accounting for the number of design variables present in our study and the runtime associated with a single design evaluation from start to finish (being ~2.5 hours in duration here based on resources utilized). The most time-consuming process in the workflow is

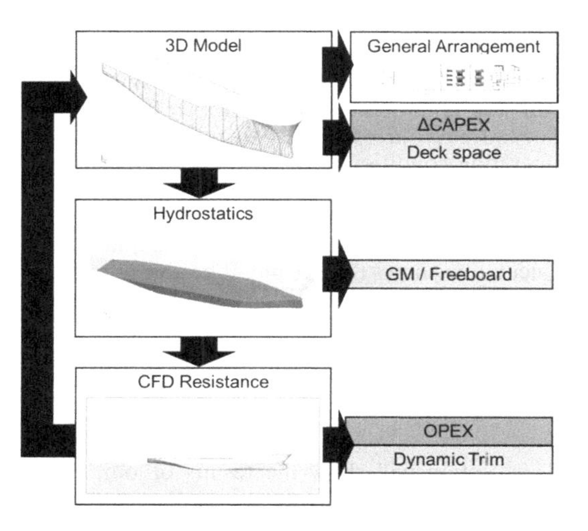

Figure 12. Design study process automated by HEEDS. Objectives in green and constraints in yellow.

the CFD resistance calculation, (limited only by allocated computing hardware resources). As such, we chose to execute in parallel two design evaluations for the resistance calculation, using a resource set of two Linux workstations. This halved our turnaround time, since we could perform twice as many CFD resistance simulations as we would if we only had a single Linux workstation available to us. With this, our study was completed within 12 days. More importantly however, SHERPA adapts its search strategies to find the best possible design alternatives in the allotted time budget. With more resources allocated to the study, either more evaluations could be performed, or the turnaround time could be reduced further.

3.6 *Computational requirements*

To complete the project in less than two weeks, only a single Windows Laptop plus a resource set of two Linux desktop workstations with each 18 cores (Intel(R) Xeon(R) CPU E5–2650 v3 @ 2.30GHz), was used. HEEDS resided on the Windows Laptop, managing the execution of NX CAD locally, whilst also managing the execution of Simcenter STAR-CCM+ remotely using a resource set (a HEEDS specific feature which allows for different design evaluations to be distributed on separate machines, and executed at the same time). So the design space exploration was parallelized (two designs running simultaneously), as were the CFD calculations (18 cores per design).

HEEDS is able to execute different software on different machines with different operating systems. While for this study the simulation was executed on a laptop and two Linux desktops, it is clear that using a larger cluster will significantly reduce the turnaround time for these design space exploration studies. By using a larger cluster we can run the CFD simulations on more cores. It also allows to run several more designs in parallel. Having more hardware resources also allows to investigate more operating conditions or perform additional simulations for strength. HEEDS is designed to allow the user flexibility to allocate resources based upon their need and available resources. A quick hand calculation shows that we might be able to perform 500 design iterations on a 256 core cluster in 3 days if we run 4 designs in parallel.

4 RESULTS

This section will show the results of our design space exploration study:

- Objective: Minimize both ΔCAPEX and OPEX (find trade-off)
- Subjected to: 4 Constraints (Table 2)
- By modifying: 8 variables (Table 1)

4.1 *Baseline design*

The main particulars for the baseline design, the starting point of this study, are listed in Table 3.

The baseline design satisfies the constraints and the total resistance is 258.8 kN. Looking at the wave pattern in Figures 13 and 14 it can be seen that there is still room for improvement; by improving the hull shape we can reduce the wave making resistance. The engine power required for propulsion is 3298 kW which gives an OPEX of $2.54M. Since this is our baseline case the ΔCAPEX is set to $0.

4.2 *Best designs and exploration*

Of the 300 designs evaluated in this study, 235 designs were feasible, 1 design was unfeasible, and 64 were error designs. The feasible designs all satisfied the performance constraints of Table 2, while the single unfeasible design did not satisfy our GM constraint. The 64 error designs were either a result of the CAD failing to successfully regenerate for the given design variable combination (41/64), or

Table 3. Main particulars baseline design.

Main particular	Value
L_{pp} [m]	115
B [m]	25
R_b [m]	2.25
$R_{b,aft}$ [m]	4.00
Z_{aft} [m]	6.50
T_{cwl} [m]	7.00
Design displacement [mt]	15000
β_{bow} [deg]	65
L_{fwd} [m]	57
$L_{fwd}2$ [m]	30
L_{aft} [m]	40

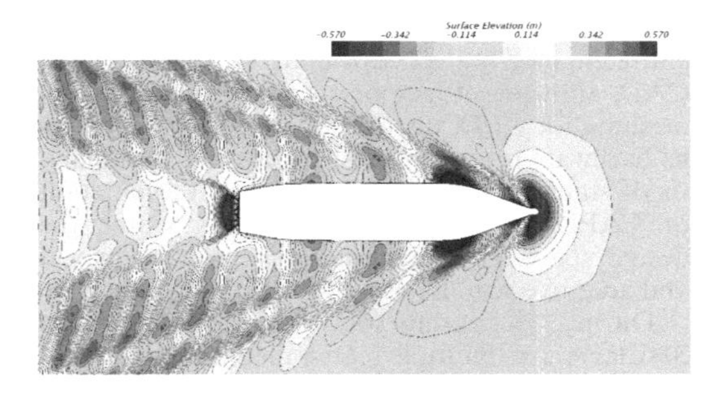

Figure 13. Result showing free surface elevation (top view).

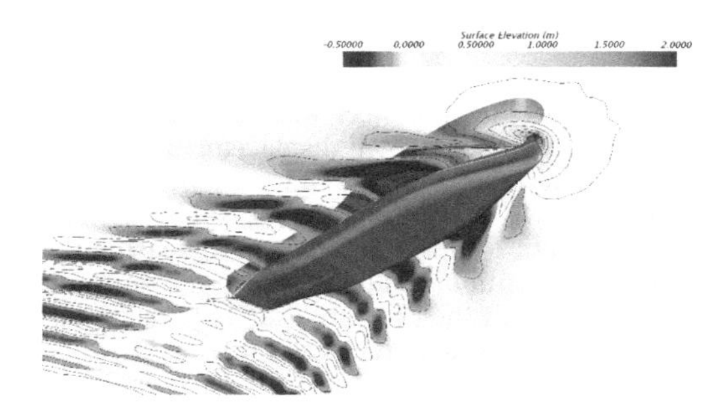

Figure 14. Result showing free surface elevation (3D view).

the hydrostatic and manufacturing requirements not being met. Recall that it was described in the context of Figure 12 that the lengthy CFD computations would only be performed if the deck space, GM and Freeboard requirements were satisfied for a given design. If these computed values did not satisfy these requirements, it was decided to skip the CFD computation associated with that design and treat the design as an error. This scenario accounts for the other 23 non CAD regeneration related error designs. It is important to note that SHERPA efficiently handles error designs, so parameterized models need-not be perfect. As the CAD robustness model showed a 75% success rate for CAD regeneration, it was deemed suitable for a design space exploration. We see here that during the design space exploration, 86% of the designs actually regenerated, and only 14% of the designs failed the hydrostatic and manufacturing requirements. This is typical of a design space exploration with HEEDS, whereas SHERPA will focus-in on promising regions of the design space and spend less time in poor producing regions, resulting in higher success rates than predicted from the CAD robustness runs which evenly sample the entire design space.

Looking at Figure 15, which highlights the trade-off between ΔCAPEX and OPEX in our OSV concept, we can easily identify the Pareto front and best set of designs that meet our goals.

The goal of a Pareto optimization is to identify a family of solutions that perform well in all of our objectives. When objectives compete, such as ΔCAPEX and OPEX here, an optimal family of designs will be found, called a Pareto front.

Visually, this can be explained by looking at Figure 15. Graphically, the goal is to get as far to the lower left corner of the graph as possible (representing lower OPEX performance and lower ΔCAPEX performance). However, we can see that we indeed have two conflicting objectives; for the best designs in blue, a decrease in ΔCAPEX leads to an increase for OPEX (and vice-versa). We can also see that all designs on the Pareto front have a negative ΔCAPEX with most designs having a lower OPEX than our starting baseline design. The designs identified on the Pareto front with a higher OPEX than the baseline, however, have a more significant reduction in the CAPEX. So this should be considered in our evaluation of what design best meets our needs for the given application. Clearly this design space exploration has provided us with multiple design options based on our needs and requirements, allowing us to trade-off ΔCAPEX and OPEX depending on other factors and design considerations.

In Figure 16 we can see a parallel plot which provides greater understanding of relationships between variables, constraints and performance goals. Each design evaluation generates a line that links the specific variables and responses used for the design variant (x-axis represents the given parameter, y-axis the span of variants). Overlaying these lines for the feasible solutions can highlight common paths or parameter values that meet per-

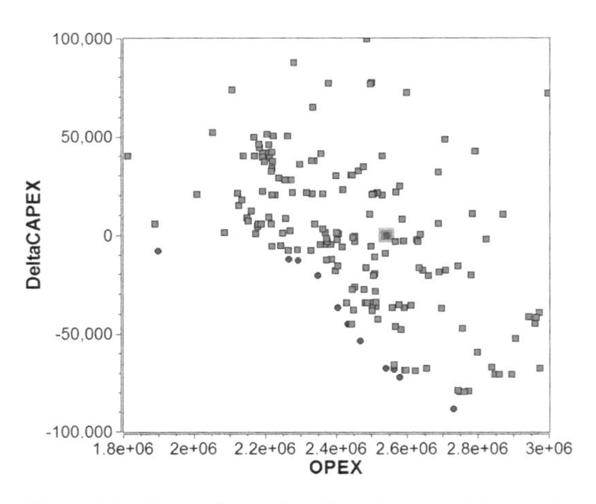

Figure 15. Pareto front plot. Non-dominated designs in blue circles which form the Pareto front, all other non-error design in red squares. Baseline design marked by large square.

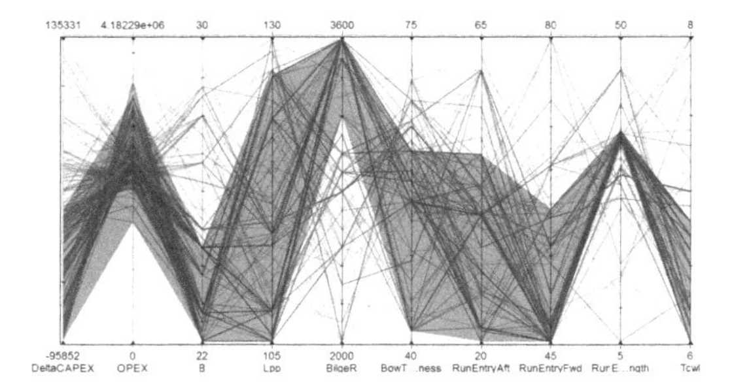

Figure 16. Parallel plot. Non-dominated designs are within the blue band.

formance requirements. The banding of lines can also indicate result sensitivity. The band of lines in blue in Figure 16 are all designs which form the Pareto front. For the beam of the vessel (B on the plot) there is only a narrow band of lines, indicating a consistent beam among the designs on the Pareto front. This could be an indication that this parameter value for beam significantly affects performance and should be kept small for both a low ΔCAPEX and OPEX.

Similarly, and even more prominently, the parameter value for the run entry forward shoulder length has an even narrower band, indicating that high performing vessels all have a very similar bow shape. The result of this bow shape is an interaction effect between the bow wave system and the fore shoulder wave system to minimize the wave making resistance. It makes sense that this parameter value has a very narrow band since we are only investigating one vessel speed and therefore the bow shape will be optimized for this vessel speed.

Conversely, for the length between perpendiculars (L_{pp}) there is a wide spread among the variants along the Pareto front. However, dividing the designs with low ΔCAPEX and low OPEX in two separate sets we can identify a clear trend for the length between perpendiculars, see Figures 17 and 18.

From Figure 18 it can be seen that to design for a low CAPEX, a short (small L_{pp}) bulky vessel is better whereas to design for low OPEX, a long (large L_{pp}) streamlined vessel is better. This can also be seen in Figures 19 and 20 (respectively the low CAPEX design and low OPEX design as highlighted in Figure 17). Both designs, although being narrow, satisfy the criterion for deck space by having a long after body and by having an optimized bow shape for a low pressure resistance and thus low OPEX. The design in Figure 19 is clearly short, minimizing the steel weight and thus the CAPEX. To keep sufficient displacement the draft is larger, increasing the frontal area of the submerged hull which increases the pressure drag. The design in Figure 20 is much longer, increasing the CAPEX, but with a small draft the transom is out of the water and the pressure drag is reduced significantly. The wetted area is slightly larger which

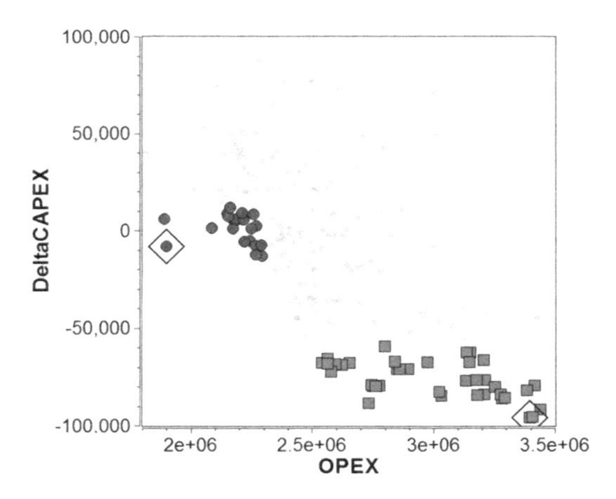

Figure 17. Selection of the low ΔCAPEX (red square) and low OPEX (blue circle). For each selection a design is highlighted (black diamond).

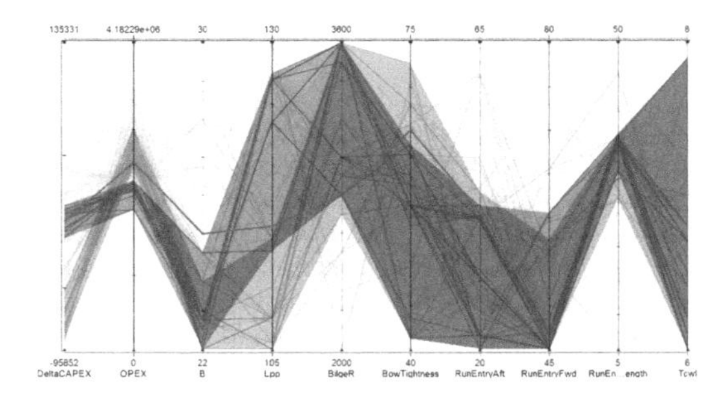

Figure 18. Parallel plot. Low ΔCAPEX designs are within the red band. Low OPEX designs are in the blue band.

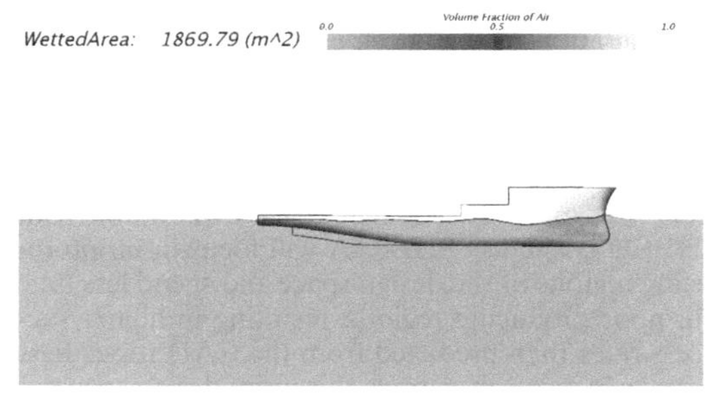

Figure 19. Wetted dynamic surface of a low CAPEX design.

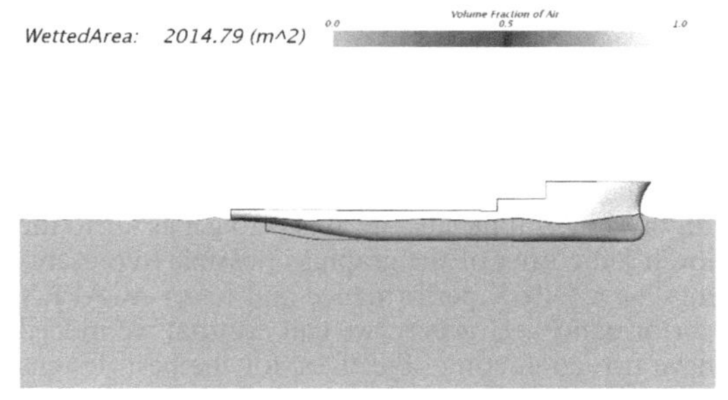

Figure 20. Wetted dynamic surface of a low OPEX design.

increases the resistance drag due to friction, but this increase is small compared to the reduction in pressure drag.

One concept over the other may be better for a given application or region of the globe, or better for scenarios that weren't considered here (e.g. for a DP operation where wind, waves and currents are all perpendicular to the vessel a short vessel can be much more energy efficient). The design exploration output and the insight we can gain from the data (here 300 vessel concepts) allows us to make informed decisions and have many design alternatives for consideration.

5 CONCLUSION

By using one Master Model and automating the coupling with simulation, an efficient work-flow is created which ensures results will not be out of sync with the design, and that the models are in a state to be used with design space exploration to find innovative concepts.

This allows for simulation to drive the design, increasing performance potential and pursuing innovative solutions. For the example case of the OSV we linked CAD with CFD and found a series of better vessel designs with a tradeoff between ΔCAPEX and OPEX. Furthermore, the design study gave us clear insight into the trends among our competing objectives, and multiple design candidates for further consideration.

In the example in this study we only linked CAD with CFD, but in principle it is possible to set up a more complex workflow which can also include FEA analysis or 1D subsystem analysis to assess the total performance of a vessel design. The approach itself can therefore be truly holistic.

The Siemens PLM software tools used in this study, NX, Simcenter STAR-CCM+, and HEEDS, allowed us to setup this automated workflow and design space exploration without the need for expert IT support or knowledge of optimization algorithms or methodologies, it is therefore easy to implement by anyone (e.g. the Naval Architect). Furthermore it allowed us to perform the design space exploration within our allocated time budget and hardware resource allotment, which can be greatly improved by using more computational power in the form of a cluster.

Using Siemens PLM software the modern design approach for ship design, to drive innovation through simulation, can bring huge benefits to the marine sector by lowering project costs and turnaround time as well as more efficient and cost-effective vessels.

6 NOMENCLATURE

Symbol	Definition
CAD	Computer-Aided Design
CAE	Computer-Aided Engineering
CAPEX	Capital Expenditure
CFD	Computational Fluid Dynamics
DFBI	Dynamic Fluid Body Interaction
DP	Dynamic Positioning
FEA	Finite Element Analysis
GM	Distance between VCG and metacentric height
IMO	International Maritime Organization
L_{pp}	Length Between Perpendiculars
OPEX	Operational Expenditure
OSV	Offshore Supply Vessel
PLM	Product Lifecycle Management
R&D	Research & Development
RANS	Reynolds-Average Navier-Stokes
VCG	Vertical Center of Gravity
VOF	Volume of Fluid

REFERENCES

Chase N., Rademacher M., Goodman E., Averill R. & Sidhu R. 2009. *A Benchmark Study of Multi-Objective Optimization Methods.*

HOLISHIP. HOLIstic optimisation of SHIP design and operation for life cycle. URL: *http://www.holiship.eu/*. retrieved Feb 26, 2018.

Papanikolaou A. 2010. Holistic ship design optimization. *Computer-Aided Design*, 42(11), November 2010, 1028–1044.

Stachowski T-H & Kjeilen H, Digitread AS, Norway 2017. Holistic ship design—how to utilise a digital twin in concept design through basic and detailed design. *International Conference on Computer Applications in Shipbuilding 2017*, Singapore.

Transport & Environment. Road to Paris: A climate deal must include aviation and shipping. *URL https://www.transportenvironment.org/road-paris-climate-deal-must-include-aviation-and-shipping*. retrieved Feb 26, 2018.

Marine Design XIII – Kujala & Lu (Eds)
© 2018 Taylor & Francis Group, London, ISBN 978-1-138-34076-3

Smart design of hull forms through hybrid evolutionary algorithm and morphing approach

J.H. Ang & V.P. Jirafe
Sembcorp Marine Ltd., Singapore

C. Goh
University of Glasgow, UK

Y. Li
School of Computer Science and Network Technology, Dongguan University of Technology, China
Faculty of Engineering, University of Strathclyde, UK

ABSTRACT: Digitalisation of ship design, construction and operation are gaining increasing attention in the marine industry as ships become eco-friendlier and smarter. 'Smart designs' can be applied under this digital revolution as an intelligent design process that is highly automated and collaborative with 'smart manufacturing' and 'smart ships' through-life. Focusing on smart design, we introduce a Hybrid Evolutionary Algorithm and Morphing (HEAM) approach which combines an evolutionary algorithm with an efficient morphing-based shape variation approach to optimise and automate the hull form design process. By combining the process of design exploration, geometry modification and performance evaluation, it enables highly automated design process where new hull forms are created, compared and analysed so as to reduce the overall design cycle and produce more optimal designs. This paper a) introduces a framework on how smart design process can be connected with smart manufacturing and smart ships to form into through-life smart shipping network, b) describes the proposed HEAM approach to optimise and automate the hull form design process and c) provides result of the HEAM approach to demonstrate the design efficiency and performance improvement.

1 INTRODUCTION

In the face of stiff competition, cost reduction and globalisation, marine industry today requires a quantum leap in the entire process of ship design, construction and operations to keep abreast and differentiate itself in the marketplace that is fast moving towards digital technologies. With the arrival of Industry 4.0 and artificial intelligence 2.0, digitalisation of ship design, construction and operation are gaining increasing attention in the marine industry as ships become eco-friendlier and smarter. However, the emphasis of 'smart' design has been lacking in comparison to 'smart' manufacturing and a 'smart' ship. In particular, how can we further automate the ship design process and integrate with smart manufacturing and smart ship considering the entire ship lifecycle? With industry 4.0, manufacturing is now moving towards more intelligent system or machineries that are highly connected. In the context of shipping, vessels are also becoming smarter with more automated systems and begun to move towards unmanned or fully autonomous vessel, with the world's first autonomous container feeder vessel YARA birkeland to be launched in 2018. Smart design is hereby proposed as an intelligent design process that is highly automated and collaborates closely with smart manufacturing and smart ships throughout the entire product lifecycle. By connecting up smart design with smart manufacturing and smart ships, important information can be shared seamlessly across entire lifecycle of a ship and becomes a fully integrated through-life smart shipping network.

As marine industry moves towards eco-friendlier and energy-efficient ships, the design and optimisation of hull forms continue to play an important role to help reduce fuel consumption and carbon dioxide emission. Traditional method of ship design and hull form optimisation requires many manhours by ship design firm and shipyards using the 'trial-and-error' approach, which is inefficient and does not guarantee optimum designs. While latest simulation based design methods and tools help to automate some of these processes, they still require considerable human input and success at end result depends heavily on the designer's experience and knowledge.

Considering smart design, we introduce an innovative concept which aim to address the above issues by automating the hull form design process with minimum user interference and yet not compromising the quality of the results. This is achieved by combining evolutionary algorithm with efficient shape variation approach known as morphing and hereby proposed as hybrid evolutionary algorithm and morphing (HEAM) approach.

The focus of this paper is to introduce the concept of smart design as well as HEAM approach which possesses the potential to improve design efficiency and produce more optimal hull forms. Section 2 describes the concept of smart design which can link up smart manufacturing and smart ship operation process considering entire lifecycle. Section 3 proposes the HEAM approach which combines evolutionary algorithm with morphing to automate the hull form design and optimisation process. Section 4 provides the results of HEAM concept, followed by discussion and conclusion in section 5.

2 SMART DESIGN OF SHIP AND HULL FORM

Hull form design and optimisation is an important topic in the marine industry due to more stringent environmental regulations and reduction of operation cost due to fuel consumption. An efficient hull form will help to reduce resistance acting on the vessel and thereby reducing fuel consumption and emission to the environment. As the marine industry moves toward digitalisation, it is essential to further automate the design process and connect up with other lifecycle processes to achieve fully automated smart design. An illustration of the various stages of hull form design developments are provided in Figure 1 as below.

From above figure, it started off with traditional method of hull form design which is based primary on 'trial-and-error' where ship designers create the initial hull form from scratch or modify from existing proven hull designs (stage 1). This manual method is extremely time consuming and only allows few design variation and testing. Since the introduction of computer into ship design, simulation-based design (SBD) approach became dominant as it accelerates the design process by semi-automating the shape variation and optimisation process and validating the performance using computational fluid dynamics (stage 2). While SBD method helps to automate some of the design process, they still require considerable human input and the result often depends heavily on the designer's experience and knowledge. This method is also isolated and does not usually consider external feedbacks such as manufacturing or ship operation. With the development of digitalisation and artificial intelligence, we can further automate the design process so as to reduce the iterative process and free up the designer's time for more critical task. This process is proposed under HEAM approach (stage 3) and will be covered in next section. Subsequently, the end goal is to fully automate the design process and connect up the entire product lifecycle process to also include ship construction and operation. This can be achieved via through-life smart shipping network (stage 4) which will be further elaborated below.

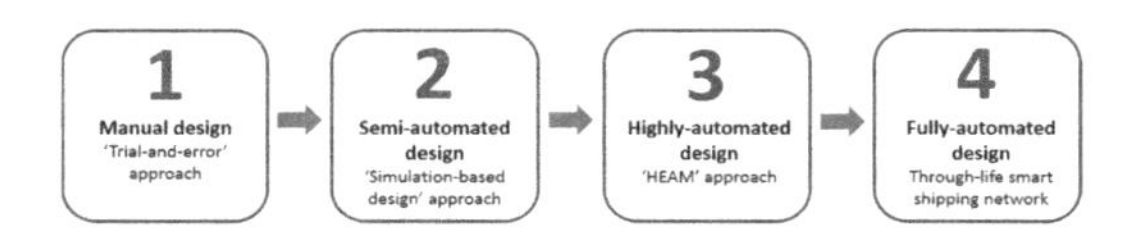

Figure 1. Development towards fully automated design process.

2.1 *Related works in simulation-based hull form design*

Simulation-based designs (SBD) are used widely for performing numerical optimisation and evaluation of hydrodynamic performance of the hull form. Most simulation-based hull form design optimisation consists of three key processes—firstly, (1) the hull shape is linked to a design exploration function to search systematically for optimal design. Next, (2) the geometry modification function will change the shape of the hull to create new designs. Following which, (3) the new shape generated will be evaluated on its performance function. This process will continue to iterate until the stopping criteria or most optimal hull design is achieved. An illustration of simulation based hull form design optimisation process is provided in Figure 2.

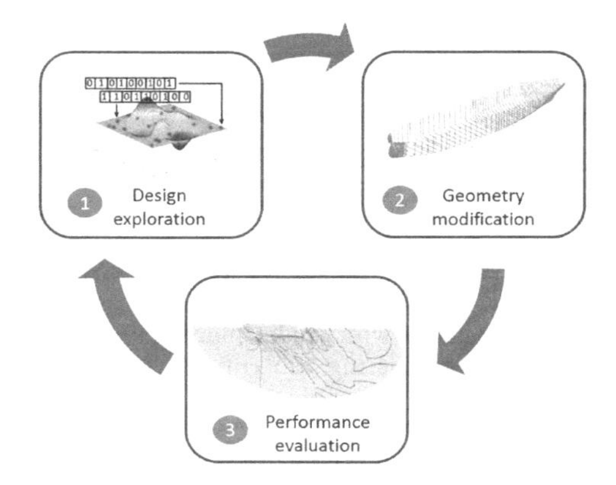

Figure 2. Simulation-based hull form design optimisation process.

While the key steps or process are somewhat similar, there are many different methodologies that are applied in hull form optimisation. Design exploration, also known as optimiser, is a parametric optimisation process where key design parameters such as shapes of the hull are modified in an iterative loop to produce a set of optimal hull forms at the end of the optimisation process. Some recent examples of design exploration methods applied to hull form optimisation include Simplex of Nelder and Mead (Jacquin et al. 2004, Kostas et al. 2014), Sequential quadratic programming (Park et al. 2015, Berrini et al. 2017) and evolutionary algorithm such as genetic algorithm (Baiwei et al. 2011, Kim 2012) and particle swarm optimisation (Tahara et al. 2011, Xi Chen et al. 2014).

Geometry modification plays an important role in ensuring the hull shape can be easily manipulated to form new shapes in order for the optimiser to investigate and evaluate. The key challenge here is to ensure every new shape generated must be smooth and of feasible design. There are 2 main approaches used to modify hull geometry—direct modification and systematic variation. Direct modification changes the hull geometry by adjusting the hull coordinates manually using control points through curve or surface representations. While this method is highly flexible, it requires large number of control points to represent the shape and hence not very efficient for modifying the entire hull shape. Examples of direct modification includes Beizer curve, non-uniform rational basis spline (NURBS) and T-splines (Kostas et al. 2014). On the other hand, systematic variation modifies the hull shape using a function which considers global hull parameters (e.g. block coefficient, Cb) or series of local hull representation. This method is particularly useful for global modification which enables the entire hull form to be transformed more efficiently. However, shape changes are somehow more restricted and not very flexible as compared to direct modifications. Some recent examples of systematic variation methods applied in hull form optimisation include parametric modification (Saha & Sarker 2010, Brizzolara & Vernengo 2011) and free-form deformation (Campana et al. 2013).

Performance evaluation assess each candidate solution produced from the optimiser based on the objective function. The most important performance parameters that are influenced by shape of the hull include resistance and sea-keeping behavior and hence selected as key objective functions in most hull form design optimisation applications. For evaluation of resistance, Computational fluid dynamic (CFD) are used extensively in hull form optimisation which had been proven as an effective means to simulate the fluid flow around vessel. Examples of CFD methods used for resistance evaluation include potential flow (Nowacki 1996) and Reynolds Averaged Navier-Strokes Equation—RANSE (Tahara et al. 2006, Zha et al. 2014). For sea-keeping analysis, there are several numerical methods which include strip theory, unified theory, green function method, etc. (Bertram 2000).

2.2 *Related works in smart design, construction and operation*

With the development of digitisation and big data, it enables ships to become more connected and smarter. However, most ship lifecycle process now are rarely connected in reality. To illustrate this, we can look at the lifecycle and key milestone of a typical ship's lifetime as provided in Figure 3 below.

From above figure, while the life-cycle processes are progressive and closely linked, information are not transferred interchangeably from one process to the other. As an example, once shipbuilder completes construction and ship owner takes over ownership of the vessel, they do not share operational information back to the shipbuilder which are useful for them to monitor the actual performance and use the information to improve on its subsequent designs. As such, it should be recognised that full digitalisation of ship life-cycle cannot be realised without elevating each process into more connected and automated process which considers the entire value chain.

Considering life-cycle for ships, we look at how these key processes can be elevated into smart design, construction and operation. Smart designs are relatively new concept and not explored widely in particular ships or hull form design. It can be defined as an intelligent design process which is highly automated and ability to collaborates closely with smart manufacturing and smart ship or operation throughout the entire lifecycle. Some early works on smart design by authors include (Ang et al. 2017a). Other smart design application related to ship includes smart ship system design for electric ships (Chalfant et al. 2017).

Smart construction or manufacturing is currently a high interest topic that is being driven under the advent of industry 4.0 (i4). I4, also known as forth industrial revolution, aims to merge

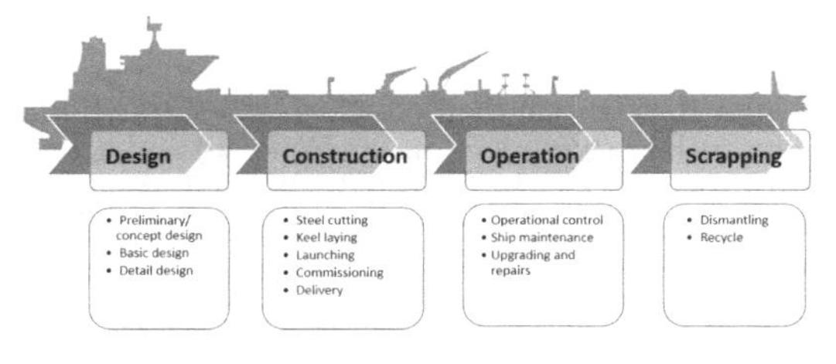

Figure 3. Typical ship life-cycle and key processes.

the real and physical space through cyber-physical system. It provides a platform to transform traditional segregated manufacturing process into fully connected manufacturing system. Basic components and enabling technologies of i4 includes internet of things, collaborative robots, cybersecurity, cloud computing, additive manufacturing and big data analytics. I4 or smart factory concept are increasingly adopted and implemented in high tech manufacturing and aviation industry. There are currently very few applications of i4 in shipbuilding. One recent work done is a study of smart pipe system for shipyard (Paula et al. 2016).

Smart operation or commonly known as smart ships is increasingly in demand as ship builders aims to build more efficient and smarter vessels. Smart ships can be defined as vessels that are highly connected through the use of big data for real time monitoring and controls so as to enhance operation performance. Smart ships are driven with promises to reduce operation cost and improve safety and believed to revolutionise ship design and operation. There are several works done recently on smart ship which includes one that considers the design of control of power and propulsion system for smart ship (Geertsma et al. 2017) and another that consider smart ships in general (Jan 2017). Smart scrapping or decommissioning of ships are not considered here due to its short duration comparing entire lifecycle but might be worth to look into in future works.

2.3 *Through-life smart shipping network*

As mentioned in the beginning, the emphasis of smart design has been lacking in comparison to smart manufacturing and smart products. In particular, how is smart design going to integrate with smart manufacturing and smart product when considering entire product lifecycle? One promising solution is a framework that connects and creates a feedback loop to link up smart manufacturing and smart product to smart design. By connecting up smart design with digital manufacturing and smart operations into a unified digital model, important information can be shared seamlessly across entire product lifecycle of a ship and becomes a fully integrated through-life smart shipping network, as introduced in (Ang et al. 2017a) and illustrated in Figure 4.

By closing the loop between ship operation and design through smart product, useful through-life data such as ship operating environment and actual performances can be collected, analysed and feedback into smart design process for producing more optimum future designs. In addition, smart design can be further enhanced by combining design automation process and digital product model under i4. By linking digital product model into the automated design process, we can provide an automated feedback loop from smart product back to design to improve the design performance of future vessels.

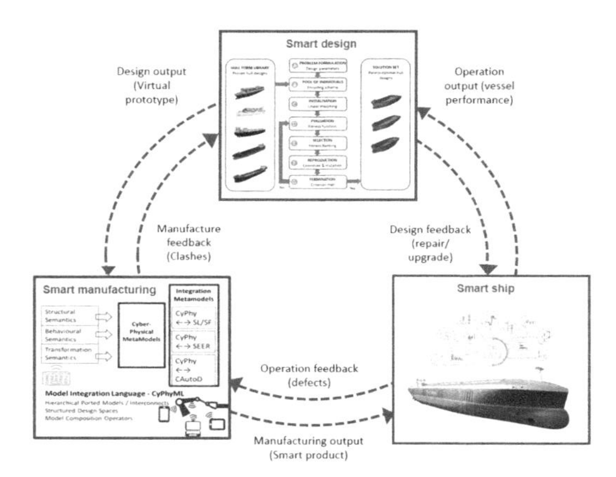

Figure 4. Through-life smart shipping network.

3 HYBRID EVOLUTIONARY ALGORITHM AND MORPHING APPROACH

Considering the issue in current simulation based hull form optimisation with respect to the lack of efficient shape manipulation and robust optimisation techniques to automate the hull form design process and goal to elevate to smart design, a hybrid evolutionary algorithm and morphing (HEAM) approach was proposed by authors in (Ang et al. 2017b). The proposed methodology integrates evolutionary algorithm and curve morphing to automate the hull form design optimisation and elevate into smart design.

3.1 *Evolutionary algorithm*

Evolutionary algorithms (EA) are a group of generic population-based meta-heuristic optimisation techniques that are widely used in many different applications due to its ability to solve complex problems and produce a set of globally optimal solutions. Among various EA methodologies, one of key methods used in hull form optimisation is genetic algorithm (GA). GA was first developed by (Holland 1975), which is a nature-inspired search heuristic method based on Darwinian Theory of natural selection and the 'survival-of-the-fittest' principle. GA works on the principle of 'genes' and 'chromosomes' as illustrated in Figure 5.

Through the use of genetic operators namely selection, crossover and mutation, information represented by genes are exchange between these chromosomes over a number of iteration, typically with

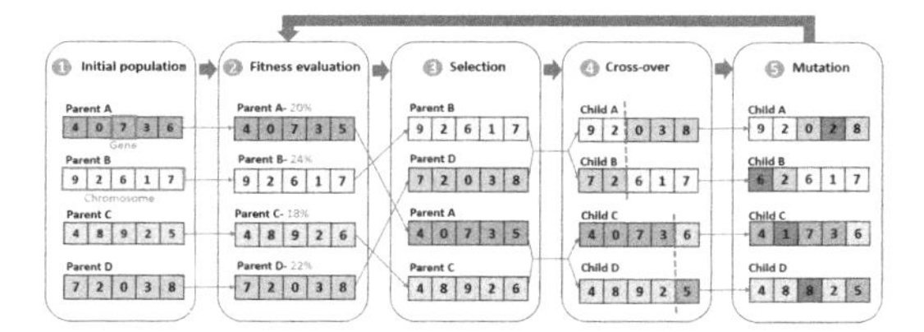

Figure 5. Genetic algorithm working principle.

the fittest solutions replacing the weaker ones and eventually leading to a set of optimal solutions. The key feature of using GA for hull form optimisation is the ability to generate many new hull design combinations, search very huge solution space and subsequently narrow down to a few optimal designs.

3.2 *Curve morphing*

Metamorphosis, also known as morphing, is a technique used widely in the animation industry to generate a sequence of images that smoothly transform a source to another target image. It is also applied in computer graphic and industrial design to compute a continuous transformation from a source to another target shape. Morphing can be a very useful tool for the designer to modify, manipulate, transform the shape or geometry of the design in pursuit to improve the design attributes such as performance, quality, aesthetic, etc. Morphing can be catergorised into 2 main types—two-dimensional (2D) or three dimensional (3D). 2D morphing consist of image morphing and curve morphing and 3D morphing include surface morphing and volume morphing. In ship application, (Tahara et al., 2006) applied morphing using 3D patch model from NAPA to transform a ship hull model into another target model. (Kang & Lee 2010) applied 3D mesh-based surface morphing to generate intermediate hull models between two parent vessels.

Since the beginning of shipbuilding and subsequent introduction of computer-aided design (CAD), 2D offset table remains the most fundamental representation of ship's hull form. Hence until today, it is still used as the basis for designer to model and modify the hull design. The advantage of using 2D hull lines from offset table are it is simple to represent the entire shape of the hull and easy to modify the hull form by adjusting the lines. In this paper, we apply curve morphing based on 2D hull lines to transform the shape of hull through interpolation and extrapolation between the hull lines of two or more hull forms.

Using morphing equation:

$$M(t) = (1-t) \times R_0 + t \times R_1 \quad (1)$$

where $M(t)$ is the morphed shape, t is the morphing parameter, R_0 denotes the source shape and R_1 the target shape.

From above equation, we can see when $t = 0$, $M(t)$ is also equal to 0 and hence the morphed shape is equivalent to source shape R_0. Likewise, when $t = 1$, $M(t) = R_1$ which is the target shape. To illustrate the concept, by using hull lines provided from the body plan of source and target vessels, we can morph and generate large number of intermediate shapes just by changing the morphing parameter (t). Other than interpolating between the source and target vessel, we can also extrapolate beyond the 2 hull lines to create new 'extended' lines. As an example, we take one hull line each from ship A (source) and ship B (target) at station 0.5 for both vessels. By applying curve morphing equation, we can generate interpolated and extrapolated curves as illustration in Figure 6.

It can observe by applying only one morphing parameter (t) constantly across all transverse stations, we can effectively morph or create the entire hull form between the source and target vessels. Key feature of this curve morphing approach is the ability to capture complex shapes such as hull form using minimal design variables, which in this case is represented by morphing parameters (t).

3.3 *Hybrid evolutionary algorithm and morphing approach (HEAM)*

By combining the advantages of GA—ability to search for best global solution—and that of morphing—ability to generate smooth intermittent shapes from the combination of two or more hull form designs, we can now potentially create a wide range of hull form designs with improved efficiency and thereby finding the most optimal hull form. An overview of the proposed HEAM concept is provided in Figure 7.

The HEAM concept comprises 7 main components, namely (a) problem formulation, (b) pool

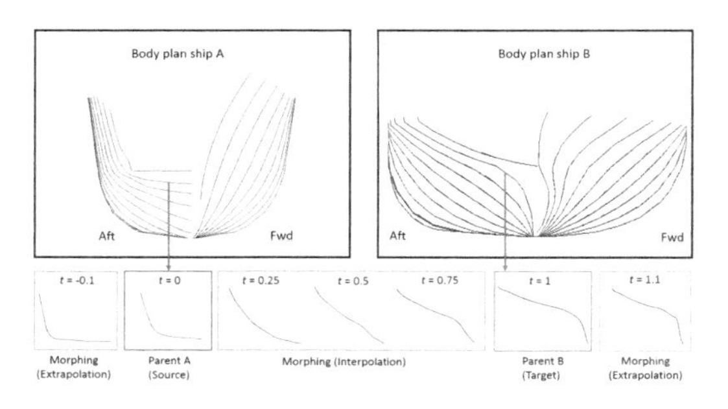

Figure 6. Curve morphing through interpolation and extrapolation at station 0.5.

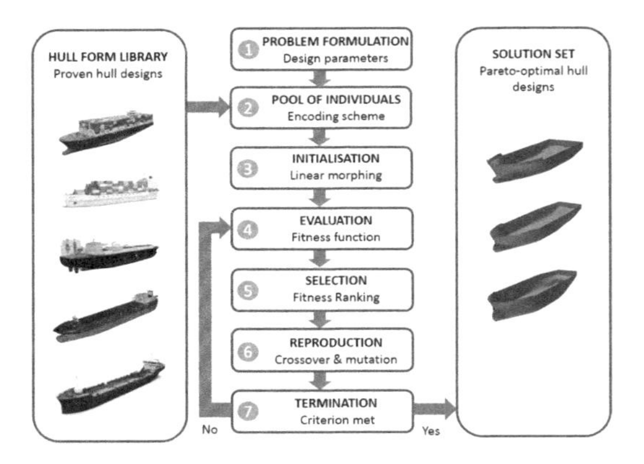

Figure 7. Hybrid evolutionary algorithm and morphing approach.

of individuals, (c) initialisation, (d) evaluation, (e) selection, (f) reproduction and (g) termination.

3.3.1 *Problem formulation*

Before any optimisation process, it is important to first specify the design parameters which include ship type, principle dimensions as well as objective functions. Depending on the number of existing hull designs in hull form library, they can be catergorised by different ship types and selected to form the initial hull designs based on the design requirement. Principle dimensions such as length between perpendiculars (Lpp), beam (B), draft (T) would need to be specified as per design requirement. In this HEAM approach, we can scale up or down the existing vessels from hull form library to meet the design requirement by applying linear transformation to modify the hull form according to desired length, beam and draft. Depending on vessel type, objective functions relating to hull form optimisation may include reducing resistance and seakeeping motion for vessels.

3.3.2 *Pool of individuals*

The next step of HEAM is to create the first pool of individuals and map them into unique encoding scheme. In ship design process, this can be obtained from existing hull forms from the hull form library or create from scratch. The advantage of using existing designs is the assurance of their performance which are validated to meet design objective and helps to shorten the design cycle, although the improvements are often incremental. Another alternative is to model a new hull form from scratch which will allow more freedom of design thereby allowing the creation for more innovative hull form designs. For the proposed HEAM approach, real-value chromosomes using morphing parameters (t) which captures the ship's geometry in X and Y planes according to their respective frame or stations across Z planes, as illustrated in Figure 8. This provides a simple yet direct representation of the ship geometry which allows the hull shape to be transformed easily by changing the morphing parameters (t) at various station locations along the entire vessel.

For initial population, first vessel model (parent A) will be assigned morphing parameter $t = 0$ and second vessel model (parent B) will be assign $t = 1$. More vessel models can be included using same arrangement to increase the variety of shapes and hence increasing the search space to achieve more optimal designs.

Under this HEAM approach, we introduce three applications of curve morphing by incorporating into above encoding scheme—i) constant morphing, ii) linear morphing and iii) varying morphing. Constant morphing is applied to morph the entire hull form using same morphing parameters within the encoding scheme. For example, by applying the same morphing parameters (t) across the entire length of vessel, we can generate large number of intermittent designs as demonstrated in Figure 9.

Other than interpolating between 2 parent vessels, we can also extrapolate beyond the parent vessels using morphing method to create more 'new' designs. Linear morphing is used during crossover and mutation function to smoothen the curve by applying gradual morphing parameters (t) between two parent models. Varying morphing is used when we combine both constant and linear morphing or when multiple vessel types are combined together.

3.3.3 *Initialisation*

Prior to morphing operation, we need to prepare the hull coordinates for all the parent vessels so as to ensure the hull form corresponds to each other. Considering no two offset tables are identical in

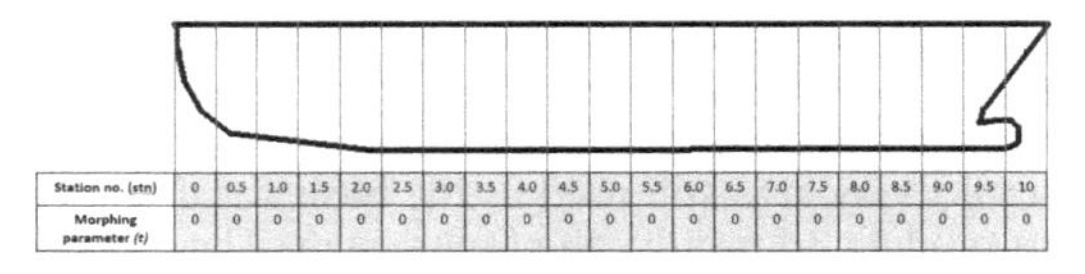

Figure 8. Encoding scheme using real value chromosome $(t = 0)$.

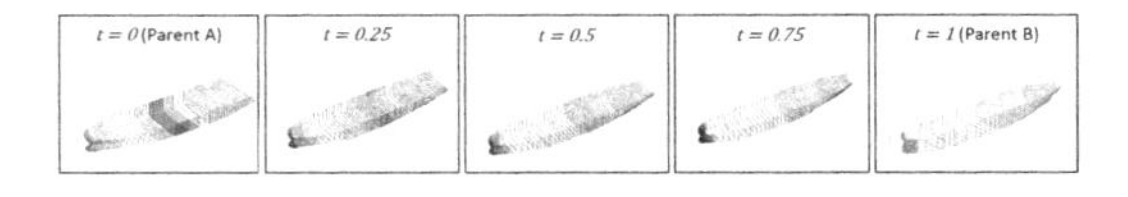

Figure 9. Two parent vessels and intermediate vessels created by parallel morphing.

terms of number of coordinate points, we perform correspondence so as to create same number of points across all section curve for the parent vessels which are to be morphed. This can be done using cubic spline interpolation to create additional points at different interval of the section curve. Cubic spline interpolation is a piecewise continuous curve which passes through each of the values in a table of points, which is represented in the following equation:

$$S_i(x) = a_i(x - x_i)^3 + b_i(x - x_i)^2 + c_i(x - x_i) + d_i; \quad for\ x \in [x_i, x_{i+1}] \qquad (2)$$

where $S(x)$ denotes the spline and $[x_i, y_i]$ represents a table of points for $i = 0,1,\ldots,n$ for function $y = f(x)$

As an example, section curve X = 0 for vessel A contain 7 points and same curve for vessel B contain ten points in their respective offset tables. In order to morph the section curve at X = 0 for both vessel A and B, we need to create ten equal points on section curve X = 0 for both vessel using cubic spline interpolation, as per illustrated in Figure 10 below.

3.3.4 *Fitness evaluation*

In order to measure the 'fitness' of the parent vessels, we need to evaluate the performance of each hull form design based on the objective function. In most hull form optimisation, the objective function will be to reduce hull resistance and ship's motion. For both objectives, the aim is to minimise the cost function as follow:

$$Min\, f(\chi),\ \chi \in X \qquad (3)$$

where f is the vector of design objectives, χ is vector of design variables and X is the feasible design variable space.

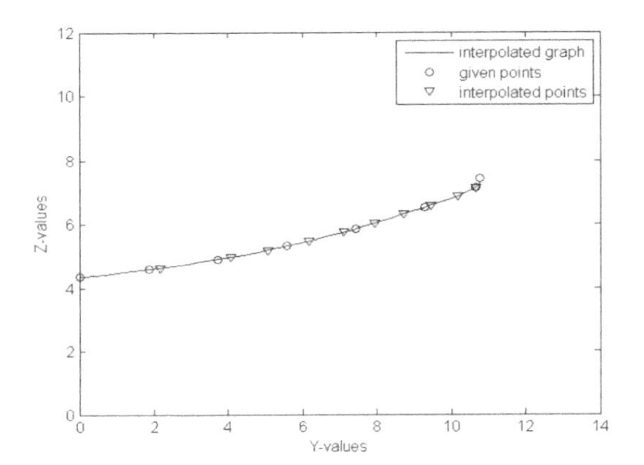

Figure 10. Curve correspondence using cubic spline interpolation.

At this stage, we will need to translate the 2D hull geometry into 3D surfaces by mapping the offset table into hull surfaces using surface generation method such as NURBS. The 3D surfaces will then be panelised and the resistance can be evaluated using numerical methods such as potential flow or Reynolds Averaged Navier-Strokes Equation (RANSE). For motion analysis, strip theory can be used which are available in most hydrodynamic analysis tools. Under this HEAM approach, we proposed the candidate design solutions should be assessed using low-fidelity CFD method potential flow for resistance analysis. This is in view of the large number of candidate solution to be evaluated and potential flow are preferred due to its efficiency and fairly good estimation during early ship design. High fidelity CFD method such as RANSE can be applied at later stage to validate the optimal design.

3.3.5 *Selection*

In GA, selection is a process of selecting which solution will be used in reproduction for generating new solutions. The principle is to always select the good solutions in order to increase the chance to obtain better individuals. For proposed HEAM approach, we apply roulette wheel selection which probalisatically selects individuals based on their performance for next round of reproduction.

3.3.6 *Reproduction—crossover and mutation*

Crossover is an important operator in GA where chromosomes of two parents are combined to form new chromosome of the child. Principle of this operator is to create new individuals by mixing the good genes of their parents and subsequently lead to fitter individuals. In this HEAM approach, we apply linear morphing to combine two or more existing hull form (parents) to generate new hull form designs (child) through interpolation to create smooth intermediate curves between the two parents, as illustrated in Figure 11 below.

On top of morphing two different hull form together in one hull concept, we can also join multiple hull forms (three or more) using varying morphing by simply applying linear morphing at intermittent stations between different vessels. For combining vessels which are vastly different in term of sizes, rescaling through linear transforma-

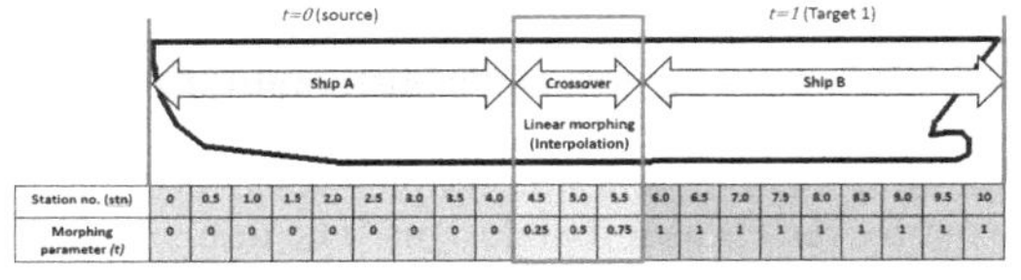

Figure 11. Crossover between ship A aft and ship B forward body through linear morphing (interpolation).

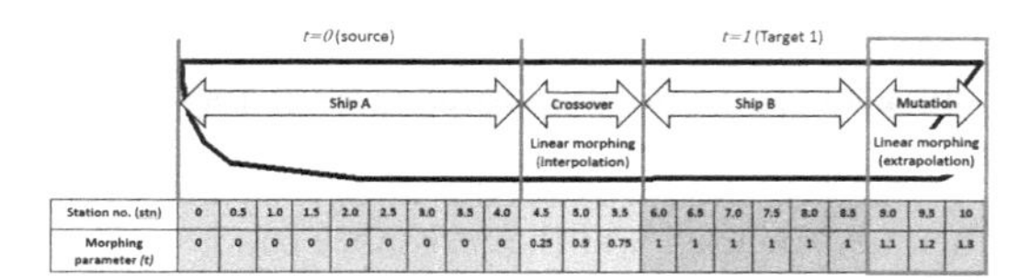

Figure 12. Mutation at random stations through linear morphing (extrapolation).

tion can be applied to reduce or enlarge the hull form to match the other vessel.

Mutation is the next reproduction process within GA where new genes are created in random to produce a new genetic structure which helps to introduce new elements into the population. In this HEAM approach, we apply linear morphing at random station through extrapolation to create the chromosome of new solutions as illustrated in Figure 12 below.

While mutation is entirely random in nature, linear morphing can help to reduce the possibility of infeasible designs such as unsmooth surface or odd shape generated during the optimisation process.

3.3.7 *Termination criterion and solution set*

Once all solution designs are ranked and termination condition are met, the iteration will stop and provide the results identifying the pareto optimum design. In this HEAM approach, the termination condition can be based on total number of iterations or terminate if there no further improvement after 10 iterations. Ultimately, the designer should decide the termination criterion based on number of initial designs available in hull form library and lead time for this initial design process.

4 RESULTS AND DISCUSSIONS

To demonstrate the flexibility and benefits of curve morphing and proposed HEAM approach, we performed two case studies to – i) optimise the hull form of container vessel using two parent models through constant and linear morphing and ii) 'split' and combine three different vessel types using varying morphing.

4.1 *Hull form optimisation of container vessel using HEAM approach*

In this first case study, we applied multi-objectives optimisation using HEAM approach to produce the hull form of a new container vessel using two existing vessels as initial designs. The design objective is to reduce total resistance and seakeeping motion of a container ship with principle dimension of 185 m length between perpendicular (Lpp), 32 m beam, 9 m draft and design speed of 20 knots. The principle dimensions for two parent container vessels are provided as follow.

From the two parent vessels selected, we applied correspondence to match the number of coordinate points and performed linear morphing (interpolation and extrapolation) to form the initial population. After creating the initial designs, the designs are evaluated based on their total pressure resistance and maximum heave response function. In this study, the performance of each candidate design is evaluated using potential flow method and strip theory in NAPA program and it took less than five minutes to evaluate one hull form design using standard quad core workstation. Prior to reproduction, we performed rescaling using linear transformation of parent vessel B to meet the design criteria set in this case study. The crossover and mutation points were then selected randomly through the HEAM program written in Matlab. This randomness helps to generate more novel candidate designs which may not been considered by the designer if carried out manually. The preliminary results of the hull form generated from HEAM program are provided as below.

From the above results generated from first run of the HEAM program, we can already see some improvement for the new vessel (child) created from crossover and mutation process as compared to two existing parent vessels. In case of crossover process, linear morphing was applied by combining parent A (aft) and parent B rescaled (forward). The next step of HEAM approach is the mutation process where extreme forward of the vessel

Table 1. Principle dimensions for two existing parent vessels.

Principle dimensions	Container A	Container B
Lpp	185 m	202.1 m
Beam (B)	32 m	32.2 m
Draft (T)	9 m	10.5 m

Table 2. Principle dimensions and results of candidate solutions generated using HEAM program.

Vessel	Lpp (m)	B (m)	T (m)	TP* (kN)	PR+ (deg/m)
Parent A	185	32	9	75.72	1.0933
Parent B rescale	185	32	9	61.34	1.0788
Crossover	185	32	9	60.26	1.0744
Mutation	185	32	9	60.17	1.0672

*Total pressure resistance.
+Maximum pitch response function.

are mutated through extrapolation from parent A to parent B rescale using morphing parameter $(t) = 1.1$ to generated the child vessel. As compared to parent vessel A and B (rescaled), child vessel achieves an improvement of 20.5% and 1.9% respectively in total resistance pressure. In terms of pitch motion, the child vessel obtained an improvement of 2.38% and 1.07% as compared to parent vessel A and B (rescaled) respectively. While the improvement is considered minor comparing stronger parent B (rescaled) and child vessel, it should be noted this result is generated based on only one run of the program. The potential of this method would be realised with more iterations of the HEAM program, which will be presented in our next study. The overall wave generated are compared between parent B (rescaled) and child vessel and illustrated in Figure 13 below.

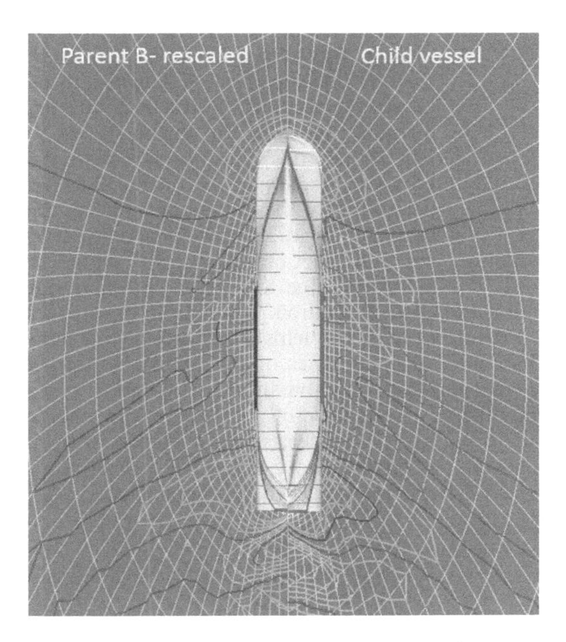

Figure 13. Wave profile comparing vessel parent B—rescaled and child vessel.

Table 3. Principle dimensions for three different types of sea-going vessels.

	Aft body	Mid-body	Forward body
Dimensions	Container	Bulk carrier	Oil Tanker
Lpp	202.1 m	215 m	314 m
Beam (B)	32.2 m	36 m	58 m
Draft (T)	10.5 m	15 m	20.92 m

4.2 *Combination of three different vessel types using two-point crossover and varying morphing*

In the next case study, we demonstrate here the possibility of combining three different vessel types into one hull form through two-points crossover and varying morphing. This unique feature allows the segregation of hull form into three main sections – stern section, mid-section and bow section, thereby enable the creation of more innovative hull form combination through mix-and-match process within the HEAM approach. Firstly, we select three different vessel types – a container vessel, one bulk carrier and an oil tanker. The principle dimensions for three vessels are provided as below

Next, we performed rescaling using linear transformation to resize the vessels to the same principle dimension, which is the same as oil tanker in this example. Following which, the three vessels are 'split' into three separate sections and 'join' together using multi-target morphing. The combined vessels are illustrated as below in Figure 14.

We can see from above example the versatility of two-points crossover and varying morphing methodology which can combine three very different vessels in terms of function and size. This is very useful when applied in HEAM program which can potential create many more different types of hull form combinations and more innovative designs. It is recognised one key limitation of using existing

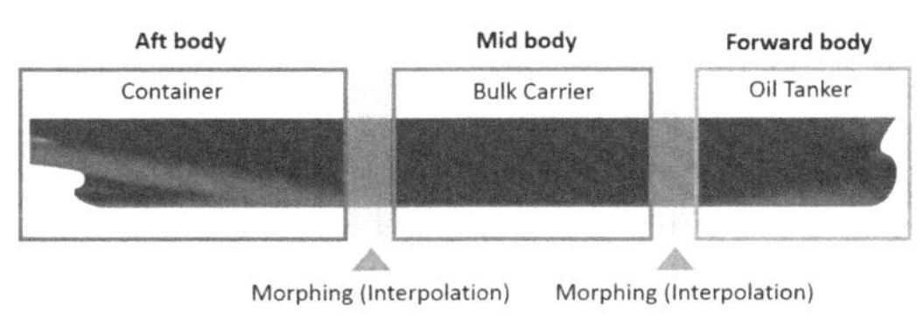

Figure 14a. Combination of three different vessel types using multi-target morphing.

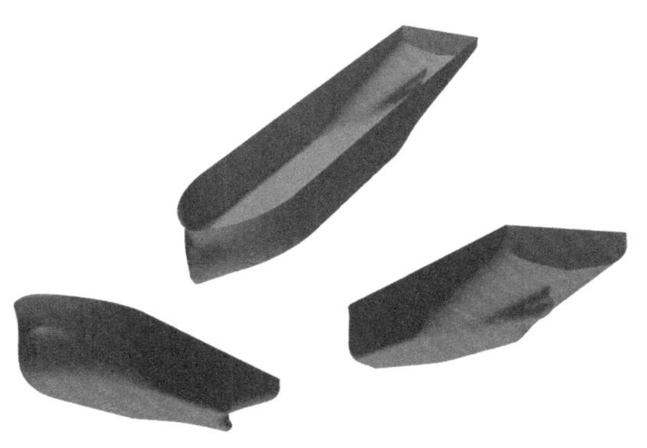

Figure 14b. Isometric view of 3 combined vessels.

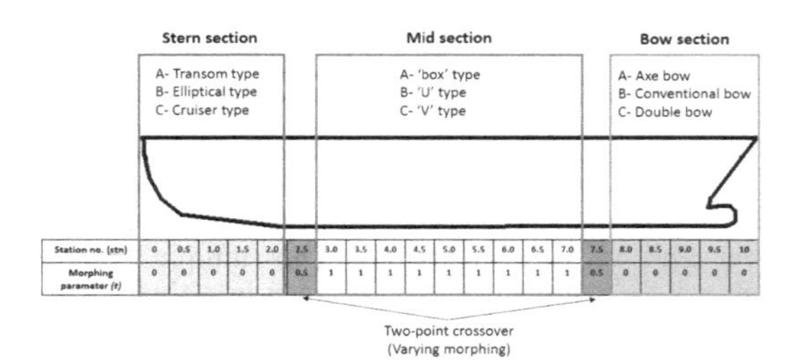

Figure 15. Possibility of mix-and-match of different designs using two-points crossover and varying morphing.

vessel design within this HEAM approach is the new designs are closely linked to the parent and lack of freedom to create more innovative designs. To overcome this limitation, designer can also select from existing vessels or create from scratch their own designs for HEAM to morph and generate new designs. For example, designer can choose between different design types of hull form in way of three main sections of the vessel—stern, mid and bow section as illustrated in Figure 15.

Subjected to the availability of the different hull form designs in the hull form library, this varying moprhing function will help to increase the solution space and thereby provides another function within HEAM or additional 'tool' for designers to create more innovative hull form.

4.3 *Discussions*

As compared to existing hull form design process, there are several benefits of this HEAM approach. Firstly, it starts from a pool of existing proven hull forms designs as compared to improvement from only one 'sister-ship' hull designs, which enables the exploration of wider solution spaces and generating more optimal hull forms. Second, by performing morphing via interpolation and extrapolation between these proven designs, it enables automated geometry modification without any designer input. It should be noted the concept of HEAM approach is not to replace the role of designer but instead to supplement and provide as an additional tool for designer to examine a wide variety of existing hull form and narrow down to few optimal designs very quickly before working on the design further. Thirdly, by incorporating morphing parameter into encoding scheme within HEAM, we can effectively transform the entire hull shape with as little as one design variable—morphing parameter, *t*, thereby increasing the overall efficiency of geometry modification and optimisation process. Finally, HEAM allows 'mix-and-match' of hull forms through unique crossover and mutation functions which enables the creation of more innovative hull form. As demonstrated in the results section, this approach has the flexibility to combine 3 very different vessels whereby increasing the solution space for more optimal hull. It should be noted HEAM approach is not limited to only ship hull form. It can also be extended to other related design applications such as aerospace and automobile main body design.

In order to realise the full benefits of HEAM and to be applied to ship design process, there are still several issues that needs to be addressed and overcome. Firstly, it is widely known within the industry automating the hull form design process is already very challenging in its own sense. In fact, there are no known procedure or tools that are able to automatically translate hull coordinates to surface and meshing for evaluation. In this study, we are only able perform the translation from morphed points to surfaces manually using NAPA modeling software and feed the results back to HEAM program. Secondly, one can observe from second case study that the principle dimensions of the three vessels were adjusted drastically as the result of re-scaling in order for all the vessels to join together. Hence, this may not represent the actual hull form characteristic as compared to the original vessel. Thirdly, the full benefits of GA are best exploited in applications with huge data set or solution space. In our example, we are limited to only few hull forms due to manual surface modeling and CFD evaluation. Finally, while crossover function within GA helps to create new designs by combining the genes of two parent vessels, it is recognised the combination of two good genes might leads to creation of a bad gene with poorer performance. Nevertheless, more studies will be carried out on the above issues and we hope to address them in our subsequent publication.

5 CONCLUSION

In this paper, we presented two concepts—through-life smart shipping network and smart design through HEAM approach. These two concepts provide the direction towards digitalised and smart shipping where design, construction and operation of ships can become more integrated and efficient. Through-life smart shipping network combine the key lifecycle processes into an integrated network where useful information can be exchanged at different phases in an collaborative manner. HEAM approach combines morphing and GA to generate series of new hull designs in a more automated manner and perform 'intelligent' search to narrow down to more optimal designs. By incorporating curve morphing concept into unique encoding scheme, we utilise GA operators such as crossover and mutation function to transform the hull shape through constant morphing, linear morphing and varying morphing. Two case studies are applied to optimise the hull form of container vessel through HEAM approach consisting constant and linear morphing and also combined three different parts of vessel into one single hull using varying morphing. Through computational intelligence and connected lifecycle network, it is envisioned this smart HEAM approach will help to improve the overall efficiency and ability to produce more efficient and smarter ships in the near future.

REFERENCES

Ang, J.H., Goh, C., Saldivar, A.F. & Li, Y. 2017a. Energy-efficient through-life smart design, manufacturing and operation of ships in an industry 4.0 environment. *Energies* 10, 610.

Ang, J.H., Goh, C., Jirafe, V.P. & Li, Y. 2017b. Efficient hull form design optimisation using hybrid evolutionary algorithm and morphing approach; *Conference on Computer Applications and Information Technology in the Maritime Industries, 3–4 October 2017*, Singapore.

Berrini, E., Mourrain, B., Duvigneau, R., Sacher, M. & Roux, Y. 2017. Geometric model for automated multi-objective optimization of foils. *VII. International Conference on Computational Methods in Marine Engineering, May 2017*, Nantes, France.

Bertram, V. 2000. *Practical Ship Hydrodynamics*. Butterworth-Heinemann.

Brizzolara, S. & Vernengo, G. 2011. Automatic optimisation computational method for SWATH, *Int. J. Mathematical Models and Methods in Applied Sciences.*

Campana, E.F., Serani, A. & Diez, M. 2013. Ship optimization by globally convergent modification of PSO using surrogate based newton method, *Engineering Computations.*

Chalfant, J., Langland, B., Rigterink, D., Sarles, C., McCauley, P., Woodward, D., Brown, A. & Ames, R. 2017. Smart Ship System Design (S3D) Integration with the Leading Edge Architecture for Prototyping Systems (LEAPS), *Electric Ship Technologies Symposium*, IEEE.

Chen, X., Diez, M., Kandasamy, M., Zhang, Z.G., Campana, E.F. & Stern, F. 2014. High-fidelity global optimization of shape design by dimensionality reduction, metamodels and deterministic particle swarm, *Engineering Optimization.*

Feng, B.W., Hu, C.P, Liu, Z.Y, Zhan, C.S & Chang, H.C. 2011. Ship resistance performance optimization design based on CAD_CFD, *International Conference on Advanced Computer Control, 18–19 January 2011*. Harbin, China.

Geertsma, R.D., Negenborn, R.R., Visser, K., Hopman, J.J. 2017, Design and control of hybrid power and propulsion systems for smart ships: A review of developments, *Applied Energy* 194: 30–54.

Holland, J.H. 1975, Adaptation in Natural and Artificial Systems, *Univ. of Michigan Press.*

Jacquin, E.; Derbanne, Q., Bellevre, D., CORDIER, S., Aleslessandrini, B., Roux, Y. 2004. Hull form optimisation using free surface RANSE solver, *25th Symp. Naval Hydrodynamics*, St.John's.

Jan, O.R. 2017, Towards Shipping 4.0, *International Conference on Smart Ship Technology*, 24–25 January 2017, London, UK.

Kang, J.Y. & Lee, B.S. 2010, Mesh-based morphing method for rapid hull form generation, *Computer-Aided Design.*

Kim, H. 2012, Multi-Objective Optimization for Ship Hull Form Design, *George Mason University.*

Kostas, K.V., Ginnis, A.I., Politis, C.G. & Kaklis, P.D. 2014, Ship-Hull Shape Optimization with a T-spline based BEM-Isogeometric Solver, *Comput. Methods Appl. Mech. Eng.*

Nowcki, H. 1996, Hydrodynamic design of ship hull shapes by methods of computational fluid dynamics, *Progress in Ind. Math. at ECMI*: 232–251.

Park, J.H., Choi, J.E. & Chun, H.H. 2015, Hull-form optimization of KSUEZMAX to enhance resistance performance, *Int. J. Nav. Archit. Ocean Eng.*

Paula, F., Diego, N., Tiago M.F., Manuel, A.D. and Miguel, V. 2016, Smart pipe system for a shipyard 4.0. *Sensors,* 16, 2186.

Saha, G.K., Sarker, A.K. 2010, Optimisation of ship hull parameter of inland vessel, *Int. Conf. Marine Technology,* Dhaka.

Tahara, Y.; Tohyama, S. & Katsui, T. 2006, CFD based multiobjective optimization method for ship design, *Int. J. Numerical Methods in Fluids* 52: p. 28.

Tahara, Y., Peri, D., Campana, E.F., Stern, F. 2011, Single and multiobjective design optimization of fast multihull ship, *J. Mar. Sci. Techn.*

Zha, R.S., Ye, H.X., Shen, Z.R. & Wan, D.C. 2014, Numerical study of viscous wave making resistance of ship navigation in still water, *J. Marine Sci. Appl.* 13: 158–166.

Marine Design XIII – Kujala & Lu (Eds)

Hull form resistance performance optimization based on CFD

Baiwei Feng & Haichao Chang
Key Laboratory of High Performance Ship Technology (Wuhan University of Technology), Ministry of Education, Wuhan, China
School of Transportation, Wuhan University of Technology, Wuhan, China

Xide Cheng
Key Laboratory of High Performance Ship Technology (Wuhan University of Technology), Ministry of Education, Wuhan, China

ABSTRACT: The combining of Computational Fluid Dynamics (CFD) and optimization techniques is a significant method for automatic multi-scheme selection in hull form optimization, which can improve the design level of energy efficient ship. This paper focuses on hull surface automatic modification and its application in hull form optimization. A new hull surface automatic modification method is proposed, which based on the Radial Basis Function (RBF) interpolation method. The method are applied to hydrodynamic optimization of Series 60. Model tests of original and optimized model are carried out at Wuhan University of Technology (WHUT) for validations. Both of numerical optimization and model tests results show that the proposed method can significantly improve optimization efficiency.

1 INTRODUCTION

The energy efficiency design index (EEDI) was proposed by International Maritime Organization (IMO) in 2009. Ship resistance performance optimization is an effective method for meeting the EEDI requirements. With the rapid development of computer technology and computational fluid dynamics (CFD), hull form optimization based on simulation has become a research hotspot. Harries (2001), Peri and Campana (2003, 2004, 2005), Yang (2011), and Tahara (2004, 2006) have combined numerical simulation techniques and optimization algorithms, realised the integration of computer aided design (CAD) and CFD, successfully completed the optimization design of hull lines and obtained an optimal hull form with improved hydrodynamic performance. Xu et al. (2001) traced the development of a classified optimization procedure consisting of five levels in order to complete the optimization of ship hull lines, and demonstrated that CFD code is applicable for the resistance performance optimization of ship hull lines. Abt et al. (2003) presented a parametric approach based on FRIENDSHIP-Modeler and SHIPFLOW. Campana et al. (2003, 2004, and 2005) investigated a variable-fidelity approach for speeding up the optimization process using a free surface RANS in a multi-objective design problem. Tahara et al. (2004) employed a finite-difference gradient-based approach for stern and sonar dome optimization. Peri et al. (2005) introduced a design optimization method based on large-scale numerical simulations for complex engineering optimization problems involving highly computationally expensive objective functions and nonlinear constraints. Zhang et al. (2009) optimised the hull form with minimum wave-making resistance based on CFD, by combining the Rankine source method with nonlinear programming (NLP). Peri et al. (2016) presented an approach for robust optimization of a bulk carrier's conceptual design, subject to uncertain operating and environmental conditions, and a particle swarm optimization algorithm was applied for the global minimisation process, minimising the expectation and standard deviation of the unit transportation cost.

Grigoropoulos et al. (2010) proposed an improved hull form optimization scheme for calm water resistance and seakeeping based on FRIENDSHIP-Modeler, FRONTIER and EASY. Yang et al. (2011) developed an efficient and effective hull surface modification technique for CFD-based hull form optimization in order to achieve both local and global hull form modifications by combining the two approaches. Kim et al. (2010) developed a new methodology for hydrodynamic optimization of a TriSWACH, which considers not only the side hull positions, but also its shape.

In order to satisfy the requirements of hydrodynamic performance, the hull form is somewhat complicated. When using a non-uniform rational B-spline (NURBS) to represent the hull surface accurately, the number of control points must be extremely large, and it is very difficult in hull surface modification. Therefore, several hull surface modification methods have been explored, and can be divided into two categories, as follows. 1) Morphing method based on the initial hull form (Tahara 2006; Feng 2010). This method can guarantee that the new hull surface is faired when the initial hull form is smooth. However, the limitation of this method is that it cannot consider the deformation of local hull lines. 2) Direct modification of the control points based on the Bezier patch and free-form deformation methods. This approach can be used for geometric reconstruction of the entire hull form, but the design variables are relatively more important and the control points must be carefully selected. Hull surface deformation is achieved using the initial ship's NURBS control points as design parameters in the above methods.

This paper focuses on the hull surface modification method. An optimization tool is developed with the hull surface modification module. The enhanced hull optimization tool is applied to the resistance performance optimization of Series 60 vessel, and model tests are carried out in order to verify the method's validity.

2 HULL SURFACE MODIFICATION METHOD

The hull surface modification method plays a vital role in ship hydrodynamic performance optimization, and should meet the following requirements.

1. Design parameters should be as few as possible.
2. The modification range of the hull form should be sufficiently large.
3. The local modified hull surface should have a fairing connection with other parts.
4. The optimal hull form should be suitable for manufacturing.

In this paper, the radial base interpolation technique is used to modify the hull surface, based on the NURBS hull surface representation.

2.1 *Hull surface modification method based on RBF interpolation*

The radial basis function (RBF) for automatic modification was briefly introduced by Kim (2008). The RBF is a type of real-valued function that is symmetric along the radial direction, the value of which depends only on the Euclidean distance between any point X and the centre of point X_i (Buhmann 2004). The basis function is described as follows:

$$\phi(\|X - X_i\|) \qquad i = 1,2,\cdots,n, \tag{1}$$

where X_i is the centre of the RBF; ϕ is the variable Euclidean distance.

When RBF interpolation is used during hull form modification, the function can be described as follows (Morris et al. 2008):

$$S(X) = \sum_{i=1}^{n} \lambda_i \phi(\|X - X_i\|) + p(X) \tag{2}$$

where $S(X)$ represents the displacement of point $X = (x, y, z)$ on the hull surface; $p(X)$ is used to recover the translation and rotation and has the specific form $p(x) = c_1 x + c_2 y + c_3 z + c_4$; n is the number of control points; and $\|X - X_i\|$ is the Euclidean distance between two points.

In this case, the 3D Wendland's function with compact support features is adopted as a basis function:

$$\phi(\|X\|) = \begin{cases} (1 - \|X\|)^4 (4\|X\| + 1), & 0 \le \|X\| \le 1 \\ 0, & \|X\| > 1 \end{cases} \tag{3}$$

This equation guarantees that the constructed system matrix is a positive definite matrix. The equation coefficients λ_i and c_i are obtained by changing the control point coordinates (Morse et al. 2001):

$$S(X_i) = f_i, i = 1,2,\cdots,n \tag{4}$$

where f_i is the control point displacement.

Additional requirements include the following (Buhmann 2000):

$$\sum_{k=1}^{n} \lambda_k X_k^T = 0; \sum_{k=1}^{n} \lambda_k = 0 \tag{5}$$

The values of coefficients λ_j and c_i can be obtained by solving the linear system:

$$\begin{pmatrix} f \\ 0 \end{pmatrix} = \begin{pmatrix} M & q \\ q^T & 0 \end{pmatrix} \begin{pmatrix} \lambda \\ c \end{pmatrix} \tag{6}$$

where

$\lambda = [\lambda_1, \lambda_2, \cdots, \lambda_n]^T$, $c = [c_1, c_2, c_3, c_4]^T$, $f = [f_1, f_2, \cdots, f_n]^T$.

$$q = \begin{pmatrix} x_1 & y_1 & z_1 & 1 \\ x_2 & y_2 & z_2 & 1 \\ \vdots & \vdots & \vdots & \vdots \\ x_n & y_n & z_n & 1 \end{pmatrix} \tag{7}$$

All of the unknown coefficients in Eq. 2 can be determined by solving the above equations. Then, new coordinates for all unknown points can be obtained by RBF interpolation.

3 RESISTANCE PERFORMANCE OPTIMIZATION

The hull form optimization process is illustrated as follows (Fig. 1):

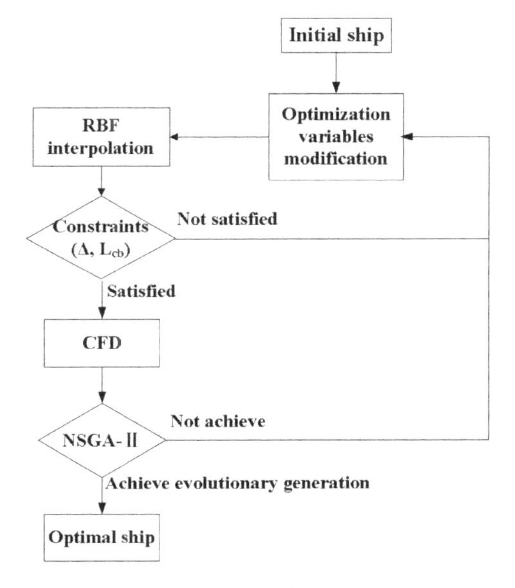

Figure 1. Hull form optimization process.

1. The RBF interpolation method described in the previous section is applied in order to realise hull surface deformation.
2. The hydrostatical properties of each new hull form, such as displacement and longitudinal position of the buoyancy centre, are calculated. If the constraint conditions are satisfied, the wave-making resistance calculation is carried out by CFD software. Otherwise, variables are modified again in order to create a new one.
3. The non-dominated sorting genetic algorithm (NSGA-II) is used to drive the entire optimization process. If it reaches the maximum generation, the process ends; else, the design variables are modified in order to repeat the above process.

CFD analysis can be performed using either potential or viscous flow theory, and potential flow theory is selected in this paper. The Shipflow software is used as the CFD solver.

3.1 *Description of optimization problem*

The Series 60 model is accredited as a standard ship by the ITTC organisation for model testing. The main elements are displayed in Fig. 2 and Table 1.

In order to save optimization time, a step-by-step optimization process is adopted. Firstly, the bow shape of the Series 60 is optimised, where the optimization objective is minimised wave resistance. Secondly, on the basis of the bow shape optimization, optimization of the tail shape is continued, where the optimization objective is minimum total resistance.

Figure 2. Series 60.

Table 1. Main elements of Series 60 vessel model.

L_{pp} (m)	L_{wl} (m)	B_{wl} (m)	T (m)	C_b	C_m	∇(m^3)	S (m^2)
3.048	3.101	0.406	0.163	0.6	0.977	0.1214	1.6

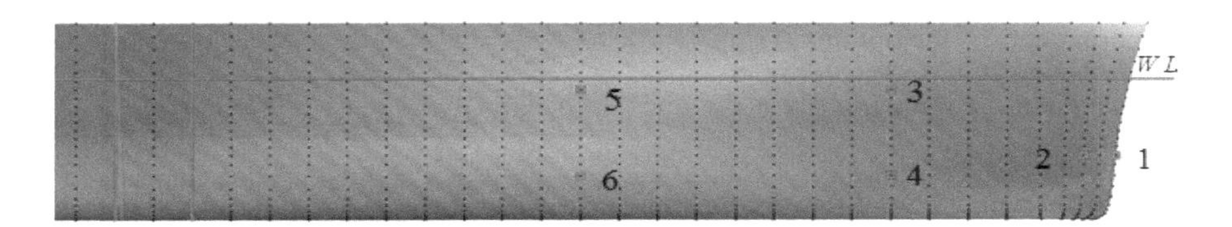

Figure 3. Variable points.

Table 2. optimization variables and range.

Optimization variables		Upper limit	Lower limit	Initial value
1	P1-X	–0.03	–0.05	–0.04
2	P1-Z	0.094	0.07	0.08
3	P2-Y	0.035	0.02	0.03
4	P3-Y	0.075	0.059	0.07
5	P4-Y	0.088	0.07	0.08
6	P5-Y	0.148	0.126	0.14
7	P6-Y	0.16	0.14	0.15

3.2 *Bow shape optimization (step 1)*

1. Optimization objectives:
The key point is to reduce resistance within a certain speed range. Therefore, the optimization objective is defined as follows:

$$\min f_{obj} = R_w \ , \ F_N = 0.27..$$

2. Optimization variables:
The RBF interpolation method is adopt to modify the hull surface. Firstly, constraints are applied to the deck on the side, line of the bottom and stern profile. According to our experience, six variable points are selected, as shown in Fig. 3, wherein variable point 1 changes along the *X* and *Z* directions to control the length and height of the bulbous bow. Furthermore, variable points 2 to 6 change along the *Y* direction, wherein variable point 2 controls the width of the bulbous bow, variable points 3 and 5 are arranged along the waterline to control the inflow angle size and the inflow section shape on the waterline, and variable points 4 and 6 can have a significant impact on the bilge line.
3. Constraints:
   1. Displacement: $\Delta_{opti} \geq 0.99\Delta$.
   2. Longitudinal location of centre of buoyancy: $\frac{|Lcb - Lcb_{opti}|}{Lcb} \leq 0.01$.
   3. Only the geometry of the bow profile is changed; the length, breadth, draft and tail shape of the ship are unchanged.

where Δ, *Lcb* and *S* are the displacement, longitudinal distance from the buoyancy centre to amidships and the wet surface area of the initial ship, respectively, while Δ_{opti}, Lcb_{opti} and S_{opti} are the corresponding parameters of the optimised ship.
4. Optimization algorithms: According to optimization experience, population size is 50, and hereditary algebra is 50.
5. Results analysis:
Compared with the initial ship, the wave-making and total resistance of the optimal ship decrease by 27.5% and 7.68%(Fig. 4 and Table 3). It can be seen from Fig. 4 that a new bulbous bow is generated by variable point 1 and 2, and it is slightly upturned, similar to the SV-shaped bulbous bow. The bow may produce significantly beneficial wave interference at high speed. Because there is no bow for the Series 60, so the variable point 1 and 2 also changes the inlet flow length. The inlet flow length are increased, which can reduce wave-making resistance. The variable point 3 to 6 changes the inlet flow angle, which makes the inlet flow angle decreases. It is beneficial factors in reducing wave-making resistance.

3.3 *Tail shape optimization (step 2)*

1. Optimization objectives:
Based on the bow shape optimization results, the tail shape is optimised in step 2. The optimization objectives are defined as follows:

$$\min f_{obj} = R_t \ , \ F_N = 0.27$$

2. Optimization variables:
The variable point numbers are selected as 7 to 12 (Fig. 5). All the variable points move along the *Y* direction and the constraint conditions are the deck at side, line of the bottom and stern profile. The upper and lower limits and initial value of each variable point are displayed in Table 4.
3. Constraints:
   1. Displacement: $\Delta_{opti} \geq 0.99\Delta$.
   2. Longitudinal location of centre of buoyancy: $\frac{|Lcb - Lcb_{opti}|}{Lcb} \leq 0.01$.
   3. Only the geometry of the bow profile is changed; the length, breadth, draft and tail shape of the ship are unchanged.

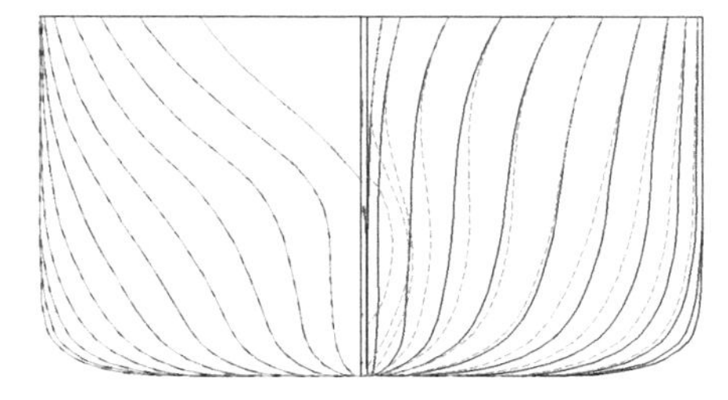

a. Transverse line comparison

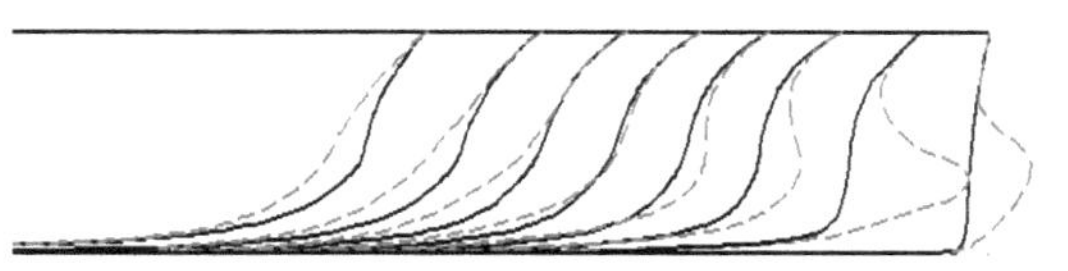

b. Longitudinal line comparison

Figure 4. Bow shape comparison (initial: solid line, step 1 optimization: dotted line).

Table 3. Parameters comparison.

	$S(m^2)$	Displacement	*Lcb*	*Rw*(N)	*Rt*(N)
Initial ship	1.6	121.38	1.59	2.471	9.103
Optimization (step 1)	1.613	122.9	1.578	1.792	8.404
Variation	0.81%	1.25%	–0.75%	–27.5%	–7.68%

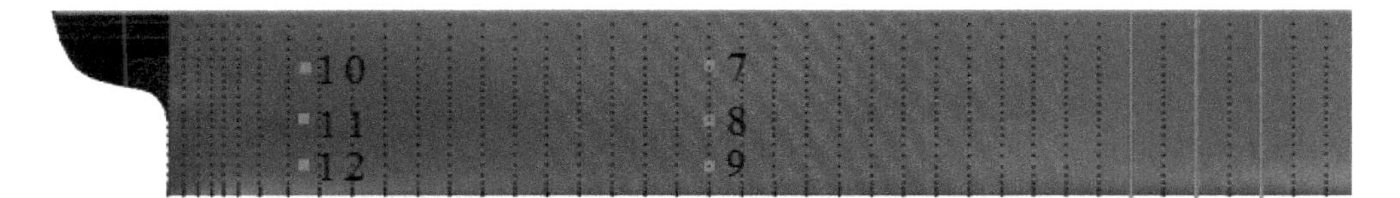

Figure 5. Variable points.

Table 4. Optimization variables and range.

Optimization variables		Upper limit	Lower limit	Initial value
1	P7-Y	0.166	0.15	0.16
2	P8-Y	0.125	0.098	0.114
3	P9-Y	0.086	0.07	0.079
4	P10-Y	0.195	0.167	0.186
5	P11-Y	0.159	0.136	0.151
6	P12-Y	0.134	0.115	0.127

In the above, Δ, *Lcb* and *S* are the displacement, longitudinal distance from the buoyancy centre to amidships and the wet surface area of the initial ship, respectively, while Δ_{opti}, Lcb_{opti} and S_{opti} are the corresponding parameters of the optimal ship.

4. Optimization algorithms: According to optimization experience, population size is 50,and hereditary algebra is 50.
5. Results analysis:

The total resistance of optimal model (S60 step 2) reduced by 10.64% compared to the initial ship at the design speed.

It can be seen from Fig. 6 that the bulbous bow of the optimised ship is slightly upturned and the longitudinal profile is S-shaped, similar to the SV-shaped bulbous bow, which may produce significantly beneficial wave interference at high speed. The transverse section area curve of the optimised ship is slightly concave after the position of 0.5 times the length, so the ship bow is thinner, the inlet flow angle decreases and the inlet flow length increases, which are beneficial factors in reducing wave-making resistance. It can be concluded from Table 5 that the stern optimization result is the scheme of the wet surface area with a small increase under the displacement constraints, so that too much friction resistance cannot be added, and the optimised stern can be positioned more inward than that of the initial ship. Furthermore, the flow length increases slightly, which is beneficial to the reduction of viscous pressure resistance.

The optimised ship resistance is calculated with the Shipflow tool during the optimization process. Computations are performed on a 2 CPU PC (Intel Core 3.10 GHZ, 4 GB RAM). The resistance curve trend of the two conditions is almost the same, as illustrated in Fig. 7. During the bulbous bow optimization stage, the total resistance decreases significantly at high speed (approximately 7%) and increases by approximately 3% before reaching the design speed ($Fr = 0.27$). This is because the bulbous bow increases the wet surface area, which results in higher friction resistance. At low speed, the wave-making resistance accounts for a smaller proportion; therefore, the total resistance increases. When the speed is increased above the design speed, the wave-making resistance component increases and the bulbous bow effect becomes obvious, which reduces the total resistance. Following optimization of the bow profile, it can be seen that, compared to the results of the first optimization step, the resistance performance is improved slightly when the speed is below the design speed. The optimization of the stern shape

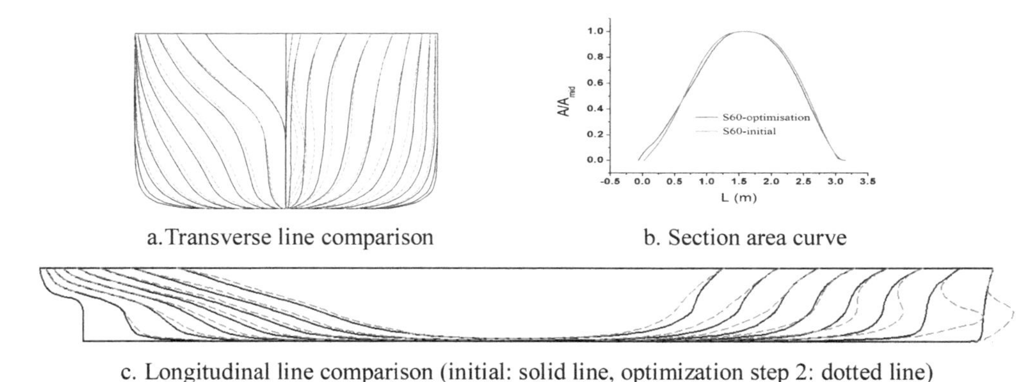

a. Transverse line comparison

b. Section area curve

c. Longitudinal line comparison (initial: solid line, optimization step 2: dotted line)

Figure 6. Contrast between optimised and initial hull forms.

Table 5. Parameters comparison.

	Wet surface area	Displacement	*Lcb*	*Rt*(N)
Initial ship	1.6	121.38	1.59	9.103
Optimization (step 2)	1.607	120.25	1.567	8.134
Variation	0.439%	−0.937%	−1.44%	−10.64%

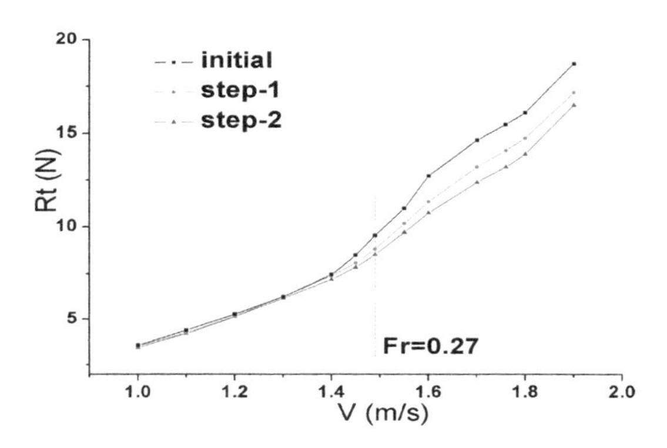

Figure 7. Calculation results by shipflow.

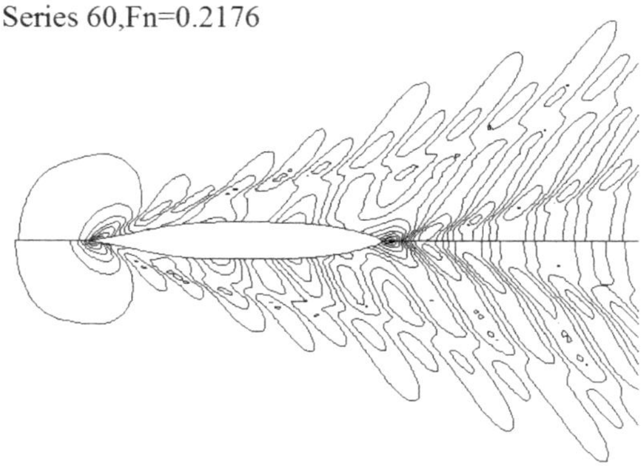

Series 60,Fn=0.2176

Figure 8. Waveform comparison at $Fr = 0.2176$ (initial: above; optimization: below).

decreases the viscous pressure resistance and total resistance across the entire speed range, compared with the initial ship.

Figs. 8 to 12 represent the waveform and side longitudinal wave comparison charts of the initial and optimised ship at three typical speeds. When $Fr = 0.2176$, as shown in Fig. 9, in addition to the reduction of the first crest in the optimised ship bow, the wave amplitude in other positions is increased to varying degrees, which indicates that the effect of the bulbous bow on reducing wave-making resistance is not obvious at low speed but is obvious at medium or high speeds. In the side longitudinal wave diagram, the first wave crest height in the bow decreases obviously and the wave crest height in other positions also decreases significantly, which suggests that the wave energy decreases so that the work done on wave-making resistance by the ship is reduced and wave-making resistance performance is improved.

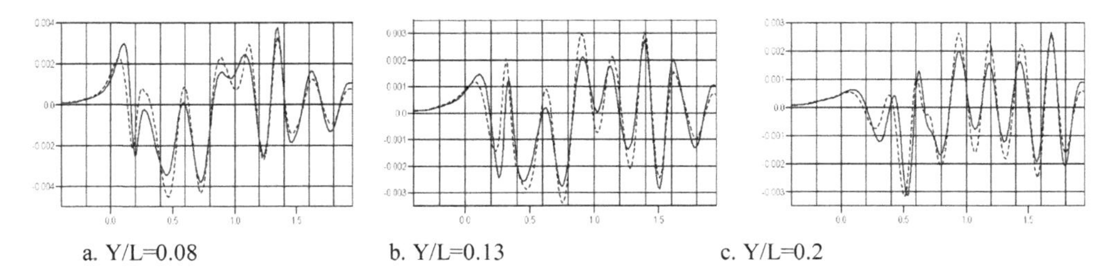

a. Y/L=0.08 b. Y/L=0.13 c. Y/L=0.2

Figure 9. Comparison of side longitudinal wave height at $Fr = 0.2176$ (initial: solid line; optimization: dotted line).

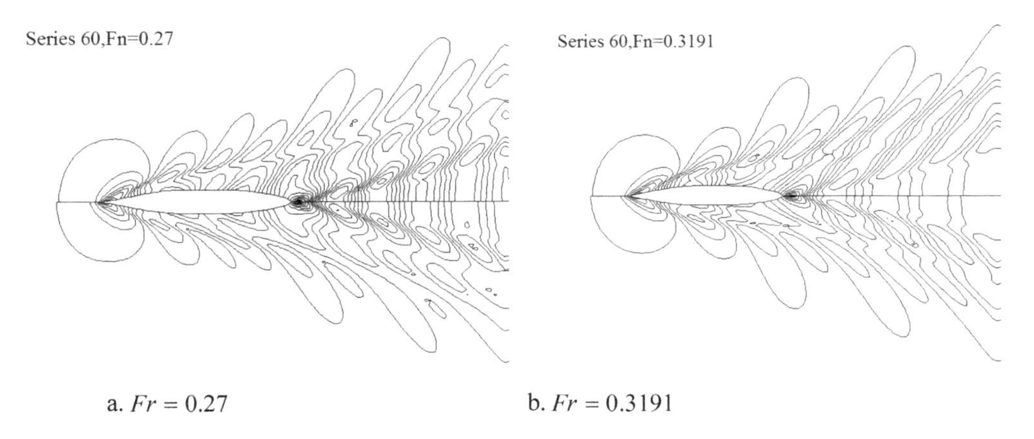

a. Fr = 0.27 b. Fr = 0.3191

Figure 10. Waveform comparison (initial: above; optimization: below).

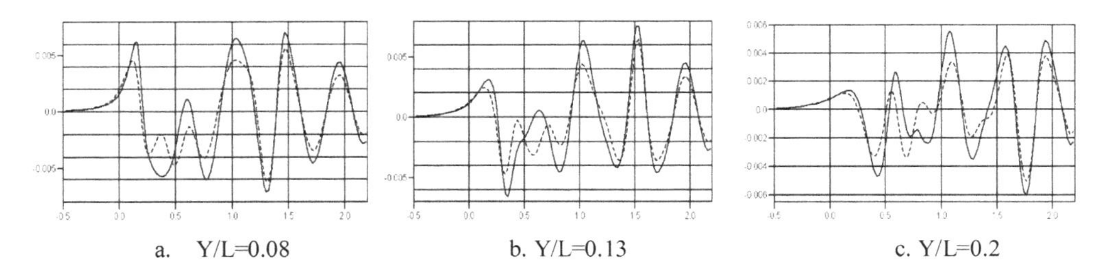

a. Y/L=0.08 b. Y/L=0.13 c. Y/L=0.2

Figure 11. Comparison of side longitudinal wave height at $Fr = 0.27$ (initial: solid line; optimization: dotted line).

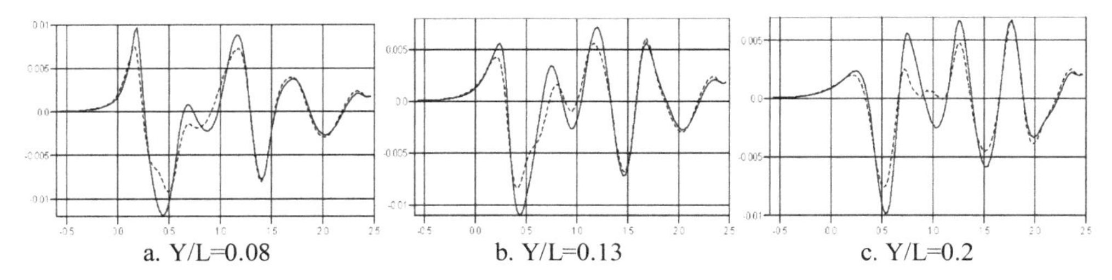

a. Y/L=0.08 b. Y/L=0.13 c. Y/L=0.2

Figure 12. Comparison of side longitudinal wave height at $Fr = 0.3191$ (initial: solid line; optimization: dotted line).

4 CONCLUSIONS

This article focuses on hull form resistance performance optimization based on RBF and CFD. The optimization results of Series 60 showes that the hull form modification method based on radial basis interpolation can provide different hull forms. A new bow are generated and the stern part is deformed. And the optimization process of two steps are valid. Although it only optimize the total resistance at design speed, the total resistance reduces significantly at higher speed.

Future studies will mainly concentrate on the following two aspects:

1. Thus far, variable control points of the bow have been uniformly arranged on the hull surface; therefore, there must be certain control points with little effect on hydrodynamic performance. For this reason, we intend to conduct a sensitivity analysis on these control points in accordance with the site. It can improve the effect and quality of hull form optimization,if the points with a greater influence on optimization results are selected.
2. Hull form optimization is time-consuming; therefore, the approximation technique are essential to save optimization time by means of establishing an approximate model to replace high-precision CFD calculations.

REFERENCES

Abt, C., Harries, S., Heimann, J., Winter, H. 2003. From redesign to optimal hull lines by means of parametric modeling. *In: 2nd intl. conf. computer applications and information technology in the maritime industries.*

Buhmann, M.D. 2000. A New Class of Radial Basis Functions with Compact Support. *Mathematics of Computation*, 70: 307–318.

Buhmann, M.D. 2004. *Radial Basis Functions: Theory and Implementations*. Cambridge: Cambridge University Press.

Campana, E.F., Peri, D., Tahara, Y., Stern, F. 2006. Shape optimization in ship hydrodynamics using computational fluid dynamics. *Computer Methods in Applied Mechanics and Engineering*, 196(1): 634–651.

Diez, M., Peri, D., 2010. Robust optimization for ship conceptual design. *Ocean Engineering*, 37(11): 966–977.

Feng, B.W., Liu, Z.Y., Zhan, C.S., Chang, H.C. 2010. Parameterization modeling based on hull form morphing method. *Computer Aided Engineering*,19(4): 3–7.

Grigoropoulos, G.J.& Chalkias, D.S., 2010. Hull-form optimization in calm and rough water. *Computer-Aided Design*, 42(11): 977–984.

Harries, S., Valdenazzi, F., Abt, C., & Vivani, U. 2001. Investigation on optimization strategies for the hydrodynamic design of fast ferries. *The 6th International Conference on Fast Sea Transportation*, UK:. Southampton.

Kim, H.Y., Yang, C., Löhner, R. et al. 2008. A practical hydrodynamic optimization tool for the design of monohull ship, *18th ISOPE,* Canada:Vancouver.

Kim, H. & Yang, C. 2010. A new surface modification approach for CFD-based hull form optimization. *Journal of Hydrodynamics*, 22(5): 520–525.

Luo, X., Fu, W., Guan, P., Zhang, C. 2011. Algorithm for dynamic delaunay triangulation of simple polygon, *Journal of Jian University (Natural Science)*, 32: 1–6.

Morris, A.M., Allen, C.B., Rendall, T.C.S. 2008. CFD Based Optimization of Aerofoils using Radial Basis Functions for Domain Element Parameterization and Mesh Deformation, *International Journal for Numerical Methods in Fluids*, 58: 827–860.

Morse, B.S., Yoo, T.S., Rheingans, P., Chen, D.T., Subramanian, K.R. 2001. Interpolating implicit surfaces from scattered surface data using compactly supported radial basis functions, *In Shape Modeling International 2001,* 89–98, Italy: Genova.

Peri, D., & Campana, E.F. 2003. Multidisciplinary design optimization of a naval surface combatant, *Journal of Ship Research*, 41(1): 1–12.

Peri, D., Campana, E.F., Dattola, R. 2004. Multidisciplinary design optimization of a naval frigate, *The 10th AIAA/ISSMO Symposium on Multidisciplinary Analysis and Optimization*, Albany: NY.

Peri, D., & Campana, E.F. 2005. High-fidelity models and multi-objective global optimization algorithms in simulation-based design, *Journal of Ship Research*, 49(3): 159–175.

Peri, D. 2016. Robust Design Optimization for the refit of a cargo ship using real seagoing data, *Ocean Engineering*, 123:103–115.

Piegl, L. 1989. Modifying the Shape of Rational B-splines, Part 1: Curves. CAD, 21: 509–518.

Tahara, Y.,Stern, F., Himeno, Y. 2004. Computational fluid dynamics-based optimization of a surface combatant, *Journal of Ship Research*, 48(4): 273–287.

Tahara, Y., Tohyama, S., Katsui, T. 2006. CFD-based multi-objective optimization method for ship design, *International Journal for Numerical Methods in Fluids*, 52: 449–527

Wen, W., Yang, Y., Yu, X. 2000. An algorithm for Delaunay triangulation realized with visual C language, *Journal of North China Electric Power University*, 27: 4–8.

Wu, Z.M. 2002. Radial Basis Function Scattered Data Interpolation and the Meshless Method of Numerical Solution of PDEs, *Journal Of Engineering Mathematics*, 19(2): 10–11.

Xu, L. & Wang, Y.Y. 2001. The fine optimization of ship hull lines in resistance performance by using CFD approach, *Practical Design of Ships and Other Floating Structures*, 1: 59–65.

Yang, X., Pei J., Zhang, Z. 2008. Compact support analysis of radial basis function in image local elastic transformation, *Journal of electronics and information technology*, 30:12–16.

Yang, C., & Kim, H. 2011. Hull Form Design Exploration Based on Response Surface Method. Proceedings of the *Twenty-first International Offshore and Polar Engineering Conference*. USA: Hawaii,.

Yang, C., Kim, H., Huang, F. 2015. Hydrodynamic optimization of a triswach, *Journal of Hydrodynamics*, 26(6): 856–864.

Zhang, B., Ma, K., Ji, Z. 2009. The optimization of the hull form with the minimum wave making resistance based on Rankine source method, *Journal of Hydrodynamics Series B*, 21(2): 277–284.

Marine Design XIII – Kujala & Lu (Eds)
© 2018 Taylor & Francis Group, London, ISBN 978-1-138-34076-3

Development of an automatic hull form generation method to design specific wake field

Yasuo Ichinose & Yusuke Tahara
National Maritime Research Institute, Japan

ABSTRACT: In this paper, we propose a design method of hull form, called wake field design system, that enables arbitrary control of wake field. The proposed method is formed by uniting the data mining method of hull-form and wake-field database with the hull-form blending (morphing) method. In comparison with conventional studies which identify a hull form by characteristic parameters, the system identifies a hull form by unique number in hull-form database, which gives high degree of freedom and scalability to the proposed design system. In addition, we demonstrate the validation of the proposed method by leave-one-out cross validation method. The result shows the great agreement of generated hull form by the proposed system and correct hull form. Furthermore, relevant knowledge is revealed by the proposed simple datamining method (based on k-nearest neighbor algorithm), which is seen to be very useful for this complex database analysis.

1 INTRODUCTION

A wake field yielded by a given hull form is deeply related to improvement of propeller efficiency and reduction of propeller cavitation. In recent years glowing social demands for reduction of underwater ship radiated noise in terms of mariners' comfort and environmental protection, the international concern over underwater ship radiated noise has risen. To reduce underwater ship radiated noise, propeller cavitation which is one of the main source of the noise should be reduced. One of the conventional countermeasures against severe cavitation is an increase in propeller development area in the view of propeller design. However, the problem with this approach is that propeller efficiency goes down according to an increase in propeller development area ratio, and major improvement cannot be expected unless the wake field generated by the hull form is improved. Improvement in the shape of the wake field distribution at the propeller plane will lead to improvement of propeller efficiency and cavitation performance. Therefore, integrated design of hull form and propeller is very important to improve the total performance. From this point of view, some studies have focused on interference between hull and propeller. However, little study has been done to design method to obtain hull form which generate a specific wake field. That is because it is difficult to solve the inverse problem from the Navier-Stokes equation in the complex wake field generated by the hull form, which has a strong nonlinear phenomenon accompanied by a three-dimensional separation. For this reason, the conventional research focuses only on a method of summarizing wake field for propeller design[1], and the wake field design is carried out only by trial and error based on the empirical rule of experienced engineers. In this paper, we propose a design method of a hull form that enables control of wake field, called wake field design system. We demonstrate the validation of the proposed method by leave-one-out cross validation method.

2 WAKE FIELD DESIGN SYSTEM

The wake field design system proposed in this paper consists of the following three technical elements.

1. Hull-form and flow-field database.
2. Wake field distribution analysis method based on the k-nearest neighbor method.
3. Hull-form generation by hull-form blending (morphing) method.

Unlike the conventional analysis[2] of the hull-form database, which organized hull form by characteristic parameters of hull form (for example, the inclination angle of the sectional area curve), the proposed design method identifies a hull form by unique number (ID) in hull-form database, which gives high degree of freedom and scalability to the proposed method. Apart from this paper, the survey of hull-form parameters that have a strong influence on wake field is going to be conducted and will be reported in near future.

The outline of the wake field design system is shown in Figure 1. The input of this system is arbitrary wake distribution on propeller plane, and the system out pus a hull form generates nearest wake distribution to the input one. The process of the system is explained step by step in following.

First, given arbitrary wake distribution on propeller plane, which composed by velocity vector (u, v, w) is transformed into polar coordinate system of 10 points on the radial direction and 36 points on the circumferential direction:

$$\overline{q_{ri\theta j}} = \left(u^q_{ri\theta j}, v^q_{ri\theta j}, w^q_{ri\theta j} \right) \quad \text{so that } i = 1,2,\cdots,10, \quad j = 1,2,\cdots,36 \tag{1}$$

Here, the data in a propeller boss is excluded from the analysis target.

Next, the Euclidean distances (d_k) between a wake distribution on the database $\left(\overline{p_{ri\theta j\,k}}\right)$ and the input wake distribution $\left(\overline{q_{ri\theta j}}\right)$ is evaluated by equation (2).

$$d_k\left(\overline{p_{ri\theta j\,k}}, \overline{q_{ri\theta j}}\right)^2 = \sum_{\substack{1\le i\le 10\\ 1\le j\le 36}} \left(u^{pk}_{ri\theta j} - u^q_{ri\theta j}\right)^2 + \sum_{\substack{1\le i\le 10\\ 1\le j\le 36}} \left(v^{pk}_{ri\theta j} - v^q_{ri\theta j}\right)^2 + \sum_{\substack{1\le i\le 10\\ 1\le j\le 36}} \left(w^{pk}_{ri\theta j} - w^q_{ri\theta j}\right)^2 \tag{2}$$

By using the Euclidean distance obtained by equation (2) as an evaluation function, multiple similar wake distributions (three in this study) are identified from the hull-form and flow-field database. Then a hull from which generates nearest wake distribution to the input one is generated by hull-form blending method by selected hull forms with similar wake fields.

The hull-form blending (morphing) method[3] is a method of generating a hull form from some basic hull forms, and its effectiveness for design is confirmed in hull-form optimization in our previous research activities. In the present method, three selected hull forms with wake distributions

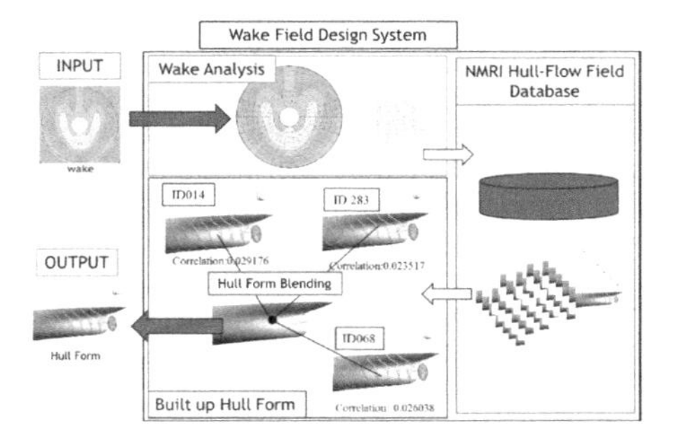

Figure 1. Overview of a wake field design system.

close to input one, are represented by resolvable discrete surface points $\left(\vec{P_1} \sim \vec{P_3}\right)$ – For example, grid points on hull surface in CFD -, and hull form $\left(\vec{P}\right)$ which generate wake field close to input one is obtained by the operation of equation (3).

$$\vec{P} = \left(a_1\vec{P_1} + a_2\vec{P_2} + a_3\vec{P_3} \right) \quad \text{so that } \sum_i^n a_i = 1 \tag{3}$$

where, the weighting factor (a_i) is obtained from the Euclidean distances (d_1, d_2, d_3) of the selected hull-form and flow-field data by the expression (4).

$$a_1 = \frac{2}{3} - \frac{d_1}{d_1+d_2+d_3}, \quad a_2 = \frac{2}{3} - \frac{d_2}{d_1+d_2+d_3}, \quad a_3 = \frac{2}{3} - \frac{d_3}{d_1+d_2+d_3} \tag{4}$$

Note that the Scipy[4] which is a Python-based ecosystem of open-source software is used for analysis of the database (calculation of Euclidean distance) in anticipation of carrying out data mining to hull-form and flow-field database in the future.

3 VALIDATION OF EFFECTIVENES OF THE PRESENT METHOD

As a validation of the effectiveness of the present method, leave-one-out cross validation method is conducted for hull-form and flow-field database of 749-gross-ton-type-domestic-general cargo ship[3]. Leave-one-out cross validation method is often used in statistics to estimate how accurately a model will perform in practice. Based on this method, we extract one sample in the database and rebuild the model of the database without the sample. Then we predict the sample as a target with rebuilt model, and show how accurately the model will predict the target hull-form in practical aspects. To simplify the validation activity, we focus only on the axial flow velocity in wake distribution at propeller plane in equation (1), (2).

The target database has 358 hull-forms and CFD-flow-filed data with different ship length (L_{PP}), breadth (B), draft (d), block coefficient (C_B), and sectional area curves in the range in Table 1. Figure 2 shows the design space of target database. All hull forms in the database is produced by systematic deformation method with a base hull forms. The method keeps frame lines of a base hull form and moves the frame lines of a base forms to the corresponding length-directional position on the sectional area curve which is obtained by $1 - C_P$ method, which is generally used to adjust C_B in practical design. All hull forms in the target database is characterized by the design parameters ξ, η, ζ of [0, 1], which is corresponding to L_{PP}, B, d as shown in Figure 3.

CFD-flow-filed database of 358 hull-forms is produced by a structured grid based Reynolds-averaged Navier-Stokes solver, NEPTUNE[5] developed at NMRI. The turbulence model used in the present work is modified Spalart-Allmaras (MSA) one—equation model without wall function. This solver and turbulence model are normally used for hull form design at NMRI to estimate model-scale flow filed. The calculation grids of basic hull forms without any appendages at full-load and even-keel-were generated with HO topology, 1.8 million cells ($i \times j \times k = 174 \times 128 \times 80$: both sides) at model-scale as shown in Figure 4. The minimum spacing normal to wall is set to be $y^+<1.0$. The effect of free surface is considered to be small and ignored in all cases. The wake distributions at propeller planes in the database is very varied as shown in Figure 5.

Leave-one-out cross validation method is conducted for the hull-form and flow-field database of 749-gross-ton-type-domestic-general cargo ship as a validation of the effectiveness of the present method. The target hull-form data which generates

Table 1. Principal dimensions in target database.

L_{PP}	73.0 m–83.0 m
B	12.8 m–14.5 m
d	3.8 m–4.8 m
C_B	0.68–0.80

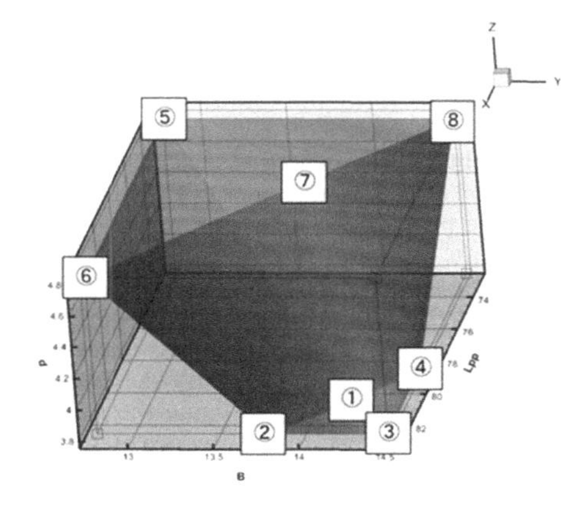

Figure 2. The design space of target database.

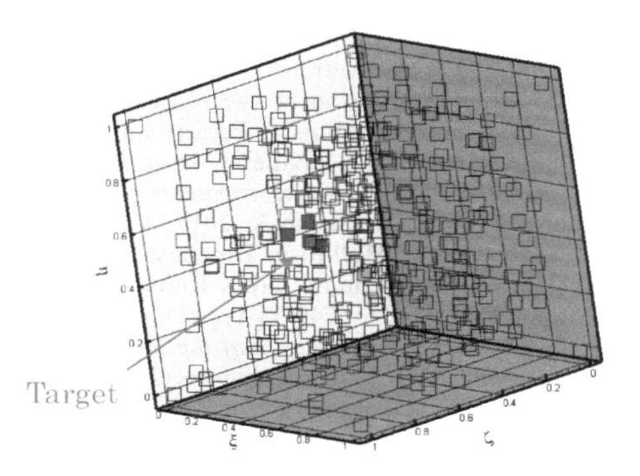

Figure 3. Overview of parameterized target database.

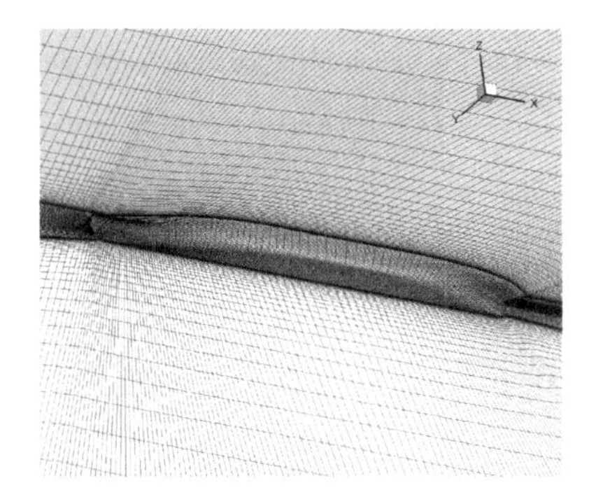

Figure 4. An example of grid of CFD.

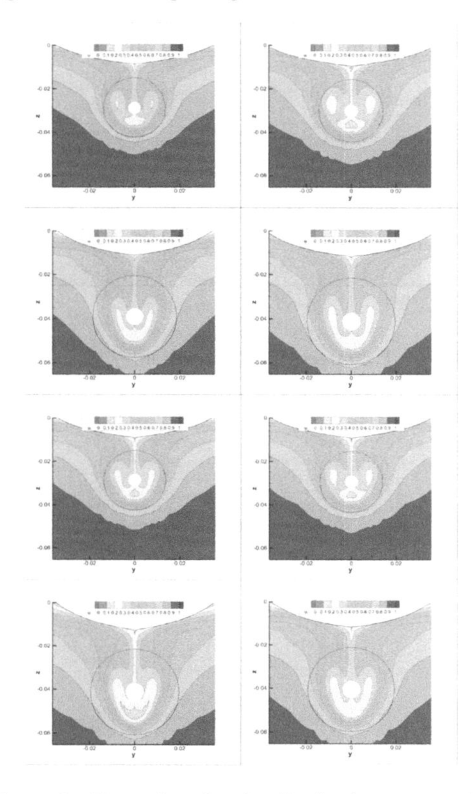

Figure 5. Examples of wake distribution at propeller plane in target database.

wake distribution shown in Figure 6 is left from the target database. The target wake distribution is input to the present wake field design system and the system outputs three similar hull forms. Table 1 shows Euclidean distance and the design parameters of selected hull forms in the target database, which is also shown in Figure 3 – red: target, blue: ID 283, 68, 14 (The ships, the Euclidean distance of which is close to target ship within 3rd place in the database and which is used for hull-form blending.) green: ID

Table 2. The Euclidean distance and design parameter in the target database.

ID	d_k	ξ	η	ζ
Target	–	0.3773	0.5093	0.6158
283	0.023517	0.3978	0.4868	0.5914
68	0.026038	0.3881	0.5826	0.6188
14	0.029176	0.3011	0.5337	0.6481
203	0.081283	0.5122	0.3685	0.5064

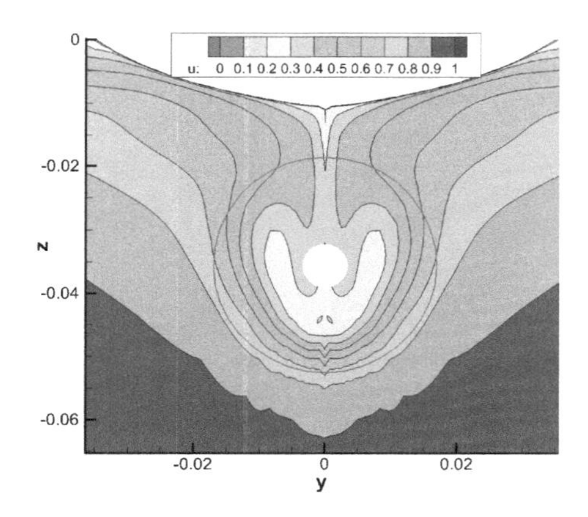

Figure 6. The target wake distribution.

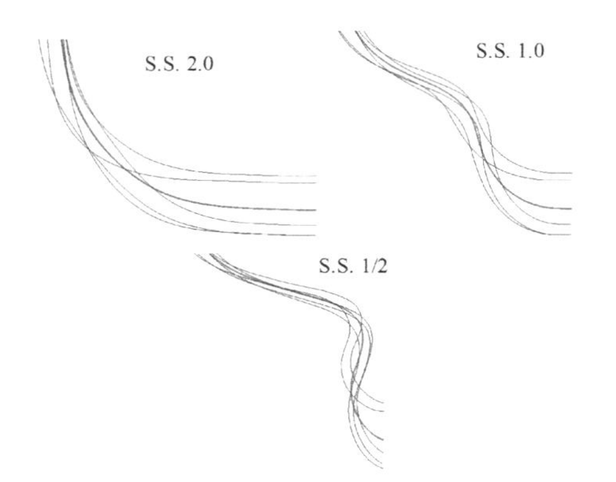

Figure 7. Comparison of frame lines. (red: target, blue: output of the system, black: samples in database).

203 (The ship for reference, the Euclidean distance of which is placed in 4th.) – . It is evident from this validation that there is a correlation between the Euclidean distance obtained from the wake distributions and the design parameters of the hull forms. Figure 4 shows the frame lines of the hull form outputted by the present system (blue), target hull form (red) and sample hull forms in the database. As Figure 7 indicates, the hull form newly outputted by the present wake field design system (blue) agrees with the target hull form very well, which is very useful for practical hull form design. (blue and red lines cannot be distinguished because they are overlapped in Figure 7).

4 CONCLUSION

In the present paper, we propose a design method of hull form that enables arbitrary control of wake flow, called wake field design system, and validate effectiveness of the method by leave-one-out cross validation method. The proposed method is formed by uniting the datamining method of hull-form and wake-filed database and the hull-form blending (morphing) method. In comparison with conventional studies which identify a hull form by characteristic parameters of hull form, the system identifies a hull form by unique number in hull-form database, which gives high degree of freedom and scalability to the proposed design system. The result shows the great agreement of generated hull form by the proposed system and target hull form. Furthermore, relevant knowledge is revealed by the proposed simple datamining method (based on k-nearest neighbor algorithm), which is seen to be very useful for this complex database analysis. We plan to expand the database, verify the extrapolation of the database and investigate the hull form parameters that have a strong influence on the wake.

REFERENCES

Mori, M. 1997. *Hull-form design* (in Japanese), Tokyo: Ship technology association.

Kanai, T. 2000. Application of the Neural Network to Estimate of Ship's Propulsive Performance and Hull Form Optimization (in Japanese), *Transactions of the West-Japan Society of Naval Architects* (99): 1–11.

Ichinose, Y. Tahara, Y. & Kume, K. 2017. A Construction and Evaluation of Hull-form Database for Domestic Vessels with Regulation on Gross Tonnage—Development of a Prototype for 749GT-type Domestic General Cargos – (in Japanese). *Journal of the Japan Society of Naval Architects and Ocean Engineers* (26): 51–32.

Nobuyuki, Hirata. & Takanori, Hino. 2000. An Efficient Algorithm for Simulating Free-Surface Turbulent Flows around an Advancing Ship, *Journal of the Society Naval Architect*. Japan (185): 1–8.

Marine Design XIII – Kujala & Lu (Eds)

Hull form optimization for the roll motion of a high-speed fishing vessel based on NSGA-II algorithm

Dan Qiao, Ning Ma & Xiechong Gu
State Key Laboratory of Ocean Engineering, Shanghai Jiao Tong University, Shanghai, China
Collaborative Innovation of Center for Advanced Ship and Deep-Sea Exploration, Shanghai Jiao Tong University, Shanghai, China
School of Naval Architecture, Ocean and Civil Engineering, Shanghai Jiao Tong University, Shanghai, China

ABSTRACT: The roll motion for fishing vessels in beam seas with zero speed deserves to be a significant criterion for ship safety evaluation. High-speed fishing vessels are prone to the surf-riding and broaching in adverse quartering and following seas with potential safety implications. The possibility of improving roll motion responses for the ITTC-A2 high-speed fishing vessel is investigated delicately through ship hull form optimization. During the process, the ship hull surfaces are reconstructed accordingly using the Radial Basis Function (RBF) interpolation technique through adjustments of a limited number of RBF control points. The statistical roll motion responses at irregular waves are numerically predicted through the general hydrodynamic software based on 3-D potential theory. The surf-riding characteristics in following seas are studied in detail through the IMO recommended level 2 vulnerability criteria for surf-riding/broaching. In addition, the calm water resistance along with the optimization process is monitored by the corrected Holtrop Method. The optimum hull form is obtained through the application of NSGA-II genetic algorithm, which is effective and robust for multi-objective optimization problems. The results indicate that the proposed ship hull form optimization method is an efficient approach for improving roll motion responses with only limited impacts on surf-riding and calm water resistance performances for high-speed fishing vessels and deserves to be generalized into a conventional method for improving general ship performances at initial design stage.

1 INTRODUCTION

With the rapid development of the shipping industry, the Simulated Based Design (SBD) has accomplished rapid growth due to the increasing optimization theories and algorithms, computing power and Computational Fluid Dynamics (CFD) techniques with higher orders of accuracy. SBD is a systematical combination of ship hull transformation methods, optimization theories and algorithms, techniques in computation and calculation of hydrodynamic characteristics.

Gammon (2011) conducted the hull form optimization of the 148/1-B fishing boat based on resistance, seakeeping and stability performances through the Multi-Objective Genetic Algorithm (MOGA). Pecot et al. (2012) introduced the work for a monohull fishing vessel on ship hull shape optimization for multiple objectives including resistance and seakeeping performances. The Parametric CAD modeling, frequency-domain seakeeping codes and extensive CFD calculations are installed into the optimization procedure. Koutroukis et al. (2013) conducted a systematical multi-objective optimization for the Required Freight Rate (RFR) and the Energy Efficiency Design Index (EEDI) for a container ship through Non-Dominated Sorting Genetic Algorithm II (NSGA-II) based on the design concept of Eco-Ship. Pinto et al. (2007) worked on the optimization design of the bulbous bow shape of an oil tanker with the objective function related to the total resistance calculated by potential theory and empirical formulas. Huang et al. (2015) combined both the local and general geometric reconstruction of the ship hull in avoiding the deterioration of resistance performance outside the considered speed range, and carried out a multi-objective ship hull form optimization of S60 ship aiming at the wave making resistance and seakeeping indexes at 3 different Froude numbers. Wang et al. (2011), Huang et al. (2013) and Luo et al. (2016) carried out the multi-objective ship hull optimization targeting respectively at a pentamaran and a large container ship, which emphasis on the resistance and seakeeping performance. Wang et al. (2017) performed a systematical ship hull form optimization targeting at the wave making resistance for a 10,000TEU large container ship, where the optimization algorithms combined NSGA-II, Non-Linear Programming by Quadratic Lagrangian (NLPQL) and the Kriging surrogate model together, which are assembled in iSight optimization platform.

The ship hull form reconstruction technique is required to guarantee the fundamental smoothness, continuity and rationality of the ship hull surface while generating as more intermediate ship types as possible. Basically, mainstream ship hull form reconstruction techniques can be roughly categorized into two types: the global surface reconstruction, represented by parameterized modeling method (Meng & Xu, 2002), Lackenby transformation method (Huang, 2012) and shifting method (Kim, 2009), as well as the local surface reconstruction, represented by free-form deformation approach (Sederberg & Parry, 1986), Bezier Patch method (Goshtasby, 1989) and RBF (Kim, 2009). The former type reconstructs the hull surface through adjusting the hull form parameters reflecting the hull surface features, and the latter type actualizes the deformation of the ship hull by adjusting the spatial positions of control points.

The optimization algorithm occupies the core position since the quality of the algorithm determines the efficiency and the accessibility of the optimization procedure. The gradient-based algorithms require strictly on the continuity and the differentiability of the objective function, thus obtains higher calculation efficiency but more difficulties in finding global optimum solutions. SQPM method, steepest descent method and variable gradient algorithm are typical representatives. Besides, the random searching algorithms avoid strict requirements upon objective functions and thus attach more convenience in finding global optimum solutions. Particle swarm optimization algorithm, simulated annealing algorithm and genetic algorithm are outstanding examples. Genetic algorithm is a widely-used probabilistic global search algorithm based on natural selection and genetic variation according to the theory of biological evolution. Srinivas & Deb (1994) put forward the Non-Dominated Sorting Genetic Algorithm (NSGA) with higher convergence rate. Subsequently, Deb et al (2002) proposed the concept of NSGA-II algorithm with significantly better capability performances, remarkable extensibility and robustness.

Fishing vessels independently navigate in operation areas far away from the base. Due to operation restrictions during fishing and hauling periods, the combinations of wind, wave and hauling directions are often in disadvantageous states, which becomes easier to trigger severe rolling motions and even capsizing. Meanwhile, fishing vessels commonly navigate at the Froude number larger than 0.3, which makes the ship easier to be overtaken by waves and accelerated to sail in wave celerity, thus gives rise to stability failures such as surf-riding and broaching. This paper pays close attention to the inherent roll motion response for ITTC-A2 high-speed fishing vessel in beam seas with zero speed, and such seakeeping performances emphasize detailed adjustments of local ship hull surface. As a result, the paper selects the RBF method for ship hull form reconstruction and NSGA-II as the governing optimization algorithm. The seakeeping motion responses, surf-riding and broaching assessment, as well as calm water resistance confirmation are carried out through 3-D potential theory, IMO recommended level 2 vulnerability criteria and the Holtrop method with minor corrections, respectively.

2 THEORETICAL BASIS

2.1 *NSGA-II Algorithm and the construction of the convex hull*

The objective conflict is a frequent phenomenon for multi-objective problems, which indicates the optimum solution that minimizes all objective functions barely exists (Tong 2009). As a result, the optimum solution for multi-objective optimization is just a set of non-dominated solutions named Pareto optimality (Censor 1977).

NSGA-II algorithm (Deb et al. 2002) is an efficient and robust approach in seeking for the Pareto optimality of a multi-objective optimization problem with better distribution of solutions and better degrees of convergence. With the application of fast non-dominated sorting algorithm, the overall complexity decreases while preserving the best individuals as much as possible. The selection of the elite strategy, the establishment of the mating pool between parental and progeny generations and the acquisition of the free populations significantly improve the applicability of the algorithm. The fundamental principles of the NSGA-II algorithm can be found in Luo's research (Luo et al. 2016).

The verification of the surf-riding/broaching and calm water resistance performances can be easily interfered by the random iteration steps that are not representative enough to reflect the general variation tendency. To strategically select representative intermediate iteration steps after the optimization iteration procedure, the paper chooses the iteration points on the lower boundary of the convex hull containing all iteration points as the characteristic iteration points. The convex hull of a set P of points inside an affine space over the reals is the smallest convex set that contains P, which may be visualized as the shape enclosed by a rubber band stretched around all the iteration points (Berg et al. 2008). As a result, every two adjacent characteristic iteration points and their connection line are able to control all the iteration results between them, which is capable of reflecting the variation tendency with strong representativeness.

2.2 *RBF Ship hull form reconstruction*

For the ship performance optimization, a local modification of hull surface is applied through Radial Basis Function (RBF) (Kim & Yang 2010), and such a transformation method can be easily approached by arranging control points around the ship hull. The interpolation function $S(\boldsymbol{X})$, which describes the displacement of the points $\boldsymbol{X}=(x,y,z)$ either on the hull surface or in the entire domain can be approximated by the sum of series of Radial Basis Functions as Equation 1:

$$S(\boldsymbol{X})=\sum_{j=1}^{N}\lambda_j\phi\left(\left\|\boldsymbol{X}-\boldsymbol{X}_j\right\|\right)+p(\boldsymbol{X}) \tag{1}$$

where $\boldsymbol{X}_j=\left(x_j,y_j,z_j\right)$ is the center of the radial basis function at which the interpolation function $S(\boldsymbol{X})$ is clearly given. In addition, $p(\boldsymbol{X})$ refers to a polynomial, N refers to the number of control points (centers) and ϕ refers to a given Radial Basis Function with respect to the Euclidean distance $\|\boldsymbol{X}\|$.

The basis function $\phi(\|\boldsymbol{X}\|)$ is chosen according to the solvability of the interpolation equation system, the stability of solutions, the time complexity and the balance of interpolation effect globally and locally. As a result, Wedndland's C^2 function, a kind of three-dimensional compact support Radial Basis Function with 2nd order continuity is chosen, which is shown in Equation 2 (De Boer et al. 2007):

$$\phi(\|\boldsymbol{X}\|)=(1-\|\boldsymbol{X}\|)^4(4\|\boldsymbol{X}\|+1) \tag{2}$$

The $p(\boldsymbol{X})$ can be defined into a linear polynomial to recover the affine transformation (rotation transformation, shear transformation and scaling) and translation transformation, which is shown in Equation 3:

$$p(\boldsymbol{X})=c_1+c_2x+c_3y+c_4z \tag{3}$$

The coefficients $\lambda_j(j=1,2,3,\ldots,N)$ in Equation 1 and $c_j(j=1,2,3,4)$ in Equation 3 can be determined according to the interpolation condition and the additional boundary condition shown in Equation 4:

$$\begin{cases} S\left(\boldsymbol{X}_j\right)=f_j & j=1,2,3,\ldots,N \\ \sum_{j=1}^{N}\lambda_j p\left(\boldsymbol{X}_j\right)=0 & j=1,2,3,\ldots,N \end{cases} \tag{4}$$

where $f_j(j=1,2,3,\ldots,N)$ demonstrate the discrete known value of the displacement on the boundary (control points). Finally, the values of the coefficients $\lambda_j(j=1,2,3,\ldots,N)$ and $c_j(j=1,2,3,4)$ are obtained through solving the following linear system:

$$\begin{gathered} \begin{pmatrix} \boldsymbol{f} \\ 0 \end{pmatrix}=\begin{pmatrix} \Phi & P \\ P^T & 0 \end{pmatrix}\cdot\begin{pmatrix} \boldsymbol{\lambda} \\ \boldsymbol{c} \end{pmatrix} \\ s.t.\,\Phi_{i,j}=\phi\left(\left\|\boldsymbol{X}_i-\boldsymbol{X}_j\right\|\right) i,j\in[1,N] \\ P_{i,j}=p_j\left(\boldsymbol{X}_i\right) i\in[1,N], j\in[1,4] \\ \boldsymbol{f}=\left[f_1,f_2,f_3,\ldots,f_N\right]^T \\ \boldsymbol{\lambda}=\left[\lambda_1,\lambda_2,\lambda_3,\ldots,\lambda_N\right]^T \\ \boldsymbol{c}=\left[c_1,c_2,c_3,c_4\right]^T \end{gathered} \tag{5}$$

Two types control points are required for modifying the hull surface expressed by a discrete triangulation, which can be either on or off the hull surface (Kim & Yang 2010). The fixed control points, whose spatial location is pre-determined to keep the hull surface near them unchanged and to meet with distinctive design requirements. The movable control points are used as the design variables where the movements of hull surface at certain degrees of freedom can be allowed. Such variables are determined by the given optimization algorithms in minimizing the objective functions. More details can be found from Rendall & Allen (2008) and Morris et al. (2008).

2.3 *Surf-riding/broaching and the vulnerability criteria*

In adverse sea states, ships are turned to sail in longitudinal waves to avoid stability failure modes induced by transverse wind and wave loads. However, severe longitudinal waves give rise to the situations with maneuverability or stability failure. In following and quartering seas, ships tend to be overtaken by waves and accelerated to sail in wave celerity with the combined action of the wave surge force, the propeller thrust and the ship resistance. Such a phenomenon is named as the surf-riding, which is regarded as the precedent of broaching (Yu et al. 2016). Surf-riding and broaching usually occurs on fishing vessels with large Froude numbers (Mata-Álvarez-Santullano et al. 2014).

The level 2 vulnerability criteria for the surf-riding and broaching (IMO, SDC2/WP.4 2015) takes the critical Froude number Fn_{cr} corresponding to the second threshold of surf-riding as the key parameter, which indicates that the surf-riding is inventible with the Froude number no less than Fn_{cr}. Subsequently, the critical number of revolutions of the propeller corresponding to the second threshold of surf-riding, n_{cr}, has been put forward, which can be numerically calculated and evaluated by a simple quadratic equation (IMO, SDC4/5/1/Add.2 2016) shown by Equation 6.

$$2\pi \cdot \frac{T_e(c_i, n_{cr}) - R(c_i)}{f_{ij}} + 8a_0 n_{cr} + \sum_{j=1}^{5} p_j a_j = 0 \quad (6)$$

where $T_e(c_i, n_{cr})$ refers to the propeller thrust, $R(c_i)$ refers to the ship resistance at the wave celerity c_i, f_{ij} refers to the amplitude of the wave surging force. $T_e(c_i, n_{cr})$ is approximated into a quadratic polynomial with respect to c_i and n_{cr}. $R(c_i)$ is approximated with the 5th power polynomial. Specifically, f_{ij} can be calculated with the wave information and the geometric information of each station of the ship. Based on such computing methods, the critical number of revolutions of the propeller corresponding to the second threshold of surf-riding, n_{cr}, can be uniquely determined once the wave condition and the ship hull form information are explicitly given. As a result, n_{cr} is a clear vulnerability criterion with simplicity, precision and robustness.

2.4 *Holtrop method and the correction*

Holtrop regression formula represents a widely-used simple approach for estimating the resistance properties and propulsion factors (Holtrop 1977, 1978, 1982, 1984). Since it's presented based on series of regression analyses of model-scale and full-scale test data with wider ranges of parameters, the Holtrop method has become the theoretical basis of many resistance estimation programs. The detailed subdivisions for Holtrop method are shown as Equation 15.

$$R_T = R_F(1 + k_1) + R_{AP} + R_W + R_B + R_{TR} + R_A \quad (7)$$

where R_T stands for the total resistance; $R_F(1 + k_1)$ stands for the frictional resistance according to the ITTC-1957 formula with the correction of the hull form factor; R_{AP} stands for the appendage resistance, R_W stands for the wave resistance; R_B stands for the additional pressure resistance of bulbous bow near the water surface; R_{TR} refers to the additional pressure resistance due to the transom immersion and R_A represents the model-ship correlation resistance. Each component can be easily calculated by series of approximate regression formulas with respect to the systematical ship hull form information and the speed of navigation.

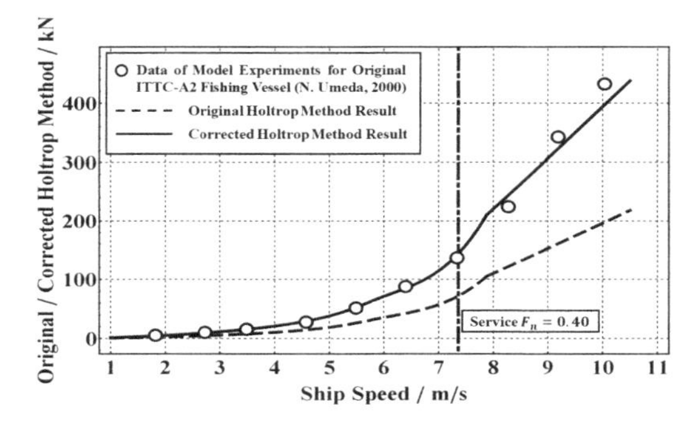

Figure 1. The comparison between original Holtrop method result, corrected Holtrop method result and data of model experiments for ITTC-A2 high-speed fishing vessel.

Nevertheless, the Holtrop method, which has been applied to ITTC-A2 high-speed fishing vessel, is provided with unsatisfying degree of agreement with the experimental data as shown in Figure 1. As a result, the formulation of the Holtrop method should be temporarily corrected according to Equation 8.

$$R_T^* = \mu \cdot R_T \quad (8)$$

where R_T^* refers to the corrected Holtrop method result for ITTC-A2 high-speed fishing vessel yet R_T refers to the original result. Parameter $\mu = 2.0075$ is the correction coefficient obtained by contrastive and regression analysis targeting at both fundamental Holtrop principles and the data of model experiments (NAOE Osaka University 2015).

The comparison analysis shown in Figure 1 indicates that the corrected Holtrop method accomplished qualified agreement with model experiment results with the Froude number up to 0.60, which can be used as an expeditious and accurate approach for the assessment and confirmation during the ship hull form optimization process.

3 HULL FORM OPTIMIZATION

3.1 *Basic information*

Based on the methods described in the previous sections, the complete ship hull form optimization tasks are carried out upon a 135GT Japanese purse seiner, ITTC-A2. The primary particulars of the ship are presented in Table 1 (NAOE Osaka University 2015).

Table 1. Primary particulars of ITTC-A2.

General ship hull – description, symbol (unit)	Value
Length overall, L_{oa} (m)	43.00
Length between perpendiculars, L_{pp} (m)	34.50
Breadth, B, (m)	7.60
Depth, D, (m)	3.07
Draught at FP, d_f, (m)	2.50
Draught at AP, d_a, (m)	2.80
Mean draught, d, (m)	2.65
Block coefficient, C_B	0.5970
Prismatic coefficient, C_P	0.7690
Water plane coefficient, C_W	0.9850
Radius of gyration-roll, k_{xx} / L_{pp}	0.108
Radius of gyration-pitch, k_{yy} / L_{pp}	0.302
Radius of gyration-yaw, k_{zz} / L_{pp}	0.302
Longitudinal center of gravity*, LCG, (m)	1.31 aft
Longitudinal center of buoyancy, LCB, (m)	1.31 aft
Longitudinal center of flotation, LCF, (m)	3.94 aft
Service Froude number, Fn	0.40
Submerged ship hull–description, symbol (unit)**	**Value**
Overall length of submerged ship hull, L_W, (m)	39.5119
Overall breadth of submerged ship hull, B_W, (m)	7.6254
Overall draught of submerged ship hull, d_W, (m)	3.1350
Water plane area, A, (m^2)	260.335
Wetted surface area, S_W, (m^2)	363.335
Submerged volume, ∇, (m^3)	413.877

*The longitudinal centers of gravity, buoyancy and flotation are all measured from the midship station.
**The submerged ship hull information is all defined according to the geometric model within MSC-Patran and HydroStar.

Table 2. Environmental wave parameters for BF-6.

Description, symbol (unit)	Value
Significant wave height, $\zeta_{W/3}$, (m)	3.0
Mean zero-crossing wave period, T_1, (s)	6.7

The ITTC two-parameter wave spectrum shown by Equation 17 is selected as the representative spectrum during the corresponding analysis of the irregular rolling of ITTC-A2 high-speed fishing vessel.

$$S(\omega) = 173 \cdot \frac{\zeta_{W/3}^2}{T_1^4} \cdot \frac{1}{\omega^5} \cdot exp\left(-691 \cdot \frac{1}{T_1^4} \cdot \frac{1}{\omega^4}\right) \tag{9}$$

To fully evaluate the significant rolling amplitude for ITTC-A2 high-speed fishing vessel under rough seas (IMO MEPC.1/Circ.796 2012), the Beaufort scale 6 (BF-6) is selected as the representative sea condition. Environmental wave parameters for BF-6 sea condition are shown in Table 2 (IMO MEPC65/INF.21 2003).

3.2 *Optimization framework and initial setup*

The ship hull optimization targeting at the significant rolling amplitude for ITTC-A2 high-speed fishing vessels in beam seas with zero speed is regarded as a typical optimization problem with the unique objective function, several monitoring variables, a couple of constraints and appropriate stopping criterion, as shown in Figure 2.

The entire optimization process is controlled by NSGA-II algorithm running on iSight optimization platform, which is effective and robust for single/multi-objective optimization problems. The input movable values of movable control points at the given degree of freedom of motion are set as the optimization variables. The RBF transformation method is utilized to conduct the ship hull form reconstruction

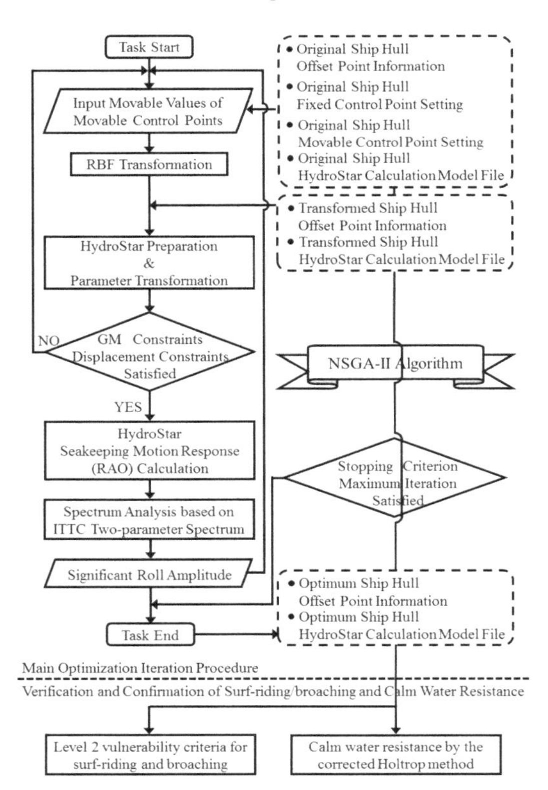

Figure 2. Optimization flow chart for significant rolling amplitude of ITTC-A2 high-speed fishing vessel.

and preparation for the seakeeping motion response calculation. The ship displacement constraints and GM constrains are employed for the fast elimination of the abnormal intermediate results to ensure the rationality of the iteration process. The seakeeping motion response calculations are carried on HydroStar based on 3-D potential theory. The maximum steps of iteration are installed as the stopping criterion, which is defined according to the population size and the number of generations in NSGA-II algorithm. The calculation, verification and confirmation of the level 2 vulnerability criteria for surf-riding and broaching, as well as the calm water resistance at service Froude number $Fn = 4.0$ are conducted after the completion of the ship hull form optimization process, where the appropriate intermediate results during the iteration should be selected strategically through constructions of the convex hull.

For the original ship hull of ITTC-A2 high-speed fishing vessel, the initial values of parameters for the optimization procedure can be shown in Table 3.

Table 3. The initial values for parameters involved into the optimization procedure.

Description, symbol (unit)	Value
Main optimization iteration procedure	
Significant rolling amplitude, $\phi_{1/3}$, (deg)	20.5465
Ship displacement, W, (t)	424.224
Natural roll period, T_ϕ, (s)	5.747
Metacentric height, GM, (m)	1.0702
Level 2 vulnerability criteria for surf-riding and broaching	
• No. 1: $\lambda / L_{pp} = 1.25, H / \lambda = 0.0504$, following sea	
Critical propeller revolutions number, n_{cr}, (rps)	5.1477
• No. 2: $\lambda / L_{pp} = 1.50, H / \lambda = 0.0396$, following sea	
Critical propeller revolutions number, n_{cr}, (rps)	5.7005
• No. 3: $\lambda / L_{pp} = 1.50, H / \lambda = 0.0504$, following sea	
Critical propeller revolutions number, n_{cr}, (rps)	5.5580
• No. 4: $\lambda / L_{pp} = 1.50, H / \lambda = 0.0600$, following sea	
Critical propeller revolutions number, n_{cr}, (rps)	5.4629
• No. 5: $\lambda / L_{pp} = 1.75, H / \lambda = 0.0504$, following sea	
Critical propeller revolutions number, n_{cr}, (rps)	5.9545
Calm water resistance by the corrected Holtrop method	
Calm water resistance at service Fn, R_T, (kN)	143.591

The significant rolling amplitude $\phi_{1/3}$ is regarded as the unique minimization optimization target, yet the variation tendency of the ship displacement W and the natural roll period T_ϕ also deserve meticulous analysis during optimization iteration. In terms of the numerical evaluation through IMO recommended level 2 vulnerability criteria for surf-riding and broaching, the critical number of revolutions of the propeller corresponding to the second threshold, n_{cr}, is highly valued. In addition, the calm water resistance at service Froude number $Fn = 4.0$, R_T, is the characteristic parameter of the resistance performance of ITTC-A2 high-speed fishing vessel.

3.3 *Case 1: Optimization with the water plane unchanged*

The Case 1 is designed to evaluate the effect of the variation of ship hull form in the middle part of the ship with particularly strict restrictions on the bow shape, stern shape, nodes on water plane and longitudinal section in center plane. Such an optimization scheme is relatively conservative, which is more applicable to the engineering demand. The distribution of the fixed and movable control points along with the original ship hull of ITTC-A2 high-speed fishing vessel are shown in Figure 3a and Figure 3b.

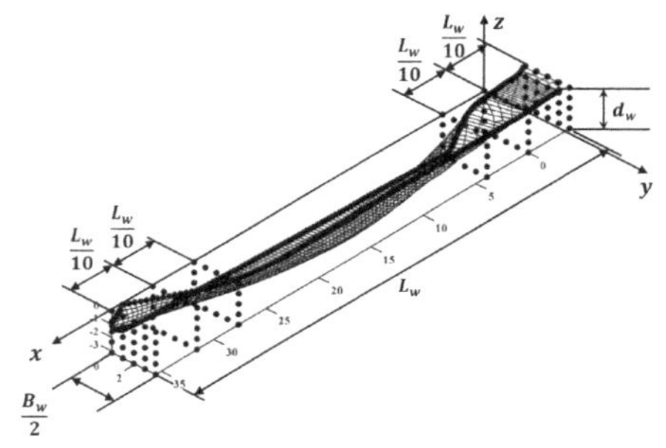

a. Fixed control points for Case 1

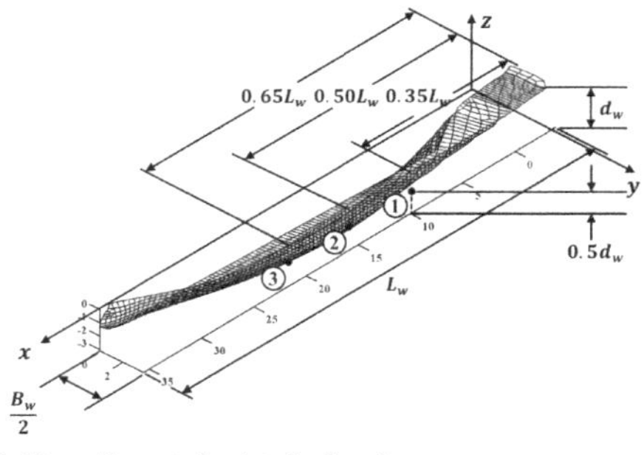

b. Moveable control points for Case 1

Figure 3. Control points distribution along with the original ship hull of ITTC-A2 high-speed fishing vessel for Case 1.

Algorithm parameters, the stopping criterion, the range of variation for movable control points at each degree of freedom, ship displacement and *GM* constraints for Case 1 are shown in Table 4.

The algorithm parameters and the stopping criterion are defined to control the overall time consuming and computing recourse occupation, and the range of variations for movable points are indented to maintain a relatively reasonable geometry of ship hull surface instead of uneven surface and excessive deformation. The limitation for variation of the ship displacement *W*, which is related to the ship hull form, resistance and economic efficiency, is controlled strictly within ±2% for the general similarity of ship hull form schemes. The lower bound of *GM* value has been prescribed as 80% of the initial value to prevent the excessive deterioration of the initial stability performance.

The results of the ship hull form optimization for Case 1 are shown in Table 5.

The history sequence chart of the optimization procedure and the contrast diagram of the ship hull lines between the initial and the optimum ship hull form for Case 1 can be shown in Figures 4 and 5.

Figure 4a demonstrates that the optimization objective $\phi_{1/3}$ has been gradually modified and it appears mild after about 200 iteration steps. Figure 5 obviously indicates that the ship hull form in the middle part becomes plumper with the bow, stern, water plane and longitudinal section in center plane strictly controlled. The increase of the plumpness of the ship hull and the changes of ship hull form lead to larger ship displacement and natural roll period, which can be clearly visualized by Figure 4b and 4c.

The convex hull of iteration results and its boundary points for Case 1 are shown in Figure 6.

For the chosen characteristic iteration points, the corresponding results for the level 2 vulnerability criteria for surf-riding and broaching, as well as the calm water resistance at service Froude number $Fn = 4.0$ are shown in Figure 7.

Table 4. Settings of initial parameters and constraints in Case 1.

Description		Lower bound	Upper bound
Population size		30	
Number of generations		30	
Maximum iteration steps		930	
Movable point ①	ΔZ_1	–0.10	+0.10
Movable point ②	ΔZ_2	–0.10	+0.10
Movable point ③	ΔZ_3	–0.10	+0.10
Ship displacement, *W*		415.739 *t*	432.708 *t*
Metacentric height, *GM*		0.85616 *m*	—

Table 5. Results of ship hull optimization for Case 1.

Description		Outcome	Variation
Iteration time consumption*		49.6689 *h*	—
Run counter of optimum $\phi_{1/3}$		777/930	—
Movable point ①	ΔZ_1	+0.092445	—
Movable point ②	ΔZ_2	+0.096026	—
Movable point ③	ΔZ_3	–0.022123	—
Roll motion amplitude $\phi_{1/3}$		19.4257 *deg*	–5.455%
Ship displacement, *W*		432.704 *t*	+1.999%
Natural roll period, T_ϕ		5.966 *s*	+3.811%

*The iteration time consumption is based on the computing device with 3.00 GHz of CPU, 8.00 GHz of RAM and Microsoft Windows XP.

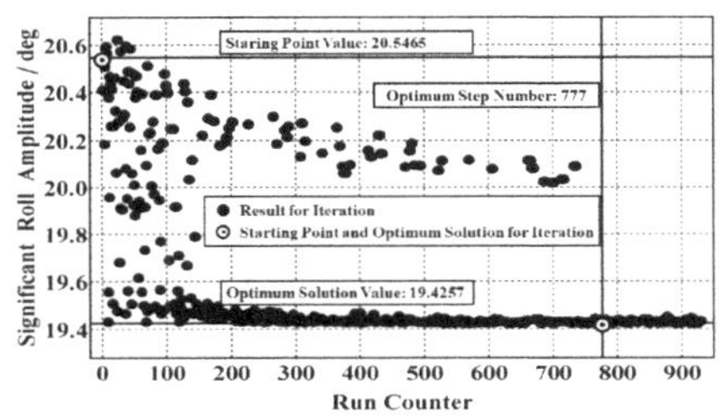

a. History sequence chart of $\phi_{1/3}$ for Case 1

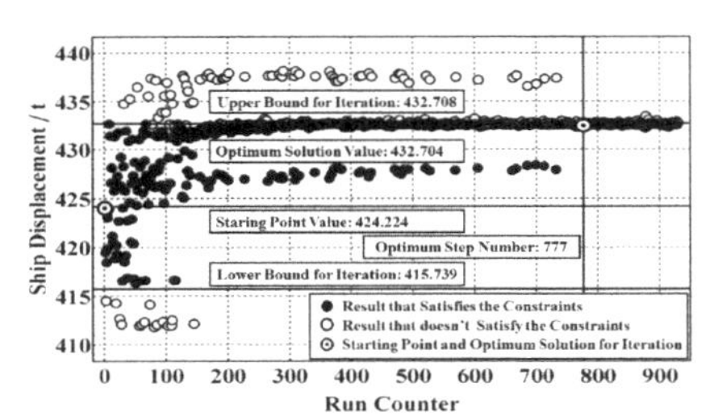

b. History sequence chart of *W* for Case 1

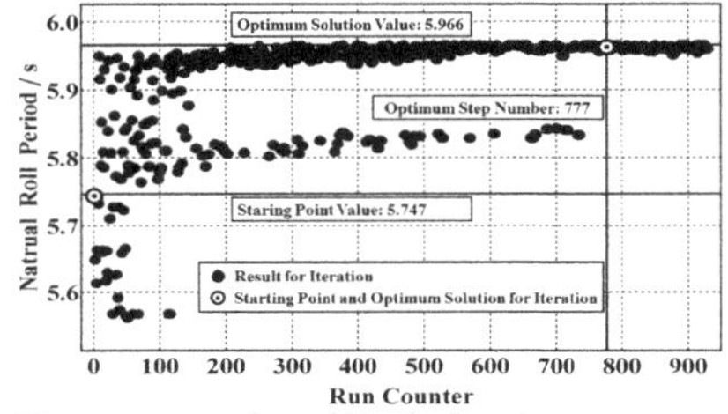

c. History sequence chart of T_ϕ for Case 1

Figure 4. History sequence charts of the optimization procedure on ITTC-A2 high-speed fishing vessel for Case 1.

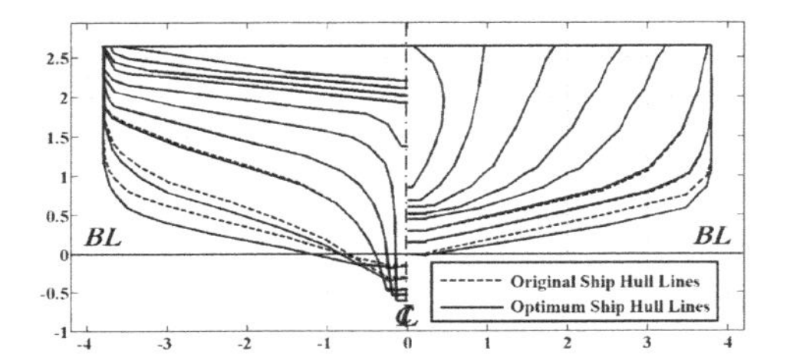

Figure 5. The contrast diagram of the ship hull lines between the initial ship hull form and the optimum ship hull form of ITTC-A2 high speed fishing vessel for Case 1.

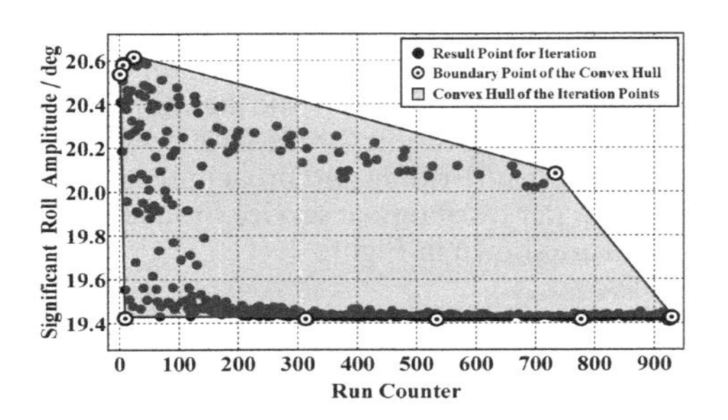

Figure 6. The convex hull of the iteration points and its boundary points for Case 1.

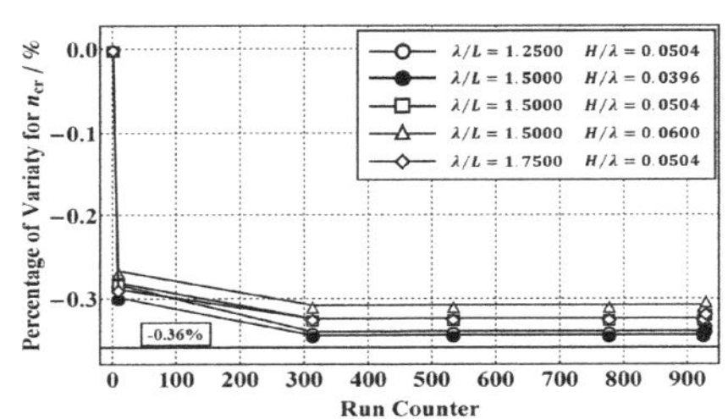

a. The variation tendency of n_{cr} for case 1

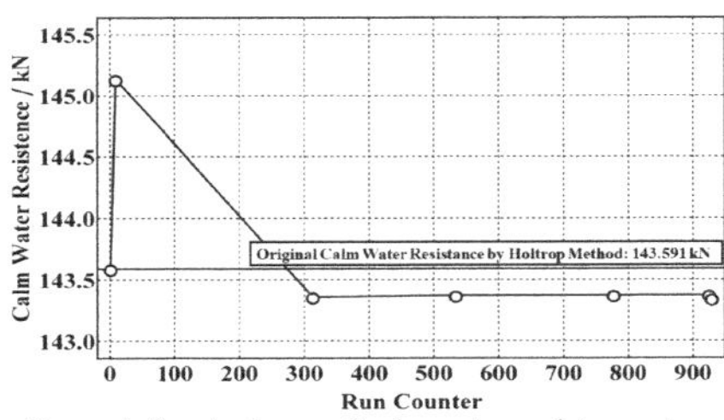

b. The variation tendency of calm water resistance at service Froude number $Fn = 4.0$, R_T, for Case 1

Figure 7. The variation tendency corresponding to the characteristic iteration points for the level 2 vulnerability criteria for surf-riding/broaching and the calm water resistance at service Froude number $Fn = 4.0$ for Case 1.

According to Figure 7, the critical number of revolutions of the propeller corresponding to the second threshold of surf-riding, n_{cr}, show a slight drop along with the optimization iteration. However, the decline has been controlled within –0.36%, which is still far away from the serious impacts on the intact stability. Fortunately, Figure 7b implies that the calm water resistance at service Froude number $Fn = 4.0$ decrease slightly to a percentage of –0.152%, which indicates that the calm water resistance performance has not been seriously affected by the increase of ship hull plumpness due to the ship hull form optimization procedures.

3.4 *Case 2: Optimization with the water plane changed*

The Case 2 is designed to evaluate the effect of the variation of ship hull form in the middle part of the ship with more rigorous restrictions on the bow shape, stern shape and longitudinal section in center plane Comparatively, the restrictions on the water plane has been replaced by series of movable control points. The transverse shifts of control points are allowed, therefore, more remarkable changes of the ship hull surfaces could be generated. The distribution of the fixed and movable control points along with the original ship hull of ITTC-A2 high-speed fishing vessel are shown in Figure 8a and Figure 8b.

Algorithm parameters, the stopping criterion, the range of variation for movable control points at each degree of freedom, ship displacement and GM constraints for Case 2 are shown in Table 6.

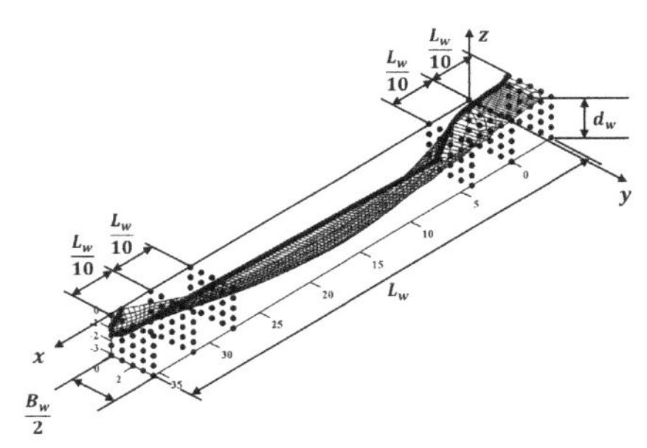

a. Fixed control points for Case 2

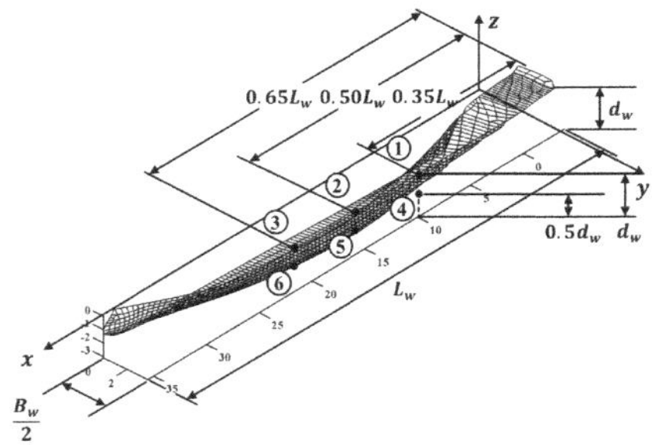

b. Movable control points for Case 2

Figure 8. Control points distribution along with the original ship hull of ITTC-A2 high-speed fishing vessel for Case 2.

Compared to Case 1, The limitation for variation of the ship displacement W in Case 2 is controlled within ±5%. Such an arrangement leaves enough margin for ship hull transformation during optimization procedures, and simultaneously, the general similarity of ship hull form schemes during optimization is basically guaranteed.

The results of the ship hull form optimization for Case 2 are shown in Table 7.

The history sequence chart of the optimization procedure and the contrast diagram of the ship hull lines between the initial and the optimum ship hull form for Case 2 can be shown in Figures 9 and 10.

The optimization history sequence charts in Case 2 demonstrate quite featured characteristics

Table 6. Settings of initial parameters and constraints in Case 2.

Description		Lower bound	Upper bound
Population size		30	
Number of generations		30	
Maximum iteration steps		930	
Movable point ①	Δy_1	–0.06	+0.06
Movable point ②	Δy_2	–0.06	+0.06
Movable point ③	Δy_3	–0.06	+0.06
Movable point ④	Δy_1	–0.06	+0.06
	Δz_1	–0.10	+0.10
Movable point ⑤	Δy_2	–0.06	+0.06
	Δz_2	–0.10	+0.10
Movable point ⑥	Δy_3	–0.06	+0.06
	Δz_3	–0.10	+0.10
Ship displacement, W	403.013 t	445.435 t	
Metacentric height, GM	0.85616 m	—	

Table 7. Results of ship hull optimization for Case 2.

Description		Outcome	Variation
Iteration time consumption*		81.8836 h	—
Run counter of optimum $\phi_{1/3}$		928 / 930	—
Movable point ①	Δy_1	+0.057481	—
Movable point ②	Δy_2	–0.049708	—
Movable point ③	Δy_3	–0.024582	—
Movable point ④	Δy_4	+0.045247	—
	Δz_4	+0.042230	—
Movable point ⑤	Δy_5	+0.048900	—
	Δz_5	+0.099410	—
Movable point ⑥	Δy_6	+0.048109	—
	Δz_6	+0.004159	—
Roll motion amplitude $\phi_{1/3}$		17.4357 deg	–15.140%
Ship displacement, W		441.561 t	+4.087%
Natural roll period, T_ϕ		6.646 s	+15.643%

*The iteration time consumption is based on the computing device with 3.00 GHz of CPU, 8.00 GHz of RAM and Microsoft Windows XP.

compared to Case 1. Figure 9a displays the variation tendency of the objective function and it turned out to be more difficult to reach the optimum solution since the optimization results appear mild after about 500 iteration steps, which is larger than the 200 iteration steps appeared in Case 1. Nevertheless, the optimization outcome shows a remarkable promotion percentage in Case 2. Simultaneously, along with the significantly improved optimization outcome, the ship displacement W turns out to be larger and closer to the artificial upper bound. The natural rolling period T_ϕ shows an obviously larger amplification, which becomes even closer to the synchronous rolling point at BF-6 representative sea condition.

Compared to Figure 5 for Case 1, Figure 10 for Case 2 demonstrates more significant deformation schemes. The hull surface near the water plane is narrowed at the fore subarea yet widened at the aft subarea in the middle part of the ship. The hull surface in the intermediate part along vertical direction embodies the tendency of outward expansion

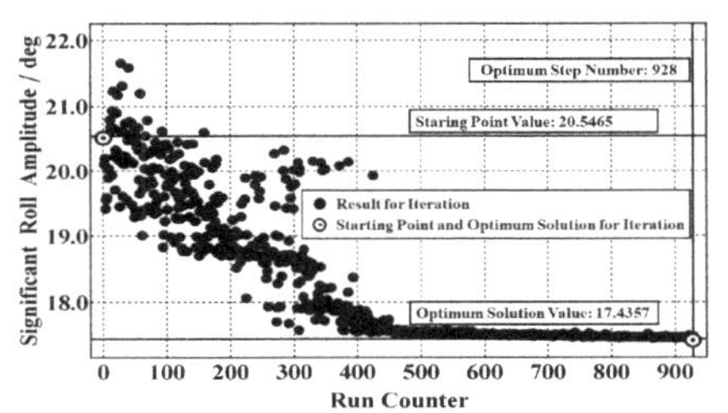

a. History sequence chart of $\phi_{1/3}$ for Case 2

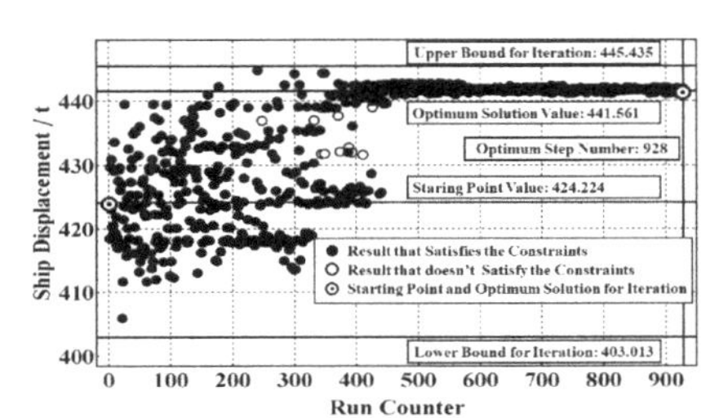

b. History sequence chart of W for Case 2

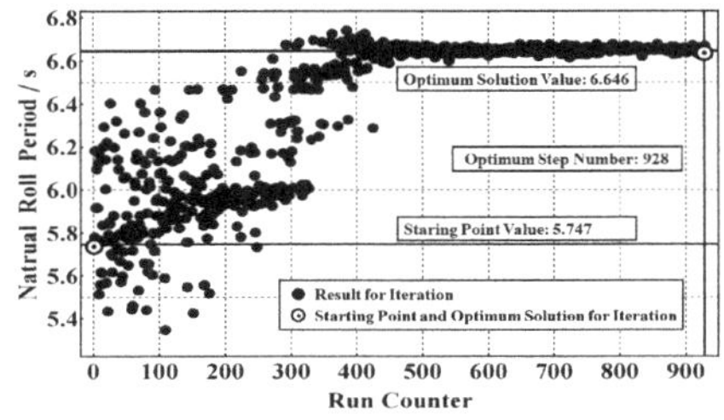

c. History sequence chart of T_ϕ for Case 2

Figure 9. History sequence charts of the optimization procedure on ITTC-A2 high-speed fishing vessel for Case 2.

and downward extension. Consequently, the ship hull deformation is controlled strictly by the artificially-defined range of variation for movable control points at each degree of freedom to ensure the general uniformity and smoothness of the hull surface for the engineering demands.

The convex hull of iteration results and its boundary points for Case 2 are shown in Figure 11.

For the chosen characteristic iteration points, the corresponding results for the level 2 vulnerability criteria for surf-riding and broaching, as well as the calm water resistance at service Froude number $Fn = 4.0$ are shown in Figure 12.

In comparison to Case 1, The critical number of revolutions of the propeller corresponding to the second threshold of surf-riding, n_{cr}, also show a slight drop along with the optimization iteration. Certainly, the decline has been controlled within –0.70%, which is also still far away from the serious impacts on the intact stability. Figure 12b shows that the calm water resistance at service Froude number $Fn = 4.0$ increases slightly during the optimization iterations. However, the calm water resistance performance corresponding to the optimum ship hull form increases only to the percentage of 0.039%, which demonstrates no obvious deterioration with the roll motion response modified prominently.

3.5 *Comparison analysis between Case 1 and 2*

The comprehensive performance of the ship hull form optimization procedure with the considera-

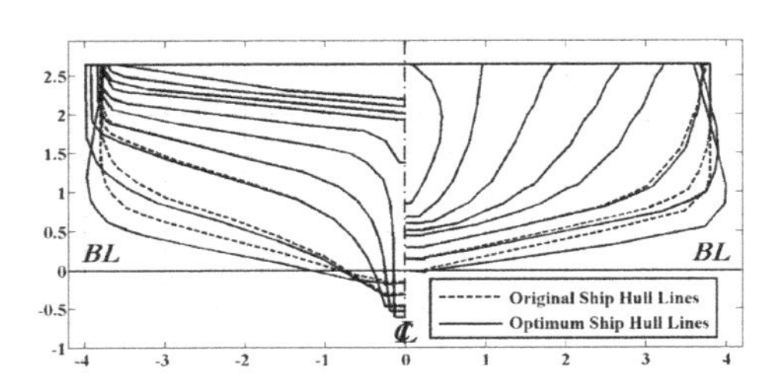

Figure 10. The contrast diagram of the ship hull lines between the initial ship hull form and the optimum ship hull form of ITTC-A2 high speed fishing vessel for Case 2.

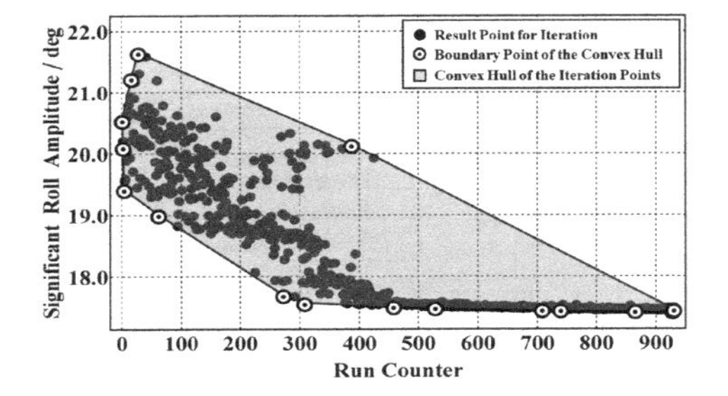

Figure 11. The convex hull of the iteration points and its boundary points for Case 2.

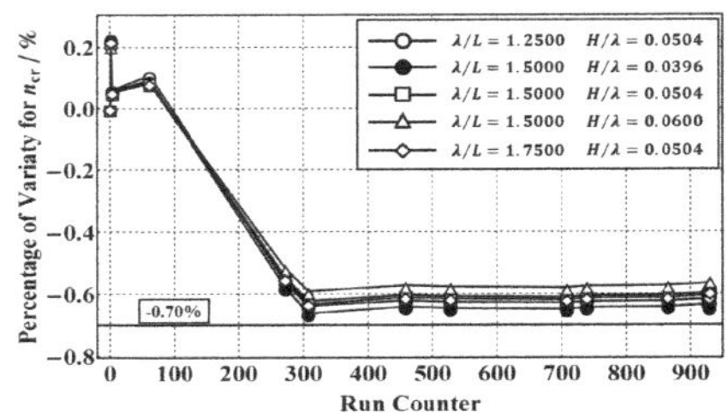

a. The variation tendency of n_{cr} for Case 2

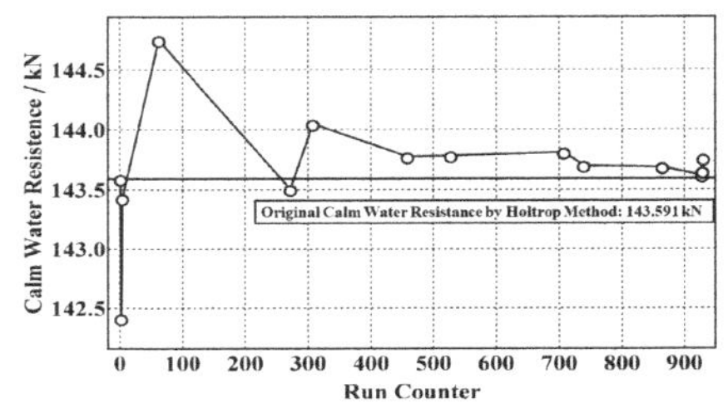

b. The variation tendency of calm water resistance at service Froude number $Fn = 4.0$, R_T, for Case 2

Figure 12. The variation tendency corresponding to the characteristic iteration points for the level 2 vulnerability criteria for surf-riding/broaching and the calm water resistance at service Froude number $Fn = 4.0$ for Case 2.

tions of computing resource consumption, objective function, monitoring variables, as well as the verification and confirmation of the level 2 vulnerability criteria for surf-riding/broaching and the calm water resistance at service Froude number $Fn = 4.0$ demonstrates both definite generalities and some sort of difference.

Firstly, NSGA-II algorithm, combined with the RBF ship hull transformation technique has been proved to have abundant accuracy, efficiency and robustness. The iterations accomplished their convergences quickly and promptly. The optimization outcome targeting at the roll motion response for ITTC-A2 high-speed fishing vessel in beam seas with zero speed has reached –5.455% and –15.140% respectively for Case 1 and Case 2, which deserve to be qualified in conforming to the engineering application. Along with the optimization procedure, there are many monitoring variables considered. In both cases, the ship displacement W grows and gradually approaches the artificial upper bound indicating the higher plumpness and economic efficiency. Although the increasing natural roll period T_ϕ gradually approaches the synchronous rolling point, the roll motions for both cases are still effectively weakened by the increase of rolling damping.

Secondly, the level 2 vulnerability criteria for surf-riding and broaching deserves to be a simple and efficacious mathematical model in evaluating the intact stability failure corresponding to the second threshold. The critical number of revolutions of the propeller where the surf-riding is inevitable,

n_{cr}, in both cases demonstrates only weak drop within only -1%. Performance on the level 2 vulnerability criteria for surf-riding and broaching proves that the ship hull optimization attached barely limited influence on the intact stability behavior for ITTC-A2 high-speed fishing vessel, which fully conforms to the engineering practice.

Thirdly, the corrected Holtrop method is proved to be with qualified precision and practicability. Although the calm water resistance at service Froude number $Fn = 4.0$ corresponding to the optimum results decreases for Case 1 but increases for Case 2, the ranges of variation are not significant enough for both cases. With remarkable optimization on rolling motion performance and the accompanying obvious increase on ship displacement and body plumpness, such slight deterioration on calm water resistance performance embodies strong engineering applicability.

Technically, the technique of contributing fixed and movable control points appropriately can effectively avoid the numerical divergence during RBF transformation and accomplish satisfying ship hull surface reconstruction. Except for placing the fixed control points on the nodes based on the grid model generated and visualized by professional software tools, the chessboard type and shell-frame type of fixed control points that uniformly distributed within the transverse planes are also effective in controlling the ship hull surface with the consistency and smoothness guaranteed. Such types of fixed control points are shown in Figure 3a and Figure 8a.

The differences of optimization results from Case 1 and Case 2 are basically originated from different strategies installed for the ship hull transformation. Case 1 represents a relatively conservative strategy that controls the water plane tightly and only enables the movable control points to shift along vertical direction. Such strategy brings about relatively smooth ship hull transformation without probable remarkable distortions. Case 2 deserves to be the

Table 8. Comparative analyses between optimization variation results of ship hull optimization for Case 1 and Case 2 (in percentage).

Symbol	Case 1	Case 2
$\phi_{1/3}$	−5.455%	−15.140%
W	+1.999%	+4.087%
T_ϕ	+3.811%	+15.643%
n_{cr}*	Deteriorated, −0.36%	Deteriorated, −0.70%
R_T	−0.152%	+0.039%

*The percentage of deterioration for n_{cr} stands only for a general range with the consideration of all combinations of wave steepness H/λ and wave length-ship length ratio λ/L_{pp} in following seas mentioned in this research. Such combinations of parameters can also be found in Figure 7a and Figure 12a.

demonstration of a relatively radical strategy that removes restrictions on the water plane and enables the transverse shifting of all movable control points. Although the deformation of the ship hull form presents to be more significant, the general consistency and smoothness, as well as the intact stability and calm water resistance performance can be basically guaranteed. Strategically, a narrowed water plane combined with an outward expanded and downward extended intermediate body probably leads to more significant ship hull form optimization outcomes.

4 CONCLUSION

In this paper, a systematical and complete ship hull form optimization method targeting at the improvement of the roll motion with the monitor of the surf-riding/broaching and calm water resistance performance applied to ITTC-A2 high-speed fishing vessel is proposed.

The NSGA-II algorithm governs the optimization procedure carried in the iSight optimization platform with platitudinous generalization, efficiency and robustness. Such a ship hull form optimization method accomplished satisfying outcomes on improvements of ship roll motion response with only limited impacts on surf-riding and calm water resistance performances. The construction of convex hulls for iteration points represents an efficient approach for selecting appropriate iteration point embodying the general variation tendency. The utilization of the convex hull is an important indicator of the degree of iteration convergence and efficiency.

The paper suggests the fixed control points to be placed on the nodes of longitudinal section in center plane and placed as the chessboard type and shell-frame type within transverse planes around the bow and stern. In contrast, this paper recommends a relatively radical strategy that removes the restrictions on water plane and enables the transverse shifting for movable control points for the convenience for the optimization of roll motion.

A narrowed water plane combined with an outward expanded and downward extended intermediate body along vertical direction are effective for the roll motion optimization with the intact stability and calm water resistance also qualified for comforting to engineering applications. Constraints should be set appropriately leaving enough room for ship hull form transformation but avoiding abnormal geometric distortions that violates the engineering demands.

ACKNOWLEDGEMENT

This research is funded and supported by the China Ministry of Education Key Research Project

KSHIP-II Project (Knowledge-based Ship Design Hyper-Integrated Platform) No. GKZY010004. Authors are also thankful for Mr. Dianfei Wang, Mr. Zhenping Huang, Mr. Shaoze Luo, Mr. Gangcheng Wang, Mr. Si Chen and Mr. Chengyuan Ma for their outstanding contributions and expertise assistance in many ways.

REFERENCES

Berg, M.D., Cheong, O., Kreveld, M.V., et al. 2008. *Computational Geometry: Algorithms and Applications-Third Edition.* Berlin Heidelberg: Springer-Verlag.

Censor, Y. 1977. Pareto optimality in multiobjective problems. *Applied Mathematics & Optimization* 4(1): 41–59.

De Boer, A., Van der Schoot, M.S., & Bijl, H. 2007. Mesh deformation based on radial basis function interpolation. *Computers & Structures* 85(11): 784–795.

Deb, K., Pratap, A., Agarwal, S., et al. 2002. A fast and elitist multiobjective genetic algorithm: NSGA-II. IEEE Transactions on Evolutionary Computation 6(2): 182–197.

Gammon, M.A. 2011. Optimization of fishing vessels using a Multi-Objective Genetic Algorithm. *Ocean Engineering* 38(10): 1054–1064.

Goshtasby, A. 1989. Correction of image deformation from lens distortion using Bezier Patches. *Computer Vision, Graphics, and Image Processing* 47(3): 385–394.

Holtrop, J. 1977. Statistical analysis of performance test results. *International Shipbuilding Progress* 24(270).

Holtrop, J., & Mennen, G.G.J. 1978. A statistical power prediction method. *International Shipbuilding Progress* 25(290).

Holtrop, J. 1982. An approximate power prediction method. *International Shipbuilding Progress* 335(335): 166–170.

Holtrop, J. 1984. A statistical re-analysis of resistance and propulsion data. *International Shipbuilding Progress* 31(363): 272–276.

Huang, F., Wang, L., & Yang, C. 2015. Hull form optimization for reduced drag and improved seakeeping using a surrogate-based method. In: *Proceedings of The Twenty-fifth International Ocean and Polar Engineering Conference. International Society of Offshore and Polar Engineers.*

Huang, J.F. 2012. The hull variation and optimization by the improved Lackenby method. *Ship & Ocean Engineering* 41(4): 54–57.

Huang, Z.P., Ma, N., & Gu, X.C. 2013. Studies of collaborative optimization of pentamaran main hull and demihulls based on motions and added resistance in waves. *Ship Engineering* 35(3): 6–9.

IMO MEPC.1/Circ.796, 2012. Interim guidelines for the calculation of the coefficient f_w for decrease in ship speed in a representative sea condition for trail use. *London, UK.*

IMO MEPC65/INF.21, 2013. Air pollution and energy efficiency—draft alternative interim guidelines for the submission for the coefficient f_w for decrease in ship speed in a representative sea condition. *Submitted by China.*

IMO SDC2/WP.4, 2015. Development of second generation of intact stability criteria report of the working group (Part 1). *London, UK.*

IMO SDC4/5/1/Add.2, 2016. Finalization of second generation intact stability criteria. *Submitted by Japan.*

Kim, H. 2009. Multi-objective optimization for ship hull form design. *Doctoral Dissertation, George Mason University, USA.*

Kim, H., & Yang, C. 2010. A new surface modification approach for CFD-based hull form optimization. *Journal of Hydrodynamics Ser. B* 22(5): 520–525.

Koutroukis, G., Papanikolaou, A., Nikolopoulos, L., et al. 2013. Multi-objective optimization of container ship design. *Developments in Maritime Transportation and Exploitation of Sea Resources: IMAM 2013*: 477–489.

Luo, S.Z., Ma, N., Hirakawa, Y., et al. 2016. Experimental and numerical study of wind drag of large containership by using open wind test in towing tank. *Journal of Shanghai Jiao Tong University* 50(3): 389–394.

Mata-Álvarez-Santullano, F., & Souto-Iglesias, A. 2014. Stability, safety and operability of small fishing vessels. *Ocean Engineering* 79: 81–91.

Meng, X.X. & Xu, N. 2002. A survey of the research works on parametric design. *Journal of Computer-aided Design and Computer Graphics* 14(11): 1086–1090.

Morris, A.M., Allen, C.B., & Rendall, T.C.S. 2008. CFD⊠ based optimization of aerofoils using radial basis functions for domain element parameterization and mesh deformation. *International Journal for Numerical Methods in Fluids* 58(8): 827–860.

NAOE Osaka University, 2015. Sample ship data sheet: ITTC A2 fishing vessel. Available at: http://www.naoe.eng.osaka-u.ac.jp/imo/a2.

Pecot, F., Yvin, C., Buiatti, R., et al. 2012. Shape optimization of a monohull fishing vessel. In: *12th International Conference on Computer and IT Application in the Maritime Industries, Liege*: 7–18.

Pinto, A., Peri, D., & Campana, E.F. 2007. Multiobjective optimization of a containership using deterministic particle swarm optimization. *Journal of Ship Research* 51(3): 217–228.

Rendall, T.C.S., & Allen, C.B. 2008. Unified fluid–structure interpolation and mesh motion using radial basis functions. *International Journal for Numerical Methods in Engineering* 74(10): 1519–1559.

Sederberg, T.W., & Parry, S.R. 1986. Free-form deformation of solid geometric models. *ACM SIGGRAPH Computer Graphics* 20(4): 151–160.

Srinivas, N., & Deb, K. 1994. Muiltiobjective optimization using nondominated sorting in genetic algorithms. *Evolutionary Computation* 2(3): 221–248.

Tong, J. 2009. Expressing and obtaining Pareto solutions in multi-objective optimization problem. *Master Dissertation, Wuhan University of Science and Technology, China.*

Wang, D.F., Ma, N., & Gu, X.C. 2011. Study on multi-objective optimization of pentamaran demihull layout based on genetic algorithm. *Ship Engineering* 33(5): 56–59.

Wang, G.C., Ma, N., & Gu, X.C. 2017. Research on the energy efficiency design index optimization of ultra large container ship. *Ship Science and Technology* 39(5): 60–64.

Yu, L.W., Ma, N., & Gu, X.C. 2016. On the mitigation of surf-riding by adjusting center of buoyancy in design stage. *International Journal of Naval Architecture and Ocean Engineering* 9(3): 292–304.

Propulsion equipment design

Marine Design XIII – Kujala & Lu (Eds)
© 2018 Taylor & Francis Group, London, ISBN 978-1-138-34076-3

The journey to new tunnel thrusters, the road so far, and what is still to come

N.W.H. Bulten
Wärtsilä Propulsion, Drunen, The Netherlands

ABSTRACT: Bow thruster design has not been in the spotlights for a long time. The increasing size of cruise vessels has led to a demand for larger, more powerful units, with strict requirements on the noise and vibration side. Due to these conflicting requirements it has only been possible to make a proper step ahead, by proper research. In the paper 'the journey towards the new generation tunnel thrusters' will be described, where increased power ratings have been achieved in combination with reductions of noise and vibrations. With the aid of full scale RANS CFD flow simulations, the inflow characteristics of the thrusters have been visualized and, based on that understanding, enhancement of the inflow has been established which contributes to clear reduction in load variations during a revolution of the propeller. The outlook of the developments which will come in the near future take into account an even wider design envelope, where performance in maneuvering mode will be addressed too. Based on all new insights it has become clear that the journey has not ended yet.

1 INTRODUCTION

Design of propulsion systems has been focused for a long time on the aspects of main propulsion. Research on increased fuel efficiency and energy saving devices has been brought to the attention of the maritime industry in various places (Mewis & Guiard, 2011; Zondervan *et al*, 2011; Schuiling, 2013).

The following step, in improving the operational aspect of the vessels, is in the direction of improved maneuverability of the vessels. Usage of azimuthing thrusters for main propulsion has introduced a new method for maneuvering the vessel, though it still acts on the aft part of the vessel (Toxopeus & Loeff, 2002). So, there is still quite some resemblance with propulsion systems based on shaft-line driven propeller with rudder. In order to get better control over the bow of the vessel, the design of the tunnel thrusters has been revisited.

At the authors company a new series of tunnel thrusters has been developed based on the operational demands of today. In order to reach the demanding targets, the use of modern tools is essential to come to the optimal balance between mechanical design and hydrodynamic performance of the unit. Another important aspect is the integration of the tunnel into the vessel structure. The interfacing between the hull and the tunnel has proven to be of significant importance for the inflow uniformity to the thrusters. In case the flow is significantly disturbed, e.g. due to severe flow separation, load fluctuations will be a direct consequence, which are experienced through vibrations and noise.

2 MARKET-PULL AND TECHNOLOGY PUSH

2.1 *Market pull*

The increase of the ship size has led to an increased demand for generation of side forces to control the vessel. This trend of increasing ships can be seen in the merchant sector (container vessels), and in the leisure market (cruise vessels). For both operations, the maneuverability in ports, and the time required to maneuver are an important aspect. The dimensions of the super structures above the waterline, are subjected to the wind forces which can lead to large side forces acting on the vessel. The side way movement of the vessel towards the quay, often denoted as crabbing mode, require sufficient side force to accelerate the vessel into a sideway direction and to counteract the wind forces acting on the vessel. Increase of the vessel sizes evidently lead to increased side thrust requirements.

Due to the growth of the vessels, the relative size of the port entrance has actually been reduced. Port entrances which could be made with shorter, more slender vessels in the past, become more challenging for the larger vessels. The conventional tunnel thruster concept is known for its reduced performance, when the vessel is sailing at low speed (around 5 knots). In order to improve the maneuverability of the vessel, the side force generation needs to be improved significantly.

Therefore, the market requests a maximum thrust for given ship draft in crabbing mode and possibly a new concept to improve the performance of the tunnel thruster, when sailing forward during maneuvering.

2.2 *Technology push*

The market demands are challenging for the development of new tunnel thrusters. However, the usage of modern analysis tools and methods provides new ways to reach the targets. Usage of Computational Fluid Dynamics (RANS-CFD) in maritime industry, has made good progress over the last decade and, with the visualization of the occurring flow phenomena, the understanding has been increased significantly. An example of the mesh of the tunnel thruster configuration with the hull geometry, the tunnel entrance details and the thruster unit, including the propeller blades is shown in Figure 1. Details of the applied CFD method have been presented before (Bulten, 2015). For the current research one of the key elements is the implementation of the actual propeller blade geometry with its transient moving mesh solution approach.

An important aspect of the use of CFD is the elimination of scale effects in case the actual full scale geometry is analyzed. In the rare case, that thruster units are tested at model scale, often units with propeller diameters in the range of 10–20 cm are used. The Reynolds numbers for those units are significantly lower than actual installations, and therefore the model scale flow phenomena are not representative for the actual thrusters (Bulten & Stoltenkamp, 2017).

The CFD simulations allow to make time-dependent analyses of the load fluctuations on a single blade during a revolution. The load fluctuations can be coupled directly to the observed inflow pattern.

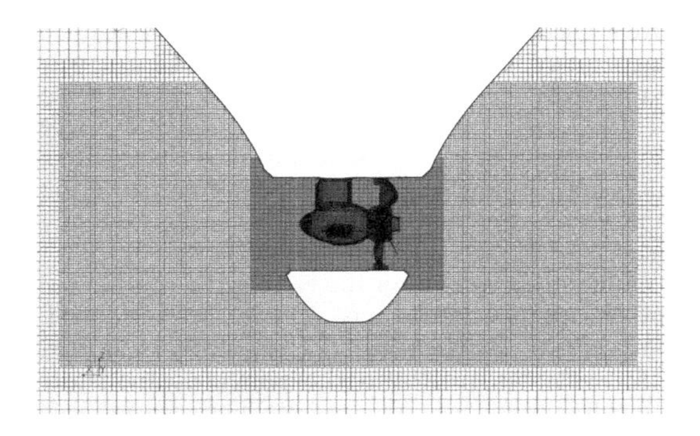

Figure 1. RANS CFD mesh for tunnel thruster analyses, including propeller blades and hull geometry.

In case of flow separation, low inflow velocities are observed, which results in increased loading. The origin of the flow separation is often a too sharp edge in the geometry, which can be detected easily from the CFD results. Another aspect which can be addressed when reviewing the calculated pressure distributions is the development of cavitation patterns on the thruster.

As part of the development of the thruster series, full scale CFD simulations have been made for the determination of the thruster performance (thrust/power ratio).

2.3 *Environmental limitations*

The solution to increase the side force generation, which has been successfully applied so far, was to install more power and/or larger installations on the vessels. On the Oasis-class for example, 4 units of 5500 kW, with 4.0 meter propellers have been installed in the bow. A simple sketch is shown in Figure 2. With the current dimensions of the tunnel thrusters the limits of the units related to the applied vessel draft, become into play. These environmental limitations will be explained in more detail in this section.

The position of the tunnel thruster is bounded by two criteria: the distance of the tunnel center line with respect to the baseline, which is about 1.0–1.5 diameter and the distance to the water surface. The minimum height above the baseline is needed to create sufficient tunnel length for the unit, given a V-shaped bow section.

On the other hand, the distance between the tunnel and the waterline (denoted with WL-margin) should be sufficient to avoid air suction into the tunnel. A calculation method, for the required water height above the tunnel, is provided by Brix (1993), where the water height is based on the diameter and the generated thrust, which is related to the applied power:

$$\frac{e}{D} = \frac{0.70}{D}\sqrt{\frac{P \cdot \frac{T}{P}}{\rho \cdot g \cdot \pi \cdot D}} \tag{1}$$

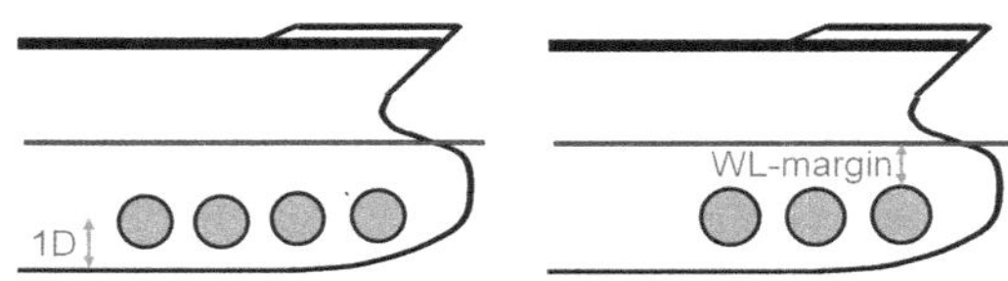

Figure 2. Comparison of lay-out with 4 and 3 tunnel thrusters.

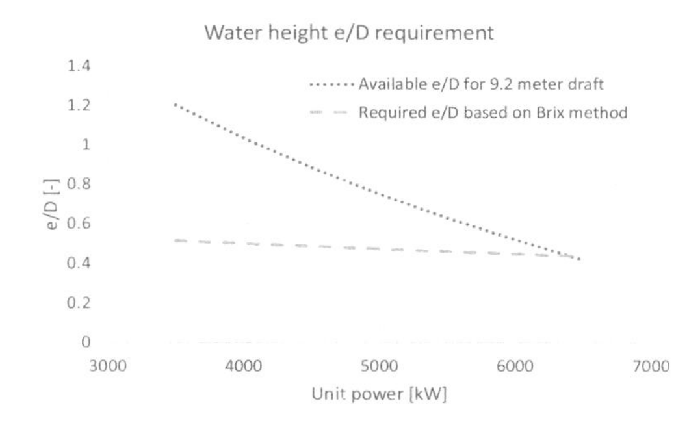

Figure 3. Comparison of required and available water height for vessel with 9.2 meter draft.

where e/D = dimensionless water height; P = engine power; T/P = thrust-power ratio; ρ = water density. The thrust-power ratio T/P is typically about 0.15. Moreover, the relation between the tunnel diameter and the unit power, can be derived from the power-density of the thruster. As a first estimate the power-density of the tunnel thruster series can assumed to be constant. As a consequence, the unit diameter is directly related to the installed power. For a given vessel draft and target power-density, the required and available dimensionless water height e/D values can be calculated. Figure 3 shows an example for a vessel operating at 9.2 meter draft.

The available distance e/D decreases with the installed power, due to the increase of the unit diameter. This evaluation shows that there is a theoretical physical limit in the allowed power per thruster unit, for given vessel draft, which is found at the intersection of both lines. When some additional margins are taken into consideration, it can been understood that the units operating at 5500 kW and 9.2 meter draft are already close to the limits.

This quick evaluation shows that 22 MW can be installed in the bow of a vessel when 4 units are selected. The alternative solution of 3 more powerful units, as shown in Figure 2, will not be feasible due to the increased size of the units and the reduced waterline-margin. In case even more side thrust is required, selection of larger units will neither be the solution. Therefore it can be concluded that the given draft of a vessel in combination with required thrust forces has become a critical design point.

3 DESIGN CHALLENGES OF TUNNEL THRUSTERS

The challenge in the tunnel thruster design is mainly found in the fact that the unit has to operate in both directions. Normal main propulsion units, and azimuth thrusters, have a preferred operating direction for which the performance can be optimized. The tunnel thruster is an exception, and it requires a more or less symmetrical solution.

One way to reach the thrust to both directions is the use of a controllable pitch propeller (CPP), which can be set either in ahead or astern pitch (as shown in Figure 4). With constant propeller shaft speed, the magnitude and direction of the thrust can be controlled. Second alternative is the use of a fixed pitch propeller (FPP) where the magnitude and direction of thrust is governed by the rotational speed and direction of the propeller.

From a hydrodynamic point of view, the design of tunnel-thruster propellers may seem simple. However, due to the requirement for two-way operation, there are some major constraints in the propeller design: (i) for controllable pitch propellers, the blades are so-called flat plates, without any pitch distribution, (ii) whereas for a fixed pitch propeller, there will be no camber used in the profile sections to keep the profile sections symmetrical.

In order to achieve the best hydrodynamic performance of the installation, a large size thruster would be selected. But, given the limited time of operation, the impact of the tunnel openings on the vessel drag in transit condition needs to be considered too. A sub-optimal tunnel thruster unit can contribute to an overall reduced fuel consumption figure for the complete vessel. A second aspect, which needs to be considered, is the generation of

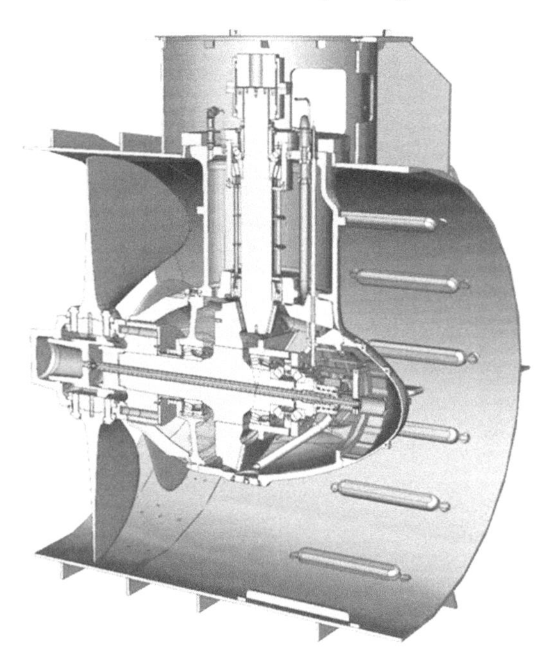

Figure 4. Mechanical layout of tunnel thruster with bevel gearbox and controllable pitch propeller.

noise and vibrations. In the cruise vessel segment, some thrust efficiency can be sacrificed to reduce the vibration levels which lead to improved comfort on board. This focus, on improved comfort on board may become a topic for other vessels too.

4 INTEGRATED DESIGNS

4.1 *Mechanical and hydrodynamic integration*

The fact that the drive line of a tunnel thruster contains a bevel gear set (as shown in Figure 4), results in conflicting design preferences. From a mechanical point of view, a relative low torque and high rotational speed of the gear wheels, is preferred. On the other hand, the hydrodynamic demands are aiming at a low rotational speed and consequently higher torques to absorb all power. Reduced propeller RPM is in general beneficial for the propulsive efficiency.

Another important aspect in modern tunnel thruster is the noise level. Noise reduction can be achieved with reduced power density and reduced blade tip speeds, as shown in Figure 5. These diagrams have been based on measurements by the Delft Institute of Applied Physics (TPD) (Buiten & Regt, 1983).

In order to reduce the power density, the propeller diameter needs to be increased. In case the tip speed is supposed to be reduced as well, the propeller rotational speed has to be reduced quite significantly to compensate for the increase of the propeller diameter. Design of a low noise tunnel thruster requires therefore a properly balanced design, where the hydrodynamic requirements, and the gear wheel selection, are matched to transfer the total power to the propeller. The shaft torque is determined by the propeller rotational speed, which is related to the desired tip speed.

A second design feature, which is important for the propeller noise, is the cavitation behavior. The amount of cavitation can be controlled to a certain extend with the selection of large propeller blade area ratios (BAR). For CPP blades, the maximum blade area is limited by the fact that the blades should be able to pass each other when changing pitch from ahead to astern (or in case of tunnel thrusters from port to starboard thrust).

In Table 1 a comparison is made of the main hydrodynamic parameters of a modern tunnel thruster compared with an existing unit. This table shows that the modern unit can absorb 20% more power, which can be translated into approximately 31% more side thrust. Due to the selection of the propeller diameter the power density of the unit is even reduced. Also the tip speed has been reduced significantly which could be achieved with the selected bevel gear set with increased gear ratio. Even with 20% extra power on the unit the overall noise level will be lower. The impact of the power increase on the noise level will be cancelled by the positive contributions from reduced power density and tipspeed.

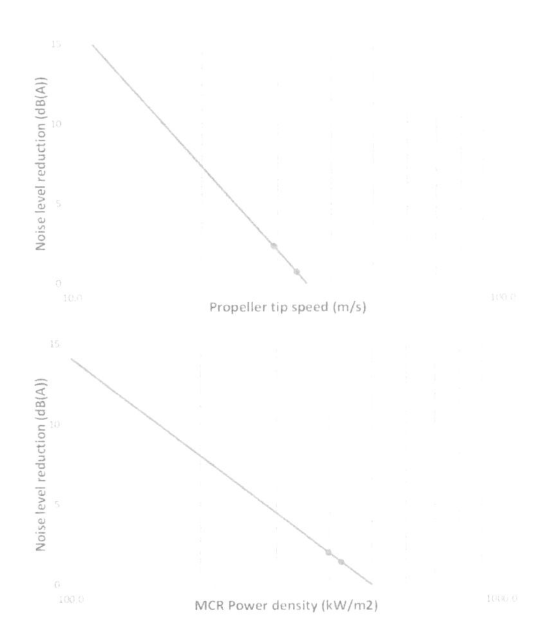

Figure 5. Impact of propeller tipspeed and power density on noise levels.

4.2 *Fixed pitch vs controllable pitch propellers*

The power absorption, and thus the thrust generation of a propeller, can be controlled in different ways. Variation of the propeller rotational speed has a direct impact on power absorption for propellers with fixed and controllable pitch. A propeller with controllable pitch has a second control option with the variation of the pitch deflection angle in operation. A tunnel thruster, with controllable pitch propeller, can thus be operated at constant propeller shaft speed over the complete range of demanded thrust values from + 100% to −100%

Table 1. Comparison of main hydrodynamic parameters of new and existing tunnel thruster units.

	New WTT-40	Existing CT300	Ratio
Power [kW	3600	3000	120%
Diameter [m]	3400	3000	113%
Prop RPM [1/min]	165	211	78%

(or port to starboard). Operation at a constant speed has a large advantage for the layout of the driveline concept where a constant RPM E-motor can be used.

In case a propeller with fixed pitch is used, then the RPM needs to be controlled with a Variable Frequency Drive (VFD), to drive the unit to either +100% RPM or –100% RPM.

Detailed transient simulations, of the performance of a tunnel thruster with fixed pitch and controllable pitch propellers in operation, have been presented before (Boletis *et al*, 2015a, 2015b). This analysis showed that the performance of a fixed pitch propeller exceeds the performance of a controllable pitch propeller by about 7% at full power (see Figure 6). This is among others attributed to the smaller propeller blade tip clearance with the tunnel.

At part load, the differences between the two configurations become even more pronounced. Figure 6 shows the example case, where the thrust demand is 50%. For such condition, the unit with fixed pitch propeller can run at 70% of the maximum RPM, absorbing about 35% power. The unit, with controllable pitch propeller, remains at 100% RPM and it absorbs 44% power. In this situation the difference in power consumption has increased to more than 25%.

The differences in operation are not only beneficial for the fuel consumption, but even more so on the noise generation. This is driven by two phenomena which are both in favor of the fixed pitch propeller: (1) reduced propeller RPM and thus tip speed at part load, (2) improved cavitation behaviour due to the differences in radial pitch distribution. The impact of RPM has been shown in Figure 6. Reduction to 70% of the tip speed reduces the noise level with 4-5 dB(A).

The constraints on the radial pitch distribution will be explained in more detail. The hydrodynamic blade design of a tunnel thruster propeller is very much determined by the requirement of bi-directional operation. The blades for controllable pitch propellers are designed with zero pitch over the span from hub to tip. As a consequence, the pitch ratio increases linear from hub to tip in case a certain pitch deflection angle is applied. This is shown in Figure 7 for two different pitch deflection angles of 15° and 30°.

The differences in radial pitch distribution compared to a fixed pitch propeller are obvious. From this diagram it also becomes clear that the hydrodynamic design of a fixed pitch propeller has much more flexibility with respect to radial pitch distribution. This flexibility can be used to vary the radial loading distribution and consequently the cavitation behavior of the propeller. Reduction of the cavitation has a positive impact on the noise levels, which can be added to the earlier mentioned 4-5 dB(A).

4.3 *Tunnel hull integration*

Even in case a tunnel thruster unit is selected with low tip speed and fixed pitch propeller, there is still a risk for unwanted high noise and vibration levels. This is related to another critical issue for optimum overall performance of the tunnel thruster: the uniformity of the inflow to the units. In case the inflow velocity to the unit varies significantly during a revolution, the forces will fluctuate which will result in overall vibration levels. This phenomenon is of course known from propellers which are operating in a ship wake field. For tunnel thrusters often the inflow field is assumed to be quite uniform. Detailed flow simulations with CFD, have indicated that the impact of the actual hull shape, and the local details near the tunnel interfacing, are important. In order to get better insights in the occurring flow phenomena and the dependence between main hull parameters and the risks for flow separation, a parametric 3D CAD model of a tunnel geometry has been developed.

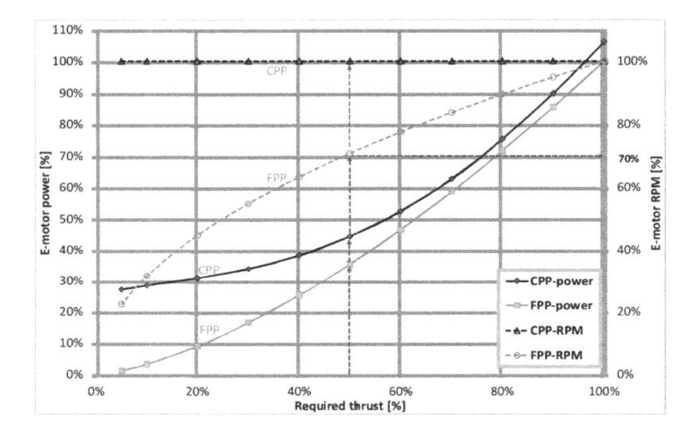

Figure 6. Performance comparison between tunnel thruster with fixed pitch and controllable pitch propeller.

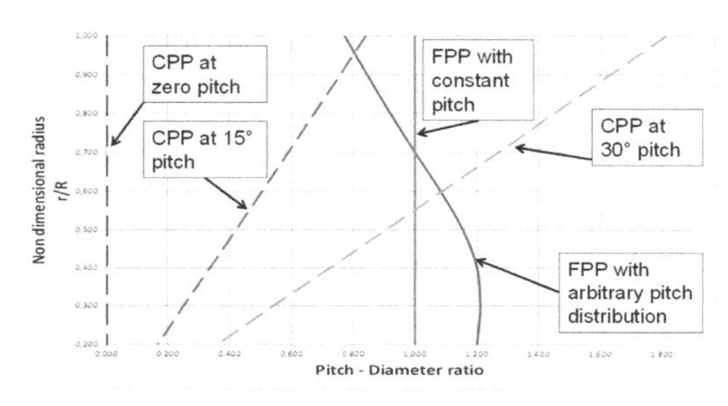

Figure 7. Radial pitch distributions for CPP and FPP installations, which reflect the typical pitch distributions of CP-blades depending on the pitch deflection angle.

The main lessons learned from the research are that V-shaped hulls are more prone to flow separation in the lower segment due to the stronger change in flow direction when the water enters into the tunnel. Since the flow separation near the entrance is the key issue, which needs to be avoided, proper rounding of sharp edges (indicated with the arrow in the picture) is necessary. Boletis *et al.* (2015a) have shown that proper rounding of the tunnel entrance can reduce the load fluctuations by a factor 2. As a consequence the cavitation behavior improves significantly, which results in lower noise and vibration levels.

Recent research has learned that the proper rounding not only improves the noise and vibration levels, but also the thrust generation. The alignment of the flow in the tunnel, leads to a larger flowrate at given input power which results in improved thrust.

Figure 9 shows the streamlines of two tunnel variants, with chamfer and rounded interface between the hull and the tunnel. Although the flow looks quite similar, there are differences in the flow rate through the tunnel. Due to this difference in flow rate, the total side thrust generation differs where the configuration, with aligned flow, resulted in more thrust. In both cases, some flow recirculation can be observed aft the propeller hub which is a consequence of the radial pitch distribution of a propeller with controllable pitch blades (as shown in Figure 7). This region of recirculating flow is not a main concern for performance or noise.

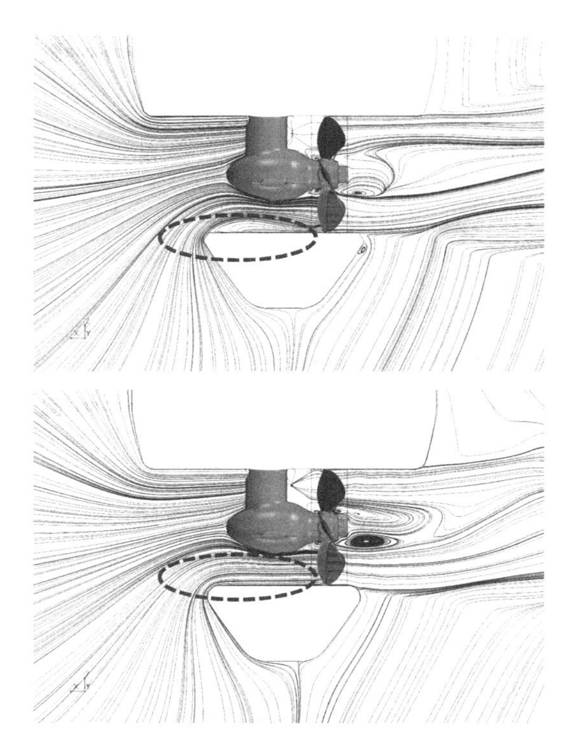

Figure 9. Flow lines for tunnel entrance geometry with chamfer and rounded interface.

4.4 *Concluding remarks*

The research so far, on the occurring flow phenomena of tunnel thrusters, has shown a number of points of attention to reach proper performance. First of all, the design of the thruster unit should be a balanced design taking the hydrodynamic and mechanical demands into account. This means that the power density and propeller tip speed of the units are selected in such way that propulsive performance is in line with market demands, and that noise levels are at minimum level. On the other hand, the selected propeller shaft RPM and torque should be delivered by the bevel gear set without compromises on structural integrity. Selection of a fixed pitch propeller means extra efforts on the driveline side with a VFD. The extra costs provide lower power consumption of the units in full load and even more in part load. Also the noise levels will be lower which is in general appreciated by the cruise vessel passengers. Finally, the importance of the interface between the hull geometry and the tunnel cannot be underestimated. In case the geometry has sharp edges, with flow separation, the load fluctuations will create unexpected cavitation behavior and consequently higher noise and vibration levels. As long as the local interface details between the tunnel and the hull are reviewed in the early design phase with dedicated CFD simulations, the risks for flow separation can be mitigated easily.

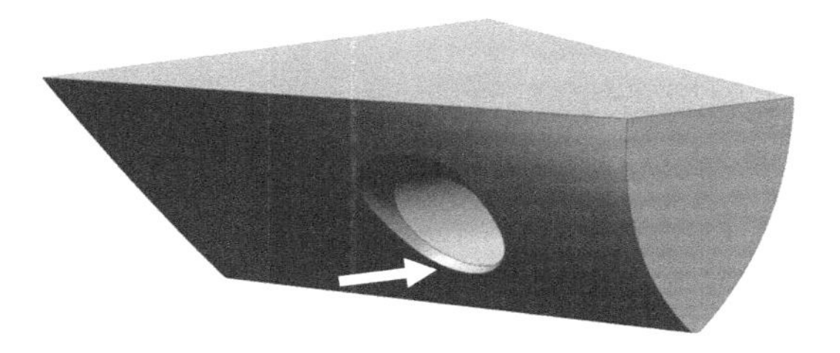

Figure 8. Example of 3D CAD geometry for parametric studies.

5 NEW DEVELOPMENTS

5.1 *Improved maneuvering capability*

In the previous section the design of the tunnel thruster unit is discussed with respect to the performance in crabbing mode which means side way movement of the ship from or to the quay. Once the vessel has moved sufficiently far from the quay, the maneuvering out of the harbor starts. This happens at relatively low forward ship speed when compared to the actual transit sailing speed.

It is known from the conventional tunnel thruster designs, that the performance at around 5 knots forward speed is dropped significantly. This is shown in the sketch of Figure 10, which has been taken from Baniela (2009).

The area of positive pressure on the entry side (marked with 2) and the area with negative pressure on the exit side (marked with 1), are both contributing to the reduced side thrust generation at forward speed. Moreover, the flow rate through the tunnel is affected by the ship's speed, which results in reduced thrust forces on the unit.

The decay in performance, due to forward ship speed, is presented in Figure 11, where the thrust/power ratio is shown as function of the jet-velocity ratio. This jet-velocity ratio increases with increasing forward speed.

For the crabbing mode, when there is no forward speed, a typical T/P-ratio for a tunnel thruster is about 0.13 (including all thruster-hull interaction effects). When the ship speed increases to about 5 knots, the T/P-ratio drops to about 0.05. The theoretical line, shown in the figure, indicates that the jet-velocity needs to be increased for given ship speed to reduce the velocity ratio V_{ship}/V_{jet}. To improve the performance in maneuvering mode, the velocity of the jet downstream of the thruster needs to be increased. This can be achieved with a reduced opening of the exit area of the tunnel.

5.2 *Concept solution*

The idea, to reduce the exit area of the jet flow, seems simple to achieve as long as a single direction operation is considered. This concept has been successfully applied for example in all water-

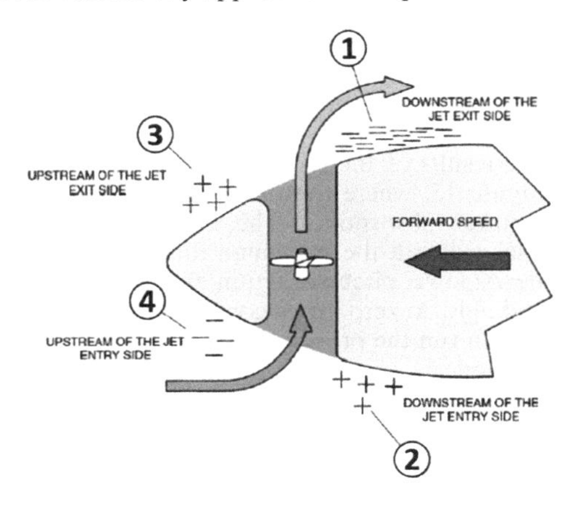

Figure 10. Pressure distribution around bow section for active tunnel thruster operating in forward ship speed (from Baniela, 2009).

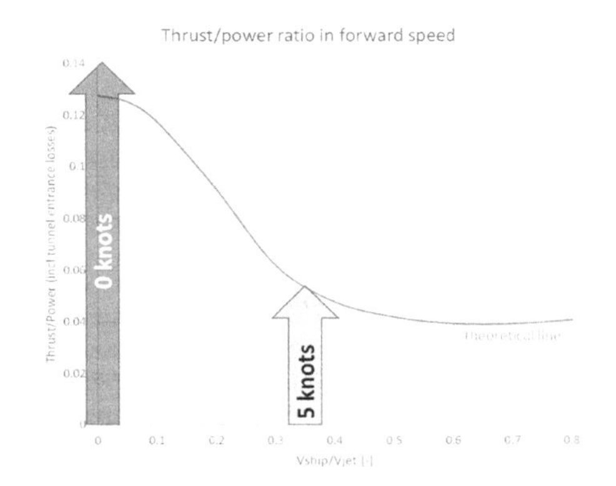

Figure 11. Thrust/power ratio for conventional tunnel thruster units for crabbing and maneuvering mode.

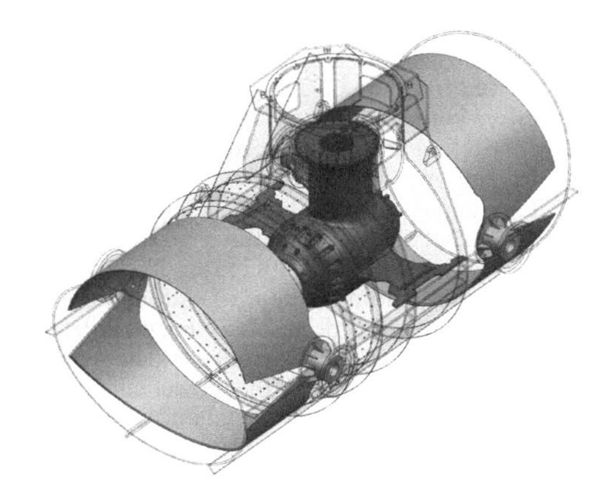

Figure 12. 3D-CAD model of tunnel thruster with adjustable outflow devices.

jet propulsion systems. Since the tunnel thrusters need to operate in both directions, a device is needed which can adjust the exit area depending on the flow direction. Depending on the flow direction, and thus the thrust direction, one side of the tunnel is either the inflow or the outflow side. Figure 12 shows a 3D sketch of the device to manipulate the outflow area which is an adjustable device to operate in both directions.

The two halve round plates on the jet exit side are rotated along the pivot axis in order to create a kind of elliptical outflow area. Then two halves on the inlet side are positioned in such way that they form a flush shape with the rest of the tunnel. As long as the two halves are positioned flush with the tunnel walls, the performance of the tunnel thruster is similar to a conventional unit so there is no performance penalty in crabbing mode.

The concept for crabbing and maneuvering mode is shown in Figure 13, where the device on the right side is located at the jet exit side. For prove of the hydrodynamic concept, a number of variables have been determined which need to be investigated. The initial main parameters of interest are the angle of closure, the length of the closing device and the actual power absorption of the unit which is governed by the propeller pitch deflection.

5.3 *Prove of hydrodynamic concept*

In order to evaluate the actual effectiveness of the concept a set of CFD simulations has been made of the device. Before the actual simulations of the hull with tunnel and thruster unit have been performed, first a series of validation calculations have been executed. Details of the work are reported by Chadha (2017).

The combination of the forward ship speed, and the jet out of the thruster, creates a complex flow regime which is also known as jet in a cross-flow. Figure 14 shows a 3D surface plot, which has been derived from the CFD simulations. Due to the interaction between the two main flows, a set of counter rotating vortices is formed. Around the jet near the wall opening, a horse shoe vortex is generated. The validation work, for the jet in a cross flow, has shown that the main features of the complex flow could be reproduced with reasonable accuracy. Based on these findings, the CFD simulations for the adjustable outflow device, have been made.

The hydrodynamic performance is based on the side thrust generation and the power absorption of the installation. For CPP installations there is a rule-of-thumb that increased pitch deflection leads to higher power absorption. It has been found that

Figure 13. Adjustable outflow device in crabbing mode (top) and maneuvering mode (bottom).

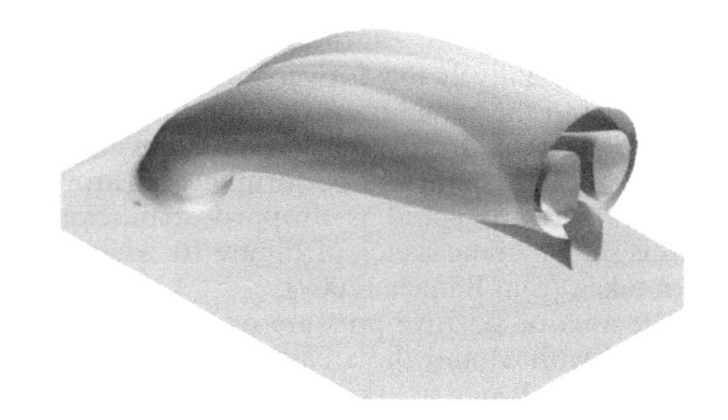

Figure 14. CFD result of jet in a cross flow, where the counter rotating vortex pair is visualised.

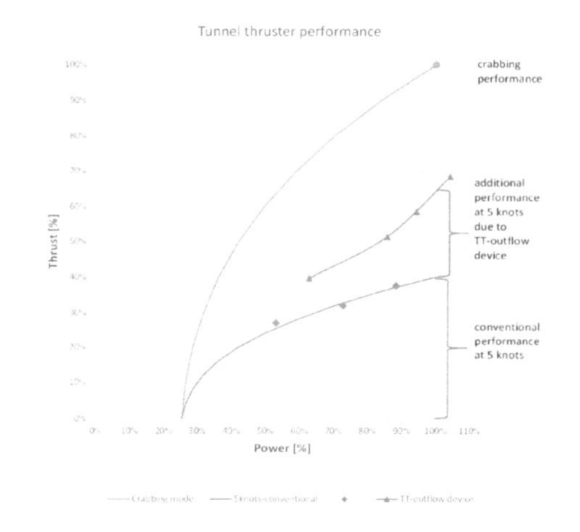

Figure 15. Thrust as function of power for crabbing and maneuvering mode without and with TT-outflow device.

this rule is not always applicable for tunnel thrusters, when operating in forward ship speed. In order to get a good overview of the occurring phenomena in the tunnel when the vessel is moving forward with 5 knots speed, a range of pitch deflections has been analyzed.

The results of the CFD simulations are shown in Figure 15, where the thrust is plotted against the power absorption. The results have been normalized with the maximum thrust in crabbing mode. At lower pitch deflection, the thrust reduces accordingly. At zero thrust condition, power is still needed to run the propeller based on the constant RPM mode.

In 5 knots forward speed the conventional tunnel configuration provides about 40% of the maximum thrust force. The different points along the line are derived from the CFD simulations at different blade pitch deflections. With the new adjustable outflow device activated, the thrust generation at 5 knots forward speed increases to about 70%. This is a clear improvement of thrust availability and it brings the usage of the tunnel thruster in forward speed into the picture.

5.4 *Other developments*

The initial CFD results, with the first generation of the closing device geometry, show significant performance improvements to proceed with the design and optimization process of the device.

Another point of attention will be the development of the actuation mechanism of the closing device. The design of the actuating mechanism will address the additional complexity of the limited space near the tunnels. The forces acting on the plates are derived from the full scale CFD simulations which give a good indication of the steady and transient forces on the device in the different operating conditions.

6 CONCLUSIONS

The initial focus on the development of the tunnel thruster series has been on the performance in crabbing mode (no forward ship speed). The main design criterion has been to optimize the thrust generation in both directions. In order to achieve this, the streamline of the thruster housing has been taken into consideration in all CFD simulations. In order to keep the noise levels under control, dedicated units have been designed with lower tip speeds and lower power densities. These design features make the unit attractive to the cruise market, where both efficiency and noise levels are of importance.

In order to get optimum integration of the tunnel thruster in the vessel, the local geometry details at the hull-tunnel interfacing need to be designed with care. With the current CFD capabilities, the flow analysis of the geometry can be made within acceptable time frames within the vessel design phase.

It has been acknowledged that the performance of a conventional tunnel thruster at 5 knots forward speed is quite disappointing which disqualifies the unit often for usage. In order to improve the thrust generation at low forward speed, a device has been designed which adjusts the exit area of the tunnel, and therefore creates a stronger jet. The side thrust force, generated with this device, is significantly improved compared to the conventional configuration. With about 70% of the maximum thrust in crabbing mode, the use of a tunnel thruster to improve maneuvering capabilities of the vessel seems logical. It can be concluded that the adjustable outflow device contributes to regaining control over the vessel bow section at low speeds.

ACKNOWLEDGEMENTS

The author wants to thanks Ms Iulia Oprea for the detailed CFD simulations of the tunnel thrusters and Mr Navneet Chadha for his MSc-thesis work on the jet in a cross flow.

REFERENCES

Baniela, S. 2009. The Performance of a Tunnel Bow Thruster with Slow Speed Ahead: A Revisited Issue. *Journal of Navigation*, 62(4), 631–642.

Boletis, E., De Lange, R., Bulten, N., 2015. Impact of propulsion system integration and controls on the vessel DP and manoeuvring capability, *10th Conference on Manoeuvring and Control of Marine Craft*, Copenhagen, Denmark.

Boletis, E., Drost, A., Bulten, N.,2015. Advancement in propulsion technology, *NAV2015 conference*, Lecco, Italy.

Brix, J.,1993, Manoeuvring technical manual, Seehafen verlag, Hamburg, Germany.

Buiten, J. and Regt, M.J.A.M. de, 1983. Handboek voor de lawaai beheersing aan boord van schepen (in Dutch), *TPD report No. 208.431*, Delft, The Netherlands.

Bulten N, Stoltenkamp, P, 2017, Full scale CFD: the end of the Froude-Reynolds battle, *Fifth International Symposium on Marine Propulsion SMP'17*, Espoo, Finland.

Bulten N., 2015, Transient blade load determination in behind ship based on CFD, *SNAME Propeller and Shafting conference*, Norfolk VA, USA.

Chadha, N. 2017, Numerical investigation of hydrodynamic interaction of jet by bow thruster with cross-flow, *MSc-thesis Technical University Eindhoven*, The Netherlands.

Ligtelijn, J.T., 2007. Advantages of different propellers for minimising noise generation, *3rd International Ship Noise and Vibration Conference*, London, UK.

Ligtelijn, J.T., Otto, R., 2013. Minimization of noise and vibration from transverse propulsion units, *2nd IMarEST Ship Noise and Vibration Conference*, London, UK.

Mewis, F, Guiard, T,, 2011, 'Mewis Duct—New Developments, Solutions and Conclusions', *Second International Symposium on Marine Propulsors SMP'11*, Hamburg, Germany.

Schuiling, B. 2013. 'The Design and Numerical Demonstration of a New Energy Saving Device' 16th Numerical Towing Tank Symposium (NuTTS), 141–146.

Toxopeus, S.L Loeff, G.B, 2002, Manoeuvring aspects of fast ships with pods, *HIPER'02 3rd International Euro-Conference on High-Performance Marine Vehicles*, Bergen, Norway.

Zondervan, G-J, Holtrop, J, Windt J, 2011, Van Terwisga, T, On the Design and Analysis of Pre-Swirl Stators for Single and Twin Screw Ships, *Second International Symposium on Marine Propulsors SMP'11*, Hamburg, Germany.

Marine Design XIII – Kujala & Lu (Eds)
 ISBN 978-1-138-34076-3

Study on the hydrodynamic characteristics of an open propeller in regular head waves considering unsteady surge motion effect

Wencan Zhang, Ning Ma, Chen-Jun Yang & Xiechong Gu
State Key Laboratory of Ocean Engineering, Shanghai Jiao Tong University, Shanghai, China
Collaborative Innovation of Center for Advanced Ship and Deep-Sea Exploration, Shanghai Jiao Tong University, Shanghai, China
School of Naval Architecture, Ocean and Civil Engineering, Shanghai Jiao Tong University, China

ABSTRACT: Based on CFD simulations, this paper conducts a quantitative study of the time-averaged and transient hydrodynamic performance of the propeller experiencing unsteady surge motion in regular head waves. As an approximate approach for simulating the actual surge motions in waves, this study utilizes the overset grid approach for simulating the rotation and the forced surge motions of the propeller and the VOF (Volume of Fluid) method for free surface tracking. The paper emphasizes on an effective method of investigating the effect of blade area ratio and pitch ratio on the hydrodynamic performance of propellers in waves. The numerical results indicate that, the unsteady surge motion can impose a negative or positive impact on the thrust and torque in different time periods within a wave period, and decreases the thrust and torque averaged over a wave period; when a propeller experiences the surge motion, its thrust and torque would increase with the blade area ratio and the pitch ratio, more rapidly the latter. The presented simulation approach and investigation method can be further improved and applied to propeller design and evaluation of ship propulsive performance in waves accordingly.

1 INTRODUCTION

In recent years, a huge amount of research has been conducted upon ship propulsive performance in waves, with rather impressive achievements. However, previous research places emphases on wave-added resistance, relatively neglecting wave-induced fluctuation and deterioration of propeller performance, which is highly responsible for the deterioration of ship propulsive performance. When a ship navigates in the real sea, the propeller can sometimes experience drastic horizontal and vertical motions due to the wave-induced ship motions such as surge, pitch and heave, which can dramatically change the inflow to the propeller, and seriously deteriorate its hydrodynamic performance; also, for ships in moderate sea conditions not experiencing drastic relative horizontal motions, in which case the propeller is a certain distance away from the free surface, the surge motion is the key factor that contributes to the fluctuation and deterioration of propeller performance. From a practical viewpoint, the drastic load fluctuation on the propeller blades in waves may cause serious strength problems on propeller and shafting, and sometimes even serious hull vibration that might possibly put the ship in danger. To sum up, it is of great significance to thoroughly analyze the propeller's hydrodynamic performance in waves.

Over the years, some researchers have conducted theoretical, numerical and experimental researches on the propeller's propulsive performance in waves.

In respect of theoretical research, Tao et al. (1991) calculated the open-water hydrodynamic performance of a propeller in waves based on the Green Function Method. Liang et al. (2006) calculated the loss of propeller thrust and torque in waves. In respect of numerical research, Yu (2008) studied open-water hydrodynamic performance of a propeller in waves based on the CFD software FLUENT. Lee et al. (2010) investigated propeller performance under seaway wave condition using a CFD method. Queutey et al. (2014) numerically evaluated the instantaneous flow field around podded propellers. In respect of experimental research, Naito and Nakamura (1979) firstly conducted tests under forced heave and pitch conditions. Cao (1988) conducted experiments on the hydrodynamic performance of propeller with variant submergence depths in waves. Tao et al. (1991) conducted an experiment for an open-water propeller with pitch and heave motions in regular waves. Politis (1999) performed open propeller tests in both regular and irregular waves, with different submergence depths, advance coefficients and wave parameters. Guo et al. (2012) experimentally studied the hydrodynamic performance of a propeller in regular waves. Dong et al. (2013) experimentally researched the effect of submergence depth and wave height on hydrodynamic characteristics of a propeller in regular head waves. Zhao et al. (2017) conducted experimental and numerical research on unsteady hydrodynamic performance of an open propeller.

In recent years, CFD has been proven to be highly accurate and effective in capturing and analyzing complicated flow physics. For an accurate prediction of propeller performance in waves, more research based on CFD is required. However, to our knowledge, not much research based on CFD simulations has been conducted on the quantitative analysis of propeller performance in waves. The reason is possibly that the nonlinear effects of the free surface can be drastic especially when the propeller emerges from the free surface, which greatly increases the difficulty of accurate simulation.

The present work conducts an investigation on the deterioration and fluctuation of the hydrodynamic performance of propellers experiencing large-amplitude surge motion in regular head waves, and on the effect of geometric parameters such as propeller's blade area ratio and pitch ratio on hydrodynamic performance of propeller in waves considering unsteady surge motion effect. A summary of the present research work is given as follows.

First, in order to validate the present numerical method and to lay a foundation for subsequent research, simulations were conducted on the hydrodynamic performance of an open propeller in regular head waves without forced surge motions, and compared with the experimental data in Cao (1988) with satisfactory agreement. Second, based on the validated numerical method, simulations were conducted on propeller hydrodynamic performance in regular head waves with/without forced surge motion in regular head waves, in order to evaluate the influence of surge motion on propeller hydrodynamic performance. Third, through comprehensive adjustment of propeller geometric parameters, such as the blade area ratio and the pitch ratio, the effect of geometric parameters on hydrodynamic performance of open propeller with forced surge motion in regular head waves is investigated.

2 NUMERICAL METHOD

2.1 *Governing equations*

Assuming that the flow is incompressible, the continuity equation and the RANS equations are written in tensor form as

$$\frac{\partial}{\partial x_i}\bar{u}_i = 0 \tag{1}$$

$$\frac{\partial \rho\left(\bar{u}_i\right)}{\partial t} + \frac{\partial}{\partial x_j}(\rho \bar{u}_i \bar{u}_j) = -\frac{\partial p}{\partial x_i} + \frac{\partial \tau_{ij}}{\partial x_j} + \rho f_i \tag{2}$$

where ρ is the density of water, p is the static pressure, τ_{ij} is the shear stress, and ρf_i is the gravity component on the ith direction.

The Realizable k–ε model is adopted for turbulence closure. In this model, the transport equations for the turbulent kinetic energy (k) and its dissipation rate (ε) are written as

$$\frac{\partial(\rho k)}{\partial t} + \frac{\partial(\rho k u_i)}{\partial x_i} = \frac{\partial}{\partial x_j}\left[\left(\mu + \frac{\mu_t}{\sigma_k}\right)\frac{\partial k}{\partial x_j}\right] + G_k - \rho\varepsilon \tag{3}$$

$$\frac{\partial(\rho\varepsilon)}{\partial t} + \frac{\partial(\rho\varepsilon u_i)}{\partial x_i} = \frac{\partial}{\partial x_j}\left[\left(\mu + \frac{\mu_t}{\sigma_\varepsilon}\right)\frac{\partial \varepsilon}{\partial x_j}\right] + \rho C_2 \frac{\varepsilon^2}{k + \sqrt{\nu\varepsilon}} \tag{4}$$

where

$$\varepsilon = \frac{\mu}{\rho}\overline{\left(\frac{\partial u_i'}{\partial x_k}\right)\left(\frac{\partial u_i'}{\partial x_k}\right)} \tag{5}$$

In Equation (3), G_k is the generation term of turbulent kinetic energy (k) induced by the average velocity gradient.

The dynamic viscosity coefficient of turbulence μ_t is defined as

$$\mu_t = \rho C_\mu \frac{k^2}{\varepsilon} \tag{6}$$

where the variable C_μ is defined as

$$C_\mu = \frac{1}{A_0 + \frac{A_S U^* k}{\varepsilon}} \tag{7}$$

and the coefficients in (7) are defined as

$$U^* = \sqrt{S_{ij}S_{ij} + \tilde{\Omega}_{ij}\tilde{\Omega}_{ij}} \tag{8}$$

$$\tilde{\Omega}_{ij} = \Omega_{ij} - 2\varepsilon_{ijk}\omega_k \tag{9}$$

$$\Omega_{ij} = \overline{\Omega_{ij}} - \varepsilon_{ijk}\omega_k \tag{10}$$

$$A_0 = 4.04 \tag{11}$$

$$A_s = \sqrt{6}\cos\phi \tag{12}$$

$$\phi = \frac{1}{3}\cos^{-1}\left(\sqrt{6}W\right) \tag{13}$$

2.2 *Numerical modeling approach*

2.2.1 *Solver options*

The commercial CFD code STAR-CCM 11.06.010 is used to simulate the incompressible, viscous, two-phase (water and air) flow. The governing equations are solved in integral form using a finite volume method (FVM). Implicit unsteady Euler schemes are used for temporal discretization, which are 1st-order for calm-water simulations and 2nd-order for regular-wave simulations, respectively. The free surface elevations are tracked by using the Volume of Fluid method (VOF), and the overset grid method is used for simulating the rotation and the forced surge motions of the propeller, as an approximate approach for simulating actual surge motions in waves.

2.2.2 *Comutational domain and meshing*

Figure 1 illustrates the setup of the computational domain. The whole domain is a rectangular parallelepiped, with its measurements in x, y, and z directions set as 2.5 times wave length, 1.5 times wave length, and 2 m, respectively. The distance is 1.5 times wave length between the propeller and the outlet boundary, and one times wave length between the propeller and the inlet boundary. The rectangular in the middle of the domain represents the "overlapping area" where the mesh is refined to ensure the accuracy of the overset grid method, with its measurement in x, y, and z directions set as 0.1 m, 0.08 m, and 0.2 m, respectively. The small cylinder lying within the "overlapping area" that encompasses the propeller is the "rotating region", the diameter of which is 1.1 times propeller diameter and the height 0.9 times propeller diameter. The incoming flow points to the negative x-axis direction.

The mesh generation method is very important for accurate computation of hydrodynamic performance of propeller experiencing forced surge motion. In this paper, the mesh is generated to ensure:

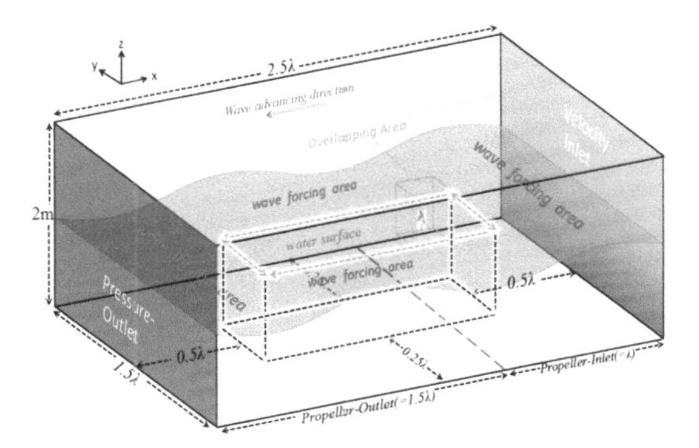

Figure 1. Setup of the computational domain.

1. An adequate distribution of cells in the near-propeller area both beneath and above the calm-water surface (See "near-propeller refinement area" in Figure 2 and Figure 3), in order to accurately capture the complicated flow field and the deformation of free surface near the propeller;
2. An adequate vertical distribution of cells in the wave-height area (See "water surface refinement area" in Figure 2 and Figure 3). In the present simulations, 20 grid cells are distributed vertically in the wave-height area;
3. An adequate and even distribution of grid cells on the propeller surface (See Figure 4).
4. An adequate mesh refinement and similar densities of grid cell distribution for both the rotating region and overlapping area to ensure a satisfactory accuracy of the overset grid method (See Figure 2 and Figure 3);
5. Smooth transition of grid density among different meshing areas.

Figure 2 presents the mesh generation of the whole computation domain. Figure 3 presents the zoom-in view of the mesh generation of the whole computation domain. Figure 4 shows the mesh generation on the propeller surface & of the rotating region.

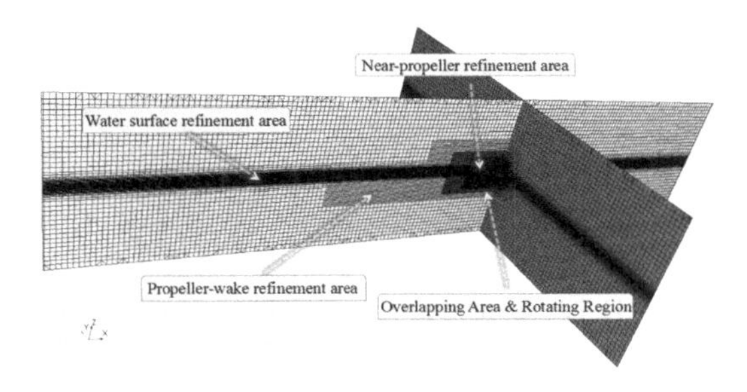

Figure 2. Mesh generation of the whole computational domain for regular-wave simulations.

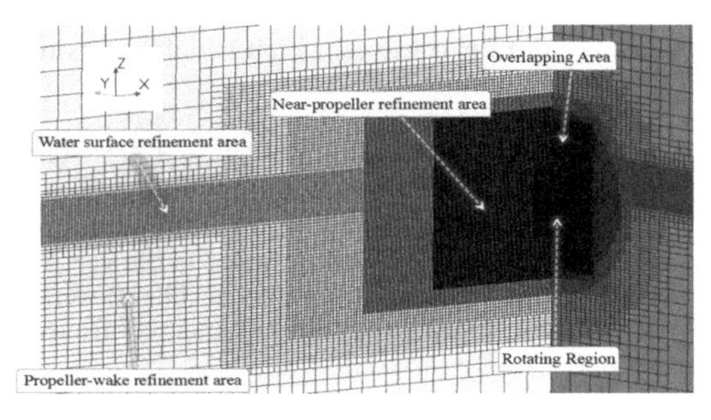

Figure 3. Zoom-in view of mesh generation in the vicinity of rotating region.

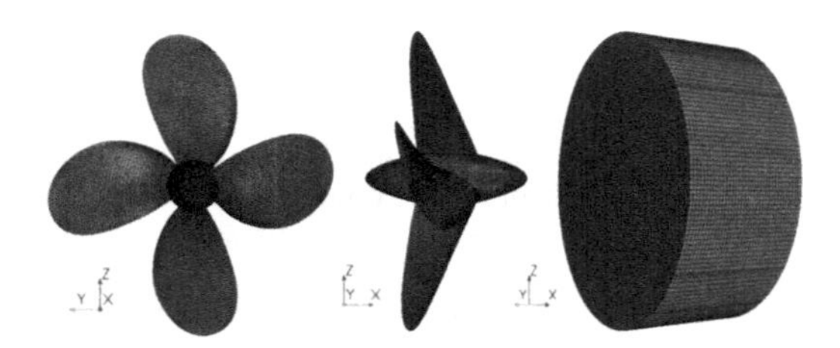

Figure 4. Mesh generation on the propeller surface & of the rotating region.

2.2.3 *Definition of hydrodynamic coefficients*

Thrust coefficient:

$$K_T = \frac{T}{\rho n^2 D^4} \tag{14}$$

Torque coefficient:

$$K_Q = \frac{Q}{\rho n^2 D^5} \tag{15}$$

Efficiency:

$$\eta_0 = \frac{K_T}{K_Q}\frac{J}{2\pi} \tag{16}$$

where T and Q represent the thrust and torque of the propeller. The advance coefficient of the propeller is defined as

$$J = \frac{V_A}{\mathrm{n D}} \tag{17}$$

where V_A is the advance speed of the propeller in the conventional sense. In regular head waves, the advance speed of the propeller is defined as

$$V'_A = V_A + V_h \tag{18}$$

where V_h is the horizontal velocity of water particles in regular head waves, which is defined as the following form

$$V_h = a\omega\cos(Kx - \omega t)\mathrm{e}^{Kz} \tag{19}$$

where a is the wave amplitude, ω is the wave frequency, K is the wave number, and z is the vertical distance from the calm water level. Therefore, in regular head waves, propeller advance coefficient is defined as

$$J' = \frac{V'_A}{\mathrm{n}D} \tag{20}$$

In this paper, however, the time-averaged value of the advance speeds in regular head waves is used to define the advance coefficient. Therefore, $V_A=V_A'$, $J=J'$.

3 VALIDATION OF SIMULATION METHOD

3.1 *Propeller geometry and working conditions*

In this paper, a four-bladed MAU-series propeller, named MAU4-6075, is used for our numerical simulations. Table 1 shows the main particulars of the propeller model.

The geometric model of MAU4-6075 is shown in Figure 5.

Table 2 presents all the conditions of numerical simulation in the present study.

In Table 2, SDR stands for the submergence depth ratio, which is defined as the ratio of H_S to D, where H_S is the submerged depth of propeller shaft axis from the calm water surface, and D is propeller diameter.

Table 1. Main particulars of propeller model MAU4-6075.

Parameter	Unit	Symbo	Value
Diameter	mm	D	130
Hub diameter	mm	d	19.5
Pitch at 0.7R	mm	P	97.5
Hub ratio	–	d/D	0.15
Pitch ratio	–	P/D	0.75
Expanded area ratio	–	A_E/A_O	0.60
Number of blades	–	Z	4
Rotating direction	–		Right-handed
Rotational speed	r/s	n	24

Figure 5. Geometric model of MAU4-6075.

Table 2. Conditions of numerical simulation.

Surge	SDR	J	Wave parameters
Without	1.0808		a = 0.0397 m τ = 1.5425s
With	1.0	0–0.5	a = 0.08 m τ = 1.5078s

Table 3. Experimental results in calm water when SDR = 1.0 and J = 0.1, Cao (1988).

SDR	J	K_T	10 K_Q
1.0	0.1	0.3075	0.3658

Table 4. Cell numbers of the three cases with different mesh resolutions.

Mesh resolution	Total	Rotating zone	Stationary zone
Coarse	1416475	617853	798622
Medium	2014796	871934	1142862
Fine	2874365	1235376	1638989

In the "wave parameters" column, a and τ represent wave amplitude and wave period, respectively.

3.2 *Grid and time-step independency study*

Independency studies are performed on grid cell number and time-step size including verification and validation. Two scalars are defined as follow,

$$\varepsilon_{fine} = \left(K_{T\&Q,fine} - K_{T\&Q,coarse}\right) / K_{T\&Q,coarse} \times 100\% \quad (21)$$

$$\lambda_{fine} = \left(K_{T\&Q,num} - K_{T\&Q,\exp}\right) / K_{T\&Q,\exp} \times 100\% \quad (22)$$

Using a parameter refinement ratio $r_i = \sqrt{2}$, numerical simulations are carried out for the case of SDR = 1.0 and J = 0.1 in calm water, and validated against the experimental data of Cao (1988) in the same condition, as listed in Table 3.

First, for grid independency verification and validation, three cases with different mesh resolutions named "Coarse", "Medium", "Fine" are simulated. The maximum number of inner iterations, as well as the time-step size, were kept constant as 28 and 2.8×10^{-4} s, respectively. The cell numbers for the three cases are given in Table 4.

Table 5 presents the results of grid independency verification and validation, where G1, G2, and G3 represent "Coarse", "Medium", and "Fine" grids, respectively. The differences in K_T and 10 K_Q between G2 and G3 are very small, which implies that the numerical results are hardly influenced by mesh resolution when the total cell number reaches the "Medium" standard.

Table 5. Results of grid independency verification and validation.

G	K_T	10 K_Q	ε(K_T)	ε (10 K_Q)	λ (K_T)	λ (10 K_Q)
1	0.2864	0.3402	–	–	8.42%	8.2%
2	0.2919	0.3458	1.92%	1.65%	5.07%	5.47%
3	0.2928	0.3467	0.31%	0.26%	4.78%	5.22%

Table 6. Results of time-step size independency verification and validation.

T (10^{-4} s)	K_T	10 K_Q	ε(K_T)	ε (10 K_Q)	λ(K_T)	λ (10 K_Q)
2.8	0.2919	0.3458	–	–	5.07%	5.47%
2	0.2972	0.3515	1.81%	1.62%	3.35%	3.91%
1.4	0.2978	0.3524	0.2%	0.26%	3.15%	3.66%

Table 7. Experimental open-water performance of B4-6075, J = 0.1, Bernitsas et al. (1981).

J	K_T	10 K_Q
0.1	0.2728	0.3118

Next, the time-step size independency verification and validation are conducted using the "Medium" mesh resolution (G2). The results are presented in Table 6, where it is seen that the time-step size plays a negligible role in influencing the numerical results once it is equal to 2×10^{-4} s or smaller.

Using the same grid resolutions and time-step sizes, Tables 8 and 9 present the results of verification and validation for a conventional open-water propeller without considering the free surface. The propeller selected here is B4-6075, a 4-bladed B-series propeller with a blade area ratio of 0.6 and a pitch ratio of 0.75. Table 7 shows the experimental open-water performance of the propeller at J = 0.1, Bernitsas et al. (1981).

Comparing Tables 8 and 9 with Tables 5 and 6 respectively, it is clear that the errors in RANS-predicted propeller hydrodynamic performance without the free surface are slightly smaller than those with the free surface, which is reasonable since the two cases are of different complexity both physically and numerically. The present method is adequately precise for simulating free surface effects on propeller hydrodynamic performance.

Table 8. Results of grid independency verification and validation for B4-6075.

G	K_T	10 K_Q	$\varepsilon(K_T)$	ε (10 K_Q)	$\lambda(K_T)$	λ (10 K_Q)
1	0.2836	0.3236	–	–	3.96%	3.65%
2	0.2783	0.3195	1.86%	1.68%	2.02%	2.47%
3	0.2776	0.3179	0.28%	0.26%	1.76%	2.22%

Table 9. Results of time-step size independency verification and validation for B4-6075.

T (10^{-4} s)	K_T	10 K_Q	$\varepsilon(K_T)$	ε (10 K_Q)	$\lambda(K_T)$	λ (10 K_Q)
2.8	0.2783	0.3195	–	–	2.02%	2.47%
2	0.2772	0.3175	1.61%	1.47%	1.75%	1.91%
1.4	0.2768	0.3169	0.12%	0.2%	1.65%	1.86%

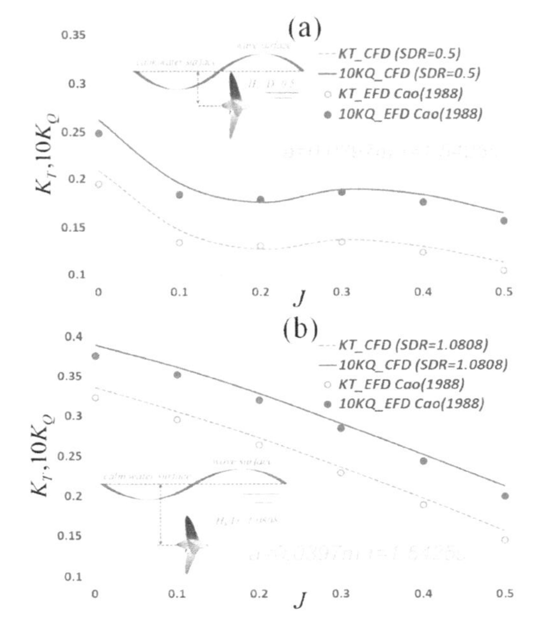

Figure 6. Comparison of CFD and EFD results for propeller MAU4-6075 in regular head waves without surge motion. (a) SDR = 0.5; (b) SDR = 1.0808.

3.3 *Accuracy validation of numerical method*

To validate the accuracy of the present numerical method, numerical simulations of propeller hydrodynamic performance in regular head waves without surge motion are conducted at two SDRs, 0.5 and 1.0808, and over a range of advance coefficient J, 0 to 0.5. The wave amplitude $a = 0.0397$ m, and the wave period $\tau = 1.5425$s.

The comparison of present numerical results and the experimental ones by Cao (1988) is shown in Figure 6 for the propeller MAU4-6075. It can be observed that, the hydrodynamic performance of the propeller is seriously compromised by propeller racing, especially in the low advance coefficient zone. Satisfactory agreement was achieved between the numerical and experimental results in the computed range of J, which indicates that the present numerical method is quite accurate and robust in dealing with propeller racing and strongly nonlinear phenomena of the free surface. Furthermore, the CFD results of K_T and 10 K_Q are both slightly higher than EFD ones, which is believed to be caused by the inability of resolving the thin ventilating vortices with the VOF scheme used.

4 PROPELLER PERFORMANCE WITH FORCED SURGE MOTION

4.1 *Setting propeller's forced surge motions and waves*

As mentioned above, the purpose of setting propeller's forced surge motions is to simulate the wave-induced surge motions of ship at real sea. Therefore, a ship must be selected as the "hypothetical" ship on which the propeller is assumed to be installed, and this ship's surge motion at a certain wave frequency and wave height must be calculated first and then applied to the open-water propeller; in addition, the phase relation between forced surge motion and regular head wave must be considered.

For this purpose, a 10,000 TEU container ship is selected as the "hypothetical" ship, and the surge motion of it within a certain range of regular head wave frequency is calculated using Hydrostar.

Figure 7 shows the RAO curve of surge motions of the full-scale 10,000 TEU container ship within a range of regular head wave frequency computed by Hydrostar. In this paper, the RAO of surge

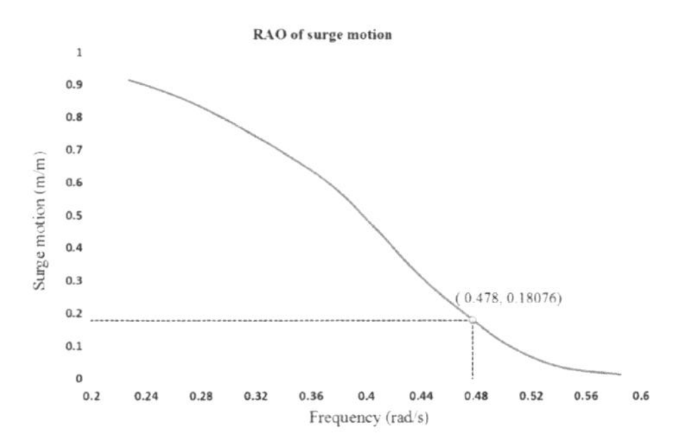

Figure 7. The RAO curve of surge motions of the 10,000TEU container ship, computed by software Hydrostar.

motion at a wave frequency of $\omega_s = 0.478$ rad/s (corresponding to a wave length $\lambda_s = 269.68$ m, approximately 0.85 times full-scale ship length L_s) is selected, which equals to 0.18076 m/m.

Assuming that the model-scale wave amplitude is $A_m = 0.08$ m, then the model-scale amplitude of surge motion equals to $S_{am} = 0.18076$ m/m × 0.08 m = 1.446 cm.

Second, the phase relation between forced surge motion and regular head wave must be considered. According to Hydrostar, the phase angle of surge motion at a wave frequency of $\omega_s = 0.478$ rad/s is $\varphi = 82.3167°$.

Then the model-scale wave frequency ω_m and wave period T_m can be calculated as follow,

$$\omega_m = \omega_s\sqrt{\lambda} = 0.478 \times \sqrt{76} = 4.1671\text{rad/s}$$

$$T_m = 2\pi / \omega_m = 1.5078\text{s}$$

where the scale ratio $\lambda = 76$.

According to the above analysis, the x-component of forced surge motion velocity can be written as $u = 0.01446\cos\left[2\pi(t/1.5078 + 82.3167/360)\right]$.

The expression above indicates that the forced surge motion of propeller can be divided into three time periods within a model-scale wave period T_m:

Period 1: (t = nT_m + 0 s~t = nT_m+0.03218s, n = 0, 1, 2, ...), this is the time period in which the propeller moves along the positive x-axis from the initial state of simulation to the first peak of surge motion;

Period 2: (t = nT_m + 0.03218s~t = nT_m + 0.78608s, n = 0, 1, 2, ...), this is the time period in which the propeller moves along the negative x-axis from the first peak position, to the first trough position of surge motion;

Period 3: (t = nT_m + 0.78608s~t = (n + 1)T_m, n = 0, 1, 2, ...), this is the time period in which the propeller moves along the positive x-axis from the first trough position, and then returns to the initial position.

Finally, the wave crest is at the ship's centre of buoyancy at the initial state of simulation. According to Yu et al. (2015), the L_{BP} of the 10000TEU container ship is 320 m; also, the distance between the centre of buoyancy and the aft-perpendicular, and the distance between the aft-perpendicular and the propeller, are 155.37 m and approximately 8 m, respectively. Therefore, the full-scale distance between the propeller and the centre of buoyancy is approximately 147.37 m. At the scale ratio of 76, the distance between the propeller and the centre of buoyancy (the wave crest) is approximately 1.9391 m at model scale when t = 0. The instantaneous profile of incident waves at the initial state of simulation is shown in Figure 8.

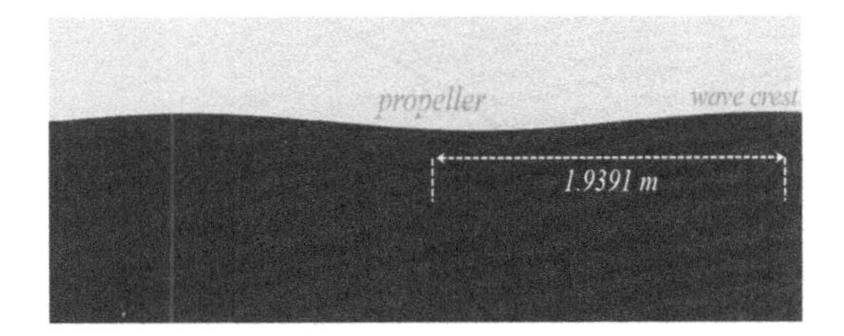

Figure 8. Free surface profile at initial state of simulation.

4.2 *Time-averaged hydrodynamic performances of propeller with/without surge motion effect*

By means of the present numerical method, the hydrodynamic performances of propeller MAU4-6075 with and without the surge motion in regular head waves are investigated. The SDR is kept constant as 1.0, and the advance coefficient J is selected as 0, 0.1, 0.2, 0.3, 0.4 and 0.5 and taken as time-averaged value. The physical time span of each simulation is 4.5234 s, which is three times the model-scale wave period T_m.

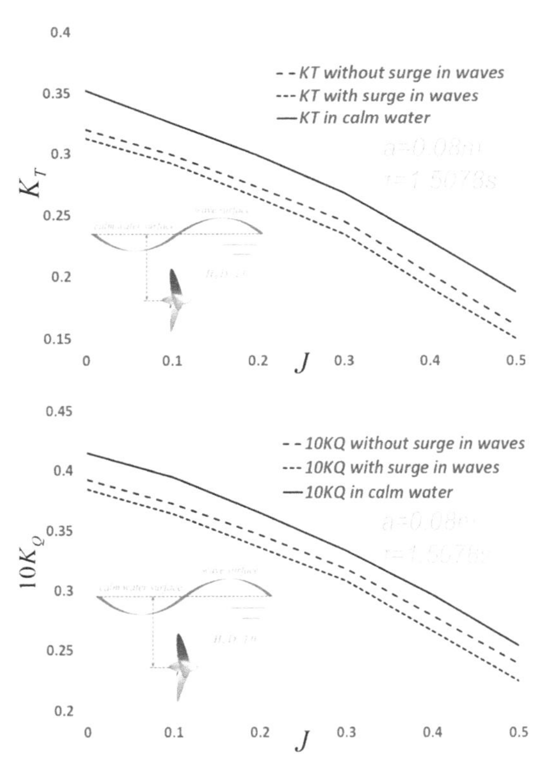

Figure 9. Time-averaged simulation results for time period 1.

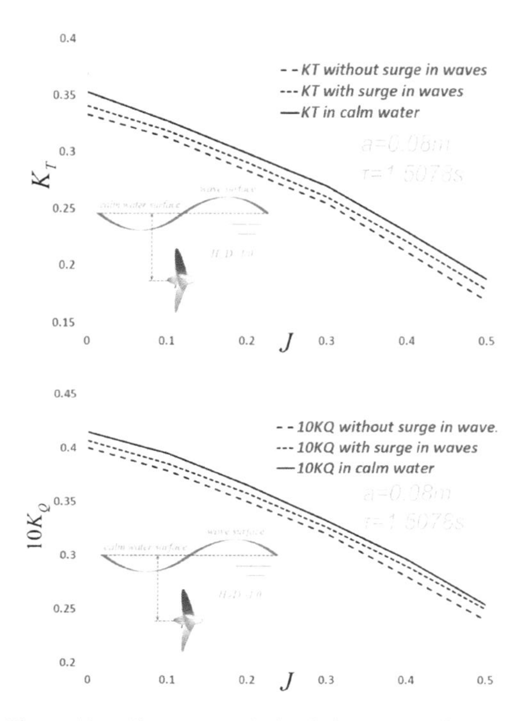

Figure 10. Time-averaged simulation results for time period 2.

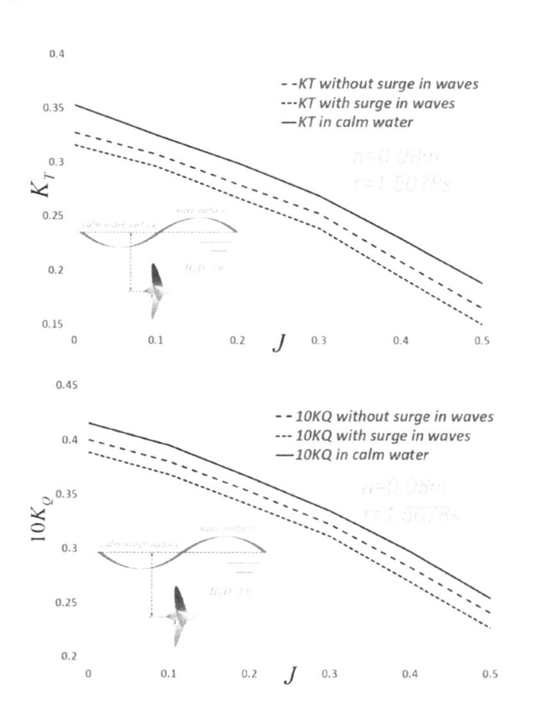

Figure 11. Time-averaged simulation results for time period 3.

The time-averaged simulation results within the third wave period (T_3 = 3.0156s~4.5234 s), with and without the surge motion in regular head waves, as well as the calm water results at SDR = 1.0, are presented in Figures 9, 10, and 11 for the three time periods 1, 2, and 3 respectively.

The results in Figures 9 through 11 indicate that,

1. when SDR = 1.0, the time-averaged thrust and torque in regular head waves are both smaller than their counterparts in calm water over the entire wave period, whether there is the surge motion or not;
2. the surge motion causes a decrease in propeller thrust and torque in time periods 1 and 3,

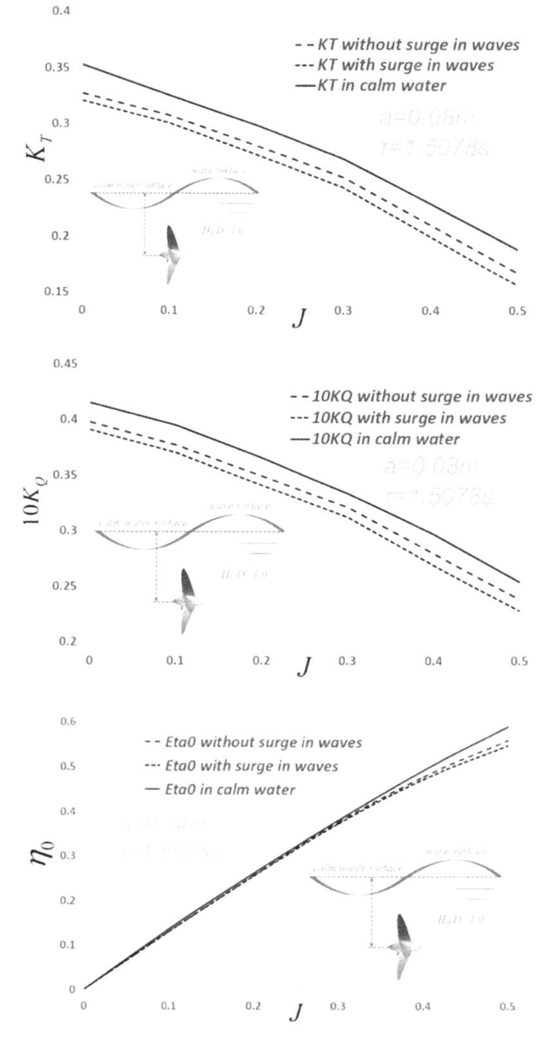

Figure 12. Time-averaged simulation results over the entire wave period.

however an increase in time period 2. This is because the propeller accelerates along the positive x-axis direction in time periods 1 and 3, so the advance speed increases. The situation is just the opposite in time period 2. Somehow, the decrease in thrust and torque due to the surge motion is most significant in time period 3.

Table 10. Coefficient of difference quantity of K_T under variant advance coefficient J in the three time periods.

J	P 1	P 2	P 3	Overall
0	−1.18	0.82	−1.45	1.22
0.1	−1.67	1.17	−1.92	−1.73
0.2	−2.37	1.45	−2.68	−2.46
0.3	−2.89	1.95	−3.23	−3.02
0.4	−3.58	2.38	−3.87	−3.81
0.5	−4.33	3.03	−4.68	−4.59

Table 11. Coefficient of difference quantity of 10 K_Q under variant advance coefficient J in the three time periods.

J	P 1	P 2	P 3	Overall
0	−1.02	0.68	−1.32	−1.08
0.1	−1.48	0.96	−1.69	−1.57
0.2	−2.07	1.21	−2.37	−2.29
0.3	−2.56	1.67	−2.95	−2.87
0.4	−3.24	2.01	−3.54	−3.59
0.5	−4.02	2.53	−4.31	−4.39

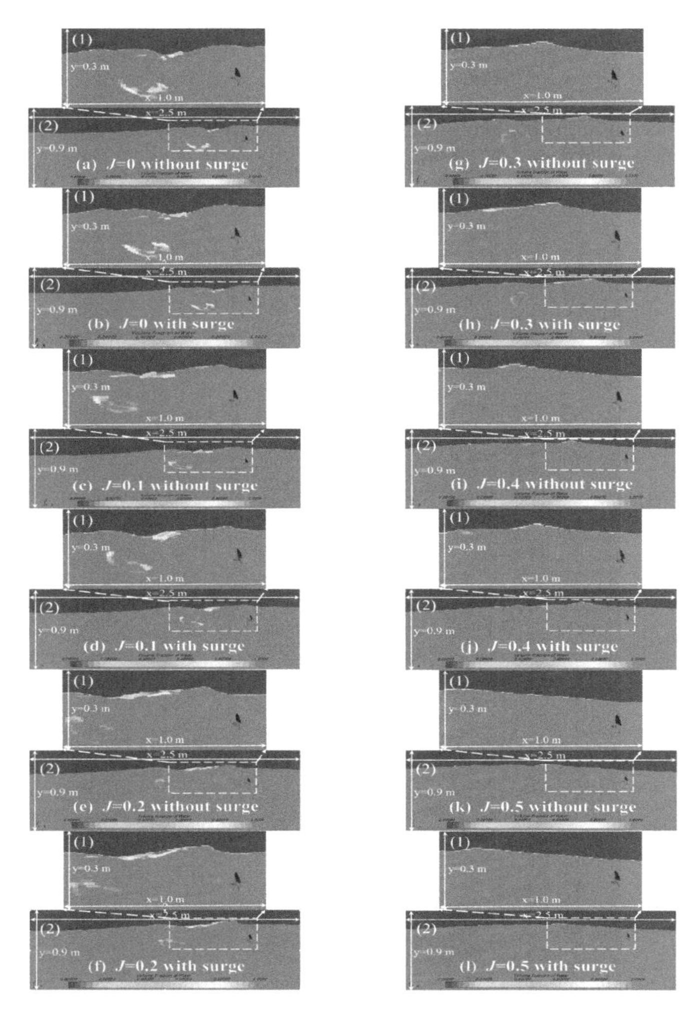

Figure 13. (a)–(l). Comparison of simulated free surface profiles with and without forced surge motion, J = 0~0.5.

Figure 12 shows the thrust and torque averaged over the entire third wave period, as well as the propeller efficiency derived from the time-averaged forces. It is observed that the forced surge motion decreases propeller thrust, torque and efficiency in the entire third wave period; also, regular head waves imposes a negative effect on propeller's hydrodynamic performance when SDR = 1.0, as mentioned above.

Table 10 and Table 11 respectively present the coefficient of difference quantity of K_T and 10 K_Q under variant advance coefficient J in the three time periods Period 1, Period 2 and Period 3 within the third regular wave period (T_3 = 3.0156 s~4.5234 s), which is defined as $(K_{T(WITH\ SURGE)}/\ K_{T(WITHOUT\ SURGE)} - 1) \times 100\%$ for Table 10, and $(10\ K_{Q(WITH\ SURGE)}/\ 10\ K_{Q(WITHOUT\ SURGE)} - 1) \times 100\%$ for Table 11. It can be observed from the two tables below that, in time period P1 and P3, forced surge motions have negative effect on K_T and 10 K_Q; in time period P2, this effect is positive. In the whole regular wave period, forced surge motions have negative effect on K_T and 10 K_Q.

The propeller-induced nonlinear phenomena of the free surface, such as rolling and breaking, are one of the reasons for the deterioration and fluctuation of the hydrodynamic performance of propeller near the free surface in waves. It is reasonable to conjecture that the forced surge motion may cause some difference in the propeller-induced nonlinear phenomena, and this difference may contribute to the deterioration of propeller's hydrodynamic performance.

Therefore, in order to better understand the influence mechanism of forced surge motions on the hydrodynamic performance of open propeller in regular head waves, Figure 13 (a)–(l) shows the simulated free surface profile with/ without forced surge motions at each advance coefficient J. For each of the pictures (a) to (l) in Figure 13, the smaller picture placed on the upper side represents the simulated free surface in the focused part, i.e. the propeller and the near-propeller area, which is marked with a white dashed rectangle in the larger view of the simulated free surface form placed downside. It can be observed from the following pictures that, for the propeller with forced surge motion, the nonlinear phenomenon of the free sur-

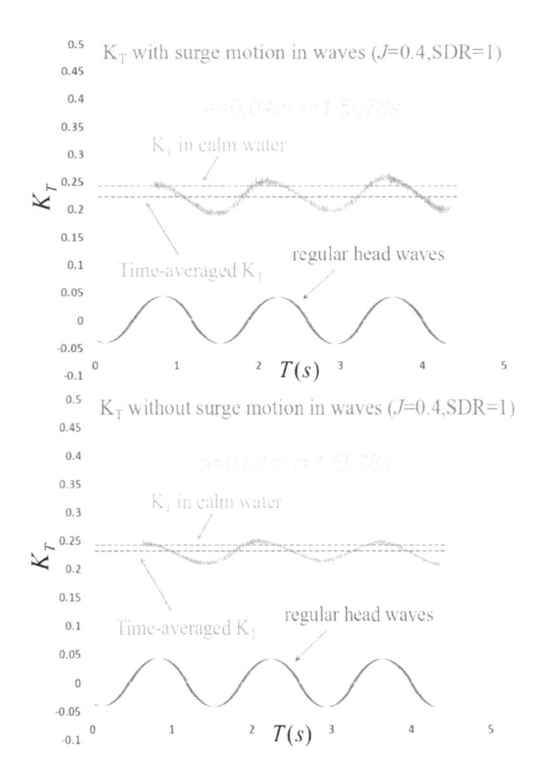

Figure 14. Comparison of instantaneous K_T in regular head waves, with and without forced surge motion. SDR = 1.0, J = 0.4.

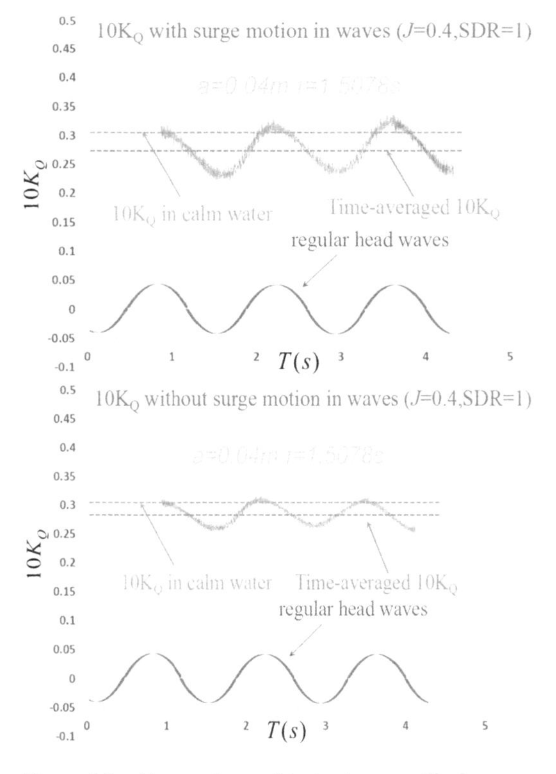

Figure 15. Comparison of instantaneous K_Q in regular head waves, with and without forced surge motion. SDR = 1.0, J = 0.4.

face becomes slightly more obvious than that without forced surge motion; this slight obviousness of the nonlinear phenomenon of the free surface with forced surge motions might to some extent be the reason for the negative impact of forced surge motion on the propeller's hydrodynamic performance, since more energy of the propeller is dissipated with the free surface nonlinear phenomenon.

4.3 *Thrust and torque fluctuations with/without surge motion effect in regular head waves*

The analysis in the preceding section focuses only upon the effects of the surge motion on propeller performance in a time-averaged sense. A further investigation is conducted on the unsteady features of propeller thrust and torque in regular head waves. For propeller MAU4-6075 working at SDR = 1.0 and J = 0.4, the effect of the forced surge motion on instantaneous thrust and torque coefficients are presented in Figures 14 and 15, respectively. It can be observed from the results that,

1. the thrust and torque both fluctuate at the wave frequency irrespective of the surge motion;
2. the fluctuation amplitudes of thrust and torque are both larger when the surge motion exists, simply because of the additional variations in the advance speed caused by the surge motion.

Obviously, the large amplitude of force fluctuations is associated with large variations of the attack angle and hence the blade loading, which can induce unsteady cavitation or even cause possibly serious strength problems in extreme motion/submergence conditions.

5 EFFECTS OF GEOMETRIC PARAMETERS ON PROPELLER PERFORMANCE WITH SURGE MOTION

In this section, the effects of blade geometric parameters on propeller hydrodynamic performance with forced surge motion are investigated numerically. The blade area ratio A_E/A_O and the pitch ratio P/D are considered.

5.1 *Blade area ratio*

As shown in Figure 16, three MAU-type propellers are selected. The blade area ratios (A_E/A_O) of them are 0.45, 0.59 and 0.73, respectively; and the pitch ratio (P/D) is 0.75 for all of them.

The numerical simulations are conducted at J = 0.1, 0.3 and 0.5. Figure 17 shows the time-averaged thrust and torque coefficients. According to the simulation results, the thrust and torque both increase significantly with blade area ratio; by increasing the blade area ratio from 0.45 to 0.73, the propeller thrust and torque can increase by approximately 20% at J = 0.1, and by approximately 25% and 35% at J = 0.3 and J = 0.5, respectively. It can be concluded that the influence of blade area ratio on the propeller's thrust and torque is significant at all three advance ratios J, and becomes even more significant as J increases.

Figure 18 present the nephogram of static pressure on blade face and blade back of the two propellers considering forced surge motion in regular head waves, with A_E/A_O respectively equal to 0.45 and 0.73. The advance coefficient J = 0, and the physical time is 4.5234s (3 times the wave period). It can be observed that, the pattern of static-pressure distribution is very similar for the two propellers with different A_E/A_O; for these two propellers, the maximum and minimum values of static pressure on the propeller blade, and the area of high-pressure and low-pressure zone, are approximately the same. It is the authors' conjecture that the blade area ratio might not have obvious effect on the pattern of static-pressure distribution on the propeller blades under this working condition.

5.2 *Pitch ratio*

As shown in Figure 19, three MAU-type propellers are selected. The pitch ratios (P/D) of them are 0.6,

Figure 16. Geometric model of the three MAU-type propellers having different blade area ratios.

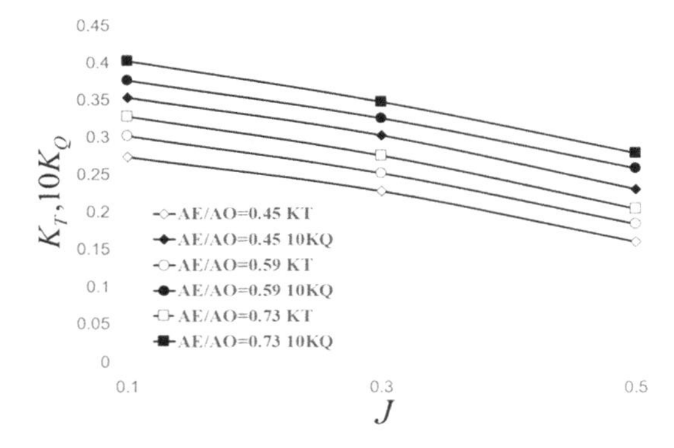

Figure 17. Time-averaged thrust and torque coefficients for the propellers with different blade area ratios.

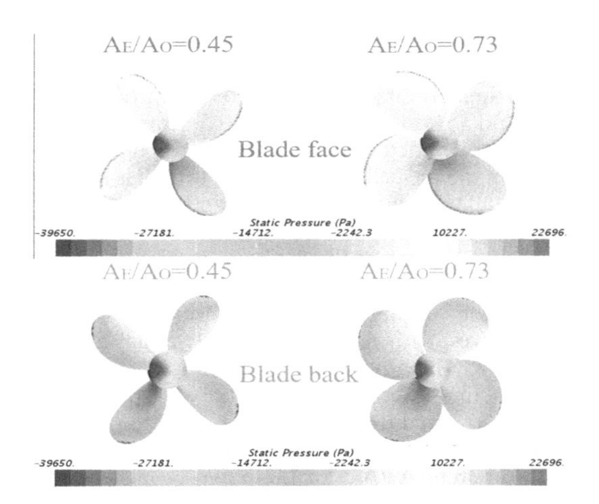

Figure 18. Nephogram of static pressure on blade face & back of the two propellers with different A_E/A_O and the same P/D.

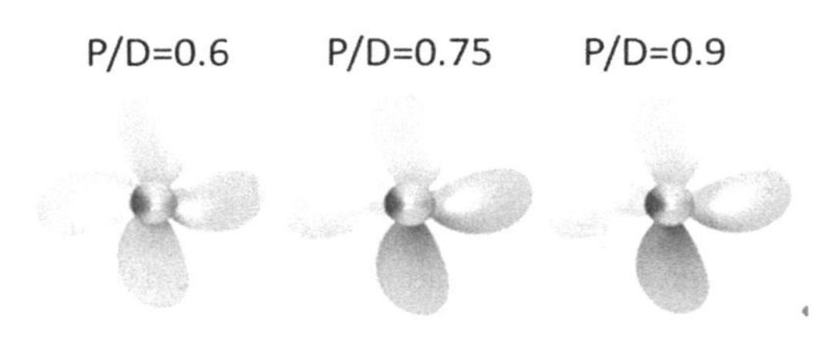

Figure 19. Geometric model of the three MAU-type propellers having different pitch ratios.

0.75 and 0.9, respectively; and the blade area ratio (A_E/A_O) is 0.59 for all of them.

The numerical simulations are conducted at J = 0.1, 0.3 and 0.5. Figure 20 shows the time-averaged thrust and torque coefficients. According to the simulation results, similar to the case of blade area ratio, the thrust and torque both increase significantly with the pitch ratio; however, unlike the case of blade area ratio, by increasing the pitch ratio from 0.6 to 0.9, the increase rate of propeller thrust and torque seems to stay constant with the advance ratio J, which is 25% approximately at all three advance ratios. It can be concluded that, the influence of pitch ratio on the propeller's thrust and torque is significant at all three advance ratios J, but remains relatively constant as J increases.

Figure 21 present the nephogram of static pressure on blade face and blade back of the two propellers with P/D respectively equal to 0.6 and 0.9 at J = 0. It can be observed that, the pattern of static-pressure distribution is not the same for the two propellers with different P/D; for the propeller with P/D = 0.9, the high-pressure area on the leading edge, as well as the low-pressure area on the blade

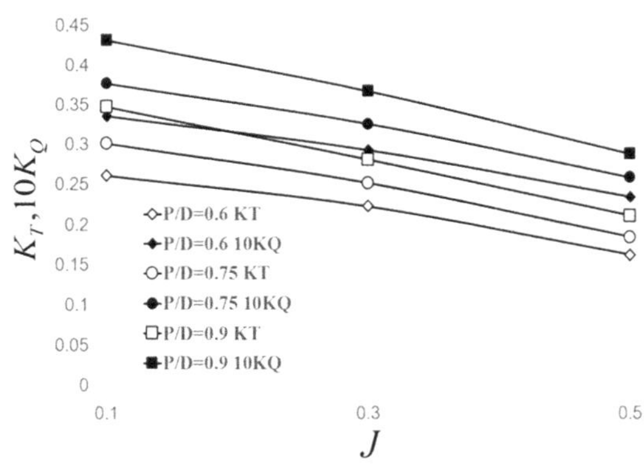

Figure 20. Time-averaged thrust and torque coefficients for the propellers with different pitch ratios.

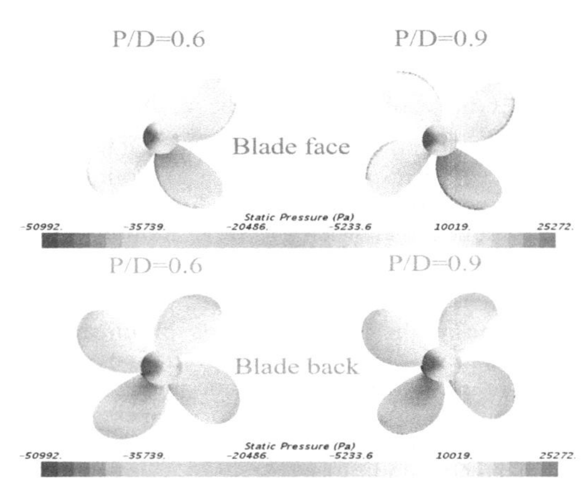

Figure 21. Nephogram of static pressure on blade face & back of the two propellers with different P/D and the same A_E/A_O.

tip, is more obvious than that with P/D = 0.6. It is the authors' conjecture that the reason for the positive correlation between the pitch ratio and the propeller thrust is that pitch ratio affects the pattern of static-pressure distribution.

6 CONCLUSIONS

The following conclusions can be drawn:

1. The numerical method applied in this paper has been found to have satisfactory accuracy and robustness when dealing with propeller racing and free-surface deformation.
2. When SDR = 1.0, regular head waves impose a negative effect on propeller's hydrodynamic performance.

3. In a time period long enough (for example in T_3), the surge motion reduces the time-averaged thrust and torque of the propeller in regular head waves. However, in smaller time periods (for example time periods 1, 2 and 3), the surge motion can reduce or increase propeller thrust and torque, depending on whether the propeller inflow is accelerated along the positive or negative direction of the *x*-axis due to the surge motion.
4. In regular head waves, the thrust and torque both fluctuate at the wave frequency irrespective of the surge motion, though the surge motion would make the fluctuation amplitudes increase.
5. When a propeller experiences the surge motion, its thrust and torque would increase significantly with the blade area ratio and the pitch ratio; it is conjectured that unlike pitch ratio, the blade area ratio might not have very obvious effect on the pattern of static-pressure distribution on the propeller blades.

Towards more comprehensive and practical evaluation of wave and ship motion effects on propeller performance, the combined effects of pitch and heave motions as well as the effects of ventilation and cavitation will be studied in the future.

ACKNOWLEDGEMENT

This research is supported by the China Ministry of Education Key Research Project "KSHIP-II Project" (Knowledge-based Ship Design Hyper-Integrated Platform): No. GKZY010004. The authors are also grateful to Mr. Dan Qiao, Mr. Si Chen, Dr. Han Liu and Mr. Chengyuan Ma for their kind help in this study.

REFERENCES

Bernitsas M.M., Ray D., Kinley P. K_T, K_Q and efficiency curves for the wagningen B-series propellers. May, 1981.

Cao M. L. The open-water characteristics of a propeller with varying depth of shaft and in waves [J]. *Journal of Shanghai Jiao Tong University*, 1988, 22(3): 27–35. (in Chinese).

Dong G. X., Li J. P., He H. M. A study on propeller open-water hydrodynamic performance in regular head waves [C]. *Proceedings of ship hydrodynamics academic conference*, 2013. (in Chinese).

Guo C. Y., Zhao D. G., Wang C., et al. Chang Xin. Experimental research on hydrodynamic characteristics of propeller in waves[J]. *Journal of Ship Mechanics*, 2012.9, 9(16).

Lee S. K., Yu K., Chen H. C., et al. CFD simulation for propeller performance under seaway wave condition [C]. *Proceedings of the Twentieth (2010) International Offshore and Polar Engineering Conference*, Beijing, China, 2010.

Liang Q. C., Wan L. A method for calculating the hydrodynamic changes of propeller in waves [J]. *Ship Science and Technology*, 2006, 28(4): 32–35. (in Chinese).

Naito S., Nakamura S., et al. Open-water characteristics and load fluctuations of propeller at racing conditions in waves [J]. *Journal of Kansai Soc of Naval Arch.*, Japan, 1979(172): 51–63.

Politis G.K. Ventilated Marine Propeller Performance In Regular And Irregular Waves; An Experimental Investigation [J]. *Transactions on Modelling and Simulation*, 1999.

Queutey P., Wackers J., Leroyer A., et al. Dynamic Behaviour of the Loads of Podded Propellers in Waves: Experimental and Numerical Simulations [C]. *Proceedings of the ASME 2014 33rd International Conference on Ocean, Offshore and Arctic Engineering, OMAE2014*, June 8–13, 2014, San Francisco, California, USA.

Tao Y. S., Zhang F., Feng T.C. A study on propeller thrust and torque increase in regular waves [J]. *Shipbuilding of China*, 1991(1): 47–57. (in Chinese).

Yu X. Numerical simulation of hydrodynamic performance of propeller with heave motion in waves [D]. Harbin Engineering university, 2008, Harbin, China.

Yu Y. H., Chen J. L., Yang H. Design of construction method of 10000TEU ultra-large container carrier [J]. *Jiangsu Ship*, 2015. (in Chinese).

Zhao Q. X., Guo C.Y., Su Y.M., et al. Study on unsteady hydrodynamic performance of propeller in waves. *Journal of Marine Science and Application*, 2017, 16(3).

Marine Design XIII – Kujala & Lu (Eds)

Application of CAESES and STARCCM + for the design of rudder bulb and thrust fins

Fan Yang, Weimin Chen, Xiaojun Yin & Guoxiang Dong
Shanghai Ship and Shipping Research Institute, Shanghai, China

ABSTRACT: With environmental concerns becoming the most important issue which the industry is facing today, there exists a strong demand for Installing Energy Saving Devices (ESDs) for both existing and manufacturing ships. Although the energy saving effect of post-ESDs are not as obvious as pre-ESDs, the structures of post-ESDs, such as rudder bulb and rudder fins, are much simpler and easier to be fixed on the ship. This paper focuses on designing a combined ESD with rudder bulb and thrust fins for a larger bulk carrier by incorporating the parameterized modeling software CAESES and STARCCM+. Genetic algorithm NSGA-II is involved to optimize the ESD intelligently. After calculating hundreds of schemes, an optimal scheme which has more than 1% energy saving effect is obtained and the result is validated by model test in SSSRI's towing tank.

1 INTRODUCTION

There are many factors leading to the deep concern about energy saving devices (ESDs) such as environmental awareness, fluctuating fuel prices, international regulations and the competitions between enterprises. As downstream ESDs, the production process of rudder bulb and thrust fin could be relatively simple. Because of this, many scholars and engineers have been studying the working principles and design methodologies.

Guo and Huang (2006a, 2006b) used lifting surface method, panel method and non-linear vortex lattice method for the calculation of propeller and rudder with additional thrust fins respectively. Through comparing the results with and without additional thrust fin, it gave the working principle of the additional thrust fin. Guo and Huang (2007) used panel method for the calculation of the unsteady performance of propeller and rudder with additional thrust fin. The influence of ship hull was given as unsteady flow on the disc of the propeller. Li (2009) used lifting surface method to analyze the impact of different diameters of rudder bulb on the hydrodynamic performance of propeller-rudder-bulb system. Guo and Hu (2010) used FLUENT to calculate the interaction among rudder, propeller and thrust fin. Liu (2011) studied the impact of parameters such as the diameter of rudder bulb, the span-chord ratio and installing angle of thrust fin based on lifting surface method. He (2012) investigated the influence of rudder bulb on propulsive system based on lifting surface method, the results showed the optimum matching point between rudder bulb diameter and propeller diameter. For the rudder bulb length, the energy saving effect improved when the length is less than a value. Shen and Liu (2013) used CFD method to design and optimize a rudder bulb for a full form ship by changing bulb diameter and profile manually, the results indicated that the energy-saving for rudder bulb is around 2%. Wang (2013) used FLUENT to optimize the energy saving effect of rudder bulb and thrust fin under propeller by changing rudder bulb diameter and thrust fin's position. Sun (2016) optimized the thrust fin which was installed on a twisted rudder with rudder bulb by changing the installation position, the span-chord ratio and the installation angle based on CFD method, the system efficiency was improved by 1.2% finally. In the FP7 project GRIP S. Coache and M. Meis (2016) used STARCCM+ to optimize a rudder bulb for a bulk carrier by changing 7 parameters manually, but unfortunately they failed.

With the development of parametric modeling tools, CFD software and calculation capabilities, it becomes more and more popular to combine these three resources for the optimization of propellers and energy-saving devices. J. Park and M. Peric from Hyundai Heavy Industries (2015) have optimized the propeller for a merchant ship by using CAESES and STARCCM+. In the FP7 project GRIP, Yan and Streckwall (2016) from HSVA and H.J. Prins (2016) from MARIN also used the similar method to design an upstream ESD for a bulk carrier and gained obvious energy saving effect. In the HSVA's Website, they claimed to use a parametric modeling approach to design an unconventional rudder bulb which has a more obvious effect than the conventional one.

This paper focuses on designing rudder bulb and thrust fin which is a combined ESD for a larger full bulk carrier by using CAESES and STARCCM+. The working process will be outlined. Finally, the 1.8% energy saving effect of the optimized ESD in model scale was validated by model test conducted by Shanghai Ship and Shipping Research Institute.

2 COMPUTATIONAL METHOD

The state-of-the-art CFD software STARCCM+ was applied for evaluating the ESD's energy saving effect in this paper. The detail methods used by this software are explained below.

2.1 *Governing equations*

In this paper, software STARCCM+ is used to simulate the resistance, open water and self-propulsion. This software uses finite volume method to solve the transport (continuity and momentum) equations (Choi et al., 2010). In terms of the unsteady incompressible fluid, the governing equations adopted here is the Unsteady Reynolds-Average Naiver-Stokes (URANS) equations. The equations are a mass conservation equation and a momentum conservation equation.

Control equations are listed as follows:

The equation of continuity:

$$\frac{\partial u_i}{\partial x_i} = 0 \tag{1}$$

Momentum equation:

$$\frac{\partial u_i}{\partial t} + \frac{\partial u_i u_j}{\partial x_j} = \frac{\partial p}{\partial x_j} + \frac{1}{\mathrm{Re}}\frac{\partial}{\partial x_j}\left(\frac{\partial u_i}{\partial x_j} + \frac{\partial u_j}{\partial x_i}\right) + \frac{\partial}{\partial x_j}\left(-\overline{u'_i\ u'_j}\right) \tag{2}$$

where ui is average Reynolds velocity components, xi is the component of coordinate, $R_e = (UL)/\nu$ is Reynolds number, υ is kinematic viscosity coefficient, $-\overline{u'_i\ u'_j}$ is the Reynolds stress item.

2.2 *VOF and turbulence model*

VOF model is used to simulate the wave making, the second order discrete format and reasonable Sharpening Factor are used to obtain more accurate wave making. Multiphase Segregated Fluid model was used to solve conservation equations for mass, momentum, and energy for air and water phases. Implicit unsteady solver was used due to the transient nature of the VOF scheme.

Since the SST k-ω turbulence model inherits the advantages of k-ω model in the near-wall region and k-epsilon model in the far field, meanwhile the model increases the transverse dissipation derivative term, the transport processes of turbulence shear stress are considered, it can be used to calculate the flow with adverse pressure gradient and the flow around the airfoil. Following Du's (2012) and Yang's (2013, 2016) study on resistance and self-propulsion calculation for some bulk carriers and ESDs using CFD software, the SST k-ω turbulence model is used to close the control equations. SIMPLE algorithm is adopted to solve the coupled equation of velocity and pressure.

3 ESD HYDRODYNAMIC DESIGN AND OPTIMIZATION PROCEDURE

It is common that a strong low-pressure core exists behind a conventional propeller hub cap which causes thrust deduction by pulling the propeller hub cap. Rudder bulb can fill the low-pressure area and provide good rectification for the down streaming flow. There is still lots of energy cannot be absorbed by a propeller which flows away with the wake. Thrust fin can provide additional thrust by recovering some rotational energy behind the propeller. According to the energy saving mechanism, we try to design a rudder bulb and thrust fin for a bulk carrier.

The design procedure of the rudder bulb and thrust fin is made of three main steps, as illustrated in Fig. 1:

1. parametric modeling rudder bulb and thrust fin by using CAESES.
2. optimization for the rudder bulb and thrust fin in the open water condition.
3. evaluation for the optimal one by self-propulsion simulation.

In the following, each step will be explained in detail.

3.1 *Geometry parametric modeling*

The geometry of a rudder bulb can be basically described by parameters as follows: Radius of front Rudder Bulb, Maximum Radius, Position of M.R., Profile of the Rudder Bulb, Length of rudder Bulb,

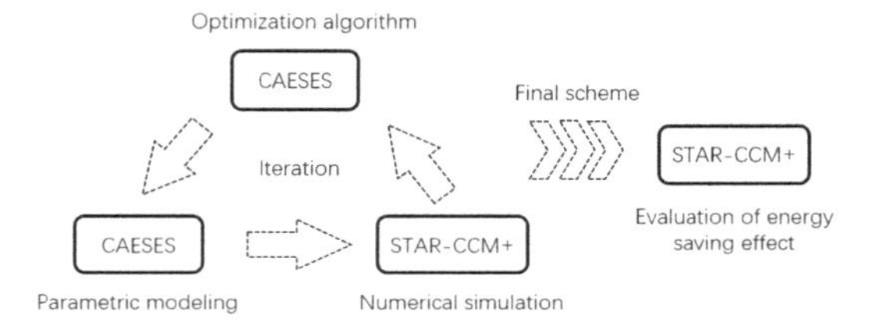

Figure 1. ESD design and optimization process.

The Gap between hub and R.B. as shown in the Fig. 2. B-spline curve is applied for describing the profile of the Rudder Bulb in our study. The shape of the rudder bulb is shown in Fig. 3.

The geometry of a thrust fin can be basically described by parameters as follows: section profile, chord, camber, angle of attack, x position of tip, x position of root, span, as shown in Fig. 4.

A NACA 4-digit cambered profile has been used for the thrust fin design, the value of camber and chord can be changed separately at the tip and root of the fin. These two parameters for the fin in each side are the same to reduce the number of parameters.

Although the extrusion path can be defined arbitrarily, to make the optimization process much simpler, we take a straight line as the path. However, considering the difference of the wake especially the velocity direction in the port and starboard sides, the general angle of attack can be varied in each side.

The pair of thrust fin is on the same horizontal plane with the propeller shaft, but the X position of tip and root can be changed. The span can also be changed. The parameters of the thrust fin are shown in Table 1, the shape of the thrust fin is shown in Fig. 5.

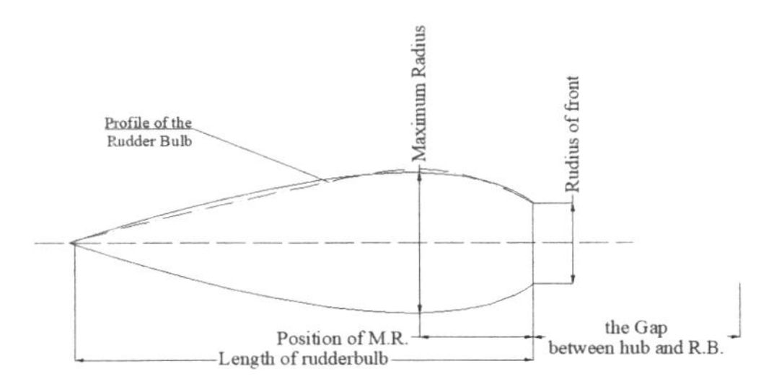

Figure 2. The parameters of rudder bulb.

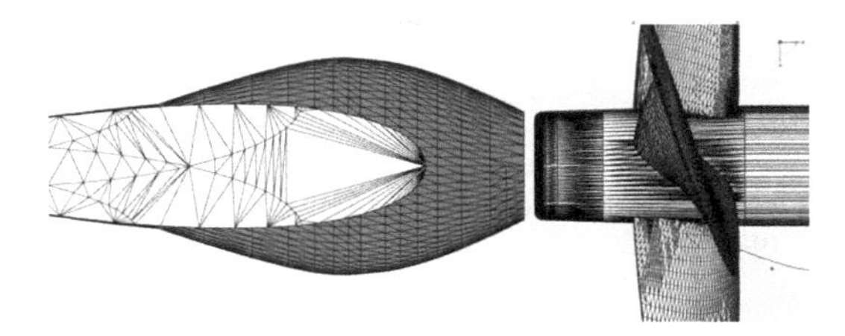

Figure 3. The shape of the rudder bulb.

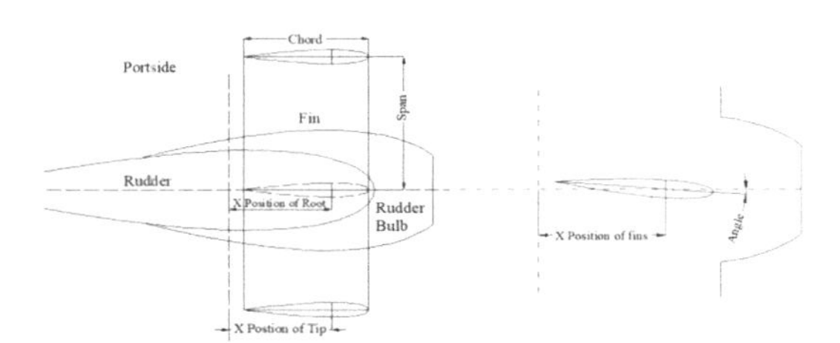

Figure 4. The parameters of thrust fin.

3.2 *optimization of rudder bulb in the open water condition*

In this paper, we try to optimize the rudder bulb in the open water condition at first. The optimization workflow combined the CAESES and STARCCM+ is shown in Fig. 6.

As mentioned before, the rudder bulb is parametric modeled by CAESES. After that the rudder bulb is installed on the rudder behind the propeller virtually.

Table 1. The parameters of thrust fin.

NO.	The Port side	The Starboard
1	chord	
2	camber	
3	Angel of attack	–
4	–	Angel of attack
5	X position of tip	
6	X position of root	
7	Span	

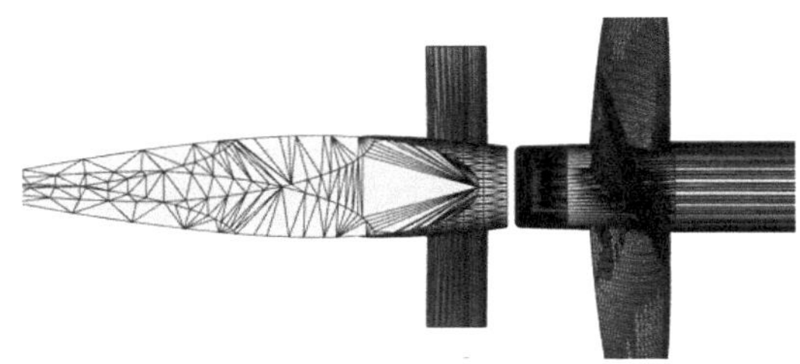

Figure 5. The shape of the thrust fin.

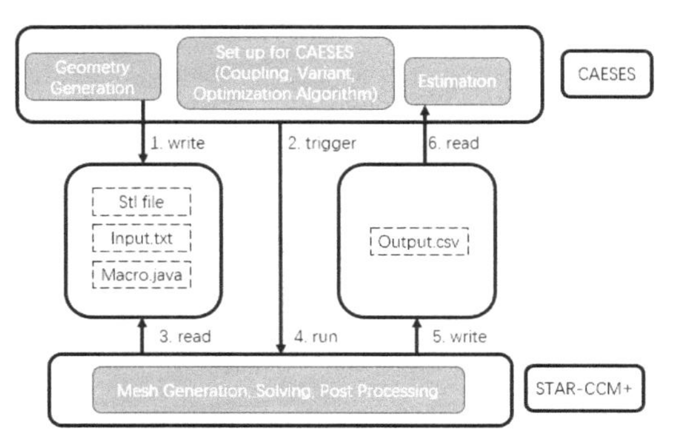

Figure 6. Optimization workflow integrating with CAESES and STARCCM+.

Ideally, self-propulsion evaluation and analysis in full scale should be needed to evaluate different ESD designs and by comparing the predicted delivered power of the ship with and without an ESD, the best ESD can be selected. But it is time consuming in optimization stage using this method, fortunately rudder bulb is a kind of downstream ESD, it would not influence the upstream flow, for simplification, here the hydrodynamic performance of the whole system inclined propeller, rudder with and without rudder bulb is estimated in the model scale and the open water condition by STARCCM+.

Applying genetic algorithm NSGA-II, we try to find the ESD which will maximize system thrust and minimize system torque. Hundreds of schemes have been evaluated, according to the optimization criterion, the best scheme is selected for further optimization for the thrust fin.

Fig. 7 shows the relationship between the width of rudder bulb and thrust of the system, there exists an optimal width of rudder bulb, too large or too small will reduce the thrust of the system. Fig. 8 shows the relationship between the width of rudder bulb and torque of the system, to obtain smaller torque, diminution of the width of rudder bulb is necessary. With the optimization process progressing, propulsion efficiency of the whole system was increased, as shown in Fig. 9.

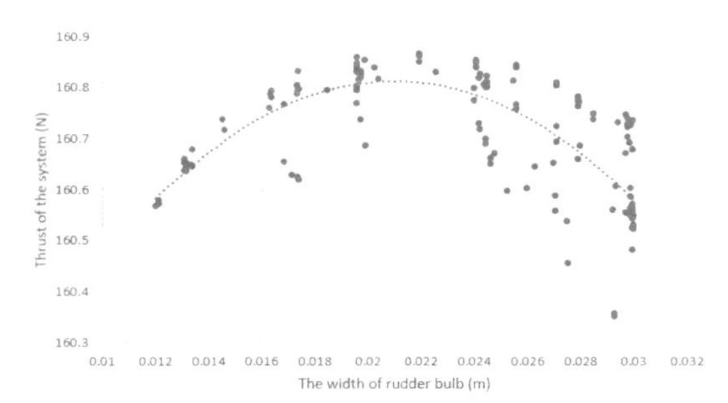

Figure 7. The relationship between the width of rudder bulb and thrust.

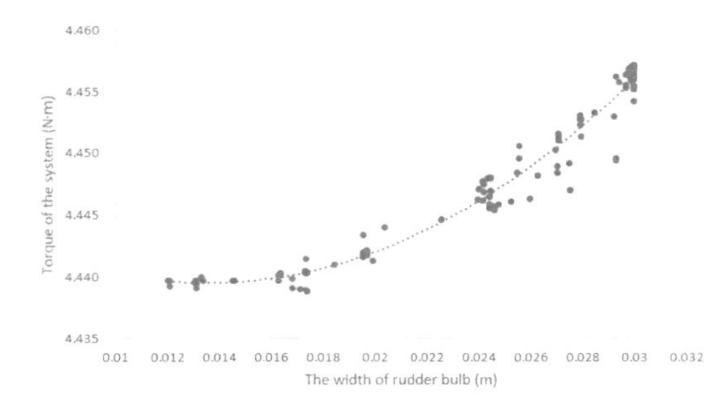

Figure 8. The relationship between the width of rudder bulb and torque.

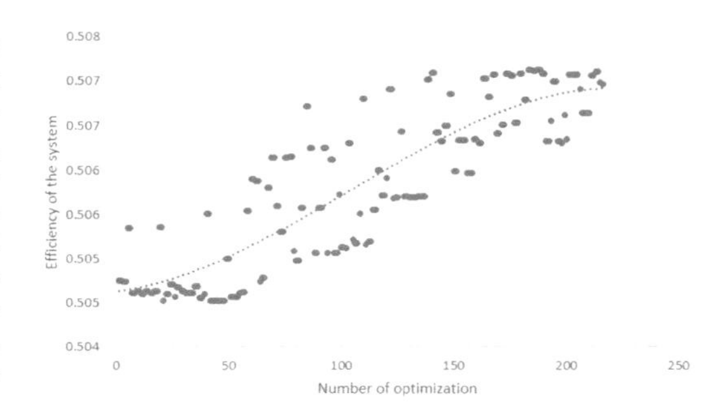

Figure 9. The relationship between the number of optimization and efficiency.

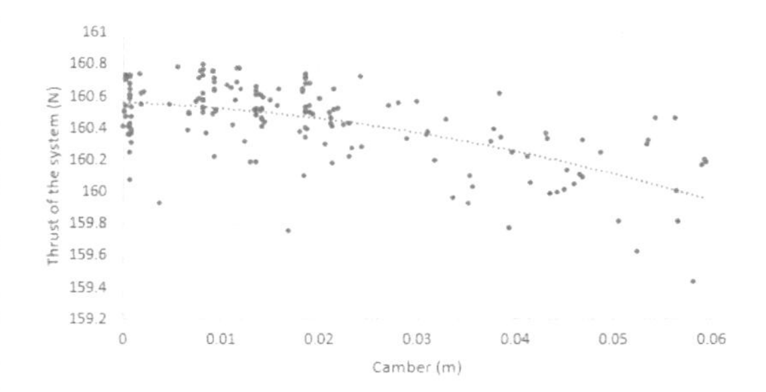

Figure 10. The relationship between camber and thrust.

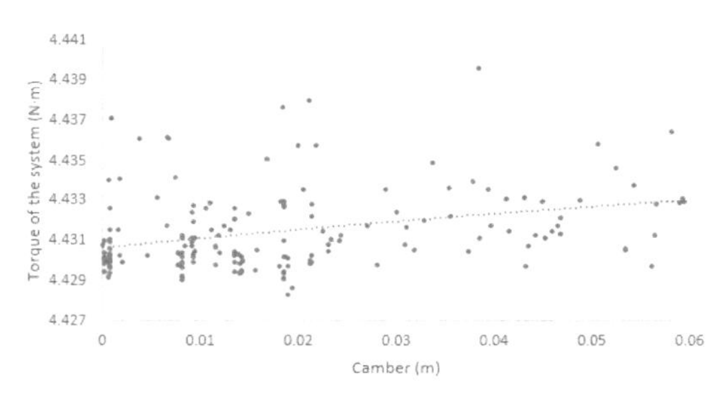

Figure 11. The relationship between camber and torque.

3.3 *Optimization of thrust fin in the open water condition*

The thrust fin is also parametric modeled by CAESES and installed on the combination of rudder and optimized rudder bulb. In the same fashion, with the control of NSGA-II, hundreds of schemes have been evaluated in open water condition. Fig. 10 and Fig. 11 show the influence of camber on the thrust and torque of the system. Fig. 12 shows that with the number of optimization increasing, a better ESD can be found. But when the number of optimization is greater than 125, it is hard to obtain a more effective ESD. Finally,

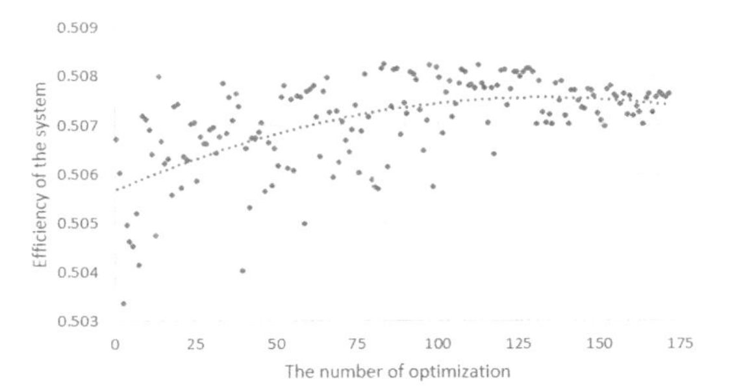

Figure 12. The relationship between the number of optimization and efficiency.

according to the same optimization criterion, the suitable scheme is selected for the final evaluation

4 EVALUATION BY SELF-PROPULSION SIMULATION

4.1 *Estimation process of energy saving effect*

The key to estimate the energy saving effect of an ESD by CFD is the numerical simulation of self-propulsion. In fact, the movement of a ship always makes waves, so the simulation need to consider the free surface which means it requires a calculation of multiphase flow, but it will take a long time to complete a calculation which is not suitable for the engineering application. Considering the position of the combined ESD which is below the surface in a certain depth, it can be assumed that the waves will not influence its hydrodynamic performance. Self-propulsion simulation can be done by using single phase calculation method for evaluating the energy saving effect of an ESD. Because the rudder bulb and thrust fin has very little influence on the resistance which is proved to be true in the next section, the estimation of the energy saving effect of the ESD can be made by comparing the propulsion efficiency with and without the ESD finally.

First of all, the open water performance of the propeller and the resistance of ship model R_M can be obtained respectively by carrying out the open water simulation and model ship resistance test simulation. Then simulating self-propulsion test of model ship by changing three different rotation rates, the resistance, thrust and torque in three different rotations can be obtained.

According to formulas,

$$\Delta = T + F_D - R_M \tag{3}$$

$$F_D = \frac{1}{2}\rho_m S_m v_m^2 \left(c_{fm} - c_{fs} \right) \tag{4}$$

Curve $\Delta \sim n$ can be built and the corresponding rotation rate n at self-propulsion point can be obtained when $\Delta = 0$. According to curve T-n, Q-n, the corresponding thrust T and torque Q_B can be obtained, then thrust coefficient K_T and torque coefficient K_{QB} can be calculated. According to the thrust identity method, the advance coefficient J, torque K_{Q0} and open water efficient η_{0m} can be obtained from open water characteristics curves.

According to

$$\eta_{Rm} = K_{Q0} / K_{QB} \tag{5}$$

Relative rotation efficiency can be obtained.

According to

$$t_m = \frac{T - R_m + F_D}{T} \tag{6}$$

$$w_m = 1 - \frac{JnD}{V_m} \tag{7}$$

Thrust deduction and wake fraction can be obtained, and then hull efficiency can be calculated. According to

$$\eta_d = \eta_0 \cdot \eta_r \cdot \eta_h \tag{8}$$

The propulsion efficiency η_D can be obtained finally.

Based on this process, the propulsion efficiency without the ESD η_{D0} and the propulsion efficiency with the ESD η_D can be obtained respectively.

The energy saving effect of the ESD can be evaluated by (9) in model scale.

$$\left(\frac{\eta_{Dpss}}{\eta_{D0}} - 1 \right) \times 100\% \tag{9}$$

4.2 *Computational domain and mesh*

In this paper, the origin of coordinate system is defined as the point of intersection among central fore and aft plane, water plane and zero station, the direction of X axis is from stern to bow and the direction of Z axis is vertical upward. The size of computational domain has a certain impact on the calculated results, if the domain is too small, the velocity of flow around the ship would not be the set value due to blockage effect. Considering the calculation time, the domain is expanded as large as possible, -3 LPP < X < 3 LPP, -2 LPP < Y < 2 LPP, -1.5 LPP < Z < 0.5 LPP. Slide mesh approach is used to simulate the rotation of a propeller. The computational domain is shown in Fig 13. In the STARCCM+ software, the inlet boundary condition is set as velocity inlet, the outlet boundary condition is set as pressure outlet, the other sides are set as symmetry plane, the boundary conditions are also shown in Fig. 13.

To some extent, the accuracy of calculation results is affected by the style, quality and quantity of the mesh. Reviewing the development and application of CFD method in recent years, especially the calculation of resistance and self-propulsion, the improvement of the calculation accuracy is often accompanied by upgrading mesh generation methods or some meshing relative techniques. Good quality of mesh can help to accelerate the convergence of calculation and improve the accuracy of the calculation.

The trimmed cell mesher and prism layer mesher methods provided by STARCCM+ are used to generate grid. The trimmed cell mesher provides a robust and efficient method of producing a high-quality grid for both resistance and self propulsion mesh generation problems. The prism layer mesh model is used with a core volume mesh to generate orthogonal prismatic cells next to the surfaces of ship, rudder and propeller. This layer of cells is necessary to improve the accuracy of the flow solution. With these robust methods, a high quality of grid can be obtained.

According to the experiences and the similar cases which have done before, we set the base size of the mesh as 1.67% of the Length of Perpendiculars which can not only guarantee the accuracy of results but also reduce the simulation time. Considering the complex geometries and the high velocity gradient flow areas, such as the bow and stern, as well as the free surface area, lots of grids have been distributed in these regions to improve the mesh quality and computation accuracy. The total number of grids is around 3 million. The grids at the bow and stern are shown in Fig. 14.

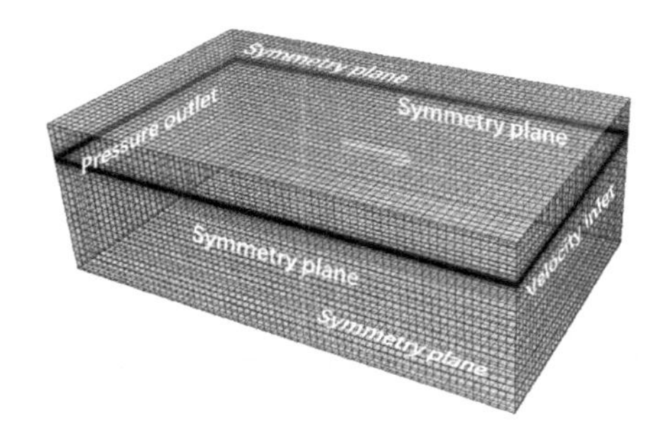

Figure 13. The computational domain and mesh.

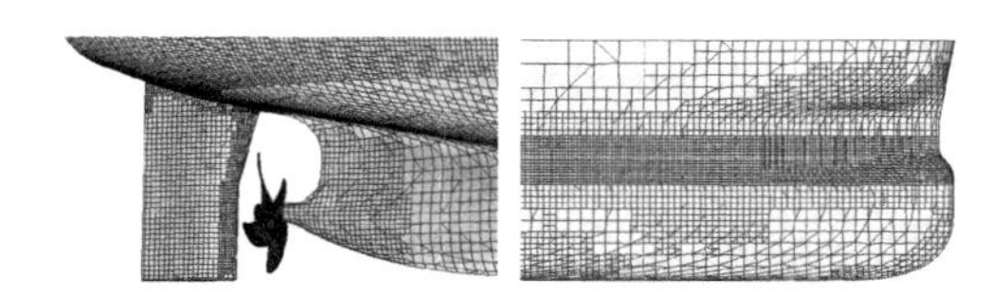

Figure 14. Local mesh at bow and stern.

4.3 *The calculation results*

Calculation results of resistance, self-propulsion performance at design speed with and without the ESD, are given in Table 2 and Table 3. According to Table 2, the influence of the ESD on the clam water resistance is very little and can be neglected. According to Table 3, this combined ESD can improve the propulsion efficiency by 1.3% in model scale. Fig. 15 shows the wave making by the bulk carrier in resistance simulation condition. Figure 16 shows the pressure contour of the rudder and the streamline behind the propeller close to the ESD.

Table 2. The influence of the final scheme to the resistance.

ESD	Resistance (N)
without	33.80
with	33.88
differ	0.2%

Table 3. Self-propulsion calculation results in model scale.

ESD	n (rps)	KT	10 KQ	ETA
without	7.185	0.1652	0.1968	0.8009
with	7.160	0.1666	0.1963	0.8115

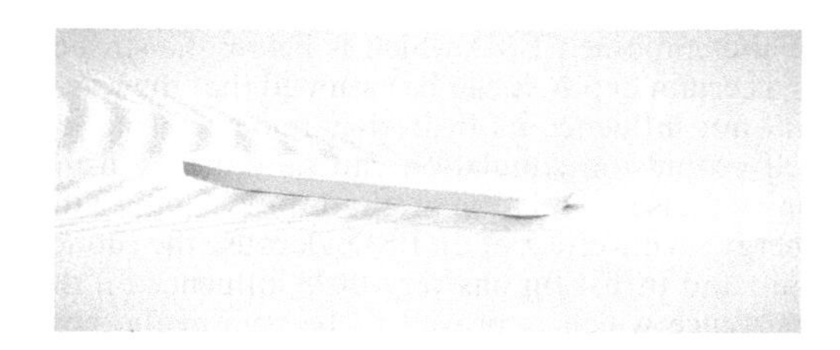

Figure 15. The wave making in resistance condition.

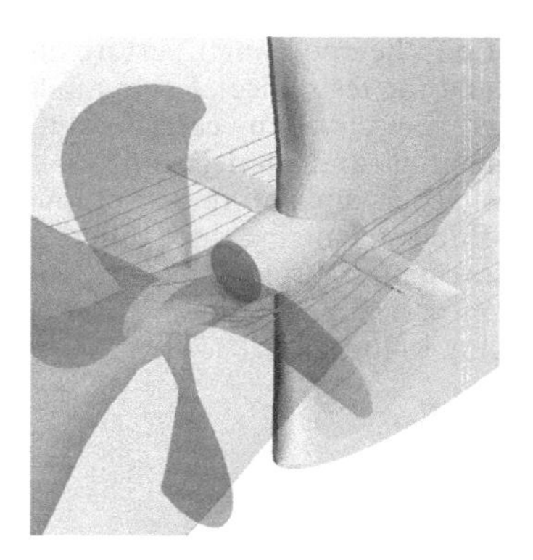

Figure 16. The flow situation in self-propulsion condition.

Table 4. The energy saving effect of the optimal ESD based on model test data.

ESD	Propulsion efficiency	Delivered power (kW)
Without	0.7505	16479.2
with	0.7639	16222.3
Differ	1.8%	−1.6%

5 MODEL TEST VALIDATION

In order to validate the evaluation of the energy saving effect of the combined ESD by CFD method, we conducted a series of model tests in SSSRI, which has already obtained EEDI certificates from BV classification society and Lloyd's Register. According to the model test results, the ESD can increase the clam water resistance by about 0.1% which is consistent with the CFD result. Table 4 shows the difference of propulsion efficiency with and without the ESD in model and the difference of delivered power with and without the ESD in full scale. According to the data, the ESD can improve the propulsion efficiency by 1.8% in model and reduce the delivered power by 1.6% in full scale.

6 CONCLUSIONS

This paper focuses on designing a combined ESD with rudder bulb and thrust fin for a larger full bulk carrier. Using the parameterized modeling software CAESES to model the rudder bulb and thrust fin respectively, using STARCCM+ to simulate the performance of the propulsion system including propeller, rudder and the ESD in open water condition. We try to find the ESD which can maximize system thrust and minimize system torque. There are many parameters can influence the performance of propulsion system, so genetic algorithm NSGA-II is involved to search the optimal scheme automatically and intelligently. After calculating hundreds of schemes, an optimal scheme obtained at the current stage. More than 1% energy saving effect of this final ESD is evaluated in self-propulsion condition by CFD methods and validated by towing tank tests.

REFERENCES

Choi, J.E. & Min, K.S. & Kim, J.H. & Lee, S.B. & Seo, H.W. 2010. Resistance and propulsion characteristics of various commercial ships based on CFD results. Ocean Eng. 37(7): 549–566.

Coache S. & Meis M. 2016. ESD design for a validation bulk carrier. International Shipbuilding Progress 63: 211–226.

Du, Yunlong & Chen, Xiaping & Chen, Changyun. 2013. Numerical simulation and comparison of performance for a bulk carrier with two different hull lines. Journal of hydrodynamics. 28:5.

Guo, Chunyu & Hu, Wenting. 2010. Using RANS to Simulate the Interaction and overall Performance of Propellers and Rudders with Thrust Fins. Journal of Marine Science Application, 9: 323–327.

Guo, Chunyu & Huang, Sheng. 2006a. Research on the rudder with additional thrust fins used by non-linear Vortex-Lattice method. Journal of Huazhong University of Science and Technology, 34(6): 87–89.

Guo, Chunyu & Huang, Sheng. 2006b. Research on performance of rudder with additional thrust fin by panel method. Journal of Harbin Engineering University, 27(4): 501–504.

Guo, Chunyu & Huang, Sheng. 2007. Numerical simulation of unsteady flow around propeller and rudder with additional thrust fin. Journal of Harbin Engineering University, 28(3): 259–262.

He, Miao. 2012. Design and hydrodynamic performance simulation of integrative energy-saving propulsion system. [D] Harbin: Harbin engineering Univ.

Li, Xin. 2009. A hydrodynamic analysis on energy-saving rudder ball behind propeller. [D] Harbin: Harbin engineering Univ.

Liu, Yebao. 2011. Study of Energy-Saving Rudder-Bulb-Fin Combination Behind Propeller. [D] Harbin: Harbin engineering Univ.

Park, Jeong-yong & Peric, Milovan. 2015. An Optimization Process for Propeller Design and its Application, Proceedings of the Twenty-fifth International Ocean and Polar Engineering Conference.

Prins, H.J. & Flikkema M.B. & Schuiling B. 2016. Green retrofitting through optimization of hull-propulsion interaction—GRIP, 6th Transport Research Arena.

Shen, Hailong & Liu Yebao. 2013. Study of Optimization Design and Energy Saving Effect Prediction of Rudder Bulb Based on CFD Technique. Advanced Materials research, Vols. 724–725: 986–989.

Sun, Yu & Su, Yumin. 2016. Research on Influence of Rudder-Bulb-Fin Parameters on Hydrodynamic Performance of Twisted Rudder. Journal of Ship Mechanics 1007–7294 09–1071–12.

Wang, Ying. 2013. The numerical analysis of propeller-rudder-ball-fin system's hydrodynamic performance. [D] Harbin: Harbin engineering Univ.

Yan, Xiong-Kaeding & Heinrich, Streckwall. 2016. ESD design and analysis for a validation bulk carrier, International Shipbuilding Progress 63 (2016/2017): 137–168.

Yang, Fan & Chen, Xiaping & Chen, Changyun. 2014. The influence of design parameters of a pre-swirl stator on its performance PAAMES2014.

Yang Fan & Dong, Guoxiang & Chen, Hao. 2016. The selection of different hull line schemes based on numerical simulation methods PAAMES2016.

Marine Design XIII – Kujala & Lu (Eds)
© 2018 Taylor & Francis Group, London, ISBN 978-1-138-34076-3

Design verification of new propulsion devices

Xun Shi, Jun Song He, Yao Hua Zhou & Jiang Li
Shanghai Rules & Research Institute, China Classification Society, Shanghai, China

ABSTRACT: As the ship's navigation areas are expanding, a variety of hydrological conditions put forward higher requirements on ship propulsion system from different aspects, such as strength, environmental protection, energy efficiency, draught and so on, which makes new propulsion devices constantly emerging. How to ensure the safety of each device in operation is a problem that every verifier needs to solve. In recent years, China Classification Society (CCS) has conducted research on various propulsion devices and propulsion methods relying on national research resources, and has preliminarily formed a verification system for different propulsion devices. This paper takes polar propeller, semi-submerged propeller and straight blade propeller as examples to introduce the verification system and verification tool, aiming at providing reference for the relevant organizations in this industry.

1 INTRODUCTION

As the demands for commercial trade and resource development increase, the ship's navigation areas are gradually expanding. In addition to meeting the basic requirements of speed, the complex use environment and water temperature conditions have put forward higher requirements on the ship propulsion system from different aspects such as strength, environmental protection, energy efficiency and draught, which make more and more new propulsion devices constantly appear. Therefore, how to guarantee the safety of new propulsion devices in operation is a problem that has to be faced.

In recent years, China Classification Society has cooperated with industry and carried out a series of research on new propulsion devices, including the brand new structure design of the new propulsion device which is different from the traditional propulsion system, the particularity of the load conditions, potential structural strength issues, and so on. Based on these studies, China Classification Society has developed strength calculation and verification software for special load conditions and formed a verification system for the safety assessment of new propulsion devices. At present, the relevant research results have been successfully applied to the design verification and safety assessment of several new propulsion devices. This paper introduces the verification system of new propulsion devices and the strength calculation and verification software for special load conditions, which will help to the design and development of more new propulsion devices in the future.

2 TYPE OF NEW PROPULSION DEVICES

In general, according to the work mode the new propulsion devices can be divided into the traditional design meeting special requirements, and the non-traditional design with the function of propeller and rudder.

2.1 *Traditional design meeting special requirements*

This kind of propulsion device is not different from the traditional propulsion system in form, layout and supporting systems. But because of the special work environment or other reasons, the propulsion device must meet the special strength or vibration requirements in addition to the general design requirements. The polar propeller is a typical representation of this type of propulsion devices.

Sea ice and low temperature are the main challenges for the design and build of polar ships. Ice can be great potential damages to the structure and power system of ship, such as hull, propulsion system and rudder. The propeller of the polar ship is usually exposed to the stern of the ship and immersed in shallow water, and the suction effect of the propeller in ice conditions can cause the hard ice in entrainment flow in ship stern, so the leading edges and tips of the propeller will inevitably collide with the ice. Therefore, the strength of the polar propeller blades and the vibration of shaft system caused by the collision require special consideration.

2.2 *Non-traditional design*

Unlike the traditional design of propeller and rudder separation but cooperation to achieve propulsion and steering, many new propulsion devices are combinations of propeller and rudder, which achieve propulsion and steering simultaneously. This kind of design includes water jet, semi-submerged propeller (as shown in Figure 1) and straight blade propeller (as shown in Figure 2) and so on.

The straight blade propeller device is used for low speed ship. The thrust of straight blade propeller device generated by the blades which inserted vertically into the water rotating around the fixed center and blades rotating around the axis of the device. The straight blade propeller is a two-in-one design of the blade and rudder, which does not require additional supporting equipment. The semi-submerged propeller and water jet, which are mainly used for high speed ships, are equipped with powerful hydraulic steering gears, and steering by adjusting the direction of the thrust.

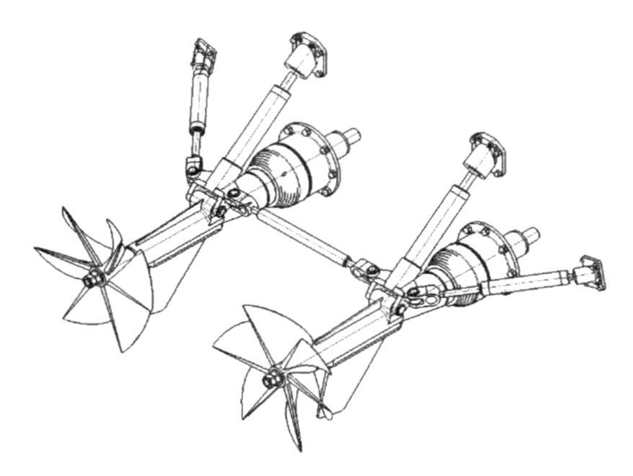

Figure 1. Semi-submerged propeller.

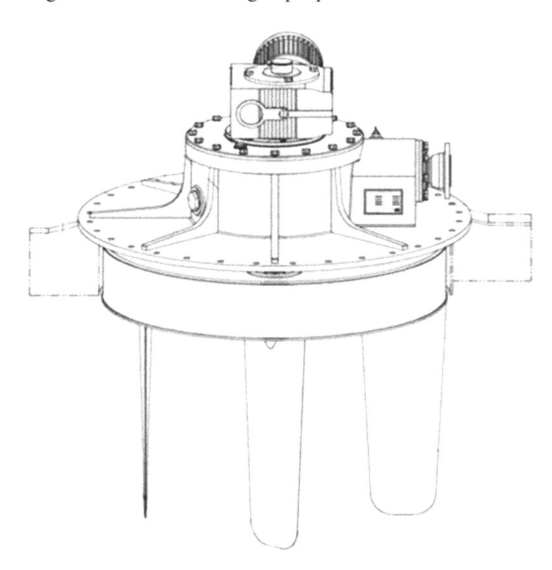

Figure 2. Straight blade propeller.

3 VERIFICATION OF NEW PROPULSION DEVICES

3.1 *General requirements*

As a kind of propulsion, new propulsion devices need to meet the basic requirements of the ship propulsion system first, including strength requirements, operation requirements, control requirements, safety alarm requirements and so on. These requirements are detailed in International convention for the safety of life at sea (SOLAS), International Association of Classification Societies (IACS) and rules of classification societies.

3.2 *Special requirements for new propulsion devices*

3.2.1 *Strength*

The strength of the main load-bearing components of the propeller devices, such as blade, shaft, flange, bracket, rudder, and other restraining members, needs to be verified, both limit condition and fatigue condition should be considered. For non-traditional design propulsion devices, special consideration should be given to the propeller load and stress analysis. The verification should include but not be limited to description of method to determine blade loading, description of method selected for stress analysis, ahead condition based on propulsion machinery's maximum rating and full ahead speed, astern condition based on the maximum available astern power of the propulsion machinery and crash astern operation, fatigue assessment, allowable stress and fatigue criteria. The blade which acts as rudder and propeller simultaneously should also be verified in the limit steering condition.

3.2.2 *Steering and control*

The propulsion devices shall have sufficient strength, capacity and necessary support system to provide effective thrust and steering for the ship under all operating conditions. When the device is the main propulsion and steering gear of the ship, two sets should generally be equipped to ensure that the failure of one of them will not render the other one inoperative. Any single failure of the steering gear, control system or power supply will not result in a complete loss of the ship's steering capability. When a single fault occurs, each steering gear should have additional capabilities to set to return to and lock in the center position. If the steering gear cannot return to the center position, locked in the current position can be accepted.

3.2.3 *Monitoring and alarm*
Most of the possible failures that may lead to the performance reduction or failure of the steering control system should be automatically monitored. The monitoring should include but not be limited to power failure, AC and DC circuit grounding fault, closed-loop system circuit failure including command and feedback loop, data communication failure, programming system failure, deviation between the steering order and the feedback, low lubricating oil level in gearbox, high lubricating oil temperature in gearbox, hydraulic blocking and other failure alarm. The navigation bridge should be equipped with the visual and auditory alerts of all the failures. When the failure occurs, the steering gear should be able to maintain the current rudder position.

3.2.4 *Product inspection*
The mechanical parts and electric control parts of the non-traditional design new propulsion devices should be inspected respectively. For the first product, type test of mechanical parts should be carried out according to the approved test program, which should include at least appearance inspection, dimension check, tightness test, idling test, functional test, durability test and connection check. The single product inspection should include at least appearance inspection, dimension check, tightness test, idling test, functional test and connection check. Type test of electric control parts should include at least functional test, environmental condition test and electromagnetic compatibility test. The single product inspection should include at least all functional tests, voltage test and insulation resistance measurements.

4 STRENGTH CALCULATION SOFTWARE

The working condition of the traditional propeller is stable. The strength requirements can be verified by checking the thickness of the blade. Unlike traditional propeller, the new propulsion devices such as polar propeller and semi-submerged propeller often need special hydrodynamic load analysis and finite element strength calculations due to the complicated loading conditions. In order to accurately evaluate the blade strength of polar propeller and semi-submerged propeller and improve the efficiency and accuracy of verification, China Classification Society developed the finite element strength calculation software for polar propeller and semi-submerged propeller.

The software solves the key problems in propeller blade modeling, finite element mesh division and accurate force loading. The software implements one-click automatic operation from geometric modeling, mesh generation, defining material properties, defining load conditions to submit model calculations, which provides great convenience for finite element strength verification of special blades.

4.1 *Blade geometric modeling*

The propeller is a complex irregular space geometry. To model the space geometry of propeller, the parameters and offsets of propellers need to be converted into space coordinates at first.

Conventional propeller drawings generally include propeller side view, front view and expended outline. The side view and the front view are respectively the lateral and forward projection of the propeller, and the expended outline determines the shape of the blade surface. The section shape of the propeller blade at different radius is determined by the table of offsets.

The coordinate system required for blade geometric modeling is a cylindrical coordinate system(R, θ, z), and the original point O is the intersection of the middle line of the blade and the center line of the propeller hub. As shown in Figure 3, the z axis points to the blade surface along the center line of the propeller hub, R is the

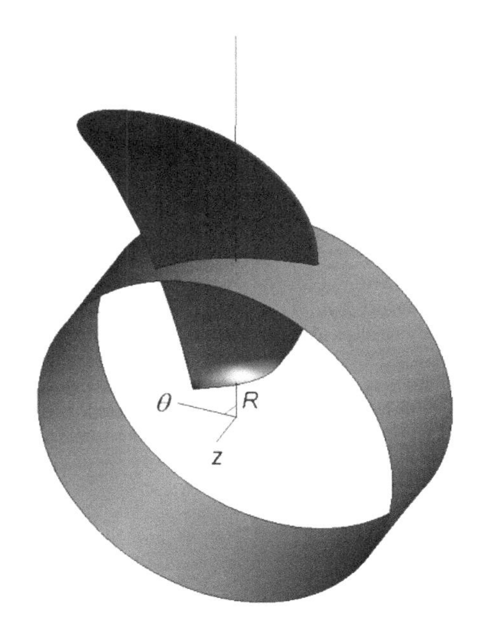

Figure 3. Coordinate system for blade geometric modeling.

direction of the radius, and θ starts from the vertical direction and satisfies the right-hand rule with the z axis. When the cylindrical surface intersects with the propeller blade, a curved section will be obtained, and the shape of the blades on each section can be drawn according to the table of offsets.

The propeller modeling needs to be able to represent the geometric shape and thickness of the blade, as well as the rounding at the root of the blade. The blade geometry can be generated by connecting the sections at each radius. After completing the modeling of the hub, the geometric modeling of propeller can be accomplished by connecting the blade and hub with a smooth surface according to the radius of the rounding. The basic process of modeling is shown in Figure 4.

The above is the basic method of propeller modeling. The software can also carry out special treatment to the fillet of leading edge and trailing edge and the ridge line on the semi-submerged propeller blade surface, so as to realize the full automatic modeling.

4.2 *Finite element mesh division*

As the propeller blade is mainly subjected to bending loads, the unsteady stress distribution will be produced along the whole blade thickness. In the stress analysis of blade, the conventional stress analysis methods are beam theory and coarse shell elements. The beam theory can deal with warping stress, and the coarse shell elements can roughly express the thickness change of the blade. However, the results obtained by these two methods cannot achieve the accuracy required in strength verification of polar propeller.

In order to simulate the composite three-dimensional stress state of the structure and achieve acceptable accuracy, a ten node tetrahedron space unit is adopted for the propeller mesh division. The thinnest area of the blade should be divided into two units at least in the direction of the blade thickness, and the shape of the unit should be kept in good condition. This method can guarantee the accuracy of the stress calculation, ensure that the local peak stress near the blade root region has a good accuracy, and provide reliable stress results for ultimate strength analysis and fatigue assessment. The sketch map of mesh division of the blade is shown in Figure 5.

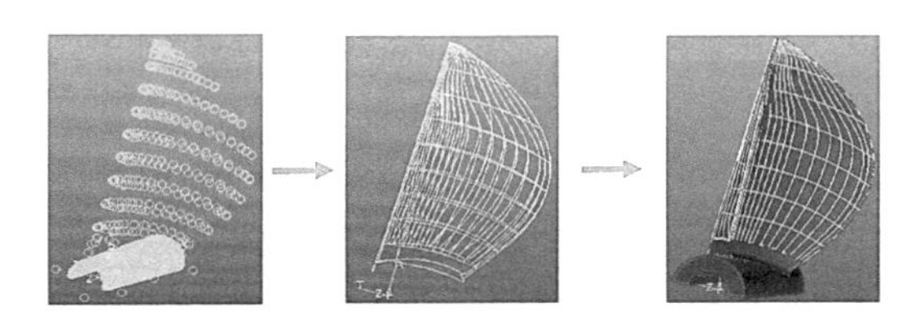

Figure 4. Basic process of propeller modeling.

4.3 *Load condition*

Accurate loading is an important guarantee for the accuracy of software calculation. The determination of load conditions needs to take full consideration of the environment and characteristics of the new propulsion device. The curved surface structure of the blade makes it more complex to ensure that the force is evenly distributed on the blade surface. Therefore, special attention should be paid to loading loads in certain areas.

Take the polar propeller as an example. For polar propeller strength verification, the load requested by IACS is the force F, not the pressure value. When the method of loading in finite element calculation is to use the force F divided by the total area S of the action surface, the pressure P is obtained, as Equation 1:

$$P = \frac{F}{S} \tag{1}$$

where, S is the total area of the action surface, and it is also the summation of the area of each small unit:

$$S = \sum_{i=1}^{N} s_i \tag{2}$$

These small units are not strictly on the same plane, but in a slightly curved surface. Then the resultant force of the pressure acting on these units is not strictly equal to F, but is slightly less than a value of F. Therefore, there is a certain gap between the calculation and the actual require-

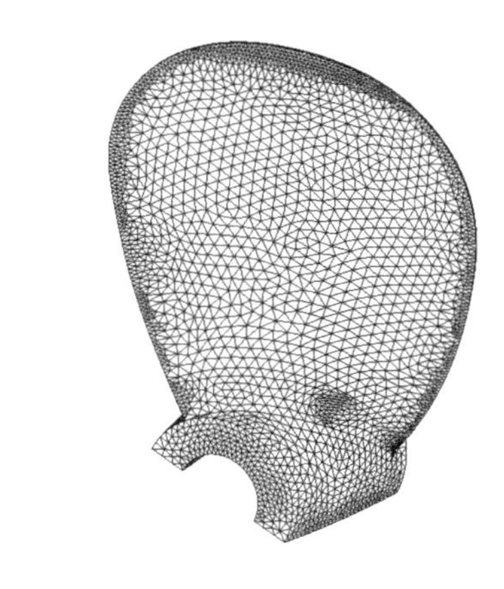

Figure 5. Sketch map of mesh division.

ment. It is necessary to consider a calculation method of pressure P, which makes the resultant force equal to F.

The software is processed as follows. Suppose the unit normal vector of s_i is $< n_{x,i}, n_{y,i}, n_{z,i} >$. Because they are unit vectors, so it has to be $n_{x,i}^2 + n_{y,i}^2 + n_{z,i}^2 = 1$.

Then, the pressure applied on each unit of the action surface P' can be calculated by Formula 3:

$$P' = \frac{F}{S'} \tag{3}$$

where,

$$S' = \sqrt{S_x^2 + S_y^2 + S_z^2} \tag{4}$$

$$S_x = \sum_{i=1}^{N} n_{x,i} s_i \tag{5}$$

$$S_y = \sum_{i=1}^{N} n_{y,i} s_i \tag{6}$$

$$S_z = \sum_{i=1}^{N} n_{z,i} s_i \tag{7}$$

Loading the calculated P' on the stressed surface, the resultant force of the pressure acting on these units is strictly equal to F. The blade accomplished loads loading is shown in Figure 6.

For semi-submerged propeller, there is no international uniform requirement on the strength of the blade. The calculation loads loading on the semi-submerged propeller blade can be the CFD calculation results or the fixed loads loading in certain areas. The speed of semi-submerged propeller is very fast and its motion is absolutely unsteady (Figure 7 shows the variation of the force acting on a semi-submerged propeller blade), combined with the characteristics of CFD calculation make it impossible to save all the load distributions of the blade at every moment, so it is very difficult to obtain the true limit load condition of the blade in a cycle through CFD calculation directly. It is generally considered to be more conservative to load the maximum load in a fixed position applying the load distribution features of CFD results or test results in the strength calculation and verification.

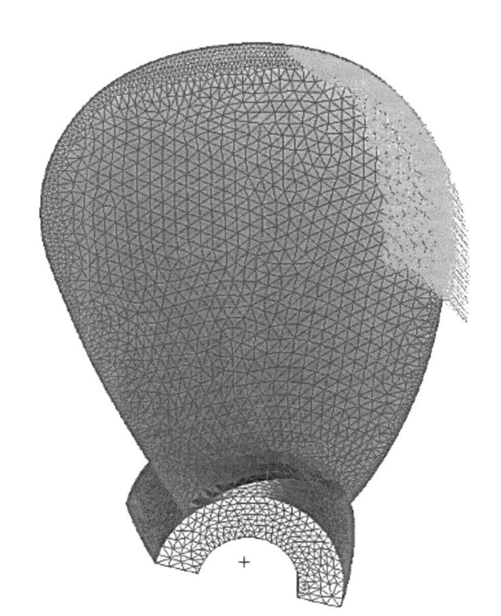

Figure 6. Loads on blade.

4.4 *Calculation and verification*

After load condition definition, the software can automatically submit the calculation cases. The results of the calculation can be imported into the software to read the stress, and the software can automatically generate the verification report according to the specification.

Figure 8 and Figure 9 respectively give examples of calculation results of polar propeller and semi-submerged propeller. By comparing with manual finite element calculation results and calculation

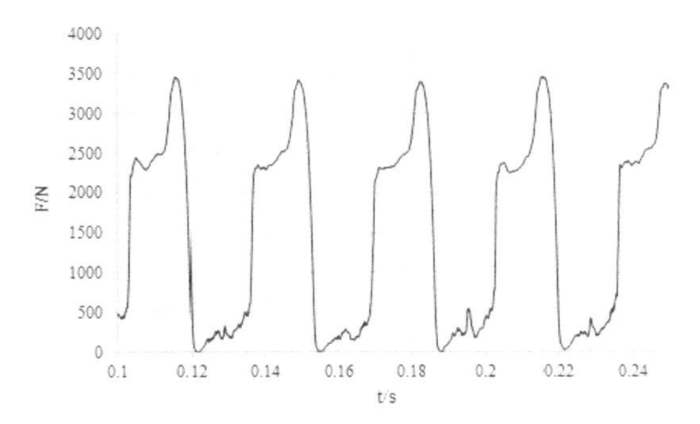

Figure 7. Variation of the force acting on a semi-submerged propeller blade.

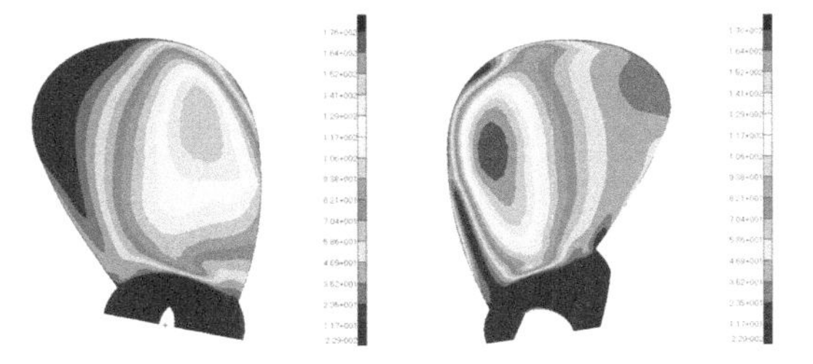

Figure 8. Calculation results of polar propeller.

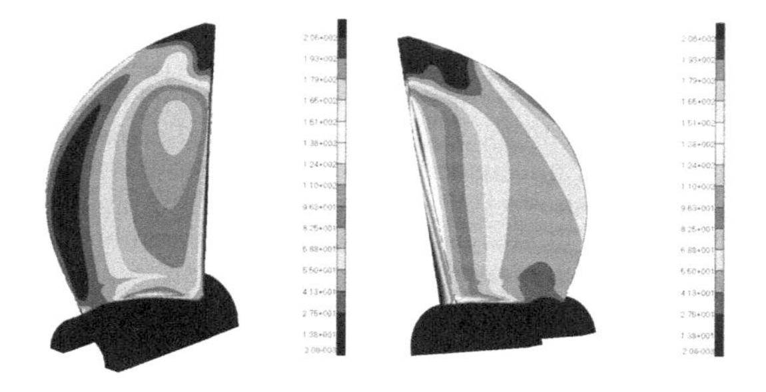

Figure 9. Calculation results of semi-submerged propeller.

results of blade strength obtained from other researches, it is found that the calculation results of the software are accurate and reliable.

5 CONCLUSION

This paper introduces the research achievements and strength calculation software of China Classification Society in the verification of new propulsion devices. At present, the corresponding methods and software have been applied to the actual verification of China Classification Society. The verified new propulsion devices have served in the real ship and achieved good economic benefits. At the same time, the strength calculation and verification software is also used in the design and development of new propulsion devices. The practical experience will help to promote the development and engineering application of more new propulsion devices.

REFERENCES

China Classification Society. 2015. Rules for Classification of sea-going steel ships. Beijing: China Communication Press Co., Ltd.

China Classification Society. 2017. Technical guidelines for straight blade propeller. Beijing: China Communication Press Co., Ltd, in press.

China Classification Society. 2018. Technical guidelines for semi-submerged propeller. Beijing: China Communication Press Co., Ltd, in press.

Gao, T. & Yang X. 2002. Simulation display and automatic manufacture of marine propeller. Ship & Ocean Engineering 145(2) 11–16.

IACS. 2007. Unified Requirements I3. London: The International Association of Classification Societies.

Sheng, Z. & Liu Y. 2004. Ship Principle. Shanghai: Shanghai Jiao Tong University press.

Navy ships

Marine Design XIII – Kujala & Lu (Eds)
© 2018 Taylor & Francis Group, London, ISBN 978-1-138-34076-3

An approach for an operational vulnerability assessment for naval ships using a Markov model

A.C. Habben Jansen, A.A. Kana & J.J. Hopman
Department of Maritime and Transport Technology, Delft University of Technology, The Netherlands

ABSTRACT: An early stage assessment of the vulnerability of systems on board naval ships needs to be carried out in order to ensure that naval ships can execute their operational scenario after hits or other damage. However, the broad scope of operational scenarios and impact levels, the interdependencies between systems and their environment, and the dynamic nature of vulnerability complicate the execution of such an analysis. While current methods mainly assess whether systems are still available after hits, the eventual question is whether the operational scenario can still be executed. An approach with a Markov chain is proposed to handle these complexities in early stage design. This paper specifically focusses on the effect of interactions between systems on vulnerability. It is shown that the fact *that* multiple systems are placed together in a layout already influences the vulnerability of the ship, regardless if they are related from a physical or logical point of view or not. Furthermore, the need for an integrated approach instead of assessing each system individually is demonstrated and quantified. The results of this methodology can be used by designers to make substantiated design choices with regard to prioritization of the different ship functions after hits.

1 INTRODUCTION

Since naval ships are designed to operate and survive in a hostile environment, one of their most important capabilities is their ability to withstand external threats. This aspect of naval ship design is known as survivability. In a naval ship context, survivability refers to "the capability of a ship and its shipboard systems to avoid and withstand a weapons effects environment without sustaining impairment of their ability to accomplish designated missions" (Said, 1995). Survivability can be subdivided in the following three stages (Said, 1995):

1. Susceptibility: The inability of a ship to avoid being damaged in the pursuit of its mission, and its probability of being hit.
2. Vulnerability: The extent to which the operational performance of the ship or system degrades after being hit.
3. Recoverability: The ability to restore the ship or system functionality by means of active response.

In some references, e.g. (Boulougouris & Papanikolaou, 2004), survivability is defined as the opposite of killability, which is the product of susceptibility and vulnerability. From this point of view recoverability is left as a separate domain, since it is by definition mainly an operational aspect, i.e. to a large extent it involves active human response. Though it has previously been shown that configurational choices can affect the recoverability as well (Piperakis & Andrews, 2012), recoverability is outside the scope of this research. Susceptibility and vulnerability, on the other hand, are mainly determined by intrinsic vessel characteristics that are fixed during the design. Though it could be argued that extensive susceptibility reduction measures reduce the need for vulnerability reduction measures, it is practically impossible to eliminate the probability that the ship gets hit or otherwise damaged. A historical overview of vulnerability incidents involving Western naval ships shows that most incidents happened in littoral waters, in situations where a hostile threat was not expected or observed, and preemptive measures where thus not taken (Schulte, 1994). These examples of incidents show that the probability of naval ships being hit or damaged is definitely present.

In order to limit the potential consequences of hits or other damage, vulnerability assessments are carried out during the design of the ship. Traditionally, vulnerability assessments focus mainly on optimizing the damage stability, such as described by e.g. (Boulougouris & Papanikolaou, 2004). Research on that topic still continues today (Boulougouris, et al., 2017). An advantage of this approach is that it can be used during all design stages. A preliminary layout with decks and bulkheads can be sufficient to provide an indication of the damage stability. In later stages the assessment can be made more detailed.

However, with the increasing number of distributed systems on board, the focus of vulnerability methods tends to shift towards system availability assessments, such as described by (van Oers et al., 2012), (Kim & Lee, 2012) and (Waltham-Sajdak, 2016). Such assessments are crucial for ensuring that distributed systems can meet their power and/or flow requirements, even in damaged scenarios. However, assessing the system vulnerability is a non-trivial effort due to the following complicating factors:

1. Operational scenario: The consequences of an external weapon hit are dependent on operational scenario, that can be broken down to the different functions that are carried out by the ship, such as propulsion and communication. For example, a hit on the radar may be of limited consequences for the propulsion, but of major consequence for the communication. The availability of such functions after a hit determines whether the operational scenario can still be carried out.
2. Impact level: Different types of external weapon hits may impose different requirements for remaining capabilities on board. For example, after a torpedo hit the priority is likely to focus on evacuating, while following an anti-tank missile hit fired from land, only minor damage is likely.
3. Interdependencies: Many distributed systems are dependent on other systems. Separately optimizing the vulnerability of individual systems may not necessarily result in the ship that is most survivable as a whole. Furthermore, the increasing electrification of naval ships may result in even stronger interdependencies between systems. Interdependencies do not only exist between systems, but also between systems and operations. A survivable system is one that is placed at a proper location, *and* connected adequately to other systems, *and* designed in a fashion that is in compliance with how it will be operated.
4. Time: The operation of a naval vessel is a highly dynamic process. The operational scenario and its associated requirements from systems vary constantly. Inherently, the vulnerability of the ship and its systems are also dependent on time.

Because of these aspects, vulnerability cannot be treated as a static phenomenon. However, current vulnerability tools are usually rather static. They assess the ship from an "engineering" point of view. That is, the vulnerability is assessed by considering the ship with a physical model that could include e.g. bulkheads, decks, machinery, and connections. Based on the configuration of these items, the vulnerability of the ship is determined by calculating the availability of systems and spaces under various damage scenarios. The situation-dependent context, for example that a naval ship has different power needs for different scenarios, is usually not taken into account. Furthermore, many vulnerability tools are analysis tools rather than design aids. Inherently, they are mostly suited for the detailed design stage. However, in detailed design it is very costly in both time and money to modify the design of the ship if that is deemed necessary based on the results of the analysis. For that reason it is sensible to have vulnerability already assessed during early design stages.

Summarizing the above, it can be concluded that a need arises for extending vulnerability assessment methods, taking into account the four situation-dependent and dynamic factors listed above, and the need for an early stage assessment. To approach this topic, this paper first looks at interdependencies between systems. The need for an integrated vulnerability assessment of systems is shown, and an early stage methodology to achieve that is introduced. Further elaborations on the operational scenario, the impact level and temporal nature of vulnerability are left for follow-on research.

The remainder of this paper is built up as follows. In Section 2 interdependencies are discussed in more detail. Furthermore, requirements for the extended vulnerability assessment methodology are formulated, based on a knowledge gap that is identified from existing methods. Subsequently, Section 3 provides a generic explanation of the method that is used to approach this problem. This involves the application of Markov theory. Several conceptual case studies are covered in Section 4 to provide an illustration of how the method could be used to do a basic vulnerability assessment on a generic layout. A case study with a mine countermeasures vessel (MCMV) in Section 5 explains how the methodology can be beneficial for a vulnerability assessment of an actual ship design. Conclusions and recommendations are given in Section 6.

2 BACKGROUND

2.1 *Effect of interdependencies on vulnerability*

Important sub-systems of ships, such as weapons or propulsion engines, often require various types of resources to operate, e.g. cooling water, chilled water, fuel, or data. Such resources are usually produced by one or multiple central sources (e.g. generators, chilled water units). A distribution network, consisting of e.g. cables, switchboards, and pipelines, distributes these resources towards the users. The systems associated with these resources are therefore known as distributed ship systems. Recent developments in instrumentation, automation, control and computer networks, such as (Janssen et al., 2016), have resulted in an increased number of interdependencies between different distributed systems on board. Due to the current interest in all-electric ships, the interdependencies between distributed systems are expected to become even stronger (Dougal & Langland, 2016). The main reason for this is that all-electric ships have a centralized power generation plant for providing both propulsion power and service power (combat systems, accommodation, etc.). This provides new challenges, not only for finding effective routings, but also for ensuring that the distributed systems can provide sufficient power for their functions, also in case of damage. Since this increasing number of interdependencies may influence vulnerability characteristics of ships, the need for an integrated vulnerability assessment is definitely present. Inter-dependencies do not only exist between systems, but also between systems and their (operational) environment. That is, systems should be designed in accordance with their expected operational use. For example, if a generator with a certain maximum output is connected to an electric propulsion motor and a high energy weapon, it is not possible to sail at high speed *and* deploy the weapon at the same time (Dougal & Langland, 2016). The vulnerability of these systems then becomes dependent on the operational scenario of the ship. For example, if the ship is sailing at high speed, the vulnerability for the task "deploy weapon" increases, because sufficient power is not available at that moment.

In order to better understand the physical and operational interdependencies between systems, a recently developed architectural framework for distributed naval ship systems is used (Brefort et al., 2017). This framework identifies three types of architecture that together form the response of a system: the physical architecture, the logical architecture, and the operational architecture. A visualization of the framework is provided in Figure 1. The physical architecture describes what a system looks like in the physical space, such as locations and dimensions of compartments and the equipment inside. The logical architecture describes how different system components are connected to each other, for example with single-line diagrams. The operational architecture describes for an operational scenario, how a system and its components are used over time. The overlap of all three architectures is the system response. This means that in order to properly assess the response of a system, the three types of architecture *all* need to be considered, because they are interdependent. In the

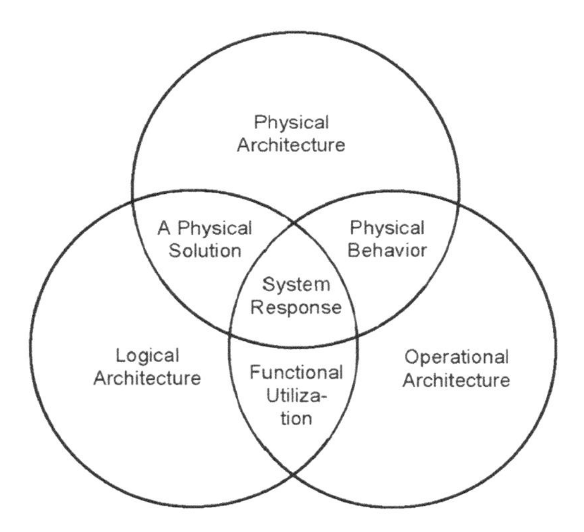

Figure 1. Visualization of the architectural framework, including physical, logical, and operational architecture for a specific operational scenario (Brefort et al, 2018).

context of the larger problem scope, as defined in Section 1, the framework also shows the need for including the operational scenario during the design and vulnerability assessment of systems. Furthermore, the framework can be beneficial for selecting the type of information that is used in the early stage design. It shows the need for combining the three types of architectures. In the early design stage, it is better to have little rough data on three architectures, than more detailed data on only one or two architectures.

2.2 *Review of existing vulnerability methods*

Various tools and methods exist for assessing the vulnerability of naval ships during the design. There are many metrics that can be used to quantify vulnerability, such as damage stability, system availability, or structural strength of blast bulkheads. Various computational tools exist for assessing this in an integrated fashion. Many of these tools have been developed with the application to a specific country in mind, such as RESIST for the Netherlands, SURVIVE for the United Kingdom, and MOTISS for the United States. These tools require rather detailed input on the ship geometry, the structural plan, or the crew composition. They provide rather high-fidelity results, but are in general mostly suited for later design stages. Since this paper focusses on system vulnerability assessments in the early design stages, several methods that deal with that topic are now briefly discussed.

Since the components and connections of distributed systems can easily be represented as a network, more examples of network-based approaches for system vulnerability assessment exist. In terms of the architectural framework, these methods cover the logical architecture, and to some extent also the physical architecture. In (Shields et al., 2016) functional complexity theory is used to describe the relation between survivability requirements and the complexity of the ship. Rather than evaluating the actual quality of designs, this method aims to predict the complexity of a design. This way, it can be studied how early stage vulnerability requirements affect the later stage complexity of the ship. In terms of the framework, this method has been identified to be in the overlap between the physical and logical architecture.

In (Kim & Lee, 2012) a vulnerability assessment methodology is proposed that identifies several critical components, and then calculates the probability of their availability based on the vulnerable area. Several probabilistic tools are compared, such as a Poisson process and a Markov chain. Both the physical and the logical architecture are considered in this methodology. Another method where physical and logical architecture are combined, is described in (van Oers et al., 2012). The ship is divided in compartments bounded by decks and bulkheads, and system components are assigned to compartments (physical architecture). In addition, a routing algorithm is added to connect the different system components (logical architecture). The routings are represented as a network. A genetic algorithm is used to minimize the total number of disabled systems after hits and the total length of all networks, which can both be seen as a metric for the system vulnerability.

Another example of an application of network theory for naval ship vulnerability is (Trapp, 2015), where network flow optimization is applied to design survivable networks. It is purely network-based and does not take into account where system components are located on the ship. Another pure network-based approach for designing survivable networks for distributed systems is proposed in (de Vos, 2014), where network topologies are generated automatically. In doing so, an extensive overview of the design space is acquired, resulting in better trade-offs between different network topologies. Where (Trapp, 2015) optimizes networks for flow distribution, (de Vos, 2014) uses several networks parameters (e.g. number of hubs, number of vertices) to steer the optimization algorithm towards a promising solution. In terms of the framework, these two methods could analyse the logical architecture. A summary of these tools is provided in Table 1. This is not intended to be a fully exhaustive list, but merely a representation of current research.

Table 1. Overview of existing methods for early stage vulnerability assessments of naval ships.

Method	Theory	Arch.
Shields	Complexity theory	P, L
Kim	Various (probabilistic)	P, L
van Oers	Networks, optimization for physical parameters	P, L
Trapp	Networks, optimization for flow	L
de Vos	Networks, optimization for network parameters	L

2.3 *Requirements for new methodology*

The situation-dependent and dynamic nature of ship vulnerability has been identified as a factor that complicates a vulnerability assessment. In terms of the framework, this factor could be considered as a part of the operational architecture. However, the overview of existing methods for assessing system vulnerability in the early design stages presented in Table 1 shows that existing methods mainly capture the physical and logical architecture. Including the operational architecture as well could potentially provide a significantly more meaningful result of a vulnerability assessment. It is then not only assessing whether systems are available after hits, but also how the damage affects the ability to carry out the operational scenario. This is particularly important for all-electric ships. Because systems on board such ships become more and more interconnected, it becomes less obvious that the availability of systems automatically implies that the operational scenario can actually be carried out. In order to include these operational architecture aspects, the following requirements for a new method need to be met:

1. Different scenarios: since vulnerability is situation-dependent, the new method needs to be able to model a broad scope of scenarios, both related to operations and the type of hit.
2. Interdependencies: The vulnerability methodology needs to be able to handle interdependencies, between systems and between systems and operations. Known interdependencies need to be modelled, and unknown interdependencies are hopefully identified.
3. Time: The interdependencies and operational scenarios mentioned above are dependent on time. For that reason, the methodology is required to handle dynamic relationships.
4. Low level of detail: The available level of detail in the early design stage is very limited. It is therefore necessary that the vulnerability approach is able to provide sufficient information with limited available data. Inherently, the

method needs to be able to handle uncertainties associated with the layout and the future operational scenarios.

3 METHOD

3.1 *Markov theory*

Based on the requirements for the new vulnerability assessment approach that have been identified in Paragraph 2.3, Markov theory will be used for further development of this methodology. It was selected because it can assess a system that moves through various conditions over time, and because it is strong in handling uncertainties within this process. More specifically, Markov theory can be used to describe the different states of a system, the probability of transferring between states, and the behaviour of a system over a longer time. These characteristics are expected to match well with the requirements. The different scenarios can be modeled with states. Temporal behaviour can be modeled with transferring between states at different time steps. Dependent on the layout and systems, several states will have a lower or higher probability of occurrence, which can be seen as an indication of interaction between systems. The level of detail stands apart from the mathematical background. In principle Markov theory can be used for any level of detail, though as usual when making models, the fidelity of the result increases with a higher level of detail.

The basis of the approach is a Markov chain, showing similarities with work that has previously been published about this topic (Kim & Lee, 2012). In addition to (Kim & Lee, 2012) this paper also highlights the need for an integrated vulnerability assessment rather than an assessment where all systems are decomposed. This is explained further in Section 4.

The basis of a Markov chain consists of a state vector and a transition matrix. Let n be the number of possible states. The states are described by the state vector **s**, which is a row vector of length n. The values of the elements of **s** represent the probability that the system is in that specific state. Hence, the sum of **s** equals 1. The initial state is represented as **s**(0). The probability of transferring between states is described by the transition matrix T. The dimensions of T are $n \times n$. Element $T_{i,j}$ denotes the probability of transferring from state i to state j. For each row in T the sum equals 1.

The behaviour of the system over time can be assessed by multiplying the state vector with the transition matrix. If t are discrete time steps and T does not change over time, the state of the system at any point in time, k, can be determined from Equation 1.

$$\begin{aligned}
t=1:&\quad s(1)=s(0)\cdot T\\
t=2:&\quad s(2)=s(1)\cdot T=s(0)\cdot T^2\\
t=3:&\quad s(3)=s(2)\cdot T=s(0)\cdot T^3\\
&\ldots\\
t=k:&\quad s(k)=s(0)\cdot T^k
\end{aligned}\tag{1}$$

3.2 *Application to vulnerability assessment*

If a basic layout of the ship with its main systems is available, a Markov chain such as described in the previous subsection can be used to do an early stage vulnerability assessment. This section describes the approach in generic terms, after which the following two sections illustrate the approach with a simple conceptual layout of the distributed systems and a basic ship layout.

In terms of the vulnerability assessment, the states represent the availability of systems. If m is the number of systems, and each system can either be on or down, the total number of possible states is 2^m. In state 1 all systems are on, in state 2^m all systems are down, and the intermediate states contain all possible combinations of several systems being on and several systems being down.

The transition probabilities can be derived from the layout of the ship. It is assumed that a basic 2D side view with decks and bulkheads is available, such as in Figure 7, for which the analysis is described in Section 5. Hence, the ship is subdivided in compartments. Let q be the number of compartments. It is now assumed that the ship gets hit, and that the hit will disable one of the compartments, while all other compartments remain intact. The hit probability is assumed to be equal for all compartments. Hence, each compartment has a probability of 1/q to get hit. Based on which systems and routings are located inside that compartment, a transition to another state may occur.

At t = 0 all systems are assumed to be on (state 1). At t = 1 the first hit occurs in an arbitrary compartment. If no systems or routings are located in the compartment that was hit, there is no transition to another state. In all other cases, there will be a transition to another state. Which state this is, depends on which systems and/or routings are located in that compartment. The probability of transferring to another state after a hit can be derived from the number of compartments that a system occupies. It is not needed to know the exact location of the hit, as this method considers all possible hit locations. In this fashion the first row of T can be constructed. For all other rows, a similar procedure can be followed. However, for these rows it needs to be assumed that the corresponding systems are already down. The number of hits that happened previously to end up in that state does not matter. In state 2^m all systems are down. Since there is no possibility to hit any more systems, the probability of staying in that state is 1. Once the transition matrix has been built, the probabilities for each state can easily be determined for each number of hits with Equation 1.

4 CONCEPTUAL GEOMETRIC CASE STUDY

4.1 *Set-up of transition matrix*

The method presented in Paragraph 3.2 is now applied to a simple geometric problem. Consider a square grid with 3 × 3 compartments. The grid contains two different systems, system A and system B. The routings of systems A and B occupy three and four compartments, respectively. The systems are placed in the grid in various ways. In Layout 1 the routings cross each other in one compartment. In terms of the framework, this is an overlap in the physical architecture. In Layout 2 the systems share the power source, so there is an overlap in the logical architecture. In Layout 3 the systems do not overlap, and they share no components. Hence, there is no physical or logical relation between the two systems. The three layouts are presented in Figure 2.

For each layout, there are two systems, that can be either on or down. This results in four possible states for the availability of systems:

- State s_1: A and B are both on
- State s_2: only A is on
- State s_3: only B is on
- State s_4: both systems are down

Hence, s has length 4, and T has size 4 × 4. In the initial state all systems are on, so:

$$s(0) = [1 \quad 0 \quad 0 \quad 0] \tag{2}$$

The transition matrix is dependent on the layout. Layout 1 is considered first. The nine compartments are referred to with the abbreviations of the compass card (N, E, S, W, their intermediate directions, and center compartment C). From the initial state, the NW, NE and SW compartments can be hit without affecting system A and/or B. Hence, the probability for staying in s_1 is 3/9. If the W, E or SE compartment is hit, only system A stays on, so the probability for transferring to s_2 is 3/9. Similarly, if the N or S compartment is hit, only system B stays on, so the probability of transferring to s_3 is 2/9. If the C compartment is hit, both systems go down, so the probability of transferring to s_4 is 1/9. These numbers fill the first row of T.

The subsequent rows of T can be filled in a similar fashion. The starting state is now assumed to be s_2. In that situation only system A is on. This state implies that either compartment W, E, SE or multiple of them are already down. Returning to s_1 is not possible, because no repair of systems is assumed. Going to s_3 is also not possible, since system B is down. It is possible to stay in s_2. This happens if compartment NW, NE or SW is hit (no systems in those compartments) or if compartment W, E or SE is hit (any of those compartments has already been hit anyway). In other words, the probability for staying in s_2 is 6/9. There is also a probability to transfer to s_4, if compartment N, C or S gets hit, i.e. a probability of 3/9. The third row of T can be filled likewise. The fourth row represents the state where all systems are down. If that state is reached, it cannot be left, because under the current assumptions no repair of systems is considered. T therefore becomes as follows for Layout 1:

$$T_1 = \begin{bmatrix} 3/9 & 3/9 & 2/9 & 1/9 \\ 0 & 6/9 & 0 & 3/9 \\ 0 & 0 & 5/9 & 4/9 \\ 0 & 0 & 0 & 1 \end{bmatrix} \tag{3}$$

For Layouts 2 and 3, T can be obtained in a similar fashion. The transition matrices for Layouts 2 and 3 are given in Equations 4 and 5, respectively. Their first rows are noticeably different due to the different layouts of the systems in the grids. Thus, T is a function of the system layout.

$$T_2 = \begin{bmatrix} 4/9 & 2/9 & 1/9 & 2/9 \\ 0 & 6/9 & 0 & 3/9 \\ 0 & 0 & 5/9 & 4/9 \\ 0 & 0 & 0 & 1 \end{bmatrix} \tag{4}$$

$$T_3 = \begin{bmatrix} 2/9 & 4/9 & 3/9 & 0 \\ 0 & 6/9 & 0 & 3/9 \\ 0 & 0 & 5/9 & 4/9 \\ 0 & 0 & 0 & 1 \end{bmatrix} \tag{5}$$

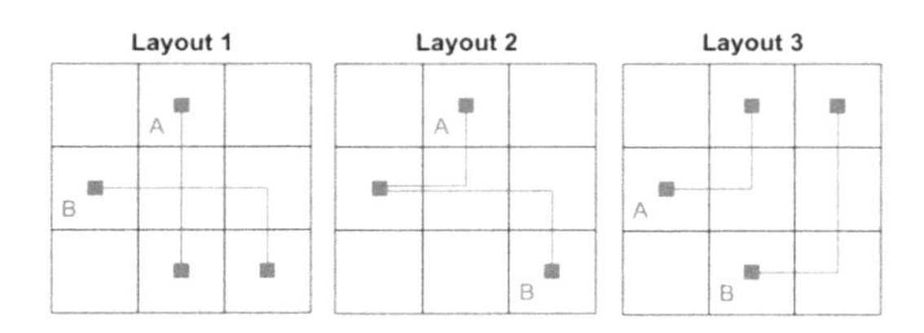

Figure 2. Three conceptual layouts for the vulnerability assessment, with a physical relation (1), a logical relation (2), and no relation (3).

The probability of being in any given state after any given number of hits can now easily be calculated by multiplying **s**(0) with T^k, where k is the number of hits.

4.2 *Decomposed approach*

In order to highlight the need for an approach where all systems are assessed together, such as the Markov-based method described above, a comparison is made with the situation where all systems are assessed individually, as would be done if the problem would be decomposed to individual systems to assess their vulnerability independently. For such an approach, the availability of the systems would merely be the product of the probabilities that each individual system is available (see Fig. 3), because in standard probability theory, the probability that multiple events happen at the same time equals the product of the probability of separate events (Dekking, Kraaikamp, Lopuhaa, & Meester, 2005). With this approach, system A has a probability of 6/9 to be available after a hit, and system B a probability of 5/9. For k hits, the probabilities for each state then become:

$$P(s_1) = (6/9)^k \cdot (5/9)^k \tag{6a}$$

$$P(s_2) = (6/9)^k \cdot (1-(5/9)^k) \tag{6b}$$

$$P(s_3) = (1-(6/9)^k) \cdot (5/9)^k \tag{6c}$$

$$P(s_4) = (1-(6/9)^k) \cdot (1-(5/9)^k) \tag{6d}$$

Since this approach does not account for the fact that the systems may overlap or share components, the probabilities in Equations 6a-d are independent of the physcal location of the systems. For reference, the mathematical expressions of the vector-matrix multiplication of the Markov approach for Layout 1 are given in Equations 7a-d. These equations are found by diagonalizing T. These equations are dependent on the transition matrix, and thus the layout. For Layouts 2 and 3 the transition matrix and equations change. They can be obtained in a similar fashion as for Layout 1, such as described above. The difference in these equations is a subtle but important property of this method, that highlights the need for an integrated approach. The implications of this are discussed further in Paragraph 4.3.

$$P(s_1) = (1/3)^k \tag{7a}$$

$$P(s_2) = -(1/3)^k + (2/3)^k \tag{7b}$$

$$P(s_3) = -(1/3)^k + (5/9)^k \tag{7c}$$

$$P(s_4) = (1/3)^k - (2/3)^k - (5/9)^k + 1 \tag{7d}$$

4.3 *Results*

The results for Layout 1 are presented in Figure 4. The solid lines represent the results of the Markov assessment. The dashed lines represent the results of the decomposed approach. The Markov approach is considered first. It can be seen that the probability of being in s_4 increases with an increasing number of hits. The probability of s_1 rapidly decreases. s_2 and s_3 also show a decreasing graph, but to a lesser extent. This is in line with what could be expected from the layout. After an increasing amount of hits, it is likely that both systems are down. In the meantime, system A is likelier to stay on for a while, since it occupies only three compartments, while system B occupies four compartments. The dashed line is the result that would be obtained if the situation would be assessed as the combination of two decomposed systems, as visualized in Figure 3. It can be seen that this does not yield the same result as the Markov assessment. That is because the decomposed approach violates a major assumption of the probabilistic equations, namely the requirement that the events are independent. For this layout this is definitely not the case, since systems A and B share a compartment.

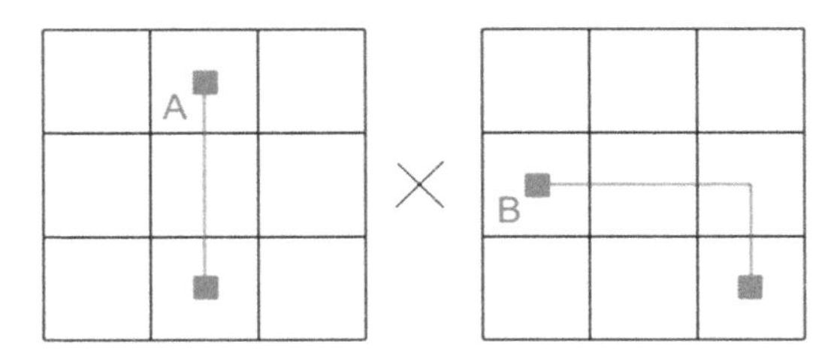

Figure 3. Layout 1 if the systems are assessed in a decomposed fashion.

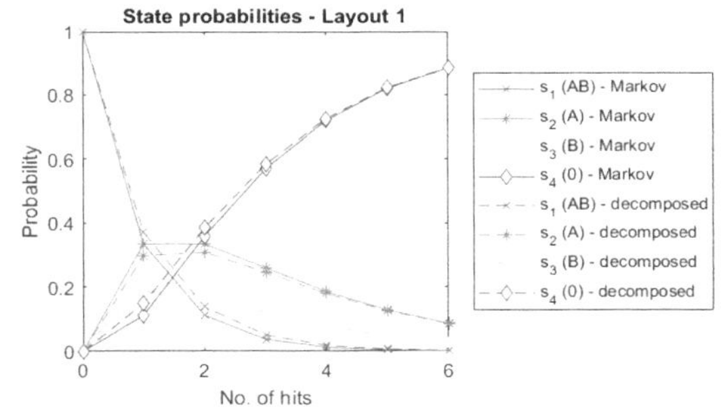

Figure 4. Results for the Markov assessment (solid) and the decomposed assessment (dashed) of Layout 1.

Assessing the vulnerability as the combination of the individual vulnerabilities of different systems is therefore not a valid method. However, for this example the differences with the Markov approach are rather small.

For Layout 2 the systems share two compartments. In one of these compartments a common source for both systems is located. In terms of the framework, there are both physical and logical relations between the systems. Figure 5 presents the result for this layout. It can be seen in this figure that the deviation between the integrated Markov approach and the decomposed approach is larger for this layout, compared to Layout 1. The probabilities that either one of the systems is on (s_2 and s_3) are overestimated with the decomposed approach. System A shares two out of its three compartments with system B. For that reason, it is in fact rather likely that system B also is down if system A is down. Hence, if the systems are assessed in the decomposed fashion, the probability that only one of the system is on is overestimated, and the probabilities that they are both on or down are underestimated.

The results for Layout 3 are presented in Figure 6. From the layout it appears that there is no relation between the systems, so at first sight, it could be expected that the interaction between the systems is small. However, the results show major differences between the decomposed approach and the integrated Markov approach. After a closer look at the layout it can be seen that only two compartments are empty. For that reason, there is a high probability that at least something goes down after a hit (s_2 or s_3). On the other hand, the probability that all systems are down (s_4) after a small number of hits is rather small, since the systems do not share sources or routings. If the systems are assessed in the decomposed fashion, there are relatively too many empty compartments, so the probability that nothing gets hit (s_1) is overestimated then. This can indeed be seen in Figure 6. The probability that everything gets hit (s_4) is also overestimated. In fact, after one hit this state cannot be reached, while the decomposed approach does not account for that. This again shows why an integrated approach is needed instead of assessing each system individually. As can be seen in the results of this layout, this also holds for situations where there are no physical or logical relations between systems.

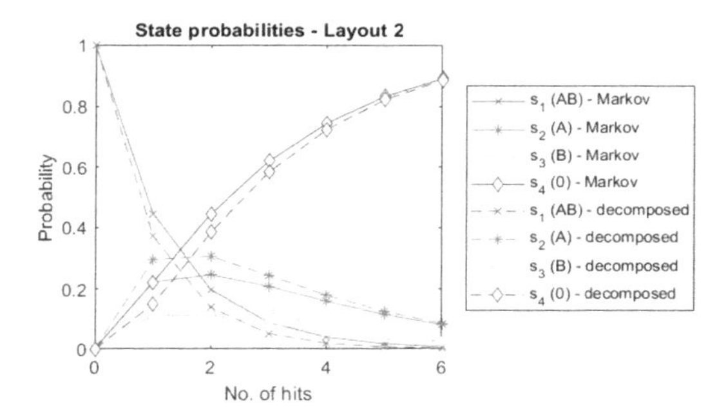

Figure 5. Results for the Markov assessment (solid) and the decomposed assessment (dashed) of Layout 2.

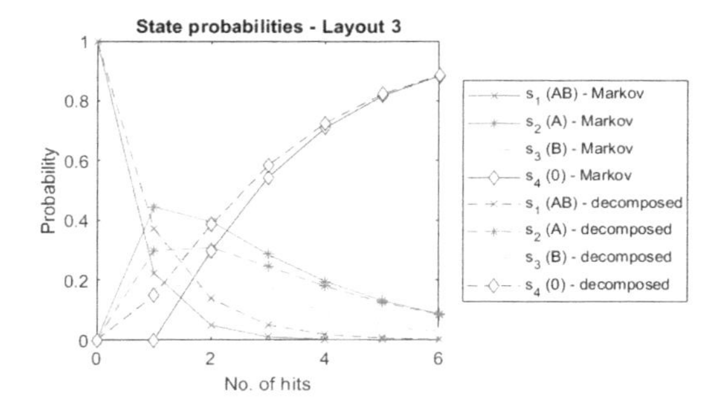

Figure 6. Results for the Markov assessment (solid) and the decomposed assessment (dashed) of Layout 3.

5 MINE COUNTERMEASURES VESSEL

5.1 *Layout of the ship*

The examples that illustrate the vulnerability assessment methodology in Paragraph 4.1 are of a conceptual nature. Representative ship and system layouts need to be introduced to investigate the practical applicability of this method in further detail. To that end, this paragraph provides the application of the methodology to a basic design of a mine countermeasures vessel (MCMV). This example is not intended to capture the complex layout and routing characteristics of a real ship, but it illustrates that the methodology can rather easily be scaled up to any desired level of representation of the ship.

The basic design comprises a 2D layout at compartment level. This design was retrieved from an existing set of designs that previously has been generated with the Packing Approach (Duchateau, 2016). In this design the main equipment for five main functions of the ship is included: propulsion; steering; navigation; RHIB launch & recovery; and UUV launch & recovery. A visualization is presented in Figure 7. In this example, power for the five main functions is provided by a generator. For each system there are one or two power sinks, such as listed in Table 2. Again assuming that functions are either available or not, the number of states is $2^5 = 32$. The probability for each state after any

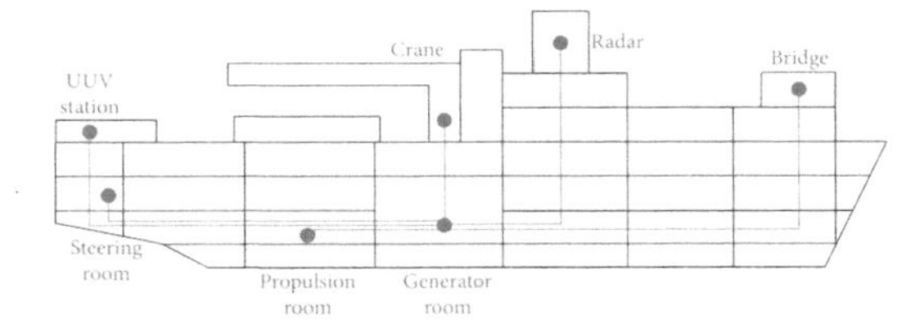

Figure 7. Layout for the MCMV used in the example of Section 5.

Table 2. Overview of main functions and systems of the MCMV.

No.	Function	Power sinks
1	Propulsion	Propulsion room
2	Steering	Steering room
3	Navigation	Radar and bridge
4	RHIB L&R	Crane
5	UUV L&R	UUV station

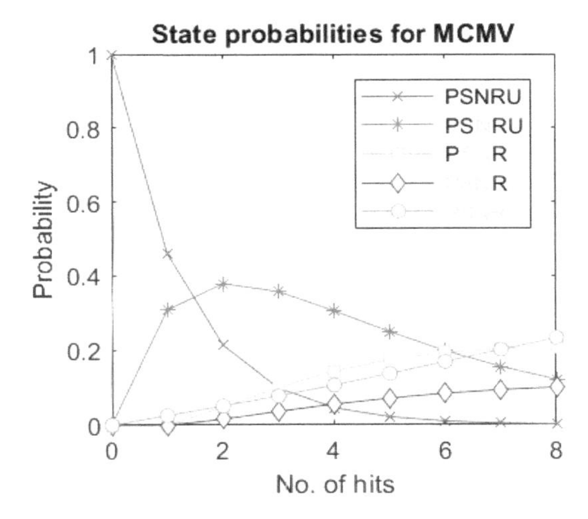

Figure 8. Results of the Markov assessment for the MCMV.

number of hits can be calculated in a similar fashion as the 3 × 3 grids.

5.2 *Results*

The results for the MCMV are presented in Figure 8. The visualization is slightly different from the figures that present the results of the 3 × 3 grids. Since these examples have shown the need for an integrated approach, the dashed lines that represent the results of the decomposed approach are left out. Furthermore, not all 32 states are plotted. Many states have a very small or even zero probability to occur, and are from a visualization perspective therefore not interesting. For that reason, Figure 8 only shows the states that have a probability of at least 10% to occur somewhere over the selected number of hits. The legend shows abbreviations of the five functions (propulsion, steering, navigation, RHIB launch & recovery, UUV launch & recovery). If an abbreviation is gray, it means that that function is not available in that state.

The state in which all systems are on shows similar behaviour to the respective states in the 3 × 3 grids: the probability for that state rapidly decreases after multiple hits. A state that is very likely to be reached instead, is the state where all functions except for navigation are on. This is because navigation requires two systems (radar and bridge) that have rather large routings because of their physical locations. These large routings make the navigation system vulnerable. For larger number of hits, it is likely that, apart from the navigation, more functions become unavailable. Steering and UUV launch & recovery have a high probability to become unavailable for larger numbers of hits, while RHIB launch & recovery is likely to stay available for a larger number of hits.

This information can be used by designers in prioritizing the tasks that the ship should still be able to perform after it has been hit. Traditionally, the functions float, move, and fight are identified. After a hit has taken place, 'float' usually has a higher priority than 'move', while that function in its turn has a higher priority than 'fight'. The functions of the MCMV can roughly be subdivided in 'move' (propulsion, steering, navigation) and 'fight' (RHIB and UUV launch & recovery). Note that in the context of an MCMV 'fight' refers to taking countermeasures against mines rather than actual firing.

The MCMV in the example quickly loses the function 'navigation', which is part of 'move', while UUV launch & recovery, and especially RHIB launch & recovery, which are both part of 'fight' have a high probability to remain available. This is contradictory to the traditional breakdown, in which 'move' needs to be available longer than 'fight'. For such strategic decisions, that are usually dealt with in combination with navy staff, the designers can use this procedure to make substantiated design choices with regard to prioritization of the different ship functions.

6 CONCLUSION AND RECOMMENDATIONS

This paper has introduced an approach for an early stage vulnerability assessment of naval ships. The approach is eventually intended to handle the complex nature of vulnerability, that manifests itself in the dependency on the operational scenario, the impact level, interdependencies, and time. Of these four aspects, this paper elaborates on interdependencies between systems. An approach with a Markov chain is proposed to evaluate the availability of major systems after one or more hits. Several case studies associated with this method show that there is a specific need for an integrated approach to assess the availability of systems after hits. Assessing the availability of systems individually is not sufficient, not even if they do not share routings or components. The fact *that* they are placed together in one layout already means that they are dependent on each other.

The examples presented in this paper are of a conceptual nature. For further incorporation of the complexity of vulnerability in an early stage assessment, several additions are actively being pursued. As mentioned in the paper, it is not sufficient to know if systems are still available after hits. The eventual question that needs to be answered, is whether the desired operational scenario can still be executed. Therefore, the availability of systems needs to be related to functions that are part of the operational scenario. Due to the dynamic nature of the operational scenario, a temporal analysis needs to be included. Furthermore, it is recommended that the results from this vulnerability assessment are fed back into the design. In this way, the results of the assessment can actually be used to generate less vulnerable designs in the early design stage, in addition to knowing why certain designs perform better or worse than others. This can also comprise machine learning and data analysis of the results of this methodology from a large set of designs.

Apart from these recommendations that contribute to the development of the approach, several practical recommendations are proposed to improve the quality of the results of the approach. More realistic ship models are needed for this. This can comprise, but is not limited to, including redundant components and routings, improving the hit probabilities of the compartments, modelling multi-compartment hit scenarios and including the effect of blast bulkheads.

ACKNOWLEDGEMENTS

Funding for this research is provided by Ms. Kelly Cooper from the United States Office of Naval Research (ONR) under grant no. N00014-15-1-2752, and is gratefully acknowledged. Furthermore, the authors would like to thank the Defence Materiel Organisation (DMO) of the Dutch Ministry of Defence for their in-kind contribution to this research.

REFERENCES

Boulougouris, E., & Papanikolaou, A. (2004). Optimisation of the Survivability of Naval Ships by Genetic Algorithms. In *Proceedings of COMPIT'04*. Siguenza, May 2004.

Boulougouris, E., Winnie, S., & Papanikolaou, A. (2017). Assessment of Survivability of Surface Combatants after Damage in the Sea Environment. *Journal of Ship Production and Design*, *33*(2), 156–165.

Brefort, D., Shields, C., Sypniewski, M., Goodrum, C., Singer, D., Habben Jansen, A., Duchateau, E.A.E., Pawling, R.G., Droste, K., Jaspers, T., Sypniewski, M., Goodrum, C., Parsons, M.A., Kara, M.Y., Roth, M., Singer, D.J., Andrews, D.J., Hopman, J.J., Brown, A. & Kana, A.A. (2018). An Architectural Framework for Distributed Naval Ship Systems. *Ocean Engineering 147, 375–385.*

Dekking, F., Kraaikamp, C., Lopuhaa, H., & Meester, L. (2005). *A Modern Introduction to Probability and Statistics —Understanding Why and How* (Springer). London.

de Vos, P. (2014). On the application of network theory in naval engineering—Generating network topologies. In *International Naval Engineering Conference*. Amsterdam, May 2014.

Dougal, R., & Langland, D. (2016). Catching it early—Modeling and simulating distributed systems in early stage design. *SNAME Marine Technology*, 63–69.

Duchateau, E. (2016). *Interactive evolutionary concept exploration in preliminary ship design.* Delft University of Technology, Delft.

Janssen, J., Butler, J., Worthington, P., Geertsma, F., & den Hartog, M. (2016). Autonomous, adaptive, aware: DINCS. In *Proceedings of INEC*. Bristol, April 2016.

Kim, K.S., & Lee, J.H. (2012). Simplified vulnerability assessment procedure for a warship based on the vulnerable area approach†. *Journal of Mechanical Science and Technology*, *26*(7), 2171–2181.

Piperakis, A., & Andrews, D. (2012). A comprehensive approach to survivability assessment in naval ship design. *International Journal of Maritime Engineering*. Trans RINA, Vol 154, Part A2, Apr-Jun 2012.

Said, M. (1995). Theory and Practice of Total Ship Survivability for Ship Design. *Naval Engineers Journal*, *107*(3), 191–203.

Schulte, J. (1994). *An analysis of the historical effectiveness of anti-ship cruise missiles in littoral warfare.* Naval Postgraduate School.

Shields, C., Sypniewski, M., & Singer, D. (2016). Understanding the relationship between naval product complexity and on-board system survivability using network routing and design ensemble analysis. In *Proceedings of PRADS2016*. Copenhagen, September 2016.

Trapp, T. (2015). *Shipboard Integrated Engineering Plant Survivable Network Optimization.* Massachusetts Institute of Technology.

van Oers, B., van Ingen, G., & Stapersma, D. (2012). An integrated approach for the design of survivable ship services systems. In *Proceedings of the International Naval Engineering Conference (INEC)*. Edinburgh, May 2012.

Waltham-Sajdak, J. (2016). Hard lessons—Evolving the design of survivable naval ship systems. *SNAME Marine Technology*, 40–46.

Marine Design XIII – Kujala & Lu (Eds)

Early stage routing of distributed ship service systems for vulnerability reduction

E.A.E. Duchateau
Maritime Systems Department, Defence Materiel Organisation, Netherlands Ministry of Defence, The Hague, The Netherlands

P. de Vos & S. van Leeuwen
Department of Maritime and Transport Technology, Delft University of Technology, Delft, The Netherlands

ABSTRACT: Warships rely on multiple distributed systems to function and perform their tasks. Vulnerability is a major driver in the design of these distributed systems. Aspects such as: the vessel and system component arrangement; the redundancy of system components; and the topology and routing of cables, ducts, piping and shafts to a large extent determine the ability of the ship to remain operational after incurring damage. The topology and routing of the connections within and between these distributed systems is often only addressed in later stages of the ship design, at which time changes to the general arrangement of the vessel and the distributed system components become difficult and costly. Earlier insight into the interrelations between the system's vulnerability, the distributed system routings, and the vessel and system arrangements is thus deemed a critical step in designing robust distributed systems and ultimately more resilient warships. To address this, we propose a novel method for the early stage routing of ship distributed systems for vulnerability reduction. The approach uses network descriptions of the vessel layout and the distributed system topology to model both the routing and vulnerability problems. The paper outlines the modelling approach as well as the algorithms used. The usefulness of the approach as an early stage design tool is demonstrated by using it to design the routing for multiple (e.g. electrical, chilled water) distributed systems of a notional warship.

1 INTRODUCTION

1.1 *Warship survivability and vulnerability*

Modern ships, and warships in particular, rely on multiple distributed ship service systems to function and perform their tasks and missions. These systems produce and distribute electricity, fluids (fuel, hydraulics, and chilled water), air, and data required to operate and function all other systems of the ship, such as: engines, cranes, and in case of warships, weapon and sensor systems. For warships these distributed ship service systems are a vital element for operating and fighting the ship. Hence they must be made resilient to damage incurred from internal and external factors such as fires, flooding, and weapon effects. In warship design this design aspect is part of the topic of survivability.

Survivability in the naval ship context refers to (Said 2009): "The capability of a ship and its shipboard systems to avoid and withstand a weapons effect environment without sustaining impairment of their ability to accomplish designated mission." Survivability can be further subdivided into the topics of: susceptibility (avoiding damage), vulnerability (resisting the effects of damage), and recoverability (recovering/repairing from damage). Although in warship design much attention is devoted to measures which prevent damage incurred from weapon effects (susceptibility), historical analysis of incidents involving anti-ship cruise missiles shows that, even for defendable warships, weapon hits cannot be avoided (Schulte 1994). Even so, a warship with a well thought out general arrangement, systems design, and well trained crew is able to survive longer and recover more quickly. This paper focusses on the vulnerability of the ship and its systems.

1.2 *Vulnerability in warship design*

In the vulnerability of a warship, design aspects such as: the ship subdivision layout (arrangement of bulkheads, blast bulkheads, and decks); the ship and system component arrangement; the redundancy of systems and their components; and the topology and routing paths of cables, ducts, and piping all play a large role. Hence, these aspects deserve specific attention during all phases of the design of a warship. Nonetheless, this paper focusses on the early stage (or concept) design phase as it is here where large and defining design decisions, regarding most of the above mentioned design aspects, are made.

In recent decades many advances have been made in the field of early stage warship design tools (Andrews et al. 2012). These novel tools allow increases in the level of detail of the ship design, while remaining flexible enough to perform broad yet thorough concept exploration studies required for naval requirements elucidation (Andrews 2011). The focus of these tools has mostly been towards allowing quick variations to the subdivision layout, arrangement of spaces, large systems and pieces of equipment, and the overall balancing of the design in terms of space, weight, power (Pawling 2007, Takken 2008, van Oers 2011a). The goal of those quick and broad design variations is to gain insights into size and cost drivers, and evaluate the impact of requirement changes (van Oers 2011b, Duchateau 2016).

In the tools mentioned above the design of distributed ship systems mostly plays a secondary role. Larger pieces of equipment (e.g. diesel engines, generator sets, weapons and sensors) are often placed and arranged as individual items, but mainly to address sizing of spaces and overall ship arrangement aspects (e.g. integrated topside design, propulsion system arrangement). Smaller equipment is often aggregated into their own functional spaces (e.g. electronic switchboard spaces or pump rooms). The general assumption being that sufficient space is available for these smaller items (van Oers 2011a). The required space and volume is often taken from experience with past vessels. During this process often design guidelines and "rules-of-thumb" are applied to ensure that critical systems (and their spaces) are made redundant, and are sufficiently separated within the vessel (Brown 1987, Cerminara & Kotacka 1990).

However, at the early stages interconnections between these spaces and distributed systems, such as, cables, piping and ducts, are not thought out (or only to a limited degree as in Andrews et al. 2004). The actual topology and routing of the connections within and between distributed systems are often only addressed in later stages of the ship design, at which time changes to the general arrangement of the vessel and the distributed system components become difficult and costly. This may lead to a sub-optimal routing of system connections, ultimately resulting in a more vulnerable vessel than could have been achieved. Thus, earlier insight into the interrelations between the ship's vulnerability, the distributed system routings, and the vessel and system arrangements is deemed a critical step in designing resilient distributed systems and warships.

1.3 *Distributed systems routing and topology*

One reason why vulnerability, and systems routing specifically, is often only addressed in later design stages is the relatively high level of detail that is required by many dedicated vulnerability analysis and routing tools. Some example of vulnerability analysis tools are: SURVIVE (QinetiQ 2017), RESIST LITE (TNO 2017). Although the examples provide excellent means of analysing in detail the effects of weapon impacts, blast effects, and fragmentation on the ship's systems vulnerability, they are less suited for the early design stages, where a quicker analysis is required to keep up with the constantly changing design concepts and requirements. Moreover, these detailed tools usually require a pre-defined systems design, topology, and routing as input to the vulnerability analysis. More recently, Goodfriend & Brown (2017) developed a vulnerability assessment specifically targeted at the early design stages. Their method however, also requires a pre-defined systems design, and does not take into account the physical routings of system interconnections.

There are several examples of automated routing tools in literature, yet these are mostly targeted at the more detailed routing of pipes during the detailed engineering phases the ship design, (e.g. see Park & Storch 2002, Asmara 2013). Nonetheless, there are only a few examples of routing tools dedicated for the early stage design of warships. Shields et al. (2017) developed a 2D routing approach specifically targeted at gaining insight into routing connections during early stage design, while van Oers et al. (2012) present a 2D routing approach including a vulnerability assessment. Although this approach can vary both the distributed system topology and routing, only variations of the shortest path are possible.

The system topology itself is another reason why it is difficult to address systems routing in early design. The topology needs to be known before connections can be made and routed through the ship's compartments. Nowadays, the system topology is often only decided upon after relatively detailed load balances have been set up for the different distributed ship service systems. However, as stated earlier, the design of distributed ship systems (including setting up and solving said load balances) often play a secondary role during early stage ship design. Therefore a method is needed to pre-determine system topology before automated routing can be applied.

A number of possibilities exist:

- using a pre-set system topology using design rules;
- copying earlier realised systems (e.g. from similar vessels);
- parametrising design variables for system topologies and setting values for these parameters based on earlier realised systems (de Vos 2014).

More recently, de Vos & Stapersma (2018) introduced a framework for node categorisation and developed an automatic system topology generator. In turn, van Leeuwen (2017) developed and implemented a vulnerability metric based on system connectivity as an objective function in the automatic topology generator of de Vos. This same metric is also used in this paper (see Step 5 in Section 2).

However, what is still lacking in the literature is an approach targeting the problem of distributed systems routing and the accompanying vulnerability analysis. To address this, we propose a novel method for the early stage routing and vulnerability assessment of ship distributed systems. The approach uses network descriptions of the vessel layout and the distributed system connections to model both the routing and vulnerability problems. To estimate vulnerability, a depth-first search is used to determine the connected state of critical users in a damaged network condition. For the routing problem, a k-shortest path algorithm (Yen 1971) coupled to a genetic search algorithm (Deb et al. 2002) is used to optimise the routing of system connections for vulnerability and total connection length (a cost driver).

The remainder of the paper outlines the proposed approach as well as the algorithms and techniques used. The usefulness of the routing and vulnerability analysis tool as an early stage design aid is further demonstrated in a test-case involving the routing design and analysis of a notional offshore patrol vessel.

2 PROPOSED APPROACH

The proposed routing and vulnerability assessment approach builds upon several other methods and tools. It should work at a level of detail appropriate to the early stage warship design process. As such, it must integrate easily into existing workflows used by designers at that stage. For example, refer to van Oers et al. (2017) for a description of the warship concept exploration and definition phase at the Netherlands Ministry of Defence's Defence Material Organisation (note that other nations and companies often follow a similar process).

The workflow of the proposed approach comprises the following steps (see Fig. 1):

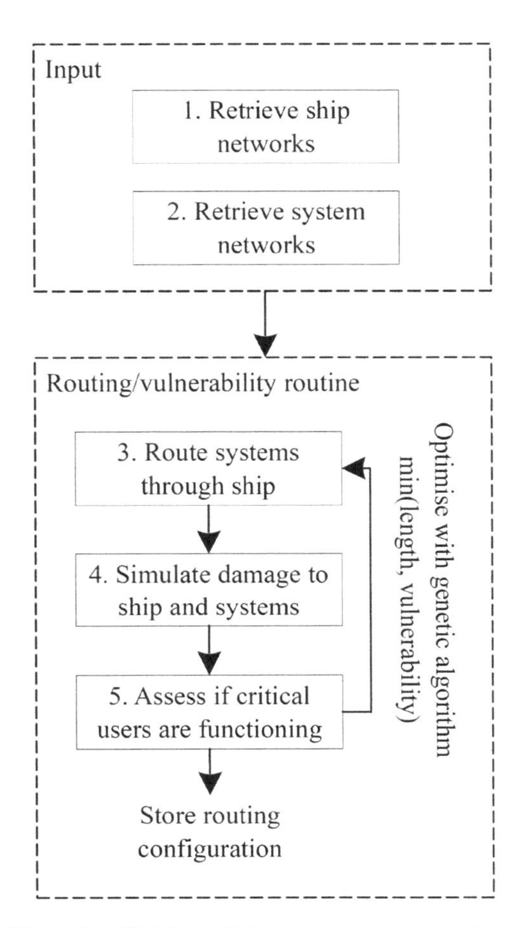

Figure 1. Workflow of the proposed routing and vulnerability assessment approach.

1. Translate the ship's 3D general arrangement and subdivisions information into a set of networks describing the routable (and unrouteable) paths and compartment adjacency and subdivision information.
2. Model the distributed systems and their logic as a network of nodes (systems) and edges (connections such as pipes, cables, ducts).
3. Route all relevant and distributed connections between components defined in Step 2 through the network describing the ship arrangement and routable paths created in Step 1.
4. Simulate one (or multiple) hits on the ship's hull and superstructure and assess which compartments and thus nodes within the networks of Step 1 are damaged.
5. Assess which components and connections are directly and indirectly damaged and thus no longer function as a result of the damaged compartments found in Step 4.

Steps 3–5 may be repeated within a genetic algorithm in an attempt to find a Pareto-set of solutions minimising both the overall routing configuration length and the distributed system vulnerability.

2.1 *Step 1: Describing the ship arrangement*

Most early stage design tools have some form of 3D description to describe the hull shape, decks and bulkheads, main spaces, and sometimes even system components (depending on the desired level of detail of the design tool or problem). Although this 3D description is an excellent visual reference for naval architects and designers working on the model, it is not well suited for use within mathematical models. Ongoing work at the University of Michigan, University College London, and Delft University of Technology (Gillespie 2012, Rigterink 2014, Collins et al. 2015, Brefort et al. 2018) has shown the potential of using network descriptions as a way of mathematically describing various parts of the ship.

Already, van Oers et al. (2012) and Rigterink (2014) explored the potential of networks describing the adjacency between compartments as a simple mathematical description of the vessel's arrangement. In this case, network nodes describe spaces or rooms and network edges describe the direct adjacency to other spaces. For example, if space A and B are directly above each other (separated by a deck) there would be an edge between nodes A and B in the adjacency network. This seemingly simple network description of all the spaces, their arrangement, and adjacency relations within the ship is very powerful as it can capture many different types of relationships.

Examples of such relations, which are relevant for the proposed routing and vulnerability assessment approach, are:

1. Arrangement adjacency of subdivision compartments: A is adjacent to B either above, below, forward or aft. This network describes how all subdivision compartments are arranged with respect to each other (where a subdivision compartment is bounded by two structural or blast bulkheads and two decks).
2. Damage adjacency of subdivision compartments. For example, subdivision compartment A is adjacent to B but they are separated by a blast bulkhead blocking or mitigating the effects of an explosion in A to extend towards compartment B. In this case the existence of an edge in the network relates to the propagation of damage between the network nodes.
3. Routing adjacency of ship spaces and compartments: space A is adjacent to B and a connection can be routed between them. This network describes if a routed connection between adjacent compartments is possible. This is useful as not all adjacent compartments automatically allow connections to be routed between them. For example, it is impractical to route electrical cables through tanks containing liquids, or it may be undesirable to route certain piping through ammunition spaces.

Figure 2 provides examples of the three network types described above. Note that the example of the routing adjacency network shows that this type of network description can change depending on the type of distribution network and connection that is under consideration. In the example, the network does not allow routing through the lower tank compartments (e.g. for electrical cables). Each type of distribution network can thus be assigned its own dedicated routing adjacency network indicating which parts of the ship

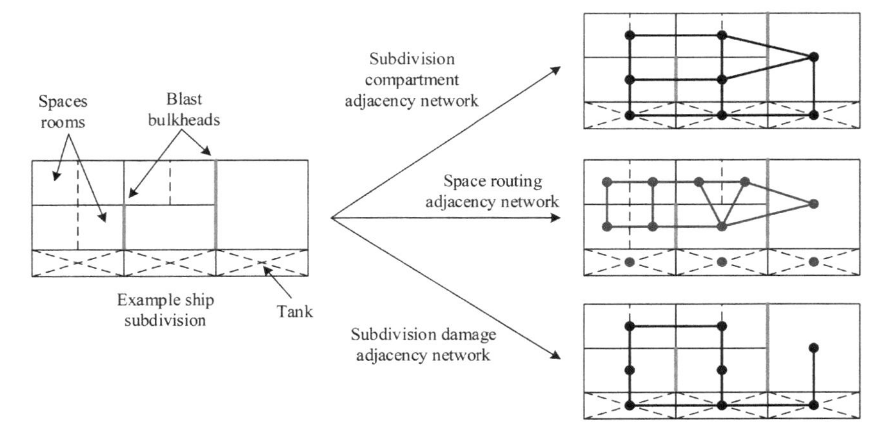

Figure 2. Describing the ship subdivision arrangement as different networks depending on the type of adjacency information to be captured.

are allowed (or must be avoided) when attempting to route connections.

2.2 *Step 2: Modelling distributed systems*

Distributed systems are modelled using the proposed framework of de Vos (2014). In his framework definition system components (e.g. generator-sets, pumps, switchboards, valve-chests, etc.) are modelled as network nodes, while connections between components (e.g. pipes, ducts, cables) are modelled as directed or undirected network edges. De Vos distinguishes three types of system nodes: users, suppliers, and hubs.

Users solely consume a resource while suppliers solely supply a resource within a network type. For example, a diesel generator set is a consumer (and thus user) within the fuel distribution system network but a supplier within the electrical distribution system network. Similarly, a warship sensor (radar) is a user of chilled water within the chilled water system network (for cooling purposes) and a supplier of sensor data within the sensor data network. Hubs neither supply nor consume resources but distribute resources from suppliers to users. For example, a switchboard may distribute electrical current at a certain voltage (e.g. 440V) to multiple end users. Note that a single component (e.g. the diesel generator-set) is thus both a user (fuel) in one network as well as a supplier in another network (electricity).

Using the above mentioned (user-hub-supplier) scheme, different types of distributed systems (e.g. fuel, chilled water, electricity, data) can each be described as a single network with a system adjacency matrix describing both the network components as well as their connections. Figure 3 and 4 give a simple example of this implementation. In the example eight components are connected in two distributed system types (a 440V electrical distribution network and a simple chilled water distribution network). Two diesel generator sets (DG1 and DG2) supply electricity at a voltage of 440V to two switchboards (SWB1 and SWB2). The switchboards distribute the electricity towards two critical users (CU1 and CU2) and two chilled water plants (CWP1 and CWP2). In addition, the two switchboards have an undirected crosslink. These, in turn, also supply chilled water (CW) towards the two critical users.

Though the given example of Figure 4 is rather simplistic, more complex networks representative of distributed systems found on board of ships can easily be modelled as shown in the test-case later in this paper (see Fig. 7).

2.3 *Step 3: Routing connections*

Steps 1 and 2 together described:

- where the distributed system components are located within the vessel's arrangement;
- how these system components are interconnected (and interdependent on one another) and by which network types;

	DG1	DG2	SWB1	SWB2	CWP1	CWP2	CU1	CU2
DG1	0	0	1	0	0	0	0	0
DG2	0	0	0	1	0	0	0	0
SWB1	0	0	0	1	1	0	1	0
SWB2	0	0	1	0	0	1	0	1
CWP1	0	0	0	0	0	0	1	0
CWP2	0	0	0	0	0	0	0	1
CU1	0	0	0	0	0	0	0	0
CU2	0	0	0	0	0	0	0	0

Figure 3. Example distributed system adjacency matrix for eight components.

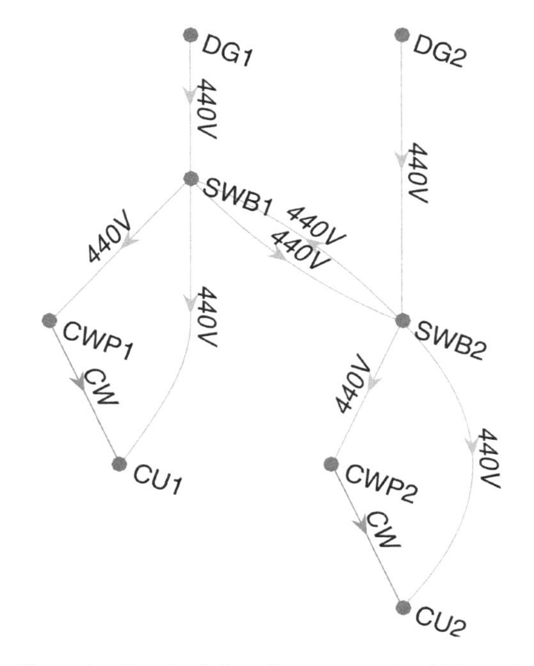

Figure 4. Graph of the adjacency matrix of Figure 3. Red edges show connection in the 440V electrical distribution system while the blue edges show connections within the chilled water system. Note that the dual connection between SWB1 and SWB2 is physically a single cable cross-link.

- the routing adjacency networks describing how connections between system components can be routed through the vessel's general arrangement;
- the damaged adjacency networks used to determine the extent of damage (Step 4).

The next step is to determine the routing configurations of all interconnection between system components. For this a k-shortest path algorithm is used to find and route different possible paths for each connection of the distributed system network as described in Step 2 (Yen 1971). The algorithm identifies not only the shortest path but also $k-1$ other paths in increasing order of path length. This allows alternative paths for connections to be found which can decrease vulnerability by taking alternative, and possible longer, routes through the ship arrangement.

Once k-paths have been identified for each connection within the distributed system network, the complete routing configuration can be assembled by choosing different combinations of paths for each connection. Paths can be chosen manually or automatically with the use of a genetic algorithm (Deb et al. 2002). In the latter case, the genetic algorithm chooses combinations of paths that configure the shortest and least vulnerable routing configuration. Algorithm 1 gives the pseudo code for the adopted approach. Refer to Step 4 and 5 in Fig. 1 for more information on assessing the vulnerability of the routing configuration.

```
Input: paths (connections) in the distributed systems
foreach path in the system do
    Retrieve path info (start node, end node,
      distribution network type);
    if path start = path end then
        No path needed, set path length = 0;
    else
        Find k-paths from start to end in this routing
          network type ;
        Store found paths and their length;
    end
    if number of paths found > 1 then
        Choose a path using the GA;
        Store chosen path for this routing
          configuration;
    else
        Store the found path for this routing
          configuration;
    end
end
Return set of chosen paths and their length for the
  routing configuration;
```

Algorithm 1. Pseudo code for finding possible routing paths for all distributed system network connections, the result is one possible routing configuration for the entire set of distributed systems and their connections.

The proposed routing approach is an improvement of the 2D routing approach used by van Oers et al. (2012) which could only identify variations of the shortest path. Shields et al. (2017) also applied a k-shortest path approach for identifying different routing configurations for system connections. They opted for a non-optimisation approach in search for general insights into different routing configurations. Our approach however, uses optimisation to identify Pareto-front solutions giving a trade-off between overall routing configuration length and vulnerability. Exploring routing solutions along and near the resulting Pareto-front gives designers more insight into routing bottlenecks, such as, poorly chosen equipment/space locations, the use of blast bulkheads, and the chosen system topology and redundancy. In addition, the overall routing configuration length (which is available per network type) may provide valuable cost estimation information.

2.4 *Step 4: Simulating damage*

Once a routing configuration has been selected (Step 3) the next step is to simulate damage to the ship and distributed systems caused by a weapon hit. The network description of the subdivision compartments and the routing network provide a quick and scalable approach. Damage is simulated using five steps:

1. Select one or multiple nodes in the damaged subdivision adjacency network as the primary hit compartments.
2. Find the adjacent subdivision compartment nodes if the damage extent is large. For a small damage extent only one compartment may be assumed as hit and damaged. Note that a blast bulkhead is represented by a removed edge (see Fig. 5).
3. Find routing nodes (system components) and paths (system connections) that are within or are routed through the damaged subdivision compartments.
4. Create a damaged system matrix with removed edges and nodes representing the damaged components and connections for this damage case.

The vulnerability assessment (see Step 5) is relatively fast. Thus, when a single or double hit damage case is simulated, it is usually possible to assess hits to all possible compartments of the vessel.

For multi-hit damage cases with more than two hits however, the number of possible hit combinations quickly increases with:

$$C_k(n) = \binom{n+k-1}{k}, \quad (1)$$

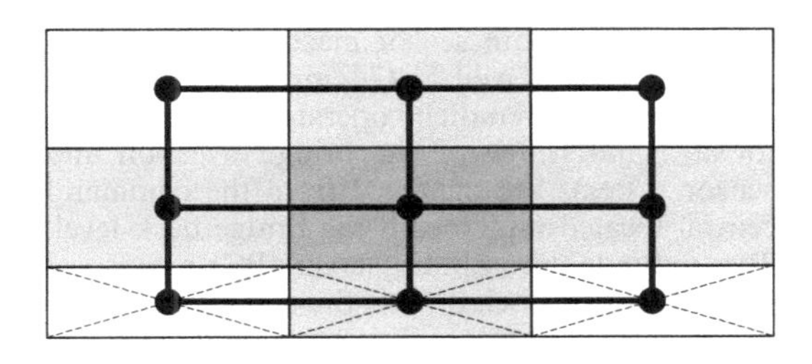

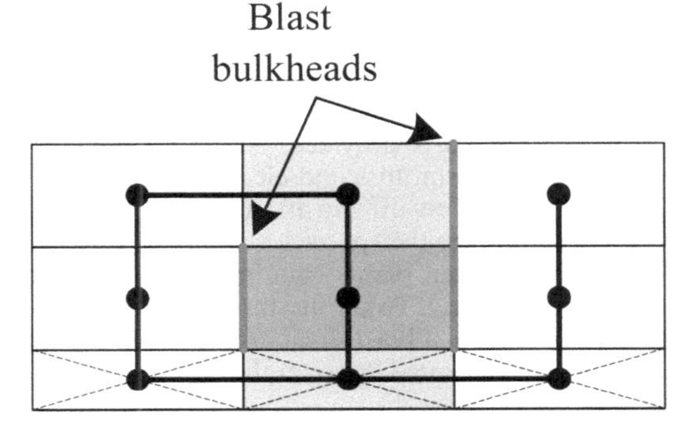

Figure 5. Using the subdivision damage adjacency matrix to determine the damage extent. The red compartment (centre) is directly hit and damaged, the adjacent compartments are indirectly damaged. Note the effect/modelling of blast resistance bulkheads (bottom).

where, n is the number of compartments and k the number of hits to simulate. For a 30 compartment ship a three-hit case already gives 4960 scenarios. In such cases, a randomly chosen sub-set may be evaluated in order to simulate the expected vulnerability score.

For the OPV test-cases presented later in this paper the individual hit probability of each subdivision compartment is set equal. That is, a compartment low down in the ship (e.g. a fuel tank) thus has the same hit probability as a higher compartment containing for instance the bridge or PC server rooms. Both this equal hit probability and the inclusions of all subdivision compartments in the analysis can of course be debated. It is highly dependent on the type of threat that is considered for the damage scenario. For example there will be a large difference between an airborne projectile and a subsea mine or torpedo. Nonetheless, more elaborate and alternative hit distributions and scenarios can be modelled using the same approach as presented above.

2.5 *Step 5: Assessing vulnerability*

The last step of the proposed approach deals with assessing the vulnerability of the routed system topology given the simulated damage to the ship and the distributed system components and connections (step 4). The assessment is based on an approach developed by van Leeuwen (2017). It uses the distributed systems framework of de Vos (Step 2) in combination with a depth-first search to check whether a critical user(s) in the network is connected. That is, it is fully supplied by all its required resources, such as: electricity, chilled water, and data. By checking one, or multiple, critical user(s) the vulnerability of the current distributed systems and their routings can be assessed. Finally, by simulating multiple hit and damage scenarios as described in step 4, the chance of losing one or more critical users may be estimated.

Algorithm 2 shows the pseudo code for the depthfirst search algorithm that checks the connectivity of a critical user (*system node*). A problem that arises when applying this network search method, is the existence of self-loops in the distributed system network. That is, distributed systems are often interdependent. For example, a switchboard panel (SWB) distributing electricity from a diesel generator set (DG) to a chilled water plant (CWP) may itself depend on chilled water for cooling purposes. The basic depth-first search algorithm will then run into an endless loop.

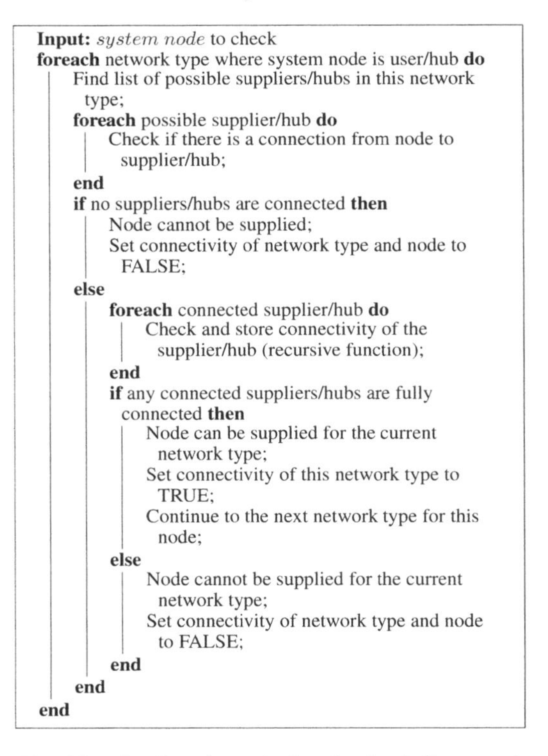

```
Input: system node to check
foreach network type where system node is user/hub do
    Find list of possible suppliers/hubs in this network
      type;
    foreach possible supplier/hub do
        Check if there is a connection from node to
          supplier/hub;
    end
    if no suppliers/hubs are connected then
        Node cannot be supplied;
        Set connectivity of network type and node to
          FALSE;
    else
        foreach connected supplier/hub do
            Check and store connectivity of the
              supplier/hub (recursive function);
        end
        if any connected suppliers/hubs are fully
          connected then
            Node can be supplied for the current
              network type;
            Set connectivity of this network type to
              TRUE;
            Continue to the next network type for this
              node;
        else
            Node cannot be supplied for the current
              network type;
            Set connectivity of network type and node
              to FALSE;
        end
    end
end
```

Algorithm 2. Pseudo code for the depth-first search tree algorithm which assesses the connectivity of a system node and its required supplies in a damaged distributed system network.

This problem is solved by keeping track of the actual search tree of nodes that has already been checked. Whenever a system node and supplying node is under investigation, a cross-check is made as to whether this node-pair has been checked before (which is an indication of a self-loop in the network). Thus, when in the example mentioned above the combination of SWB $\leftarrow$ CWP is checked for the second time, the algorithm stops this branch of the search tree automatically as this connection was already checked earlier.

3 OFFSHORE PATROL VESSEL TEST-CASE

To demonstrate the working of the proposed approach, a test-case was devised using a notional offshore patrol vessel (OPV) design. The goal of the exercise is to maximise the availability (and thus minimize the associated vulnerability) of the main 76 mm gun. Note that, because the availability of multiple systems can be assessed simultaneously it is also possible to assess functional capabilities of a vessel combining multiple, often redundant, systems as opposed to a single system as was done for this test-case (e.g., self-defence, propulsion, or anti-air warfare). The following sections elaborate on the OPV design, the distributed systems modelled, the problem set-up, and gives an indication of the type of output the approach provides.

3.1 *Vessel arrangement*

The test-case notional OPV has a traditional small combatant design (see Fig. 6). The OPV has one large main sensor mast located at midship, a medium calibre (76 mm) gun forward, and a 30 mm remotely operated machine gun forward and below of the bridge. Weapon and sensor systems are managed from the command central located underneath the bridge deck level. Two redundant computer server (PC) rooms are located in the vessel, one on the command central deck (below the bridge) and one aft and below the helicopter hangar (see red block in Fig. 6 below the helicopter hangar).

For propulsion and power generation machinery a centrally located, but split two compartment fore and aft engine room configuration (see green blocks in Fig. 6) is provided. These contain the main propulsion motors and electric power diesel generator sets. Forward and aft of the engine room compartments are the spaces for auxiliary systems (e.g. chilled water plants, pump rooms, electric switchboards, etc.). In an attempt to reduce vulnerability further, a blast resistant transverse bulkhead is placed located between the two propulsion machinery rooms.

3.2 *System description*

To function properly, the 76 mm main gun requires several inputs. These are: chilled cooling water, 440V electricity, and targeting data. These are provided by the main distributed systems on the vessel: the 440V electricity distribution network, the chilled water network (CW), and the data distribution network (DataShoot). Table 1 lists all simulated system components and their associated role in the different networks (user, hub, or supplier). Figure 8 gives an overview of the applied distribution network.

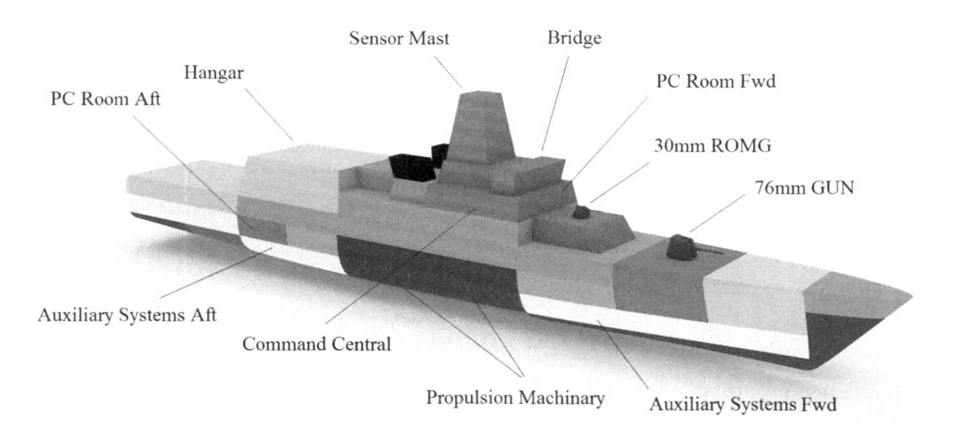

Figure 6. OPV general arrangement and main system locations.

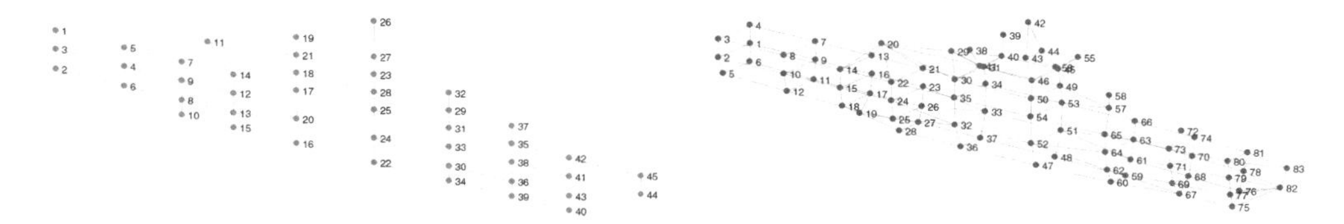

Figure 7. OPV subdivision compartment adjacency network (left) and routing compartment adjacency network (right). Note the blast bulkhead edges that are missing between the two engine room compartments in the subdivision adjacency network.

Table 1. List of OPV system components and associated distribution network types. The 76 mm gun is considered the critical system in the vulnerability analysis.

	Network in which system component is		
System component	User	Hub	Supplier
4 × Diesel-generator sets	–	–	440V
2 × Main Switchboard (aft & fwd)	–	440V	–
2 × Tranformer (SB & PS)	440V	–	690V
2 × Electric Motor (SB & PS)	690V	–	–
3 × Load Centre (aft, mid & fwd)	–	440V	–
2 × Chilled Water Plant (aft & fwd)	440V	–	CW
2 × Chilled Water Hub (aft & fwd)	–	CW	–
Sensor Mast (radar)	440V; CW	–	DataSens
2 × PC Server Room (aft & fwd)	440V; CW	DataSens	
Command Central (CIC)	440V; CW; DataSens	–	DataShoot
30 mm Gun	440V; CW; DataShoot	–	–
76 mm Main Gun (critical user)	440V; CW; DataShoot	–	–

Figure 8. Graph of the OPV distributed systems network. The 76 mm gun is located in the bottom left of the figure, the 4 diesel generator sets in the top right of the figure. All distribution networks are shown: 440V (orange), 690V (blue), chilled water or CW (yellow), data (purple and light blue).

3.3 *Test-case setup*

Initially two test-cases, each with a different damage scenario are evaluated. The first assumes a small single compartment damage, the second a larger multicompartment damage where the directly adjacent subdivision compartments are also considered lost (refer to Step 4 in Section 2). Both test-cases assume a single hit to the vessel. The OPV has 45 subdivision compartments resulting in 45 different possible hit locations. So, for each routing configuration selected by the genetic algorithm, 45 hit cases and possibly damaged distributed system networks are evaluated.

The two test-cases take roughly a day to set-up, run, and interpret, where the required steps are:

1. Translating the 3D CAD model into the required subdivision and routing compartment adjacency networks. This process is fully automated using scripts in the CAD software.
2. Defining the distributed system components, their location (taken from the 3D CAD model of the ship's arrangement), and the connections between them.
3. Running the routing approach for the test-cases. Running each case takes 20 minutes on a standard laptop PC.
4. Interpreting the results and possibly adjusting the model to run additional cases.

3.4 *Output and results*

Figures 9–12 display the two extreme Pareto solutions found for both the single and the multi compartment damage test-cases (i.e. the shortest least vulnerable and the least vulnerable shortest routing configurations). The side-view plots only show the chilled water network (yellow lines) and indicate which subdivision compartments when hit result in loss of functionality of the 76 mm gun.

Furthermore, the results indicated that this chilled water distribution network, and specifically the chilled water connection between the forward chilled water hub (CWP HUB 2) and the command central (CIC) highlighted in black in Figures 10 and 12, is a bottleneck. Note that the CIC in this case is only connected by a single source of chilled water and that this chilled water is required for cooling the processing equipment. In both the multi and the single compartment damage scenario the shortest path solution routes this CIC chilled water connection down three decks towards the forward engine room compartment and then forward towards the second chilled water hub. The less vulnerable path has a slight detour and goes down two decks before turning forward and down towards the chilled water hub. In both cases this second routing option results in a lower vulnerability score as a hit to forward

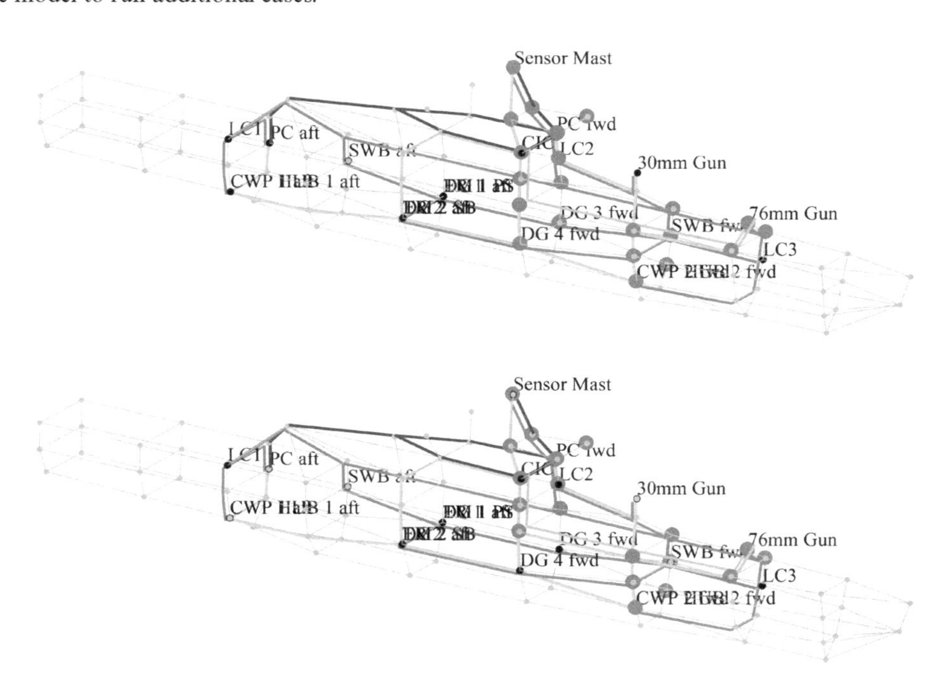

Figure 9. Single small hit compartment damage case. Both the shortest path Pareto solution (top) and lowest vulnerability solution (bottom) are shown. The colours of the routing paths correspond to the different network types (see Fig. 8). The red nodes indicate nodes which if damaged cause the main gun to be unavailable.

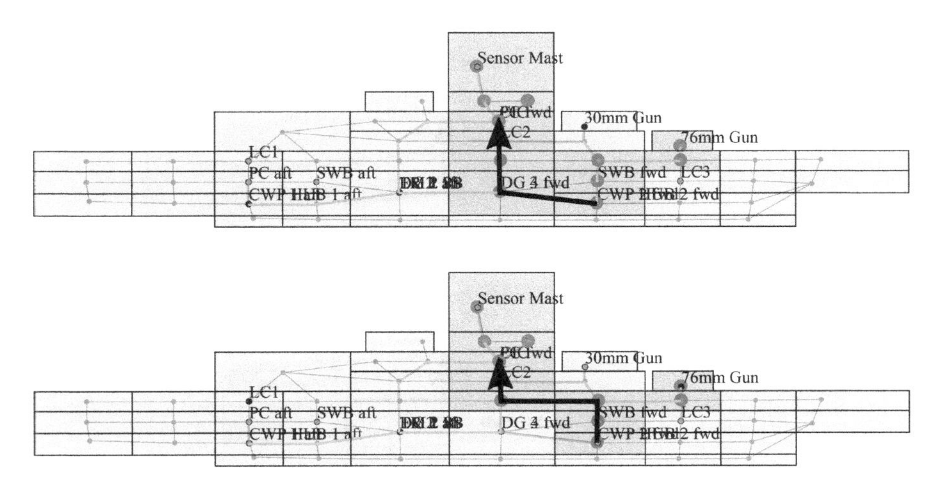

Figure 10. Single small hit compartment damage case. Both the shortest path solution (top) and lowest vulnerability solution (bottom) are shown. Only the chilled water distribution system connections are displayed in this side view.

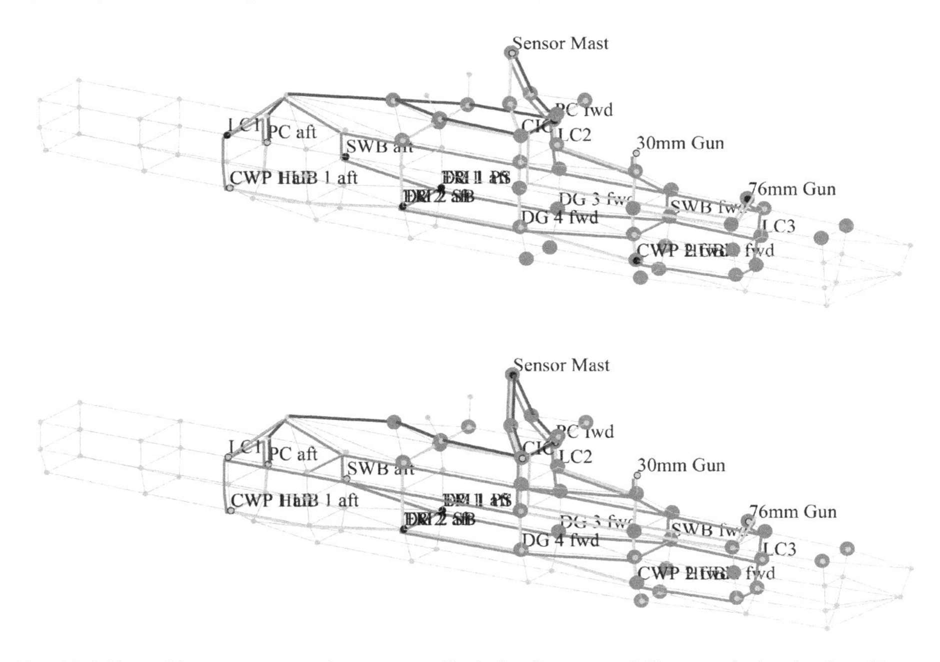

Figure 11. Multi large hit compartment damage case. Both the shortest path Pareto solution (top) and lowest vulnerability solution (bottom) are shown. The colours of the routing paths correspond to the different network types (see Fig. 8). The red nodes indicate nodes which if damaged cause the main gun to be unavailable.

engine room no longer cuts the chilled water connection towards the CIC.

For the single compartment damage test-case a second evaluation was made using a slightly altered distributed network topology. In an attempt to decrease vulnerability, an additional chilled water connection between the aft chilled water hub (CWP HUB 1) and the CIC, as well as an additional data connection (shown in blue) between the CIC and the 76 mm gun, was added. The least vulnerable routing results of this third test-case are shown in Figure 13. The highlighted black arrows in the side-view plot show the forward and aft redundant chilled water connection towards the CIC. The light-blue lines show the redundant data connection from the CIC towards

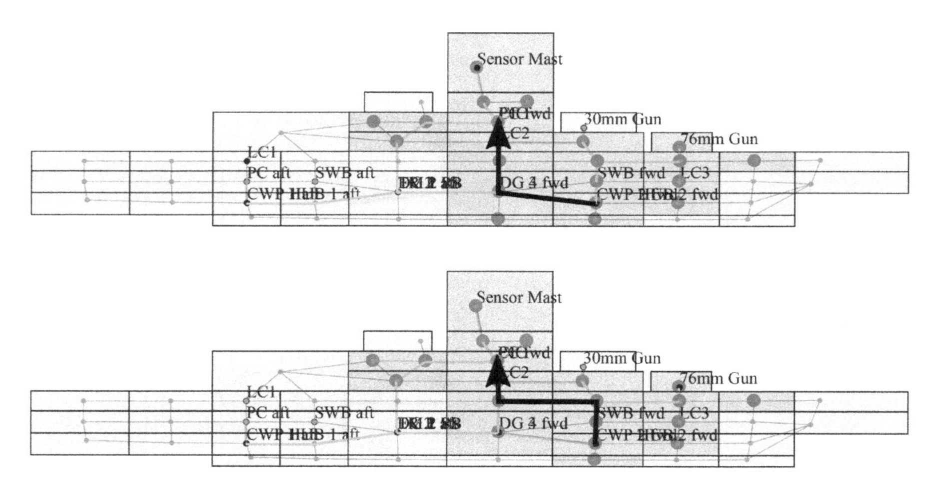

Figure 12. Multi large hit compartment damage case. Both the shortest path solution (top) and lowest vulnerability solution (bottom) are shown. Only the chilled water distribution system connections are displayed in this side view.

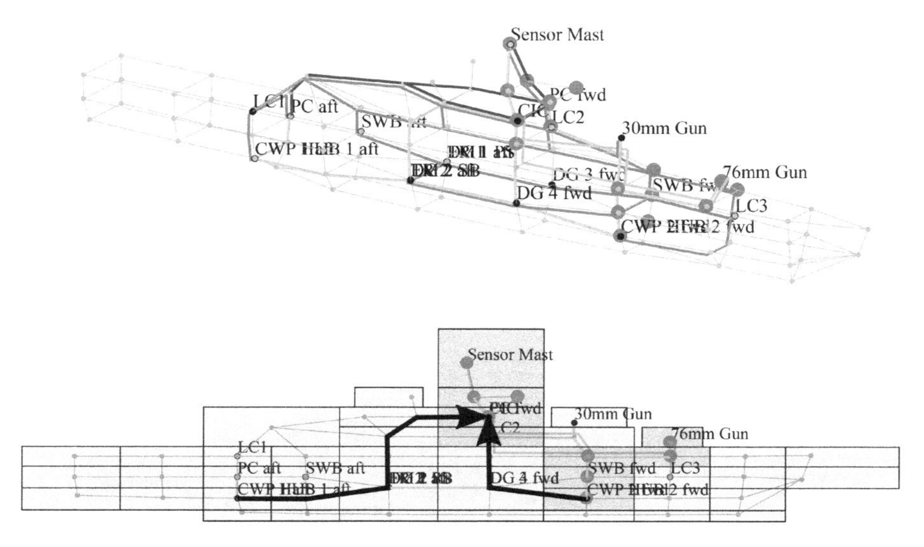

Figure 13. Single small hit compartment damage case with an additional chilled water connection to the CIC (highlighted in black). The blue lines show the now redundant data connections from the CIC towards the 76 mm gun. Note the subdivision compartment above the forward engine room which in Fig. 10 still caused a malfunction of the 76 mm gun.

the 76 mm gun. Note that they are routed through different compartments. Adding these two connections increased the total routing length from 924 to 1034 meter but cleared an additional compartment from the hit locations. This resulted in a further reduction in the probability of losing the main gun as one additional compartment was removed from the critical hit locations (compare Figs. 10 and 13).

In the case of the multi compartment damage scenario test-case, adding the additional connections had no effect on the vulnerability outcome. Although naturally the total routing configuration length did increase. Here, a different solution is required. For example, by adding additional blast/fragment resistant bulkheads or by adding an additional chilled water hub. Although these options where not further pursued for this paper, it is relatively easy to adjust the model and run the analysis.

4 CONCLUSIONS

This paper presents an approach to the early stage routing and vulnerability assessment of distributed ship service systems. It is specifically targeted towards the early stage design of warships where insight into the impact of several design aspects on the ship's vulnerability is wanted. This not only assists designers in improving the ship and system designs, but also aids the discussion of setting requirements with respect to distributed systems design and integration. Herein aspect such as: the ship and its systems arrangement; the chosen system topology; and, especially, the routing of distributed systems play a prominent role.

The approach implements several types of networks and network algorithms to model the ship arrangement, distributed system topology, and system logic. A k-shortest path algorithm generates many possible routing configurations while a depth-first search tree approach is used to assess if the distributed system and its critical components still function. This is combined within a genetic algorithm which optimises the complete routing configuration of the distributed ship systems for minimal connection length and vulnerability.

Two test-cases involving an offshore patrol vessel demonstrated the working of the approach and indicated how it could be used as an early design aid. The routing configuration tool proposed alternative (longer) connection paths for specific distribution network types in order to reduce vulnerability of operating the main 76 mm gun. Moreover, by carefully studying the results it was possible to further reduce vulnerability by adding additional connections within the defined distributed system network topologies.

The ability to identify and resolve bottlenecks (e.g. as identified in the OPV test-case) in the distributed system topology, the ships' general arrangement, and the system component locations is considered a valuable outcome of the approach. This should further facilitate the dialogue between ship and system designers, and aid in setting-up and evaluating requirements related to system and ship vulnerability early on.

5 FUTURE WORK

The approach presented in this paper is workinprogress. It is currently under further development both at the Defence Materiel Organisation as well as Delft University of Technology. As such, several improvements and recommendations are envisioned. These are:

- Improvements to the k-shortest path algorithm. The factor k can be linked to the length of the shortest path for that network connection eliminating the need to find k paths for every connection. For example, if we only need a connection between two adjacent compartments, there is no need to examine 1000 different options, whereas a connection from the foremost compartment to aft most compartment of vessel has many different possibilities. Using graph theory we can calculate the total number of possible paths with a predefined length (Fiol & Garriga 2009), this would allow us to make a quick estimate for the range of the factor k to use in the k-shortest path algorithm.
- Vulnerability of multiple systems. The OPV test-case considered only a single critical system (the main gun) whereas it is also possible to analyse the combined vulnerability of multiple systems at once. For example, maximise the availability of all weapon systems, or investigate different groups of systems associated to capabilities of the ship (e.g. anti-air warfare, anti-surface warfare, missile defence). Ultimately, the desired post-hit availability of systems or groups of systems depends on the operational use concept and scenario.
- Including the propulsion related systems. Although this paper focussed on the traditional distribution networks, according to de Vos (2014) the mechanical drive train of the propulsion system can also be thought of as a distribution of power. In this case the components are: electric motors, diesel engines, gearboxes, and clutches, and the connections are physical rotating shaft lines. This opens up possibilities for analysing the vulnerability of more complex drive train configuration including both a combination of electric motors (and thus also the electrical distribution network) and diesel engines as often found in newer CODELAD configurations.
- Modelling of data distribution systems. The test-case OPV example contains a small and simplistic data distribution network example. When multiple sensors, server rooms, and weapon systems are involved this quickly increases in complexity. In addition, the information flow through data (network) cables consists of multiple data types (e.g. raw sensor data, processed sensor data, command decision data, or weapon control data) which often "flow" through the same physical cables in both directions. This distinguishes the data distribution from the other traditional non combat system oriented distribution systems (e.g. electricity, liquids, or air). Hence, it requires a different modelling approach.

DISCLAIMER

The opinions presented are the personal opinions of the authors and the authors alone. Specifically, they do not represent any official policy of The Netherlands Ministry of Defence, the Defence

Materiel Organization, or the Royal Netherlands Navy. Furthermore, the results presented here are for the sole purpose of illustration and may not have an actual relation with any past, current or future warship procurement project at the Defence Materiel Organization.

REFERENCES

Andrews, D.J. (2011). Marine requirements elucidation and the nature of preliminary ship design. *International Journal of Maritime Engineering 153*, A23–A39.

Andrews, D.J., E.A.E. Duchateau, J.W. Gillespie, J.J. Hopman, R.G. Pawling, & D.J. Singer (2012). State of the art report: Design for layout. In *Proc. 11th Int. Marine Design Conf. (IMDC)*, Volume 1, Glasgow, UK, pp. 111–137.

Andrews, D.J., A. Greig, & R.G. Pawling (2004). The implications of an all electric ship approach on the configuration of a warship. In *INEC 2004: Marine Technology in Transition*. IMarEst.

Asmara, A. (2013). *Pipe routing framework for detailed ship design*. Ph. D. thesis, Delft University of Technology.

Brefort, D., C. Shields, A. Habben Jansen, E.A.E. Duchateau, R.G. Pawling, K. Droste, T. Jasper, M. Sypniewski, C. Goodrum, M.A. Parsons, M.Y. Kara, M. Roth, D.J. Singer, D.J. Andrews, J. Hopman, A. Brown, & A.A. Kana (2018). An architectural framework for distributed naval ship systems. *Ocean Engineering 147*, 375–385.

Brown, D.K. (1987). The architecture of frigates. In *Proc. Warship '87: Int. Conf. on ASW*, Volume I, London, UK. RINA. Cerminara, J. & R.O. Kotacka (1990). Ship service electrical systems designing for survivability. *Naval Engineers Journal 102*(5), 32–36.

Collins, L., R.G. Pawling, & D.J. Andrews (2015). A new design approach for the incorporation of radical technologies: Rim drive for large submarines. In *Proc. 12th Int. Marine Design Conf. (IMDC)*, Providence, RI, USA.

Deb, K., A. Pratap, S. Agarwal, & T. Meyarivan (2002). A fast and elitist multiobjective genetic algorithm: NSGA-II. *IEEE Trans. on Evolutionary Computation 6*(2), 182–197.

Duchateau, E.A.E. (2016). *Interactive evolutionary concept exploration in preliminary ship design*. Ph. D. thesis, Delft University of Technology.

Fiol, M.A. & E. Garriga (2009). Number of walks and degree powers in a graph. *Discrete Mathematics 309*(8), 2613–2614. Gillespie, J. (2012). *A network science approach to understanding and generating ship arrangements in early-stage design*. Ph. D. thesis, University of Michigan, Ann Arbor, USA.

Goodfriend, D. & A.J. Brown (2017). Exploration of system vulnerability in naval ship concept design. *Journal of Ship Production and Design 309*(8), 2613–2614.

van Leeuwen, S.P. (2017). Estimating the vulnerability of ship distributed system topologies. Master's thesis, Delft University of Technology.

van Oers, B. (2011a). *A packing approach for the early stage design of service vessels*. Ph. D. thesis, Delft University of Technology, Delft, The Netherlands.

van Oers, B.J. (2011b). Designing the process and tools to design affordable warships. Technical report, NATO-RTO-MP AVT-173 Workshop on virtual prototyping of affordable military vehicles, Sofia, Bulgaria.

van Oers, B.J., E. Takken, E.A.E. Duchateau, R.J. Zandstra, J.S. Cieraad, W. van den Broek de Bruijn, & M. Janssen (2017). Warship concept exploration and definition at the netherlands defence materiel organisation. In *ASNE Design Sciences Series Workshop on Set-Based Design*.

van Oers, B.J., G. van Ingen, & D. Stapersma (2012). An integrated approach to design, route and evaluation of resilient ship service systems. In *11th International Naval Engineering Conference (INEC)*, Edinburgh, UK.

Park, J.-H. & R.L. Storch (2002). Pipe-routing algorithm development: case study of a ship engine room design. *Expert Systems with Applications 23*(3), 299–309.

Pawling, R.G. (2007). *The application of the design building block approach to innovative ship design*. Ph. D. thesis, University College London.

QinetiQ (2017). SURVIVE. www.qinetiq.com.

Rigterink, D.T. (2014). *Methods for Analyzing Early Stage Naval Distributed Systems Designs, Employing Simplex, Multislice, and Multiplex Networks*. Ph. D. thesis, University of Michigan.

Said, M.O. (2009). Theory and practice of total ship survivability for ship design. *Naval Engineers Journal 107*(4), 191–203.

Schulte, J.C. (1994). *An analysis of the historical effectiveness of anti-ship cruise missiles in littoral warfare*. Ph. D. thesis, Naval Postgraduate School, Monterey, CA, USA.

Shields, C.P.F., D.T. Rigterink, & D.J. Singer (2017). Investigating physical solutions in the architectural design of distributed ship service systems. *Ocean Engineering 135*(3), 236–245.

Takken, E. (2008). Concept design by using functional volume blocks with variable resolution. Master's thesis, Delft University of Technology, Delft, The Netherlands.

TNO (2017). RESIST LITE. www.tno.nl.

de Vos, P. (2014). On the application of network theory in naval engineering: Generating network toplogies. In *12th International Naval Engineering Conference and Exhibition (INEC)*, London, UK.

de Vos, P. & D. Stapersma (2018). Automatic topology generation for early design of on-board energy distribution systems. *Ocean Engineering*. under review.

Yen, J.Y. (1971). Finding the k shortest loopless paths in a network. *Management Science 17*(11), 712–716.

Offshore and wind farms

Marine Design XIII – Kujala & Lu (Eds)
© 2018 Taylor & Francis Group, London, ISBN 978-1-138-34076-3

An innovative method for the installation of offshore wind turbines

Prabhu Bernard & Karl Henning Halse
Department of Ocean Operation and Civil Engineering, NTNU Ålesund, Norway

ABSTRACT: Emergence of offshore wind turbine as an extremely promising renewable energy resource has transformed the future of marine applications and ship design. The objective of this paper is to present a smart and efficient mechanism for installing an offshore wind turbine at water depths more than 200 meters without relying on existing high lift offshore cranes. The integration of this mechanism into the design of a construction vessel in order to meet the future market demand is tested and verified in this project. The paper is primarily structured around three phases. A design phase in which a new offshore wind turbine deployment mechanism is developed by considering manufacturability and material selection. An analysis phase that ensures the structural stability of the concept using NX Nastran and a simulation phase that verifies the kinematic stability of the design using NX motion simulation.

1 INTRODUCTION

1.1 *Background*

Among several sources of renewable energy resources available today, Offshore Wind Turbines (OWT) have proved to be extremely promising for the future of humanity. Currently there are several OWT installations around the world. Installation of these wind turbines are extremely expensive and time consuming owing to the complexities involved in erecting a huge structure out in the ocean. European offshore wind energy statistics shows the erection of only 182 wind turbines during the first half of 2016 in 13 wind farms operated by four countries [1].

Meteorological studies have proved that the average wind speed in the offshore area is less turbulent and more intense when compared to the onshore sites due to low surface shear. Installing a wind turbine out in offshore location has always been a challenge and it remains to be one. A smart and efficient way of installing a 1000-ton wind turbine safely and efficiently in the middle of the ocean is still a prominent research question. This project aims in developing an innovative concept design for the installation of offshore wind turbines, which could surpass the current limitations in the installation time and contribute towards a better future. Reducing the installation time could have a significant impact on the capital investments required for the project. The dynamic stability and structural integrity of the concept developed in this project is later verified using motion simulation and structural analysis with the help of a CAD (Computer Aided Design) software, Siemens NX.

1.2 *Problem definition*

Most of the wind turbines in operation today are on top of monopiles driven deep in to the seabed for a stable and reliable operation. The installation of these heavy wind turbines are often carried out in an effective manner using jack up vessels which are rooted to the sea bed for increased stability during assembly process.

In order to reap the full benefits of an offshore wind turbine, the installation needs to happen far into the ocean where there is a more reliable source of wind speed. However, installing a wind turbine on a foundation far from the shore has its limitations due to the depth of the seabed and the unpredictability of the sea states.

The concept developed in this paper is inclined more towards the installation of an offshore wind turbine far into the ocean on top of a floating foundation. Usage of fixed jack up vessels cannot be an option in this scenario owing to the depth of the seabed, which in this case is more than 200 meters. This also questions the reliability of existing technology in achieving a safe and quick installation. The innovative concept developed in this paper has the possibility of carrying a number of pre-assembled wind turbines on a catamaran vessel for installation.

Huge catamaran installation vessels could provide more stability while carrying out the operation presented in the paper. The main objective behind installing a pre-assembled wind turbine on a floating foundation rather than carrying out a part-by-part installation out in the sea is to reduce the overall installation time. A quick and safe installation out in the sea require calm waves and wind conditions, which are available only at rare intervals during a 24-hour period. The mechanism developed in this paper is also equipped with advanced motion compensation systems and lifting mechanisms, which could significantly reduce the time and cost involved in the process and carry out installations in higher sea states.

1.3 *Scope of work*

After evaluating the current investments and projections in the field of offshore energy, it is evident that the total installation of wind turbines needs to increase in order to reduce the consumption of fossil fuels that induces global warming. To achieve the clean energy target projected by the European countries, a paradigm shift is required.

The current industrial focus is towards, increasing the production of fully assembled wind turbines. Providing safety and stability during the transportation and installation. Developing new technologies that could significantly reduce the installation time and increasing the operability at higher sea states by improving motion compensation systems. Scope of the work takes the implications of the above factors into consideration.

1.4 *Motivation*

The motivation behind choosing the project has two major aspects, an environmental aspect and a technical aspect. The use of fossil fuels has a huge negative impact on the humanity and any effort to reduce it is a step closer to saving our planet. A shift from fossil fuels to renewable energy could save our future generations from the ill effects of global warming. Introducing this concept design ensures a smooth transition to renewable energy. This in turn boosts the generation of electricity using wind power and encourage the decommissioning of existing coal power plants, which poses a constant threat to humanity.

The technical aspect of choosing the project is to reduce the huge amount of capital investment required for chartering the installation vessels. As per the statistics presented in the report published on offshore wind cost reduction [2], it shows that a significant 26% of the whole installation time is due to weather downtime. The weather downtime is expensive due to the cost involved in hiring the installation vessels. In order to carry out a successful wind turbine installation out in the sea, significant wave height (Hs) is an important factor. For example, installation of a monopile required for mounting a wind turbine requires a significant wave height less than 1.5 m. However, a sea state higher than 1.5 m dominates the North Sea sites for almost 40% time of the year.

Increasing the installation capabilities from a significant wave height of 1.5 m to 2.5 meter could reduce the weather downtime to less than one third of the existing time. The floating vessels with highly efficient dynamic positioning system used currently for this purpose have expensive charter rates per day. The operating cost for a large installation vessel with highly advanced dynamic positioning system is around 280,000 USD a day as per the reports published by The Crown Estate [3]. With this huge investment at stake, reducing weather downtimes could have a significant impact on the trade and strengthen our drive towards clean energy.

In addition to using a catamaran vessel for the installation of offshore wind turbine, the project focuses on developing a fully functional motion compensation system to assist the installation. This could further increase the possibilities of installing the wind turbines at a much higher significant wave height, thereby reducing the downtime due to weather conditions further more. The most important challenge while designing a motion compensation system for heavy lifting is to provide a safe and stable operation for the installation at these high sea states. Verifying the dynamic stability of the concept in this paper ensures a safe and stable operation.

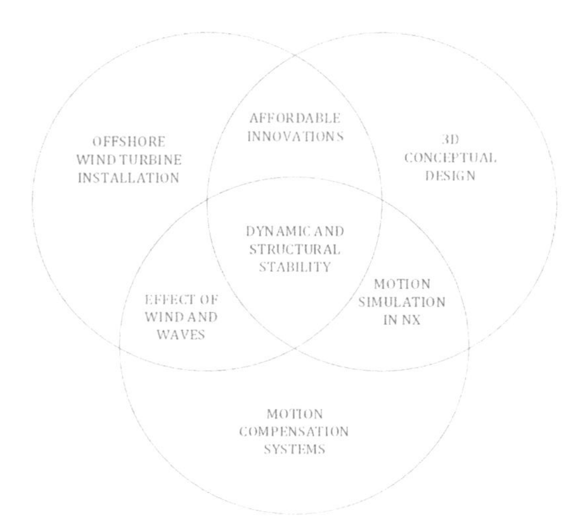

Figure 1. Scope of work.

2 CONCEPT DESIGN OVERVIEW

2.1 *Methodology*

This part of the paper provides an overview of the concept developed, which involves the end-to-end methodology followed by a systematic detailing of the entire installation process.

The project has mainly three phases, which structure the whole process followed. A design flow chart connect these phases (See Figure 2).

The flow chart explains the steps followed during the completion of the project. It also summaries every step which needs to be validated with the respective analysis to be able to proceed to the next level.

Design of Lifting cage for transferring wind turbine on to foundation: Transportation of the wind turbines to the installations location is usually as separate pieces. In this project, vertical transportation of the wind turbine as a complete assembly in a vessel is the main highlight. However, for transferring the turbine assembly from a vessel to the foundation requires utmost precision. First step in the process was to design a special lifting and deploying cage for this purpose.

Design of lifting hook inside the lifting cage: The lifting hooks designed in this project is an attachment coming along with the lifting cage. The hooks works on hydraulic actuation and supports the weight of the tower during lifting process. Hook is a critical part since it takes up the entire weight of the turbine. Hence, a separate structural analysis is required to ensure the stability of the hook as well as the stability of the lifting cage along with the hook.

Design of the motion compensated platform and support structures: The motion compensated platform assists in compensating the pitch, roll and heave motion. However, this platform has to be at a height almost half of that of the turbine, which is around 50 meters from the base to maintain the balance of the system. In order to achieve that, a tall support structure acts as the base of the motion compensated platform, which also compensate for the sway motion. Finite element analysis verifies the structural stability of the main platform.

Design of sway compensated platform: In order to compensate the movement in the sway direction, the entire motion compensated platform is on top of steel rollers controlled by a rack and pinion system. This will assist in achieving a safe and secure installation by compensating the sway motion.

Design of Yaw compensated platform: Dynamic positioning system of the vessel and the tunnel thrusters helps in compensating the yaw motion. However, in order to achieve high precision during installation, a rotating platform driven with the help of helical gears and electric motors are used.

Figure 3 shows a 3D model of the concept during the initial stages of development. It highlights the major components involved in the design as mentioned in the flowchart.

Introduce prescribed motion to the vessel: After completing the design and analysis phase, motion simulation module in NX analyzes the kinematic stability of the joints. A prescribed motion is set to move the vessel in the direction of heave, pitch and roll. This is to study the response of the joints to these motions.

Develop a motion simulation model with all joints intact: Plotting the motion response from the joints with respect to the relative motion of the wind turbine assembly ensures a perfectly synchronized motion simulation model.

The steps 2, 4 and 6 discuss about structural analysis. A detailed analysis ensures the structural stability of the concept [4]. However, part of the project that involves the results of structural analysis is not included in this paper.

2.2 *Concept development*

The proposal for the project started with two sketches that combines the initial design idea. These when tweaked into a working model after checking the feasibility of the kinematics behind the movements gives the final design of the motion compensation system.

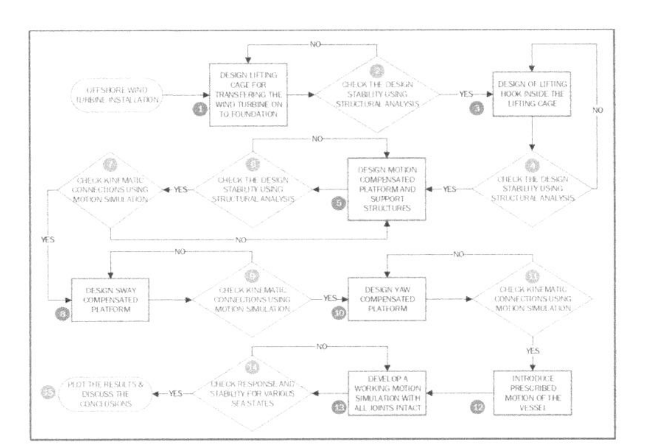

Figure 2. Design flow chart of the process involved.

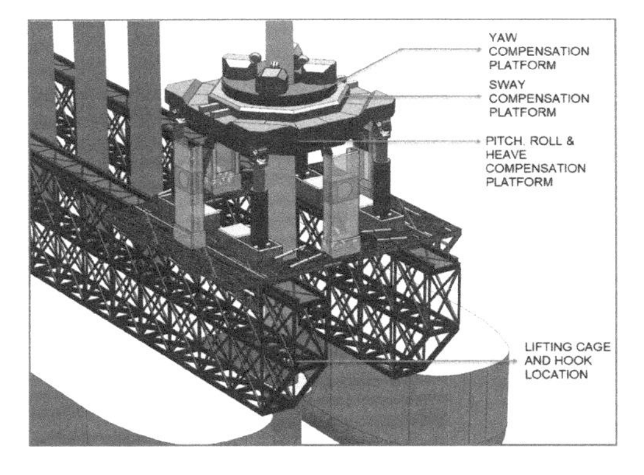

Figure 3. Concept during the initial stages of the design in NX.

The sketches shown in Figure 4 were among the very first ideas for this concept. This arrangement can slide on rails to assist precise installation. A platform mounted on four hydraulic cylinders and support links compensate for the pitch, roll and heave motion. A split divides the entire platform in to two sections controlled by hydraulic cylinders and support mechanisms. The initial step of the design process was to ensure the motion stability of the concept. Verifying several options and its response to wave motion as shown in Figure 5 helped in finalizing a stable concept shown in Figure 6.

The final concept selected as shown in Figure 6 has joints at the center of the platform in order to prevent the entire structure from collapsing due to the movement of the vessel in response to waves. The motion simulation helps in identifying all the possible movement combinations made by the joints and links during the wave motion. The result shows that the chosen arrangement with two pin joints, a rotating link and a sliding link on all four corners are essential for the stability of the design. However, Figure 6 shows support links only on two sides. The final complete design has the support links on all four sides considering the fact that the platform has to support the weight of an entire wind turbine assembly weighing more than 1000 tons.

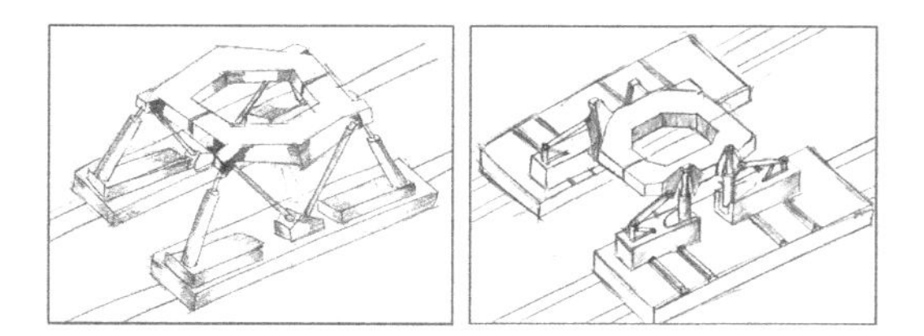

Figure 4. Concept sketch.

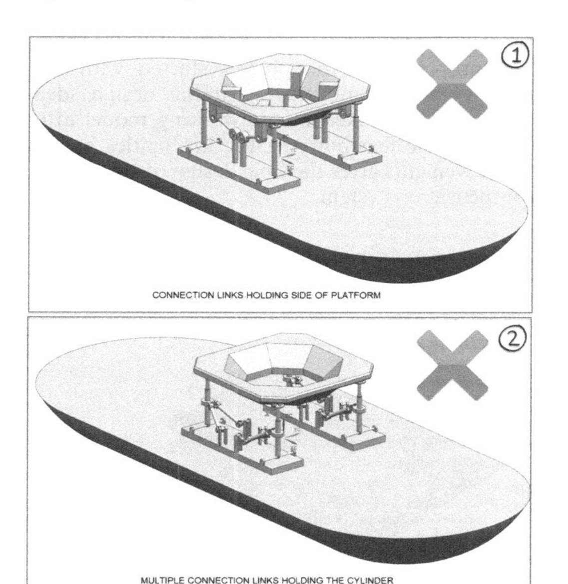

Figure 5. Joint response verified using NX motion simulation.

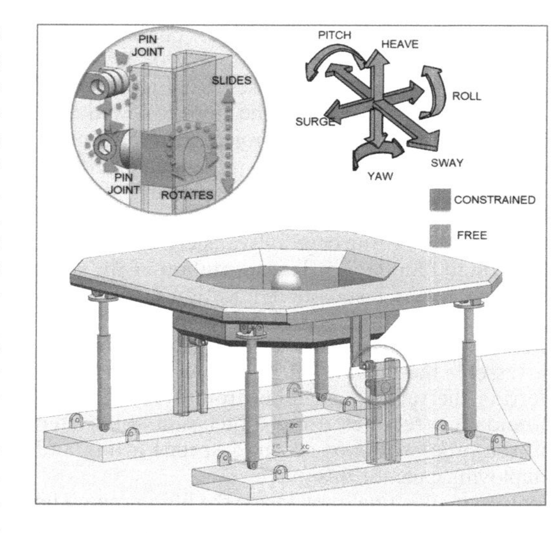

Figure 6. Motion compensation platform with three DOF.

Degrees of freedom of the motion compensation platform achieved by the hydraulic cylinders and the universal joint is as shown in Figure 6. The universal joints on top of the cylinder rod compensates the roll and pitch motion of the platform. Spherical joints at the bottom end of the cylinder housing allows the hydraulic cylinders to have a slight angular movement to allow freedom of movement while compensating roll motion.

The motion compensation platform discussed in Figure 6 can actively compensate only the pitch, roll and heave motion of the vessel. The sway, surge and yaw movements are compensated by other additional mechanisms which are discussed further in the paper. There are rotating platforms controlled using electric motors as well as sliding platforms controlled by hydraulic actuators to assist the compensation of sway, surge and yaw movements, which is explained further in the coming chapters.

3 DESIGN PROPOSAL

The detailed design of the concept is a 3D model developed using Siemens NX and rendered using KeyShot. The most important factor considered in the entire design process is that, every single component used in the whole assembly are standard cross sections of steel that are available in the European market. The forged and welded components in the assembly also completes the test for structural stability using finite element analysis.

The whole installation process is a series of steps illustrating the complex mechanisms involved in the concept design. This is a summarized version of the entire process.

Step 1: Arrangement of Wind turbines on vessel
The turbine assembly considered in the concept is around 135 meters tall and weighs close to 1000 tons. The turbine assemblies are stacked vertically on the vessel (See Figure 7) using specially designed mounts and tie rods. The assumed vessel dimension used for the installation is of a length of 270 meters and a width of 100 meters. The increased width and multi hull design will contribute towards lowering the center of gravity of the whole vessel and thus improve the stability. This is approximately 75% in size to the dimension of the catamaran vessel, Pioneering Spirit [5]. The main intention behind this approach is to utilize a vessel of approximately the same size to implement this installation process, which will minimize the capital investment required for the entire project. The OWT foundation will align in the slot between the hulls and the turbine assembly comes on top of it during installation. A major simplification in this paper is that the OWT foundation is floating. To have a more precise installation, it requires a connection between the floating foundation and the vessel. However, the design of this connection, which requires extensive further research, is not included in this paper.

Support rail structures that runs along the entire length of the vessel slides the base platform, which moves in between wind turbines. This platform houses the components required for the heave compensation system and the equipment required for lowering the turbine assembly on top of the transition piece.

Step 2: Opening and closing of the Base Platform
The base platform designed can switch between wind turbines during installation. The four hydraulic cylinders used for motion compensation are on a rolling platform (Cylinder Platform) that can move in and out as the base platform open and closes. The open and close conditions of the base platform is as shown below (See Figure 8). This step is required to ensure an easy switch between turbines after installation. The movement of the entire cylinder platform is with the help of steel rollers on a defined track. The actuators used for initiating the motion is a rack and pinion system, which drives along a tooth on both the sides.

Step 3: Opening and closing of support structures
Four hydraulic cylinders help in the up and down movement of the motion compensated platform. The main heave compensated platform has a weight of around 2000 tons. After the installation, this platform retracts back to the initial position of the cylinder. At this position, additional hydraulic actuated support brackets contribute towards supporting the weight of the entire platform and associated mechanisms on top of it. However, for safety purpose, these supports structure opens up during motion compensation. Once closed, the weight of the heave compensated platform transfer along the four support structures and the four hydraulic cylinders. This will allow safe retraction of the platform by ensuring equal weight distribution along 8 points. There are four additional guide mechanism, which can take the weight of the struc-

Figure 7. Concept designed in NX and rendered in KeyShot.

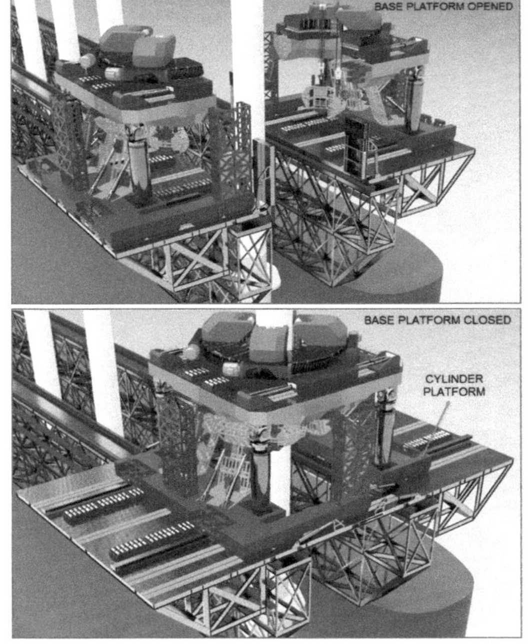

Figure 8. Open and close condition of the base platform.

ture. However, this is a redundant system used to ensure more safety and stability during operation. The platform supported by support structure is as shown below (See Figure 9).

Step 4: Pitch and roll movements of the platform

The pitch and roll motion of the platform is achieved with the use of four main components as shown in Figure 10. Four universal joints that connects the main hydraulic cylinder to the base of the motion compensated platform, four fixed links with multiple pin joints along with four slide links and four rotating links that constraints the motion of the platform within the prescribed limits. These components arranged in a specific pattern achieves the perfect motion compensation.

There are eight points on the heave compensated platform where these links are connected. In fact, only two of the fixed link connections are enough to constraint the motion. However, four connections are there to support a combined load of around 3000 tons. As the motion compensation happens, the sliding link will move up and down along the vertical guides and will restrict the motion beyond a certain limit. This will allow the

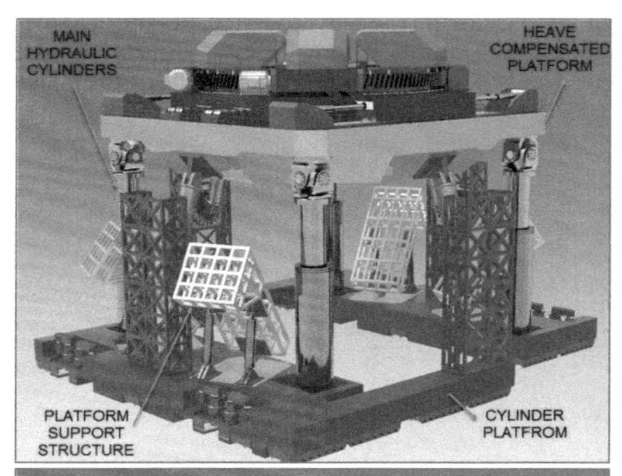

Figure 9. Open and close condition of support platform.

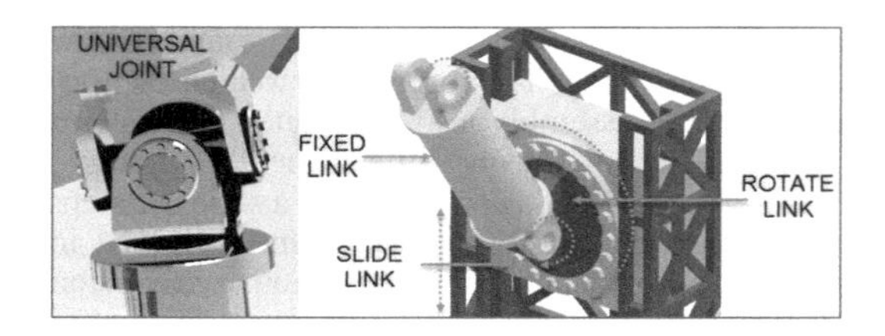

Figure 10. Four main components for motion compensation.

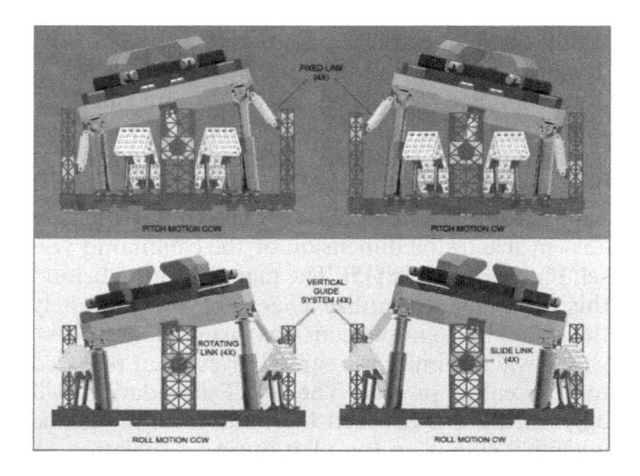

Figure 11. Pitch and roll motion limitation of the platform.

main hydraulic cylinders to stay upright during motion compensation. Swivel bearings used at the base of cylinder housing enables roll motion. The pitch and roll motion of the platform constrained to a specific limit is shown below (See Figure 11).

Step 5: Sway and Yaw movements of the platform

While installing the wind turbine on top of the transition piece, some fine adjustments in the direction of yaw and sway in inevitable. A new mechanism developed, obtains sway and yaw compensation on top of the heave compensation platform as shown in Figure 12. Sway platform moves with the help of eight hydraulic cylinders. These cylinders propel the platform to move 3.5 meters in both directions. The entire platform rests on top of steel rollers and rubber tracks to assist the smooth motion compensation. Yaw platform moves with the help of a helical gear system driven by four electrical motors. The yaw platform houses a winch system, which is required to lift and lower the wind turbine.

Step 6: Lowering of the Lifting cage and hook

A lifting cage is a structure that can lift a fully assembled wind turbine from the vessel and lower it down to a transition piece. Four electric winches assist in lifting and lowering which makes the installation a little flexible as shown in Figures 13 and 14.

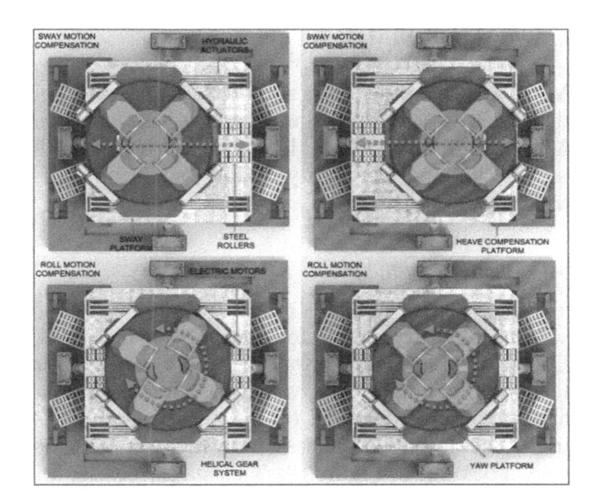

Figure 12. Sway and yaw motion limitation of the platform.

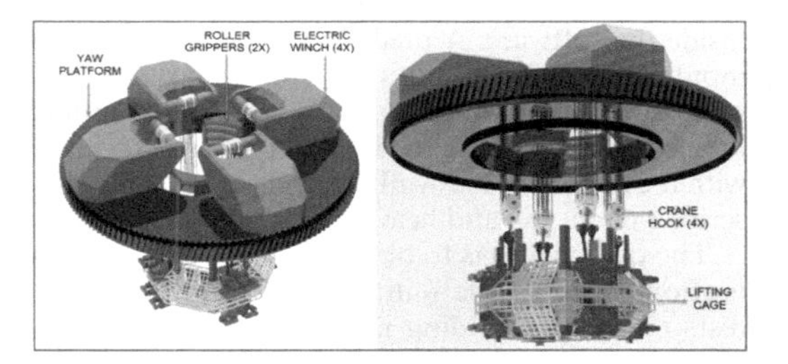

Figure 13. Lifting cage, crane hook and electric winches.

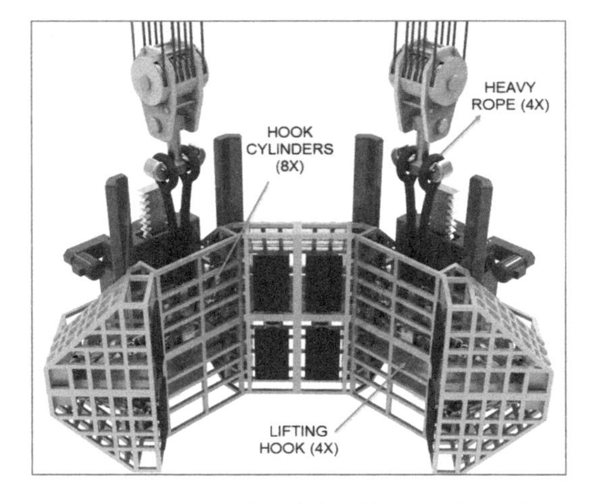

Figure 14. Cross section of the lifting cage for clarity.

There are two hydraulic actuated roller grippers installed inside the yaw platform, which can extend and retract while gripping the sides of the turbine assembly. These grippers contribute towards constraining and supporting the turbine walls during the process of lifting and lowering. Without the grippers, there is a high chance of turbine colliding with the platform walls due to the wind conditions offshore. Rollers, which are horizontally arranged, works only during lifting and lowering operation. This yaw platform is there to assist in the precise turning of the turbine assembly during installation.

Design of the crane hook is in the form of a typical hook used in heavy lifting operation. Heavy ropes connects the lifting hooks with the yaw platform and the electric winches. Hooks extends and retracts with the help of hydraulic cylinders. Lifting cage can split into two sections after installation.

Step 7: Lifting and lowering the turbine assembly

In order to lift the turbine assembly, the lifting cage has to open with the help of a rack and pinion system attached to it. The important part of lifting comes when the upper half of the cage aligns to the base of the wind turbine as shown in Figure 15. After the alignment, the lifting hooks are closed. This secures the turbine assembly inside the upper half of lifting cage. This will then allow the ropes and the crane hook to take the entire load of the wind turbine as well as the lifting cage. The distribution of the lifting points are in such a way that the load distributes equally on all four hooks.

During lowering, the lower half of the lifting cage aligns with the transition piece on top of the OWT foundation and initiates the lowering process. The hydraulic grippers inside the cage activates and takes up the weight of the wind turbine assembly on 16 hydraulic cylinders. The hooks then open and the rack and pinion system slowly lowers the assembly between the lower and upper half of the lifting cage. The ropes move into slacking mode and are relieved from the weight of the turbine (See Figure 16). The entire weight of the turbine is on the 16 cylinders until the rack and pinion installs it on foundation.

Step 8: Detaching the cage after installation

Once the installation is complete, the lifting cage pulls itself back to the top by keeping it in the closed condition. This will not cause interference

Figure 15. Lifting of turbine assembly from vessel.

Figure 16. Lowering of the turbine assembly to the foundation.

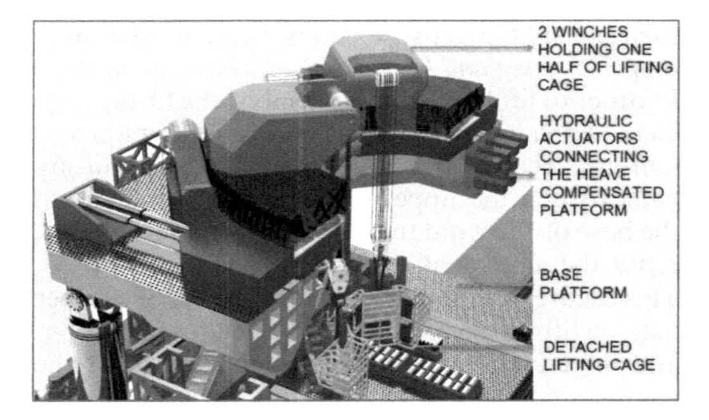

Figure 17. One-half of lifting cage assembly after detachment.

with the turbine assembly owing to the tapered structure of the turbine tower. Once it reaches a height of almost up to 3 meters below the heave compensation platform. The Cage is detached into two and pulled back to either sides of the installed turbine. This project does not involve the detachment methods used. This can either be a hydraulic cylinder arrangement, which locks and detaches the cage once it reaches the required height or a manually operated locking mechanism, controlled by operators on top of the platform and foundation. The two electrical winches pulling each half of the lifting cage shares distributed load of the entire lifting assembly. There are also hydraulic actuators that moves inside one-half of the platform to improve the stability of the structure once in closed condition. These are as shown in Figure 17. The lifting cage once detached lowers down to the base platform floor and rest there for inspection before starting the installation of the next turbine.

All the steps mentioned in this design proposal is the overview of the various stages involved in the installation process. The 3D models shown in the figures are created using the software Siemens Nx and rendered using the software keyshot for a clear and precise presentation of the steps invovled in this concept.

4 MOTION SIMULATION

Verifying the kinematic stability of the concept developed is an important part of the project. Motion simulation module in NX helps in verifying the kinematic stability of the joints and links used in the concept.

4.1 *Setting up the model*

This module has the ability to verify the motion of the various joints in response to the motion of the vessel. Due to the immense computing power required for the motion simulation to solve the kinematic linkages, a simplified model of the entire assembly is used. The simplified final assembly made in NX is as shown below (See Figure 18). This makes it easier for the solver to compute the link movements.

In order to set up the model for motion simulation, a whole assembly restructuring is required. Components installed changes to links and joints inside the software. A predefined point receives the input pitch and roll motions of the ship. Another point super imposed on top of it receives the inputs for heave motion due to the restrictions associated with the software. This will in effect simulate the prescribed pitch, roll and heave motion of the vessel.

The objective was to obtain heave, roll and pitch motion on the same joint. Spherical joints which rests on another sliding joint serves this purpose. This is a work around done to get the simulation to run in the solver. There are four spherical joints, on the base of the main cylinder housing, which allows the cylinder to tilt sideways. The cylinder rods connected to the housing uses slider joints with limits. The next part is the universal joint connected to the top of the cylinder rod using four revolute joints, which allows the rotation. The other end of the universal joint connects four locations to the top platform as a revolute joint that allows rotation along a specific axis. The fixed link for the platform support connects revolute joints to another four

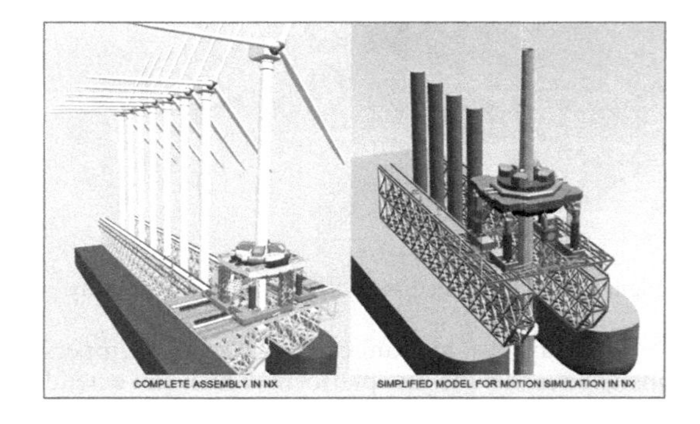

Figure 18. Simplified model for motion simulation.

locations on the heave compensated platform. The bottom end of the fixed link connects to the rotating component as a revolute joint at four locations. The rotating component connects to the slider as a revolute joint at four locations. The slider component connects to the cylinder support platform using a sliding joint at four locations. The sway motion and the surge motions are provided for the both the platforms to enable the compensation to happen in the respective directions. The heave motion applied to the spherical joint has an amplitude of 1.25 meters allowing a wave compensation of around 2.5 meters. A roll and pitch motion are applied as drivers to vessel which gives 1.5 degrees amplitude each allowing a total wave compensation of 3 degrees. Graphs plotting the motion of the vessel needs markers added to a specific location and sensors attached to it. The sensors plot determines the heave in mm as well as pitch and roll in degrees. The pitch and roll applied as a combination of sine wave and cos wave gives realistic vessel motion effects. A detailed explanation about the connection is inside the thesis report [4].

4.2 *Implementing the relations*

The point marked "1" takes the prescribed input for the roll motion (See Figure 19). It is an assumed point where the pitch, roll and heave motion for the vessel is applied. The figure shows how the entire assembly compensates the pitch motion. D1 is the distance from the center point of the ship to the base of the first cylinder and the one behind it. Similarly, D2 is the distance to the base of the second cylinder and the one behind it. The "θ" is the pitch applied to the center of the vessel. In this case, an amplitude of 1.5 degrees. After applying that motion, a sensor placed at the center point captures the θ. This is used to finds the X1 and X2 (See Figure 19).

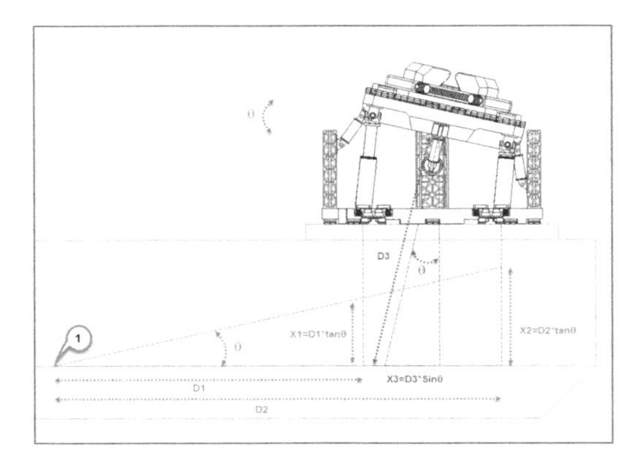

Figure 19. Trigonometric relations behind pitch and surge.

These are displacements applied to the first two cylinders and the next two cylinders respectively. This will allow the motion compensation for the pitch motion. While compensating for the pitch, the design of the platform will initiate a surge movement along the horizontal axis. The dimension D3 in Figure 19 helps in compensating the surge motion. D3 is the distance from the base to the center point of the platform, which is a constant. Finding the X3 will provide the distance that the surge platform should move to compensate the surge motion. Thus the pitch and surge motions are compensated by the platform.

The prescribed roll motion of the vessel is applied to an assumed center point marked as "1" (See Figure 20). The "ϕ" is the roll applied to the center of the vessel, 1.5 degrees amplitude. In this case the known values are D1 and D2 which represents the length of the platform which is a constant. The ϕ is captured as usual by placing a sensor at the center point. The distance X1 is the distance that the first two cylinder needs to displace in order to compensate for the roll motion. Similarly, a negative value is applied to the other two cylinders to compensate for the roll at the other end.

The Design of the platform is made in such a way that there will be a slight movement of sway once the roll compensation is achieved. This can be corrected by subtracting the value of X2 from the cylinder displacement (See Figure 20). The

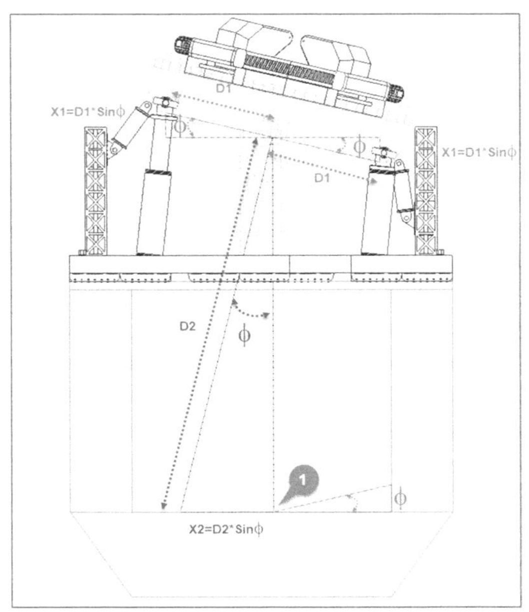

Figure 20. Trigonometric relations behind roll and sway.

output values are continuously updated as the input is derived from a sensor. These values gives a real time simulation result by keeping the platform steady in response to the vessel motion.

A detailed description of the heave motion is not provided since it is a direct vertical movement applied to the point and can be easily compensated by lowering the cylinders to the exact same distance. A simulation for the Yaw motion is not included in this project, there is a yaw platform on top which can directly compensate for the yaw motion of the ship. The trigonometric relations behind compensating all the 5 degrees of freedom is explained. The resulting real time simulation results are explained further.

4.3 *Discussion of results*

There are several ways of implementing pitch, roll and heave compensation for this system. The primary idea would be to give the prescribed motion to the vessel and calculate the force required for each cylinder by gathering inputs from various sensors placed in and along the joints. This in itself requires a more detailed study of the control algorithm which is required for running the simulation. Motion simulation module in NX has immense potential to implement real time inputs from matlab and Simulink or even from an external PLC's from which real time the control signals can be connected. However, this project does not go into the details of implementing the control logics for the motion simulation. The main objective behind this project is to show the kinematic stability of the structure using prescribed motion given as inputs to the joints.

In this project, the sway and surge compensations are activated by applying driver to both the joints and by giving inputs to compensate for the motion.

The result for the roll and pitch is obtained as plot of time against angular displacement. The red line shows the prescribed input roll and pitch motion with an amplitude of 1.5 degrees and the blue line shows the response of the platform obtained from a sensor placed on the platform. The real time motion simulation shows a blue marker over the plot, which shows the real time positioning of the vessel with respect to roll and pitch motion. (See Figure 21) which is a replication of the direct output screen from NX which gives the real time response to the simulation along with the 3D model.

The result of the heave motion is obtained as a plot of time against displacement. The red line shows the prescribed heave movement of the vessel at an amplitude of 1.25 meters and the blue line shows the response of the platform achieving the heave compensation.

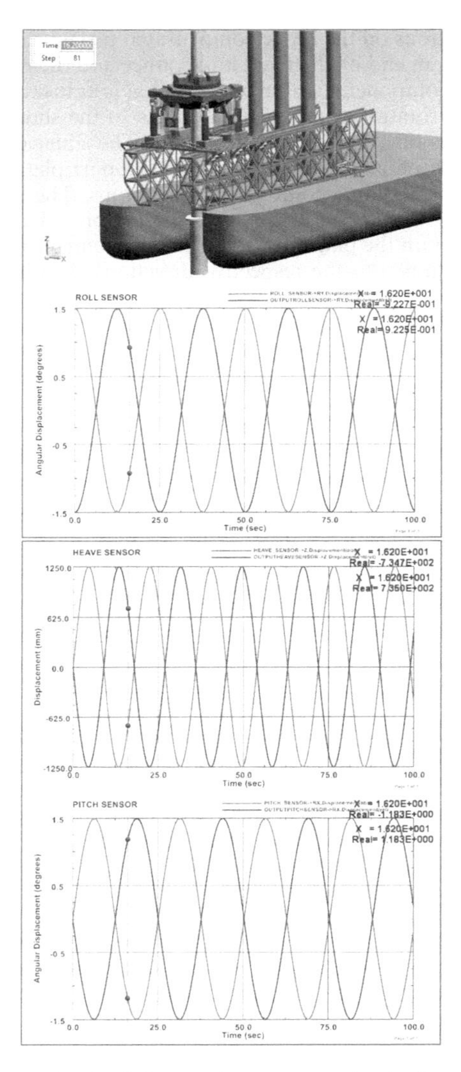

Figure 21. Real time motion compensation in Siemens NX.

The simulations results are obtained by placing sensors on the motion compensated platform and by plotting the relative motion of the platform with respect to that of the point at which the prescribed motion of the vessel is applied. The plotted result has heave, pitch and roll compensated. The yaw motion of the platform is not considered for motion simulation.

5 CONCLUSION

The paper thus discuss the implementation of an innovative concept for the installation of offshore wind turbines. The overall focus is towards the presentation of a new and improved method for installing wind turbines in water depths more than 200 meters without relying on existing heavy lift cranes.

The paper also validates the motion stability of the concept and presents a simulation result validating the kinematic stability in response to prescribed wave motions. This concept in its entirety requires extensive further research in order to develop a working model. Few of the major areas that needs further research is included in the thesis paper [4].

This concept upon implementation could save huge capital investments in the field of offshore wind turbine installations and ensure a significant return of investments over the years. Moreover, it strengthens the drive of humanity towards green energy and serves as a bright initiative in our journey towards cleaner and sustainable future.

REFERENCES

[1] The European offshore wind industry (2016): https://windeurope.org/wp-content/uploads/files/about-wind/statistics/WindEurope-mid-year-offshore-statistics-2016.pdf

[2] Offshore wind cost reduction pathways Technology work stream. (2012), (May): https://www.thecrownestate.co.uk/media/5643/ei-bvg-owcrp-technology-workstream.pdf

[3] Study, P. (n.d.). Offshore Wind Cost Reduction: http://www.we-at-sea.org/wp-content/uploads/2015/01/offshore-wind-cost-reduction-pathways-study.pdf

[4] Bernard, P. (2017). Master ' s degree thesis: http://hdl.handle.net/11250/2462438

[5] Pioneering spirirt, World's Largest catamaran installation vessel. [Online]. Available: https://allseas.com/equipment/pioneering-spirit/, accessed on 03-12-2016

Marine Design XIII – Kujala & Lu (Eds)
© 2018 Taylor & Francis Group, London, ISBN 978-1-138-34076-3

Loads on the brace system of an offshore floating structure

T.P. Mazarakos & D.N. Konispoliatis
Laboratory for Floating Structures and Mooring Systems, Division of Marine Structures, School of Naval Architecture and Marine Engineering, National Technical University of Athens, Athens, Greece

S.A. Mavrakos
Laboratory for Floating Structures and Mooring Systems, Division of Marine Structures, School of Naval Architecture and Marine Engineering, National Technical University of Athens, Athens, Greece
Hellenic Center for Marine Research, Anavyssos, Greece

ABSTRACT: The paper investigates the shear forces and bending moments on the braces and pontoons of a semi-submersible platform suitable for offshore applications under wave and wind loads. The responses of the platform which is composed of an array of floating vertical cylindrical bodies are calculated from the analytical solution of two boundary value problems, namely, the diffraction and the motion-radiation problems. They have been obtained through an analytical solution method using matched axisymmetric eigenfunction expansion formulations. In doing this, the hydrodynamic interactions among the cylindrical bodies that form the buoyant elements of the semi-submersible platform is of particular importance since the hydrodynamic characteristics of each cylindrical member in a multi-body configuration may differ from the ones obtained for the same body considered as isolated due to hydrodynamic interaction phenomena. In that context, comparisons between the loads on the braces and pontoons with and without taking into consideration the interaction phenomena are presented.

1 INTRODUCTION

Over the past two decades nations worldwide have been made considerable efforts and advances in exploiting the offshore wind and wave energy. Besides the numerous concepts for offshore wind energy installed in shallow water sites, the main challenge in offshore renewable energy devices is to build a structure capable to withstand the challenging environmental conditions while being financial competitive with regard to other types of energy sources exploitation systems. In the case of floating supporting structure, meaning the floater and the moorings, the cost of steel plays a very important role in overall cost of the system. In this context, it is evident that in order to control the construction cost of the steel structure, the design of the braces and pontoons of the platform is very challenging towards keeping the steel weight as low as possible. In addition, the resultant loads due to wave impact with the structure are essential in the design of connection equipments and behaviors of the braces and pontoons.

In order to evaluate the connection loads, the responses of the floating structure should be determined. The main difference between an isolated submerged cylinder and an array of such bodies, merging the floaters of a floating supporting structure, is the hydrodynamic interaction phenomena between the array's members. Each member scatters waves which excite the remaining bodies, which in turn respond to this excitation and scatter waves contributing to the excitation of the initial body and so on. The importance of taking into account these multiple scattering effects in predicting the exciting wave loads and wave induced motions for offshore supporting platforms and systems has already experimentally been confirmed and numerically predicted by several investigators (Mavrakos & Koumoutsakos, 1987; Mavrakos, 1991; Konispoliatis & Mavrakos, 2016; Konispoliatis et al., 2016; Mazarakos et al., 2014a; 2014b; 2017). Therefore, in the present analysis the interactions among the floaters are determined and their effect on the loads of the brace system are estimated.

There are many numerical methods dealing with the hydrodynamic analysis of arbitrary shaped floating bodies in the presence of regular waves. In the present contribution, the examined bodies were assumed rigid and the hydrodynamic interaction phenomena among the floaters of the floating structure have been taken into account through the physical idea of multiple scattering (Twersky, 1952; Okhusu, 1974, Mavrakos & Koumoutsakos, 1987; Mavrakos, 1991). Besides the multiple scattering approach, direct matrix inversion methods have been presented in the literature as well to solve the hydrodynamic interaction problem among arrays of bodies for which the diffraction solution is known (Kagemoto & Yue, 1986, Mazarakos, 2010). This method has been also used by Siddorn & Eatock Taylor (2008) in order to elaborate an exact algebraic method for the diffraction and independent radiation by an array of truncated cylinders.

In exploiting the multiple scattering approach the method of matched axisymmetric eigenfunction expansion as it was implemented for truncated vertical cylinders (Miles & Gilbert, 1968; Black et al., 1971; Yeung, 1981) has been used, producing corresponding hydrostatic stiffness-, added mass- and radiation damping- matrices formulating the structure's motion equation. For the analysis of a TLP designed concept the equation of motion is modified by employing the mooring stiffness matrix. Results concerning the shear forces and bending moments on the braces when the structure is moored are discussed as a function of the angle and the wave frequency of the incoming wave to the floater array.

2 EXAMINED SUPPORTING PLATFORM AND SITED LOCATION

2.1 *Description of the floating structure*

The examined floating system has been proposed for supporting the DTU 10 MW Wind Turbine (WT), which is a variable-speed variable-pitch controlled WT (see also subsection 2.2). The structure encompasses an array of three hydrodynamically interacting cylindrical floaters at the corners of a triangular arrangement and a central cylindrical body supporting the WT (see Figures 1, 2). The tower of the wind turbine is cantilevered at an elevation of 10 m above the sea water level (SWL) to the top of the central column of the platform. The draft of the platform is 20.0 m (see Figure 2). The notation concerning the cylindrical bodies; braces and pontoons is depicted in Figure 2. The braces and the pontoons have cylindrical shape with diameter 1.6 m. A summary of the platform's geometry is given in Table 1.

Normal steel (i.e. density 7.850 tn/m^3; Yield stress 235 MPa; Young's modulus 2.1E5 MPa) is applied to all the pontoons and braces. The position of each member of the floating supporting platform is tabulated by Table 2.

The mass, including ballast, of the floating platform is 6289 tn. This mass was calculated such that the combined weight of the rotor-nacelle assembly, tower, platform, plus the applied TLP pretension and the weight of the mooring system in water, balances with the buoyancy (i.e. weight of the displaced fluid) of the platform in the static equilibrium position in still water. The centre of

Figure 1. Group of four cylindrical bodies connected as a unit, forming an integrated floating supporting platform.

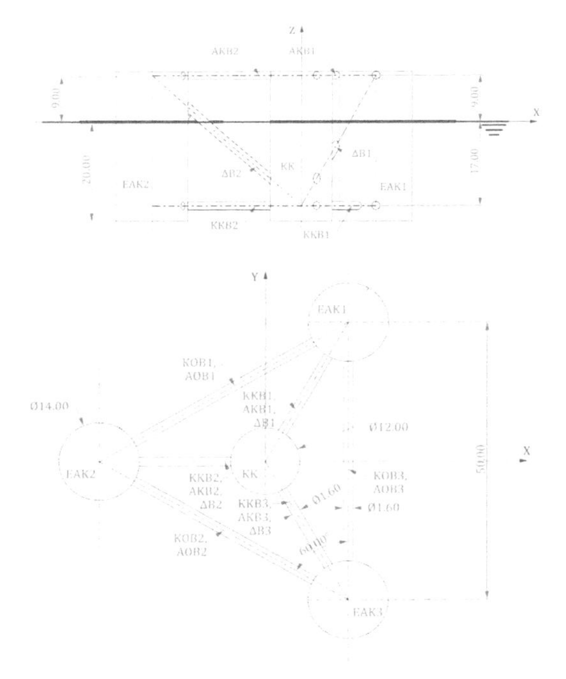

Figure 2. Side and top (down) view of the platform.

mass (CM) of the floating platform, including ballast, is located at 11.59 m along the platform centre line below the SWL. The roll and pitch inertias of the floating platform about its CM are

6.385E6 tn.m^2 and 6.385E6 tn.m^2 about the platform x-axis and y-axis respectively, while the yaw inertia of the floating platform about its centre line is 1.170E7 tn.m^2.

To secure the platform's position, the floating structure is moored with a TLP mooring system of three tendons spread symmetrically about the platform Z - axis. The fairleads (body-fixed locations where the mooring tendons are mounted to the platform) are located at the base of the offset columns at a depth of 20.0 m below the SWL. The anchors are located at a water depth of 200 m below the SWL. Each of the 3 mooring tendons has an unstretched length of 180.0 m, a diameter of 1.2192 m, wall thickness 0.0422 m, an equivalent mass per unit length of 1.224 tn/m, an equivalent submerged weight in fluid per unit length of 1.067 kN/m. The pretension of each tendon is 18838 kN. Table 1 summarizes these properties.

Table 1. Floating platform and mooring properties.

Spacing between columns	50.00 m
Draft of the structure	20.00 m
Elevation of main/offset column above SWL	10.00 m
Diameter of central column	12.00 m
Diameter of outer floaters	14.00 m
Diameter of pontoons and cross braces	1.60 m
Length of outer pontoons	36.00 m
Length of inner pontoons	15.87 m
Length of cross braces	28.77 m
Cross brace and pontoons connection point below SWL	17.00 m
Cross brace and pontoons connection point above SWL	7.00 m
Number of tendons	3
Tendon length	180.0 m
Tendon diameter	1.2192 m
Tendon mass per unit length	1.224tn/m
Submerged mooring line weight per unit length	1.067 kN/m
Pretension of each tendon	18838 kN
Tendon stiffness k_{xx} of each tendon	104 kN/m
Tendon stiffness k_{xx} of each tendon	173533 kN/m.

2.2 *WT specifications and aerodynamic loading*

In the corresponding frequency domain formulation, the contribution of the WT is projected on the 6 floater motion. This is carried out in the context of Hamiltonian dynamics with gravity and aerodynamics being the external forcing. The Blade Element Momentum theory defines the aerodynamic loading as a function of the operational conditions and the motions of the floater that change the effective angles of attack seen by the blades. By linearizing all terms with respect to the case of zero floater motions (static equilibrium), added mass, damping and stiffness matrices are defined which contribute the WT's aerodynamic, inertial—gyroscopic and gravitational effect on the floater (Mazarakos et al., 2014a).

The height of the selected WT's tower is 105.63 m, the CM of the WT is 46.7 m and the

Table 2. Floating platform geometry.

Name	Initial	Beginning coordinates	Ending coordinates
1. Central column	KK	(0,0,–20)	(0,0,10)
2. Column 1	EAK1	(14.43,25,–20)	(14.43,25,10)
3. Column 2	EAK2	(–28.87,0,–20)	(–28.87,0,10)
4. Column 3	EAK3	(14.43,–25,–20)	(14.43,–25,10)
5. Delta Upper Brace 1	AOB1	(8.36,21.5,10)	(–22.8,3.5,10)
6. Delta Upper Brace 2	AOB2	(–22.8,–3.5,10)	(8.36,–21.5,10)
7. Delta Upper Brace 3	AOB3	(14.43,–18,10)	(14.43,18,10)
8. Delta Lower Brace 1	KOB1	(8.36,21.5,–17)	(–22.8,3.5,–17)
9. Delta Lower Brace 2	KOB2	(–22.8,–3.5,–17)	(8.36,–21.5,–17)
10. Delta Lower Brace 3	KOB3	(14.43,–18,–17)	(14.43,18,–17)
11. Y Upper Brace 1	AKB1	(3,5.196,10)	(10.93,18.93,10)
12. Y Upper Brace 2	AKB2	(–6,0,10)	(–21.87,0,10)
13. Y Upper Brace 3	AKB3	(3,–5.196,10)	(10.93,–18.93,10)
14. Y Lower Brace 1	KKB1	(3,–5.196,–17)	(10.93,18.93,–17)
15.Y Lower Brace 2	KKB2	(–6,0,–17)	(–21.87,0,–17)
16. Y Lower Brace 3	KKB3	(3,–5.196,–17)	(10.93,–18.93,–17)
17. Cross Brace 1	ΔB1	(3,5.196,–11.6)	(10.93,18.93,9.2)
18. Cross Brace 2	ΔB2	(–6,0,–11.6)	(–21.87,0,9.2)
19. Cross Brace 3	ΔB3	(3,–5.196,–11.6)	(10.93,–18.93,9.2)

Table 3. Environmental conditions on the 50-year contour surfaces with maximum U_w or maximum H_s.

Condition	U_w	H_s	T_p
Conditions with maximum U_w	18.82 m/s	5.44 m	10.20 s
Conditions with maximum H_s	18.76 m/s	5.48 m	10.28 s

mass of the WT is 1200 tn. The roll and pitch inertias of the latter about its CM are 1.089E7 tn.m^2 and 1.089E7 tn.m^2 respectively, while the yaw inertia about its centre line is 9.757E4 tn.m^2.

2.3 *Climate conditions at the installation location*

In this subsection the design environmental conditions in the installation location (at the Aegean Sea with coordinates 35.34°N, 26.80°E) are presented (Mazarakos et al., 2017).

Following Soukissian et al. (2015), which have estimated the contour surfaces with return period 50 years for the specific location, the maximum wind speed U_w, and the maximum significant wave height H_s with the corresponding associated values of H_s; T_p (i.e. peak period); and U_w; T_p; respectively, are presented in Table 3.

3 DESCRIPTION OF THE HYDRODYNAMIC MODELLING METHOD

We consider that the group of four floating cylindrical bodies is excited by a plane periodic wave of amplitude $H/2$, frequency ω and wave number k. It is assumed small amplitude, inviscid, incompressible and irrotational flow, so that linear potential theory can be employed. A global Cartesian coordinate system O-XYZ with origin on the sea bed and its vertical axis OZ directed positive upwards is used. Moreover, four local cylindrical coordinate systems (r_q, θ_q, z_q), $q = 1,2,3,4$ are defined with origins on the sea bottom and their vertical axes pointing upwards and coinciding with the vertical axis of symmetry of each body.

The fluid flow around each body (cylinder) $q = 1,2,3,4$, which can be described by the potential function $\Phi^q(r_q, \theta_q, z; t) = \mathrm{Re}\{\phi^q(r_q, \theta_q, z) \cdot e^{-i\omega t}\}$ can be decomposed, on the basis of linear modeling, as (Mavrakos & Koumoutsakos, 1987; Mavrakos, 1991):

$$\phi^q = \phi_0^q + \phi_7^q + \sum_{p=1}^{4} \sum_{j=1}^{6} \dot{\xi}_{j0}^p \cdot \phi_j^{qp} \tag{1}$$

Here, ϕ_0^q is the velocity potential of the undisturbed incident harmonic wave; ϕ_7^q is the scattered potential around the q body, when it is considered fixed in waves; ϕ_j^{qp} is the motion-dependent radiation potential around the body q resulting from the forced oscillation of the p body in j direction with unit velocity amplitude, $\dot{\xi}_{j0}^p$, ($j = 1, 2, \ldots, 6$). It holds $\dot{\xi}_j^p = \mathrm{Re}\{\dot{\xi}_{j0}^p \cdot e^{-i\omega t}\} = \mathrm{Re}\{-i\omega\xi_{j0}^p \cdot e^{-i\omega t}\}$ and the subscript j stands for surge ($j = 1$), sway ($j = 2$), heave ($j = 3$), roll ($j = 4$), pitch ($j = 5$) and yaw ($j = 6$) modes of motions, respectively.

The potentials ϕ_j^l ($l \equiv q$ or $l \equiv qp$; $j = 0, 1, \ldots, 6, 7$; $p, q = 1, 2, 3, 4$) are solutions of Laplace's equation in the entire fluid domain and satisfy proper conditions at the free surface and on the sea bed. Furthermore, the potentials have to fulfill kinematic conditions on the mean body's wetted surface. Finally, a radiation condition must be imposed which states that disturbance propagation must be outgoing.

The unknown potential functions involved in Equation 1 can be established throughout the method of matched axisymmetric eigenfunction expansions by subdividing the flow field around each body in coaxial ring shaped fluid regions. In each of those regions different series expansions of the velocity potentials are derived. These are solutions of the Laplace equation in each fluid region and are selected so that the kinematic boundary condition at the horizontal walls of the body, the linearized condition on the free surface, the kinematic condition on the sea bottom and the radiation condition at infinity are satisfied. The various potential solutions are then matched by continuity requirements of the hydrodynamic pressure and radial velocity along the vertical boundaries of adjacent fluid regions, as well as by fulfilling the kinematic conditions at the vertical walls of the body (cylinder). The method has been extensively described in the past (Mavrakos & Koumoutsakos, 1987; Mavrakos, 1991; Konispoliatis & Mavrakos, 2016) therefore it is no further elaborated here.

The hydrodynamic interaction phenomena among the members of the multi-body configuration have been taken into account by properly superposing the incident wave potential and the propagating and evanescent modes that are scattered and radiated by the array elements, thus exact representations of the total wave field around each body of the array can be obtained. This method has been in details presented in (Mavrakos, 1991) to solve the diffraction, the motion-dependent radiation problems for an interacting array of vertical axisymmetric bodies.

The gravitational, inertial and aerodynamic loads that the WT contributes on the floater dynamics are defined in the simplest possible way; with respect to gravitational and inertial loading, the wind turbine is modeled as a collection of concentrated masses, namely the masses of the blades, the nacelle, the hub and the tower. By assuming the WT rigid (defining only its inertial characteristics) only the 6 degrees of freedom corresponding to the floater rigid motions remain (Mazarakos et al., 2014a). Applying the Lagrange equations, the dynamic equations for the 6 degrees of freedom, can be reduced in the following linearized form:

$$\sum_{j=1}^{6}\left[-\omega^2(M_{i,j}^{WT}+\frac{i}{\omega}B_{i,j}^{WT})+C_{i,j}^{WT}\right]\cdot \mathrm{q}=Q \tag{2}$$

Here q denotes the vector of the 6 WT motions projected to the motions of the floating supporting structure and Q the generalized external loads (aerodynamic, gravity). The left side of the above equation contains gravity, buoyancy as well as the aerodynamic part that corresponds to the reference state, while the mass matrix $M_{i,j}^{WT}$ includes the WT inertia (including the gyroscopic effects due to the rotation), the damping matrix $B_{i,j}^{WT}$ includes the WT damping due to rotation and aerodynamics and finally the stiffness matrix $C_{i,j}^{WT}$ includes the contribution from both aerodynamics and gravity.

The investigation of the dynamic equilibrium of the forces acting on the floating structure leads to the following well-know system of differential equations of motions, describing the couple hydro-aero-elastic problem of the investigated floating supporting structure in the frequency domain (Mazarakos et al., 2014b, Mazarakos & Mavrakos, 2017):

$$\sum_{j=1}^{6}\begin{bmatrix}-\omega^2(M_{i,j}+M_{i,j}^{WT}+\\ +A_{i,j}+\frac{i}{\omega}B_{i,j}+\frac{i}{\omega}B_{i,j}^{WT})\\ +C_{i,j}+C_{i,j,mooring}+C_{i,j}^{WT}\end{bmatrix}\cdot\xi_{j0}=f_i \tag{3}$$

for $i = 1,2,\ldots,6$

The superscript WT corresponds to physical quantities associated to the wind turbine. Moreover, $M_{i,j}$ and $C_{i,j}$ are elements of the mass and stiffness matrix of the floating structure, $A_{i,j}$, $B_{i,j}$, and $C_{i,j,mooring}$ represent its 6 by 6 added mass, damping, and mooring line stiffness matrices, respectively. f_i^T is the six by one vector that contains the hydrodynamic exciting forces on the floating supporting structure, and ξ_{j0}, is the motion displacement of the entire system at the j–th direction with respect to the global co-ordinate system G.

4 CALCULATION OF THE SHEAR FORCES AND BENDING MOMENTS AT CONNECTION POINTS

For the calculations of the shear forces and bending moments, three simple steps can be followed:

1. The global motions of the structure are calculated through Eq. (3). To calculate the motion components of the structure along an intersection point $\boldsymbol{r} = (x,y,z)$, the following relations are used:

$$\begin{aligned}s_1 &= \xi_1 + z\xi_5 - y\xi_6\\ s_2 &= \xi_2 + x\xi_6 - z\xi_4\\ s_3 &= \xi_3 + y\xi_4 - x\xi_5\\ s_4 &= \xi_4\\ s_5 &= \xi_5\\ s_6 &= \xi_6\end{aligned} \tag{4}$$

2. The shear force along an intersection point $\boldsymbol{r} = (x,y,z)$, counterbalance the sum of the diffraction, radiation and inertia forces exerted on the considered structural member. They are obtained by considering the dynamic equilibrium at the structural member intersection to the rest of the structure (with or without the interactions), $\boldsymbol{F}_i = (F_{xi}, F_{yi}, F_{zi})$, where $i = 1,\ldots 19$ (number of bodies, see Table 2).
3. The bending moment along an intersection point $\boldsymbol{r} = (x,y,z)$, counterbalance the sum of the diffraction, radiation and inertia moments exerted on the considered structural member. They are obtained by considering the dynamic equilibrium at the structural member intersection to the rest of the structure (with or without the interactions), $\boldsymbol{M}_i = (M_{xi}, M_{yi}, M_{zi})$, where $i = 1,\ldots 19$, plus the cross product of the force exerted on each body of the configuration at its distance from the point of intersection: $\boldsymbol{M}_{int} = \boldsymbol{r}x\boldsymbol{F}_i + \boldsymbol{M}_i$

5 NUMERICAL RESULTS

In the present section numerical results corresponding to the exciting loads on the structure; the shear forces and bending moments at the floater – brackets joints, at the mooring structure connections and at the tower – floater connection, are presented in form of figures. The presented results were obtained using the in – house developed computer code HAMVAB (Hydrodynamic Analysis of Multiple Vertical Axisymmetric Bodies, Mavrakos, 1995) software in FORTRAN programming language. The above software which relied on analytical representations of the velocity potential around each cylinder—type body of the array was preferred in the present contribution against other available numerical tools (3D panel methods, e.g. WAMIT, or finite elements methods) applicable to general 3D geometries since by keeping the same accuracy with them it is usually less CPU time—consuming (Mavrakos & Koumoutsakos, 1987; Konispoliatis & Mavrakos, 2016), hence representing an efficient alternative tool in the early design phases of such type of arrays. The number of interactions between the bodies of the array were taken equal to 7 and the azimuthal modes m = ±7, as it was found that the results obtained for those values were correct to an accuracy of within 1% (Konispoliatis & Mavrakos, 2016).

5.1 *Exciting wave loads*

In the below figures the exciting forces and moments acting on the entire supporting structure are presented versus the wave frequency with and without taking into consideration the interaction phenomena between the bodies of the platform for wave heading angle 0° (the angle is propagating along the positive x-axis, see also Figure 2).

More specifically in Figures 3–5 the surge (Fx); heave (Fz); pitch (My) exciting forces and moments are depicted, respectively. The dash line corresponds to the exciting loads taking into account the interaction effect and the dot line without.

It can be obtained from the below figures that the interaction phenomena seem to have minor effect on the Fz forces, while there are discrepancies in the Fx, My due to the interactions between the cylinders. Moreover, the values of the horizontal forces and moments are overestimated for most of the examined wave frequencies without taking into consideration the interaction phenomena, highlighting the importance of the determination of the interactions between the bodies of the platform.

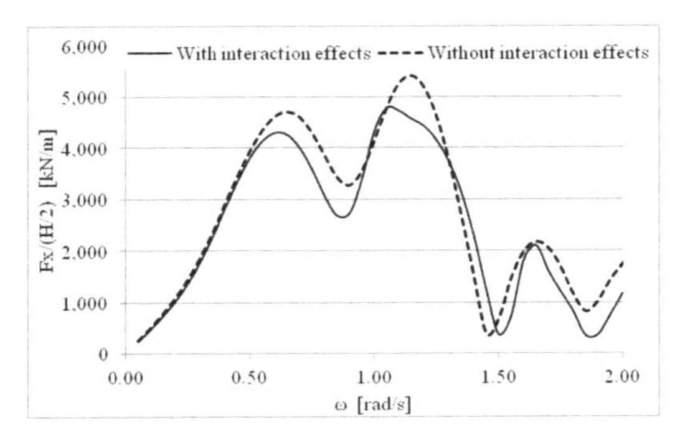

Figure 3. Fx exciting forces on the entire supporting structure versus the wave frequency.

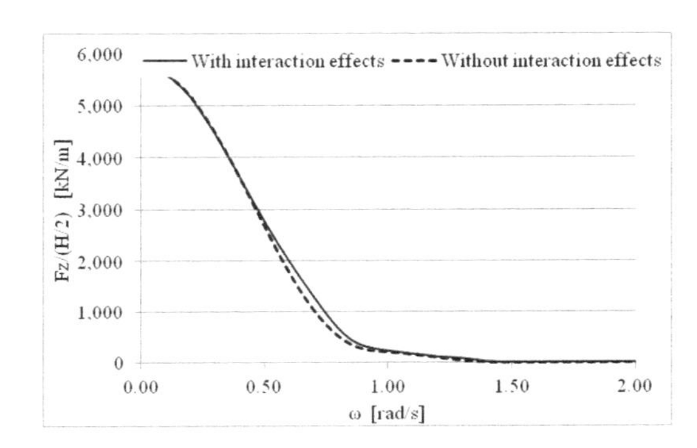

Figure 4. Fz exciting forces on the entire supporting structure versus the wave frequency.

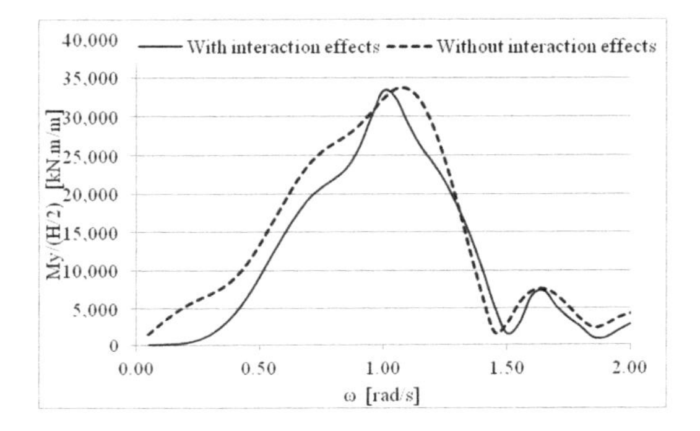

Figure 5. My exciting moments on the entire supporting structure versus the wave frequency.

5.2 *Shear forces and bending moments at operational conditions*

In the below subsection the shear forces and bending moments at several connection points are presented. In Figures 6–8 the corresponding forces and moments at the WT's connection points (i.e. base of the WT's tower, 10 m above SWL) are depicted, respectively. In addition, in Figures 9, 10 the surge and heave shear forces at the three floaters—tendons connection points (i.e. at the bottom of each floater, –20 m below SWL, see Figure 2) are presented. As mentioned above the latter figures are compared to each other with/without considering the interaction effects. Here the wind speed is assumed equal to 18 m/s thus the WT's mass, damping and stiffness coefficients were properly taken into consideration, and the wave angle equals to 0°.

It is depicted that at low wave frequencies (i.e. ω<0.5rad/s) the interaction phenomena between the bodies of the array do not affect the acting loads on the structure. On the other hand, as the value of the wave frequency increases the influence of the interaction effects is evident. However it isn't clear an overestimation or underestimation

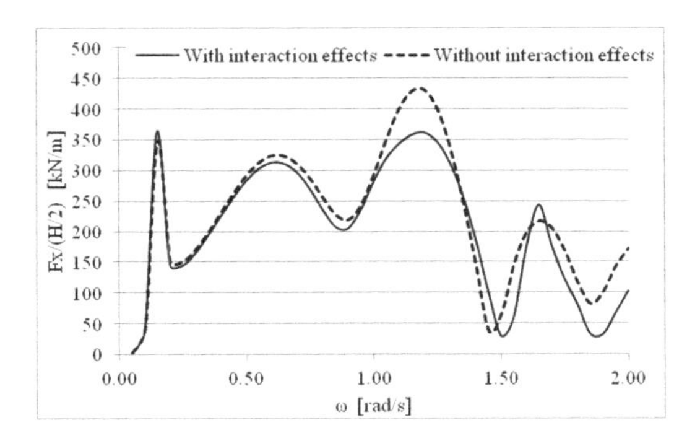

Figure 6. Fx shear forces at the base of the WT tower versus the wave frequency, for wave angle heading 0° wind velocity 18 m/s.

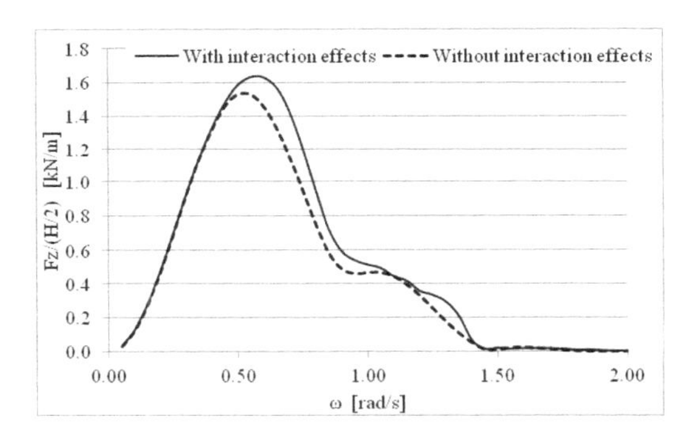

Figure 7. Fz shear forces at the base of the WT tower versus the wave frequency, for wave angle heading 0° wind velocity 18 m/s.

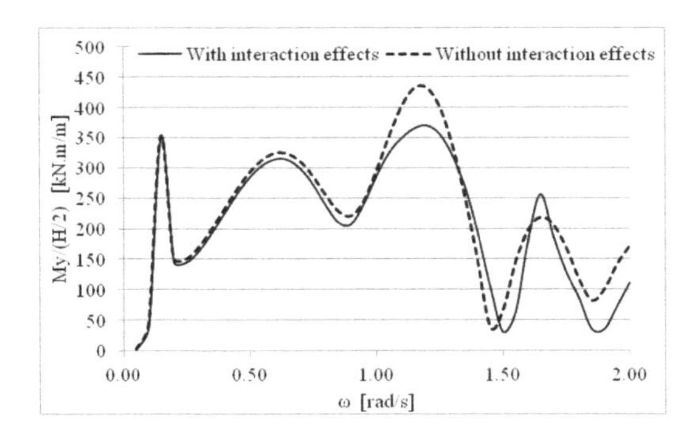

Figure 8. My bending moments at the base of the WT tower versus the wave frequency, for wave angle heading 0° wind velocity 18 m/s.

of the loads values at all the examined frequencies with interaction considerations.

In addition, as far as the vertical shear forces at the moorings connection points is concerned it can be seen that due to the high vertical axial stiffness of the mooring lines, the Fz presents higher values than

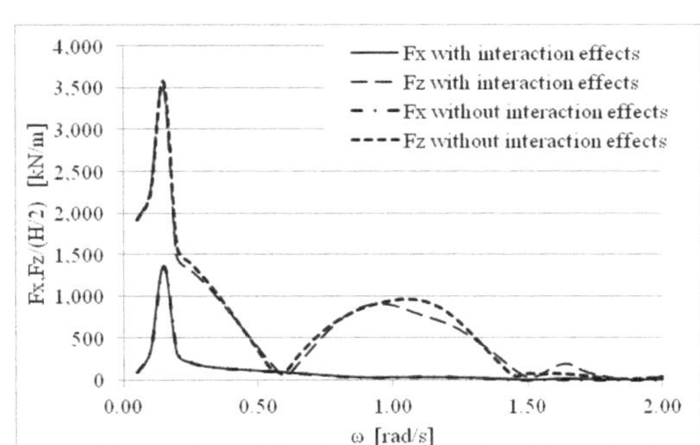

Figure 9. Fx and Fz shear forces at the 1st mooring—structure point versus the wave frequency, for wave angle heading 0° and wind velocity 18 m/s.

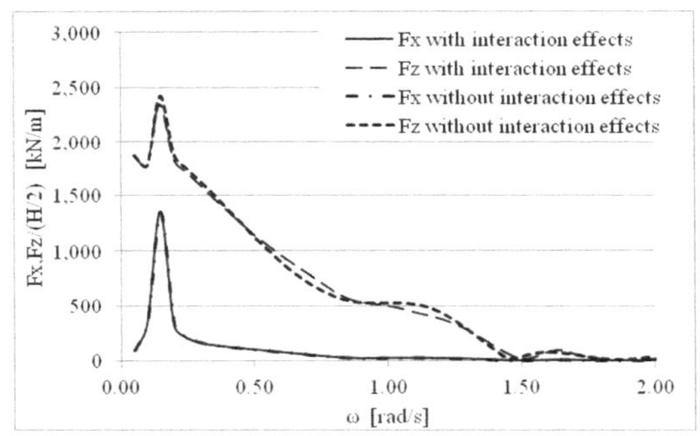

Figure 10. Fx and Fz shear forces at the 2nd mooring—structure point versus the wave frequency, for wave angle heading 0° and wind velocity 18 m/s.

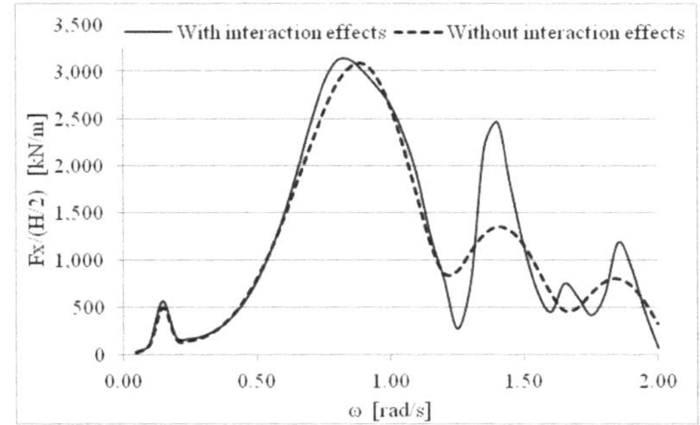

Figure 11. Fx shear forces at the 1st floater—lower delta brace joint versus the wave frequency, for wave angle heading 0° wind velocity 18 m/s.

the corresponding Fx loads for the interaction- and no interaction- condition. Moreover, the interaction effects seem not to affect the corresponding shear loads at floaters—tendons connection points.

In the below figures the shear forces and bending moments at the floater—bracket joint are presented. More specifically in Figures 11–12 the acting loads (i.e. Fx, Fz) between delta lower brace 1 and floater 1 are depicted, respectively. In addition, in Figure 13 the corresponding bending moments My are presented. The below figures are compared to each other with/without considering the interaction effects; for the corresponding moored floating supporting structure; for wind speed 18 m/s and wave impact angle 0°.

As mentioned above for the shear forces and bending moments at the base of the WT tower, the interaction phenomena seem to have significant effect on the acting loads at the brace joints only for large wave frequencies. The latter are the values where considerable variations between the interaction- and no-interaction- condition occur.

5.3 *Shear forces and bending moments at extreme conditions*

For the extreme values of the shear forces and bending moments calculations, the Jonswap Spectrum (DNV-RP-C205) with peakness factor equal to 3.3 is considered (Mazarakos et al. 2017).

Results for the significant values of shear forces and bending moments at the base of the WT, the 1st lower delta brace joint and the mooring lines 1, 2, 3, respectively are presented in Tables 4 and 5.

Table 4. Significant values for shear forces and bending moments at the WT base and lower delta brace joints.

Name	Fx(sign) [kN]	Fy (sign) [kN]	My(sign) [kNm]
Base of the WT	0.8949E3	0.33437E1	0.89386E3
1st Lower Delta Joint	0.5484E4	0.14573E4	0.49549E5

Table 5. Significant values—Mooring lines.

Name	T(sign) [kN]
Mooring 1	0.18658E4
Mooring 2, 3	0.19633E4

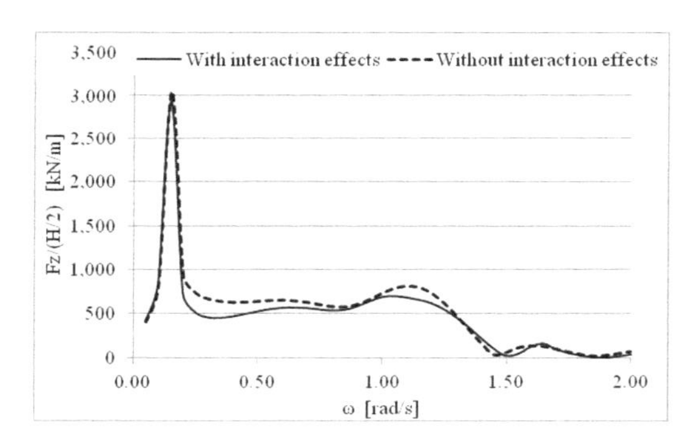

Figure 12. Fz shear forces at the 1st floater—lower delta brace joint versus the wave frequency, for wave angle heading 0° wind velocity 18 m/s.

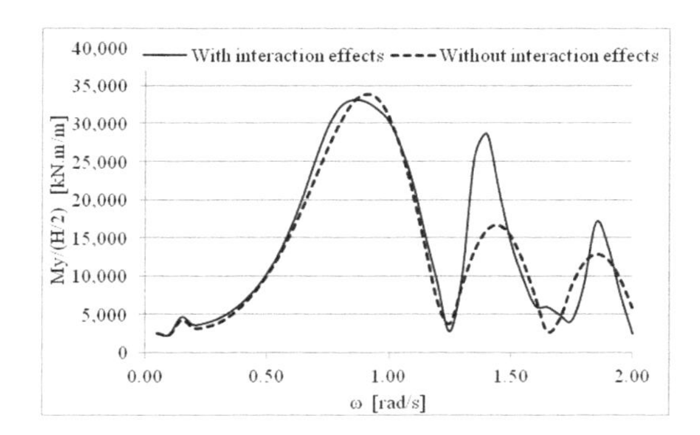

Figure 13. My bending moments at the 1st floater—lower delta brace joint versus the wave frequency, for wave angle heading 0° wind velocity 18 m/s.

6 CONCLUSIONS

A TLP floater for the DTU 10 MW WT has been presented encompassing three cylinders.

For this design, RAO's of the complete system have been calculated using a frequency domain solution, considering the WT's contribution on the floater's degrees of freedom.

The following conclusions were drawn:

1. Both methods (i.e. with or without taking into consideration the interaction effects) capture the acting loads for low wave frequencies values. On the other hand for larger wave frequencies the discrepancies between the two methods are notable.
2. Due to the high pretension in the structure, the exciting Fz forces are identical, while there are small differences in the exciting Fx forces and My moments due to the interactions between the cylinders.
3. Also the values of the horizontal shear forces at the moorings connection points are very low in relation with the corresponding vertical loads, due to the high axial vertical stiffness of the mooring lines.
4. From the presented analysis it hasn't been clear if the interaction phenomena between the devices overestimate or underestimate the acting loads at all the examined wave frequencies. Therefore, it would be interesting to study in further works how the distance between the bodies of the structure in relation with the wavelength

influence the acting loads on the structure for both methods (i.e. with or without the interaction effects).

ACKNOWLEDGMENTS

This research has been partially financed by the European Union, Horizon 2020, the E.U. Framework Program for Research and Innovation, Research Fund for Coal and Steel, Program: JABACO: Development of Modular Steel Jacket for Offshore Wind Farms (2015–2018).

Part of this research has been co-financed by the by the European Union, Horizon 2020, the E.U. Framework Program for Research and Innovation, Research Fund for Coal and Steel, Program: REFOS (709526): Life—Cycle Assessment of a Renewable Energy Multi—Purpose Floating Offshore System.

REFERENCES

Black, J.L., Mei, C.C., Bray, M.C.G., 1971. Radiation and scattering of water waves by rigid bodies, *J. Fluid Mech.* **46**, 151–164.

DNV, 2007. Recomanded Practice – Environmental Conditions and Environmental Loads, DNV-RP-C205, Det Norske Veritas.

Kagemoto, H., Yue, D.K.P., 1986. Interactions among multiple three – dimensional bodies in water waves: an exact algebraic method, *J. Fluid Mech.* **166**, 189–209.

Konispoliatis, D.N., Mazarakos, T.P., Mavrakos, S.A., 2016. Hydrodynamic analysis of three-unit arrays of floating annular oscillating-water-column wave energy converters, *Applied Ocean Research*, vol. 61, 42–64.

Konispoliatis, D.N., Mavrakos, S.A., 2016. Hydrodynamic analysis of an array of interacting free–floating oscillating water column devices, *Ocean Eng.* 111, 179–197.

Mavrakos, S.A., Koumoutsakos, P., 1987. Hydrodynamic interaction among vertical axisymmetric bodies restrained in waves, *Applied Ocean Research*, **9**, 128–140.

Mavrakos, S.A., 1991. Hydrodynamic coefficients for groups of interacting vertical axisymmetric bodies, *Ocean Eng.* **18**, 485–515.

Mavrakos, S.A., 1995. Users Manual for the Software HAMVAB, School of Naval Architecture and Marine Engineering, Laboratory for Floating Structures and Mooring Systems, Athens, Greece.

Mazarakos, T.P., 2010. Second – Order Wave Loading and Wave Drift Damping on Floating Marine Structures, Ph.D. Thesis, School of Naval Architecture and Marine Engineering, Division of Marine Structures, Laboratory of Floating Structures and Mooring Systems, National Technical University of Athens, Greece, 1–272,

Mazarakos, T.P., Manolas, D., Grapsas, T., Mavrakos, S.A., Riziotis, V.A. and Voutsinas, S.G., 2014a. Conceptual design and advanced hydro-aero-elastic modeling of a TLP concept for floating wind turbine applications, *RENEW 2014, 1st International Conference on Renewable Energies Offshore,* Lisbon, Portugal.

Mazarakos, T.P., Mavrakos, S.A., Konispoliatis, D.N., Voutsinas, S.G., Manolas, D., 2014b. Multi- pupose floating structures for offshore wind and wave energy sources exploitation, *COCONET Workshop for Offshore Wind Farms in the Mediterranean and Black Seas,* Anavyssos - Greece.

Mazarakos, T.P., and Mavrakos S.A., 2017. Experimental Investigation On Mooring Loads And Motions Of A TLP Floating Wind Turbine, *Special Session on Offshore and Marine Renewable Energy: Conversion and Transmission, Twelfth International Conference on Ecological Vehicles & Renewable Energies, EVER 2017,* Grimaldi Forum, Monaco.

Mazarakos, T.P., Konispoliatis, D.N., Mavrakos, S.A., 2017. Hydrodynamic loading and fatigue analysis of an offshore multi-purpose floating structure for offshore wind and wave energy sources exploitation, *International Maritime Association of the Mediterranean*, Lisbon, Portugal.

Miles, J., Gilbert, F., 1968. Scattering of gravity waves by a circular dock, *J. Fluid Mech.* **34** (No.4), 783–793.

Okhusu, M., 1974. Hydrodynamic forces on multiple cylinders in waves. University College London, London, Int. Symposium on the Dynamics of Marine Vehicles and structures in Waves.

Siddorn, P., Eatock Taylor, R., 2008. Diffraction and independent radiation by an array of floating cylinders, *Ocean Eng.* **35**, 1289–1303.

Soukissian, T., Adrianopoulos, K., Mpoukovalas, G., 2014. Report on the estimation of the design environmental conditions (in Greek). Technical Report No. D1.1, Program POSEIDON (2014), Greek General Secretariat for Research and Technology.

Twersky, V., 1952. Multiple scattering of radiation by an arbitrary configuration of parallel cylinders, J. Acoust. Soc. Am. **24** (No.1).

Yeung, R.W., 1981. Added mass and damping of a vertical cylinder in finite—depth waters, *Appl. Ocean Res.* **3** (No.3), 119–133.

WAMIT, Inc, http://www.wamit.com/manual.htm/ "User Manual", 2016.

Marine Design XIII – Kujala & Lu (Eds)

Downtime analysis of FPSO

M. Fürth & J. Igbadumhe
Charles V. Schaefer, Jr. School of Engineering and Science, Stevens Institute of Technology, Hoboken, USA

Z.Y. Tay
Engineering Cluster, Singapore Institute of Technology, Singapore

B. Windén
University of Southampton, Southampton, UK

ABSTRACT: Hydrocarbons remain the most important fuel in today's world despite recent depreciation in their economic value and continuous developments in the use of alternative sources of energy such as solar power. The Organization for Economic Cooperation Development (OECD) predicts a bounce back in the value of hydrocarbons in coming years which implies that there will be more exploration, offshore drilling and production of hydrocarbons worldwide. Floating Production Storage and Offloading (FPSO) vessels have been widely used for these operations due to their cost effectiveness especially in remote areas. The oil produced from an FPSO can be transported to the mainland by the use of tankers and the offloading process from FPSO to tanker can be affected by weather conditions. Hence, a rigorous downtime analysis for the offloading operation has to be conducted to ensure crew safety and to maximize production. A simplified method for analysing the weather downtime on a turret moored FPSO and the tandem offloading process is presented in this paper and the workability for both vessels is presented.

Keywords: FPSO; Downtime; Weather; Shuttle Tanker; Offloading Operation; Wave Height

1 INTRODUCTION

Floating Production Storage and Offloading (FPSO) units will continue to remain relevant in the oil and gas industry because of their ability to operate in deep waters and offering earlier exploration and production services. The recently published annual report on FPSO units by Offshore Magazine (2017) showed an increase of FPSOs by 5 percent from 2015 to 2016 and approximately 24 percent increase from 2016 to 2017, despite the recent depreciation in the economic value of oil and gas. The annual report also showed that 26 percent of FPSO's are operating in Brazil, 24 percent in Asia, 21 percent in West Africa, 10 percent in the North Sea, 5 percent in the Gulf of Mexico, 4 percent in the Norwegian Sea and 10 percent in other locations. Depending on the economic situation, an FPSO can be designed to operate in all types of location and all possible weather condition or, the FPSO can be designed to operate in selected locations and situations where certain downtime is acceptable. The elements that play a key role in FPSO offloading operations are weather, the FPSO itself, the shuttle tanker, hawser and the judgement of the shuttle tanker operator.

Numerous studies have shown that the Gulf of Mexico and the North Sea have extreme weather conditions particularly in the winter that affect the operations of FPSO's while West Africa has benign weather conditions. A good example was given by Orji & Woodward (2015) which showed that FPSOs would further experience downtime in benign weather locations when sea swells, characteristic long waves and wind seas occur at the same time. Anundsen (2008) also highlighted that even though West Africa is dominated by calm weather conditions, occurrence of swells can cause floating vessels to oscillate in such a way that operations would have to be put on hold.

The process of offloading hydrocarbon from an FPSO to a shuttle tanker typically takes 3 to 5 days. Rodriguez et al, (2009) summarized the basic FPSO offloading process from a shuttle tanker's point into five stages: tanker approaches FPSO, connects to FPSO with a hawser and connect loading hose, transfer hydrocarbon from FPSO to shuttle tanker, disconnect hose and hawser and then depart to onshore receiving station. With such a tight schedule in offloading activities, predicting and analysing downtime is significant. Generally, most FPSO's are designed to allow tanker connection at a significant wave height of 4.5 meters and maintain position at significant wave height of 5.5 meters (Meyer and Huglen, 2003).

Over the last 20 years, recommended typical maximum significant wave heights to be used for design has increased in many parts of the world. This is shown in the comparison between the values given by Centre for Marine and Petroleum Technology (CMPT) (1998) and DNV (2013) in Table 1. In addition, Harrison & Wallace (2005) highlighted that since the 1980s, there has been a 2% yearly rise in mean wave height due to changes such as the increase in the UK winter wind speed over the past years and carbon emission. Harrison & Wallace also stated that evidence has shown alterations in the wind and wave climates over the years, and the UK sea level is expected to increase by 36 cm by 2080. Hence, frequent prediction and analysis of weather downtime is beneficial in optimizing the FPSO offloading operations.

Offshore downtime analysis has been conducted in the past by a large number of researchers. Wal & Boer (2004) discussed two methods for analysing downtime. The first method is based on a wave scatter diagrams and the second method is based on the duration of the job being carried out using scenario simulations. The scatter diagram approach employs the use of wave scatter diagrams which is simple, easy to understand and thus it is widely used and provides an overview of operational limitations relative to weather conditions but this does not account for sudden changes in probabilities when random waves are generated. On the other hand, the scenario simulations method applies the long term sea state time records and provides stochastic variation of project durations.

An example of the downtime analysis of a spread moored FPSO system evaluated using a static calculation of the Dynamic Positioning Shuttle Tanker (DPST) capacity during offloading operation in Brazilian waters was presented by Corrêa *et al.* (2013). 8-year environmental conditions that covered current, wind, local- sea and swell were used in the downtime calculation. Static wind forces were obtained using the coefficients recommended by the Oil Companies International Marine Forum (OCIMF), wave forces were obtained using the Joint North Sea Wave Project (JONSWAP) wave spectrum and the static current forces were also obtained using non-dimensional coefficients defined by surge forces, sway forces and yaw moments. The environmental mean forces due to these wave, wind and current components were then evaluated for the DPST under fully loaded or ballasted condition. The thrust allocation algorithms were then used to obtain the mean force on each thruster of the DPST. A safe operation was defined as when the DPST system was able to allocate thruster force to counteract the environmental forces and when the environmental force directed towards the stern was less than 22 KN. Correa *et al.* (2013) also showed that increasing the DPST weathervane safe angles could decrease the offloading downtime, though such decisions are subjected to a comprehensive risk analysis. According to Correa *et al.* (2013), the offloading from the bow or stern of an FPSO has negligible effect as due to the large variation of wave-wind conditions along the year and in addition, the DPST is able to avoid these conditions by pushing towards the FPSO.

Cueva *et al.* (2009) also performed a time domain downtime analysis for offloading operations by comparing a DP Shuttle Tanker (DPST) with a non-DP Shuttle Tanker (ST) in the Gulf of Mexico by using six different scatter tables to define the environmental conditions. With both the DPST and ST both having the same hull dimensions, the downtime was obtained as the summation of all the probabilities of occurrence under which the DPST and the ST were outside its safe operation zone also known as the green zone. The analysis was performed for both ballasted and fully loaded conditions for the DPST and the ST. It was observed that the loaded condition experienced higher downtime for both the DPST and the

Table 1. Maximum significant wave height and wave period for design.

	CMPT (1998)		DNV (2013)	
Location	Max Sig. Wave Height (m)	Sig. Wave Period (sec)	Max Sig. Wave Height (m)	Sig. Wave Period (sec)
Norwegian Sea			16.5	17.0–19.0
Northern North Sea	15	17	15.0	15.5–17.5
West of Shetlands/ North Sea	18	20	14	15–17
Gulf of Mexico (Hurricane)	13	16	15.8	13.9–16.9
Gulf of Mexico (Winter storm)	–	–	7.3	10.8–12.8
West Africa (swell)	4	17	4.8	15.6
Brazil	7	14	8.0	13.0
Philippines/ South China (non-typhoon)	11	15	7.3	11.1
South China (typhoon)	–	–	13.6	15.1

ST, but the difference in the downtime between the DPST and ST was negligibly small.

Previous downtime analysis by various researchers on offshore operations for floating and fixed production facilities generally tend to perform analysis for the maximum operational wave and/or current (Cueva *et al.*, 2009). The method provided in this paper aims to differ from other methods by evaluating the motions of the FPSO and the shuttle tanker under scatter environmental data in the operation area (green zone), incorporating an interpolation scheme and combing environmental conditions and operational limits into workability.

2 THE FPSO SYSTEM, SHUTTLE TANKER AND ENVIRONMENTAL CONDITIONS

This paper focuses on a turret moored FPSO and a dynamic positioned tanker, connected by tandem at the stern and located in the Gulf of Mexico. The aim is thus to perform an offloading downtime analysis for the considered vessels placed adjacent to each other under the environmental conditions of Gulf of Mexico (GOM) which is a hurricane prone environment. The general environmental conditions for the Gulf of Mexico are shown in Table 2 (Meyer and Huglen, 2003). It is assumed that the FPSO is required to maintain position for all extreme weather conditions including the 10-year hurricane condition.

The FPSO vessel in this analysis is an internal turret moored vessel at an approximate water depth of 2,000 meters which must be continuously operational for 20 years and subjected to a sustained wind of up to 27 m/s, a current speed up to 1.2 m/s, a wave height of up to 6.0 m and a peak wave period of 10 seconds. The assumed characteristics of the FPSO is detailed in Table 3. The internal turret FPSO was chosen for this analysis because the aforementioned annual report on FPSO units published by Offshore Magazine (2017), showed that most of the FPSO's operating in the GOM were internal turret moored. Since the FPSO is turret moored, it will weathervane to the wave, wind and current directions.

The assumed DP shuttle tanker is based on a Panamax ship with assumed double hull main characteristics. The characteristics of the shuttle tanker are presented in Table 4.

Table 2. General environmental conditions of the GOM.

Design condition	Loop Curr.	10-year Winter Storm	100-year winter storm	10-year Hurr.	100-year Hurr.
Wave height (m)	3.8	5.9	8	8.5	13.4
Period (Sec)	9	11.05	13	13	14.9
Wind speed (m/s), 1 hr	15	20	23	27	38.6
Current Speed (m/s)	2.13	0.6	0.5	1	1

Table 3. FPSO characteristics.

Parameter	Value	Unit
Length LOA	253	M
Beam B	42	M
Depth	21	M
Draft T	16	M
L/B ratio	5.55	
B/T ratio	1.97	
Block coefficient Cb	0.93	
Water area coefficient Cw	0.95	
Section area coefficient Cm	0.97	
Length of waterline LWL	221.35	M
Storage capacity	2,000,000	barrels
Displacement	105,411.95	Ton
Displaced volume	471,398.4	M
Individual Cargo tank length	30	M
Individual Cargo tank beam	16	M
Individual cargo tank depth	19	M
Vertical Centre of Gravity VCG	6.0	M
Roll Radius of Gyration K11	1.5	M
Fraction of Critical roll damping (zeta)	0.2	

Table 4. Shuttle tanker characteristics.

Parameter	Value	Unit
Length LOA	185	M
Beam B	27	M
Depth	17	M
Fully loaded Draft T	12.2	M
Ballasted Draft	6	M
Block coefficient Cb	0.83	
Water area coefficient Cw	0.95	
Section area coefficient Cm	0.97	
Length of waterline LWL	178.38	M
Storage capacity	1,500,000	Barrels
Displacement (fully loaded)	44,909.8	Ton
Displacement (ballasted)	21297.3	Ton
Individual Cargo tank length	20	M
Individual Cargo tank beam	17	M
Individual cargo tank depth	15	M
Vertical Centre of Gravity VCG	8.84	M
Roll Radius of Gyration	9.3	M
Fraction of Critical roll damping (zeta)	0.2	

3 METHODOLOGY

Figure 1 shows the flow chart for performing an offloading downtime analysis of the FPSO and DP shuttle Tanker.

The NORDFORSK (1987) criteria for acceleration and roll motion for vessels under heavy manual work was used as the operational limit for both the FPSO and the tanker. These criteria are given as RMS values which allows for direct comparison with the obtained response spectrum (last step in Figure 1).

The Response Amplitude Operator (RAO) was obtained using the Seakeeeping Performance Program (SPP) by Parsons, Li and Singer (1998). The SPP is a probabilistic analysis involving the use of strip theory where the Lewis transformation is used to represent each hull section subjected to long-crested seas. This makes it possible to calculate the non-dimensional motion response in heave, pitch and roll with just the section area, design waterline and draft of the vessel. Strip theory was used because only requires hull properties for the stations, thus reducing a 3D problem to a 2D problem, results obtained can be converted to three dimension by performing integrations over the length of the FPSO (Journée and Adegeest, 2003). Three dimensional program was not used because of its high demand in computer resources and the need for rigorous detailing of the grid for both the vessel and the free surface (Milgram, 2007). In initial design studies speed comes before accuracy to some degree, before finalising the design a more in depth method is recommended.

Relevant vessel characteristics, presented in Table 3 and Table 4 are applied in obtaining the RAO. Both vessels are assumed to have a zero speed and encountering a head sea since the turret moored FPSO is always weathervaning and the tanker is tandem connected to the FPSO. 19 wave frequencies are considered in the analysis ranging from 0.48 radians per second to 1.25 radians per second in order to cover the periods available in the used wave statistics given by Hogben and Lumb, (1986).

The GOM environment is defined by wave scatter diagram environmental data adopted from Hogben and Lumb (1986), with significant wave heights ranging from 1 to >14 m and the zero crossing wave period from 3 to >12 seconds. The wave scatter diagram represents the percentage of joint probability of occurrence of the significant wave height and the zero crossing wave period combination. Based on the wave scatter diagram by Hogben and Lumb (1986), eight different scatter tables; one for each direction according to Table 5 was created using random waves.

Figure 2 presents the annual random waves generated at 135 degrees (waves coming from the

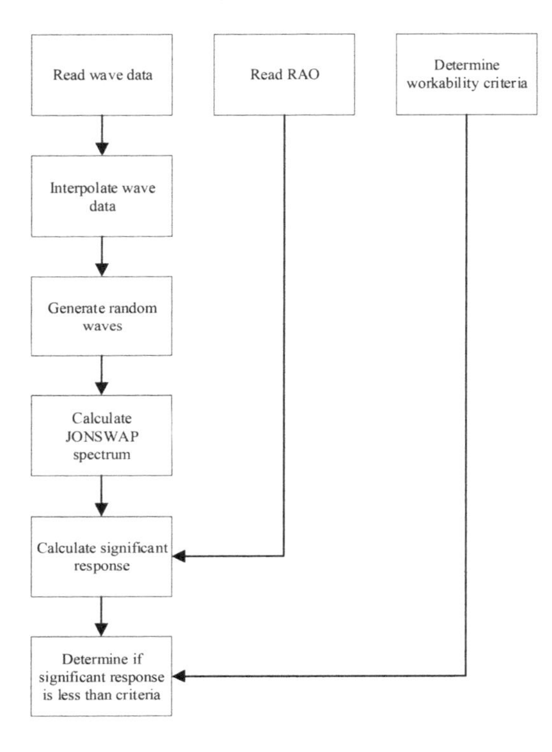

Figure 1. Downtime analysis summary.

Table 5. Probabilities of selected conditions.

	Annual (%)	Dec–Feb (%)	Mar–May (%)	Jun–Aug (%)	Sept–Oct (%)
315 degrees (NW)	4.93	7.30	4.70	3.04	4.63
0 degrees (NN)	9.82	14.6	8.42	3.68	12.42
45 degrees (NE)	18.19	20.19	15.84	12.84	23.86
90 degrees (E)	3.20	3.25	3.05	3.88	3.03
135 degrees (SE)	29.28	24.30	29.32	33.40	30.39
180 degrees (SS)	3.47	2.79	2.94	5.45	2.63
225 degrees (SW)	9.81	9.07	11.05	12.19	6.99
270 degrees (W)	19.41	16.79	23.45	23.07	14.38

south-east) conforming to the probability given in Table 5 using an interpolation scheme. The random waves were generated within the limits of the scatter data using the linear wave theory and MATLAB. First a grid corresponding to the wave periods and heights in the area was created, where extra care must be taken for low wave heights. In order to get a smoother distribution of the probability for waves with a wave height of 0 to 0.5 m the probability was linearly interpolated. Since the data by Hogben and Lumb (1986) is rather course; the probability is given for an increment of one second and the wave height for increments of one-meter linear interpolation was used to get a finer distribution of waves. MATLAB's random function was used to create waves with random wave height and period within the limits of the wave scatter data. The probability of the generated waves corresponded to the probability given by Hogben and Lumb (1986).

Random waves fitted to the same scatter diagram without using an interpolation scheme is presented in Figure 3. It can be seen that the probabilities for the constructed diagrams in Figures 2 and 3 correspond to the probabilities observed by Hogben and Lumb (1986) in Table 5 when a sufficiently large N is considered. The interpolation scheme, as applied in Figure 2, is used to get smaller statistical bins for random wave generation in order to avoid the sudden change in probability between bins (defined by Hogben and Lumb as a 11×15 grid) as observed in Figure 3. Without the interpolation scheme, the individual wave bins are clearly visible among the random generated waves as seen in Figure 3.

It is noted that the interpolation grid has to be kept at reasonable size and not be too fine. Using a too fine interpolation grid causes more rounding truncation resulting in fewer than the specified number of waves being created. In this case truncation contours become visible in the scatter diagram which is not correct as seen. It is noted that the scatter diagram at the top in is generated using the same interpolation as in Figure 2.

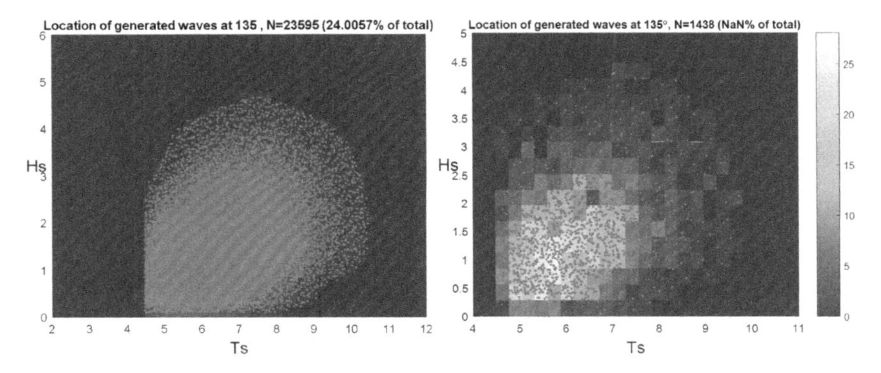

Figure 2. Random waves generated using interpolation, (Left) 100 000 waves total. (Right) 10 000 waves total. Each generated wave is marked with a red dot and the color bar shows an intensity map of waves in each region.

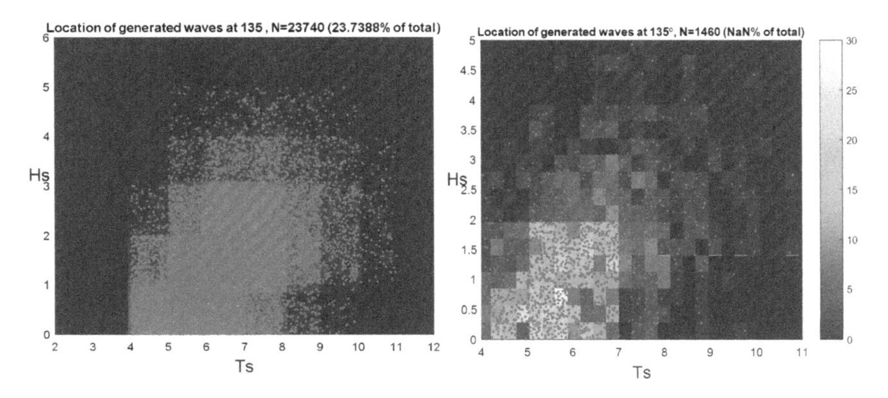

Figure 3. Random waves generated without interpolation, (Left) 100 000 waves total. (Right) 10 000 waves total. Each generated wave is marked with a red dot and the color bar shows an intensity map of waves in each region.

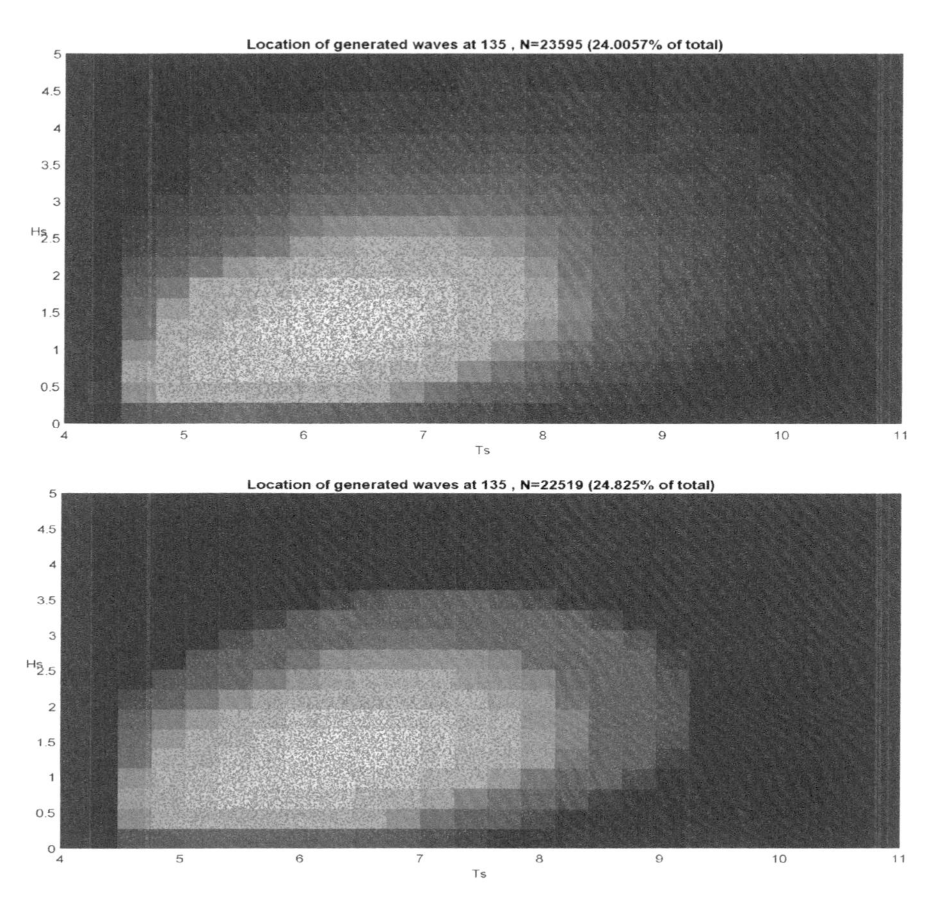

Figure 4. Random waves generated with a suitable interpolation grid (top) and a too fine interpolation grid (bottom). Each generated wave is marked with a red dot and the color shows an intensity map of waves in each region. 100 000 waves in total are generated.

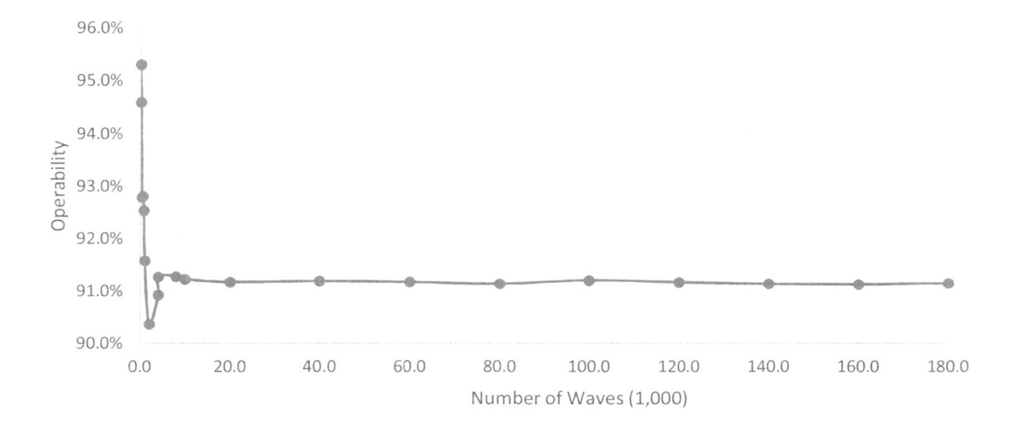

Figure 5. Convergence of operability.

The wave spectrum and the response spectrum were generated for each sea state corresponding to individual cells of the scatter diagram (red dots in Figures Figure 2 to.) The JONSWAP wave spectrum was used for the seaway characteristics and it was reversely constructed using the significant wave height and peak period according to Lee and Bales (1980). The peak period is assumed to be approximately 1.4 times the zero crossing period as given by the scatter diagram. The JONSWAP spectrum was then used to obtain the response spectrum by multiplying it with the square of the RAO for each seastate. The significant response was obtained from the response spectrum using the area under the graph. The workability is defined as the percentage of conditions where the significant response does not exceed the operating limit criteria whereas the downtime is defined as the conditions where the significant responses exceed the specified operating limit.

3.1 *Convergence of operability*

The specific characteristics of the generated waves is random (but conforming to the statistics of Hogben and Lumb (1986)) hence, the calculated operability of the vessel would not be constant even though the same case was studied. Too few waves also give large statistical variations. To obtain a convergence of operability level for the vessels, the number of random waves generated was varied for all directions as shown in Figure 5. This provides good judgement on the number of waves that should be used in analysing the downtime and workability. Figure 5 shows that there is variation in the operability up to when the number of waves reaches 120,000 and convergence is assumed thereafter. The effect of the number of waves towards the operability level has been studied up to 600,000 and the results does not vary more than 1%.

4 RESULTS AND DISCUSSION

The percentage workability and downtime of the FPSO and the shuttle tanker was evaluated by considering 120,000 random waves. Both ballast and fully loaded conditions were considered for the tanker. The vessels' respective workability annually and for different seasons of the year are as shown in Table 6. The workability and downtime are presented in terms of the two degrees of freedom; Heave and Pitch as they are the dominant degrees of freedom when considering head seas. The results in Table 6 for waves generated in all direction show that the winter season (December to February) has the highest downtime due to the severe sea state while the summer season (June to August) has the lowest downtime for both the FPSO, ballasted tanker and loaded tanker. The loaded DP shuttle tanker shows a slightly higher downtime compared to the ballasted tanker which was also as observed by Cueva et al., (2009)

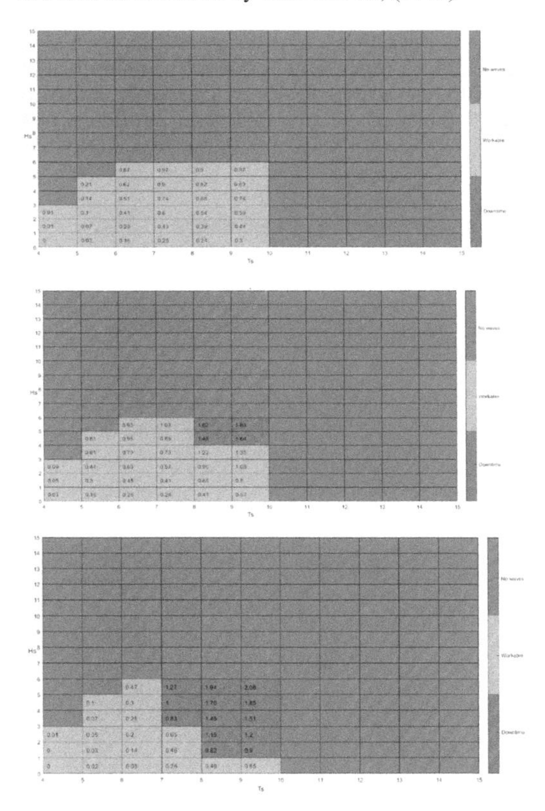

Figure 6. Workability and downtime due to heave for FPSO (top), Ballasted tanker (middle) fully loaded tanker (bottom) for waves generated from the NW from December to February.

Table 6. Workability percentages in all directions.

Vessel	Annual	Dec–Feb	Mar–May	Jun–Aug	Sept–Nov
FPSO	91.15%	84.12%	88.84%	97.20%	85.53%
Ballast Tanker	83.33%	74.32%	80.74%	93.82%	81.44%
Full Tanker	83.14%	74.27%	80.54%	93.68%	77.76%

Figure 6 and Figure 7 show that the workability and downtime for the fully loaded tanker for heave and pitch motions, respectively. The waves are generated from the South West. The results presented in Figure 6 and Figure 7 are in terms of the significant response of the vessels considered, where a red square denotes that the significant response is over the NORDFORSK (1987) criteria while a green square indicates that the response is below the criteria. Grey color is used when there are no waves within the given period and wave height.

The results obtained in the conventional method as described by Wal and Boer (2004), are solely based on wave direction and season but does not cater for operational modes which describes how a given operation contributes to the downtime. In addition, time domain downtime analysis with six different scatter tables by Corrêa *et al.*(2013), noted that each scatter table had numerous possible combinations which was not feasible to analyze. Hence, data was treated, and the number of combinations was reduced. Comparing the results by Wal and Boer (2004) and the method by Corrêa *et al.* (2013), with this paper, the results of this paper provides workability with regards to the degrees of freedom which would assist in planning offshore activities better.

The workability tables show that offloading from the FPSO to the shuttle tanker can be safely carried out in the green zones and offloading cannot occur in the red zones. It can be observed that a strong correlation exists in the occurrence of wave height and period when the downtime occurs for both FPSO and tanker where it is interesting to find that tanker will experience the same downtime wave height and period as the FPSO. Results in Figure 6 and Figure 7 also show that the tanker experienced more downtime than the FPSO due to the greater displacement in the former vessel. Also, the turret moored station keeping system has a slower response towards wave action as compared to the DP system.

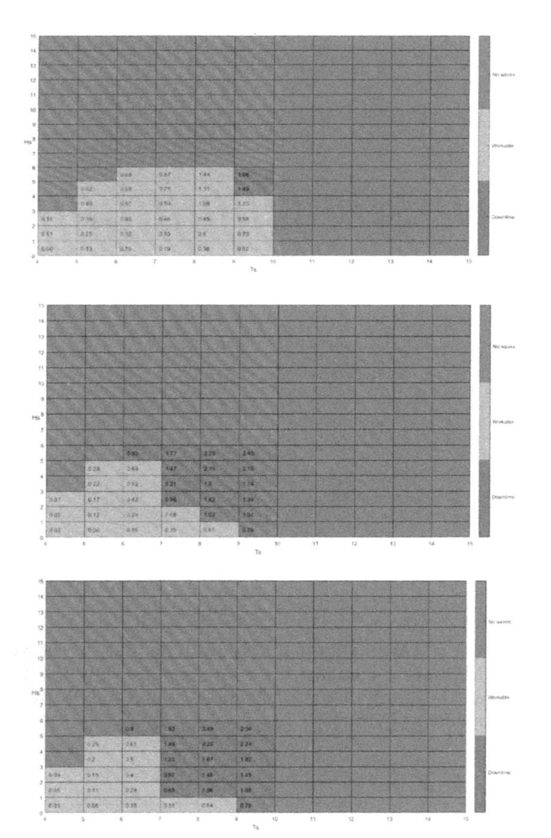

Figure 7. Workability and downtime due to pitch for FPSO (top), Ballasted tanker (middle) fully loaded tanker (bottom) for waves generated from the NW from December to February.

5 CONCLUSION AND FUTURE DIRECTIONS

This paper has presented a simplified method for analysing the weather downtime on a turret moored FPSO and the tandem offloading process by evaluating the motions of the FPSO and the shuttle tanker under scatter environmental data in the GOM. The wave scatter diagram considered was based on the actual wave statistic data whereas the industrial standard NORDFORSK criteria were adopted for the operational limits for the FPSO and DPST. The workability and downtime were presented for four different seasons in the year and the direction from which the waves were generated. The results showed that the winter seasons had the highest downtime while the summer seasons had the lowest downtime. The results presented on the workability of the shuttle tanker and FPSO under loaded and ballast condition enhanced the understanding of the behavior of FPSO under different operating sea states.

The conventional method for analyzing downtime with wave scatter diagrams as explained by Wal and Boer (2004), similar to the method presented in this paper combines the environmental conditions and the operational limits into workability. However, conventional methods do not incorporate an interpolation scheme which prevents sudden changes in probabilities when random waves are generated. In addition, conventional methods provide workability without identifying the degree of freedom that is contributing to downtime at any given time based on the wave direction and the season of the year.

This paper has excluded the hawser connection and shuttle tanker engine power but focused on workability of the FPSO and DPST under different wave heights and wave periods. Therefore, more work needs to be carried out to include location depth, current effect and shuttle tanker – FPSO hawser connection using the method described in this paper, and the method could to be tested using a model test and/or a three-dimensional Computational Fluid Dynamics program to obtain the Response Amplitude Operators (RAO). In addition, further studies of this method would also be beneficial in proposing improvements to FPSO designs.

ACKNOWLEDGEMENT

Appreciation goes to the Petroleum Technology Development Fund (PTDF) for providing funds to carry out this research, Professor Raju Datla and members of Furth Lab, Stevens Institute of Technology.

REFERENCES

Anundsen, T. (2008) *Operability comparison of three ultra-deepwater and harsh environment drilling vessels*. University of Stavanger.

Centre for Marine and Petroleum Technology (1998) *Floating structures: A Guide for Design and Analysis, Volume 1 & 2*. Edited by N.D.P. Barltrop. CMPT 1998.

Correa, D.C., de Oliveira, A.C., Tannuri, E.A. and Sphaier, S.H. (2013) 'Comprehensive Downtime Analysis of DP-Assisted Offloading Operation of Spread Moored Platforms in Brazilian Waters', in *ASME 2013 International Conference on Ocean, Offshore and Arctic Engineering*. France, p. 8. doi: 10.1115/OMAE2013-10405.

Corrêa, D.C., de Oliveira, A.C., Tannuri, E.A. and Sphaier, S.H. (2013) 'Comprehensive Downtime Analysis of DP-Assisted Offloading Operation of Spread Moored Platforms in Brazilian Waters', *Volume 1: Offshore Technology*, (August 2014), p. V001T01 A023. doi: 10.1115/OMAE2013-10405.

Cueva, M., Matos, V., Correa, S., Tannuri, E.A. and Mastrangelo, C. (2009) 'Downtime Analysis for Offloading Operation: DP X Non-DP Shuttle Tanker', in *ASME 28th INternational Conference on OCean, Offshore and AArtic Engineering OMAE*. Honolulu, Hawaii, pp. 1–10.

DNV-OS-E301, D.O.S. (2013) 'Position Mooring', (October).

Harrison, G.P. and Wallace, A.R. (2005) 'Climate sensitivity of marine energy', *Renewable Energy*, 30(12), pp. 1801–1817. doi: 10.1016/j.renene.2004.12.006.

Hogben, N. and Lumb, F.E. (1986) *Global Wave Statistics*. London.

Journée, J.M.J. and Adegeest, L.J.M. (2003) *Theoretical Manual of Strip Theory Program " SEAWAY for Windows*. Available at: www.amarcon.com (Accessed: 26 February 2018).

Lee, W. and Bales, S. (1980) *A Modified JONSWAP Spectrum Dependent Only on Wave Height and Period*. Available at: http://www.dtic.mil/dtic/tr/fulltext/u2/a090342.pdf (Accessed: 15 January 2018).

Meyer, E. a. and Huglen, Ø. (2003) 'What Dynamic Positioning Means to Floating Storage and Shuttle Tankers in the Gulf of Mexico', in *Dynamic Positioning Conference*. Houston, USA.

Milgram, J.H. (2007) 'Strip Theory for Underwater Vehicles in Water of Finite Depth', *Journal of Engineering Mathematics*, 58(1–4), pp. 31–50. Available at: http://web.mit.edu/flowlab/NewmanBook/Milgram.pdf (Accessed: 27 March 2018).

NORDFORSK (1987) *Assessment of Ship Performance in a Seaway*. Available at: http://unina.stidue.net/Universita%27 di Trieste/Ingegneria Industriale e dell%27Informazione/Nabergoj/Temporary/NORDFORSK.pdf (Accessed: 10 January 2018).

Offshore Magazine (2017) *Offshore Magazine: Oil and Gas News Covering Oil Exploration, Offshore Drilling, Drilling Rigs, Oil Industry Production & amp; more*. Available at: http://www.offshore-mag.com/maps-posters.html (Accessed: 23 November 2017).

Orji, C.U. and Woodward, M. (2015) 'Roll Motion Analysis of FPSO from Free Decay Data in Calm Sea', 3, pp. 313–324.

Parsons, M., Li, J. and Singer, D. (1998) 'Michigan conceptual ship design software environment', (November).

Rodriguez, Carmen E.P.; de Souza; Gilberto F.M.; Martins, M.R. (2009) 'Risk-Based Analysis of Offloading Operations With Fpso Production Units', *20th International Congress of Mechanical Engineering*, (January). Available at: http://www.abcm.org.br/anais/cobem/2009/pdf/COB09-2003.pdf.

Wal, Remmelt J and Boer, G. d (2004) 'Downtime Analysis Techniques for Complex Offshore and Dredging Operations', in, pp. 1–9. doi: 10.1115/OMAE2004-51113.

Production

Marine Design XIII – Kujala & Lu (Eds)
© 2018 Taylor & Francis Group, London, ISBN 978-1-138-34076-3

Prediction of panel distortion in a shipyard using a Bayesian network

C.M. Wincott & M.D. Collette
University of Michigan, Michigan, USA

ABSTRACT: The distortion incurred along a shipyard panel line is a costly factor of shipbuilding influenced by design parameters and manufacturing operations. By identifying the most impactful parameters, design for production can effectively mitigate panel line distortion. Bayesian networks can model the potential for distortion as a result of specific design parameters and process operations. Bayesian networks capture a complex web of cause-effect relationships, such as the mechanics of a structural panel's production, and break them down into a series of smaller conditional relationships that are more easily established. By breaking down a panel line and creating a series of connected networks modeling each step, the potential for distortion can be predicted from design parameters such as plate thickness and cutout locations and manufacturing decisions such as the use of automated seam welding. The series of networks allows the distortion measured upstream to update the probabilities of downstream nodes. Determining the effect of panel parameters on the distortion generated along a panel line allows informed detail design decisions that could reduce costly distortion in the shipyard. This paper explores the capability of Bayesian networks to predict such distortion through demonstration on an active shipyard panel line.

1 INTRODUCTION

The distortion incurred by a panel along a shipyard panel line has complex cause and effect relationships with the manufacturing processes of that line. Bayesian networks are an effective tool for modeling such complex relationships and, when applied to the manufacturing processes of a shipyard, can predict distortion and its effect on downstream processes. Such a network could be used to identify the most impactful design parameters and process operations to mitigate distortion and reduce costly corrective work later in production.

Shrinkage forces and resulting distortion are an unavoidable consequence of the heating and cooling cycle of welding (Blogdett, 1966). Additional manufacturing processes like cutting and fitting provide further opportunity for distortion as ship structure panels progress down a panel fabrication line. The shipyard panel line considered in this work encompasses workstations tasked with joining panels by seam welding, cutting openings and fitting inserts and attaching longitudinal stiffeners and transverse framing. The welding processes all contribute to distortion through shrinkage forces; these forces are exacerbated by too large welds and heavy rework. Thin panels gain additional distortion from sagging when they are poorly supported during transfers and storage in shipyards.

The prediction and control of distortion in a ship structure panel due to welding and the fabrication process has been approached with several methods: elastic (Wang, et al., 2013) and thermal-elastic-plastic (Wang, et al., 2015) finite element analyses, artificial neural networks and photogrammetry (Lightfoot & Bruce, 2007), and inherent strain theory (Jang, et al., 2004).

On the panel line distortion is both caused by manufacturing processes and leads to additional manufacturing processes, like welding and fitting, at later points; this identity of distortion as both cause and effect results in complex relationships that are difficult to capture and define. Bayesian networks provide a promising method for capturing these complex cause-effect relationships by breaking them down into a network of smaller causal relationships.

Bayesian networks (BN) are probabilistic networks helpful in representing causal relationships while maintaining independence and dependence relationships between entities of the domain (Kjaerulff & Madsen, 2013). BN are acyclic directed graphs of nodes and edges defining influences and independence between variables. Variables in the model are represented by nodes connected by edges that capture the direction of causality and dependence between variables. Each node of the network contains a conditional probability table (CPT) defining the relationship between the node's variable and the parent variables which directly affect its likelihood. By providing evidence of the state of a node gained by observation, BN update our beliefs of the rest of the network variables in the light of this evidence.

BN uniquely capture the ability to explain away causal beliefs through intercausal reasoning; if belief in one cause is increased, belief in another cause of the same variable is decreased.

Figure 1 provides an example of a simple predictive BN relating wet grass, rain and a sprinkler. In this example rain is a root parent node; it has no parent nodes itself but influences both the probability that the sprinkler is turned on and that the grass is wet. Without evidence provided the prior probability that rain will occur is 30%. The likelihood of the sprinkler being turned on is influenced by the occurrence of rain; if it is not raining, there is a 40% chance the sprinkler will be turned on, but if it is raining, it is much less likely, only 5%. Additionally, an occurrence of wet grass could be caused by rain or a sprinkler. However, if we know the sprinkler was on we know both that the grass is more likely to be wet and it is less likely that it rained than we would estimate from seeing wet grass alone. This is intercasaul reasoning.

In development Bayesian networks are viewed as the combination of two components: the qualitative structure and quantitative probability parameters. The structure of a network is always established first by identifying relevant variables and their relationships. Once the structure is formed, values of the parameters quantifying conditional dependences are elicited. BN can be constructed with varying levels of machine learning from data, from fully-automated to fully manual construction. Generally, both the structure and probability tables are developed from a combination of the data that the network models and expert knowledge.

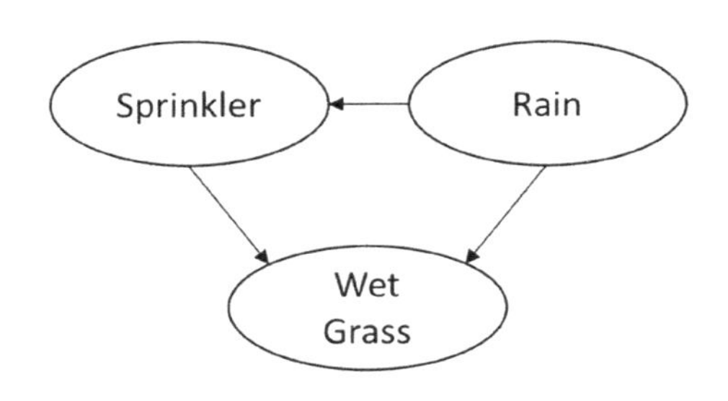

P(Rain) = (0.7,0.3)

		Rain	
		n	y
Sprinkler	n	0.6	0.95
	y	0.4	0.05

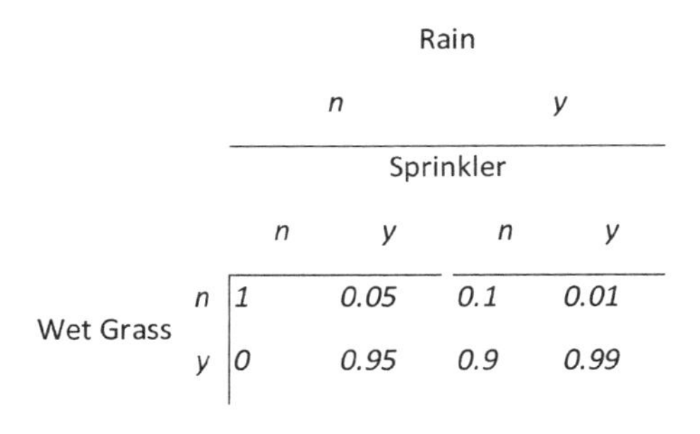

		Rain			
		n		y	
		Sprinkler			
		n	y	n	y
Wet Grass	n	1	0.05	0.1	0.01
	y	0	0.95	0.9	0.99

Figure 1. Simple Bayesian network.

Bayesian networks are beneficial to the prediction of distortion in shipyard panels lines because prediction of the final distortion can be continually updated with new evidence of what distortion has so far been incurred. Additionally, the network can be tuned to individual shipyard assembly systems by incorporating evidence as experience into the prior knowledge of the network parameters.

BN have been applied as predictive models in fields ranging from medical diagnoses to traffic pattern prediction. Marine applications of BN were first applied by Friis-Hansen (Friis-Hansen, 2000) with focus on structural reliability methods, failure detection, and risk prediction and analysis. Outside of the marine field, Bayesian networks have been extensively applied to structural health monitoring and reliability models (Straub & Papaioannou, 2015), (Straub & Der Kiureghian, 2010), (Cheung & Beck, 2009), (Mahadevan, et al., 2001), (Ayello, et al., 2014). In manufacturing, Bayesian networks have been used to diagnose the sources of variation and faults among assembly fixtures (Dey & Stori, 2005), (Jin, et al., 2012), (McNaught & Chan, 2011), (Sayed & Lohse, 2013), (Wolbrecht, et al., 2000). These manufacturing applications generally focus on the machining operations as the root cause, rather than the properties of the fabricated product.

2 METHODOLOGY

For this project, a Bayesian network was assembled to model the processes of a shipyard panel line to track distortion as an effect of plate parameters and manufacturing processes. After previous work to develop this network through automation resulted in predictions only slightly better than a polynomial regression, it was determined that a more complex structure developed with expert knowledge in addition to distortion and parameter data would be needed. The network structure was developed manually and probability relationships were derived semi-automatically through data and through expert knowledge.

Data for the network came from 221 panels produced on the panel line of an American shipyard. These panels were largely thin-plate panels, 81 percent less than 10 mm in thickness.

2.1 *Network structure development*

An initial attempt conducted to assemble a BN to predict distortion from the shipyard panel line used an almost fully-automated approach to learn the structure of the network. The data regarding the plate parameters and distortion measured at various points in the fabrication process were evaluated by the Necessary Path Condition (NPC) algorithm, a constraint-based learning algorithm, of the software, Hugin Researcher (HUGIN EXPERT A/S, 2017), used. Only inverse causality was corrected for in this initial construction attempt. The resulting network predicted distortion with 38% error when provided only the plate parameters; error decreased to 25% when intermediate distortion measurements were also introduced as evidence to the model. From the construction of this initial model, it was determined a human-expert developed network was necessary to gain greater accuracy.

The second attempt, developing the network described in this paper, relies only on expert knowledge for the development of the network's structure. Several iterations occurred in the manual development of this network. Additionally, to simplify the complex connections of the network it was separated into six sub-networks, each representing a process step of the panel line fabrication; these sub-networks have distortion as their only input and output, passing that knowledge to the next process. The outer system of this network, containing each process as an instance node is demonstrated by Figure 2.

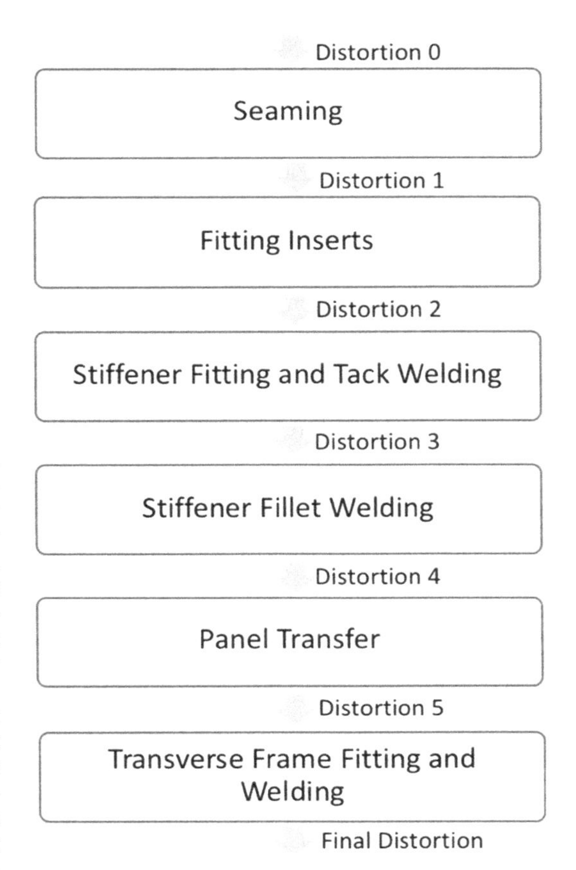

Figure 2. Full panel line network.

The initial manual construction of this network included the same nodes as the original, automatically developed network and additional variables of manufacturing operations that either prevented or corrected distortion with some coinciding labor cost. This strategy neglected to fully capture the causality relationships leading to distortion, and it was decided that additional intermediate factor variables such as heat input and shrinkage force were needed as variable nodes.

Four categories of variables are included in each process subnetwork: input plate parameters, cost operations, distortion factors and distortion measurements. Input parameters include both geometric parameters such as plate size and thickness and process parameters such as weld type. As determinations of detail design, these are known for all panels before the manufacturing process begins and are used as evidence to predict distortion levels. Cost operations include manufacturing operations which prevent or correct distortion, often at an additional labor cost. These operations included procedures such as clamping and grinding. As noted above, distortion factors are the intermediate factors which directly cause distortion such as shrinkage force and heat input. The distortion nodes capture the distortion of the panel as measured between each process step. Six of the seven major distortion nodes reflect the points during fabrication when distortion data was collected by the shipyard; distortion was not measured between stiffener fitting and fillet welding.

The edges connecting the nodes were established concurrently with the nodes by verbally working through the cause and effect relationships with subject matter experts.

2.2 *Parameter estimation*

The probabilities contained in the nodal probability tables were developed as a fusion of data and expert knowledge, depending on the amount and type of data available for each node. Because the network structure was developed and variables were identified after the distortion data was collected, quantitative data is not available for all measurable variables. Future applications of this distortion prediction method would benefit from additional data gathering.

As parent nodes, most input parameter variables were assigned distributions directly reflecting the training data from 206 panels. Many of the intermediate variables including cost operations and additional distortion factors have no collected data from this study to relate their values to those of the parent nodes. For these variables expert knowledge was used to determine the most likely outcome given parent node conditions. For children nodes with multiple parents, a CPT was developed for each parent and combined as a weighted average.

This simplifies the knowledge gathering by allowing the expert to identify more direct relationships but does assume parents can combine linearly.

CPTs were developed similarly for distortion nodes; however, distributions of measured distortion were available. The weight of each parent's influence was adjusted until the joint probability distribution of the node reflected the training data.

When available, regressions developed by outside analysis were used to evaluate for expected values of distortion factors. This approach was used for heat input (Yang, et al., 2013), shrinkage force and weld size (National Steel and Shipbuilding Company, 1993).

3 RESULTS

The effectiveness of the developed BN was analyzed given two levels of evidence. The first level includes variable states known before manufacturing begins as evidence. These variables include design decisions made on plate characteristics, like size and thickness, and manufacturing subprocesses, such as automated or manual welding. In application this level of evidence allows designers to predict the impact of detail design decisions on expected distortion. The second level of analysis also provides distortion measured between panel line processes as evidence. The results of this second level of analysis provide information on the accuracy of prediction for process subnetworks; in application, the ability to provide this evidence during manufacturing allows an engineer to make more informed design decisions based on the current level of distortion and the impact of subprocesses for distortion correction and prevention.

A total of 14 test panels were analyzed. While the design decisions of all panels are known, gaps in distortion measurement data mean that the distortion of each panel between every process step is not known. The number of test panels with measured distortion is reported for each prediction.

3.1 *Developed subnetworks*

Six subnetworks were developed to model the propagation of distortion along the shipyard panel line. Effort to minimize the number of parent nodes and node states resulted in a relatively small CPT for each child node. Ideally, variable domains are small enough to minimize the complexity of

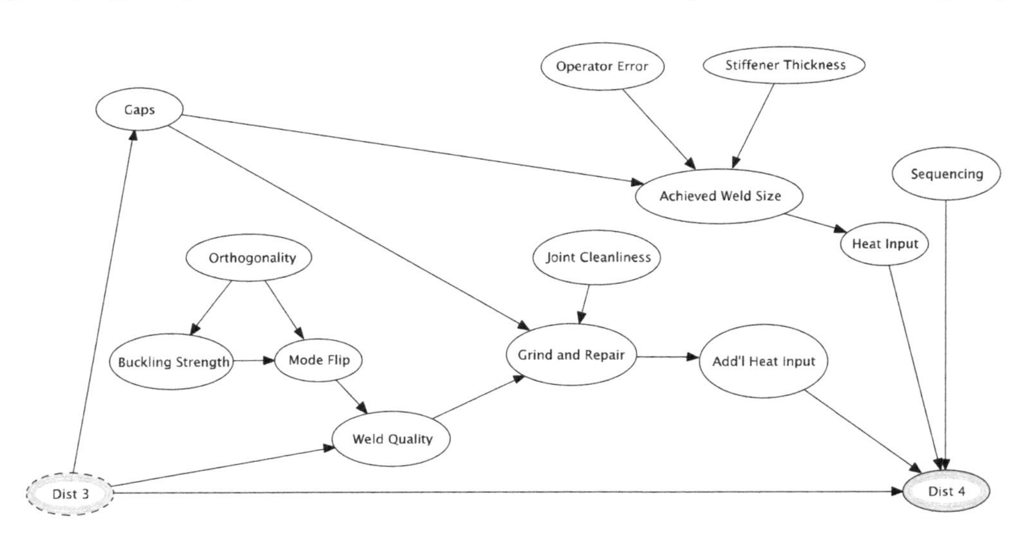

Figure 3. Bayesian subnetwork modeling stiffener fillet welding process.

eliciting probabilities and minimize the complexity of inference while providing a sufficient number of node states to accurately express relationships (Kjaerulff & Madsen, 2013). Table 1 provides a summary of the properties for each subnetwork, including the total number of nodes, total number of edges, the average number of parent nodes (number of inbound edges) and average number of node states.

A visual example of a process subnetwork is provided in Figure 1. This subnetwork models Stiffener Fillet Welding. It has one input node, Dist 3, which is the output Distortion 3 distribution from the previous process subnetwork, Stiffener Fitting and Tack Welding. The output node, Dist 4, is used to predict the distortion measured after the process and outputs a distribution to the Panel Transfer subnetwork. Two nodes are given evidence in this subnetwork: Orthogonality and Sequencing. For each panel the node state for these two variables is provided to aid in prediction of Distortion 4.

3.2 *Design decisions as evidence*

For this network, five plate characteristics and three process decisions were included as evidence on design decisions provided to the network. These eight variables are enumerated in Table 2. These variables were chosen based on data available for this project.

Values of these eight variables were provide to the network for all 14 test panels. The mean value of the distortion likelihood distribution was extracted at each process step and compared to the measured values of distortion after stiffener welding, panel transfer and frame welding. Table 3 provides a summary of the distortions predicted after three process step subnetworks and their error compared to those measured. These three process steps were the only steps with enough data available to meaningfully analyze the effectiveness of the network.

Table 1. Summary of subnetwork sizes and connections.

Process	Nodes	Edges	Average Number of Parent Nodes	Average Number of Node States
Seaming	21	32	1.52	3.33
Fitting Inserts	21	31	1.48	4.24
Stiffener Fitting and Tack Welding	14	18	1.29	3.71
Stiffener Fillet Welding	15	18	1.20	3.07
Panel Transfer	6	5	0.83	3.50
Transverse Frame Fitting and Welding	26	40	1.54	3.50

Table 2. Design decision variables provided as evidence.

Plate characteristics	Plate thickness
	Plate orthogonality
	Plate area
	Number of cutouts
	Number of inserts
Process Decisions	Elimination of force fitting procedures
	Weld sequence optimized for minimum distortion
	Use of new high rail for panel transfer and storage

Table 3. Summary of accuracy of distortions predicted provided design decision evidence.

Process step	No. test panels	Error mean percent	Error std dev percent
Stiffener Weld	6	23	21
Panel Transfer	11	25	19
Frame Weld	10	27	36

Error here is calculated as the absolute percent difference between measured and predicted values.

Distortion was predicted for these three steps at an average error of 26%. This is a significant reduction from the 38% error provided by the previously NPC developed Bayesian network and the 33% error of a polynomial regression modeled by the shipyard. Neither the previous BN nor regression model considers the process decisions in predicting distortion.

3.3 *Previous distortion as evidence*

An advantage of Bayesian networks is their ability to incorporate a varying amount of evidence. This characteristic allows us to update the prediction of future distortion as we gather measurements of distortion progression. Distortion measurement data were available, in varying amounts, for intermediate distortions measured after individual process steps along the panel line. The network's accuracy is assessed for the prediction of subsequent distortions when provided intermediate distortions measured after two processes: stiffener fillet welding and panel transfer. A summary of these accuracies is provided in Table 4.

Table 4. Summary of accuracy of distortions predicted provided distortion evidence.

Evidence process step	Predicted process step	No. test panels	Error mean	Error std dev
			percent	percent
Stiffener Weld	Panel Transfer	6	18	14
	Frame Weld	6	16	14
Panel Transfer	Frame Weld	8	32	33

Provided evidence on the level of distortion after stiffener welding, the network's accuracy increases to an average 17% error. Providing additional evidence on the distortion measured after a panel transfer increases the error in prediction of the final distortion considered, after frame welding, to 32%. In consideration of the highest initial prediction error for this processes step as well, this high level of error shows that the last subnetwork, modeling the transverse frame fitting and welding, is least effective in the prediction of distortion. This inaccuracy is likely due to a neglected cause of distortion mitigation. There is a general decrease in distortion measured before and after the framing workstation likely due to the fitting of the frames; however, there was no data available for this project on factors that would decrease distortion, such as frame spacing. The evidence provided in this step has no correlation to decreased distortion and providing additional impactful evidence would likely increase the ability of this subnetwork to predict distortion.

Generally, the effectiveness of this network is limited by the data used in learning the probability parameters and provided as evidence. Accuracy of predictions could be improved by additional panel data surrounding frame and stiffener number or spacing, insert size and thickness, and designed weld size for both training and use as evidence. For this project the lack of data is a result of the data used being non-specific to use for the BN. Ideally, the variables identified in the qualitative constriction of the network would influence the variable data collected.

4 DISCUSSION

Similar implementation of Bayesian networks for distortion prediction in a shipyard could aid in design and production decisions by predicting their influence. In the design of plate characteristics, a BN would allow engineers to evaluate the impact of a changed characteristic or process decision, input as evidence, on the distortion at each stage of fabrication.

It is also valuable in making manufacturing process decisions surrounding distortion mitigation or correction along the panel line. If distortion is measured and provided as evidence in real-time, the network provides a tool for decision making surrounding cost-linked manufacturing processes by allowing engineers to predict the most cost-effective subprocess given current distortion conditions of the panel. Similarly, well trained networks can be used to determine the overall most impactful operations for distortion mitigation.

If use continuously with a panel line, a learned network can be continuously updated with new evidence through adaptation of experience probability tables. This would take into account changing conditions on a panel line.

5 CONCLUSIONS

This work demonstrates definite predictive capability in application to distortion along a shipyard panel line. It is a useful tool for predicting the impact of design changes as the change can be easily incorporated as evidence for a specific panel or as the single component to evaluate change on overall distortion likelihood distributions. Once learned, the network can be used to assess the impact of operation changes or transitioned to an experience and adaptations model, incorporating the knowledge of each new panel evaluated.

ACKNOWLEDGEMENTS

The authors wish to acknowledge the support of the LIFT project, operated by the American Lightweight Materials Manufacturing Innovation Institute (ALMMII) for supporting this work. Furthermore, the expert knowledge of Steve Scholler, TD Huang and Michele Bustamante was vital to the development of this network.

REFERENCES

Ayello, F., Jain, S., Sridhar, N. & Koch, G., 2014. Quantitive assessment of corrosion probability – A Bayesian network approach. *Corrosion,* November, 70(11), pp. 1128–1147.

Blogdett, O.W., 1966. *Design of Weleded Structures.* Cleveland, Ohio: The James F. Lincoln Arc Welding Foundation.

Cheung, S.H. & Beck, J.L., 2009. Bayesian Model Updating USing Hybrid MOnte Carlo Simulation with Appliction to Structural Dynamic Models with Many Uncertain Parameters. *Journal of Engineering Mechanics,* 135(4), pp. 243–255.

Dey, S. & Stori, J., 2005. A Bayesian network approach to root cause diagnosis of process variations. *International Journal od Machine Tools & Manufacture,* 45(1), pp. 75–91.

Friis-Hansen, A., 2000. *Bayesian Networks as a Decision Support Tool in Marine Applications. PhD Thesis.* Denmark: Department of Naval Architecture and Offshore Engineering, Technical University of Denmark.

HUGIN EXPERT A/S, 2017. *HUGIN Researcher.* 8.5 ed. Aalborg: HUGIN EXPERT A/S.

Jang, C., Ryu, H. & Lee, C., 2004. *Prediction and control of welding deformations in stiffened hull blocks using inherent strain approach.* Toulon, France, The Fourteenth International Offshore and Polar Engineering Conference – ISOPE 2004, pp. 159–165.

Jin, S., Liu, Y. & Lin, Z., 2012. A Bayesian network approach for fixture fault diagnosis in launch of the assembly process. *International Journal of Production Research,* 50(23), pp. 6655–6666.

Kjaerulff, U.B. & Madsen, A.L., 2013. *Bayesian Networks and Influence Diagrams: A Guide to Construction and Analysis.* 2nd Edition ed. Aalborg: Springer.

Lightfoot, M. & Bruce, G., 2007. *The measurement and prediction of welding distortion using ANN and photogrammetry.* Portsmouth, United Kingdom, RINA - International Conference on Computer Applications in Shipbuilding 2007, pp. 269–273.

Mahadevan, S., Zhang, R. & Smith, N., 2001. Bayesian networks for system reliability reassessment. *Structural Safety,* 23(3), pp. 231–251.

McNaught, K. & Chan, A., 2011. Bayesian networks in manufacturing. *Journal of Manufacturing Technology Management,* 22(6), pp. 734–747.

National Steel and Shipbuilding Company, 1993. *Weld Shrinkage Study,* Bethesda, MD: Naval Surface Warfare Division, Carderock Division.

Sayed, M. & Lohse, N., 2013. Distributed Bayesian diagnosis for modular assembly systems – A case study. *Journal of Manufacturing Systems,* 32(3), pp. 480–488.

Straub, D. & Der Kiureghian, A., 2010. Bayesian Network Enhanced with Structural Reliability Methods: Application. *Journal of Engineering Mechanics,* 136(10), pp. 1259–1270.

Straub, D. & Papaioannou, I., 2015. Bayesian Updating with Structural Reliability Methods. *Journal of Engineering Mechanics,* 141(3), p. 04014134.

Wang, J., Ma, N. & Murakawa, H., 2015. An efficient FE computation for predicting welding induced buckling in production of ship panel structure. *Marine Structures,* Volume 41, pp. 20–52.

Wang, J., Rashed, S., Murakawa, H. & Luo, Y., 2013. Numerical prediction and mitigation of out-of-plane welding distortion in ship panel structure by elastic FE analysis. *Marine Structures,* 34(1), pp. 135–155.

Wolbrecht, E., D'Ambrosio, B., Paasch, R. & Kirby, D., 2000. Monitoring and diagnosis of a multistage manufacturing process using Bayesian networks. *Artificial Intelligence for Engineering Design, Analysis and Manufacturing,* 14(1), pp. 53–67.

Yang, Y.-P.et al., 2013. Uniform-Panel Weld Shrinkage Data Model for Neat Construction Ship Design Engineering. *Journal of Ship Design and Production,* 29(1), pp. 1–16.

Zhu, J. & Collette, M., 2017. A Bayesian approach for shipboard lifetime wave load spectrum updating. *Structure and Infrastructure Engineering,* 13(2), pp. 298–312.

Marine Design XIII – Kujala & Lu (Eds)

Author index

#0038 - 130618 - C0 - 246/174/27 [29] - CB - 9781138340763